Applied Analyses in Geotechnics

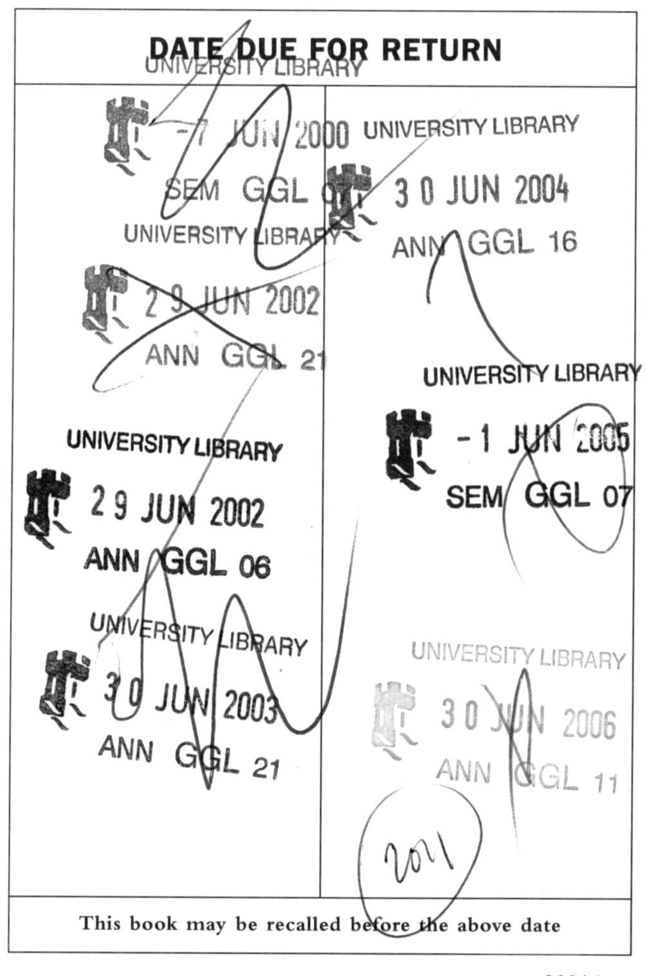

Applied Analyses in Geotechnics

Fethi Azizi
University of Plymouth, UK

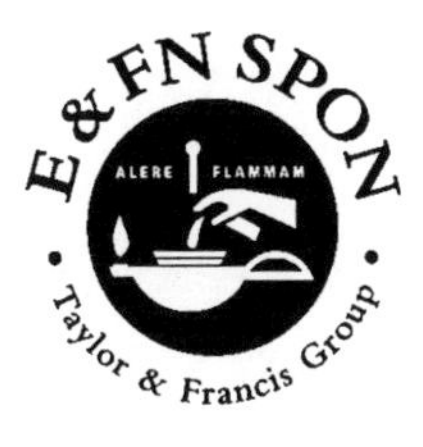

London and New York

First published 2000 by E & FN Spon
11 New Fetter Lane, London EC4P 4EE

Simultaneously published in the USA and Canada
by E & FN Spon
29 West 35th Street, New York, NY 10001

E & FN Spon is an imprint of the Taylor & Francis Group

Printed and bound in Great Britain by TJ International Ltd, Padstow, Cornwall

British Library Cataloguing in Publication Data
A catalogue record for this book is available from the British Library

Library of Congress Cataloging in Publication Data
Applied analyses in geotechnics / Fethi Azizi
p. cm.
Includes biographical references and index.
1. Engineering geology I. Title.
TA705.A97 1999
624.1'51-dc21

99-35126
CIP

ISBN 0-419-25340-8 (hb)
ISBN 0-419-25350-5 (pb)

To the memory of my Father

La raison nous commande bien plus impérieusement qu'un maître; car en désobéissant à l'un on est malheureux, et en désobéissant à l'autre on est un sot. Blaise Pascal

Contents

Preface

When writing this book, I was always guided by four things: (*a*) the book must be organised to suit different needs of undergraduate and postgraduate teaching programmes; (*b*) the content should stimulate the reader to think, thus avoiding it being a user's manual that describes procedures and methodologies; (*c*) the book must be innovative in that it should contain aspects of analyses that are only accessible through specialised books or research papers, and (*d*) the analysis related to each section should be thorough and well referenced so that the reader can concentrate on thinking and understanding.

I have also endeavoured to make the book enjoyable to read, but with respect to this subjective aspect, only your judgement matters, and so, do not hesitate to compare this book with other textbooks in terms of style and content while remembering the words of *Benjamin Disraeli '...this shows how much easier it is to be critical than to be correct'*.

I am assuming that the reader has a *working knowledge* of engineering mechanics, especially of stresses and strains, Mohr's circle representation of stresses and strains, equilibrium equations in terms of moments and forces, elasticity and plasticity.

I am aware that part of the analysis in some sections of the book may appear somewhat mathematically involved. In fact, you would be surprised to find how easy they are once you have re-read them: it is only by reading the analysis time and time again that an intellectual skill can be mastered. The book contains more than 65 worked examples and in excess of 450 illustrations which are presented in a clear and detailed way which can only enhance the understanding of the principles involved.

By the time you have read through this book, you will have hopefully understood, learned and mastered its content. In addition, you will have been able to satisfy some of your curiosity through some detailed derivations of formulae and equations which are presented as a *fait accompli* in other textbooks. You will have also noticed the rather lengthy process related to the detailed calculations of some worked examples. In practice, these calculations are almost exclusively undertaken numerically, because not only specialised software packages dealing with different themes in geotechnics are widely available, but a designer can always develop a software to suit his or her needs. So, do not despair if you find the hand calculations in some instances long and tedious because that was the feeling I had when I wrote them! However, as a learning experience, every example presents you with an opportunity to see how these details are handled and, more to the point, helps you develop your engineering judgement, in that the outcome of any computation always reflects the choice made *vis à vis* the soil parameters. Remember that geotechnics is not an exact science; rather, it can be described as an art in which the artist (*i.e.* the engineer) has to rely sometimes, if only partly, on his or her intuition (*i.e.* engineering judgement).

Moreover, you will be able to appreciate that, although in practice calculations may be undertaken in a different way, the fundamental thinking is similar to what you would have learned throughout the book. If anything, once you have assimilated the basic knowledge and principles, you will find it much easier to make the *appropriate assumptions* with confidence. This reinforces the statement made earlier about mastering an intellectual skill. So do not be put off if you do not fully understand what you have read; go through it again bearing in mind that the last reading *is* the one that matters.

This book is also meant to be useful to a postgraduate student who seeks a deeper understanding and a specialised knowledge about different subjects in geotechnics. In addition, practising engineers may find the book valuable since most of the mathematical formalisms can be forgone, provided that the assumptions of the formulation are well understood, so that it is not used in a perfunctory way as a recipe. Remember, the book is meant to make you think, then act.

Bearing in mind what *Sir Arthur Conan Doyle* once said, through *Sherlock Holmes, 'it is a capital mistake to theorise without data',* I went to great lengths to try to relate any theoretical aspects to practice, and this is reflected in the choice of the title for the book *'Applied Analyses in*

Geotechnics'. The emphasis is always on the practical side although, admittedly, some sections contain a fair amount of accessible mathematical formalism. There is no escape from it, an engineer *must* be able to understand how to translate a physical behaviour into *non trivial* mathematical equations; after all, that is what design *is* about!

I have refrained from making any suggestions on how the book may be used by colleagues to organise a course at whatever level. Had I done so, it would not only be presumptuous on my part, but it would also contradict my earlier statement regarding the different needs of different programmes.

I would like to thank the many friends and colleagues who contributed, in their own ways, to the improvement of the content of this book. In particular, I am grateful to *Clive Dalton, John Davidson, Kenneth Fleming, Henri Josseaume, David Naylor, David Rowe Bottom, Robert Whittle* and *Dennis Wilkinson* for taking the trouble to go through part or all of the manuscript and make valuable constructive comments.

Finally, on a personal note, notwithstanding the sporadic late-night frustrations or the occasional social deprivation, I have had a tremendous pleasure writing this book. I hope you will enjoy reading it.

Fethi Azizi, July 1999

List of main symbols

a	radius, area, amplitude
A	air content, activity, area, porewater pressure coefficient
A_b	base area
A_s	shaft area
B	width, porewater pressure coefficient
B'	effective width
b	width
c	apparent cohesion
c_h	coefficient of horizontal consolidation
c_u	undrained shear strength
c_v	coefficient of vertical consolidation
c_w	cohesion at soil/wall interface
C_c	compression index
C_g	coefficient of gradation
C_s	swelling index
C_u	coefficient of uniformity
C_α	slope of secondary consolidation graph
CSL	critical state line
D, d	diameter, depth, depth factor, depth of embedment
d_c, d_q, d_γ	depth factors
d_w	drawdown
$[D]$	elasticity matrix
e	void ratio, eccentricity, depth corresponding to zero net pressure
E	stiffness (elasticity modulus)
E_b	soil secant modulus
E_c	stiffness modulus of pile material
E_m	pressuremeter modulus
δE	work done by an external load
E_pI_p	flexural stiffness
f	motion frequency
F	force, factor of safety
f_o	correction factor
$\{F\}$	vector of nodal forces

g	acceleration due to gravity, intercept of Hvorslev surface, depth corresponding to zero bending moment
G	shear modulus
G^*	modified shear modulus
G_s	specific gravity
h, H	height, total head, length of drainage path, slope of Hvorslev surface
h_c	capillary rise
h_e	elevation head
h_p	pressure head
h_s	capillary saturation level
i	hydraulic gradient
i_{cr}	critical hydraulic gradient
i_e	exit hydraulic gradient
I	moment of inertia, rigidity index
I_c, I_q, I_γ	inclination factors
I_L	liquidity index
I_P	plasticity index
I_σ	influence factor
I/y	steel section modulus
J	cross modulus
k	permeability
$k_h^/$	equivalent horizontal permeability
$k_v^/$	equivalent vertical permeability
k_t	transformed permeability
K	ratio of horizontal to vertical effective stresses, bulk modulus
$[K]$	stiffness matrix
K^*	modified bulk modulus
K_a	coefficient of active earth pressure
K_e	length ratio
K_o	coefficient of active pressure at rest
K_p	coefficient of passive pressure
L	length
M	mass, moment
M_s	soil/shaft flexibility factor
m	soil creep parameter
m_α	slope constant
m_v	coefficient of volume compressibility

n	porosity
N	normal force, stability number
N_c, N_q, N_γ	bearing capacity factors
N_f	number of flow channels
N_d	number of equipotential drops
$[N]$	shape functions
NCL	normal consolidation line
OCR	overconsolidation ratio
p	mean stress, perimeter
p_e'	equivalent mean effective pressure
P_a	active thrust
P_c	creep pressure
P_l	limit pressure
P_p	passive thrust
PBP	prebored pressuremeter
$[P]$	permeability matrix
q	load, rate of flow, deviator stress
Q	load, rate of flow
Q_a	safe working load
Q_b	ultimate base resistance
Q_n	negative skin friction
Q_s	ultimate shaft friction
Q_u	ultimate loading capacity of a single pile
Q_{ug}	ultimate loading capacity of a pilegroup
R, r	radius
r_u	pore pressure ratio
Re	Reynolds number
R_o	radius of influence
RC	relative compaction
S	degree of saturation, sensitivity, spacing, total settlement
S_e	elastic shortening
s	reduced mean stress
s_c, s_q, s_γ	shape factors
SBP	self-boring pressuremeter
SBS	state boundary surface
ΔS	stress error
t	time, reduced deviator stress
T	surface tension force, torque, shear force per linear metre
T_v, T_r	time factors

u	porewater pressure, lateral displacement
u_g	pore gas pressure
U	degree of consolidation, thrust due to water pressure
$\{U\}$	vector of nodal unknowns
v	velocity, specific volume, vertical displacement
V	volume, model centrifuge in flight velocity
V_a	volume of air
V_s	volume of solids
V_w	volume of water
w	water content, weight, radial displacement
w_L	liquid limit
w_P	plastic limit
w_S	shrinkage limit
W	weight
δW	work dissipated per unit volume
z	depth
z_f	neutral depth for negative skin friction
β	soil slope behind a retaining wall
γ	unit weight
γ'	effective unit weight
δ	soil/structure friction angle
ε	strain
η	dynamic viscosity, ratio of total heads, drainage coefficient, ratio of deviator stress to mean effective stress, efficiency factor, optimum angle of failure behind a retaining wall
θ	temperature
$\Gamma, \kappa, \lambda, M$	critical state parameters
Γ_c	Coriolis acceleration
Γ_h	horizontal shaking
Γ_v	vertical acceleration
λ, Ω	wall inclinations
ν	Poisson's ratio
ξ	correction factor, strength ratio
ρ	density, sheet pile stiffness
σ	total stress
σ'	effective stress
σ_a	active pressure
σ_h, σ_3	horizontal stress

σ_p	preconsolidation pressure
σ_v, σ_1	vertical stress
τ	shear stress
τ_{mob}	mobilised shear stress
υ	angle of dilation
ϕ	angle of shearing resistance, from factor
ϕ_p, ϕ_c, ϕ_r	peak, critical and residual angles of shearing resistance
Φ	stress function, velocity potential
χ	constant
Ψ	stream function
ω	angular velocity

Conversion factors

- multiplication factors

10^9	*giga*	G
10^6	*mega*	M
10^3	*kilo*	k
10^{-3}	*milli*	m
10^{-6}	*micro*	μ
10^{-9}	*nano*	n

- length

$1\,cm = 0.3937\,in$ $\qquad$ $1\,in = 2.54\,cm$

$1\,m = 3.28\,ft$ $\qquad$ $1\,ft = 30.48\,cm$

- area

$1\,m^2 = 10.76\,ft^2$ $\qquad$ $1\,ft^2 = 929\,cm^2$

- volume

$1\,l = 1000\,cm^3 = 61.02\,in^3$ $\qquad$ $1\,in^3 = 16.388\,cm^3$

$1\,m^3 = 35.32\,ft^3$ $\qquad$ $1\,ft^3 = 0.02832\,m^3$

- mass

$1\,g = 0.0022\,p$ $\qquad$ $1\,p = 453.6\,g$

- density

$1\,g/cm^3 = 1\,Mg/m^3 = 62.43\,p/ft^3$ $\qquad$ $1\,p/ft^3 = 0.01602\,Mg/m^3$

- force

$1\,N = 102\,g = 0.2248\,p$ $\qquad$ $1\,p = 4.448\,N$

- pressure

$1\,N/m^2 = 1\,Pa$ $\qquad$ $1\,bar = 100\,kPa$

$1\,kN/m^2 = 20.89\,p/ft^2$ $\qquad$ $1\,p/ft^2 = 0.04787\,kN/m^2$

$1\,kN/m^2 = 0.1450\,psi$ $\qquad$ $1\,psi = 6.895\,kN/m^2$

- angle

$1\,rad = 57.296^\circ$ $\qquad$ $1^\circ = 0.017453\,rad$

- temperature

$^\circ C = 0.555\,^\circ F - 17.778$ $\qquad$ $^\circ F = 1.8\,^\circ C + 32$

Acknowledgements

We are very grateful to the following for granting us permission to use published material for which they are copyright holders: the *American Society of Civil Engineers (figures 8.32, 8.33, 4.24)*, the *Institution of Civil Engineers, London (figures 1.8, 4.35, 11.7)*, the *National Research Council of Canada (figures 4.28, 10.4)*, *Longman Scientific & Technical (figures 7.19, 9.14)*, the *British Research Establishment (figure 8.27)*, *Solétanche-Bachy (figure 9.8)*, *la Revue Française de Géotechnique (figures 5.4, 5.26, 6.32)*, *le Laboratoire Centrale des Ponts et Chaussées (figures 5.27, 5.31, 5.32, 9.12, 11.22, 11.31, 14.18)*, the *Transportation Research Board, Washington DC (figure 2.20)*, *Atlas Palen Franki Geotechnics, Belgium (figure 9.3)*, *Kvaerner Cementation Foundations (figure 9.44)*, *Cambridge Insitu (figures 12.18, 12.19)*, *A. A. Balkema (tables 10.1, 10.2, figure 13.1)*, *John Wiley VCH, Winheim (figure 3.45)*, *Construction Industry Research and Information Association, London (figure 9.15)*, *John Wiley, New York (figures 3.6, 7.20)*.

CHAPTER 1

Formation and physical properties of soils

1.1 Geological aspects of soil formation

Since its formation *circa* 4600 Ma (million years) ago, the planet Earth was (and still is) subjected to continuous geological changes due mainly to volcanic, seismic and climatic activities. To understand the constitution of soils, which is a consequence of these activities, it is of paramount importance to relate it to the geological evolution at the Earth crust (*i.e.* the Earth upper layer).

The basic definition of a soil is that of a material having three components, namely solid particles, air and water. The mechanical behaviour of a soil mass, for instance its response to any externally applied load, is, in large part, governed by the *size* and *proportion* of the *solid particles* it contains as will be explained. However, the question to be addressed at this stage relates to the formation of these solid particles which have originated from *rocks*; thus one needs to explore, first, the formation of rocks.

Rocks comprise one or more minerals. The behaviour of a rock in terms of hardness, nature, and chemical reaction to external agents, depends on the minerals it contains. The most common minerals in rocks are *silicates* (such as feldspar, quartz and muscovite).

There are three main types of rocks:

- ***igneous rocks*** such as granite, dolerite and basalt are formed as a result of magma cooling either on or below the surface of the Earth. Rocks which have resulted from the cooling of magma on the ground surface are called *extrusive*, the basalt blocks of the *Giant's Causeway* in *Northern Ireland* being a well known example. On the other hand, rocks made from the slow cooling of magma within the Earth crust, in other words below the ground surface, are known as *intrusive*, dolerite and granite being typical examples of this type of rocks.

- ***metamorphic rocks*** such as slate, schist and marble are products of existing rocks subjected to changes in temperature and/or pressure leading to changes in mineral composition of the parent rock. In *regional metamorphism,* pressure rather than heat represents the main agent of change, whereas in *thermal metamorphism,* the changes in mineral composition occur in a relatively short time, that is in geological terms, and are mainly generated by an increase in temperature. For example, the process of thermal metamorphism can cause a pure limestone to be re-crystallised as marble. A famous example is provided by some of the mountains in *Tuscany, Italy,* which are virtually solid marble. Slate on the other hand is a mildly metamorphosed shale.

- ***sedimentary rocks***, such as sandstone, shale and limestone, are formed through compaction of rock fragments and organic debris. These fragments are usually cemented through the action of calcite and silica. The *white cliffs of Dover, England*, provide a typical example of soft limestone (also known as chalk).

The geological process of soil formation is based on rock *weathering* which can occur either chemically when the minerals of a rock are altered through a chemical reaction with rainwater, or mechanically through climatic effects such as freeze–thaw cycles and erosion (see for instance Bell, 1993).

Soils thus formed consist of a mixture of solid particles ranging from boulders (rock fragments usually larger than $200\,mm$ in size) to clay minerals (solid particles of less than $2 \times 10^{-3}\,mm$ in size), and are usually classified according to the way they were deposited. Hence, if the present location of the soil is that in which the original weathering of the parent rock occurred, the soil is known as *residual.* If it is not the case, the soil is referred to as *transported.* The process by which the soil is transported from one location to another can be due to gravity, wind, ice or water.

- ***gravity deposits*** result generally from rock weathering at the face of a cliff that causes the rock fragments to fall to the foot of the cliff, thus forming a pile of rock deposit known as *talus.* The final position of each fragment is entirely due to gravitational forces and no other transporting agent is involved in this process depicted in figure 1.1.

-***wind deposits*** are best illustrated by sand dunes which are shaped according to the size of sand particles and the velocity of the wind.

- most of the ***glacial deposits*** were formed throughout the glacial age which, in terms of geological timescale, corresponds to the Pleistocene epoch (between 1.5 and 2 million years ago). From a mechanical viewpoint, the flow of large ice blocks is such that it can erode soils and crush rocks. The debris thus transported, known as glacial drift, is formed of soil particles and rock fragments of various sizes and shapes and can be buried at the bottom of glaciers, forming in the process what is known as *glacial tills* which can contain large boulders.

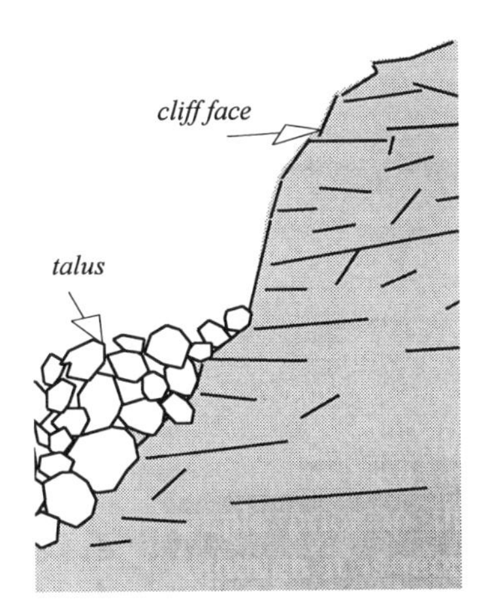

Figure 1.1: Rock weathering and gravity deposits.

- ***alluvial deposits*** are formed through a process of erosion during which the solid particles are transported by water and deposited as soon as the water velocity becomes minimal. During the process, a natural selection occurs in that the heavier grains will be deposited first and the lighter ones will travel further downstream and be deposited through sedimentation. The soil thus formed contains a layered structure.

1.2 Physical properties of soils

On the basis of the grain sizes of different soils given in table 1.1 (notice the logarithmic scale), one would expect, for instance, a clay with solid particles of a size smaller than 2×10^{-3} *mm* to be much more compressible than a sand; similarly, a gravel would be much more permeable than a silt. These logical conclusions are closely related to the soil composition, whose behaviour is dependent on the type and size of solid particles and their volumetric proportion with respect to water and air, as well as on the soil stress history (*i.e.* the way in which it was deposited and its subsequent loading-unloading cycles).

Table 1.1: Grain size corresponding to different types of soils

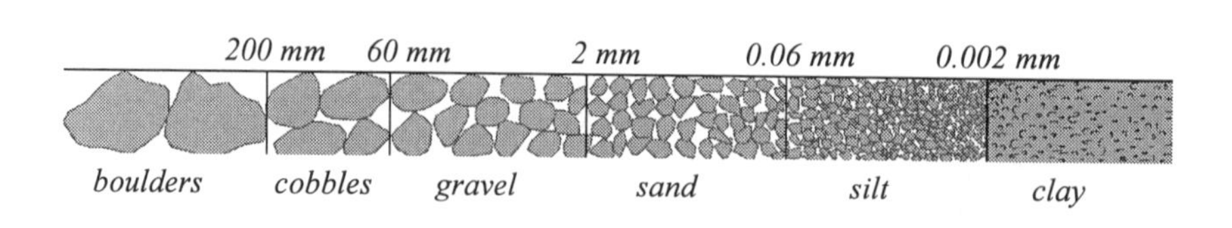

Consider the volume of soil depicted in figure 1.2*a*. Assume that a microscopic analysis revealed that water is not filling all the voids between solid particles and that some pockets of air exist within the soil matrix. Now imagine that the volume V of this *unsaturated soil* is rearranged in a (theoretical) way such that all solid particles are squeezed together so that no gap is left for water or air to fill (figure 1.2*b*). The remainder of the volume is therefore constituted of water and air, hence :

$$V = V_s + V_w + V_a \tag{1.1}$$

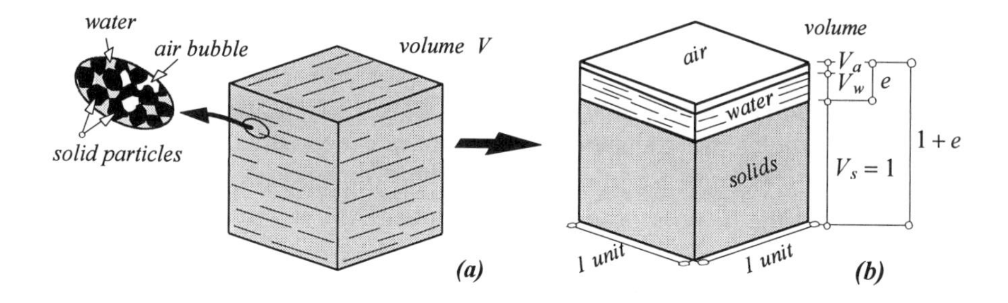

Figure 1.2: (a) Elementary volume of soil, (b) soil composition.

The volume V_s in the above equation is that of solid particles and the quantity $V_w + V_a$ corresponds to the *volume of voids* V_v. Let the total volume of soil V be selected so that V_s corresponds to one unit. While this astute choice does not affect in any way the general aspect of the following formulation, it makes it easier to handle and to establish. Thus, with reference to figure 1.2*b*, the following basic soil properties can now be defined in a straightforward way:

- *the void ratio*:

$$e = \frac{V_v}{V_s} \tag{1.2}$$

- *the specific volume*:

$$\begin{aligned} V &= V_s + V_v \\ &= 1 + e \end{aligned} \tag{1.3}$$

- *the degree of saturation*:

$$S = \frac{V_w}{V_v} \tag{1.4}$$

- *the water content*:

$$w = \frac{M_w}{M_s} \tag{1.5a}$$

$$w = \frac{V_w}{V_s}\frac{1}{G_s} \tag{1.5b}$$

$$G_s = \frac{\rho_s}{\rho_w} \tag{1.6}$$

The dimensionless parameter G_s in equation 1.6 refers to the *specific gravity* and represents the relative density of solid particles with respect to water. Typical values of G_s for soils are in the range 2.65 (sands) to 2.75 (clays). M_s and M_w correspond to the mass of solids and that of water respectively, and ρ_w refers to the density of water $\left[= 1 Mg/m^3\right]$.

Equation 1.4 indicates that a *fully saturated soil* has a volume of water identical to that of voids because no air is contained within the soil matrix, in which case the degree of saturation is $S = 100\%$. On the other hand, a totally dry soil contains no water and therefore its degree of saturation according to equation 1.4, is $S = 0\%$.

In practice, the water content is calculated from equation 1.5*a* in which the quantities M_w and M_s are measured in the laboratory in a very simple way.

For a given volume of soil, the operator needs to determine the mass of soil in its natural state M_t, then its mass M_s after being thoroughly dried in an oven at $105°C$ for long enough (usually 24 h) to ensure that all but the chemically bonded water has evaporated. The mass of water M_w is therefore $M_t - M_s$.

Also, it is easy to show that, by substituting for V_v and V_s from equations 1.4 and 1.5*b* respectively, into equation 1.2, the following expression for water content can be established:

$$w = \frac{e\,S}{G_s} \tag{1.7}$$

so that for a fully saturated soil (*i.e.* $S = 100\%$), the latter equation reduces to $e = wG_s$. However, for a totally dry soil, equation 1.7 cannot be used to calculate the void ratio since, in this case, both w and S are zero. *Typical values* for the void ratio range from 0.5 to 0.8 for sands and between 0.7 and 1.3 for clays.

Another useful property related to the void ratio and known as *porosity* is defined as the proportion of the volume of voids (that is air *and* water) with respect to the total volume (refer to figure 1.2*b*):

$$n = \frac{V_v}{V} = \frac{e}{1+e} \tag{1.8}$$

The air content in a soil mass corresponds to the ratio of the volume of air to the total volume and, referring to figure 1.2*b,* it can be seen that:

$$A = \frac{V_a}{V} = \frac{V_v - V_w}{V} = \frac{V_v}{V}\left[1 - \frac{V_w}{V_v}\right] = \frac{e}{1+e}(1-S) \tag{1.9a}$$

Substituting for S from equation 1.7, it follows that:

$$A = \frac{e - wG_s}{1+e} \tag{1.9b}$$

A relationship between the degree of saturation S of a soil and its air content A can easily be found were the void ratio in the above equation to be replaced by its value from equation 1.7, in which case:

$$S = \frac{1 - A}{1 + A/wG_s} \tag{1.10}$$

Equation 1.10 is markedly non-linear when the air content is larger than zero.

The soil *density* is defined as the ratio of the soil mass to its volume. Accordingly, the following relationships which will be most useful for soil compaction calculations (see section 1.6) are easily established using equations 1.2 to 1.6:

- *the bulk density:*

$$\rho = \frac{M_s + M_w}{V} = \frac{G_s(1 + w)}{1 + e}\rho_w \tag{1.11}$$

- *the dry density* (*zero water content:* $w = 0$):

$$\rho_d = \frac{G_s}{1 + e}\rho_w \tag{1.12}$$

- *the saturated density* ($w = e/G_s$, equation 1.7 with $S = 100\%$):

$$\rho_{sat} = \frac{G_s + e}{1 + e}\rho_w \tag{1.13}$$

A combination of equations 1.11 and 1.12 yields the relationship between the bulk and dry densities of a soil:

$$\rho_d = \rho/(1 + w) \tag{1.14}$$

Furthermore, it is most helpful during a compaction process (which will be explained shortly) to relate the dry density to the air content of the soil. Thus, rearranging equation 1.1:

$$\frac{V_s}{V} + \frac{V_w}{V} + \frac{V_a}{V} = 1 \tag{1.15}$$

Making use of equations 1.3 and 1.12, it follows that:

$$\frac{V_s}{V} = \frac{1}{1+e} = \frac{\rho_d}{\rho_w}\frac{1}{G_s}$$

On the other hand, both equations 1.2 and 1.4 can be exploited:

$$\frac{V_w}{V} = \frac{V_w}{V_v(1 + V_s/V_v)} = S\frac{1}{1 + 1/e}$$

Hence, substituting for the degree of saturation S from equation 1.7 and rearranging:

$$\frac{V_w}{V} = \frac{wG_s}{1+e} = \frac{w\rho_d}{\rho_w}$$

The last ratio in equation 1.15 corresponds to the air content (equation 1.9*a*). From whence, a straightforward substitution for the different ratios into equation 1.15 yields:

$$\frac{\rho_d}{\rho_w}\frac{1}{G_s} + w\frac{\rho_d}{\rho_w} + A = 1$$

or

$$\rho_d = \frac{G_s(1-A)}{1+wG_s}\rho_w \qquad (1.16)$$

As will be seen later, the calculation of stresses are undertaken using the appropriate soil *unit weight* instead of the density, the two being related through the acceleration due to gravity $g = 9.81\,m/s^2$. Accordingly, the *bulk unit weight* of a soil is established from equation 1.11:

$$\gamma = \rho g = \frac{G_s(1+w)}{1+e}\gamma_w \qquad (1.17)$$

γ_w being the unit weight of water $\left[= 9.81\,kN/m^3\right]$.

Both the *dry* and the *saturated unit weights* are obtained in a similar way from equations 1.12 and 1.13 respectively:

$$\gamma_d = \frac{G_s}{1+e}\gamma_w \qquad (1.18)$$

$$\gamma_{sat} = \frac{G_s + e}{1 + e}\gamma_w \tag{1.19}$$

When a soil is fully saturated (or submerged), its solid particles are subjected to a buoyancy due to the water in accordance with *Archimedes'* principle. In that case, the *effective unit weight* of the soil corresponds to the difference between its saturated unit weight and the unit weight of water:

$$\gamma' = \gamma_{sat} - \gamma_w = \frac{G_s - 1}{1 + e}\gamma_w \tag{1.20}$$

Example 1.1

A sample of sand occupying a total volume $V = 1000\,cm^3$ has a total mass $M = 1960\,g$. Once dried in the oven, the mass of the sample was reduced to $M_s = 1710\,g$. Assuming the specific gravity of the solid particles is $G_s = 2.65$, let us evaluate the different following quantities:

- *the water content:* using equation 1.5*a*, it is seen that:

$$w = \frac{M_w}{M_s} = \frac{1960 - 1710}{1710} = 14.6\,\%$$

- *the void ratio:* equation 1.2 yields $e = \dfrac{V_v}{V_s}$, where the volume of solids is calculated from equation 1.6:

$$V_s = \frac{M_s}{G_s \rho_w} = \frac{1710}{2.65 \times 1} = 645\,cm^3$$

Hence the volume of voids: $V_v = V - V_s = 1000 - 645 = 355\,cm^3$,

and therefore the void ratio:

$$e = \frac{355}{645} = 0.55$$

- *the degree of saturation:* equation 1.4 is now used:

$$S = \frac{V_w}{V_v} = \frac{1960 - 1710}{355} = 70.4\%$$

- *the bulk density:* calculated from equation 1.11:

$$\rho = \frac{M}{V} = \frac{1960}{1000} = 1.96\, Mg/m^3$$

- *the air content:* found from equation 1.9*a*:

$$A = \frac{V_a}{V} = \frac{V_v - V_w}{V} = \frac{355 - 250}{1000} = 10.5\,\%$$

If the sample of sand were saturated, then obviously the air content would be reduced to zero, implying that the volume of voids is equal to that of water: $V_v = V_w$. Accordingly, the mass of water becomes:

$$M_w = V_v \rho_w = 355 \times 1 = 355\, g$$

thus yielding a water content (for the saturated sand) of:

$$w = \frac{355}{1710} = 20.8\,\%$$

The void ratio is defined as the ratio of the volume of voids to that of solids, both of which are unchanged, and hence, the void ratio remains constant. Knowing that the degree of saturation is in this case $S = 100\%$, equation 1.7 then yields:

$$e = wG_s = 0.208 \times 2.65 = 0.55$$

which is the same value as the one calculated previously. The density of the sand, however, increases to reflect the saturation and is calculated from equation 1.11 in which the total mass is now:

$$M = M_s + M_w = 1710 + 355 = 2065\, g$$

Therefore:

$$\rho_{sat} = \frac{2065}{1000} = 2.065\, Mg/m^3$$

Example 1.2

A sample of compacted clay with a total volume $V = 7.85 \times 10^{-4} m^3$ has a moisture content $w = 15\%$, an air content $A = 8\%$, and a specific gravity $G_s = 2.7$. Required are: the degree of saturation, void ratio, porosity, and bulk, dry and saturated densities.

The volume of air is 8% of the total volume and thus :

$$V_a = 0.08 \times 7.85 \times 10^{-4} = 0.628 \times 10^{-4} m^3$$

Moreover, the volume of water as a proportion of the volume of solids can easily be calculated from equation 1.5*b*:

$$V_w = wG_s V_s = 0.15 \times 2.7 V_s = 0.405 Vs$$

Since the total volume is $V = V_a + V_w + V_s$, it follows that:

$$V = 0.08V + 1.405 V_s$$

Hence:

$$V_s = \frac{0.92}{1.405} V = 5.14 \times 10^{-4} m^3$$

and

$$V_w = 0.405 V_s = 2.08 \times 10^{-4} m^3$$

The degree of saturation is evaluated from equation 1.4:

$$S = \frac{V_w}{V_a + V_w} = \frac{2.08}{0.628 + 2.08} = 77\%$$

the void ratio and the porosity being calculated from equations 1.7 and 1.8 respectively:

$$e = \frac{wG_s}{S} = \frac{0.15 \times 2.7}{0.77} = 0.53, \qquad n = \frac{e}{1+e} \approx 0.35$$

Finally the bulk, dry and saturated densities are determined from equations 1.11, 1.12 and 1.13 respectively, in which the density of water is taken as $\rho_w = 1 Mg/m^3$:

$$\rho = G_s \frac{(1+w)}{1+e} \rho_w = \frac{2.7 \times 1.15}{1.53} = 2.03\, Mg/m^3$$

$$\rho_d = \frac{G_s}{1+e} \rho_w = \frac{2.7}{1.53} = 1.77\, Mg/m^3$$

$$\rho_{sat} = \frac{G_s + e}{1+e} \rho_w = \frac{2.7 + 0.53}{1.53} = 2.11\, Mg/m^3$$

1.3 Particle size analysis

The mechanical behaviour of a given soil depends on the size of its solid particles, on the minerals it contains, as well as on its stress history. In particular, the resistance of a soil to any applied load or the ease with which water can flow through it is governed by its *granulometry* (*i.e.* the range of solid particles it contains). Soils can be classified either as *granular* or as *cohesive*.

Granular soils such as sands and gravels have individual solid particles, most of which can be identified by sight. The strength of such materials results from the interlocking of solid particles which provides resistance through friction. Accordingly, the range of particle size present within the soil matrix and the way these particles interact are a key element to predicting the soil behaviour. The grain size distribution can be determined with the help of a technique known as *sieving*.

This old technique, which applies to granular soils (*i.e.* soils that do not contain silt or clay), has the advantage of being simple and cheap. During the standard procedure (a detailed description of which can be found in any standard laboratory testing manual), the soil is sifted through progressively finer woven-wire sieves, down to a mesh size of 31 μm. The percentage *by weight* of material passing through each sieve is then plotted on a chart representing the particle size distribution such as the one depicted in figure 1.3.

The shape of the curve thus obtained gives an indication of the distribution by weight of different sizes of solid particles within the soil. However, the graph in question does not correspond to the *true* weight distribution since the material retained in any sieve yields the weight of solid particles with an individual size exceeding that of the sieve mesh, including those (elongated) particles with only one dimension larger than the mesh size. Accordingly, the outcome of such a test depends on the time and method of operation: the longer it takes to undertake the test, the more likely that some (or in some instances all) elongated solid particles will fall through the sieve because of a change in orientation.

Assuming that these shortcomings are acceptable, then the two typical curves in figure 1.3 are indicative of different soil grading: the soil corresponding to curve *A* is referred to as *well graded* because the graph reflects relatively similar proportions of different particle sizes. For curve *B*, however, although there is no disproportion between the solid particles' size distribution, the graph is more compact than in case *A*, and the corresponding soil is known as *uniformly graded.*

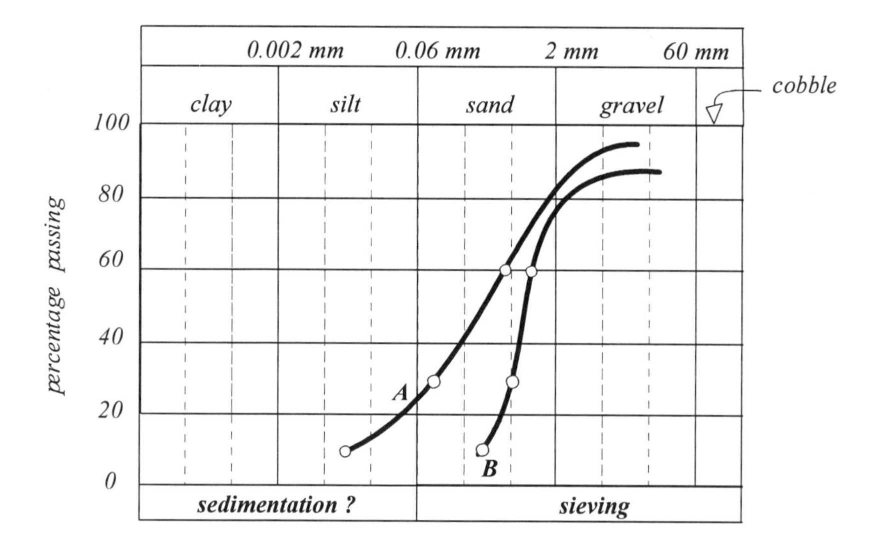

Figure 1.3: Particle size distribution for granular soils.

Moreover, if d_{10}, d_{30}, and d_{60} are the solid particle sizes corresponding to 10%, 30% and 60% respectively of percentage passing, then it is useful to calculate the two following coefficients:

- *the uniformity coefficient*: $$C_u = \frac{d_{60}}{d_{10}} \qquad (1.21)$$

- *the coefficient of gradation*: $$C_g = \frac{d_{30}^2}{d_{10}d_{60}} \quad (1.22)$$

The higher the value of C_u, the larger the range of particle sizes contained within the soil matrix. Also, the coefficient of gradation of a well graded sand is usually in the range $1 \leq C_g \leq 3$.

Example 1.3

Calculate the coefficients of uniformity and gradation of the soils corresponding to graphs A and B in figure 1.3.

For soil A, it is seen that: $d_{10} \approx 0.01\,mm$, $d_{30} \approx 0.09\,mm$, $d_{60} \approx 0.55\,mm$.
Hence: $C_u = \frac{0.55}{0.01} = 55$, $\quad C_g = \frac{0.09^2}{0.55 \times 0.01} = 1.47$.

For soil B: $\quad d_{10} \approx 0.35\,mm$, $d_{30} \approx 0.6\,mm$, $d_{60} \approx 1.1\,mm$, and

$$C_u = \frac{1.1}{0.35} = 3.1, \qquad C_g = \frac{0.6^2}{1.1 \times 0.35} = 0.94.$$

Soil A contains a wide range of particle sizes, from gravel to silt (hence the large value of C_u), none of which is predominant. This is reflected in the value of the coefficient of gradation C_g. Soil B on the other hand has a uniform grading indicated by a grading coefficient value of slightly smaller than one.

Cohesive soils have solid particles with a size smaller than 0.063 *mm*, making the sieving technique impractical. Instead, the grain size distribution of such soils can be determined using other techniques such as *sedimentation.* This traditional method, which is only (realistically) applicable to grain sizes in the range 2 to 50 μm, is based on Stokes' law relating the terminal velocity v of a solid particle with an equivalent sphere diameter D_s, falling in water of dynamic viscosity η, to its weight (read diameter D_s):

$$v = \frac{(\rho_s - \rho_w)}{18\eta} g D_s^2 \quad (1.23)$$

where g is the acceleration due to gravity, ρ_s and ρ_w are the densities of solid particles and of water respectively, and $\eta = 1.005 \times 10^{-6}\, Mg/m.s$ is the water dynamic viscosity at a standard temperature of $20^\circ C$.

Equation 1.3 can therefore be used in a straightforward way to calculate the settling time.

Example 1.4

Estimate the time t that will take a solid particle with an equivalent diameter $D_s = 1\, \mu m$ and a density $\rho_s = 2.5\, Mg/m^3$ to settle a distance $h = 1\, cm$ in water at $20^\circ C$.

Rearranging equation 1.23, it follows that:

$$v = \frac{h}{t} = \frac{(\rho_s - \rho_w)}{18\eta} g.D_s^2 \quad \Rightarrow \quad t = \frac{18\eta h}{(\rho_s - \rho_w) g D_s^2}$$

therefore:

$$t = \frac{18 \times 10^{-2}}{(2.5 - 1) \times 10^6 \times 9.81 \times 10^{-12}} \approx 12{,}200\, s$$

The time needed for this particle to settle a mere 1 cm (about 3.4 $hours$) is an indication that sedimentation is an exceedingly slow process.

Knowing that Stokes' law uses an equivalent diameter D_s which assumes that the solid particle has a regular compact spherical shape, one therefore expects an irregularly shaped normal particle to have a larger surface area than its equivalent sphere, thus causing the settling to be even slower because of the increased drag. Consequently, the time calculated using equation 1.23 is likely to deviate substantially from the actual settling time. Furthermore, theoretical and experimental evidence indicate that for particles smaller than $2\, \mu m$ in size, the gravitational settling calculated from Stokes' law is markedly affected by the Brownian movement (*i.e.* the irregular oscillations of particles suspended in water) which can induce an error of more than 20%. On the other hand, Stokes' law is no longer applicable for particles larger than $50\, \mu m$ in size, the settling being turbulent. Therefore the range 2 to $50\, \mu m$ mentioned earlier within which equation 1.23 can be applied under a strict temperature control since the water viscosity η is a temperature dependent parameter (it is useful to remember that a $1°C$ change in temperature will induce a 2% change in water viscosity).

To offset these shortcomings, other sophisticated techniques have been developed and are slowly being adopted in soil mechanics laboratories, although their use was restricted until recently to the fields of clay mineralogy and sedimentology. These techniques include *photon correlation specstroscopy* which can be applied to measure particle sizes in the range $1\,nm$ to $1\,\mu m$ (that is $10^{-9}\,m$ to $10^{-6}\,m$). Because of their small size, these particles, known as *colloids,* are subjected mainly to a Brownian motion that scatters light. The principle of the method consists of relating the diffusion of particles to the autocorrelation function of the scattered light, and more details can be found in MacCave and Syvitski (1991) and Weiner (1984).

Laser diffraction spectroscopy is another reliable technique that can be applied to measure particle sizes in the range $0.1\,\mu m$ to $2\,mm$. The method is based on the fact that the light diffraction angle is inversely proportional to particle size. The technique itself consists of passing a laser beam through a suspension and focusing the diffracted light on to a ring detector which senses the angular distribution of scattered light intensity. This distribution is then related to the size distribution of the suspension through an appropriate mathematical expression, the details of which can be found in Weiner (1979) and Agrawal and Riley (1984).

1.4 Design of soil filters

Soil filters are used to drain water seeping out of a given soil surface in a controlled way so as to prevent the erosion of the natural soil. Examples include seepage through an earth dam, behind a retaining wall or around a pumping well. To fulfil these requirements, the grain size distribution of the filter material, relative to that of the soil it supports, must include enough large particles to ensure free drainage of water, and an adequate proportion of small particles to preclude the migration of the natural soil's finer material. Also, to prevent any gaps in the grading of the filter material, it is suggested that the shape of its grain size distribution curve should be similar to that of the natural soil against which the filter is applied.

Extensive experimental studies led to the establishment of empirical rules satisfying the previously mentioned requirements. In what follows, the design rules suggested by the *Hong Kong Geotechnical Engineering Office* (1993) are adopted, other rules suggested by different authors being marginally different. Note that subscripts f and s refer to the *filter* and to the natural *soil* respectively.

1- to prevent the migration of the natural soil's fine particles through the filter:

$$d_{15f} \leq 5 \times d_{85s}$$

2- to ensure that the filter is more permeable than the soil:

$$d_{15f} \geq 5 \times d_{15s}$$

3- to ensure a good performance of the filter:

$4 \leq d_{60f}/d_{10f} \leq 20$ and

$d_{\max f} \leq 50\,mm$, ($d_{\max f}$ being the maximum particle size of the filter material)

These design criteria can be extended to include the suggestions by Somerville (1986):

4- to ensure an adequate drainage of water:

$$d_{5f} \geq 0.0750\,mm$$

5- to prevent any segregation of the filter material:

$$d_{50f} \leq 25 \times d_{50s}$$

6- were the filter to be placed against a screen mesh (the case of a well for instance),

d_{85f} should not be less than *twice* the mesh size,

7- the grading curve of the natural soil should be limited to a maximum particle size:

$$d_{\max s} \leq 19\,mm$$

Example 1.5

Consider the case of a well whose lining has maximum mesh size of 0.75 *mm*. Assuming that the soil in which the well is dug has a grading represented by curve *A* in figure 1.4, it is seen that:

$$d_{15s} = 0.07\,mm, \quad d_{50s} = 0.15\,mm, \quad d_{85s} = 0.5\,mm, \quad d_{\max s} = 2\,mm.$$

Applying the above rules to design the soil filter to be placed around the lining, the following limiting values are easily computed:

1. $d_{15f} = 5 \times d_{85s} = 5 \times 0.5 = 2.5\,mm,$
2. $d_{15f} = 5d_{15s} = 5 \times 0.07 = 0.35\,mm,$
3. $d_{\max f} = 50\,mm,$
4. $d_{5f} = 0.075\,mm,$
5. $d_{50f} = 25 \times d_{50s} = 25 \times 0.15 = 3.75\,mm,$
6. $d_{85f} = 2 \times \text{mesh size} = 2 \times 0.75 = 1.5\,mm,$
7. $d_{\max s} = 19\,mm.$

These values are then plotted in the same figure yielding the shaded area within which should lie the grain size distribution curve of the filter material.

Although any graph within the shaded area meets the filter design requirements, the one shown in bold represents a good compromise because: (*a*) it has no grading gaps, (*b*) its shape is somewhat similar to curve *A*, and more importantly (*c*) it meets the design requirement represented by rule 3. In fact, taking the bold curve as that of the filter material, it follows that:

$$\frac{d_{60f}}{d_{10f}} = \frac{2.5}{0.5} = 5$$

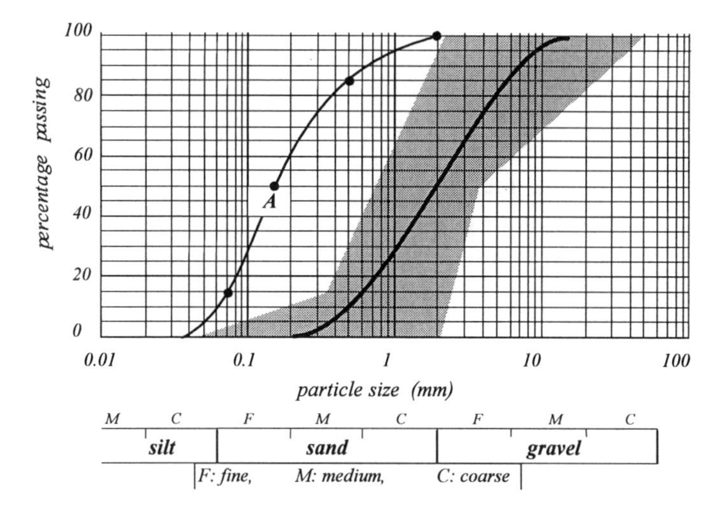

Figure 1.4: Grain size distribution graphs for natural soil and filter material.

1.5 Classification of cohesive soils

Given the small size of its solid particles, a cohesive soil is characterised by a large *specific surface* (*i.e.* the surface area per volume of solid particles) and extremely small pores in comparison with sand. In fact, taking the sand pores as a reference, the clay pores in figure 1.5 (depicting the arrangement of perfectly spherical solid particles for identical volumes of sand and clay) are in reality at least five times smaller than the ones shown in the figure.

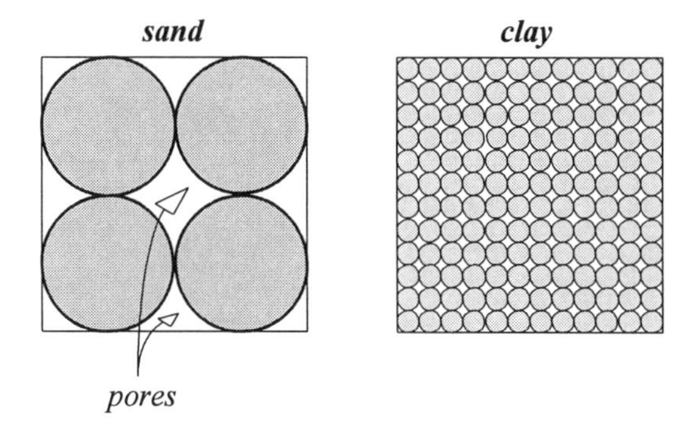

Figure 1.5: Theoretical arrangement of solid particles for sands and clays.

Accordingly, clays have a high *surface tension* which results in a high *capillary rise* as will be explained in details in section 2.2; most importantly, the velocity with which water can seep through the pores, known as *permeability* is much lower for clays than for sands.

Furthermore, the behaviour of a clay can be markedly affected by the types of mineral that it contains and their reaction to porewater. The three main groups of clay minerals are *kaolinite, illite* and *montmorillonite.* Of these three, kaolinite is the most stable *vis à vis* water so that practically no volume change occurs if water content changes. On the other hand, illite has a moderate reaction to water in that the changes in volume due to a variation in water content can result in modest swelling or shrinkage. However, montmorillonite is well known for its swelling and shrinkage properties so that a clay containing this mineral is bound to be subject to large volume changes in the event of water content variation.

The water content of cohesive soils has therefore an effect on their mechanical behaviour, so much so that the empirical system of classification, based exclusively on the moisture content and known as the consistency limits (sometimes referred to as *Atterberg* limits) is widely used to determine the type of cohesive soil. Hence, the *liquid limit* w_L corresponds to the moisture content of the soil as it changes from a plastic state to a slurry type material. The *plastic limit* w_P on the other hand is the water content when the soil changes from plastic to friable. The quantity w_S in figure 1.6 represents the *shrinkage limit* which is the moisture content at which the soil volume starts to decrease.

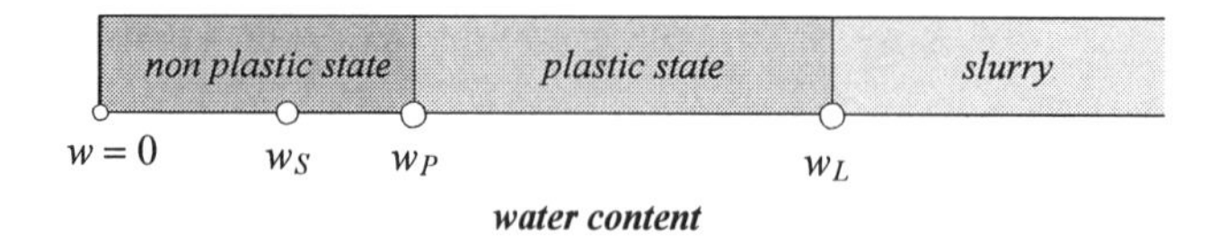

Figure 1.6 : Consistency limits for cohesive soils.

Notice that these limits are measured using a sample of soil whose fabric has been totally destroyed.

Because the liquid limit is proportional to the specific surface, it follows that the greater the liquid limit, the higher the clay content, and the more compressible the soil.

These limits are most useful when applied in conjunction with the plasticity chart (due to *Casagrande*) represented in terms of variation of the liquid limit of the soil versus its *plasticity index* defined as:

$$I_P = w_L - w_P \tag{1.24}$$

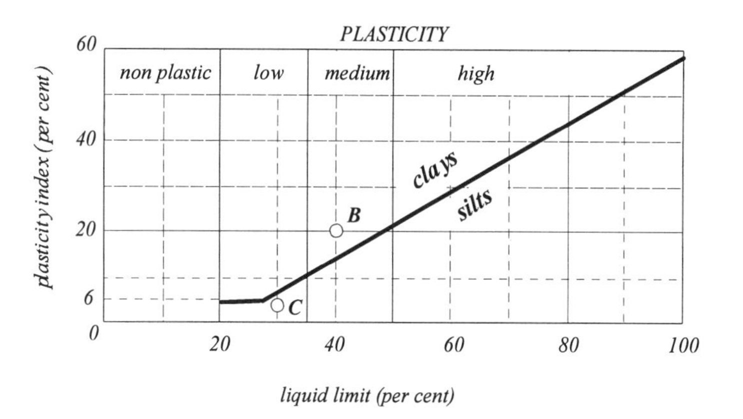

Figure 1.7: Casagrande empirical plasticity charts for soils.

The chart, shown in figure 1.7, provides a quick and accurate way of classification according to the moisture content of the soil and its plasticity index. The skewed line in the figure separates clays from silts so that, for instance, a soil with a liquid limit $w_L = 40\%$, a plastic limit $w_P = 20\%$ corresponds to a *clay of intermediate plasticity* (point B on the chart); similarly, a soil with $w_L = 30\%$, $w_P = 4\%$ is a *silt of low plasticity* (point C on the chart).

Highly plastic silts with liquid limits $w_L > 50\%$ are *organic soils* whose natural moisture content at a saturated state can be as high as 1000% (bog peat for instance) corresponding therefore to void ratios in excess of 10.

Another useful relationship between the natural *in situ* moisture content w of a soil and its plasticity index I_P is given by the *liquidity index* I_L:

$$I_L = \frac{w - w_P}{I_P} = \frac{w - w_P}{w_L - w_P} \tag{1.25}$$

It is clear from equation 1.25 that:

$I_L < 0 \quad \Rightarrow$ soil is in a non-plastic state,
$0 \le I_L \le 1 \quad \Rightarrow$ soil is in a plastic state,
$I_L > 1 \quad \Rightarrow$ soil is in a liquid state.

Finally, the *activity* of a soil, defined as:

$$A = \frac{I_P}{\%\, clay\ particles\ (< 2\,\mu m)} \tag{1.26}$$

reflects the *degree of plasticity* of the clay content of the soil. Typical A values for the three main clay minerals are as follows:

- kaolinite $\quad A = 0.5$
- illite $\quad 0.5 \le A \le 1$
- montmorillonite $\quad A > 1.25$.

From a mechanical point of view, undisturbed clays with low plasticity are characterised by small deformations when subjected to external loading. Once disturbed, a clay can potentially exhibit a marked decrease in its resistance or *shear strength,* depending on its structure. Microscopic studies of such soils, though not conclusive, tend to link the structure of natural clays to the way in which they were formed; thus a glacial till does not have the same structure as a lacustrine clay. Accordingly, once disturbed or *remoulded*, through excavation for instance, part of the natural structure will be destroyed. The effect that this structural dislocation has on the resistance of the clay can be measured by the *sensitivity*, defined as:

$$Sensitivity = \frac{strength\ of\ undisturbed\ soil}{strength\ of\ remoulded\ soil}$$

Hence, clays with a sensitivity larger than 16 are referred to as *quick clays*, and extra-sensitive clays are known to have a sensitivity in excess of 100. The structure in this case is liable to total collapse, leading to a transformation of the clay from a plastic material to a viscous liquid almost instantaneously. Quick clays can be activated by any type of shock such as vibration or earthquake activities, and are spread especially throughout Scandinavia and Canada. Bjerrum's (1954) empirical scale can be used as a guide to clay sensitivity:

sensitivity	*<2*	*2 – 8*	*8 – 16*	*>16*
clays	*insensitive*	*sensitive*	*highly sensitive*	*quick*

Bjerrum suggested that the sensitivity of the post-glacial Scandinavian clays (*i.e.* 10,000 to 15,000 year-old clays) is related to the salt content of their pores. This suggestion is supported by the graph in figure 1.8 showing a dramatic increase in sensitivity when the salt content becomes smaller than 10 *g/l*. One has to bear in mind, however, that Bjerrum's suggestion is not universal; in fact Sangrey (1972) reported that some of the highly sensitive post-glacial Canadian clays were deposited in fresh water.

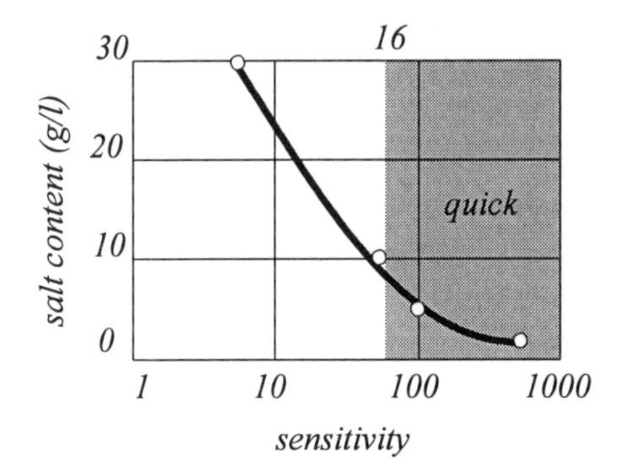

Figure 1.8: Relationship between sensitivity and salt content for Norwegian post-glacial clays (Bjerrum, 1954, by permission of the Institution of Civil Engineers, London).

1.6 Soil compaction

The process of compaction applies to *remoulded unsaturated soils*, and consists of increasing mechanically the *density* by reducing the volume of air contained within the soil matrix, without any significant change in the moisture content. This process is generally used in conjunction with fill materials behind retaining structures or during the construction of embankments and earth dams. The increase in density, generated by a reduction in the void ratio, results in a substantial increase in *shear strength* of the soil and a marked decrease in its *compressibility* as well as *permeability*. The compaction is usually measured in terms of dry density ρ_d, whose value depends on the level of compacting energy, the soil type and its natural moisture content. Accordingly, the maximum dry density $\rho_{d\max}$ that can be achieved for a given soil increases with increasing level of compacting energy.

The maximum dry density is measured in the laboratory using compaction tests. The *standard* test is undertaken on a completely remoulded soil sample with a moisture content well below its natural value (often the sample is dried prior to testing). The test itself is conducted in stages, the first of which consists of mixing the dry soil sample with a small amount of water so that it becomes damp. An extended mould is then filled with moist soil in three equal layers, each one being compacted using 25 evenly distributed blows from a rammer weighing 2.5 *kg* and falling freely from a height of 305 *mm*. Once compaction is completed, the mould extension is then removed and the soil at the top of the mould is levelled off so that the remaining volume of compacted soil is precisely $1000\,cm^3$ (1 *litre*). The bulk density ρ of the soil is then calculated by weighing the known volume of the sample. A small quantity of soil is then taken randomly from within the mould and placed in an oven for drying so that the moisture content w can be calculated. Knowing ρ and w, the corresponding dry density ρ_d is thence determined from equation 1.14:

$$\rho_d = \rho/(1 + w)$$

The sample is next removed from the mould, mixed with the remainder of the original soil to which an increment of water is added to increase its moisture content. The compaction procedure is then repeated in precisely the same way as described previously. Usually four to five points are enough to yield a compaction graph such as the ones depicted in figure 1.9, corresponding to a clay with a specific gravity $G_s = 2.7$.

The figure represents the variation of moisture content with the dry density at different energy levels:

- graph *EFG* corresponds to a standard compaction energy,
- graph *BCD* represents a higher compaction energy.

In both cases, the *maximum dry density* $\rho_{d\max}$ is achieved at a water content known as the *optimum moisture content* w_{op}, and in the case of figure 1.9, it can be seen that:

- along EFG : $\rho_{d\max} \approx 1.72\,Mg/m^3$, $\quad w_{op} \approx 15.7\,\%$,
- along BCD : $\rho_{d\max} \approx 1.845\,Mg/m^3$, $\quad w_{op} \approx 13\,\%$,

where $\rho_w = 1\,Mg/m^3$.

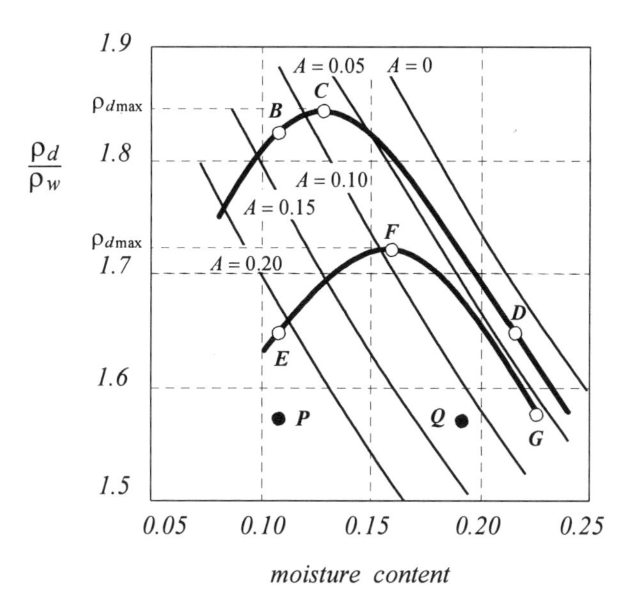

Figure 1.9: Effects of the energy of compaction on the relative dry density of a soil.

Both graphs, which are similar in shape, indicate that initially the test yields a comparatively small dry density, because at a low moisture levels the relatively dry soil tends to form in lumps that have to be crushed before any significant reduction in void ratio takes place. As the moisture content increases, so do the soil workability and energy efficiency, leading thus to a gradual decrease in void ratio and a steady increase in dry density until $\rho_{d\max}$ is reached at the optimum moisture content w_{op}. Beyond the value w_{op}, the build-up of porewater pressure within the soil matrix starts in earnest, increasing in the process the void ratio, thus decreasing the dry density of the soil.

On the other hand, the dry density equation can be expressed as follows:

$$\rho_{d\max} = \rho/(1 + w_{op}) \tag{1.27}$$

and so figure 1.9, together with equation 1.27 show clearly that the higher the compaction energy, the higher the maximum dry density, and the lower the optimum moisture content.

Example 1.6

The graphs in figure 1.9, corresponding to laboratory measurements, contain valuable information for a site engineer in charge of building an embankment, for instance, using the same material. Under such circumstances, the engineer has to make a decision as to what maximum dry density (and hence what compaction energy level) is required to minimise post-construction problems related to settlement and shear strength. If, for example, the material used corresponds to point P on the figure with a natural moisture content $w = 11\%$, the choice consists of either increasing the moisture content of the soil by about 5% and using a standard compaction energy to achieve a maximum dry density similar to that of the graph EFG, or increasing the moisture content by 3% and using a high compaction energy leading to a maximum dry density comparable to that of the curve BCD.

If, on the other hand, the material used has a natural moisture content $w = 19\%$ (point Q on the figure), the decision is more straightforward since either level of compaction energy used on site would lead to similar values of $\rho_{d\max}$, the difference being marginal.

The graphs corresponding to different air contents in the figure are calculated from equation 1.16:

$$\rho_d = \frac{G_s(1-A)}{1+wG_s}\rho_w$$

On the basis of these graphs, the air content relating to both points *C* and *F* where both optimum moisture contents occur can be estimated, then the corresponding degrees of saturation can be determined from equation 1.10:

$$S = \frac{1-A}{1+A/wG_s}$$

According to figure 1.9, point *C* with an optimum moisture content $w_{op} = 13\%$ corresponds to an air content $A \approx 8\%$. Similarly in the case of curve *EFG*, the air content corresponding to point *F* with an optimum moisture content $w = 15.7\%$ is $A \approx 9\%$. Therefore the degrees of saturation at *C* and at *F* are respectively:

$$S_C = \frac{1-A}{1+A/w_{op}G_s} = \frac{1-0.08}{1+\frac{0.08}{0.13\times 2.7}} = 75\%$$

$$S_F = \frac{1-0.09}{1+\frac{0.09}{0.157\times 2.7}} = 75\%$$

These simple calculations indicate that the optimum moisture content corresponds approximately to a constant degree of saturation, usually between 75% and 80%, regardless of the level of energy used for compaction.

1.7 Practical aspects of soil compaction

Because of the random shape of solid particles, it is not physically possible to remove all air content from within a volume of an *unsaturated soil*. In other words, it is not possible to achieve full saturation through compaction. Accordingly, the graph in figure 1.9 corresponding to $A = 0$ is just a theoretical limit. Moreover, experience shows that the *maximum dry density* achieved in ideal laboratory conditions is difficult to reproduce under field conditions and so, in practice, one rather aims at achieving a *minimum* dry density *in situ* through the use of the *relative compaction* defined as follows:

$$RC = \frac{\rho_{d(field)}}{\rho_{d\max}} \times 100\% \tag{1.25}$$

where a value $RC = 95\,\%$ or thereabouts is deemed acceptable. (*Note that* $RC = 95\,\%$ *does not imply an air content* $A = 5\%$.)

Now consider what effect the notion of relative compaction has on the *in situ* compaction of soils. If the compacted fill were cohesive, then specifying a relative compaction of, say, 98% as is the case in figure 1.10 does not constitute the only criterion for compaction. The reason is depicted in the figure, in that an $RC = 98\,\%$ can be achieved at two different values of moisture content ($w_1 \approx 12\%$, $w_2 \approx 18.5\,\%$) on each side of the optimum value $w_{op} = 15.7\,\%$.

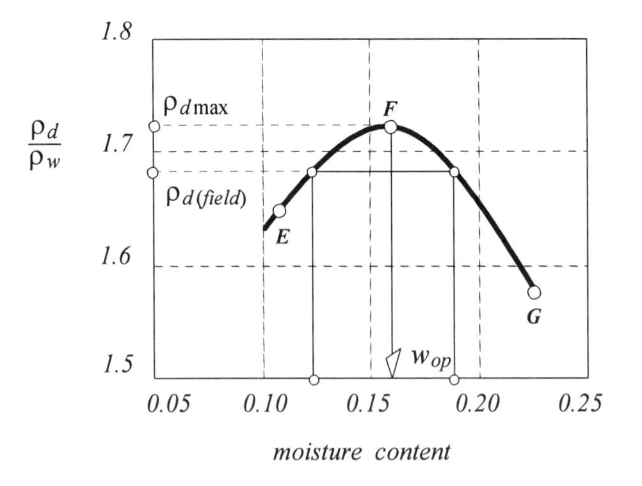

Figure 1.10: Effects of moisture content on the in situ dry density.

Consequently, the moisture content becomes a criterion for compaction and has to be specified together with the relative compaction. However, in the case of granular soils, ample experimental evidence indicates that the moisture content is not as important a criterion, and that the choice of the compaction equipment has more impact on the dry density to be achieved.

On the practical side, the fill material is usually compacted in the field in layers about 0.2 *m* thick. Usually, a number of two to four passes can achieve the required density which must be checked randomly through *in situ* tests. Moreover, the type of equipment used for compaction depends on the nature of fill and the energy level required to achieve a given relative compaction. Considering that an average family car exerts an average pressure through its tyres of around $200\,kN/m^2$, it should come as no surprise that all compaction equipment used for road building generates pressures well in excess of this value.

Thus, for example, a *smooth wheel roller* produces a contact pressure of $400\,kN/m^2$ and can be used in conjunction with all types of soils with the exception of boulder clays. When used on sands, the wheels may be vibrated to create the conditions whereby the solid particles are rearranged in an optimal way that minimises the voids, much in the way that fresh concrete is vibrated to increase its density. A *pneumatic roller* that has two or more rows of narrow tyres, whose position is alternated between consecutive rows, can generate pressures of up $700\,kN/m^2$ and can be used to compact sand or clay fills. *Sheepsfoot rollers* on the other hand are only used for clays and can produce pressures ranging from $1.5\,MN/m^2$ to $7\,MN/m^2$.

Example 1.7

A road embankment is constructed of clay fill compacted to a bulk density $\rho = 2\,Mg/m^3$, at a water content $w = 20\,\%$. With an assumed specific gravity of the solid particles $G_s = 2.7$, the following quantities can be calculated:

- *the clay porosity:* it is seen that the use of equation 1.8 necessitates the knowledge of the void ratio, which, in turn, can be evaluated from equation 1.11:

$$e = G_s(1+w)\frac{\rho_w}{\rho} - 1 = \frac{2.7 \times 1.2}{2} - 1 = 0.62$$

Therefore:

$$n = \frac{e}{1+e} = \frac{0.62}{1.62} = 0.38$$

- *the degree of saturation:* using equation 1.7:

$$S = \frac{wG_s}{e} = \frac{0.2 \times 2.7}{0.62} = 87\%$$

- *the air content:* clearly, a combination of equations 1.7 and 1.9*b* yields:

$$A = n(1-S) = 0.38 \times (1-0.87) = 4.9\%$$

- *the fill dry density:* either equations 1.12 or 1.14 yield $\rho_d = 1.67\,Mg/m^3$.

Example 1.8

A laboratory compaction test undertaken on a clay soil resulted in a maximum dry density $\rho_{d\max} = 1.65\,Mg/m^3$. The same soil, which is characterised by a specific gravity $G_s = 2.7$, was then used *in situ* to build a 7 *m* high embankment for a motorway, where it was subjected to a relative compaction of $RC = 95\%$.

Assuming that the (*in situ*) compacted clay has a bulk density $\rho_f = 1.8\,Mg/m^3$, then its optimum moisture content w_{op} can be estimated from equation 1.14:

$$\rho_f = \rho_{df}(1 + w_{op})$$

where the field dry density is calculated as follows:

$$\rho_{df} = 0.95\rho_{d\max} = 1.57\,Mg/m^3$$

Hence:

$$w_{op} = \frac{1.8 - 1.57}{1.57} = 14.6\%$$

The air content of the embankment soil can be determined from equation 1.16:

$$\rho_{df} = \frac{G_s(1-A)}{1+w_{op}G_s}\rho_w \qquad \Rightarrow \qquad A = \frac{2.7 - 1.57\times(1+0.146\times 2.7)}{2.7} = 18.9\%$$

Now the void ratio and degree of saturation are found from equations 1.9*b* and 1.10 respectively:

$$e = \frac{A + w_{op}G_s}{1-A} = \frac{0.189 + 0.146\times 2.7}{1-0.189} \approx 0.72$$

$$S = \frac{1-A}{1+A/w_{op}G_s} = \frac{1-0.189}{1+\frac{0.189}{0.146\times 2.7}} = 55\%$$

Problems

1.1 A thick layer of clay has an average void ratio $e = 0.75$ and a specific gravity $G_s = 2.7$, the water level being 2 *m* below the ground surface.

(*a*) Calculate the water content of the (saturated) clay below the water level.
(*b*) If the clay above the water table has an air content $A = 0.04$, determine its water content and degree of saturation.
(*c*) Calculate both bulk and saturated unit weights of the clay.

Ans: (*a*) $w = 28\%$, (*b*) $w = 25.2\%$, $S = 90.6\%$, (*c*) $\gamma = 18.9\ kN/m^3$, (*d*) $\gamma_{sat} = 19.3\ kN/m^3$.

1.2 Calculate the ratio γ_{sat}/γ between saturated and bulk unit weights of a clay having a void ratio $e = 1.0$, a water content $w = 25\%$, and a specific gravity $G_s = 2.7$.

Ans: $\gamma_{sat}/\gamma = 1.1$.

1.3 A cylindrical soil sample with a diameter of 100 *mm* and a height of 100 *mm* was extracted from within a thick layer of a stiff clay. The sample has a weight of 1320 *g*, and after being thoroughly dried, its weight has reduced to 1075 *g*.

Assuming the clay has a specific gravity $G_s = 2.7$, determine its:

(*a*) water content,
(*b*) void ratio,
(*c*) degree of saturation,
(*d*) air content,
(*e*) bulk and dry densities.

Ans: (*a*) $w = 22.8\%$, (*b*) $e = 0.97$, (*c*) $S = 63.3\%$, (*d*) $A = 18\%$, (*e*) $\rho = 1.68\, Mg/m^3$, $\rho_d = 1.37\, Mg/m^3$.

1.4 Refer to the graph in figure *p*1.4 obtained in the laboratory from a test on a soil in which a well is being dug.

(*a*) Determine the coefficients of curvature and grading.
(*b*) Give an estimate of the soil permeability.
(*c*) Using the criteria given in section 1.4, design a soil filter to be placed around the well lining whose maximum mesh size is 2 *mm*.

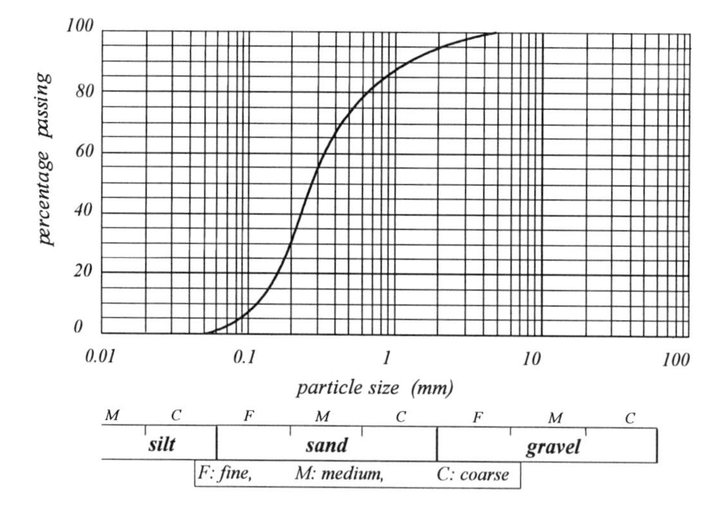

Figure p1.4

Ans: (*a*) $C_u \approx 3$, $C_g \approx 1.2$, (b) $k \approx 1.1 \times 10^{-4}\, m/s$.

1.5 A soil with 23% of clay content has an activity $A = 0.72$ and a plastic limit $w_p = 16\%$.

(*a*) Calculate its liquid limit, then use the plasticity chart in section 1.5 to classify the soil.
(*b*) In its natural state, the soil is characterised by a liquidity index $I_L = 0.75$. Determine its natural moisture content.

Ans: (*a*) $w_L = 32.5\%$, $I_P = 16.6\%$, $\Rightarrow$ *clay of low plasticity.*
(*b*) $w = 28.4\%$.

1.6 A compaction test undertaken on a sample of clay with a specific gravity $G_s = 2.7$ yielded an air content $A = 3\%$ and a maximum dry density $\rho_{d\max} = 1.65\, Mg/m^3$. Calculate the corresponding optimum moisture content.

Ans: $w_{op} \approx 22\%$.

1.7 A laboratory compaction test undertaken on clay yielded the following results:

Moisture content (%)	11	13	16	21
Bulk density (Mg/m^3)	17.75	18.92	19.84	19.48

(*a*) Plot the variation of the dry density with the moisture content, then determine the maximum dry density and the optimum moisture content.

The same clay was then used for the construction of an embankment, where it was subjected to a relative compaction $RC = 95\%$ at the optimum moisture content.

(*b*) Calculate the bulk density of the compacted fill.
(*c*) Assuming the clay fill has a specific gravity $G_s = 2.7$, determine its air content, degree of saturation and porosity.

Ans: (*a*) $\rho_{d\max} = 1.71\, Mg/m^3$, $w_{op} = 16\%$. (*b*) $\rho = 1.88\, Mg/m^3$,
(*c*) $A \approx 13.8\%$, $S = 65.2\%$, $n \approx 0.40$.

References

Agrawal, Y. C. and Riley, J. B. (1984) *Optical Particle Sizing for Hydrodynamic Based on Near-forward Scattering.* Society of Photo-Optical Instrumentation Engineers (489), pp. 68–76.

Bell, F. G. (1993) *Engineering Geology.* Blackwell Science, London.

Bjerrum, L. (1954) *Geotechnical properties of Norwegian marine clays.* Géotechnique, Vol. 4 (2), pp. 49–69.

Hong Kong Geotechnical Engineering Office. (1993) *Review of Granular and Geotextile Filters.*

MacCave, I. N. and Syvitski, J. P. M. (1991) *Principles and Methods of Particle Size Analysis.* Cambridge University Press.

Sangrey, D. A. (1972) *Naturally cemented sensitive soils.* Géotechnique, 22, pp. 139–152.

Somerville, S. H. (1986) *Control of Groundwater for Temporary Works.* CIRIA Report 113, Construction Industry Research and Information Association, London.

Weiner, B. B. (1979) *Particle and Spray Sizing Using Laser Diffraction.* Society of Photo-Optical Instrumentation Engineers (170), pp. 53–56.

Weiner, B. B. (1984) *Particle Sizing Using Photon Correlation Spectroscopy.* Modern Methods of Particle Size Analysis, Edited by H.G. Barth, Wiley, New York, pp. 93–116.

CHAPTER 2

Stress distribution within a soil mass

2.1 Effective stress principle

Consider the case of figure 2.1 where a *fully saturated* soil (*i.e.* all voids are filled with water) is subjected at its surface to a given load (which can be of any type). Thus, under *static water conditions*, the overall *total force* F, due to the weight of a selected volume V of soil, generates a *total strtess* $\sigma = F/A$, where A represents the cross-sectional area. The force F is in fact the sum of a component due to the solid particles and one due to water:

$$\begin{aligned} F &= F_s + F_w \\ &= g(\rho_s V_s + \rho_w V_w) \end{aligned} \qquad (2.1)$$

where g is the acceleration due to gravity, ρ_s is the density of solid particles and ρ_w that of water, and V_s, V_w are the volumes of solids and water respectively.

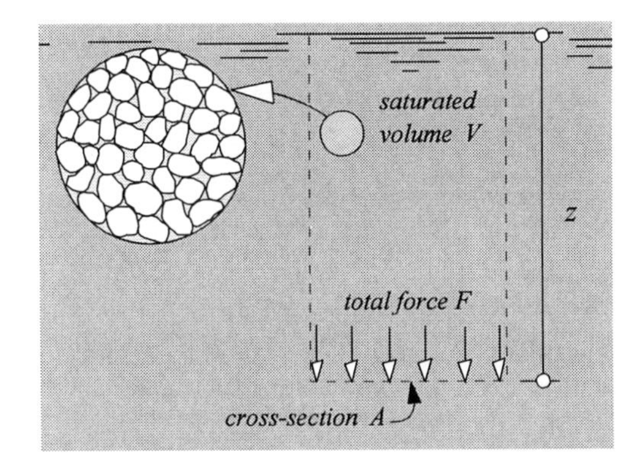

Figure 2.1: Pressure due to a loaded volume of saturated soil.

Since the total volume is $V = V_s + V_w = Az$, it is easy to establish that:

$$\frac{F}{A} = \frac{1}{A}[gV_s(\rho_s - \rho_w)] + \frac{V}{A}g\rho_w \tag{2.2}$$

Knowing that $V/A = z$, $F/A = \sigma$ and $g\rho_w = \gamma_w$, the above equation then reduces to:

$$\sigma = \frac{1}{A}[gV_s(\rho_s - \rho_w)] + z\gamma_w \tag{2.3}$$

Equation 2.3, known as *the effective stress principle,* was established by Karl Terzaghi (1936) and constitutes one of the most fundamental relationships that governs soil behaviour, the effect of which will become clearer in the following chapters.

Because the quantity $z\gamma_w = u$ represents the pressure exerted by a column of water of a height z, equation 2.3 is usually written as follows:

$$\sigma = \sigma' + u \tag{2.4}$$

where u is referred to as the *porewater pressure* and σ' represents the *effective stress.* Accordingly, the effective stress principle stipulates that the total stress existing at a depth z is the sum of the component of normal stress applied to the soil skeleton through contact between solid particles, and a porewater pressure induced in the water filling the voids. Since water cannot sustain any shear stress, both volume changes and shear strength of the soil are controlled by the effective stress.

The *actual* effective stress corresponds to the *average* normal stress generated through contact between solid particles on a given plane within the soil mass (see Mitchell (1993), for instance). However, experimental evidence indicates that equation 2.4 is highly reliable, provided that the solid particles are relatively incompressible and the overall intergranular contact area is small compared with the cross-sectional area of the corresponding volume of soil. Both requirements are fulfilled for most saturated soils.

2.2 Stresses due to self weight

As has already been established in the previous chapter, in the absence of any external loading, the stresses generated within a soil mass can be calculated using the soil unit weight γ, whose value can be estimated using any of the three equations 1.17 to 1.19, depending on the soil's degree of saturation.

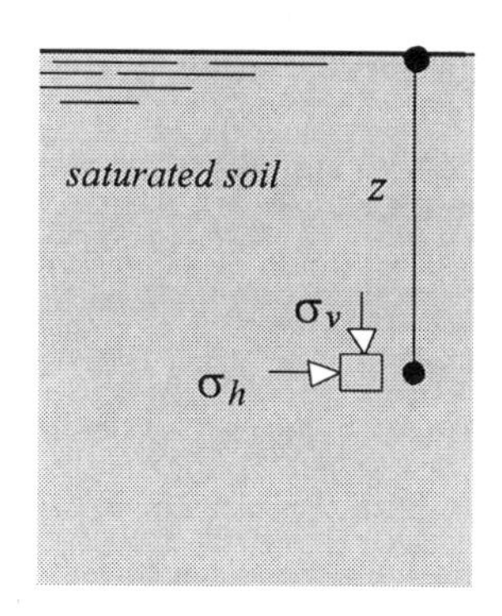

Figure 2.2: Total stress due to self weight.

Under *static water conditions*, the overall *total stresses* due to self weight, for a *fully saturated* soil, are calculated at any depth z as follows (see figure 2.2):

$$\sigma_v = \gamma_{sat}\, z \tag{2.5a}$$

$$\sigma_h = K_o\, \sigma_v \tag{2.5b}$$

where σ_v and σ_h are the vertical and horizontal stresses respectively, K_o, known as the *coefficient of earth pressure at rest,* is a constant that reflects the mode of soil deposition and its stress history, and will be discussed thoroughly in section 10.2, and γ_{sat} is the saturated unit weight of soil.

Because of the linear nature of equation 2.5*b*, the following derivations will deal exclusively with the vertical stress component σ_v, for which the corresponding effective stress is found by using the general equation 2.4:

$$\sigma_v^{/} = \sigma_v - u \tag{2.6}$$

Accordingly, the *effective vertical stress* due to the self weight of soil at a given depth depends on the porewater pressure, in other words on the level of the groundwater.

Consider the general case of figure 2.3 depicting a homogeneous soil with the groundwater at a depth z_w below the ground surface. Below water, the fully saturated soil has a *saturated unit weight* γ_{sat} that can be calculated from equation 1.18; however, above the water level depicted in the figure, a *bulk unit weight* $\gamma < \gamma_{sat}$ applies since the soil is only partially saturated.

The presence of moisture above the water level is due to *capillary rise* (much in the way water rises against gravity from the root of a tree to the branches which can be as high as 30 m or more), which in turn depends on the size of voids between solid particles, and therefore on the type of soil.

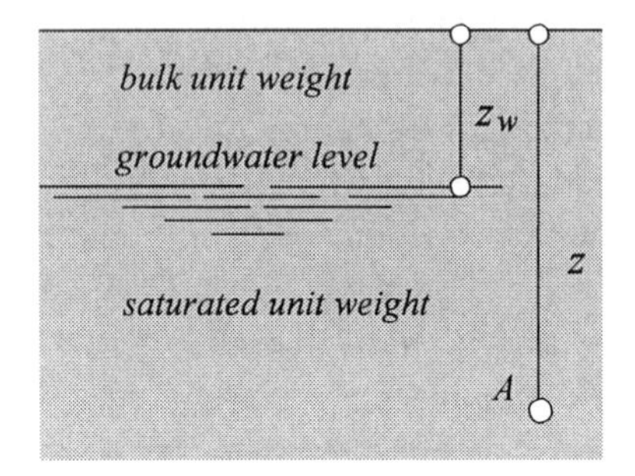

Figure 2.3: Effect of water on the unit weight of soils.

The voids between solid particles within a soil matrix are interconnected, thus forming a complex network of channels that can be thought of as capillary tubes. According to (the enlarged) figure 2.4, the capillary rise h_c of water inside a channel is proportional to the ratio of the *surface tension force* T to the "effective" channel diameter d. Hence, at equilibrium, the following equation applies:

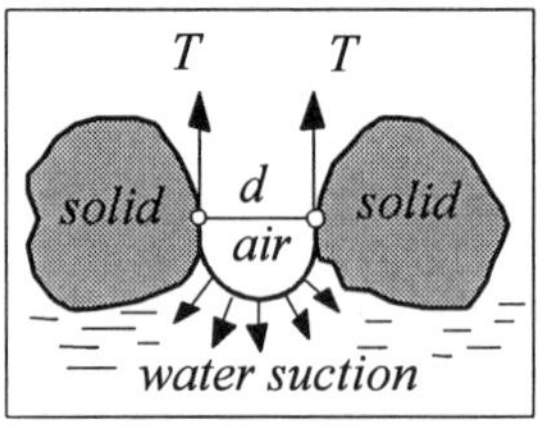

Figure 2.4: Surface tension and water suction within a soil matrix.

$$T\pi d = \frac{\pi d^2}{4} u_w \tag{2.7}$$

where $u_w = h_c \gamma_w$ is the *water suction* (*i.e.* the negative water pressure), and T represents the surface tension of the water/air interface ($T = 0.073\ N/m$ at $10°C$). Hence, substituting for u_w into equation 2.7, then rearranging:

$$h_c = \frac{4T}{d\gamma_w} \tag{2.8}$$

Equation 2.8 indicates that, in theory, a clay having an average voids size of, say, $d = 10^{-6}\ m$ will have a capillary rise of just under 30 m. In practice however, the nature of the connection of voids within a soil matrix has a limiting effect on the theoretical value of h_c calculated using the above equation. Terzaghi and Peck (1967) suggested relating the capillary rise within a soil to its effective size d_{10} (mm) [refer to section 1.3] and void ratio e as follows:

$$h_c \approx \frac{C}{e\,d_{10}} \tag{2.9}$$

C having a value between 10 mm^2 and 50 mm^2.

While equation 2.9 yields a maximum capillary rise h_c, the actual level of capillary saturation h_s can be estimated from the relationship:

$$h_s \approx h_c \frac{d_{10}}{d_{60}} \tag{2.10}$$

d_{60} being the particle size as defined in section 1.3.

Example 2.1

Consider the case of the two following soils:

(a) a clay with the parameters: $e = 1.0, \quad d_{10} = 0.7\,\mu m, \quad d_{60} = 3\,\mu m,$

(b) a sand having : $e = 0.8, \quad d_{10} = 0.2\,mm, \quad d_{60} = 0.7\,mm.$

Assuming that both soils are characterised by the same parameter $C = 10\,mm^2$, then equations 2.9 and 2.10 yield:

- for the clay:

$$h_c = \frac{10^{-5}}{1 \times 7 \times 10^{-7}} = 14.28\,m \quad \Rightarrow \quad h_s = 14.28 \times \frac{0.7}{3} = 3.33\,m$$

- for the sand:

$$h_c = \frac{10^{-5}}{0.8 \times 2 \times 10^{-4}} = 62.5 \times 10^{-3}\,m \quad \Rightarrow \quad h_s = 62.5 \times \frac{0.2}{0.7} = 17.9 \times 10^{-3}\,m$$

The corresponding water suction (*i.e.* the *negative* porewater pressure) in each case is:

$u = h_c \gamma_w = -14.28 \times 10 = -142.8\, kN/m^2$ for the clay,

as opposed to $u = -0.62\, kN/m^2$ for sand.

This simple example shows that cohesive soils such as silts and clays are characterised by capillary saturation levels that are much higher than those corresponding to granular soils.

A *partially saturated soil* occurs naturally above the groundwater level which, in a temperate climate, does not extend to any great depth. Thus, with reference to figure 2.5, the suction generated above the groundwater level can be ignored, and a unique *bulk unit weight* γ (equation 1.17) can be assumed to apply from the ground surface down to the static water level situated at a depth z_w.

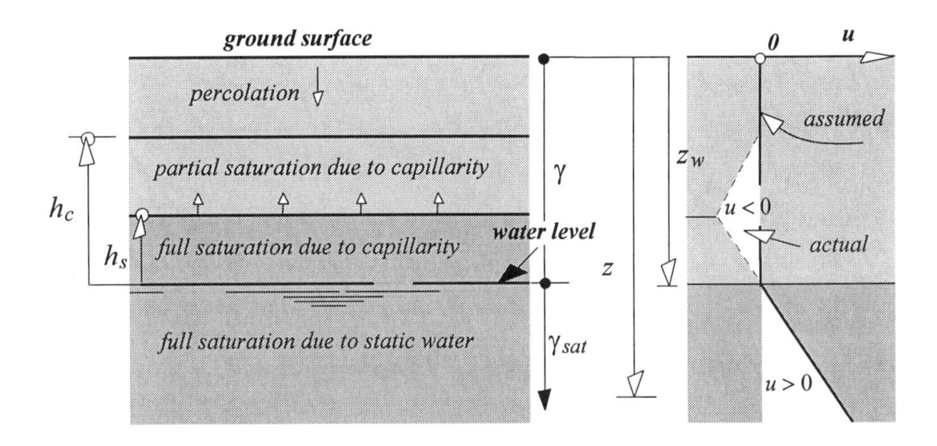

Figure 2.5: Assumption related to the use of bulk unit weight.

Therefore, the *total vertical stress* at any given depth *z*, as depicted in figure 2.5, is calculated in the following way:

$$\sigma_v = \gamma z_w + \gamma_{sat}(z - z_w) \tag{2.11}$$

the porewater pressure at the same point being:

$$u = \gamma_w (z - z_w) \tag{2.12}$$

so that the *effective vertical stress* is:

$$\sigma_v' = \sigma_v - u$$
$$= \gamma z_w + (z - z_w)(\gamma_{sat} - \gamma_w) \qquad (2.13)$$

Equations 2.11 to 2.13 imply that for a dry soil, $z_w = z$ and both *effective* and *total vertical stresses* are identical since the porewater pressure u is reduced to zero. On the other hand, when the soil is fully saturated, then $z_w = 0$ and equation 2.13 reduces to equation 2.6. These conclusions are further illustrated in figure 2.6.

In practice, the difference between saturated and bulk unit weights is typically of the order of 1 to $2\,kN/m^3$; also, the unit weight of water can be taken as $\gamma_w \approx 10\,kN/m^3$.

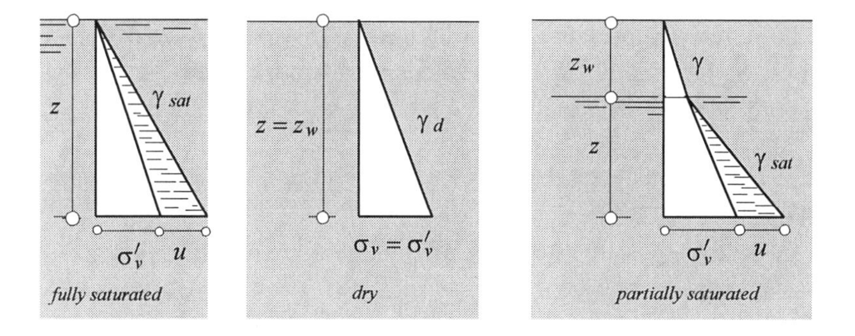

Figure 2.6 : Effects of groundwater on the stress due to self weight of a soil.

Example 2.2

A thick clay layer has a bulk unit weight $\gamma = 19\,kN/m^3$ and a saturated unit weight $\gamma = 19.5\,kN/m^3$; the unit weight of water being $\gamma_w \approx 10\,kN/m^3$. Knowing that the groundwater level is at a depth of 4 m below the surface, then the depth z_1 at which the *vertical effective stress* has a value $\sigma_{z1}' = 120\ kN/m^2$ is such that:

$$120 = 4 \times 19 + (z_1 - 4)(19.5 - 10) \quad \Rightarrow \quad z_1 = \frac{120 - 76}{9.5} + 4 = 8.63\,m$$

After the exceptional occurrence of two wet seasons, the groundwater has risen to a new level, so that its depth has dramatically reduced from the original 4 m to z_w as depicted in figure 2.7. Assuming this rise in the water level caused the effective stress (estimated at the same depth z_1 calculated above), to decrease to $\sigma'_{z1} = 90\,kN/m^2$, then the corresponding depth z_w is:

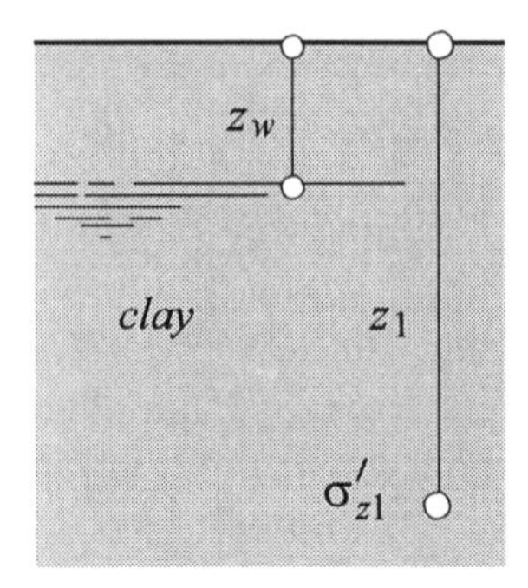

Figure 2.7: Soil conditions.

$$90 = z_w \times 19 + (8.63 - z_w)(19.5 - 10) \quad \Rightarrow \quad z_w = \frac{90 - 82}{9.5} = 0.83\,m$$

The variation with depth (up to $z_1 = 8.63\,m$) of the effective vertical stress σ'_v, the porewater pressure u and the total stress $\sigma = \sigma'_v + u$ is depicted in figure 2.8. Originally, when the groundwater level was 4 m deep, σ'_v increased linearly from $\sigma'_{v1} = 19 \times 4 = 76\,kN/m^2$ at 4 m depth, to:

$$\sigma'_{v2} = 76 + (8.63 - 4)(19.5 - 10) = 120\,kN/m^2 \quad \text{at } 8.63\ m \text{ depth}$$

The porewater pressure on the other hand varied between $u_1 = 0$ at 4 m,

to: $\quad u_2 = (8.63 - 4) \times 10 = 46.3\,kN/m^2$ at 8.63 m depth

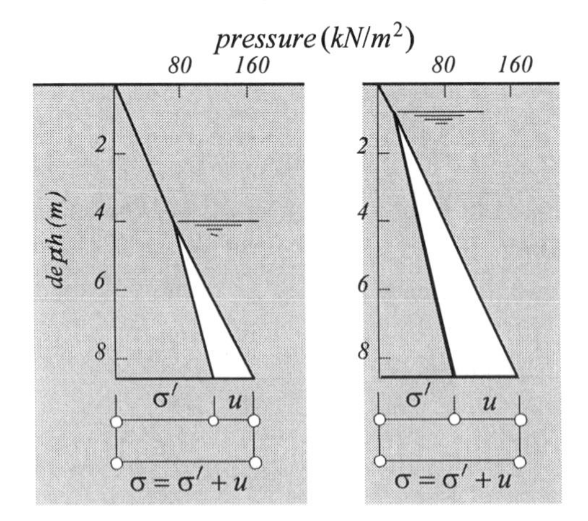

Figure 2.8: Pressure distribution with depth.

After the depth of the groundwater level has reduced to 0.83 *m*, the variation of the effective stress then reduced from:

$$\sigma'_v = 0.83 \times 19 = 15.8\, kN/m^2 \text{ at } 0.83\, m,$$

to:

$$\sigma'_v = 15.8 + (8.63 - 0.83)(19.5 - 10) = 89.9\, kN/m^2$$

while the porewater pressure increased from:

$u = 0$ at 0.83 *m*, to $u = (8.63 - 0.83) \times 10 = 78\, kN/m^2$ at 8.63 *m*

In this case, while the rise of the groundwater level has caused the porewater pressure to increase and the effective stress to decrease, the total stress (at any depth) remained virtually constant. For instance, at the depth 8.63 *m*, it is seen that:

- groundwater level 4 *m* deep: $\sigma = \sigma'_v + u = 120 + 46.3 = 166.3\, kN/m^2$,

- groundwater level 0.83 *m* deep: $\sigma = \sigma'_v + u = 89.9 + 78 = 167.9\, kN/m^2$.

The marginal difference reflects the difference between bulk and saturated unit weights of the soil.

This result, however, is very significant in that, as stated earlier, the *shear strength* of the soil is controlled by the *effective stress*. Accordingly, any increase in the porewater pressure will induce a decrease in the effective stress and therefore a decrease in the soil strength.

Example 2.3

Following a prolonged period of heavy rain, a saturated thick layer of clay was subjected to flood water as depicted in figure 2.9*a*. At its peak, the water level above the ground surface reached 1 *m*, and the corresponding pressure diagram shown in the figure indicates clearly that the water above the ground causes a uniform increase of the porewater pressure with depth. The stresses can be evaluated at any depth in a straightforward manner. For example at a depth of 4 *m*, assuming a saturated unit weight of the clay $\gamma_{sat} = 20\, kN/m^3$ and a unit weight of water $\gamma_w = 10\, kN/m^3$, it follows that:

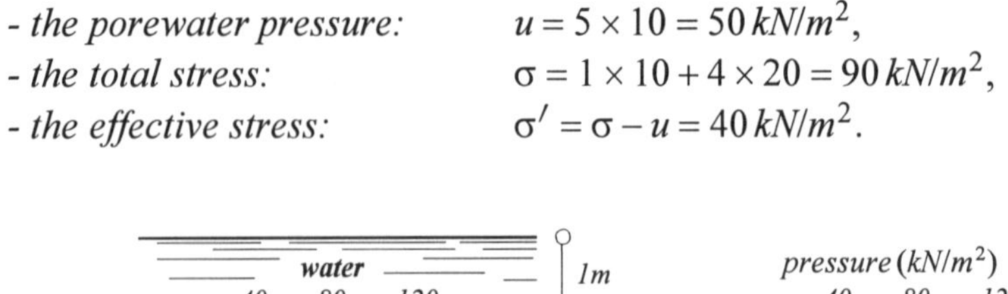

- *the porewater pressure:* $u = 5 \times 10 = 50\,kN/m^2$,
- *the total stress:* $\sigma = 1 \times 10 + 4 \times 20 = 90\,kN/m^2$,
- *the effective stress:* $\sigma' = \sigma - u = 40\,kN/m^2$.

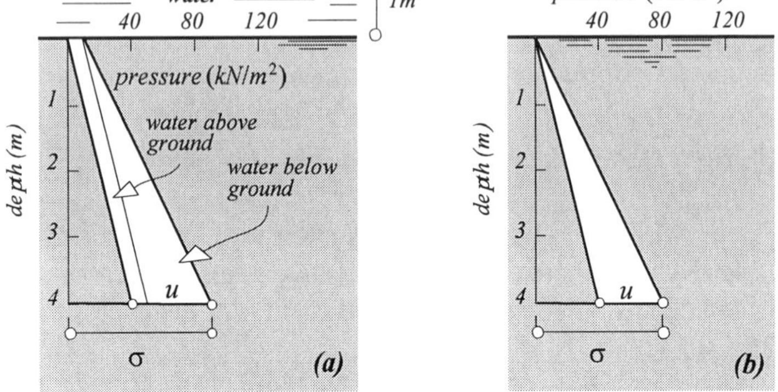

Figure 2.9: Pressure diagrams:(a) at the peak of the flood, (b) after drainage of flood water

Once the flood water is drained away leaving a fully saturated clay layer, the pressure diagram remains remarkably similar to that of figure 2.9*a* except for the porewater pressure due to the water above the ground which will then disappear as depicted in figure 2.9*b*. This means that the effective stress remains constant because the drainage of flood water decreases both the total stress and the porewater pressure in the same proportion, leaving unchanged the effective stress in the process. Accordingly at a depth of 4 *m*, it is seen from figure 2.9*b* that:

- *the porewater pressure:* $u = 4 \times 10 = 40\,kN/m^2$,
- *the total stress:* $\sigma = 4 \times 20 = 80\,kN/m^2$,
- *the effective stress:* $\sigma' = \sigma - u = 40\,kN/m^2$.

2.3 Effective stresses related to unsaturated soils

There are instances in which the void space within a soil matrix is in part filled with air or any other gas, the remaining volume of voids being filled with water. Such an *unsaturated soil* occurs naturally in a hot arid climate where the groundwater level can be situated at a great depth. Also, soils used for the construction of embankments and earth dams are usually compacted and, as such, are unsaturated. A third group is composed of marine deposits within which the decomposition of organic sediments leads to the formation of bubbles of methane gas.

Naturally, the behaviour of such soils depends on the proportion of air or gas contained within the voids, in other words on the degree of saturation. In most cases of interest in civil engineering practice, unsaturated cohesive soils have an intermediate degree of saturation (*i.e.* $S < 0.85$).

Figure 2.10 depicts the surface tension force T generated at the water/gas interface at a typical voids channel between two solid particles of an unsaturated soil. The relationship between the pore water pressure u_w, the pore gas pressure u_g and T is as follows:

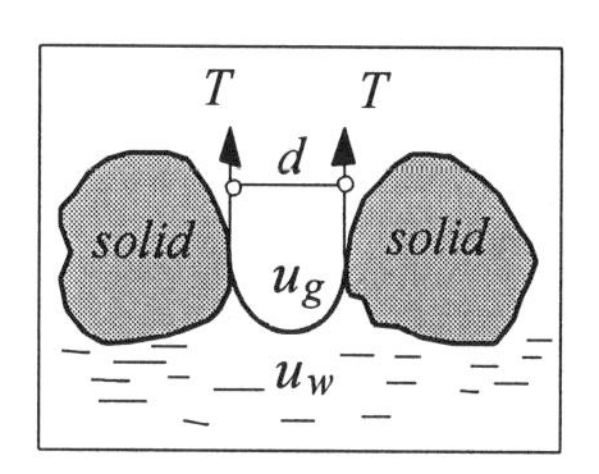

Figure 2.10: Surface tension for an unsaturated soil.

$$(u_g - u_w)\frac{\pi d^2}{4} = T\pi d$$

or

$$(u_g - u_w) = \frac{4T}{d} \tag{2.14}$$

Conclusive evidence (see Jennings and Burland (1962), for example) indicates that the behaviour of such unsaturated cohesive soils (*i.e.* with a degree of saturation $S < 85\%$), *cannot* be represented by a single effective stress variable as in the equation suggested by Bishop and Donald (1961), which is still being (unjustifiably?) advocated in some quarters for this particular type of soil. Rather, the shear strength and the void ratio are related to two stress variables, namely:

- the effective stress : $\sigma' = \sigma - u_g$,
- the soil matrix suction as per equation 2.14: $(u_g - u_w)$.

The variation of the void ratio ***e*** with these two stress variables yields a warped surface as depicted in figure 2.11, which is only used for illustration purposes. Clearly, as the suction $(u_g - u_w)$ decreases (in other words as the degree of saturation increases), the soil behaves differently depending on the magnitude of the net stress $(\sigma - u_g)$. A soil (volume) expansion occurs under low values of $(\sigma - u_g)$ (stress path *CD* in figure 2.11), as opposed to a contraction under high $(\sigma - u_g)$ values (stress path *AB*).

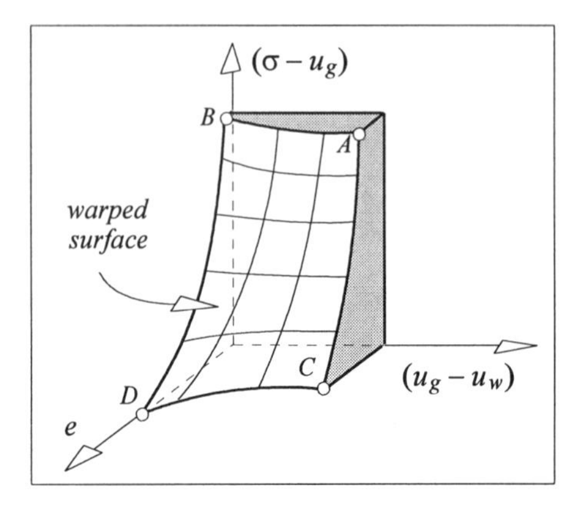

Figure 2.11: Stress–void ratio relationship for unsaturated cohesive soils (S<0.85).

The shear strength of such soils is also related to the two stress variables used in figure 2.11, and reference should be made to the paper by Fredlund (1979) in which appropriate expressions relating the void ratio and the shear strength to the above two stress variables are discussed.

For unsaturated cohesive soils with a degree of saturation limited to $0.85 \leq S \leq 1$, the presence of gas leads to the formation of gas bubbles. If the size of these bubbles is smaller than the void space between solid particles as depicted in figure 2.12, then the behaviour can be described by the Bishop and Donald (1961) equation :

$$\sigma' = (\sigma - u_g) + \chi\,(u_g - u_w) \qquad (2.15)$$

where χ represents a constant $(0 \leq \chi \leq 1)$ that depends on the degree of saturation, u_g and u_w being as defined previously. Because of the presence of gas bubbles, the porewater becomes compressible, so that a positive porewater pressure will cause the bubbles to dissolve, decreasing in the meantime the volume of water. A negative porewater pressure on the other hand will lead to the expansion of the gas bubbles and an increase in the volume of water. Accordingly, depending on the soil stress history and the level of loading applied at its surface, a positive porewater pressure will generate almost immediately a decrease in the volume of water and an increase in the intergranular stresses, and therefore an increase (in the short term) of the soil strength.

A negative porewater pressure, on the other hand, will engender an immediate expansion of the volume of voids, thus causing a decrease in the effective stresses and hence a decrease in the soil strength in the short term. With time, as the excess porewater pressure (of whatever sign) dissipates, the magnitude of the soil strength will become similar to that of an equivalent saturated soil.

This type of unsaturated soil containing small gas bubbles commonly occurs just above the capillary saturation level (refer to figure 2.5).

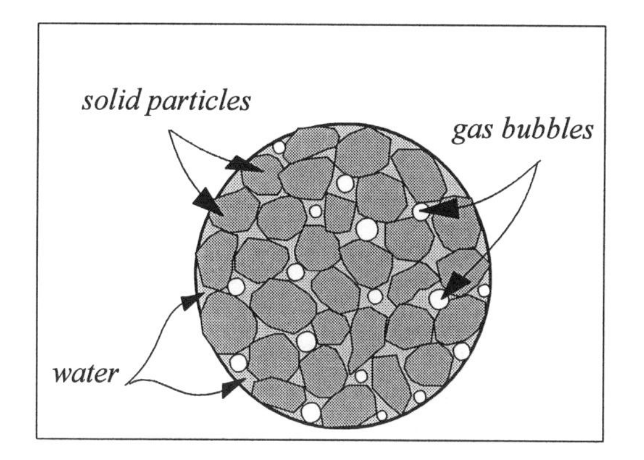

Figure 2.12: Unsaturated soil with small gas bubbles $(0.85 \le S \le 1)$.

If, under similar saturation conditions $(0.85 \le S \le 1)$, the gas bubbles are larger than the void space between solid particles as shown in figure 2.13, then the soil behaviour *can no longer* be described by equation 2.15. Rather, the soil has to be modelled as a saturated soil matrix containing large gas-filled cavities.

Such modelling can be complex in nature (see for instance Toll (1990), Wheeler (1991), Sills *et al.* (1991), Thomas and He (1998)), and relies mainly on sophisticated concepts such as critical state soil mechanics, which will be presented in chapter 6.

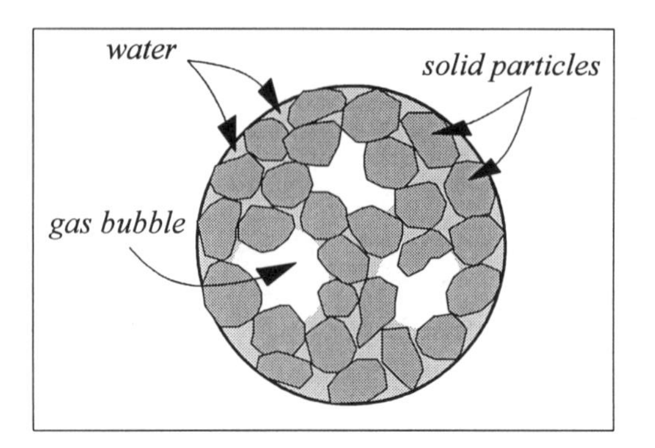

Figure 2.13: Unsaturated soil with large gas bubbles $(0.85 \leq S \leq 1)$.

It is useful to remember that this type of unsaturated soils is a common occurrence in the marine environment. Prior knowledge of their behaviour is therefore of paramount importance for foundation design of offshore structures such as oil platforms.

2.4 Stresses due to surface loading

2.4.1 Introduction

Any type of loading transmitted to a soil mass through a foundation is bound to generate a contact pressure which is three-dimensional in nature. The distribution of this pressure within the soil mass depends on the following.

(*a*) *The nature of load transmitted through the foundation*: One would expect the stresses generated by a uniform load applied at the soil surface over a relatively wide area to be different from those corresponding to a concentrated load, for instance.

(*b*) *The type of foundation transmitting the load*: The deformation of a foundation under an applied load depends on its rigidity. Thus in the case of a uniformly loaded *rigid foundation,* as depicted in figure 2.14, a uniform settlement beneath the foundation surface generates a pressure distribution whose maximum depends on the type of soil. For a clay, the maximum occurs under the foundation edges, whereas for a sand, the maximum pressure is beneath the foundation centre. In the case of a *flexible foundation*, the deformation reflecting the degree of flexibility leads to a uniform contact pressure as shown in figure 2.14.

From a practical perspective, the calculation of settlements undertaken using a pressure distribution related to a flexible foundation, are corrected to take into account any foundation rigidity when applicable (refer to section 4.6.1).

(*c*) *The soil type within which the contact pressure is generated*: Because of the three-dimensional nature of the pressure distribution, the soil can (justifiably) be considered as an infinite half-space compared with the foundation area, so that any stress analysis is not affected by the boundaries.

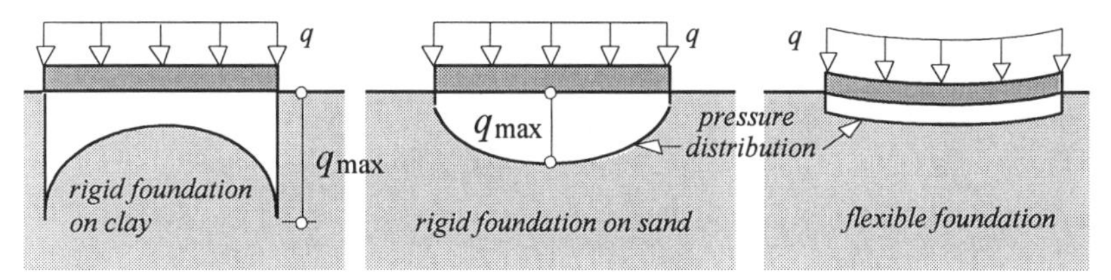

Figure 2.14: Pressure distribution within a soil mass due to a surface load.

The profiles with depth of the *vertical stress* $\Delta\sigma_z$ induced by a surface loading are of particular interest since they affect the magnitude of *settlements* as will be explained shortly in chapter 4. These profiles, which are closely related to the shape and size of the loaded area, as well as to the type of applied load, show a variable rate of *decay* of $\Delta\sigma_z$ as the depth increases, corresponding to a continuous *stress redistribution* within the soil mass. This is in sharp contrast to the *monotonous* (linear) *increase* of the vertical stress due to self weight discussed earlier. Figure 2.15 depicts a three-dimensional pear-shaped vertical stress profile under the central point of a circular loaded area of diameter *B*, transmitting a uniform pressure q. The profile corresponding to an (assumed) infinitely long strip of width *B*, subjected to an identical pressure q is no longer three-dimensional since the confinement is limited to each side of the strip; accordingly, the rate of decay of $\Delta\sigma_z$ with depth is affected by these boundary conditions.

The profile of other stress components (such as radial or shear stresses) with depth due to surface loading is equally important. For example, the design of a retaining structure requires the prior knowledge of the radial stress distribution as will be seen in chapter 10. Consequently, it is essential to develop a method by which the stress increase with depth, due to surface loading, can be calculated, if not precisely, at least with sufficient accuracy so that (with the help of a factor of safety perhaps) a safe design can be achieved.

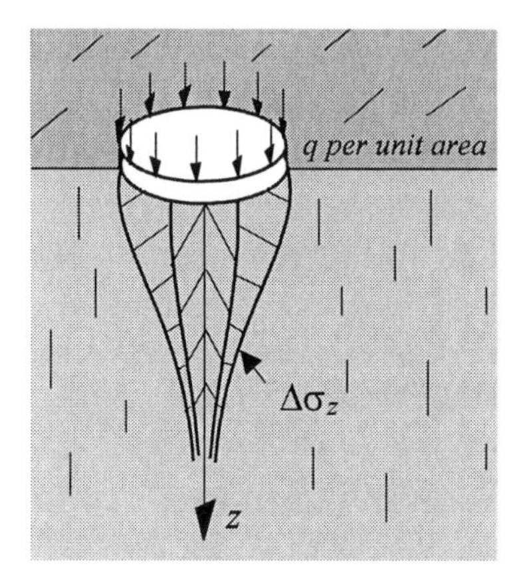

Figure 2.15: Vertical stress profile.

Natural soils, regardless of the way they were formed, are likely to be :

- *anisotropic*, in other words they have different properties in every direction of the space;
- *heterogeneous*, meaning that their stiffness varies with depth, and
- of a *restricted elastic behaviour*.

These natural conditions imply that an analytical solution to the stress distribution due to a given type of surface loading will be fraught with difficulties, the corresponding strenuous boundary conditions being very difficult to satisfy. To find a way around these difficulties (at the expense of some accuracy), the following analysis based on elastic equilibrium assumes that:

- the soil mass is a *semi-infinite* half space,
- the soil is *weightless*,
- the soil is *isotropic* and *homogeneous*, and its behaviour is *linear-elastic*,
- the load is of a *flexible* type.

Obviously, these assumptions are bound to lead to ideal stress distributions. In particular, the assumption related to elasticity is only realistic as long as the increase in stress caused by the surface load does not exceed in any significant way the yield stress as depicted in figure 2.16 in the case of a dense sand.

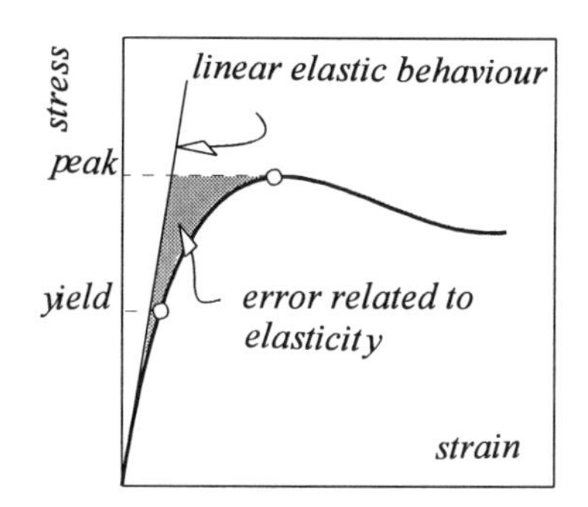

Figure 2.16: Potential errors related to the assumption of elastic soil behaviour.

Also, the assumption of a homogeneous soil can no longer apply if the soil mass is of finite thickness as in the case of a clay layer underlain by a thick layer of rock. Under such circumstances, *other* methods of analysis, mainly of a numerical type, must be applied.

Despite the restrictive nature of the previous assumptions, Burland *et al.* (1977) have shown that, in many cases, the corresponding elastic solutions to the *vertical stress* distribution compare favourably with solutions obtained using more sophisticated methods of analysis. Consequently, the following analysis, which is limited mostly to vertical stresses, must be used cautiously; the radial stress distribution (when needed) can be estimated using the coefficient of earth pressure at rest of equation 2.5*b*.

2.4.2 Elastic equilibrium of a weightless soil subject to a point load: Boussinesq solution

Consider an element of a *weightless semi-infinite soil mass,* situated at a distance $R = \sqrt{r^2 + z^2}$ from a surface load P as depicted in figure 2.17. Assuming that the soil is *homogeneous* (that is identical in composition between any two chosen points), *isotropic* and *elastic* with a modulus of elasticity E and a Poisson's ratio ν, then it can be shown (see Poulos and Davis (1974)) that the differential equation of elastic equilibrium, expressed in terms of a stress function Φ in the cylindrical co-ordinates system, is as follows:

$$\nabla^4\Phi = \nabla^2\nabla^2\Phi = 0 \tag{2.16}$$

with $$\nabla^2 = \left(\frac{\partial^2}{\partial r^2} + \frac{1}{r}\frac{\partial}{\partial r} + \frac{\partial^2}{\partial z^2}\right)$$

the stresses being related to the stress function in the following manner:

$$\sigma_r = \frac{\partial}{\partial z}\left[\nu\nabla^2\Phi - \frac{\partial^2\Phi}{\partial r^2}\right] \tag{2.17a}$$

$$\sigma_\theta = \frac{\partial}{\partial z}\left[\nu\nabla^2\Phi - \frac{1}{r}\frac{\partial\Phi}{\partial r}\right] \tag{2.17b}$$

$$\sigma_z = \frac{\partial}{\partial z}\left[(2-\nu)\nabla^2\Phi - \frac{\partial^2\Phi}{\partial z^2}\right] \tag{2.17c}$$

$$\tau_{rz} = \frac{\partial}{\partial r}\left[(1-\nu)\nabla^2\Phi - \frac{\partial^2\Phi}{\partial z^2}\right] \tag{2.17d}$$

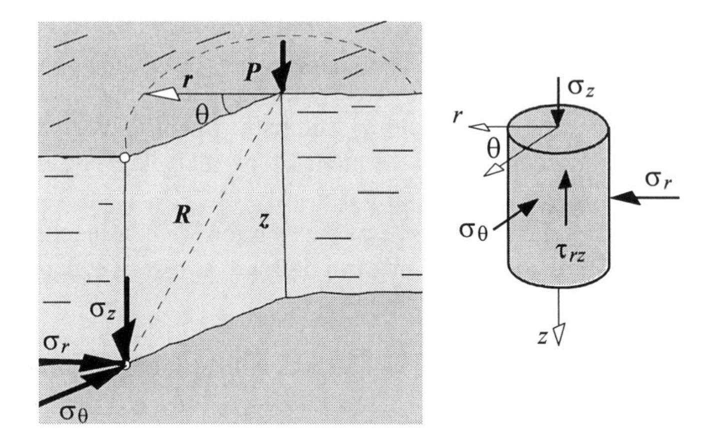

Figure 2.17: Stress field generated by a point load.

As early as 1885, Boussinesq published his *Application des potentiels à l'étude de l'équilibre et du mouvement des solides élastiques* in which he presented the solution to equation 2.16 in the case of a concentrated load P. The mathematical formulation of such an elegant solution, which is outside the scope of this text, can be found in Timoshenko and Goodier (1970), for instance. The solution resulted in the following *vertical and shear stresses*, respectively:

$$\Delta\sigma_z = \frac{3Pz^3}{2\pi R^5} \tag{2.18a}$$

$$\Delta\tau_{rz} = \frac{3Prz^2}{2\pi R^5} \tag{2.18b}$$

Analytical solutions to the same equation related to other types of surface loading have since been established, and in some cases, equation 2.16 has been solved numerically in the absence of closed form solutions. The stress distribution due to loads most commonly encountered in practice are presented in the following sections, and reference should be made to the comprehensive analysis of Poulos and Davis (1974).

2.4.3 Stress distribution due to a concentrated horizontal load

In this case, the vertical stress $\Delta\sigma_z$ and shear stress $\Delta\tau_{xy}$ engendered by the horizontal concentrated load depicted in figure 2.18 are:

$$\Delta\sigma_z = \frac{3Pxz^2}{2\pi R^5} \tag{2.19a}$$

$$\Delta\tau_{xy} = \frac{Py}{2\pi R^3}\left[\frac{3x^2}{R^3} + \frac{(1-2\nu)}{(R+z)^2}\left(R^2 - x^2 - \frac{2Rx^2}{R+z}\right)\right] \tag{2.19b}$$

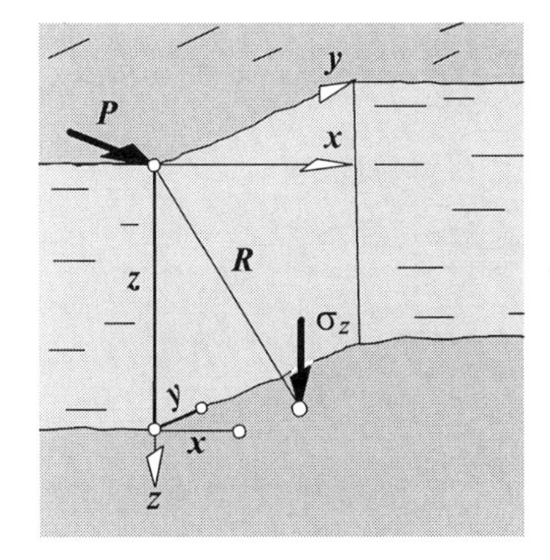

Figure 2.18: Stresses due to a horizontal point load.

2.4.4 vertical stress due to a uniformly loaded circular area

The stresses induced by this type of loading under the *centre* of the loaded area can be obtained by assuming that the vertical load applied on an elementary area such as the one depicted in figure 2.19, whose magnitude is $qr\,dr\,d\theta$, acts as a point load at the centre of the element. The stresses at any depth below the *centre* of the whole area can thence be found by integration of equations 2.19*a*, in which the quantity $qr\,dr\,d\theta$ is substituted for *P*. Knowing that:

$$R=\left[r^2+z^2\right]^{1/2} \quad \Rightarrow \quad R^5=z^5\left[1+(r/z)^2\right]^{5/2}$$

the increase in vertical stress corresponding to figure 2.19 is then calculated as follows:

$$\Delta\sigma_z=\int_0^{2\pi}\int_0^{a}\frac{3q\,r\,dr\,d\theta}{2\pi z^2\left[1+(r/z)^2\right]^{5/2}}$$

A straightforward integration yields :

$$\Delta\sigma_z=q\left[1-\frac{1}{\left[1+(a/z)^2\right]^{3/2}}\right] \tag{2.20}$$

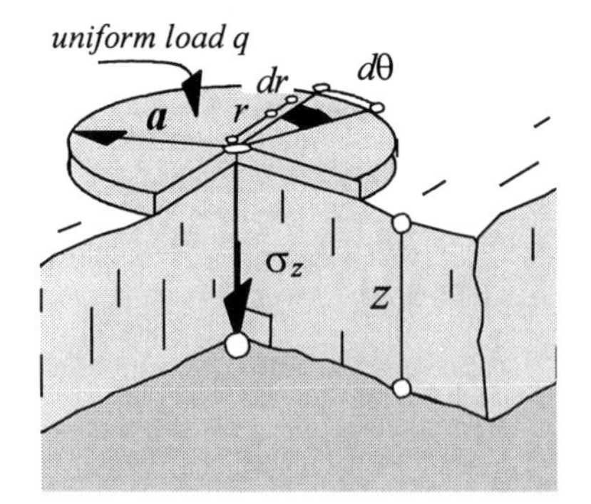

Figure 2.19: Stress increase due to a circular uniform load.

Equation 2.20 is *only* used in conjunction with the calculation of the vertical stress under the *centre* of the uniformly loaded circular area. For points outside the central axis, the analytical solution, though it exists (see Harr (1966), for example), is very complex in nature. However, graphical solutions to the stress distribution are available, and figure 2.20 represents the charts produced by Foster and Ahlvin (1954) for the evaluation of *vertical* stresses at any point beneath a uniformly loaded circular area. A more comprehensive graphical solution for this type of loading can be found in Poulos and Davis (1974), for instance.

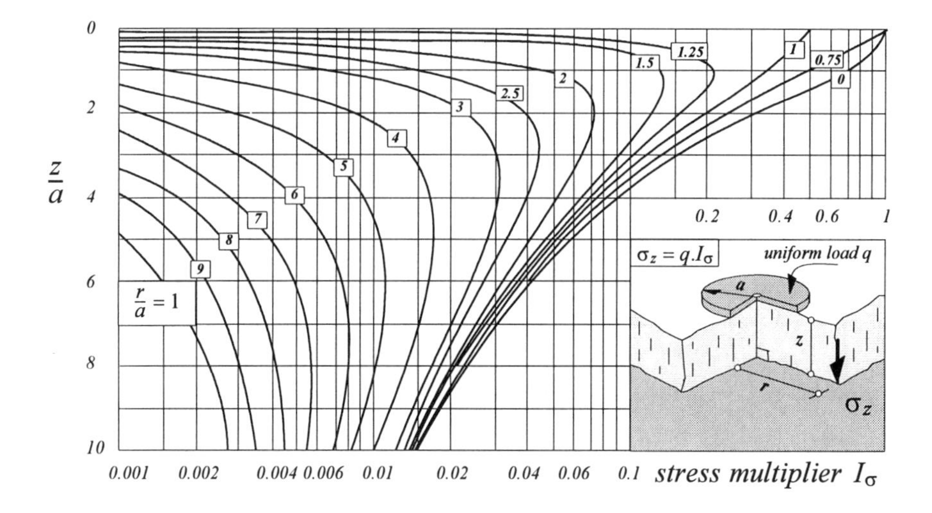

Figure 2.20: Stress multiplier for uniformly loaded circular areas due to Foster and Ahlvin (1954).

Example 2.4

A storage tank is to be built on the surface of a thick clay layer containing a pocket of softer material at a depth of 6 *m* below the ground as depicted in figure 2.21. When fully operational, the flexible foundation of the tank will be transmitting a uniform pressure q to the ground surface and, because of the nature of the soft material, it is required that the *increase* in vertical pressure (due to the tank) at point b (figure 2.21) should in no way exceed $\Delta\sigma_v = 60\,kN/m^2$. Estimate the maximum pressure q and the corresponding increase in vertical pressure at different points indicated in the figure.

First, the stress multiplier at point b is determined in the following way using the notation in figure 2.20:

$r = 6\,m$, $a = 8\,m$, $z = 6\,m$, whence, $r/a = z/a = 0.75$. Using the charts in figure 2.20, in conjunction with these ratios, it is seen that the stress multiplier $I_\sigma \approx 0.6$.

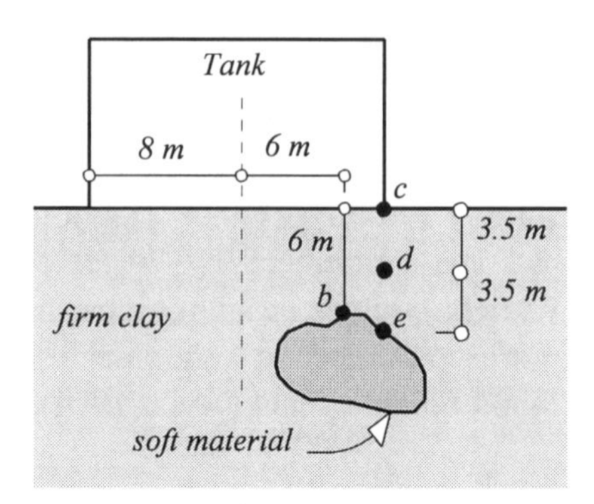

Figure 2.21: Tank dimensions.

Because of the limit imposed on the maximum vertical stress increase at b, it follows that:

$$\Delta\sigma_v = q_{max} I_\sigma = 60\,kN/m^2 \quad \Rightarrow \quad q_{max} = \frac{60}{0.6} = 100\,kN/m^2$$

Under this maximum pressure, the increase in vertical stress at points c, d, and e is (with reference to the charts in figure 2.20) such that:

- *point c:* $r/a = 1, \quad z/a = 0 \quad \Rightarrow \quad I_\sigma = 0.5 \quad \Rightarrow$
$\Delta\sigma_v = 100 \times 0.5 = 50\,kN/m^2$,

- *point d:* $r/a = 1, \quad z/a = 3.5/8 = 0.437, \quad \Rightarrow \quad I_\sigma = 0.43 \quad \Rightarrow$
$\Delta\sigma_v = 43\,kN/m^2$,

- *point e:* $r/a = 1, \quad z/a = 7/8 = 0.875, \quad \Rightarrow \quad I_\sigma \approx 0.35 \quad \Rightarrow$
$\Delta\sigma_v = 35\,kN/m^2$.

2.4.5 Vertical stress due to a uniformly loaded rectangular area

The stresses due to this type of loading under a *corner* of the uniformly loaded rectangular area can be calculated by integration of Boussinesq solutions for a concentrated vertical load (equations 2.19). Referring to figure 2.22, it is seen that the load applied to the elementary area $dx\,dy$ can be assumed to act as a concentrated load of a magnitude $q\,dx\,dy$.

Thus, substituting for the load P in equations 2.19 in the knowledge that $R=\sqrt{x^2+y^2+z^2}$, it follows that the increase in vertical stress, for instance, generated by the whole rectangular area under a corner such as the one illustrated in figure 2.22, is calculated by integration of equation 2.19*a*:

$$\Delta\sigma_z = \int_0^l \int_0^b \frac{3z^3 q\,dx\,dy}{2\pi\left(x^2+y^2+z^2\right)^{5/2}} \tag{2.21}$$

It can be shown that the calculus related to this integration (which is out of the scope of this text) yields the following solution:

$$\Delta\sigma_z = \frac{q}{4\pi}\left[\frac{2mn\,(m^2+n^2+2)\sqrt{m^2+n^2+1}}{(m^2+n^2+m^2n^2+1)(m^2+n^2+1)} + \tan^{-1}\left(\frac{2mn\sqrt{m^2+n^2+1}}{m^2+n^2+m^2n^2-1}\right)\right] \tag{2.22}$$

with $m=b/z, \quad n=l/z.$

Equation 2.22 can be rewritten as:

$$\Delta\sigma_z = qI_\sigma \tag{2.23}$$

where the stress multiplier I_σ depends on m and n, in other words on the size of the rectangle and the depth under the corner at which the stress is calculated. The corresponding graphical solution (Fadum (1948)) is reproduced in figure 2.23 in which the variables m and n are interchangeable.

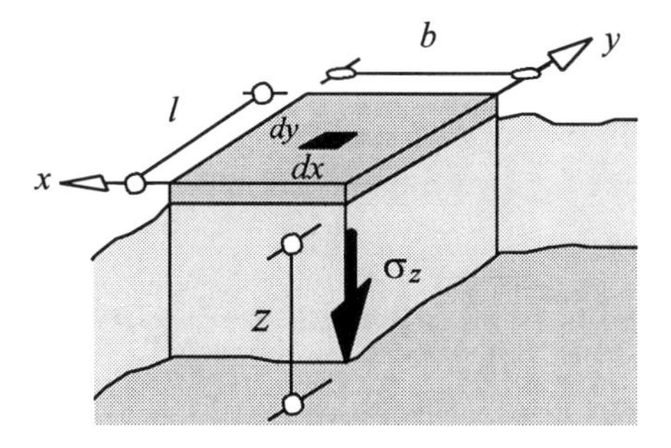

Figure 2.22: Stress due to a uniformly loaded rectangular area.

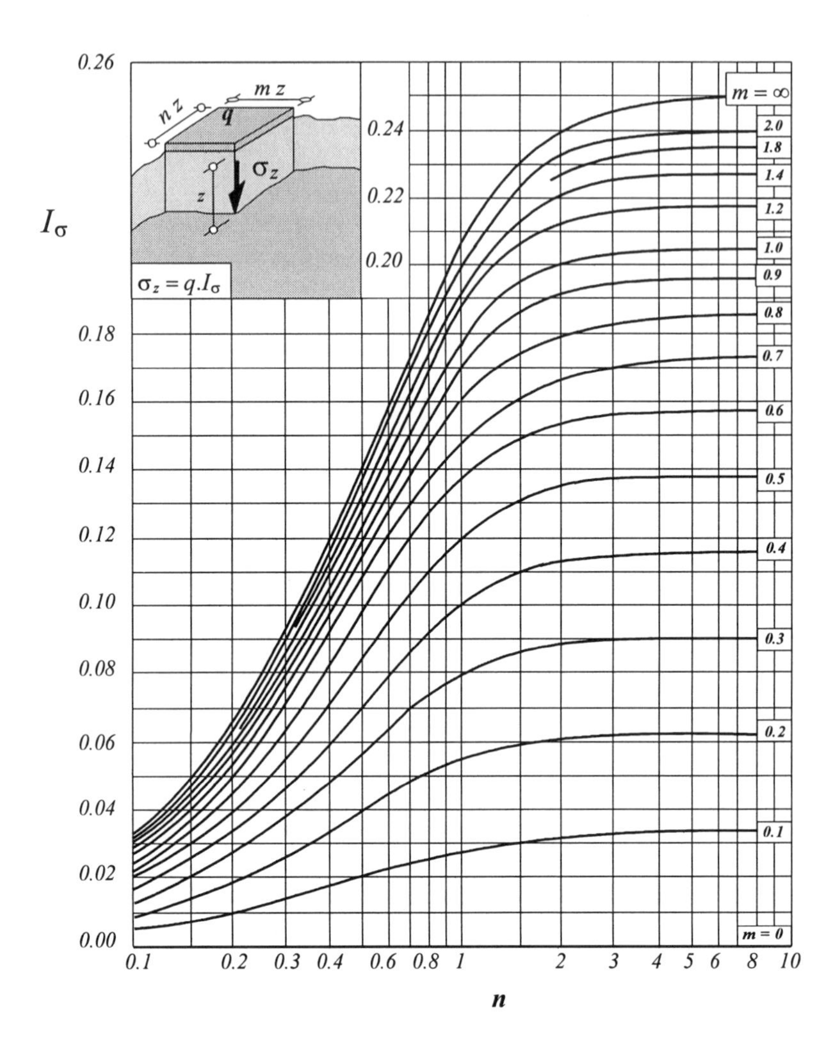

Figure 2.23: Stress multiplier for uniformly loaded rectangular areas (Fadum, 1948).

2.4.6 Vertical stress due to any shape of a uniformly loaded area

The graphical solution devised by Newmark (1942) amounts to a brilliant idea based on rearranging the expression of the vertical stress induced by a flexible circular area subjected to uniform load q and derived previously.

Using the diameter $D=2a$, it can be seen that equation 2.20 can be rewritten as follows:

$$\frac{D}{z}=2\left[\left(\frac{1}{1-\frac{\Delta\sigma_z}{q}}\right)^{2/3}-1\right]^{1/2} \qquad (2.24)$$

Accordingly, the magnitude of the vertical stress increase $\Delta\sigma_z$ at a depth z, induced by a circular area of diameter D subjected to a uniform load q, depends on the ratio D/z. For instance, a stress increase $\Delta\sigma_z=0.3q$ corresponds to a ratio $D/z=1.036$, whence the following values:

$\Delta\sigma_z/q$	0.1	0.2	0.3	0.4	0.5	0.6	0.7	0.8	0.9	1.0
D/z	0.54	0.80	1.04	1.27	1.53	1.84	2.22	2.77	3.82	∞

Significantly, these values indicate that at a given depth z, the vertical stress generated by a flexible circular area transmitting a pressure q will have a magnitude $\Delta\sigma_z=0.1q$ when the diameter of the loaded area is $D=0.54z$; similarly, an increase of $\Delta\sigma_z=0.2q$ will result from an area of diameter $D=0.8z$, and so on. Consequently, Newmark suggested *choosing* an arbitrary scale line z for which concentric circles of diameters corresponding to the ratios in the above table are then drawn (refer to figure 2.24). The circle with the smallest diameter corresponds precisely to the flexible circular area that will engender a vertical stress increase $\Delta\sigma_z$ (at a depth z) equal to 10% of the uniform load q transmitted to the soil, so that when the area of the circle is subdivided radially into, say, 20 equal elements, the vertical stress generated at the depth z by each one of these elements is:

$$d(\Delta\sigma_z)=\Delta\sigma_z/20=0.1q/20=0.005q \qquad (2.25)$$

Referring to figure 2.24, it is seen that for a given depth z, the vertical stress induced by a uniformly loaded area of *any shape* can be estimated provided that the area in question is plotted to the scale indicated in the figure, and that the point at which the stress is calculated is placed at the centre of the circles.

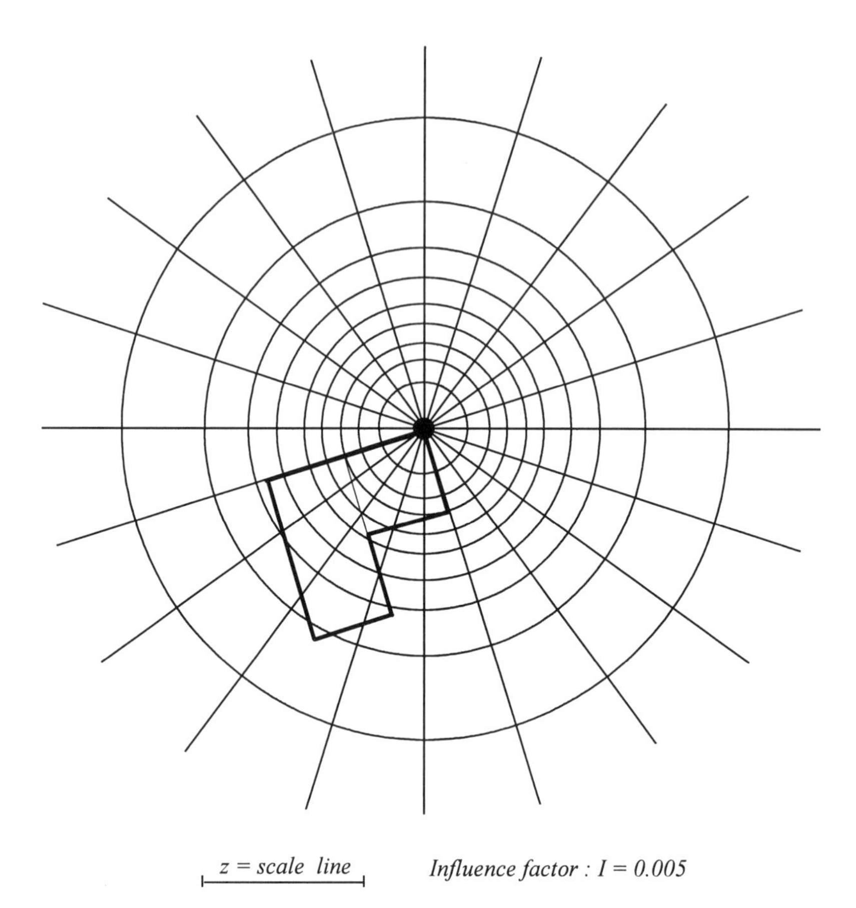

Figure 2.24: Newmark charts for the calculation of vertical stress increases due to uniformly loaded areas of any shape.

Once the number N of elements covered by the scaled area is counted (estimating partial elements), $\Delta\sigma_z$ is then calculated as follows:

$$\Delta\sigma_z = 0.005Nq \tag{2.26}$$

Example 2.5

Consider the flexible foundation depicted in figure 2.25 where the vertical stress, due to the uniform pressures q_1 and q_2, at point A situated at a depth $z = 6\,m$ needs to be evaluated. The scale to which the foundation is drawn in bold in figure 2.24 is such that the scale line in the figure corresponds to $z = 6\,m$ (*i.e.* the depth at which the stress needs to be estimated). Notice that the scaled foundation is drawn in a way that the point at which the stress is calculated (point A) is placed at the centre of the figure.

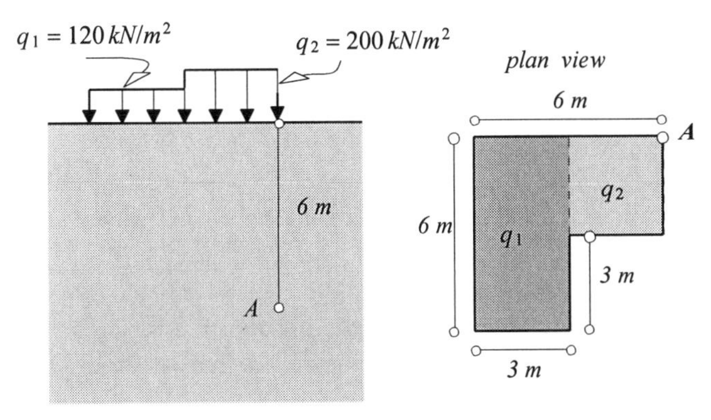

Figure 2.25: Example of a flexible, uniformly loaded foundation.

The procedure is then straightforward. It is seen that the number of elements covered by the section of foundation subjected to the uniform pressure q_1 corresponds to $N_1 \approx 10.6$, whereas the number of elements corresponding to the section subjected to q_2 is approximately $N_2 \approx 17.3$. So an increase in the (vertical) stress at point A:

$$\Delta\sigma_A = 0.005 \times (10.6 \times 120 + 17.3 \times 200) = 23.7\,kN/m^2$$

As an alternative, Fadum's charts of figure 2.23 can be used to estimate the increase $\Delta\sigma_z$ at A. In this case, it is obvious that the method, as described earlier, needs to be adjusted, in that point A needs to be at the corner of *every* rectangle or square into which the actual foundation will be subdivided. For that to happen, some *artificial* areas are needed, so that the requirement is fulfilled; in which case care must be taken so that any stress generated by whatever *artificial* area is subtracted from the total calculated value.

The procedure is best illustrated in figure 2.26, whereby the total increase in vertical stress $\Delta\sigma_{z1}$ at A is that calculated from the enlarged area $ABCD$ subjected to q_1, from which $\Delta\sigma_{z2}$ due to the artificially added area $ABEH$

is then subtracted, and to which $\Delta\sigma_{z3}$ due to the area *AGFH* subjected to q_2 is finally added. Notice that all three areas have point *A* at their top right-hand side corners.

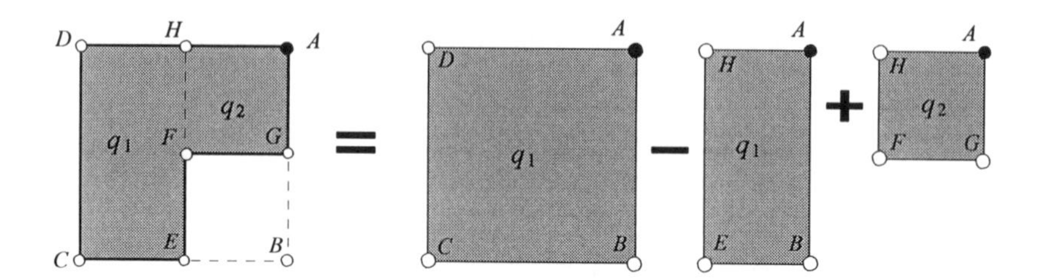

Figure 2.26: Practical aspects related to the use of Fadum charts.

Accordingly, the three stress increases at a depth $z = 6\,m$ beneath point *A* can now be estimated using the charts in figure 2.23 where it is seen that :

- *area ABCD* : $mz = nz = 6\,m, \quad z = 6\,m \quad \Rightarrow m = n = 1 \Rightarrow I_\sigma = 0.177,$

$$\Delta\sigma_{z1} = I_\sigma q_1 = 0.177 \times 120 = 21.24\,kN/m^2$$

- *area ABEH* : $mz = 3\,m, \quad nz = 6\,m, \quad z = 6\,m \quad \Rightarrow$
$m = 0.5, \quad n = 1, \quad \Rightarrow \quad I_\sigma = 0.122,$

$$\Delta\sigma_{z2} = I_\sigma q_1 = 0.122 \times 120 = 14.64\,kN/m^2,$$

- *area AGFH* : $mz = nz = 3\,m, \quad z = 6\,m \Rightarrow m = n = 0.5 \Rightarrow I_\sigma = 0.084.$

$$\Delta\sigma_{z3} = I_\sigma q_2 = 0.084 \times 200 = 16.8\,kN/m^2.$$

Hence the *estimated* vertical stress increase at point *A*:

$$\Delta\sigma_A = 21.24 - 14.64 + 16.8 = 23.4\,kN/m^2$$

A comparison with the stress increase calculated using Newmark charts shows that both methods yield remarkably similar values.

2.4.7 Stresses due to a vertical line load

Under plane strain conditions (that is when a strain component in one direction of the space is very small compared with the two other components, as in the case of an embankment, for instance), the equilibrium equation 2.16 (corresponding to zero body force), expressed in terms of the stress function Φ in the Cartesian co-ordinates system, reduces to:

$$\frac{\partial^4 \Phi}{\partial x^4} + 2\frac{\partial^4 \Phi}{\partial x^2 \partial z^2} + \frac{\partial^4 \Phi}{\partial z^4} = 0 \qquad (2.27)$$

The stresses in the two dimensional stress field (see figure 2.27) are related to Φ in the following way:

$$\sigma_x = \frac{\partial^2 \Phi}{\partial z^2}, \quad \sigma_z = \frac{\partial^2 \Phi}{\partial x^2}, \quad \tau_{xz} = \frac{\partial^2 \Phi}{\partial x \partial z}$$

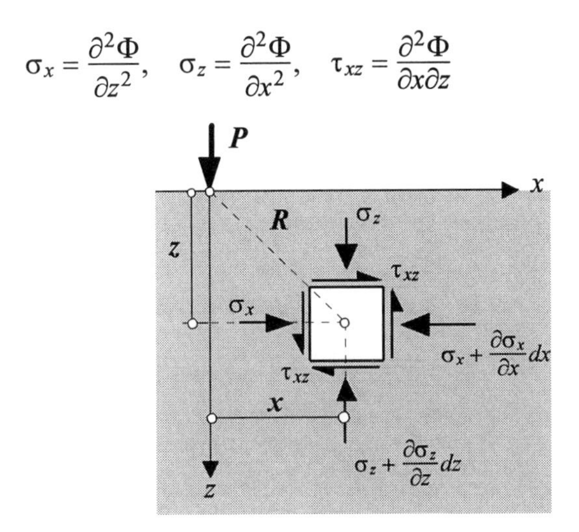

Figure 2.27: Two-dimensional stress field.

The case of a vertical line load, depicted in figure 2.28, is often referred to as the integrated Boussinesq problem, and its analytical solution can be established fairly easily by inserting the following stress function into equation 2.27:

$$\Phi = cx \tan^{-1}\left(\frac{z}{x}\right) \qquad (2.28)$$

where c is a constant. At this stage, the derivation of different stresses is easy to follow. For example, the expression of the vertical stress σ_z is:

$$\sigma_z = \frac{\partial^2 \Phi}{\partial x^2} = -\frac{2c z^3}{\left(x^2 + z^2\right)^2}$$

The constant c is evaluated from the boundary condition:

$$Q = \int_{-\infty}^{+\infty} \sigma_z \, dx = -2c z^3 \int_{-\infty}^{+\infty} \frac{1}{\left(x^2 + z^2\right)^2} dx$$

Q being the *load per unit length*. It can be shown that this integration yields the following c value: $c = -Q/\pi$. Whence:

$$\sigma_z = \frac{2Q z^3}{\pi R^4} \tag{2.29a}$$

For the shear stress increase, it is easy to establish that:

$$\tau_{xz} = \frac{2Q z^2 x}{\pi R^4} \tag{2.29b}$$

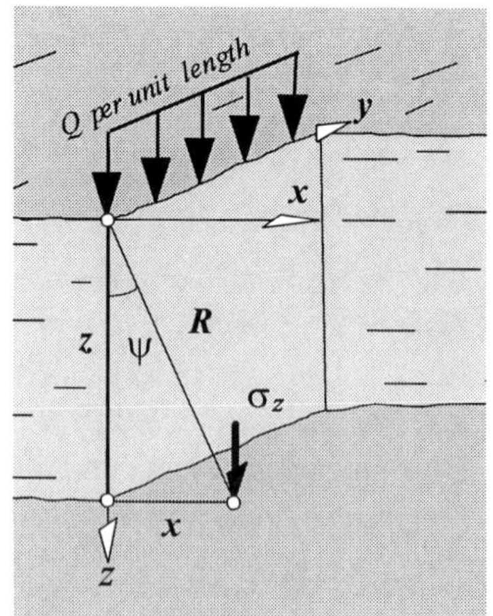

Figure 2.28: Stresses due to a vertical line load.

Example 2.6

Consider the railway line depicted in figure 2.29, used for freight, and subjected to a maximum vertical (static) load $Q = 500\ kN/m$.
To evaluate the increase in both vertical and shear stresses due to Q at points A and B, equations 2.29a and b yield the following:

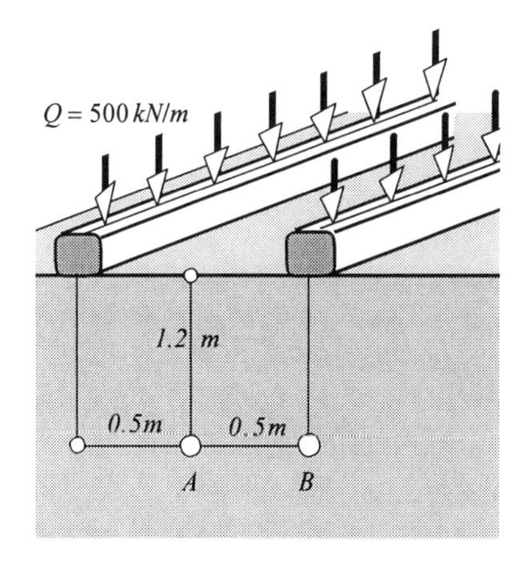

Figure 2.29: Loading conditions.

- *point A:* (remember both lines count)

$$\Delta\sigma_z = 2 \times \frac{2 \times 500 \times 1.2^3}{\pi\left(1.2^2 + 0.5^2\right)^2} = 385.2\ kN/m^2$$

$\Delta\tau_{xz} = 0$ (because of symmetry)

- *point B:* applying the superposition principle, it follows that:

$$\Delta\sigma_z = \frac{2 \times 500 \times 1.2^3}{\pi \times 1.2^4} + \frac{2 \times 500 \times 1.2^3}{\pi\left(1.2^2 + 1^2\right)^2} = 357.6\ kN/m^2$$

$$\Delta\tau_{xz} = \frac{2 \times 500 \times 1.2^2 \times 1}{\pi\left(1.2^2 + 1^2\right)^2} = 77\ kN/m^2$$

It is seen that, for the considered depth, the maximum vertical stress occurs at the central point A.

2.4.8 Stresses due to a horizontal line load

In the case of a semi-infinite soil mass subject to a horizontal line load as illustrated in figure 2.30, the derivation of different stresses results in the following:

$$\Delta\sigma_z = \frac{2Qz^2x}{\pi R^4} \qquad (2.30a)$$

$$\Delta\tau_{xz} = \frac{2Qx^2z}{\pi R^4} \qquad (2.30b)$$

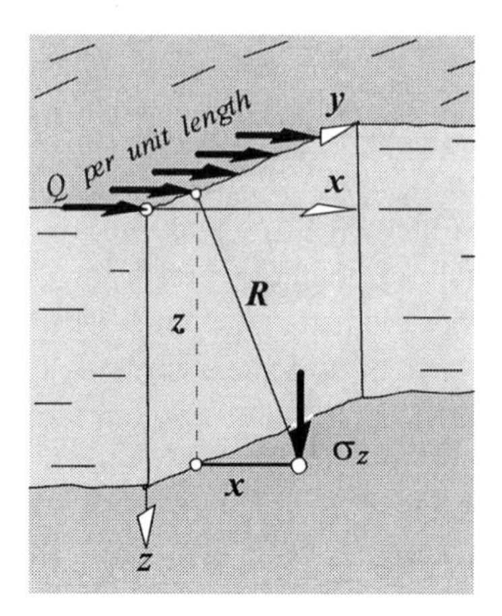

Figure 2.30: Stresses due to a horizontal line load.

2.4.9 Stresses due to a strip load: vertical uniform loading

If a strip, supposedly of infinite length, is transmitting a *vertical uniform load q* (*i.e.* a *load per unit area*) to the soil surface, then the stresses generated at any depth within the soil mass can be found by integration of equations 2.29 corresponding to a vertical line load. The integration can be made using the angle ψ indicated in figure 2.28, in conjunction with equations 2.29 :

$$\cos\psi = \frac{z}{R} \qquad (2.31)$$

Accordingly, the following integration must be undertaken between the limits indicated in figure 2.31:

$$\sigma_z = \int_{-b}^{+b} \frac{2Q}{\pi}\frac{\cos^4\psi}{z}dx \qquad (2.32a)$$

$$\tau_{xz} = \int_{-b}^{+b} \frac{2Q}{\pi}\frac{\cos^3\psi\sin\psi}{z}dx \qquad (2.32b)$$

Note that Q in the above expressions represents a *load per unit length.*

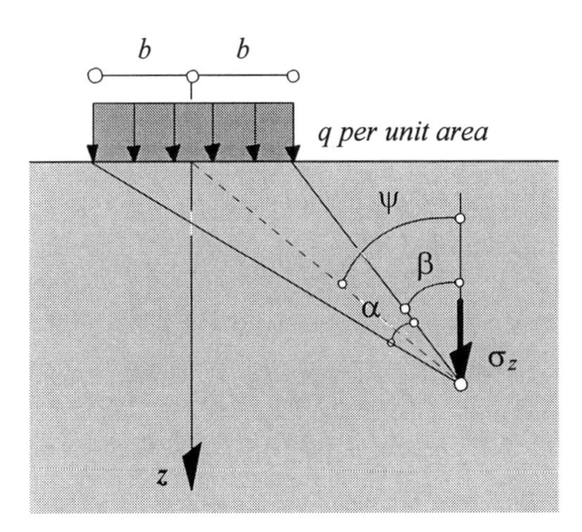

Figure 2.31: Stresses due to a vertical strip load.

Notwithstanding the somewhat cumbersome nature of the mathematical formalism, it can be shown that an integration using the angles α and β instead of ψ as depicted in figure 2.31 yields the following relationships in which q corresponds to a *load per unit area*:

$$\Delta\sigma_z = \tfrac{q}{\pi}[\alpha + \sin\alpha \cos(\alpha + 2\beta)] \qquad (2.33a)$$

$$\Delta\tau_{xz} = \tfrac{q}{\pi}\sin\alpha \sin(\alpha + 2\beta) \qquad (2.33c)$$

α and β are expressed in *radians.*

2.4.10 Stresses due to a strip load: horizontal uniform loading

The stress components at a point such as the one depicted in figure 2.32 are calculated in a way similar to that used in the previous case of vertical uniform loading:

$$\Delta\sigma_z = \tfrac{q}{\pi}\sin\alpha \sin(\alpha + 2\beta) \qquad (2.34a)$$

$$\Delta\tau_{xz} = \tfrac{q}{\pi}[\alpha - \sin\alpha \cos(\alpha + 2\beta)] \qquad (2.34b)$$

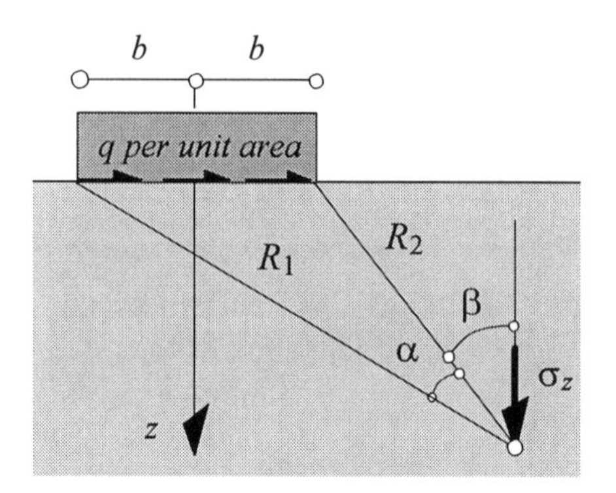

Figure 2.32: Stresses due to a horizontal strip load.

2.4.11 Stresses due to a linearly increasing vertical uniform loading

The stress increases due to a triangular area supposedly of infinite extent, at a point with co-ordinates (x,z) are evaluated as follows:

$$\Delta\sigma_z = \frac{q}{2\pi}\left[\frac{\alpha x}{b} - \sin 2\beta\right] \tag{2.35a}$$

$$\Delta\tau_{xz} = \frac{q}{2\pi}\left[1 + \cos 2\beta - \frac{z\alpha}{b}\right] \tag{2.35b}$$

Note that the uniform load varies linearly from zero to q.

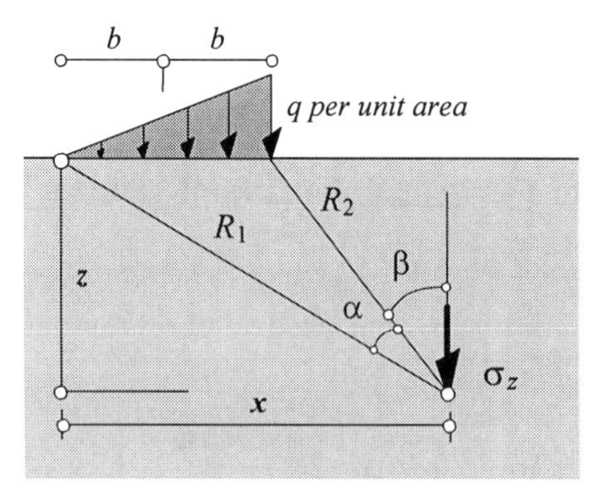

Figure 2.33: Stress field due to a triangular vertical load.

This type of loading is encountered during embankment construction whereby the total loaded area is formed of a strip in the middle and a triangular area, such as those illustrated in figure 2.33, on each side. The total stress increase due to an embankment can therefore be calculated at any point beneath the surface using the superposition principle: the contribution of the central strip is first calculated, and the contributions of both triangular areas are then added.

Example 2.7

Estimate the increase in the vertical stress $\Delta\sigma_z$ within the stiff clay layer at points A and B, due to the loading conditions generated by the embankment shown in figure 2.34.

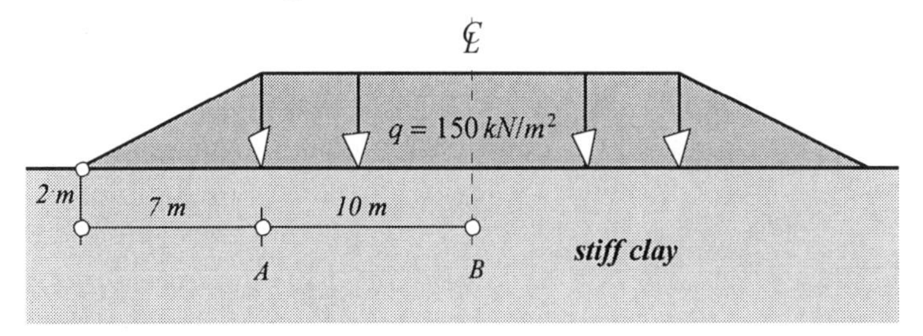

Figure 2.34: Embankment dimensions.

Referring to figures 2.34 and 2.35, and using the notation of figure 2.33, the different quantities needed to calculate the stresses at B are such that:

$$\beta_1 = 0, \qquad \alpha_1 = \tan^{-1}\left(\tfrac{10}{2}\right) = 1.373\ rad,\ \beta_2 = \alpha_1 = 1.373\ rad,$$

$$\alpha_2 = \tan^{-1}\left(\tfrac{17}{2}\right) - \alpha_1 = 0.08\ rad, \qquad R_1 = \left[2^2 + 17^2\right]^{1/2} = 17.11\ m,$$

$$R_2 = \left[2^2 + 10^2\right]^{1/2} = 10.2\ m, \quad x = 17\ m, \quad b = 3.5\ m$$

Because of symmetry, the increase in vertical stress at B can be calculated using a combination of equations 2.33a and 2.35a as follows:

$$\Delta\sigma_z = 2\left[\frac{q}{\pi}(\alpha_1 + \sin\alpha_1\cos\alpha_1) + \frac{q}{2\pi}\left(\frac{\alpha_2 x}{b} - \sin 2\beta_2\right)\right]$$

$$= 2 \times \frac{150}{\pi}\left[(1.373 + \sin 1.373 \times \cos 1.373) + \left(\frac{0.08 \times 17}{7} - \frac{1}{2}\sin(2 \times 1.373)\right)\right]$$

$$= 149.7\, kN/m^2$$

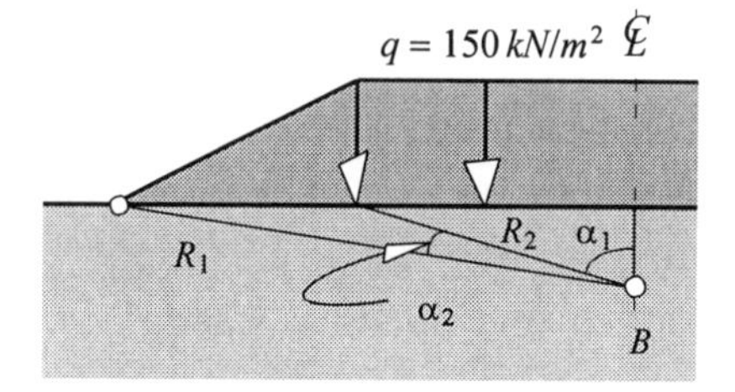

Figure 2.35: Parameters related to stress calculations at B.

It is noticeable that, because of the embankment dimensions and the central position of point B, the contribution of both triangular areas to the vertical stress increase is, under these circumstances, negligible.

Assessing the stress increase at A, figure 2.36 shows clearly that the contribution of the three areas, that is the two triangles and the central rectangle, have to be evaluated separately, then added together to yield the total stress increase. Using the angles indicated in the figure, as well as the dimensions in figure 2.34 and the notation of figure 2.33, the following quantities are easily established :

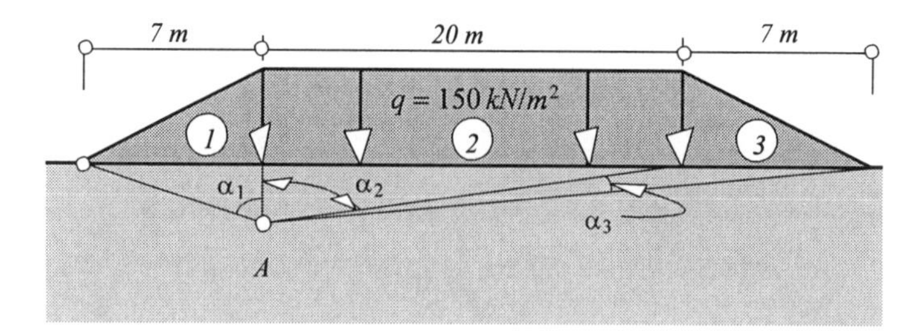

Figure 2.36: Parameters related to stress calculations at A.

- *triangle 1:* $\alpha_1 = \tan^{-1}\left(\frac{7}{2}\right) = 1.292\, rad,$

$\beta_1 = 0, \quad R_1 = \left[2^2 + 7^2\right]^{1/2} = 7.28\, m, \quad R_2 = 2\, m, \quad x_1 = 7\, m, \quad b = 3.5\, m.$

- *rectangle 2:* $\alpha_2 = \tan^{-1}\left(\frac{20}{2}\right) = 1.471\, rad, \quad \beta_2 = 0.$

- *triangle 3:* $\alpha_3 = \tan^{-1}\left(\frac{27}{2}\right) - \alpha_2 = 0.026\, rad, \quad \beta_3 = \alpha_2 = 1.471\, rad,$

$R_1 = \left[2^2 + 27^2\right]^{1/2} = 27.07\, m,$

$R_2 = \left[2^2 + 20^2\right]^{1/2} = 20.1\, m, \quad x_3 = 27\, m, \quad b = 3.5\, m.$

Hence, a combination of equations 2.33*a* and 2.35*a* yields the following increase in vertical stress at *A*:

$$\Delta\sigma_z = \frac{150}{2\pi}\left[\frac{1.292 \times 7}{3.5} - \sin(2 \times 0)\right] + \frac{150}{\pi}[1.471 + \sin 1.471 \cos 1.471]$$

$$+ \frac{150}{2\pi}\left[\frac{0.026 \times 27}{3.5} - \sin(2 \times 1.471)\right]$$

$$= 61.7 + 62.2 + 0.05 = 124\, kN/m^2$$

Because of the position of point *A,* the contribution of triangle 3 (refer to figure 2.36) to the increase in vertical stress is very small indeed.

2.5 Depth of influence and stress isobars

The stress redistribution within a soil mass induced by a surface loading depends not only on the type and magnitude of the applied load, but also on the size and shape of the loaded area as shown unambiguously in figure 2.37. It is noticeable that the depth at which $\Delta\sigma_z = 0.1q$ is about 2*B* for a circular area, increasing to around 6*B* in the case of a long strip. Consequently, it can be safely assumed that any *vertical stress increase* induced by a surface loading becomes *marginal* beyond a depth of 3*B* for a circular or a rectangular foundation (*i.e.* loaded area), increasing to about $8B$ in the case of a long strip foundation; B being the diameter or width of the loaded area.

These depths of influence delimit the area within which the vertical stress varies, so that a profile of stress isobars (*i.e.* the contours of equal stresses) can easily be found from the different stress expressions derived previously. It can be shown that the shapes of the stress isobars (also referred to as stress bulbs because of their shape) differ according to the foundation shape. Thus, those depicted in figure 2.38*a* correspond to a circular area, whereas the isobars of figure 2.38*b* relate to a long strip foundation. Both figures reflect the stress distribution discussed previously and, most importantly, indicate that under the same loading and soil conditions, the stressed area below a circular foundation having a diameter B is markedly smaller than that occurring under a strip foundation of a width B.

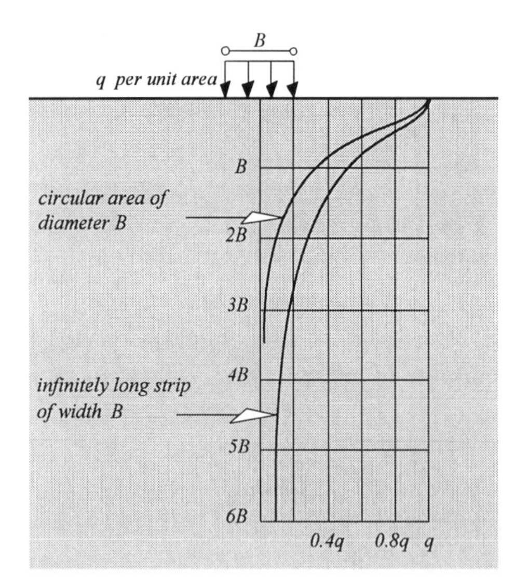

Figure 2.37: Effect of depth on the distribution of vertical stresses due to a uniform load.

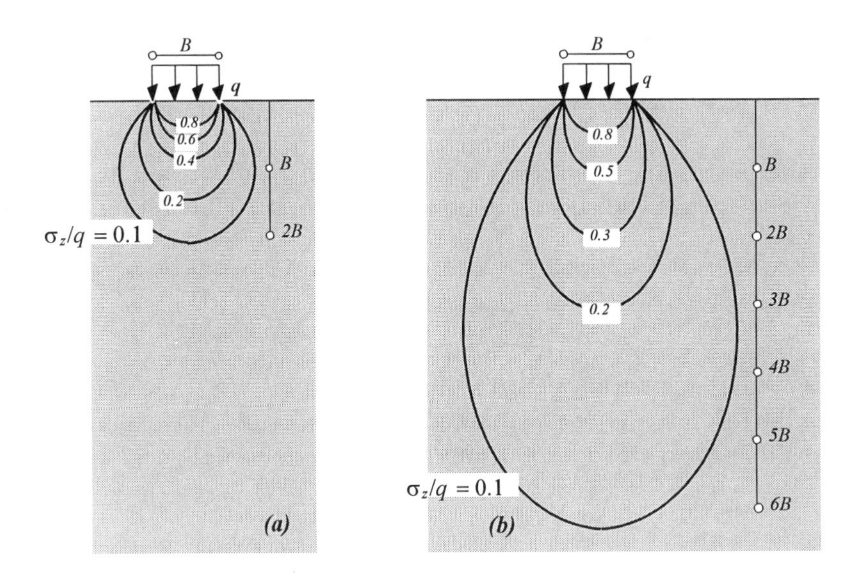

Figure 2.38: Isobars corresponding to: (*a*) *circular area,* (*b*) *long strip.*

Moreover, ample experimental evidence shows that, in the presence of adjacent loaded areas (as in the case of foundation pads for example), the interaction between the stress bulbs results in a wider and deeper stressed area of soil as depicted in figure 2.39. If the effect of a neighbouring loaded area did not exist, then the individual stress isobars would correspond to the ones represented in broken lines in the figure. However, the interaction between adjacent areas results in the spreading of stresses within the soil mass in a way that the overall stressed area is much larger than the individual ones.

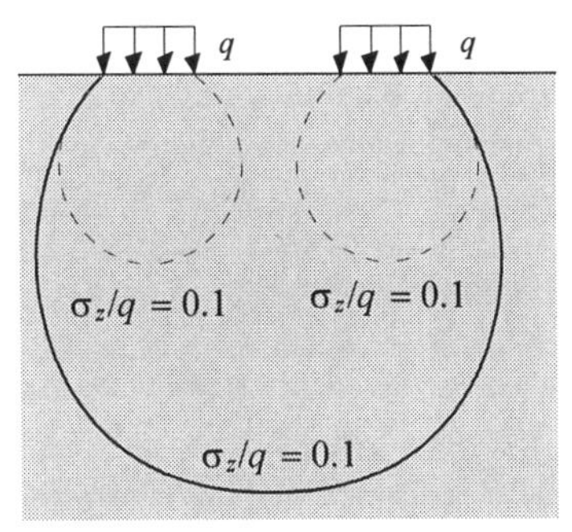

Figure 2.39: Interaction effects.

Because this stress redistribution has implications on the subsequent soil deformation (or settlement), care must be taken so as to include any interaction effect in the design.

In this respect, some empirical design rules can be used cautiously. For instance, if the individual loaded areas depicted in figure 2.40 are such that:

$$\sum B \times A \geq 0.75 \times B_1 \times A_1$$

then it is advisable to consider using the area $A_1 \times B_1$ for which the stress bulb is larger than the one that takes into account the interaction effect between individual areas, leading thus to a conservative design.

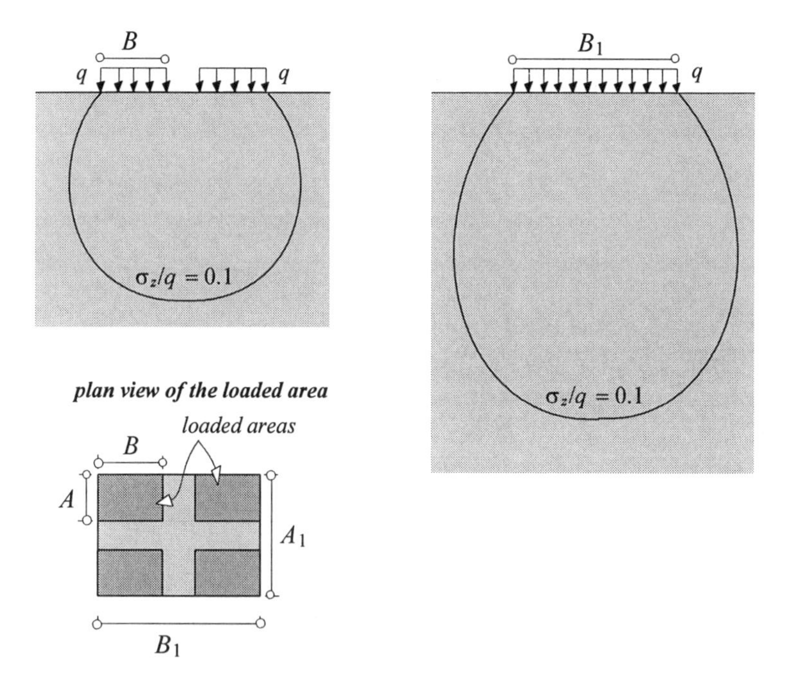

Figure 2.40: Effect of adjacent loaded areas on the stress distribution within a soil mass.

2.6 Shortcomings of elastic solutions

The solutions to the stress distribution presented earlier are based on Boussinesq analysis which *assumes* that the soil is elastic, homogeneous and isotropic. However, natural soils are usually anisotropic, heterogeneous and, most importantly, are characterised by non-linear stress–strain relationships. Experimental evidence shows that, while Boussinesq analysis can be extended to anisotropic soils, the extent of the elastic behaviour depends on the soil nature and is generally restricted to small deformations.

In particular, the assumption of elastic behaviour can no longer be justified for soils such as loose sands or young clays that have never been subjected to unloading during their geological history (known as normally consolidated clays, see section 4.1); the magnitude of the corresponding yield stresses being very low as depicted in figure 2.41.

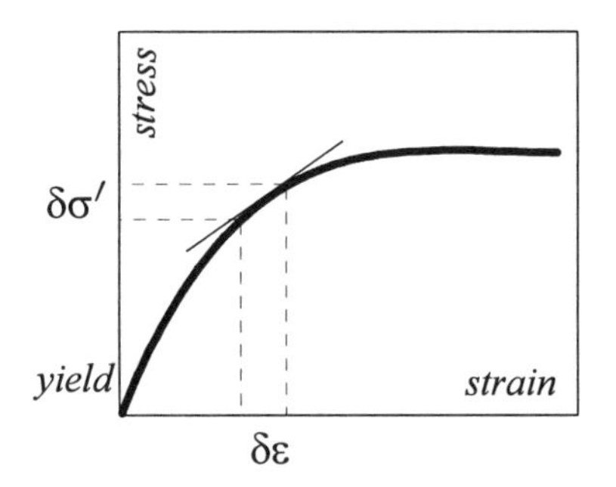

Figure 2.41: Tangent modulus for non-linear soil behaviour.

Under such circumstances, an iterative procedure is needed to evaluate the soil *stiffness* (*i.e.* the elasticity modulus) because the stress–strain relationship becomes incremental, and one solution consists of using the tangent modulus (refer to figure 2.41) defined as:

$$E = \frac{\delta\sigma'}{\delta\varepsilon} \tag{2.36}$$

This type of analysis forms the basis of numerical techniques such as finite difference or finite element methods which are exclusively undertaken using computers.

Despite their shortcomings, the solutions to the *vertical stress* distribution based on elasticity presented previously are deemed acceptable in most cases encountered in engineering practice (see Burland *et al.* (1977)). However, these solutions must be handled with great care when dealing with soils such as *loose sands* and *normally consolidated clays* (for which a comprehensive description of their behaviour is presented in section 4.1); in which case, a more sophisticated numerical modelling based on the finite elements method, for example, might be needed (see chapter 14).

Problems

2.1 A silty clay has the following characteristics:
$e = 0.94, \quad C = 15\,mm^2, \quad d_{10} = 1\,\mu m, \; d_{60} = 3.5\,\mu m$. Calculate the suction pressure (*i.e.* the negative porewater pressure) generated by the maximum capillary rise. Also, estimate the actual capillary saturation rise within the clay layer.

Ans: $u_c = -159.5\,kN/m^2, \quad h_s = 4.56\,m$

2.2 An oil tank is built at the surface of an 8 *m* thick clay layer underlain by a sandy gravel. The groundwater level is 2 *m* below the surface and the clay has a saturated unit weight $\gamma_{sat} = 20\,kN/m^3$ and a bulk unit weight $\gamma = 18.5\,kN/m^3$.

A deep excavation needs to be undertaken near the tank and, ideally, the site engineer would like his (her) team to work under dry conditions, meaning that the water table has to be lowered. The engineer then realised that the owners of the tank are adamant that the effective stress 5 *m* beneath the centre of the tank should in no way be increased by more than 20%, otherwise serious settlement problems may occur.

Estimate the depth z_w in figure *p*2.2 by which the water table can be lowered.

Ans: $z_w = 1.58\,m$

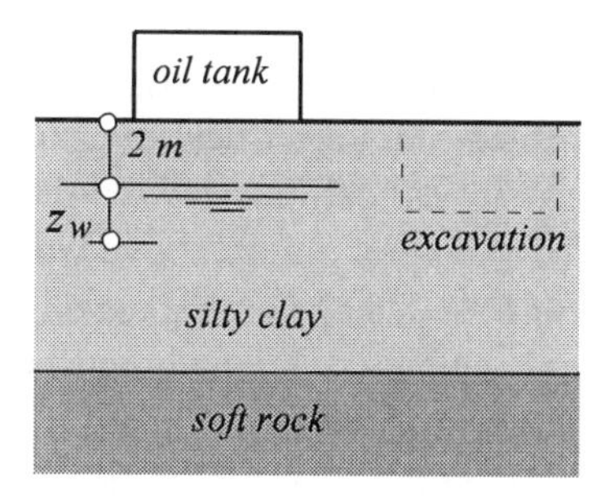

Figure p2.2

2.3 Consider the case of a 3 *m* deep lake above the surface of a 6 *m* thick layer of silty clay with a saturated unit weight $\gamma_{sat} = 20\,kN/m^3$ underlain by a stiff clay.

(*a*) Calculate the total, porewater and effective pressures (σ, u, and σ' respectively) at point *A*, 3 *m* below the ground surface.

(*b*) After an exceptionally hot summer, the lake has disappeared and the water level has receded to ground surface level. Estimate in this case the change in total, porewater and effective pressures at point *A*.

(*c*) At the peak of the dry season, the water level has further fallen to 1 *m* below the ground surface. Assuming a bulk unit weight $\gamma = 17\,kN/m^3$ applies above the water level, calculate the new values of σ, u and σ' at *A*.

(*d*) Plot the variation of σ, u and σ' across the silty clay layer corresponding to the three water levels.

Ans: (*a*) $\sigma = 90kN/m^2$, $u = 60\,kN/m^2$, $\sigma' = 30\,kN/m^2$,
(*b*) $\sigma = 60\,kN/m^2$, $u = 30\,kN/m^2$, $\sigma' = 30\,kN/m^2$,
(*c*) $\sigma = 57\,kN/m^2$, $u = 20\,kN/m^2$, $\sigma' = 37\,kN/m^2$.

2.4 Three line loads of magnitude: $Q_1 = Q_3 = 1200\,kN/m$ and $Q_2 = 2000\,kN/m$ are applied on a strip (assumed to be) of infinite length, at the surface of a very thick layer of glacial till (see figure *p*2.4). Preliminary investigations showed that the water table is at a static depth of $z_w = 5\,m$, and that, above water level, the bulk unit weight of the soil is $\gamma = 19\,kN/m^3$, increasing to $\gamma_{sat} = 20\,kN/m^3$ below the saturation line. Given the magnitude of loading, it is suggested that a thorough site investigation must be undertaken up to a depth z_i at which the increase in the vertical effective stress due to the surface loading is limited to $\Delta\sigma'_v = 0.05\sigma'_o$ beneath the centre of the loaded surface. Knowing that σ'_o represents the effective vertical *in situ* stress prior to the application of any surface loading, calculate the depth z_i.

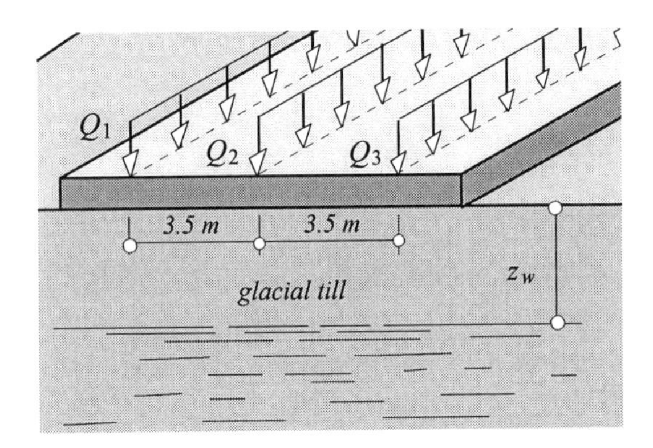

Figure p2.4

Ans: $z_i = 72.6\,m$

2.5 The base of a long retaining wall, assumed to be a long strip, is transmitting a resultant load $F = 620\,kN/m$, inclined at an angle $\beta = 12°$ to the vertical as depicted in figure *p*2.5.

(*a*) Calculate the vertical and shear stresses at point *A*, 1 *m* beneath the centre of foundation.

(*b*) Estimate the same stresses at point *B* situated at the same depth under the corner of foundation.

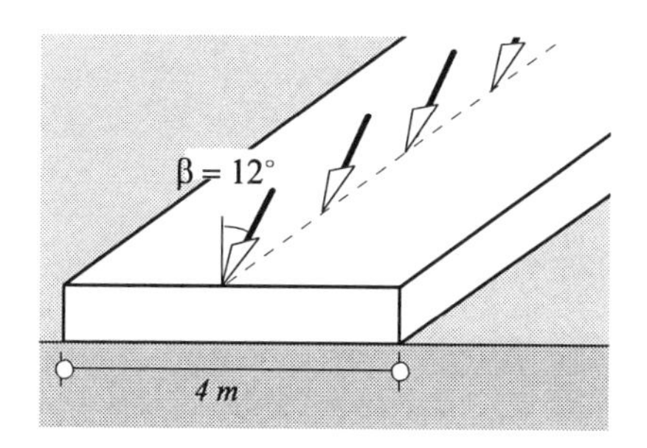

Figure p2.5

Ans: $(a)\ \tau_{xz} = 0,\ \sigma_z = 386.1\,kN/m^2,\ (b)\ \tau_{xz} = 44\,kN/m^2,\ \sigma_z = 22\,kN/m^2.$

2.6 Consider the case of the clay dam illustrated in figure *p*2.6, the clay having a saturated unit weight $\gamma_{sat} = 21\,kN/m^3$. Slope stability considerations require slopes of 1 to 3 and 1 to 2.5 on the upstream and downstream sides respectively.

Because of bearing capacity problems, the increase in vertical stress due to the weight of the dam at point *B*, 2 *m* deep as depicted in the figure should not exceed $150\,kN/m^2$.

Calculate under these circumstances the maximum height *H* of the dam.

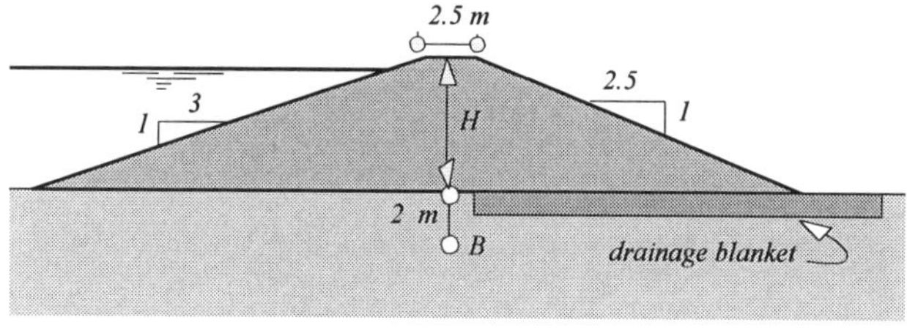

Figure p2.6

Ans: $H \approx 7.3\,m$

2.7 An oil tank with a flexible circular foundation is to be built on a layer of medium clay in the vicinity of a buried pipe as depicted in figure *p*2.7. When fully operational, the tank will transmit a net pressure $q = 250\,kN/m^2$ to the ground. To protect the pipe, it is suggested that any increase in effective vertical stress due to the tank loading at point A must be limited to $\Delta\sigma_v' = 5\,kN/m^2$.

Calculate the distance x in figure $p2.7$.

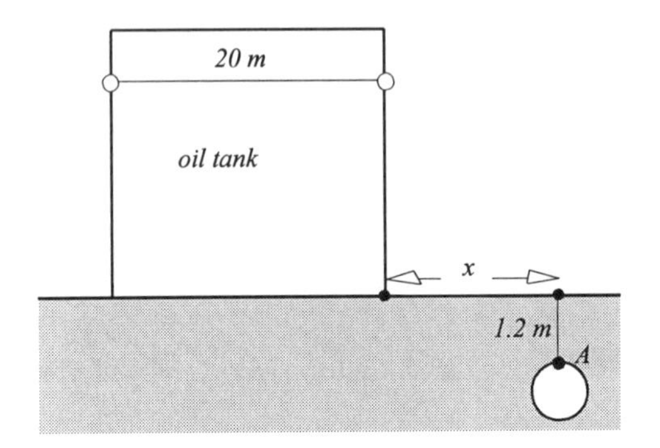

Figure p2.7

Ans: $x = 15\,m$

2.8 Consider the foundation with the shape illustrated in figure $p2.8$, transmitting a uniform pressure q to the ground.

Use Fadum's charts and estimate the ensuing increase in vertical stresses due to q at a depth $z = 1\,m$ beneath points *A, B,........, H.*

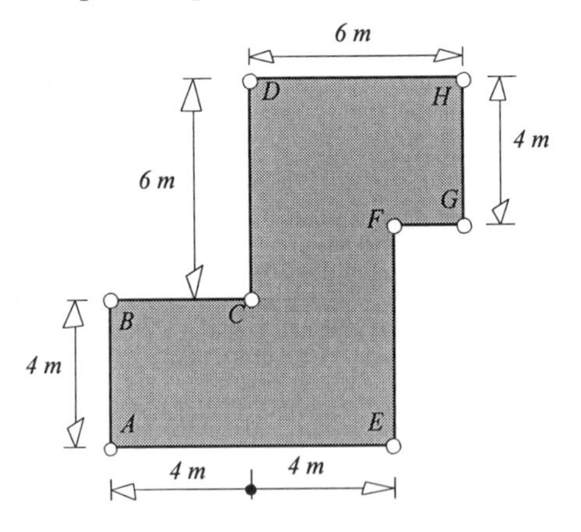

Figure p2.8

Ans: $\Delta\sigma_A \approx 0.245q$, $\Delta\sigma_B \approx 0.25q$, $\Delta\sigma_C \approx 0.724q$
$\Delta\sigma_D \approx 0.244q$, $\Delta\sigma_E \approx 0.244q$, $\Delta\sigma_F \approx 0.728q$
$\Delta\sigma_G \approx 0.247q$, $\Delta\sigma_H \approx 0.249q$.

2.9 Assume the same foundation of problem *p*2.8 is now transmitting different uniform pressures q_1 and q_2 as illustrated in figure *p*2.9.

(*a*) Use the superposition principles, and estimate with the help of Fadum's charts, for instance, the increase in vertical effective stresses at a depth $z = 1\,m$ beneath points *K, L* and *M*.

(*b*) Compare these results with those calculated in problem *p*2.8 at points *B, E* and *F*, when $q_1 = q_2$.

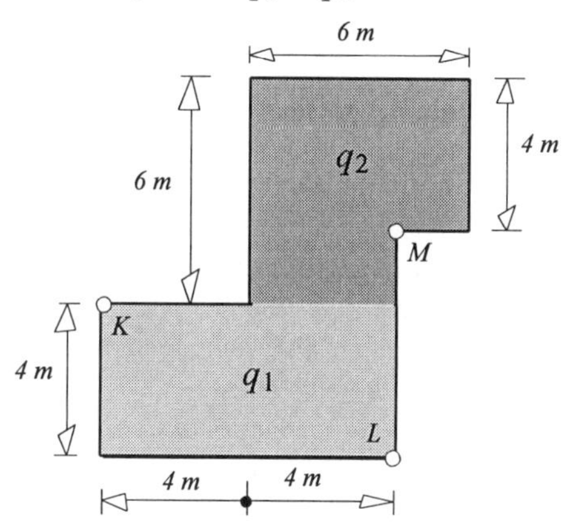

Figure p2.10

Ans: (*a*) $\Delta\sigma_K \approx 0.245q_1 + 0.005q_2$, $\Delta\sigma_L \approx 0.242q_1 + 0.002q_2$
$\Delta\sigma_M \approx 0.008q_1 + 0.72q_2$.

2.10 Use Newmark charts to estimate the increase in vertical pressures at a depth $z = 10\,m$ beneath points *K* and *M* in figure *p*2.9 when $q_2 = 2q_1 = 200\,kN/m^2$.

Ans: $\sigma_K \approx 19.5\,kN/m^2$, $\sigma_M \approx 32.5\,kN/m^2$

References

Bishop, A. W. and Donald, I. B. (1961) *The experimental study of partly saturated soils in the triaxial apparatus.* Proceedings of the 5th International

Conference on Soil Mechanics, Paris, 1, pp. 13–21.

Boussinesq, J. (1985) *Application des potentiels à l'étude de l'équilibre et du mouvement des solides élastiques.* Gauthier-Villars, Paris.

Burland, J. B., Broms, B. and De Mello, V. F. B. (1977) *Behaviour of foundations and structures.* State-of-the-art review. Proceedings of the 9th ICSMFE, Tokyo,Vol. 2, pp. 495–546.

Fadum, R. E. (1948) *Influence values for estimating stresses in elastic foundations.* Proceedings of the 2nd ICSMFE, Rotterdam, Vol. 3, pp. 77–84.

Foster, C. R. and Ahlvin, R. G. (1954) *Stresses and deflections induced by a uniform circular load.* Proceedings of Highway Research Board, Vol. 33, pp. 467–470.

Fredlund, D. G. (1979) *Appropriate concepts and technology for unsaturated soils.* Canadian Geotechnical Journal, 16, pp. 121–139.

Harr, M. E. (1966) *Foundations of Theoretical Soil Mechanics.* McGraw-Hill, New York.

Jennings, J. E. B. and Burland, J. B. (1962) *Limitations to the use of effective stresses in partly saturated soils.* Géotechnique, 12 (2), pp. 125–144.

Mitchell, J. K. (1993) *Fundamentals of Soil Behaviour.* 2nd edn, John Wiley, New York.

Newmark, N. M. (1942) *Influence charts for computation of stresses in elastic foundations.* Engineering experiment station bulletin, No 338, University of Illinois.

Poulos, H. G. and Davis, E. H. (1974) *Elastic Solutions for Soil and Rock Mechanics.* John Wiley, New York.

Sills, G. C., Wheeler, S. J., Thomas, S. D. and Gardner, T. N. (1991) *Behaviour of off-shore soils containing gas bubbles.* Géotechnique, 41 (2), pp. 227–241.

Terzaghi, K. (1936) *The shearing resistance of saturated soils.* Proceedings of the 1st International Conference on Soil Mechanics, Harvard, pp. 54–56.

Terzaghi, K. and Peck, R. B. (1967) *Soil Mechanics in Engineering Practice.* Wiley, New York.

Thomas, H. R. and He, Y. (1998) *Modelling the behaviour of unsaturated soil using an elastoplastic constitutive model.* Géotechnique, 48 (5), pp. 589–603.

Timoshenko, S. P. and Goodier, J. N. (1970) *Theory of Elasticity.* 3rd edn, McGraw-Hill, International edition, Singapore.

Toll, D. G. (1990) *A framework for unsaturated soil behaviour.* Géotechnique, 40 (1), pp. 31–44.

Wheeler, S. J. (1991) *An alternative framework for unsaturated soil behaviour.* Géotechnique, 41 (2), pp. 257–261.

CHAPTER 3

Seepage flow

3.1 Introduction

The *flow* of water within a soil mass depends, by and large, on the soil *porosity*, in other words on the space between solid particles. A flow of water within a porous medium occurs when there is an energy imbalance, in which case, water flows from the high level energy towards the low level energy. Daniel Bernoulli established that, in the case of an *incompressible* (*i.e.* constant density) *inviscid* (*i.e.* non-viscous) *fluid*, the *total energy*, often referred to as *total head* in a *steady pipe flow* remains constant regardless of the position of the point, within the flow domain, at which it is measured. Referring to figure 3.1, Bernoulli's equation is written as follows :

$$h = z + \frac{u}{\gamma_w} + \frac{v^2}{2g} = constant \tag{3.1}$$

where z is the *elevation head* above any arbitrarily chosen datum, the quantity u/γ_w represents the *pressure head* which is in fact the height to which water would rise in a standpipe inserted into the pipe as shown in the figure, the quantity $v^2/2g$ being the *velocity head*.

N.B. A steady flow means that the velocity does not vary with time, i.e. $\partial v/\partial t = 0$

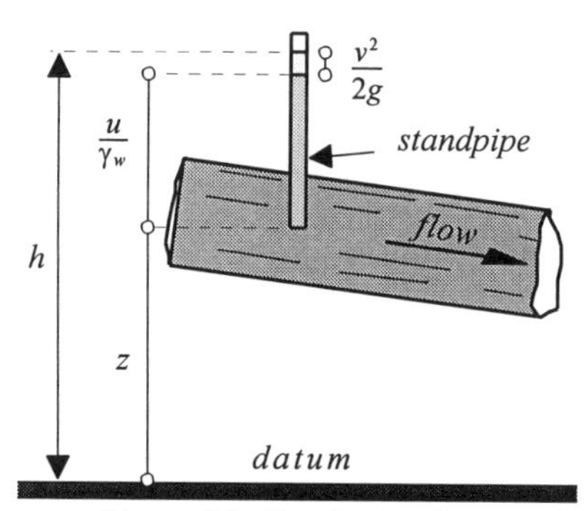

Figure 3.1: Steady pipe flow.

However, the nature of groundwater flow differs from that of a pipe flow in the sense that, for water to flow through a soil, it has to negotiate its way through the spaces existing between solid particles. This is bound to lead to a *loss of energy* through friction that must be accounted for in equation 3.1. Moreover, in most cases of interest in engineering practice, the flow of water in a *saturated soil* (the only type of soil under consideration) occurs at a steady state and at low velocities. Water therefore moves in layers or *laminae* and the corresponding flow is known as *laminar flow*. It has been established that, for a saturated soil, laminar flow conditions are maintained if Reynolds number Re is kept smaller than 10 (Bear (1972)), with:

$$Re = v\, d\, \rho/\mu \tag{3.2}$$

where v is the discharge velocity, d is the (average) diameter of soil particles, μ is the coefficient of dynamic viscosity, and ρ is the fluid density.

Under these conditions (*i.e.* soil saturation and laminar flow), an upper bound discharge velocity of 0.3 *m/s* is typical, leading to a velocity head $v^2/2g = 9 \times 10^{-2}/2 \times 9.81 = 0.0046\, m$. Considering that in all probability, the discharge velocity for most saturated sands and sandy clays will be below 0.3 *m/s*, it is clear that the velocity head can be discarded. Consequently, the total head h in equation 3.1 is no longer constant because of the loss of energy due to friction, and equation 3.1, applied to groundwater flow then becomes:

$$h = z + \frac{u}{\gamma_w} \tag{3.3}$$

where u represents the porewater pressure and z is as per figure 3.1.

The *total head loss* between any two points, due to water flow in a saturated soil is, with respect to figure 3.2:

$$\Delta h = h_1 - h_2$$

or

$$\Delta h = z_1 - z_2 + \frac{1}{\gamma_w}(u_1 - u_2) \tag{3.4}$$

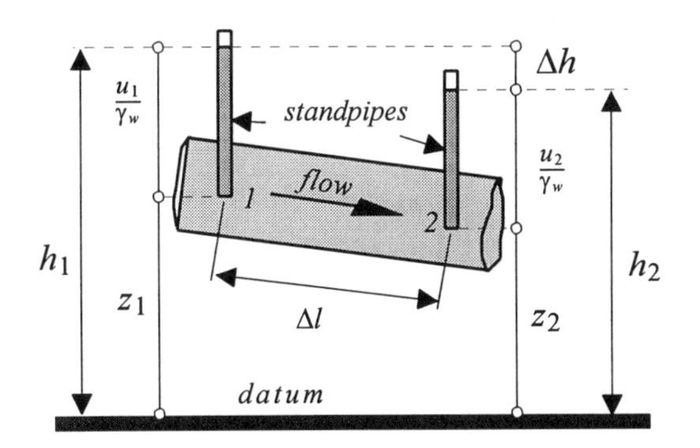

Figure 3.2: Flow in a saturated soil.

It therefore emerges that any flow rate results in a total head loss, the extent of which is related to the length of the flow path: the longer the water travels within the soil mass, the more it is subject to friction and the higher the energy loss. The opposite argument is also valid: the higher the total head loss, the longer the water has to travel within the soil mass, thus developing more friction. Whichever argument is considered, it is obvious that the flow of water within a soil mass is related to both total head loss Δh and flow path length Δl (see figure 3.2).

3.2 Darcy's law, seepage forces, critical hydraulic gradient

As early as 1856, Henri Darcy established the following simple yet powerful empirical relationship between the *rate of flow* q, the *permeability* of soil k and the gradient of flow i, better known as the *hydraulic gradient*:

$$q = A k i \tag{3.5}$$

where A is the area of flow.

Because the quantity q/A represents the flow velocity, Darcy's law is often written as:

$$v = k i \tag{3.6}$$

with the *hydraulic gradient* defined as follows (refer to figure 3.2):

$$i = \frac{h_2 - h_1}{\Delta l} = -\frac{\Delta h}{\Delta l} \tag{3.7}$$

so that at the limit, when $\Delta l \to 0$:

$$i = -\frac{\partial h}{\partial l} \tag{3.8}$$

The flow velocity v in equation 3.6 is *assumed* to be proportional to the hydraulic gradient i, thus meaning that the coefficient of permeability k of the soil mass is constant. Accordingly, the validity of (the empirical) Darcy's law depends to a large extent on the coefficient k which must be carefully determined so as to be representative of the soil mass. The practical aspects related to the measurement of k will be thoroughly discussed shortly. However, at this stage, it is worth remembering that Darcy's law can be safely applied as long as the flow remains *laminar*. This condition is best discussed in terms of Reynolds number of equation 3.2. In fact, conclusive evidence indicates that *linear laminar* groundwater flow conditions are achieved when $1 \leq Re \leq 10$. Outside these boundaries *non-linear laminar* flow occurs up to $Re \approx 100$, beyond which Darcy's law is no longer valid. Consequently, equation 3.6, though still valid, is *no longer linear* at *high flow rates* through highly porous media such as fissured rocks for which $10 \leq Re \leq 100$; similarly, there is *non-linearity* between the *velocity* and the *hydraulic gradient* in the case of groundwater flow under *very low* hydraulic gradients corresponding to $Re < 1$.

Moreover, the velocity used in equation 3.6 is in fact an apparent (or a fictitious) one since it represents the velocity over a cross-sectional area A (refer to figure 3.3) composed of solids through which water *cannot* flow and voids *via* which water seeps. Therefore, the *actual seepage velocity* v_s is n-times the apparent velocity of equation 3.6, n being the soil's porosity, hence:

$$v_s = \frac{v}{n} = v\frac{(1+e)}{e} \tag{3.9}$$

Because the porosity $n < 1$, it follows that the actual seepage velocity is larger than the artificial velocity of equation 3.6.

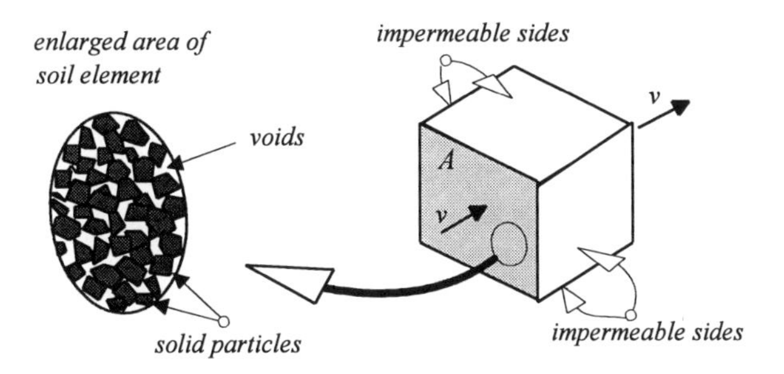

Figure 3.3: Actual seepage velocity.

The flow of water generates *seepage forces* within the soil mass, thus affecting the *effective stresses* (*i.e.* the contact pressure between solid particles). The change in effective stresses depends therefore on the orientation of these seepage forces, in other words on the direction of flow. Consider the simple flow cases represented in figure 3.4, simulating an upward and a downward water flow through a layer of an isotropic sand. In both cases, the hydraulic gradient throughout the soil is:

$$i = \frac{h}{l} \tag{3.10}$$

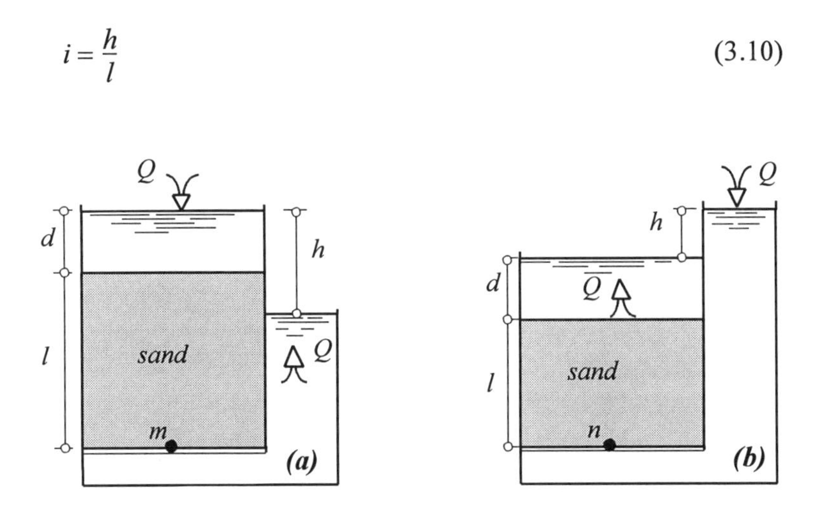

Figure 3.4: (a) Downward and (b) upward flow through an isotropic soil.

Let us now assess the stresses at the bottom of the sand layer (points *m* and *n*), starting with the downward flow (figure 3.4*a*). The *total stress* at *m* is:

$$\sigma_m = d\gamma_w + l\gamma_{sat} \qquad (3.11)$$

where γ_{sat} represents the saturated unit weight of sand and γ_w is the unit weight of water. The porewater pressure at the same point is calculated as follows:

$$u_m = (l + d - h)\gamma_w \qquad (3.12)$$

By virtue of the effective stress principle, the *effective stress* at *m* can now be found:

$$\sigma'_m = \sigma_m - u_m = l(\gamma_{sat} - \gamma_w) + h\gamma_w \qquad (3.13)$$

Using the effective unit weight of sand $\gamma' = (\gamma_{sat} - \gamma_w)$, and substituting for the hydraulic gradient from equation 3.10, the expression of the effective stress becomes:

$$\sigma'_m = l(\gamma' + i\gamma_w) \qquad (3.14)$$

For the upward flow case (figure 3.4*b*), equation 3.11 applies for the calculation of the total stress at point *n*. The porewater pressure however is now calculated as follows:

$$u_n = (l + d + h)\gamma_w \qquad (3.15)$$

therefore the effective stress at *n* is:

$$\sigma'_n = l(\gamma' - i\gamma_w) \qquad (3.16)$$

It is now becoming clear from equations 3.14 and 3.16 that the effective stresses depend on the direction of flow. This is illustrated in figure 3.5 where the *effective forces*, that is the effective stresses times the selected area of flow *A*, can be calculated at any depth on each side of the cut-off wall provided that the hydraulic gradient is known. Hence, on the upstream side (behind the wall):

$$A\sigma'_m = A z_m(\gamma' + i\gamma_w) \tag{3.17}$$

and on the exit side (in front of the wall):

$$A\sigma'_n = A z_n(\gamma' - i\gamma_w) \tag{3.18}$$

The *effective force* $A\sigma'$ exerted at any depth is therefore composed of the *effective weight of soil* $Az\gamma'$ at that depth and a *seepage force* $Azi\gamma_w$, the sign of which depends on the direction of flow as per equations 3.17 and 3.18.

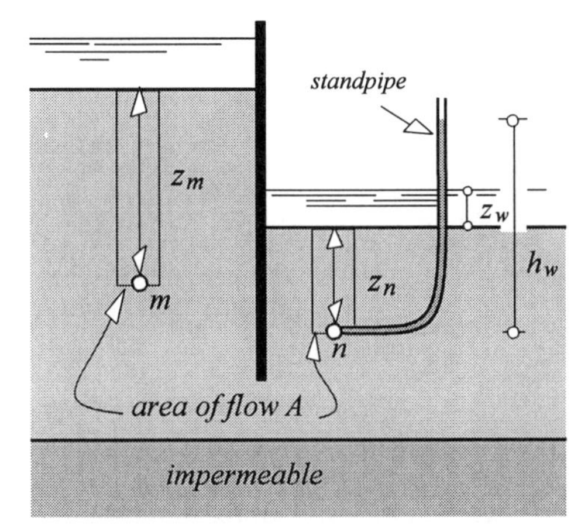

Figure 3.5: Effective stress calculations related to seepage flow.

One more important implication of the previous derivations concerns the upward flow on the downstream side of figure 3.5, represented by equation 3.18 above. This equation yields a zero effective stress when the hydraulic gradient reaches the *critical* value γ'/γ_w. Hence, using equation 1.20 established in section 1.2:

$$i_{cr} = \gamma'/\gamma_w = (G_s - 1)/(1 + e) \tag{3.19}$$

Under such conditions, there is little or no contact between solid particles, and consequently, the sand loses its entire shear strength and behaves like a liquid. This is known as *quick conditions* which are also referred to as *piping* when they develop in a localised area. The hydraulic gradient occurring in a sand must therefore be kept smaller than the critical value at all times. The margin of safety must be such that the factor of safety against quick conditions should be at least three, ideally four or more.

Notice that according to equation 3.19, $i_{cr} \approx 1$ for the vast majority of sands (loose to dense).

The expression of the *critical hydraulic gradient* of equation 3.19, derived using seepage forces and effective stresses, can also be established using the following total stress analysis. Consider the soil element represented in broken lines in figure 3.5 with a depth z_n and a flow area A. A straightforward stress analysis indicates that the *total stress* applied at the bottom of the element is as follows:

$$\sigma_n = \gamma_w z_w + \gamma_{sat} z_n$$

γ_{sat} being the unit weight of saturated soil. When the element is on the verge of failure, the above total stress must balance the porewater pressure at the element base caused by the upward flow and calculated in the following manner:

$$u = h_w \gamma_w$$

Hence, equating these two quantities, then rearranging, it follows that:

$$h_w = \left(\frac{\gamma_{sat}}{\gamma_w}\right) z_n + z_w \qquad (3.20)$$

On the other hand, the hydraulic gradient calculated at the base of the element is, with reference to figure 3.5:

$$i = \frac{h_w - (z_n + z_w)}{z_n} \qquad (3.21)$$

and thus, substituting for h_w from equation 3.20 into equation 3.21 and rearranging, it is seen that:

$$i_{cr} = \left(\frac{\gamma_{sat} - \gamma_w}{\gamma_w}\right) = \frac{\gamma'}{\gamma_w} \qquad (3.22)$$

which is identical to equation 3.19 derived previously.

3.3 Permeability of soils

3.3.1 Introduction

The coefficient of permeability k of a soil is one of the key parameters on which depends the soil response to an applied loading. The rate of soil settlement, for instance, is related to the transfer from total to effective stresses. This transfer is dependent on the rate of dissipation of the excess porewater pressure induced by the loading which, in turn, depends on the ease with which water can seep through the tortuous paths connecting the voids within the soil matrix, *i.e.* the soil's permeability. The importance of this parameter is amplified by the difficulties associated with its measurement. This is mainly due to the fact that *permeability* is *variable* from point to point within the same soil mass because of the random arrangement of solid particles in any one direction. Moreover, k is dependent on the size of solid particles, and the *crude* empirical relationship attributed to Hazen for a variety of sands suggests that k is proportional to the square of the effective size d_{10}:

$$k \approx 10^{-2} \times d_{10}^2 \qquad (3.23)$$

where k is in *m/s* and d_{10} in *mm*.

Hence there is a very wide range of *average* values of permeability associated with different types of soils as listed in the following:

gravel:	$> 10^{-2}$	m/s
sands:	$10^{-1} - 10^{-5}$	m/s
silts:	$10^{-4} - 10^{-7}$	m/s
clays:	$< 10^{-7}$	m/s

3.3.2 Laboratory measurement of the coefficient of permeability

The coefficient of permeability depends on the size of solid particles contained in a soil and on their arrangement within the soil matrix. Accordingly, it is of the utmost importance to use a *least disturbed sample* whenever the laboratory measurement of the permeability is required. Although it is accepted that some degree of disturbance is bound to occur during extraction, transportation and preparation of soil samples (especially in the case of granular soils), the operator must keep any disturbance to a minimum since a significant level of disturbance would render the measured values of permeability totally useless. Also, it can be argued that, apart from the effect of sample disturbance, the physical significance of the measured k value is largely affected by the sample size and soil structure: the smaller the sample dimensions, the less representative the permeability. Most importantly, it is well known that the vast majority of soils *in situ* are anisotropic, with a horizontal permeability far in excess of the vertical one (sometimes by three orders of magnitude). Since the flow regime applied during laboratory tests is one dimensional (*i.e.* vertical), the measured permeability is therefore expected to be well below the *actual in situ* permeability of the soil.

Moreover, the permeability of a soil is affected by the viscosity of the porewater and is therefore temperature dependent (Cedergren, 1967). In this respect, the following temperature correction can be applied cautiously:

$$k_T = \frac{k_{20}}{\xi} \tag{3.24}$$

where k_{20} represents the permeability at the standard temperature of $20°C$, k_T the permeability measured at a temperature T, and ξ is a correction factor as per figure 3.6.

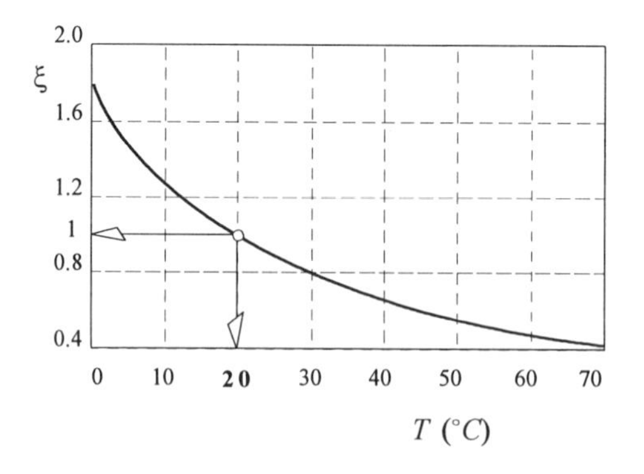

Figure 3.6: Correction related to temperature variation.

Notwithstanding these shortcomings, laboratory testing represents a useful tool that can be used to good effect to estimate the permeability of soil filters (refer to section 1.4). Two laboratory tests are widely used for permeability measurements: the falling head and the constant head tests.

(a) Falling head test

This test is suited for relatively permeable soils (such as silts). The details of the falling head permeameter are shown in figure 3.7.

According to Darcy's law, the flow rate is:

$$q = Ak\frac{h}{L} \qquad (3.25)$$

Referring to figure 3.7, it is seen that:

$$q = -a\frac{dh}{dt} \qquad (3.26a)$$

whence: $$-a\frac{dh}{h} = \frac{Ak}{L}dt \qquad (3.26b)$$

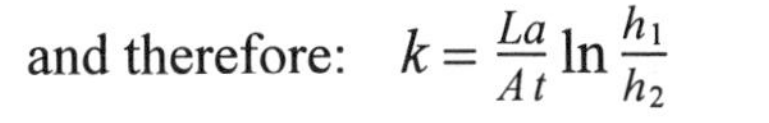

and therefore: $$k = \frac{La}{At}\ln\frac{h_1}{h_2} \qquad (3.27)$$

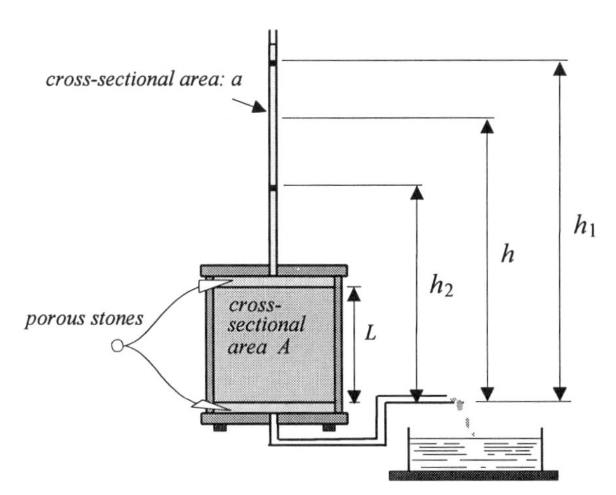

Figure 3.7: Falling head permeameter.

Example 3.1

A falling head test, undertaken on a sample of a fine silty sand 200 *mm* long and 100 *mm* in diameter, resulted in a drop of the total head $\Delta h = 400\,mm$ in a time $t = 120\,s$. Assuming the tube cross-sectional area is $a = 15\,mm^2$ and the total head $h_1 = 1200\,mm$, the permeability can then be estimated from equation 3.27:

$$k = \frac{200 \times 15}{\pi \times 50^2 \times 120} \times \ln \frac{1200}{(1200 - 400)} = 1.29 \times 10^{-3}\,mm/s$$

If the temperature in the field is only $12°C$, then the above coefficient of permeability has to be corrected according to figure 3.6 which yields a correction factor $\xi \approx 1.2$. Hence:

$$k_{12} = \frac{1.29}{1.2} \times 10^{-6} \approx 1.08 \times 10^{-6}\,m/s$$

(b) Constant head permeameter

For more permeable soils such as sands, the constant head permeameter sketched in figure 3.8 is more suited for the measurement of permeability.

According to Darcy's law, the coefficient of permeability in this case is expressed as follows:

$$k = q\frac{L}{hA} \qquad (3.28)$$

with L: length of the flow path,
h: total head loss as per figure 3.8,
A: cross-sectional area of the sample,
q: the flow rate.

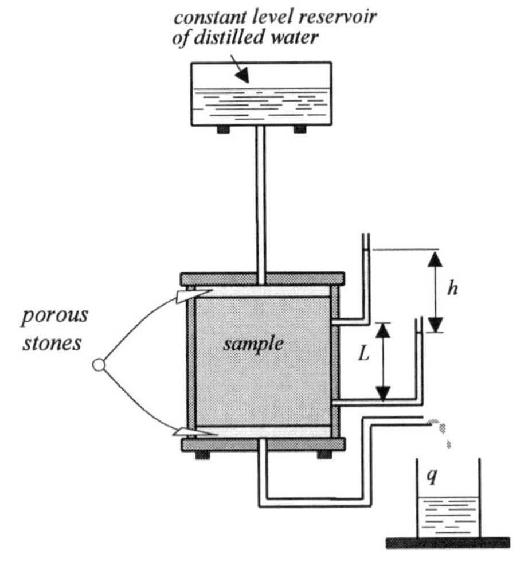

Figure 3.8: Constant head permeameter.

3.3.3 *In situ* measurement of permeability

In situ tests yield more reliable and representative values of the coefficient of permeability than laboratory tests. Not only are these tests undertaken on sites where the soil disturbance is kept to a minimum, but also they involve a larger and therefore a more representative soil volume.

The analysis of field tests is related to the type of *aquifer* (*i.e.* the permeable layer of soil) and involves *pumping* water from a well at a *steady rate* while noting the *drawdown* (or the lowering) of the *piezometric surface* (*i.e.* the surface at which the water pressure is atmospheric) at nearby observation wells.

(a) Pumping test: confined radial flow

In the case of a *confined aquifer*, the permeable stratum is sandwiched between two relatively impermeable soil layers as depicted in figure 3.9. Accordingly, for a well penetrating the full thickness of the aquifer, the flow is everywhere horizontal and the hydraulic gradient can be estimated as follows:

$$i = \frac{dh}{dr} \tag{3.29}$$

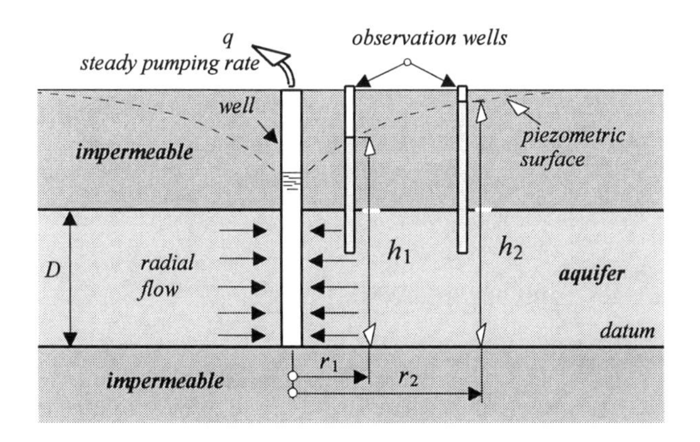

Figure 3.9: Confined radial flow.

Knowing that at a distance r from the well centre, the area through which flow occurs is $A = 2\pi rD$, and substituting for the hydraulic gradient i in Darcy's law of equation 3.5 leads to:

$$q = A\,k\,i = 2\pi\,k\,D\,r\frac{dh}{dr} \tag{3.30}$$

Hence:

$$q\int_{r_1}^{r_2}\frac{dr}{r} = 2\pi D\,k\int_{h_1}^{h_2} dh \tag{3.31}$$

and therefore:

$$k = \frac{q}{2\pi D}\frac{\ln\left(\frac{r_2}{r_1}\right)}{(h_2 - h_1)} \tag{3.32}$$

Note that when the piezometric level is above the ground surface, in other words when the water level inside a well inserted into the confined aquifer rises above the ground surface as depicted in figure 3.10, then *Artesian* conditions are said to prevail.

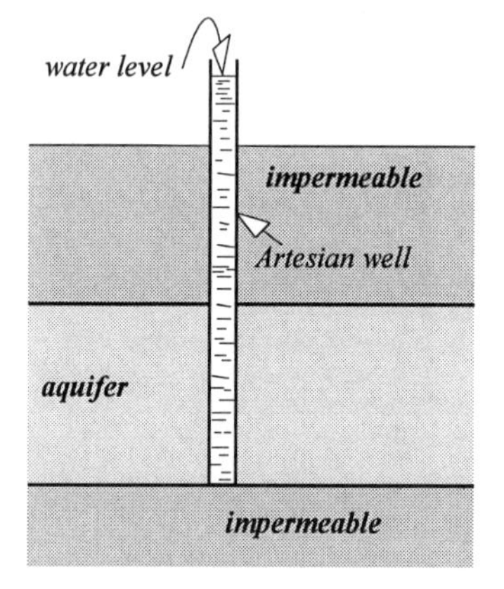

Figure 3.10: Artesian conditions.

(b) Pumping test: unconfined radial flow

For an *unconfined aquifer*, the permeable stratum is not overlain by an impermeable soil layer, and therefore the piezometric surface is the same as the *phreatic* surface or the *water table* as illustrated in figure 3.11. In such a case, the direction of flow towards a well inserted through the aquifer is not horizontal; however, the slope of the phreatic surface is very small according to Dupuit approximation, and therefore *the hydraulic gradient can still be calculated from equation* 3.29. The area of flow at a distance r from the well centre is $A = 2\pi r h$, and from Darcy's law:

$$q = A\,k\,i = 2\pi k\,h\,r\frac{dh}{dr} \tag{3.33}$$

Once integrated between the limits indicated in figure 3.11, the coefficient of permeability can then be established:

$$k = \frac{q}{\pi}\frac{\ln\left(\frac{r_2}{r_1}\right)}{\left(h_2^2 - h_1^2\right)} \tag{3.34}$$

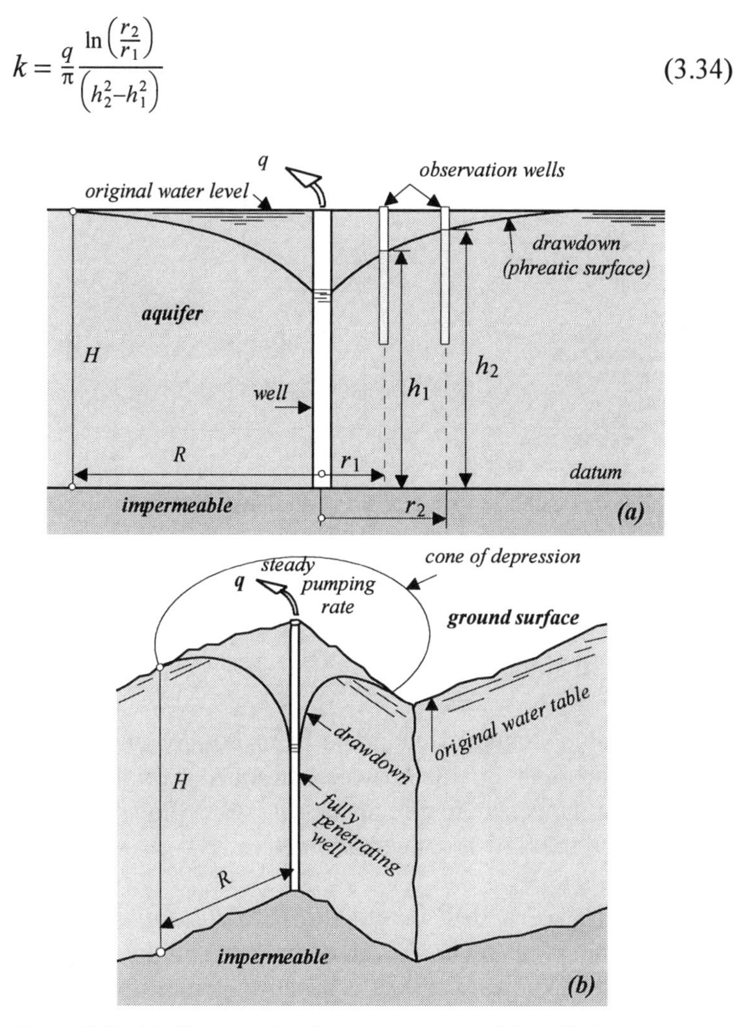

Figure 3.11: (a) Cross-sectional view of unconfined flow, (b) 3-D view.

The quantities h_1 and h_2 in equation 3.34 only represent an *approximation* of the heights to which water would rise inside the standpipes. This is due to the fact that the flow is not radial as stated earlier. Although the principles of flownets are yet to be detailed, figure 3.12 shows that water inside the standpipe nearest to the well rises a distance h_o corresponding to the level of the point of intersection of the equipotential line and the phreatic surface. In contrast, water is assumed to rise to the level h_1 in equation 3.34. However, it can be seen from the figure that the difference $(h_1 - h_o)$ is small and consequently, for all practical purposes, Dupuit assumption used in conjunction with equation 3.34 is in most cases appropriate.

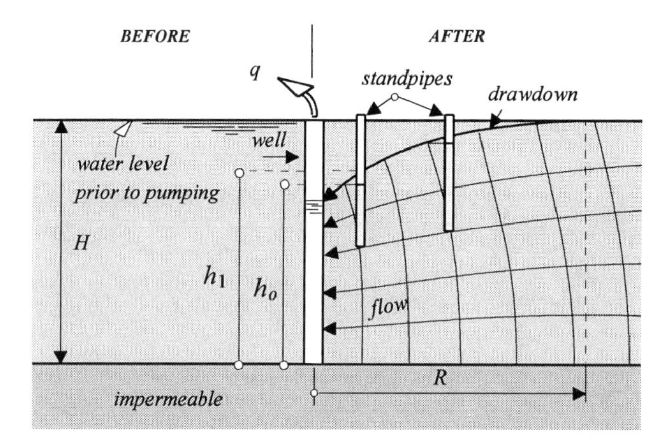

Figure 3.12: Flownet for unconfined radial flow.

(c) Pumping test: spherical flow

If the impermeable stratum is underlain by an aquifer of infinite thickness as illustrated in figure 3.13, a well can only partially penetrate the permeable layer. Under these circumstances, water flows radially towards the *well base* and the flow is spherical in nature.

Consider the case of a well driven down to *the top* of the permeable stratum. Assuming the hydraulic gradient can still be calculated using equation 3.29, and knowing that at a distance r from the well centre, the area of the hemisphere through which flow occurs is $A = 2\pi r^2$, Darcy's law then yields:

$$q = A\,k\,i = 2\pi k r^2 \frac{dh}{dr} \tag{3.35}$$

taking the well base as datum:

$$q \int_{r_1}^{r_2} \frac{dr}{r^2} = 2\pi k \int_{h_1}^{h_2} dh \tag{3.36}$$

thence the expression of k:

$$k = \frac{q}{2\pi} \frac{(r_2 - r_1)}{(h_2 - h_1) r_2 r_1} \tag{3.37}$$

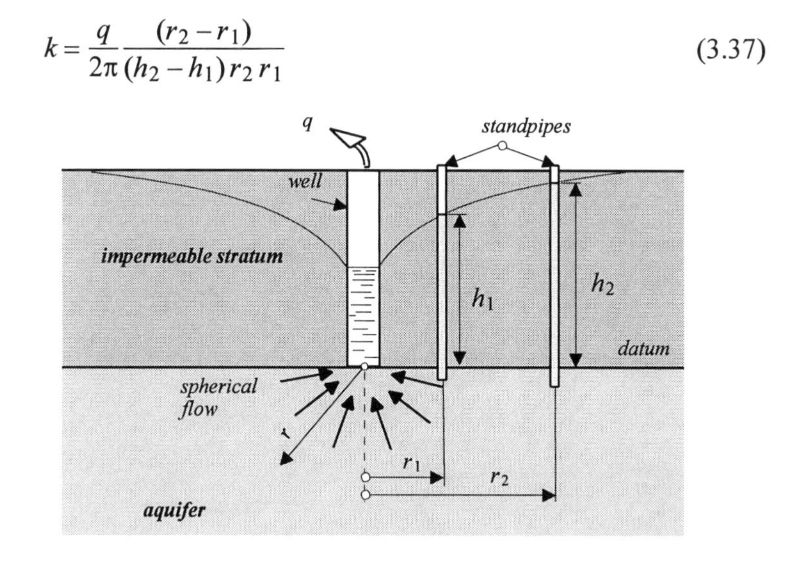

Figure 3.13: Spherical flow.

3.3.4 Design of dewatering systems

Apart from being used to yield estimates of soil permeability *in situ*, pumping wells are often used to reduce the groundwater level (in the case of unconfined flow) or the piezometric level (for confined flow) around an excavation so that work can proceed under *dry safe* conditions. An adequate reduction of the ground water or piezometric levels automatically reduces the porewater pressure applied at the bottom of excavation, thus ensuring an appropriate degree of safety against failure due to piping or to bottom heave.

The design of dewatering systems depends on the type of flow (confined or unconfined) and also on the *recharge boundaries*. The following formalism applies to the case of *radial flow* (refer to figures 3.9 and 3.11). Excavations made near a river or a lake have different recharge boundaries and should be treated accordingly (see Tomlinson (1995), Somerville (1986)).

(a) Case of confined radial flow

Consider the case of an excavation of a size $a \times b$ to be made in a layer of relatively impermeable soil underlain by a confined aquifer of sand of a known permeability k. To ensure safety *vis à vis* bottom heave failure, relieve wells (called thus because their use induces a relieve in porewater pressure at the bottom of excavation) are to be installed in sufficient numbers so that the original piezometric level is reduced by a distance d_w at the centre of excavation as depicted in figure 3.14.

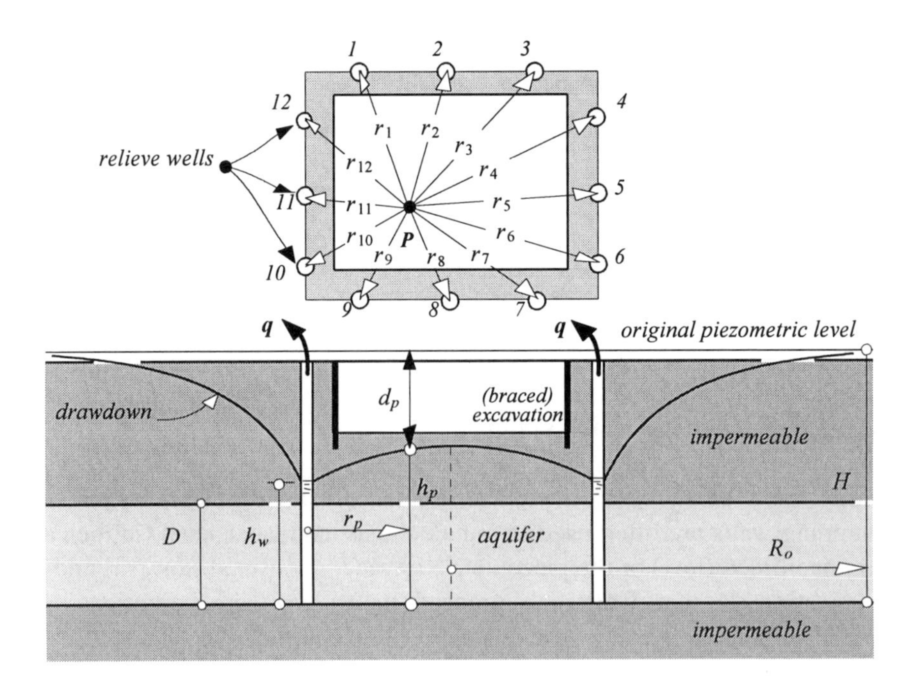

Figure 3.14: Relieve wells system for confined radial flow.

In order to determine the number of wells needed to achieve a *drawdown* d_w, one has to estimate the total quantity of water to be pumped around the excavation. To do so, the entire excavation is assumed to act as an equivalent well with an average radius R_w corresponding to the area of excavation:

$$ab = \pi R_w^2 \quad \Rightarrow \quad R_w = \left(\frac{ab}{\pi}\right)^{1/2} \tag{3.38}$$

Under these circumstances, the drawdown corresponds to :

$$d_w = H - h_w$$

Equation 3.32 can thence be rearranged to calculate the total quantity of flow that needs to be pumped :

$$Q = 2\pi Dk \frac{d_w}{\ln\left(\frac{R_o}{R_w}\right)} \tag{3.39}$$

where the quantity R_o corresponds to the *radius of influence* of the equivalent well. If the soil is assumed homogeneous and isotropic, then the flow towards the well is radially symmetrical, and the drawdown d_w decreases as the radial distance from the well centre increases. The radius of influence corresponds to the radial distance (from the well centre) beyond which the drawdown is zero (*i.e.* the radius beyond which the piezometric level remains unchanged).

On the other hand, equation 3.39 indicates that the radius R_o is not constant, since an increase in the drawdown d_w or in soil permeability k results in an increase in R_o. In practice, the radius of influence is very difficult to measure accurately. However, equation 3.39 indicates that R_o occurs within a natural logarithm term and, as such, there is no need for such a quantity to be known with great accuracy. Accordingly, in the case of plane flow within an isotropic soil, R_o can be estimated from Sichardt empirical relationship (see Somerville, (1986)):

$$R_o \approx C d_w \sqrt{k} \tag{3.40}$$

where d_w represents the drawdown (in *metres*), k is the soil permeability (in m/s), and C is a constant with a value of between 1500 and 2000 (in $[s/m]^{1/2}$) for plane flow.

Once the total quantity of flow is estimated, the number n of wells is then found by assuming that identical pumps yielding an identical discharge q under steady state conditions are used for each well. This assumption is realistic as long as the wells are equally spaced around the excavation. Under such conditions, the *drawdown* at *any* point such as P in figure 3.14 can be calculated using the superposition principle in conjunction with equation 3.32:

$$H - h_P = \frac{q}{2\pi Dk} \sum_{j=1}^{n} \ln\left(\frac{R_o}{r_j}\right) \tag{3.41}$$

with : q: flow rate from each individual well,
r_j: radial distance from point P to the centre of well j (see figure 3.14).

Example 3.2

Consider the case of a $55\,m \times 45\,m$ excavation to be made in a 20 m thick layer of firm clay underlain by a 15 m thick layer of sand as depicted in figure 3.14. Preliminary investigation indicated that artisian conditions prevail in that the original piezometric level was found to be 1 m above the ground surface. Furthermore, carefully undertaken measurements from a pumping test carried out on a well penetrating the entire sand layer revealed that the piezometric levels at two observations wells 15 m and 75 m from the pumping well were reduced by 0.98 m and 0.35 m respectively; the corresponding steady state flow rate being $q = 7 \times 10^{-2}\,m^3/s$.

Because of the substantial depth of excavation, it is suggested that a minimum drawdown $d_w = 10.5\,m$ must be achieved at the *centre* of excavation so as to avoid any potential bottom heave.

Check if this condition can be achieved using the arrangement depicted in figure 3.14, assuming the radius of influence of the entire pumped zone is $R_o = 900\,m$.

First, the permeability of the aquifer must be evaluated. According to the measurements from the observation wells, and using the stated dimensions with the bottom of the sand layer as datum, it is seen that:

- *observation well 1:* $h_1 = (15 + 20 + 1) - 0.98 = 35.02\,m, \quad r_1 = 15\,m,$
- *observation well 2 :* $h_2 = 36 - 0.35 = 35.65\,m, \quad r_2 = 75\,m.$

Hence, from equation 3.32:

$$k = \frac{q}{2\pi D}\frac{\ln\left(\frac{r_2}{r_1}\right)}{(h_2 - h_1)} = \frac{7\times10^{-2}}{2\times\pi\times15} \times \frac{\ln(75/15)}{(35.65-35.02)} \approx 1.9\times10^{-3}\,m/s$$

Next, the radius of the equivalent well is calculated from equation 3.38:

$$R_w = \left[\frac{50\times45}{\pi}\right]^{1/2} = 26.76\,m$$

The total quantity of flow that needs to be pumped is then estimated from equation 3.39:

$$Q = 2\pi Dk\frac{d_w}{\ln\left(\frac{R_o}{R_w}\right)} = 2\pi\times15\times1.9\times10^{-3}\times\frac{10.5}{\ln(900/26.76)} \approx 535\times10^{-3}\,m^3/s$$

Accordingly, the capacity of each pump is such that:

$$q = Q/12 = 44.6\times10^{-3}\,m^3/s$$

and therefore 12 pumps, each with a capacity $q = 2800\,litres/\min$ (that is $4.6\times10^{-2}\,m^3/s$) can be selected.

Now the drawdown at the *centre* of excavation (where the total head is h_c) is checked using equation 3.41:

$$H - h_c = d_w = \frac{q}{2\pi Dk}\sum_{j=1}^{12}\ln\left(\frac{R_o}{r_j}\right)$$

Referring to figure 3.15, a straightforward calculation yields the following radii with respect to the centre of excavation:

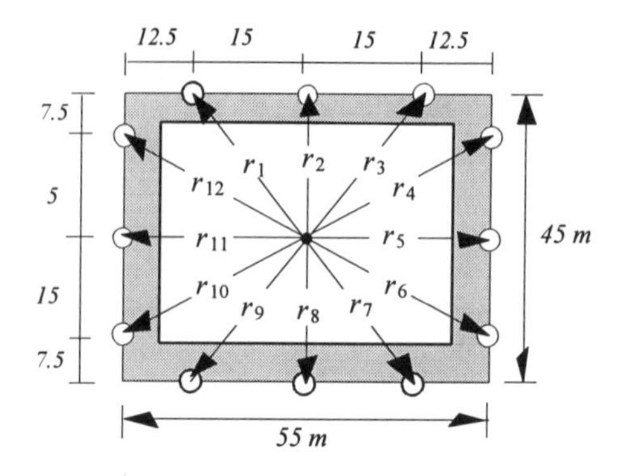

Figure 3.15: Radial distances in the case of example 3.2.

$$r_1 = r_3 = r_7 = r_9 = \left[15^2 + 22.5^2\right]^{1/2} = 27.04\,m$$
$$r_2 = r_8 = 22.5\,m$$
$$r_4 = r_6 = r_{10} = r_{12} = \left[15^2 + 27.5^2\right]^{1/2} = 31.32\,m$$
$$r_5 = r_{11} = 27.5\,m$$

Whence:

$$d_w = \frac{4.6 \times 10^{-2}}{2\pi \times 15 \times 1.9 \times 10^{-3}}[4\ln(900/27.04) + 2\ln(900/22.5)$$

$$+ 4\ln(900/31.32) + 2\ln(900/27.5)] = 10.72\,m$$

(b) Case of unconfined radial flow

In this case, a similar analysis to that used for confined radial flow, this time based on equation 3.34, results in the following equation in the case of n wells, each yielding a flow quantity q as depicted in figure 3.16:

$$H^2 - h_p^2 = \frac{q}{\pi k} \sum_{j=1}^{n} \ln\left(\frac{R_o}{r_j}\right) \tag{3.42}$$

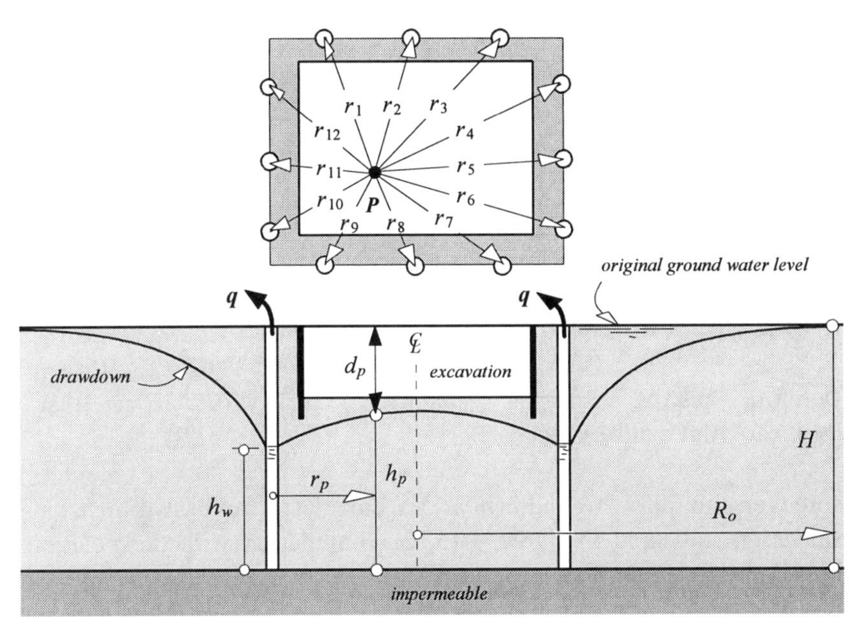

Figure 3.16: Wells system for unconfined radial flow.

Example 3.3

Assume the same excavation of example 3.2 is to be made in a 20 m thick layer of a silty sand with a permeability $k = 4 \times 10^{-4}\ m/s$, underlain by a relatively impermeable soil.

Estimate the capacity of pumps needed to achieve a minimum drawdown $d_p = 8\,m$ at point P with the co-ordinates $(-8\,m,\ -4\,m)$ with respect to the excavation centre point as depicted in figure 3.16. The original water table is at ground level, and the radius of influence of the entire pumped zone is assumed to be $R_o = 700\,m$.

The radius of the equivalent well, calculated previously is $R_w = 26.7\,m$, and therefore the total quantity of water that needs to be pumped to achieve the required drawdown $d_p = 8\,m$ at P can be estimated from equation 3.34:

$$Q = \pi k \frac{\left(H^2 - h_p^2\right)}{\ln\left(\frac{R_o}{R_w}\right)}$$

with $d_p = H - h_p \quad \Rightarrow \quad h_p = 20 - 8 = 12\,m$. Accordingly:

$$Q = \pi \times 4 \times 10^{-4} \times \frac{\left(20^2 - 12^2\right)}{\ln(700/26.76)} = 98.6 \times 10^{-3}\, m^3/s$$

With the 12 wells rearranged as in figure 3.16, the capacity of individual pumps is such that:

$$q = Q/12 = 8.2 \times 10^{-3}\, m^3/s$$

Therefore pumps with a capacity $q = 8.33 \times 10^{-3}\, m^3/s$ (that is 500 *litres*/min) can be used.

Moreover, the following radii (in *m*) are calculated from the dimensions of excavation indicated in figure 3.15, in conjunction with the position of point *P* in figure 3.16:

r_1	r_2	r_3	r_4	r_5	r_6	r_7	r_8	r_9	r_{10}	r_{11}	r_{12}
27.4	27.7	35.1	40.3	35.7	37.2	29.5	20.1	19.8	22.4	19.9	27.2

Whence:

$$\sum_{j=1}^{12} \ln\left(\frac{R_o}{r_j}\right) = \sum_{j=1}^{12} \ln\left(\frac{700}{r_j}\right) = 38.76$$

According to equation 3.34:

$$H^2 - h_p^2 = (q/\pi k) \sum_{j=1}^{12} \ln\left(R_o/r_j\right) \qquad \Rightarrow$$

$$h_p = \left[20^2 - \frac{8.33 \times 10^{-3}}{\pi \times 4 \times 10^{-4}} \times 38.76\right]^{1/2} = 11.96\,m$$

and the drawdown at point *P*:

$$d_p = H - h_p = 20 - 11.96 = 8.04\,m$$

3.4 Seepage theory

3.4.1 Governing flow equation

For the typical soil element in figure 3.17 representing the flow of an *incompressible* fluid through a *fully saturated* soil, the condition of continuity requires that the quantity of water flowing through the entrance faces of the element *must* equal that emerging at the opposite exit faces.

Using Darcy's law in the form of equation 3.8, the hydraulic gradients at the entrance faces of the element are:

$$i_x = \partial h/\partial x \tag{3.43a}$$
$$i_y = \partial h/\partial y \tag{3.43b}$$
$$i_z = \partial h/\partial z \tag{3.43c}$$

At the exit faces, the hydraulic gradients are, respectively:

$$i_x + \frac{\partial i_x}{\partial x}dx = \frac{\partial h}{\partial x} + \frac{\partial}{\partial x}(\frac{\partial h}{\partial x})dx = \frac{\partial h}{\partial x} + \frac{\partial^2 h}{\partial x^2}dx \tag{3.44a}$$

$$i_y + \frac{\partial i_y}{\partial y}dy = \frac{\partial h}{\partial y} + \frac{\partial^2 h}{\partial y^2}dy \tag{3.44b}$$

$$i_z + \frac{\partial i_z}{\partial z}dz = \frac{\partial h}{\partial z} + \frac{\partial^2 h}{\partial z^2}dz \tag{3.44c}$$

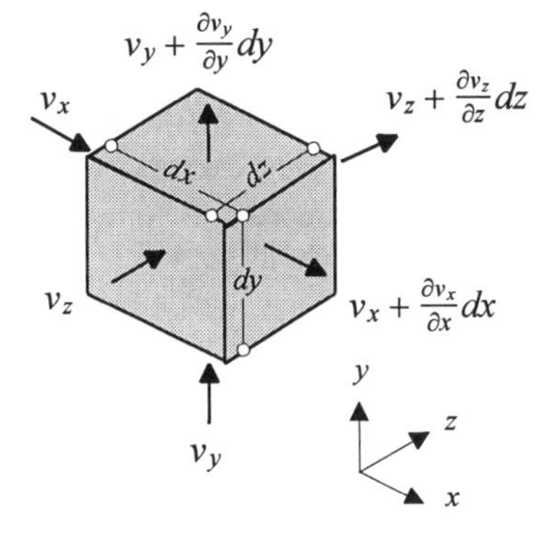

Figure 3.17: Flow through a typical soil element.

the quantity h in both sets of equations being the total head. Equating the flow quantities at the entrance to those at the exit faces leads to the *continuity equation*:

$$\frac{\partial v_x}{\partial x} + \frac{\partial v_y}{\partial y} + \frac{\partial v_z}{\partial z} = 0 \tag{3.45}$$

Introducing Darcy's law in the form of equation 3.6 in the above expression, the general governing equation of flow can then established :

$$k_x \frac{\partial^2 h}{\partial x^2} + k_y \frac{\partial^2 h}{\partial y^2} + k_z \frac{\partial^2 h}{\partial z^2} = 0 \tag{3.46}$$

This equation, represents the general expression of flow in the space (x, y, z), and indicates that the flow depends on the three components of permeability k_x, k_y, k_z.

Assuming that the flow in figure 3.17 is mainly planar, and neglecting accordingly the flow component in the z-direction, equation 3.46 can then be reduced to the following:

$$k_x \frac{\partial^2 h}{\partial x^2} + k_y \frac{\partial^2 h}{\partial y^2} = 0 \tag{3.47}$$

In practice, only few boundary value problems represented by the simple Laplacian associated with equation 3.47 have been solved analytically, principally because the boundary conditions related to the closed form solution of this equation are often difficult to satisfy. Fortunately, an approximate method of solution is at hand in that a graphical flownet, when drawn skilfully, can lead to accurate estimates of different parameters such as seepage quantities and porewater pressures. A brief description of the essential mathematical features of this technique follows.

3.4.2 Confined flow

When all boundaries of the flow domain are known *ab initio*, the flow is referred to as *confined flow*. From a mathematical view point, equation 3.47 produces families of curves, four of which are plotted in figure 3.18.

The curves referred to as *flow lines* represent the flow paths of trickles of water, whereas an *equipotential line* corresponds to a line through which the *total head* remains constant, so that if two standpipes were inserted at different locations in a way that their bases are on the same equipotential line, the water in both standpipes will rise to the same level as depicted in the figure. As one moves to the next equipotential line in the direction of flow, the total head drops by an amount Δh.

Let us examine the conditions that one *must* satisfy to obtain an acceptable flownet. With reference to figure 3.18, the slope of the flow line at the point I (where it crosses an equipotential line) is:

$$\tan\alpha = -\frac{v_y}{v_x} \tag{3.48}$$

N.B. The vectors v_x and dx are not represented in figure 3.18 for the sake of clarity.

Substituting for the velocity from equation 3.19:

$$\tan\alpha = -\frac{k_y}{k_x}\frac{\partial h/\partial y}{\partial h/\partial x} \tag{3.49}$$

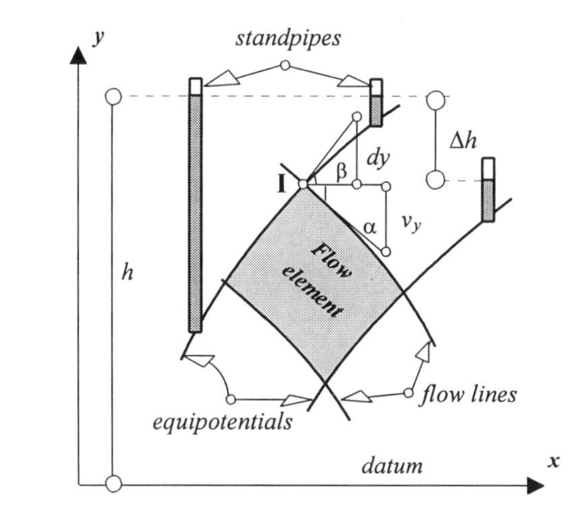

Figure 3.18: Flow and equipotential lines.

Moreover, the total head h has a constant value, with respect to any chosen datum, along any equipotential line, and in particular along the one passing through the point I. Under these circumstances, the total differential dh is zero, hence:

$$dh = \frac{\partial h}{\partial x}dx + \frac{\partial h}{\partial y}dy = 0 \qquad (3.50a)$$

alternatively :

$$\frac{dy}{dx} = -\frac{\partial h / \partial x}{\partial h / \partial y} \qquad (3.50b)$$

Furthermore, figure 3.18 shows that:

$$\frac{dy}{dx} = \tan\beta \qquad (3.51)$$

where β represents the slope of the equipotential line at I. We now, therefore, have a relationship between α and β such that:

$$\tan\beta = \frac{k_y}{k_x}\frac{1}{\tan\alpha} \qquad (3.52)$$

Three cases arise from the above equation as follows.

(a) Case of isotropic homogeneous soils

In this case, by virtue of isotropy, $k_x = k_y$ and, accordingly, equation 3.47 becomes a *Laplace* equation:

$$\frac{\partial^2 h}{\partial x^2} + \frac{\partial^2 h}{\partial y^2} = 0 \qquad (3.53)$$

equation 3.52 is thence reduced to:

$$\tan\beta - \frac{1}{\tan\alpha} = 0$$

which has a solution : $\alpha + \beta = \frac{\pi}{2}$.

This solution implies that for *isotropic homogeneous soils*:
- the flow is *independent* of the *permeability*, and
- flow lines and equipotential lines always cross at *right angles.*

Also, figure 3.19 shows the total head h decreasing continuously in the direction of the flow, indicating that *adjacent equipotential lines must never cross.* It is also clear, from the same figure, that the quantity of water per unit length, flowing within any *flow channel* (*i.e.* the space between adjacent flowlines) is constant.

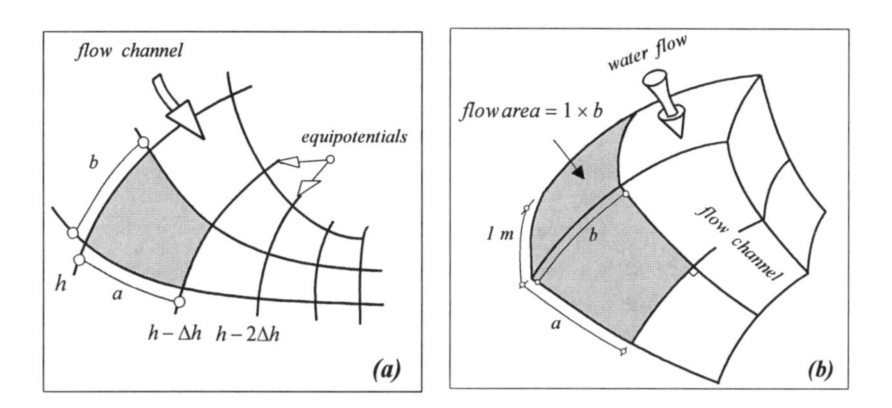

Figure 3.19 : (a) Flow and equipotential lines and (b) 3-D flow element.

Applying Dracy's law to the shaded element:

$$q = (1 \times b)\, k \frac{\Delta h}{a} \text{ per flow channel per unit length} \qquad (3.54)$$

If the flownet is drawn such that the elements of flow are *curvilinear squares*, then $b \approx a$ and:

$$q = k\Delta h \text{ per flow channel per unit length} \qquad (3.55)$$

Referring to the number of intervals between equipotential lines as N_d, the quantity Δh is then related to the total head h through the relationship :

$$\Delta h = \frac{h}{N_d} \qquad (3.56)$$

Accordingly, for a flownet having N_f flow channels, the total flow quantity per unit length is therefore:

$$Q = qN_f = k\,h\frac{N_f}{N_d} \;\; per\ unit\ length \tag{3.57}$$

Example 3.4

Figure 3.20 depicts an example of a confined flow represented by a sheet-pile cut-off wall driven in an isotropic homogeneous sand with a permeability $k = 2 \times 10^{-5}\ m/s$, underlain by an impervious material. Let us estimate the quantity of flow per unit width of the wall, then calculate the distribution of the porewater pressure along the pile, as well as the exit hydraulic gradient and the factors of safety against piping and against bottom heave.

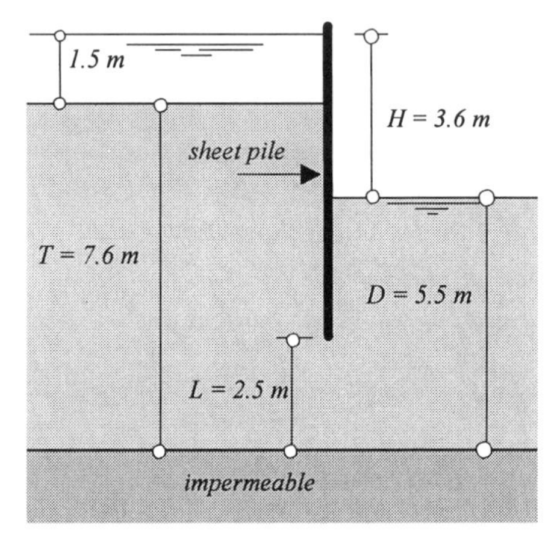

Figure 3.20: Sheet pile cut-off wall.

The corresponding flownet is depicted in figure 3.21. Notice that, in this case, the flownet has four flow channels and eleven equipotential lines labelled with their respective total heads in the figure (N_d being the number of intervals between adjacent equipotential lines).

We therefore have: $N_f = 4$, $N_d = 10$. Hence the flow quantity per unit width:

$$Q = kH\frac{N_f}{N_d} = \frac{2 \times 3.6 \times 4}{10} \times 10^{-5} = 2.88 \times 10^{-5}\ m^3/s\ per\ metre\ run$$

As for the porewater pressure, it can be estimated at any point from the flownet, since at any location, the total head as well as the elevation head (with respect to the chosen datum) can be measured. For example at point

A, the total head is: $h_A = 3.24\ m$; the elevation head (scaled from the figure) being $y = +0.4\,m$. Similarly, at B, $h_B = 0.36\,m$ and $y = -1.4\,m$. The porewater pressures at A and B are therefore:

$$u_A = \gamma_w(h_A - y_A) \approx 10(3.24 - 0.4) = 28.4\,kN/m^2$$

$$u_B = \gamma_w(h_B - y_B) \approx 10(0.36 + 1.4) = 17.6\,kN/m^2$$

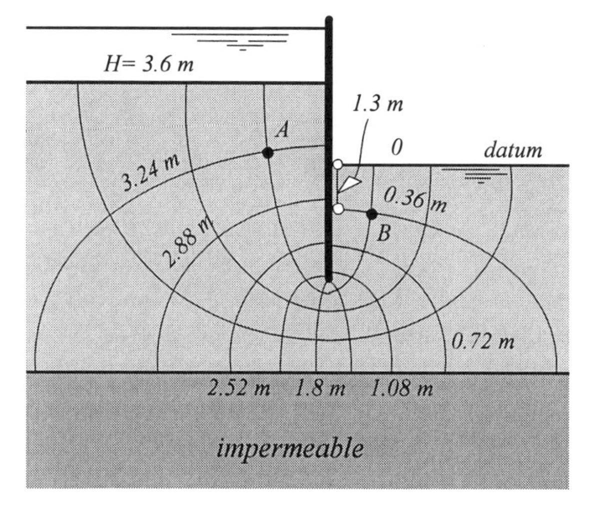

Figure 3.21: Flownet.

At the exit, the hydraulic gradient is measured from the square nearest to the pile at the top of the discharge area. From figure 3.21, the corresponding total head loss across the element is 0.36 *m,* and the flow path is 1.3 *m* long, thence:

$$i_e = \frac{0.36}{1.3} = 0.28$$

Using the critical hydraulic gradient defined earlier through equation 3.19, the factor of safety against piping (*i.e.* localised quick conditions) is:

$$F = \frac{i_{cr}}{i_e} \approx \frac{1}{0.28} = 3.6$$

Figure 3.22 illustrates the possibility of the soil mass, represented by the white element in front of the pile, failing by being carried upward by seepage forces. This may occur were the seepage forces at the element base to be larger than the submerged weight of the soil contained within the element. An adequate factor of safety is therefore essential in order to prevent this possibility from occurring. In this respect, the empirical Terzaghi procedure which consists of selecting an element with a volume $1 \times D \times D/2$, where D represents the depth of embedment as per figure 3.22, can be adopted. The factor of safety against bottom heave is therefore defined as the ratio of the submerged weight of the element to the upward seepage force exerted on its base:

$$F = \frac{(\gamma_{sat} - \gamma_w)}{u\frac{D}{2}}\frac{D^2}{2} = \frac{(\gamma_{sat} - \gamma_w)}{u} D \tag{3.58}$$

where u represents the *average* porewater pressure exerted at the element base. Thus, referring to figure 3.22, it can be seen that both the total head and elevation head at the middle of the element base are: $h = 1.08\,m$, $h_e = -3\,m$. Hence a pressure head: $h_p = h - h_e = 3.08\,m$.

Assuming that the saturated unit weight of sand is $\gamma_{sat} = 22\,kN/m^3$, and knowing that $D = 3\,m$, equation 3.58 then yields the following factor of safety:

$$F = \frac{(22-10)}{3.08\times 10} \times 3 = 1.17$$

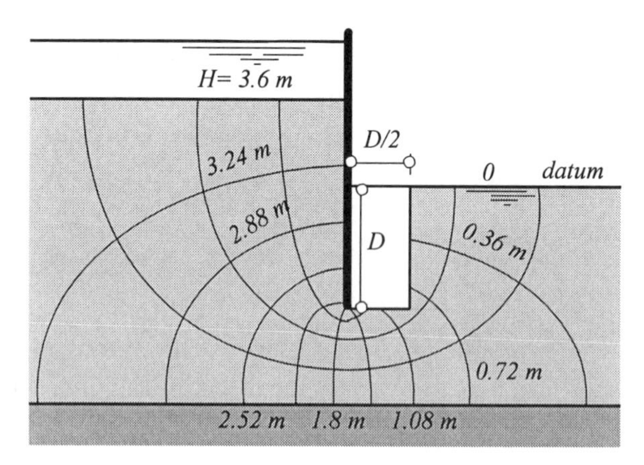

Figure 3.22: Bottom heave of an excavation.

Obviously this value is linked to that of the factor of safety against piping, whose value must be ideally 4 or more as suggested earlier.

(b) case of anisotropic homogeneous soils

Natural soils usually exhibit *anisotropic properties* since they have been, originally, laid down in layers. The values of horizontal and vertical permeabilities are, therefore, considerably *different*, and due to the formation process, the coefficient of permeability in the horizontal direction is usually higher (by as much as three orders of magnitude in some cases) than the one in the vertical direction. Consequently, the permeability of a homogeneous soil depends on the *direction of the flow*. Let us examine the three distinctive possibilities.

(b.1) The water flow is parallel to the bedding

When dealing with anisotropic homogeneous soils, if the water flows parallel to the bedding (assumed to be horizontal), the entire soil mass can be considered to be isotropic if an *equivalent horizontal permeability* is used.

Consider the case of horizontal flow represented in figure 3.23*a* where water seeps through n layers of soil, each of which having a thickness t and a permeability k. The total head loss is H, and the length of flow path is L. The figure indicates that the total flow quantity per metre width seeping through the soil is the sum of the flow passing *via* each layer:

$$Q = q_1 + q_2 + \ldots + q_n = \sum_{p=1}^{n} q_p$$

Moreover, the figure shows that all layers have the same hydraulic gradient: $i = H/L$.

Thus, in accordance with Darcy's law, the total flow quantity per unit width is:

$$Q = \frac{H}{L} \sum_{p=1}^{n} t_p k_p \qquad (3.59)$$

On the other hand, if, all other things being equal, the n layers of soil are replaced by a unique homogeneous layer of permeability k'_h as per figure 3.23*b*, with a thickness T (equal to the sum of individual thicknesses in figure 3.23*a*), then the total flow quantity seeping through the new layer becomes:

$$Q = T\frac{H}{L}k'_h \tag{3.60}$$

Manifestly, the flow quantities given by equations 3.59 and 3.60 are identical, hence the *equivalent horizontal permeability*:

$$k'_h = \frac{\sum_{p=1}^{n} t_p k_p}{\sum_{p=1}^{n} t_p} \tag{3.61}$$

where n is the number of layers and t and k are, respectively, the thickness and permeability of each individual layer.

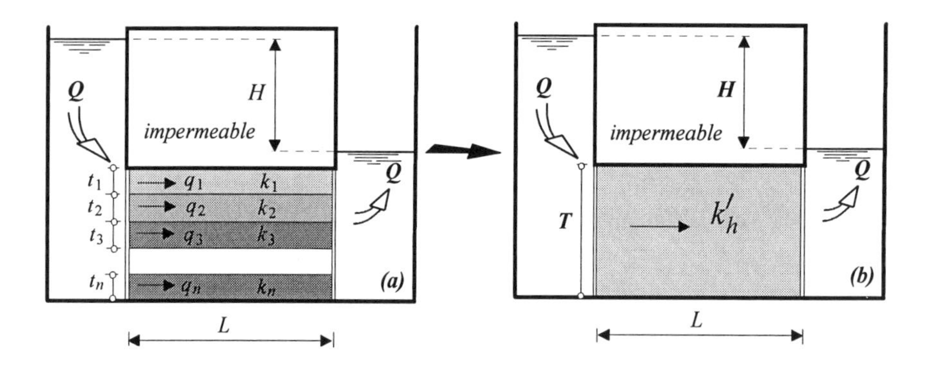

Figure 3.23: (a) Layered soil, (b) equivalent isotropic soil.

(b.2) The water flow is perpendicular to the bedding

This time, the same total flow quantity per unit width seeps through each of the n soil layers represented in figure 3.24*a* in which the bottom layer is assumed to be infinitely permeable. Thence, for the pth layer, having a thickness l_p and a permeability k_p, Darcy's law states that:

$$Q = A\,k_p\,i_p \tag{3.62}$$

where A represents the flow area which is identical to all layers.

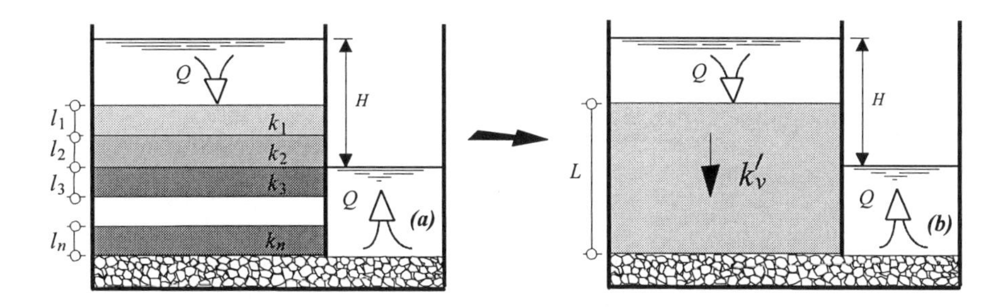

Figure 3.24: (a) Layered soil, (b) equivalent isotropic layer.

Furthermore, the thickness l_p is also the length of the flow path through the pth layer. Therefore, the corresponding hydraulic gradient is:

$$i_p = \frac{h_p}{l_p}$$

Inserting this quantity into equation 3.62 leads to the head loss occurring across the pth layer: $h_p = Q\,l_p/A\,k_p$. The total head loss can now be written as the sum of individual head losses:

$$H = \sum_{p=1}^{n} h_p = \frac{Q}{A}\sum_{p=1}^{n}\frac{l_p}{k_p} \tag{3.63}$$

Consider what happens if the n layers were replaced by a single homogeneous layer, as in figure 3.24*b*, with a total thickness L equal to the sum of individual thicknesses, and a unique permeability k'_v. All other things being equal, the same quantity of flow Q seeps through the same area A and an identical total head loss H occurs across the flow path L. Making use of Darcy's law:

$$Q = A\,k'_v\,\frac{H}{L}$$

The quantity H in equation 3.63 can now be inserted into the above expression which, once rearranged yields the *equivalent vertical permeability* k'_v:

$$k'_v = \frac{\sum_{p=1}^{n} l_p}{\sum_{p=1}^{n} l_p/k_p} \tag{3.64}$$

The latter equation represents the equivalent vertical permeability in the case of a normal flow through a soil containing n layers, each of which has a thickness l and a coefficient of permeability k.

It is clear from both equations 3.61 and 3.64 that the equivalent horizontal permeability is *larger* than the vertical one.

(b.3) The water flow is two–dimensional

In this case, the flow is governed by equation 3.47 which can be rewritten as:

$$\frac{k_x}{k_y}\frac{\partial^2 h}{\partial x^2} + \frac{\partial^2 h}{\partial y^2} = 0 \tag{3.65}$$

A closer analysis of the above equation indicates that a simple Laplacian may be engendered if an adequate change of variable is used. Let us rewrite the same equation using $X = cx$ (where c is a constant):

$$\frac{\partial^2 h}{\partial X^2} + \frac{\partial^2 h}{\partial y^2} = \frac{\partial^2 h}{\partial (cx)^2} + \frac{\partial^2 h}{\partial y^2} = 0 \tag{3.66a}$$

or

$$\frac{\partial^2 h}{\partial X^2} + \frac{\partial^2 h}{\partial y^2} = \frac{1}{c^2}\frac{\partial^2 h}{\partial x^2} + \frac{\partial^2 h}{\partial y^2} = 0 \tag{3.66b}$$

The constant c is then identified by comparing equations 3.65 and 3.66b:

$$c = \sqrt{k_y/k_x} \tag{3.67}$$

The flow is therefore dependent on the ratio of horizontal to vertical permeabilities, and the flownet associated to equation 3.65 is characterised by curvilinear rectangles of a sides ratio $\sqrt{k_x/k_y}$. Furthermore, equation 3.52 indicates that, for $k_x \neq k_y$, equipotential and flow lines do *not* cross at right angles since, at any crossing point, the tangent to the flow line does not correspond to the normal to the equipotential line as was the case for isotropic soils. As a result, the flow net is this time harder to sketch because the sides of *every* quadrangle in the net must have a ratio $\sqrt{k_x/k_y}$.

However, by choosing a co-ordinate system whereby $X = x\sqrt{k_y/k_x}$, it is possible to create an artificial domain known as *the transformed section*, where the flow is represented by a Laplacian, and where the soil can be considered isotropic. The implications are that the corresponding flownet must satisfy the less stringent conditions of an isotropic soil, *i.e.* the elements of the mesh are curvilinear squares. Also, because the soil is isotropic in the transformed section, its coefficient of permeability has a unique value k_t known as the *transformed permeability*, and the flow is, hence, independent of the permeability, the result of which are right-angle crossings between flow and equipotential lines.

The task of finding the expression of the transformed permeability k_t is greatly simplified by considering the vertical flow corresponding to figure 3.25.

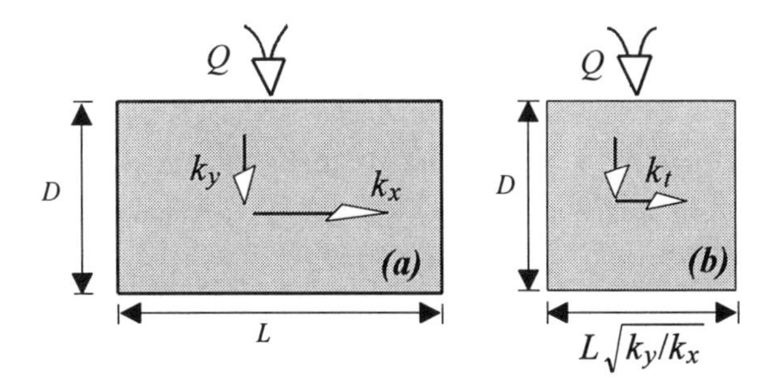

Figure 3.25: (a) Flow through natural anisotropic section, (b) flow through transformed (isotropic) section.

Starting with the natural section, the flow quantity per unit length is :

$$Q = (L \times 1)\, k_y i$$

Within the transformed section, the same quantity of flow is calculated as follows:

$$Q = \left[\, 1 \times L\sqrt{k_y/k_x}\, \right] k_t i$$

Comparing the two expressions, the transformed permeability is then established:

$$k_t = \sqrt{k_x k_y} \qquad (3.68)$$

The same expression could have been found using the flow in the *horizontal* direction. Remember, then, the hydraulic gradient changes from $i = H/L$ in the natural section to $i = H/L\sqrt{k_y/k_x}$ in the transformed section.

Example 3.5

Figure 3.26 depicts a concrete dam which is built on a layer of a homogeneous anisotropic clay having permeabilities $k_x = 16 \times 10^{-8}\, m/s$ and $k_y = 10^{-8}\, m/s$. To control seepage pressures, a cut-off sheet pile is driven on the upstream side of the dam to a depth of 4.6 *m*. Required are: the flow quantity seeping on the downstream side of the dam, the uplift force at the base of dam, the factor of safety against piping as well as the factor of safety against heave at the toe of the dam.

Because of anisotropy, a flownet needs to be sketched on a transformed section. The dimensions represented in figure 3.26 need only be reduced in the horizontal direction by a factor:

$$c = \sqrt{k_y/k_x} = \frac{1}{4}$$

Within the transformed section, the soil is considered *isotropic* with a *transformed permeability*:

$$k_t = \sqrt{k_x k_y} = 4 \times 10^{-8}\ m/s$$

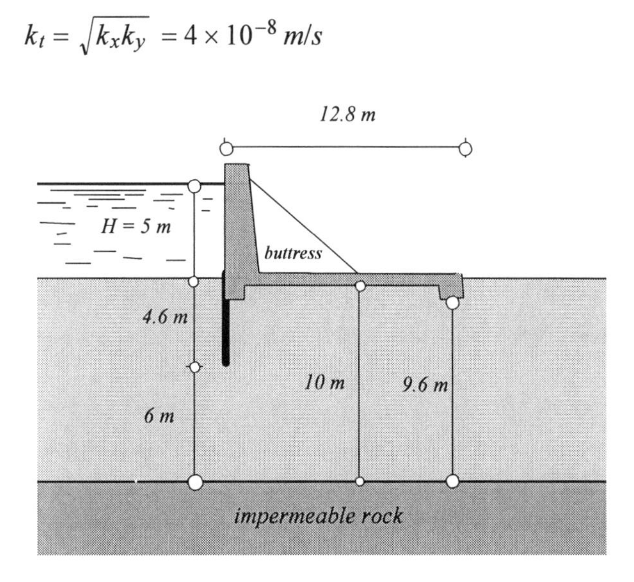

Figure 3.26: Flow through anisotropic soil.

The corresponding flownet, drawn in figure 3.27, has five flow channels ($N_f = 5$) and ten equipotential drops ($N_d = 10$).

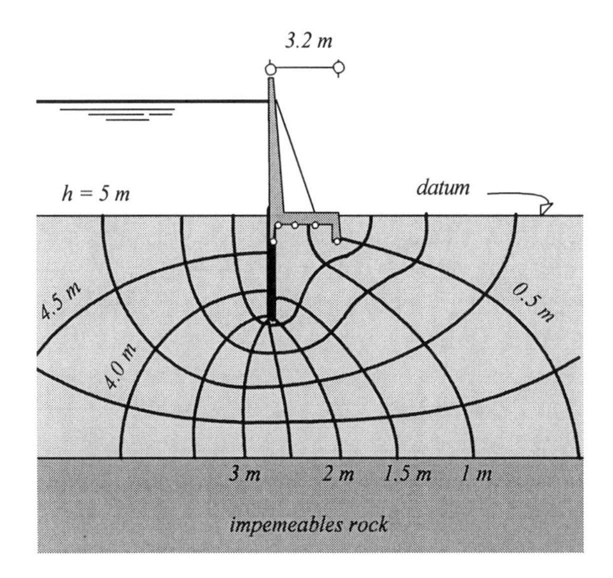

Figure 3.27: Flownet on the transformed section.

The flow quantity per unit length is therefore:

$$Q = k_t H \frac{N_F}{N_d} = 4 \times 5 \times \frac{5}{10} \times 10^{-8}\, m^3/s/m \;= 10^{-7}\, m^3/s/m$$

In order to calculate the uplift force applied to the base of the dam, one needs, first, to estimate the porewater pressure at the five selected points at the base of the dam represented by open circles in figure 3.27. The porewater pressure is calculated as follows:

$$u = \gamma_w (h - y)$$

where γ_w is the unit weight of water ($\approx 10\, kN/m^3$), h is the total head (estimated from the flownet) and y is the elevation head (scaled from the flownet) with respect to the indicated datum.

The following results correspond to the five points depicted in the figure, starting from the upstream side.

point	*h (m)*	*y (m)*	$u\,(kN/m^2)$
1	1.20	–1.0	22.0
2	1.12	–0.6	17.2
3	1.04	–0.6	16.4
4	0.9	–0.6	15.0
5	0.6	–1.0	16.0

It is clear that the porewater pressure does not vary considerably at base level (bearing in mind that a transformed section is used and that the *actual* width of the base is 12.8 *m*). A value $u \approx 17\, kN/m^2$ represents a fair average of the porewater pressure at that level and, consequently, the uplift force per unit length beneath the dam is:

$$F_{up} = u \times area = 17 \times 12.8 \times 1 = 217.6\, kN/m$$

Now that the uplift force is known, the cross-sectional area of the dam in figure 3.27 can be chosen so that the total weight W per metre run of the dam can be calculated, securing in the process a factor of safety against uplift:

$$F = W/F_{up}$$

typically in the range 1.2 to 1.5.

The exit hydraulic gradient i_e is estimated from the upper flow square nearest to the toe of the dam. This square is roughly 1.1 *m* long, and through it, the head drops by 0.5 *m*. Thus:

$$i_e = 0.5/1.1 = 0.45$$

Piping, which occurs at the toe of the dam, as illustrated in figure 3.28, can cause considerable damage and, therefore, there is a need for a good factor of safety to prevent piping from occurring. The factor of safety against piping is calculated according to the following:

$$F = i_{cr}/\,i_e \approx 1/0.45 = 2.22$$

i_{cr} being the critical hydraulic gradient.

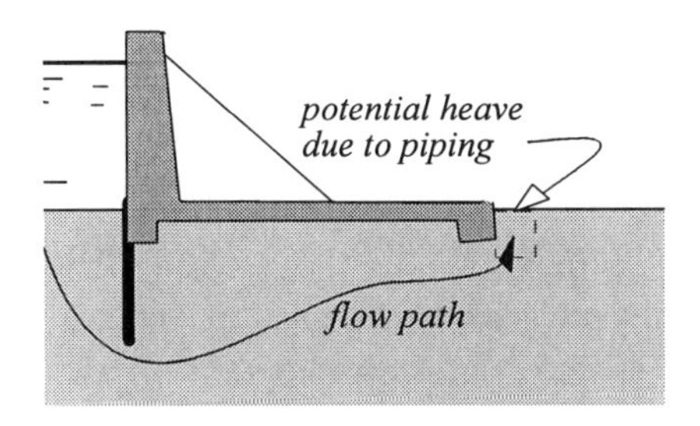

Figure 3.28: Piping related problems.

In practice, factors of safety against piping should have ideally a minimum value of 4, although some engineers would settle for smaller values. In this particular example, a larger factor of safety is needed, and this can be achieved by reducing the exit hydraulic gradient. For that to happen, the flow path must be increased so that the total head within the element of soil shown in broken lines (at the toe of the dam in figure 3.28) is substantially reduced. One solution consists of driving a sheet pile at the toe of the dam, or increasing the length of the pile used on the upstream side.

For the factor of safety against heave due to piping, equation 3.58 is applied. Considering therefore the prism shown in broken lines at the toe of the dam (figure 3.28) with dimensions of 1 *m* depth, 0.5 *m* width and 1 *m* length. The factor of safety against heave corresponds to the ratio of the *submerged weight* of the prism to the *uplift force* due to the porewater pressure at the base of the prism, that is:

$$F = \frac{(\gamma_{sat} - \gamma_w) \times 0.5 \times 1 \times 1}{u \times 0.5 \times 1}$$

Assuming that the soil has a saturated unit weight $\gamma_{sat} = 20\, kN/m^3$, and considering an average porewater pressure at the base of the prism of $13\, kN/m^2$, the ratio is therefore:

$$F = 10 \times 0.5/13 \times 0.5 = 0.77$$

This value is obviously unacceptable because it indicates that the soil at the toe of the dam will heave. To prevent this from occurring, a coarse filter can be applied in front of the dam so as to increase the effective weight of the soil mass in question.

Notice that an increase in the factor of safety against piping would mean a decrease in the porewater pressure at the base of the soil element shown in figure 3.28, therefore causing an improvement in the factor of safety against heave due to piping.

Once all calculations are undertaken, the flownet is scaled back to the natural section by multiplying all dimensions in the x-direction by a coefficient $1/c = 4$. The resulting effect is shown in figure 3.29 which represents the *actual* flownet and, most importantly, indicates how difficult it would have been to obtain such a net had a transformed section not been used in the first instance. The difficulties being twofold in that the flow elements are *not* squares and the crossings between flow and equipotential lines are *not* right angles.

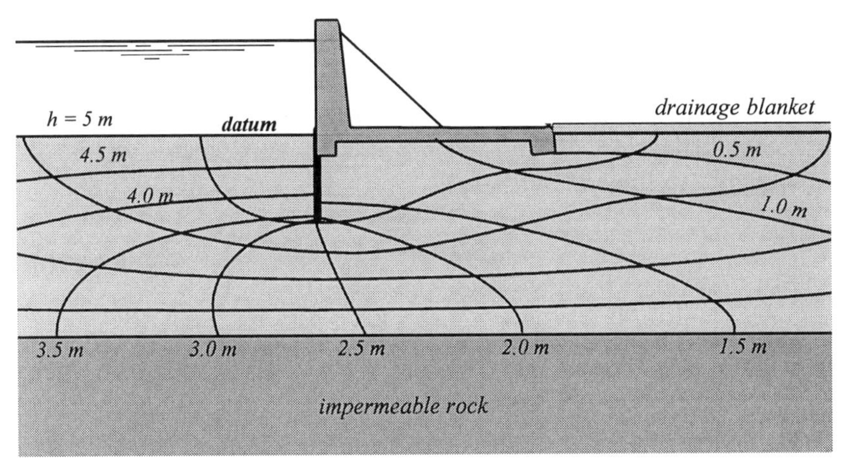

Figure 3.29: Flownet in the natural section.

The scaling back operation is performed using the flownet of the transformed section in figure 3.27. This tedious operation consists, for every crossing point between flow and equipotential lines of:

- measuring, on the transformed section of figure 3.27 the co-ordinates (X,y);
- calculating the natural co-ordinates (x,y) by multiplying X by a factor of 4 (y is not affected) and
- plotting the point (x,y) on the natural section in figure 3.26.

A flownet is then sketched on the natural section by joining smoothly all the points.

(c) Case of multi-layered soils

When dealing with non homogeneous soils (*i.e.* soils comprising layers with inherently different permeabilities), the flow patterns change according to the relative permeabilities of adjacent soil layers.

If, for instance, water flows from a high permeability layer to a lower permeability one, the energy dissipated through friction increases as a result of a steeper hydraulic gradient. At the boundary between adjacent layers, flow lines bend according to the incidence law in optics. The flownet must therefore reflect the change in permeability. As shown in figure 3.30, when water flows from one layer having a permeability k_1, to the next characterised by a permeability k_2, the shape of the flow elements changes to account for the change in permeability in such a way that the ratio between the sides of each element in the bottom layer is related to the ratio of permeabilities as follows:

$$\frac{c}{d} = \frac{k_2}{k1} \tag{3.69}$$

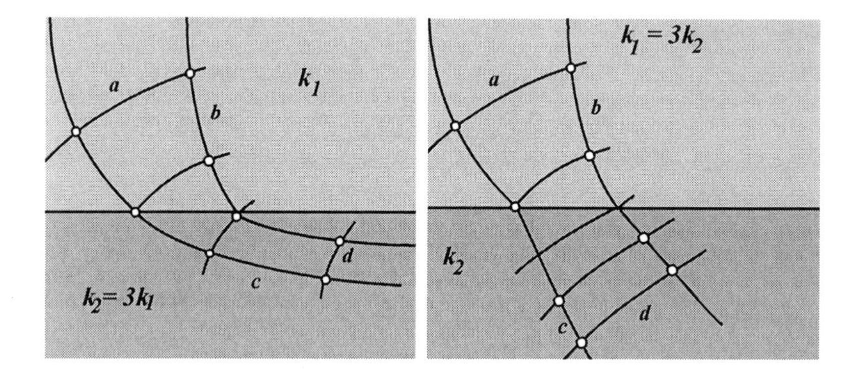

Figure 3.30: Flow through multilayered soils.

Example 3.6

Consider the dam represented in figure 3.31, built at a site characterised by two layers of isotropic clay underlain by impervious rock. The top clay layer is 2 m thick and has a permeability k_1. The bottom layer is 2.7 m thick and is three times more permeable than the overlaying soil ($k_2 = 3k_1$).

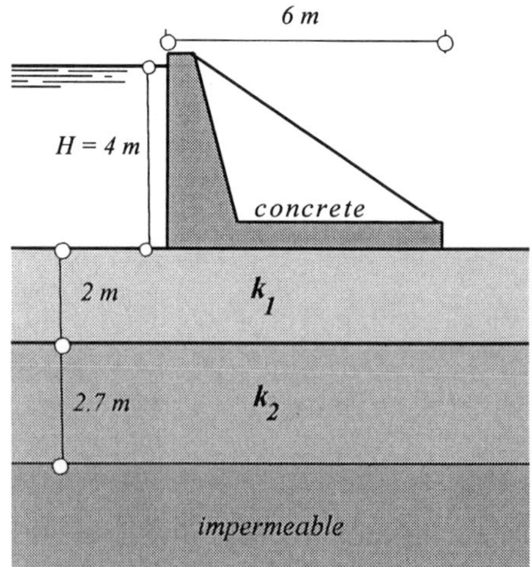

Figure 3.31: Flow through a multi-layered soil.

The corresponding flownet, sketched in figure 3.32, shows how the flow elements change from being curvilinear squares in the top layer to curvilinear rectangles in the underlying layer. In fact, the flow elements in the bottom layer have been elongated in such a way that the ratio of the sides of each rectangle is equal to $k_2/k_1 = 3$.

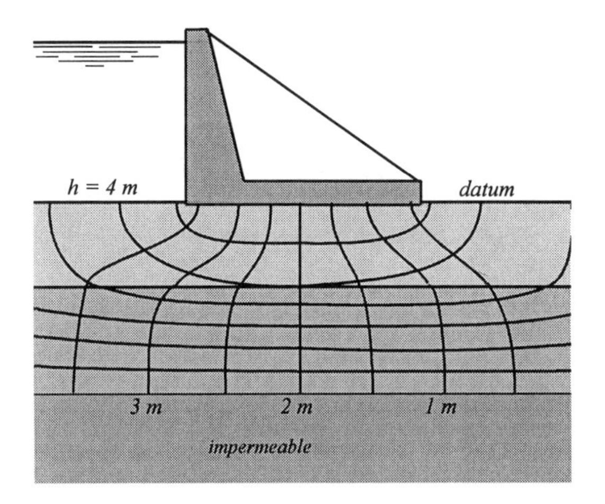

Figure 3.32: Flownet for the multi-layered soil corresponding to figure 3.31.

3.4.3 Unconfined flow

When the boundaries of flow are in part not known *a priori*, the flow is referred to as *unconfined* since it contains a *free surface* at which the pressure is atmospheric. Consider, for example, the case of seepage across an earth dam where the position of the flow patterns are, *prima facie*, hard to predict. Yet, the position of the top flowline, also known as the *phreatic surface*, is entirely defined by the fact that it represents a flowline on which the total head varies linearly with the elevation above any chosen datum. Referring to figure 3.33, the pressure head h_p (read the water pressure) at the points *A, B, C* and *D* is zero since these points are on the phreatic surface where the pressure is atmospheric. Consequently, if the equipotential line going through *A* corresponds to a total head h_A, then we can write:

$$h_p = h_A - y_A = 0$$

which shows that at *A*, the total head is equal to the elevation head y_A measured with respect to any chosen datum.

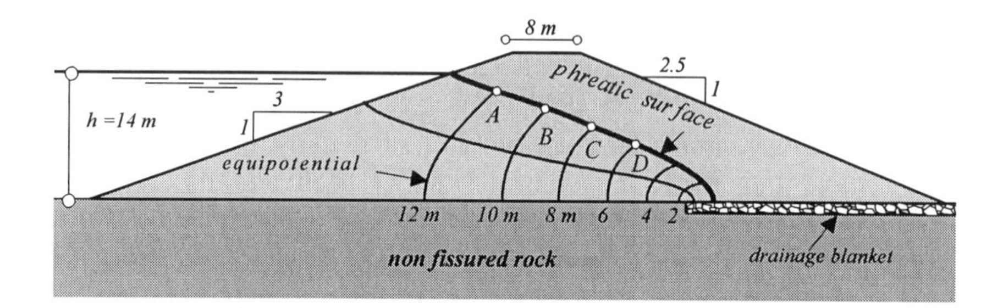

Figure 3.33: Phreatic surface for unconfined flow.

For the equipotential lines going through *B* , *C* and *D*, corresponding respectively to a total head $h_A - \Delta h$, $h_A - 2\Delta h$ and $h_A - 3\Delta h$, it follows that:

$$h_A - \Delta h - y_B = 0 \quad (3.70a)$$

$$h_A - 2\Delta h - y_C = 0 \quad (3.70b)$$

$$h_A - 3\Delta h - y_D = 0 \quad (3.70c)$$

A straightforward manipulation of equations 3.70 leads to:

$$y_C - y_D = y_B - y_C = y_A - y_B = \Delta h$$

indicating that equipotential lines drawn at *equal* intervals of total head Δh intersect the free surface at the *same* Δh in the vertical direction as shown in figures 3.34*a* and 3.34*b* where the shape of both flownets, corresponding to the same seepage problem through the same soil, is affected by the extent of the drainage blanket (*i.e.* a highly permeable drainage layer). Having this condition in mind, the problem of sketching a correct flownet then becomes a matter of practice.

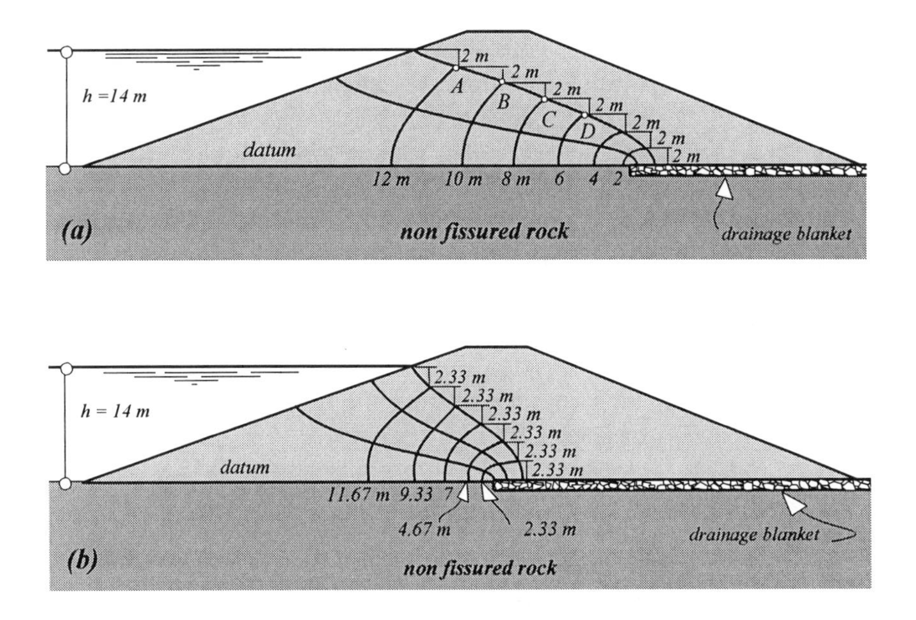

Figure 3.34: (a) Flownet construction for unconfined flow, (b) effect of the use of a wider drainage blanket.

Once an acceptable flownet is obtained, it can be used to estimate different quantities such as the porewater pressure at any location, the hydraulic gradient or the total flow quantity.

Note that aspects related to safety calculations of this type of construction, especially that concerned with the *rapid draw down conditions*, simulating a sudden decrease in the level of retained water, are presented in detail in section 7.5.4 of chapter 7.

3.5 Conformal mapping

3.5.1 Velocity potential – stream function

As mentioned earlier, only a handful of seepage problems have been solved analytically, mainly because of the difficulties arising from the boundary conditions of the flow equation which cannot always be satisfied.

However, a theoretical approach to the analysis of some seepage problems has been developed and, notwithstanding its reliance on advanced mathematical concepts such as elliptic functions and conformal mapping, the solutions obtained are none the less very elegant. The aim of this section is to use a special conformal mapping technique developed by Mandel (1951) to solve analytically seepage problems of *steady state confined flow* in *isotropic soils*, related to excavations, trenches and cofferdams. For more thorough analysis, the reader is referred to the excellent work of Pavlovsky (widely reported by Harr (1962)), Halek and Svec (1979) and Polubarinova-Kochina (1962).

Let us now consider the two-dimensional form of the continuity equation established earlier *via* equation 3.45:

$$\frac{\partial v_x}{\partial x} + \frac{\partial v_y}{\partial y} = 0 \tag{3.71}$$

If the *velocity potential* is defined as $\Phi(x,y) = -kh$, then, according to Darcy's law:

$$v_x = \frac{\partial \Phi}{\partial x} \quad \text{and} \quad v_y = \frac{\partial \Phi}{\partial y}$$

Substituting for v_x and v_y into equation 3.71 leads to the following *Laplace* equation:

$$\nabla^2 \Phi = \frac{\partial^2 \Phi}{\partial x^2} + \frac{\partial^2 \Phi}{\partial y^2} = 0 \tag{3.72}$$

The *stream function* is defined as $\Psi(x,y)$, such that:

$$v_x = \frac{\partial \Psi}{\partial y} \quad \text{and} \quad v_y = -\frac{\partial \Psi}{\partial x}$$

Substituting for v_x and v_y from the above expressions yields the *Cauchy–Reimann* equations:

$$\frac{\partial \Phi}{\partial x} = \frac{\partial \Psi}{\partial y} \quad \text{and} \quad \frac{\partial \Phi}{\partial y} = -\frac{\partial \Psi}{\partial x}$$

Moreover, using the velocity potential, it seen that:

$$\frac{\partial v_x}{\partial y} = \frac{\partial v_y}{\partial x} = \frac{\partial^2 \Phi}{\partial x\, \partial y}$$

so that when the stream function is substituted for the velocity potential in the last equation, the following Laplacian is established in a straightforward way :

$$\nabla^2 \Psi = \frac{\partial^2 \Psi}{\partial x^2} + \frac{\partial^2 \Psi}{\partial y^2} = 0 \qquad (3.73)$$

Both equations 3.72 and 3.73 indicate that the potential and stream functions are *harmonic functions*. More importantly, it can be shown that if the *complex potential* $\omega = \Phi + i\Psi$ is an *analytic function* of the complex variable $z = x + iy$, then Φ and Ψ are *conjugate harmonic functions,* meaning that curves representing constant potentials and constant stream lines intersect at right angles.

Accordingly, an analytical solution to a seepage problem can be developed *if* an appropriate complex potential can be found for which both stream and potential functions are conjugate harmonic functions. In what follows, the mathematical details of the analytical solutions developed by Mandel, relating to steady state confined seepage flow through isotropic soils, are presented.

3.5.2 Seepage along a sheet pile cut-off wall driven in a homogeneous soil layer of infinite thickness: exact solution

Consider the case of a sheet pile wall embedded in a *homogeneous* layer of soil, assumed to be of infinite thickness as depicted in figure 3.35. It is important to bear in mind that the following analysis applies equally to an *excavation* (where the water level behind the pile is below the ground) and a *cofferdam* (for which water behind the pile is above the ground surface).

In order to study the nature of seepage around the pile, Mandel considered the following mapping function:

$$z = \alpha \sinh\left(-\frac{i\omega}{\lambda}\right) + \frac{i\omega}{\lambda}\beta$$

which can be rewritten as:

$$x + iy = \alpha \sinh\left(\frac{\psi - i\Phi}{\lambda}\right) - \beta\frac{\psi - i\Phi}{\lambda} \tag{3.74}$$

where α, β and λ are positive real numbers (α > β).

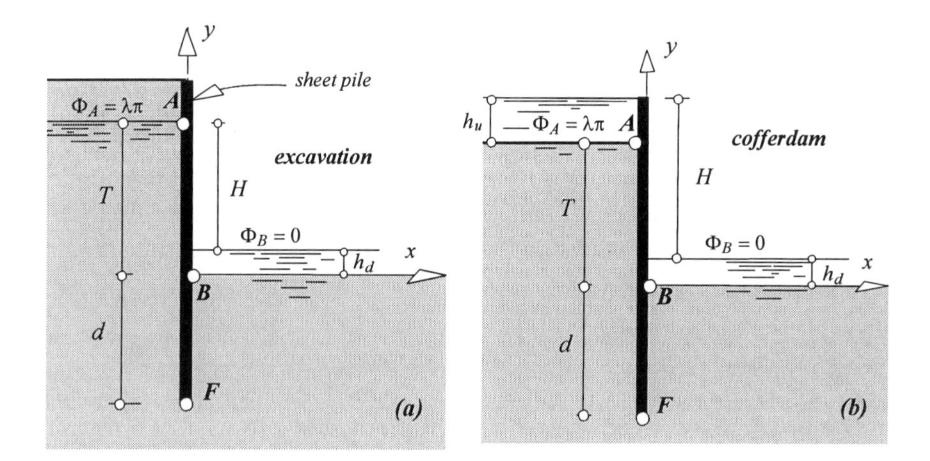

Figure 3.35: Seepage around a sheet pile cut-off wall in the case of (a) an excavation, and (b) a cofferdam.

Knowing that:

$$\sinh z = \sinh x \cos y + i \cosh x \sin y$$

equation 3.74 then yields the quantities x and y:

$$x = \alpha \sinh\frac{\psi}{\lambda}\cos\left(\frac{\Phi}{\lambda}\right) - \beta\frac{\psi}{\lambda} \tag{3.75a}$$

$$y = -\alpha \cosh\frac{\psi}{\lambda}\sin\left(\frac{\Phi}{\lambda}\right) + \beta\frac{\Phi}{\lambda} \tag{3.75b}$$

Referring to figure 3.35, it is seen that the associated boundary conditions are:

$$\Phi = 0, \qquad y = 0 \quad \text{and} \quad x \geq 0$$

$$\Phi = \lambda\pi, \quad y = \beta\pi \quad \text{and} \quad x \leq 0$$

Because both sides of the pile represent a flow line, the stream function along these frontiers (AF on the upstream side and FB on the downstream face) is therefore $\Psi = 0$. Consequently, along the pile, equations 3.75 are reduced to:

$$x = 0 \tag{3.76a}$$

$$y = \beta\frac{\Phi}{\lambda} - \alpha \sin\frac{\Phi}{\lambda} \tag{3.76b}$$

At the foot of the pile (*i.e.* point F in figure 3.35), the co-ordinate y is at its minimum and hence $\partial y/\Phi = 0$. Accordingly, if the velocity potential at F is Φ_F, then the first derivative of equation 3.76*b* with respect to Φ at the foot of the pile yields:

$$\beta = \alpha \cos\frac{\Phi_F}{\lambda} \tag{3.77}$$

Moreover, with reference to figure 3.35, it can be seen that the quantities λ and β are such that:

$$\Phi_A - \Phi_B = H = \lambda\pi \tag{3.78a}$$

$$T = \beta\pi \tag{3.78b}$$

Hence, the quantity y_F at F:

$$y_F = d = \alpha \sin\left(\frac{\Phi_F}{\lambda}\right) - \beta\frac{\Phi_F}{\lambda} \tag{3.79}$$

T is the depth of water behind the wall. When the water level is at or above ground level, T is taken as the height of the ground behind the wall, d being the depth of embedment. *Notice that at F, the quantity y_F is negative, hence the change of sign in equation 3.79 compared with equation 3.76b.*

Eliminating α between equations 3.77 and 3.79, and introducing the quantity $\beta = T/\pi$, it follows that:

$$\tan\left(\frac{\Phi_F}{\lambda}\right) - \frac{\Phi_F}{\lambda} = \pi\frac{d}{T} \tag{3.80}$$

If the ratio of the head loss occurring in the downstream side (between B and F) to the total head loss between A and B is:

$$\eta = \frac{\Phi_F - \Phi_B}{\Phi_A - \Phi_B} = \frac{\Phi_F}{\lambda\pi}$$

(since, from the boundary conditions, $\Phi_B = 0$ at $y = 0$), then equation 3.80 can be rewritten as follows:

$$\tan(\pi\eta) - \pi\eta = \pi d/T \tag{3.81}$$

Equation 3.81 is a *transcendental equation* that depends on the ratio T/d and which can be solved using, for instance, the chart represented in figure 3.36. *Notice that the chart is plotted in terms of the quantity T/d as opposed to d/T used in the equation.*

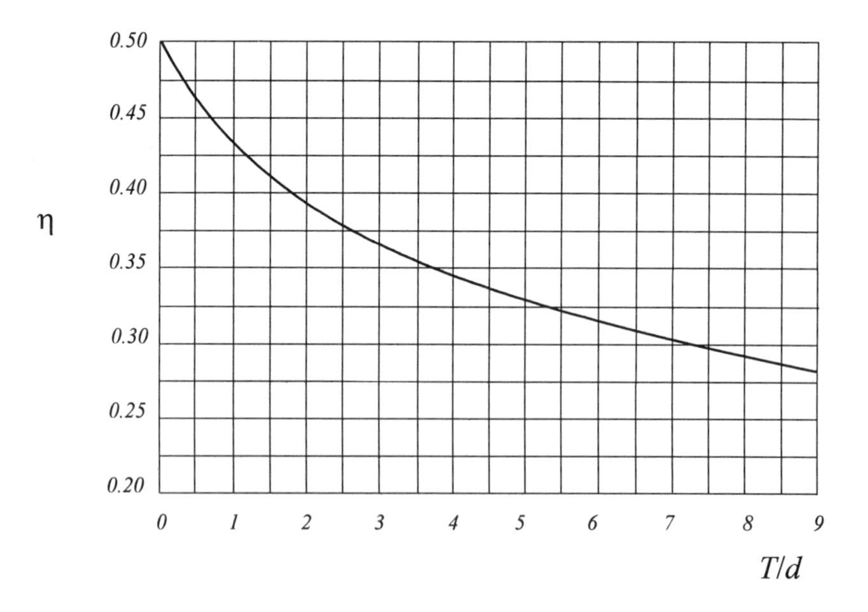

Figure 3.36: Graphical solution to equation 3.81.

The velocity potential at the foot of the pile (point F) is then:

$$\Phi_F = \eta H \tag{3.82}$$

and the expressions for the three constants can therefore be established:

$$\lambda = \frac{\Phi_A - \Phi_B}{\pi} = \frac{H}{\pi} \tag{3.83a}$$

$$\beta = \frac{T}{\pi} \tag{3.83b}$$

$$\alpha = \frac{\beta}{\cos\left(\frac{\Phi_F}{\lambda}\right)} = \frac{T}{\pi \cos(\eta\pi)} \tag{3.83c}$$

so that at any depth y along *both sides* of the sheet pile, the velocity potential can be calculated using equation 3.76b, that is:

$$y = \frac{\Phi_y}{\lambda} - \alpha \sin \frac{\Phi_y}{\lambda} \tag{3.84}$$

Notice that according to figure 3.35, y is negative between B and F on both sides of the pile. Also, the quantities Φ_F/λ and Φ_y/λ in equations 3.83c and 3.84 are expressed in *radians*.

The hydraulic gradient is thereafter determined according to the following expressions:

- on the upstream side (from A to F in figure 3.35):

$$i_u = \frac{\Phi_A - \Phi_y}{T - y} = \frac{H - \Phi_y}{T - y} \tag{3.85a}$$

- on the downstream side (from B to F):

$$i_d = -\frac{\Phi_y - \Phi_B}{y} = -\frac{\Phi_y}{y} \tag{3.85b}$$

The porewater pressure at any depth y on either side of the pile is calculated from the general expression:

$$u = \gamma_w(h - y)$$

$$= \gamma_w[(\Phi_y + h_d) - y] \tag{3.86}$$

where γ_w is the unit weight of water and $(\Phi_y + h_d)$ represents the excess head at the depth y. The quantity Φ_y is calculated from equation 3.84 and h_d corresponds to the height of the tailwater in front of the wall as shown in figure 3.35.

Example 3.7

Examine the distribution of porewater pressure along the sheet pile cut-off wall depicted in figure 3.37, and driven in a layer of isotropic sand (assumed to be infinitely thick) having a permeability k.

From the figure, it is seen that:

$h_u = 1.5\,m, \quad h_d = 0.5\,m, \quad H = 4\,m, \quad T = 3\,m, \quad d = 4\,m$

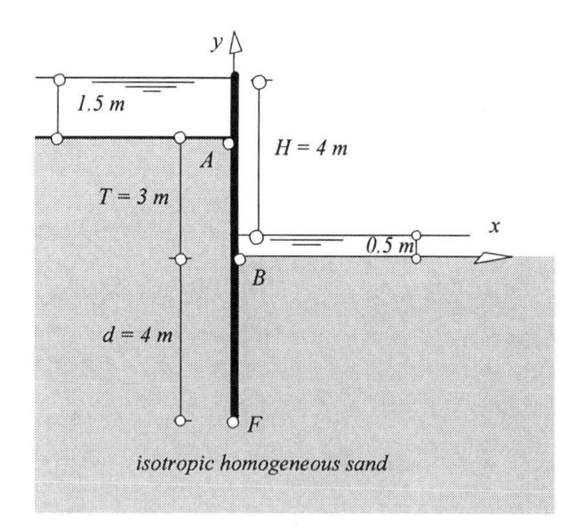

Figure 3.37: Seepage flow along a cut-off wall embedded in an isotropic homogeneous sand.

First, solve for η using equation 3.81 and figure 3.36 *(remember figure 3.36 is used in conjunction with the quantity T/d)*:

$$\tan(\pi\eta) - \pi\eta = \pi d/T = 4.189 \quad \Rightarrow \quad \eta \approx 0.443$$

The velocity potential at the foot of the pile *(point F)* is thereafter calculated from equation 3.82:

$$\Phi_F = \eta H = 1.774\,m$$

Now that the values of the velocity potential at A, F and B are known, in that:

$$\Phi_A = H = 4\,m, \quad \Phi_F = 1.774\,m \quad \text{and} \quad \Phi_B = 0$$

the coefficients λ, β and α are then determined from equations 3.83:

$$\lambda = \frac{\Phi_A - \Phi_B}{\pi} = 1.273\,m, \quad \beta = \frac{T}{\pi} = 0.955\,m, \quad \alpha = \frac{\beta}{\cos\frac{\Phi_F}{\lambda}} = 5.416\,m$$

(Φ_F/λ in radians).

Equation 3.84 can now be readily established:

$$y = 0.75\Phi_y - 5.416\sin\left(\frac{\Phi_y}{1.273}\right)$$

(the quantity $\Phi_y/1.273$ is expressed in radians).

For the sake of clarity, the above equation is first solved for the upstream side of the pile (where y varies from $+3\,m$ at A to $-4\,m$ at F, and the potential changes from $\Phi_A = 4\,m$ to $\Phi_F = 1.77\,m$), then for the downstream face for which y ranges from 0 at B to $-4\,m$ at F (with $\Phi_B = 0$).

The results, presented in tabular form, include the values of:

- the hydraulic gradient computed using equations 3.85,
- the porewater pressure according to equation 3.86.

Upstream side of the pile:

$\Phi_y(m)$	$y(m)$	i_u	$h(m)$	$u\,(kN/m^2)$
4.00	3.00	–	4.50	15
3.75	1.76	0.20	4.25	25
3.50	0.55	0.20	4.00	34
3.25	–0.57	0.21	3.75	43
3.00	–1.58	0.22	3.50	51
2.75	–2.44	0.23	3.25	57
2.50	–3.13	0.24	3.00	61
2.25	–3.62	0.26	2.75	64
2.00	–3.92	0.29	2.50	64
1.78	–4.00	0.32	2.28	63

Downstream face of the pile:

$\Phi_y(m)$	$y(m)$	i_d	$h(m)$	$u\,(kN/m^2)$
0.00	0.00	–	0.50	5
0.25	–0.87	0.29	0.75	16
0.50	–1.70	0.29	1.00	27
0.75	–2.44	0.31	1.25	37
1.00	–3.08	0.32	1.50	46
1.25	–3.57	0.35	1.75	53
1.50	–3.88	0.39	2.00	59
1.78	–4.00	0.44	2.28	63

It is worth mentioning that, in these tabulated values, the hydraulic gradient on each side of the pile increases with depth and reaches its maximum value at the foot of the pile.
The results are plotted in figure 3.39 where the hydrostatic pressure ($u = \gamma_w h_w$), as well as the porewater pressure estimated from the flownet of figure 3.38, are included. The hydrostatic pressure is plotted purely for comparison purposes. Also, notice how well the results measured from the flownet compare to the analytical solution.

The discharge area in the downstream side is of particular interest to the designer in that the hydraulic gradient should always be kept smaller than the critical hydraulic gradient i_c by an adequate factor of safety. From the downstream values tabulated earlier, the hydraulic gradient varies between $i \approx 0.29$ at the top of the discharge area and $i = 0.44$ at the foot of the pile. On the other hand, the flownet of figure 3.38 yielded a hydraulic gradient of 0.33 at the top and 0.41 around the foot of the pile (on the downstream side); values that compare favourably with those calculated analytically.

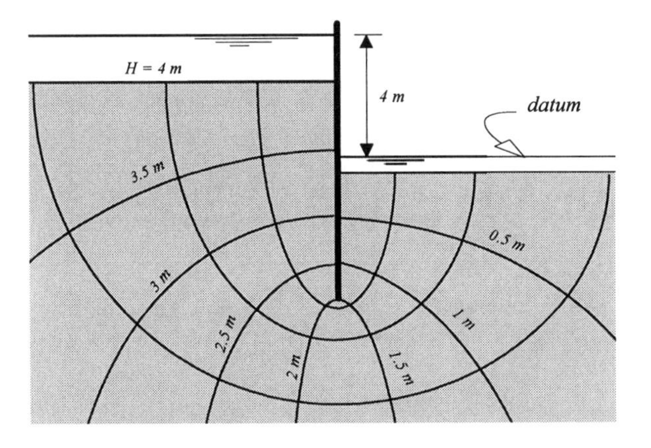

Figure 3.38: Flownet solution to example 3.7.

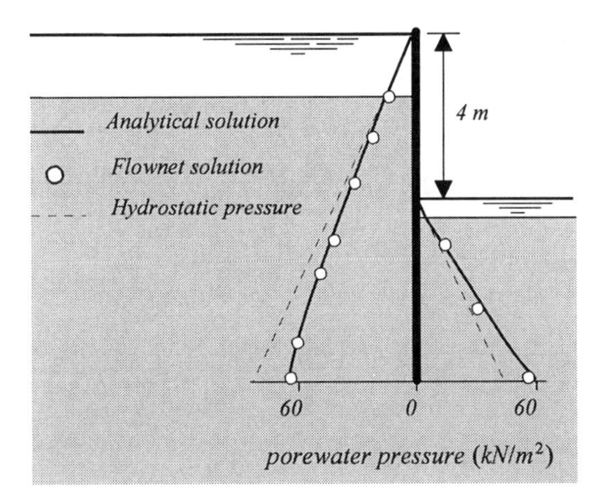

Figure 3.39: Analytical solution of the porewater pressure distribution along the pile.

3.5.3 Seepage along a sheet pile cut-off wall driven in a homogeneous soil layer of finite thickness

Consider the sheet pile driven in a soil having a permeable layer of thickness T, underlain by an impervious material as depicted in figure 3.40.

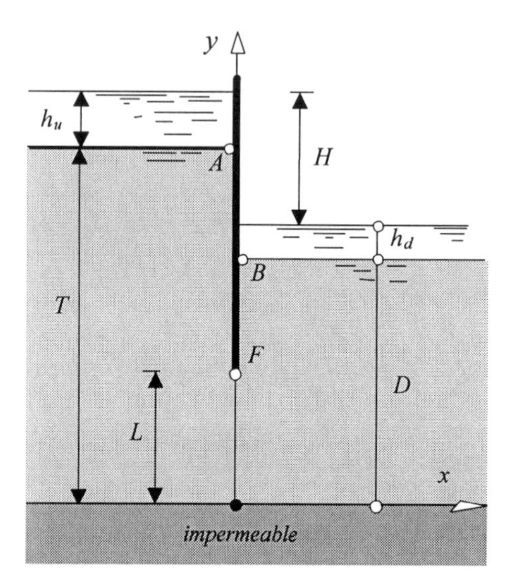

Figure 3.40: Seepage around a pile embedded in a permeable soil layer of finite thickness.

This time, Mandel chose the following mapping function:

$$z = iL \cosh \frac{\omega}{\xi}$$

which can be equally expressed in the following way:

$$x + iy = iL \cosh \frac{\Phi + i\psi}{\xi}$$

where L is the depth of the impervious layer measured from the foot of the pile and ξ is a positive real number.

It is straightforward to show that:

$$x = -L \sinh \frac{\Phi}{\xi} \sin \frac{\psi}{\xi} \tag{3.87a}$$

$$y = L \cosh \frac{\Phi}{\xi} \cos \frac{\psi}{\xi} \tag{3.87b}$$

The boundary conditions are as follows:

$$\psi = 0, \quad x = 0, \quad y \geq L,$$

$$\psi = \frac{\pi}{2}\xi, \quad y = 0 \text{ and } -\infty < x < +\infty.$$

So that in the vicinity of the pile, these conditions are virtually satisfied and amount to the following:

$$x = 0, \quad \psi = 0$$

$$y = L \cosh \frac{\Phi}{\xi} \tag{3.88}$$

Making a change of variables so that $y/L = x$ and $\Phi/\xi = \sigma$, equation 3.88 can then be rewritten as follows:

$$x = \cosh \sigma$$

or, alternatively

$$(e^{\sigma} - x)^2 - \left(x^2 - 1\right) = 0$$

which has a straightforward solution:

$$e^{\sigma} = x \pm \sqrt{x^2 - 1} \tag{3.89}$$

Taking into account the change of variables made earlier, the expression of the potential along the sheet pile is obtained by rearranging equation 3.89, so that the velocity potential at a depth y along the upstream side of the pile *(AF* in figure 3.40*)* is:

$$\Phi_y = \xi \ln\left(\frac{y}{L} + \sqrt{\frac{y^2}{L^2} - 1}\right) \tag{3.90a}$$

and along the downstream face *(FB)*:

$$\Phi_y = -\xi \ln\left(\frac{y}{L} + \sqrt{\frac{y^2}{L^2} - 1}\right) \tag{3.90b}$$

Whence the velocity potentials at A (where $y = T$), F (where $y = L$) and B (where $y = D$) are:

$$\Phi_A = \xi \ln\left(\frac{T}{L} + \sqrt{\frac{T^2}{L^2} - 1}\right) \tag{3.91a}$$

$$\Phi_F = 0 \tag{3.91b}$$

$$\Phi_B = -\xi \ln\left(\frac{D}{L} + \sqrt{\frac{D^2}{L^2} - 1}\right) \tag{3.91c}$$

On the other hand, figure 3.40 indicates that:

$$\Phi_A - \Phi_B = H$$

Making use of equations 3.91, the value of the quantity ξ can then be established:

$$\xi = \frac{H}{\ln\left(\frac{T}{L} + \sqrt{\frac{T^2}{L^2} - 1}\right) + \ln\left(\frac{D}{L} + \sqrt{\frac{D^2}{L^2} - 1}\right)} \tag{3.92}$$

Having set the relationships leading to the velocity potential at any depth along each side of the pile through equations 3.90, and having also determined its precise nature at the upstream ground level *(point A)*, at the foot of the pile *(point F)* and at the downstream ground level *(point B) via* equations 3.91, the hydraulic gradient at any level is then calculated as follows:

- on the upstream side (between A and F in figure 3.40):

$$i_u = \frac{\Phi_A - \Phi_y}{T - y} \tag{3.93a}$$

- on the downstream side (from F to B):

$$i_d = \frac{\Phi_y - \Phi_B}{D - y} \tag{3.93b}$$

The porewater pressure is subsequently calculated in the usual way, that is:

$$u = \gamma_w (h - y) \tag{3.94}$$

with h representing the total head, at the depth y, calculated as follows:

- behind the wall (upstream side):

$$h = (h_u + T) - i_u(T - y)$$

$$= \Phi_y - \Phi_A + (h_u + T)$$

- in front of the wall (downstream face) :

$$h = (h_d + D) + i_d(D - y)$$

$$= \Phi_y - \Phi_B + (h_d + D)$$

where D is the depth of the impervious layer measured from the bottom of the excavation in front of the wall and T is the thickness of the same impervious layer behind the wall; h_u and h_d being as indicated in figure 3.40.

Example 3.8

Let us re-examine the sheet pile studied previously with the help of a flownet in section 3.4.2 (example 3.4, figure 3.20). The dimensions as well as the co-ordinate systems are replotted in figure 3.41. Required : use the theoretical analysis to study the porewater pressure distribution along the pile.

Known are: $h_u = 1.5\,m$, $h_d = 0$, $T = 7.6\,m$,
$L = 2.5\,m$, $D = 5.5\,m$, $H = 3.6\,m$

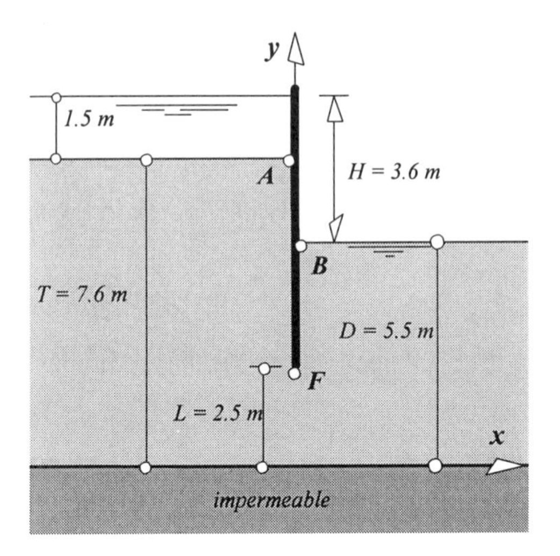

Figure 3.41: Seepage along a sheet pile embedded in a permeable soil of finite thickness.

First, use equation 3.92 to calculate the quantity ξ:

$$\xi = \frac{3.6}{\ln\left[\frac{7.6}{2.5} + \sqrt{\left(\frac{7.6}{2.5}\right)^2 - 1}\right] + \ln\left[\frac{5.5}{2.5} + \sqrt{\left(\frac{5.5}{2.5}\right)^2 - 1}\right]} = 1.124\, m$$

The velocity potentials Φ_A at ground level behind the wall, Φ_F at the foot of the pile and Φ_B at ground level in front of the wall, are thereafter computed from equations 3.91, leading to:

$\Phi_A \approx 2\, m, \quad \Phi_F = 0, \quad$ and $\Phi_B = -1.60\, m$

The sequence of subsequent calculations is then as follows:

- evaluate the velocity potential along the pile using equations 3.90,
- calculate the hydraulic gradient according to equations 3.93,
- estimate the porewater pressure with the help of equation 3.94.

The results, tabulated below are plotted in figure 3.42. Once more, the precise nature of the results calculated from the flownet in figure 3.21 is noticeable.

Behind the wall (upstream side):

$y(m)$	$\Phi_y\ (m)$	i_u	$h(m)$	$u(kN/m^2)$
7.6	2.00	–	9.10	15
6.5	1.81	0.17	8.91	24
5.5	1.60	0.19	8.70	32
4.5	1.34	0.21	8.44	39
3.5	0.97	0.25	8.07	46
3.0	0.70	0.28	7.80	48
2.5	0.00	0.39	7.10	46

In front of the wall (downstream face):

$y(m)$	$\Phi_y\ (m)$	i_d	$h(m)$	$u(kN/m^2)$
5.5	–1.60	–	5.50	0
5.0	–1.48	0.24	5.62	6
4.5	–1.34	0.26	5.76	13
3.5	–0.97	0.32	6.13	26
3.0	–0.70	0.36	6.40	34
2.5	0	0.53	7.10	46

At the discharge area, the emphasis would be on the exit gradient whose value is of paramount importance to the stability of the wall. From the above tables, the exit gradient has a value $i_e = 0.24$. The flownet of figure 3.21 yielded an exit gradient of 0.28.

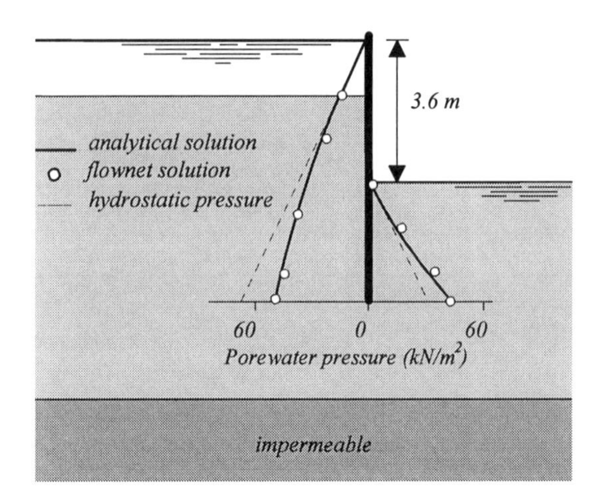

Figure 3.42: Analytical solution to the distribution of porewater pressure along the pile.

3.6 The method of fragments

Thus far, different types of two dimensional seepage problems for which the flow occurs mainly in the (x,y) plane have been considered. The method of fragments, first developed by Pavlovsky (1933), is an approximate theoretical method of solution that can be applied to any type of *confined flow* occurring within a *finite depth.*

In this method, the flow region is subdivided into sections or *fragments*, each of which is assumed to have a *straight vertical equipotential line* at its various parts. For example, in the case of seepage problem corresponding to the cofferdam in figure 3.43, the flow region is subdivided into two fragments, and the equipotential line at the foot of the pile is assumed to be vertical.

If a standpipe was inserted at each side of the pile in a way that their bases are on the same equipotential line, then water will rise to the same level inside both tubes. Manifestly, the loss of total head in fragment *I* $[h_1]$is different from the one occurring through fragment *II* $[h_2]$since the length of flow path and the nature of flow are different on both sides of the pile.

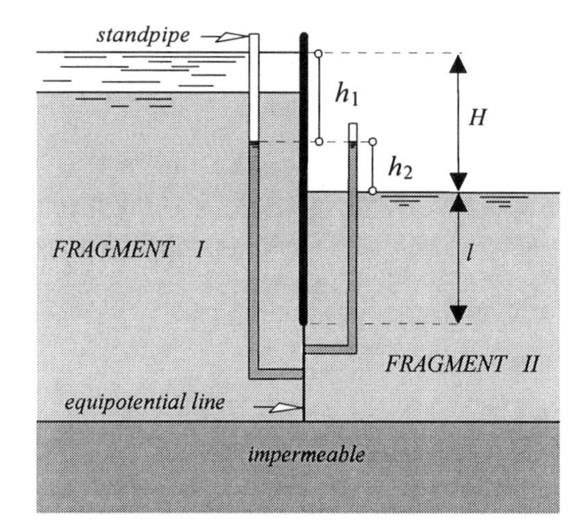

Figure 3.43: Fragments corresponding to seepage around a sheet pile.

The total head loss is such that:

$$H = h_1 + h_2$$

Furthermore, equation 3.57 giving the flow quantity per unit width, can be rearranged as follows:

$$Q = k\frac{H}{\phi} \tag{3.95}$$

where ϕ represents a (dimensionless) *form factor*. Since the same flow quantity per unit width seeps through the entire region, equation 3.95 can therefore be used *ad lib* in conjunction with h_1 and h_2, so:

$$k\frac{H}{\phi} = k\frac{h_1}{\phi_1} = k\frac{h_2}{\phi_2} \tag{3.96}$$

ϕ_1 and ϕ_2 are the form factors associated, respectively, with fragments *I* and *II*:

$$\phi_1 = h_1\phi/H \qquad \text{and} \qquad \phi_2 = h_2\phi/H.$$

Whence: $\phi = \phi_1 + \phi_2$ (3.97)

Substituting for ϕ into equation 3.95 yields the relationship between the quantity of flow Q and the form factors ϕ_1 and ϕ_2:

$$Q = kH\frac{1}{\phi_1 + \phi_2} \tag{3.98}$$

As for the *average* hydraulic gradient in the discharge area, it can be estimated from the ratio of total head loss h_2 in the discharge area (*i.e.* the total head at the foot of the pile) to the length of the flow path l indicated in figure 3.43:

$$i_{av} \approx \frac{h_2}{l} \tag{3.99}$$

with:

$$h_2 = H\frac{\phi_2}{\phi_1 + \phi_2} \tag{3.100}$$

Pavlovsky based his theoretical method on elliptic functions to establish a series of typical form factors and the mathematical details can be found in Harr (1962), for instance. To make the method more appealing to practising engineers, Davidenkoff and Franke (1965, 1966) have used the electrical analogue method to study the problem of seepage through trenches and cofferdams (problems characterised by two fragments), and have established charts from which the form factors ϕ_1 and ϕ_2 are obtained. The *modus operandi* of these charts, reproduced in figure 3.45, consists of:

- calculating, from the geometry of the problem, the dimensionless quantities d_1/T_1, d_2/T_2 and T_2/b (d_1, d_2, T_1, T_2 and b are as indicated in figure 3.44),

- reading the form factor $\phi_1 = F(d_1/T_1, T_2/b = 0)$ on the curve $T_2/b = 0$ in conjunction with the ratio d_1/T_1, and finally

- reading the form factor $\phi_2 = F(d_2/T_2, T_2/b)$ on the curve corresponding to the ratio T_2/b, in association with the quantity d_2/T_2 (use your judgement if an extrapolation is needed).

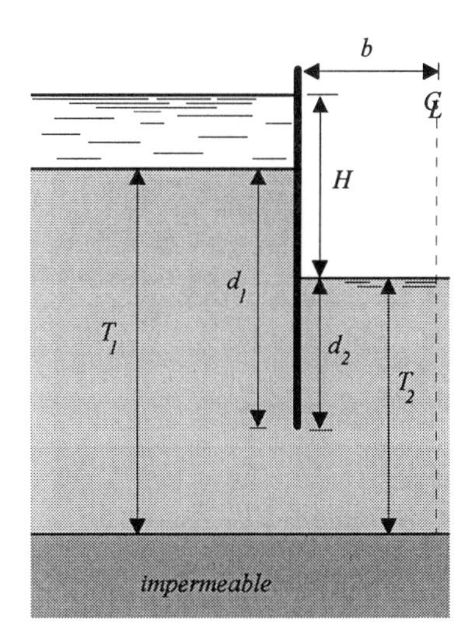

Figure 3.44: Dimensions used in conjunction with the charts in figure 3.45.

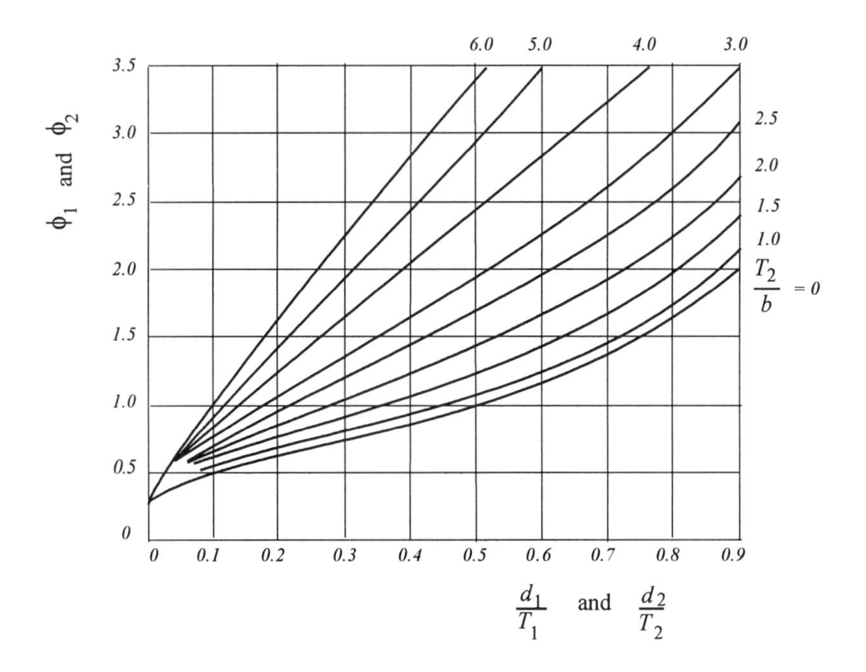

Figure 3.45: Davidenkoff and Franke form factors charts (reproduced by permission of Ernst & Sohn).

The charts can be used to estimate the total head loss in the discharge area as well as the flow quantity *per unit length of perimeter* in the three following cases:

1- Trench having a width 2b and an infinite length:

$$h_2 = H\frac{\phi_2}{\phi_1+\phi_2} \tag{3.101}$$

$$q = kH\frac{1}{\phi_1+\phi_2} \tag{3.102}$$

2- Circular cofferdam of a diameter 2b:

$$h_2 = 1.3H\frac{\phi_2}{\phi_1+\phi_2} \tag{3.103}$$

$$q = 0.8kH\frac{1}{\phi_1+\phi_2} \tag{3.104}$$

3- Square cofferdam of a side 2b:

$$h_2 = 1.3H\frac{\phi_2}{\phi_1+\phi_2} \quad \textit{(in the middle of the sides)} \tag{3.105}$$

$$h_2 = 1.7H\frac{\phi_2}{\phi_1+\phi_2} \quad \textit{(in the corners)} \tag{3.106}$$

$$q = 0.75kH\frac{1}{\phi_1+\phi_2} \tag{3.107}$$

In all cases, the *average* hydraulic gradient on the downstream side is calculated from equation 3.99.

Example 3.9

Examine the flow problem for which the planar dimensions are as represented in figure 3.46. The permeable soil layer consists of an isotropic sandy clay with a permeability $k = 10^{-6}\ m/s$.

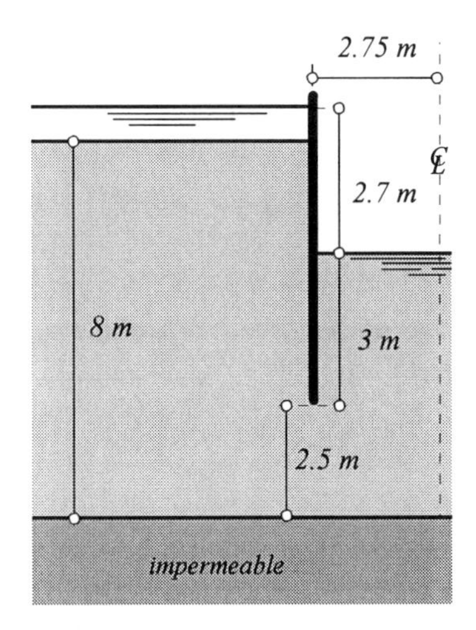

Figure 3.46: Seepage problem.

The corresponding two-dimensional flownet is sketched in figure 3.47, where it can be assumed that the equipotential line at the foot of the pile is vertical.

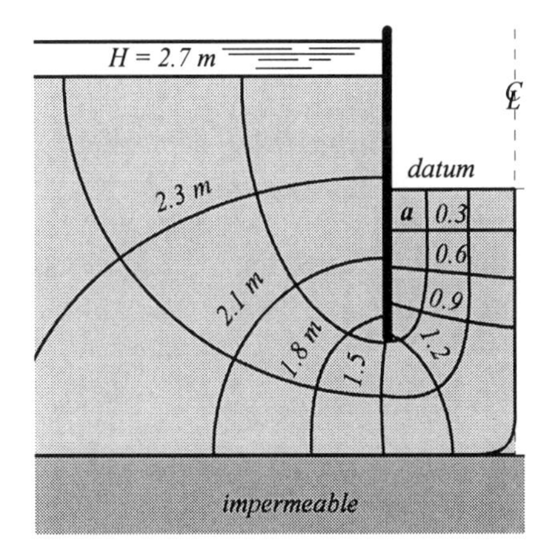

Figure 3.47: Flownet solution.

The flownet has six flow channels (only one half on the flownet is produced in figure 3.47 because of symmetry) and nine equipotential drops, thence:

$$Q = 2.7 \times \frac{6}{9} \times 10^{-6} = 1.8 \times 10^{-6}\ m^3/s/m.$$

The exit gradient is estimated from element a in figure 3.47:

$$i_e \approx \frac{0.3}{0.72} = 0.41$$

Let us now apply the Davidenkoff and Franke method. The following quantities can be calculated from figure 3.46, using the notation of figure 3.44:

$$\frac{d_1}{T_1} = \frac{5.5}{8} = 0.69, \quad \frac{d_2}{T_2} = \frac{3}{5.5} = 0.545, \quad \frac{T_2}{b} = \frac{5.5}{2.75} = 2$$

The form factors are read from the charts in figure 3.45 (note that ϕ_1 is read on the curve corresponding to $T_2/b = 0$):

$$\phi_1 = 1.35, \quad \phi_2 = 1.6$$

As drawn in the plane, figure 3.47 can be projected in the third dimension to form one of three cases which are of particular interest.

Case 1: Trench of infinite length, having a width 2b = 5.5 m

The three-dimensional flownet sketched in figure 3.48, gives a clear indication as regards the flow patterns. The quantities h_2 and q are calculated from equations 3.101 and 3.102:

$$h_2 = 2.7 \times \frac{1.6}{1.35+1.6} = 1.462\ m$$

$$q = \frac{2.7}{1.35+1.6} \times 10^{-6} = 9.1 \times 10^{-7}\ m^3/s \textit{ per metre of perimeter}$$

hence the total flow quantity on both sides of the trench:

$$Q = 2q \quad = 1.82 \times 10^{-6}\ m^3/s \textit{ per metre run}$$

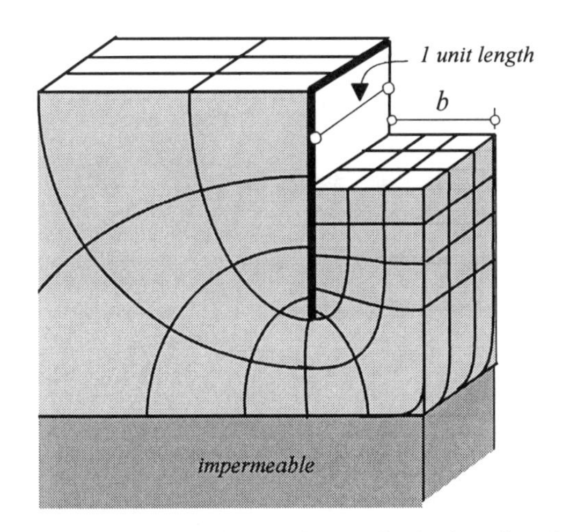

Figure 3.48: 3-D flownet in the case of a trench of infinite length.

The average gradient is estimated from equation 3.99:

$$i_{av} = \frac{h_2}{l} = \frac{1.46}{3} = 0.49$$

These results are in close agreement with those calculated from the flownet of figure 3.44:

	flownet solution	*Davidenkoff and Franke method*
$Q\,(m^3/s/m)$	1.8×10^{-6}	1.82×10^{-6}
i_e	0.41	$i_{av} = 0.49$

Case 2: Circular cofferdam of diameter 2b = 5.5 m (perimeter $2\pi b$)

A cross-section of the corresponding three-dimensional flownet is shown in figure 3.49. In this case, equations 3.103 and 3.104 are used to calculate h_2 and q:

$$h_2 = 1.3 \times 2.7 \times \frac{1.6}{1.35+1.6} = 1.90\,m$$

$$q = \frac{0.8 \times 2.7}{1.35+1.6} \times 10^{-6} = 7.3 \times 10^{-7}\ m^3/s \text{ per metre of perimeter}$$

hence the total flow quantity:

$$Q = 2\pi bq = 1.26 \times 10^{-5}\ m^3/s$$

The average gradient is in this case: $i_{av} = \frac{1.90}{3} = 0.63$

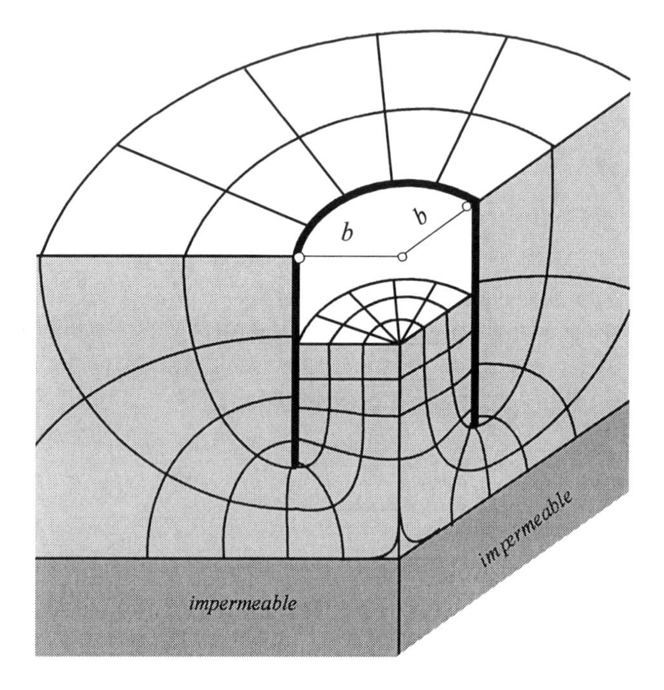

Figure 3.49: Cross-section of the 3-D flownet for a circular cofferdam.

Case 3: Square cofferdam of side 2b = 5.5 m (perimeter 8b)

Figure 3.50 depicts a cross-section of the three-dimensional flownet associated with this case. The quantities h_2 and q are computed according to equations 3.105 to 3.107:

$$h_2 = 1.3 \times 2.7 \times \frac{1.6}{1.35+1.6} = 1.90\,m \text{ (in the middle of each side)}$$

$$h_2 = 1.7 \times 2.7 \times \frac{1.6}{1.35+1.6} = 2.49\,m \text{ (in the corners)}$$

$$q = \frac{0.75 \times 2.7}{1.35+1.6} \times 10^{-6} = 6.9 \times 10^{-7}\ m^3/s \text{ per metre of perimeter}$$

so that the total flow rate is:

$$Q = 8bq = 1.52 \times 10^{-5}\ m^3/s$$

Finally the average hydraulic gradient is:

$$i_{av} = \frac{1.90}{3} = 0.63 \text{ (in the middle of each side)}$$

$$i_{av} = \frac{2.49}{3} = 0.83 \text{ (in the four corners)}$$

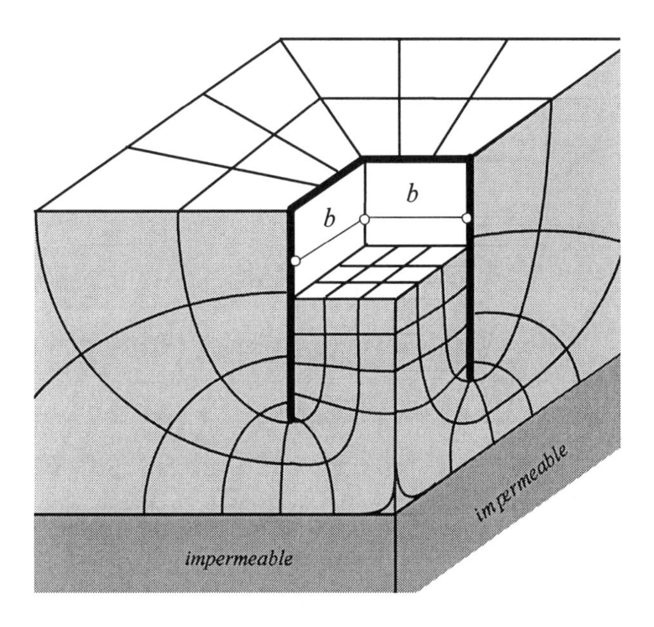

Figure 3.50: Cross-section of the 3-D flownet in the case of a square cofferdam.

Example 3.10: Asymmetric trench

Consider the problem of seepage flow beneath an asymmetric trench assumed to be of infinite length, whose dimensions are as indicated in figure 3.51. The pervious soil is an isotropic homogeneous sand of permeability *k (m/s)*.

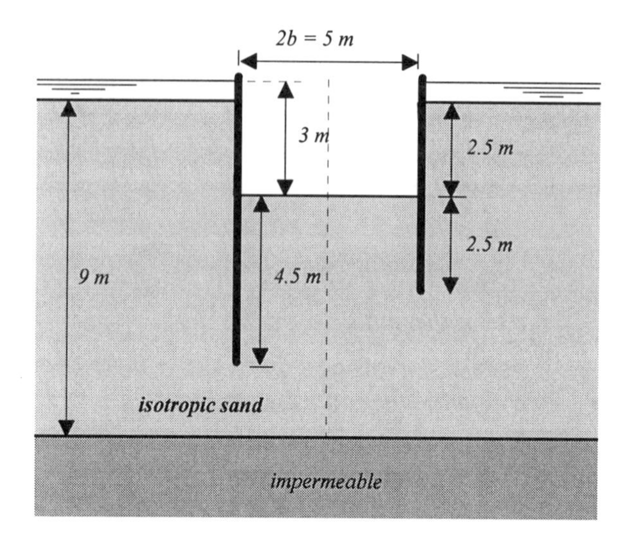

Figure 3.51: Asymmetric seepage flow.

The corresponding flownet solution, sketched in figure 3.52, reflects the asymmetrical nature of flow (five flow channels on the right-hand side and only three on the left-hand side). Based on the flownet, the flow rate can be estimated using equation 3.57:

$$Q = kH\frac{N_f}{N_d} = 3k\frac{(5+3)}{13} \approx 1.85k\ (m^3/s/m$$

With reference to figure 3.52, it is seen that the exit hydraulic gradient has different values on either side of the trench. On the left-hand side, the exit gradient calculated from element *a* in the flownet is:

$$i_e \approx \frac{0.23}{0.66} = 0.35$$

On the right-hand side, however, the exit gradient is calculated from element *b* in the flownet:

$$i_e \approx \frac{0.23}{0.46} = 0.50$$

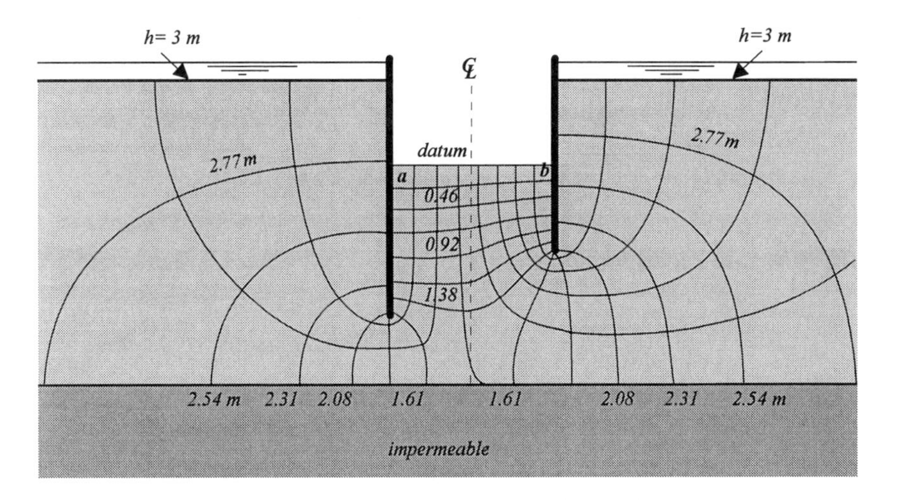

Figure 3.52: Flownet associated with asymmetric flow.

An alternative way of solving this flow problem would be to use the charts of figure 3.45. The flownet of figure 3.52 shows that both equipotential lines at the foot of the piles can be assumed to be vertical. The asymmetry dictates that the superposition of figure 3.53 is used.

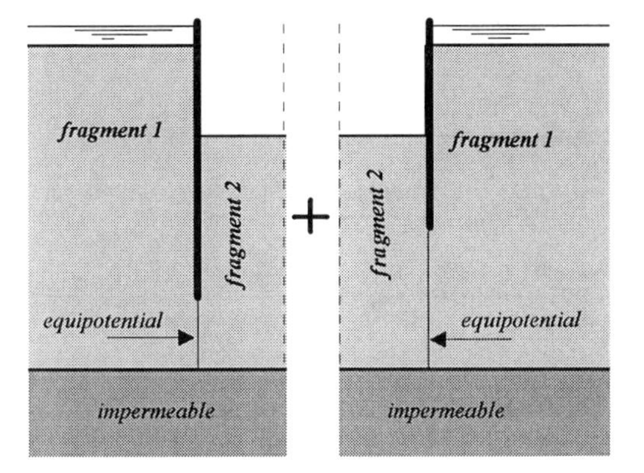

Figure 3.53: Superposition of fragments in the case of asymmetric flow problem.

The different parameters used in conjunction with the charts (refer to figures 3.44 and 3.51), as well as the ensuing results are:

- on the left-hand side of the trench:

$$b = 2.5\,m, \quad d_1/T_1 = 0.78, \quad d_2/T_2 = 0.69, \quad T_2/b = 2.6$$

Hence the form factors : $\phi_1 = 1.58$, $\phi_2 = 2.3$.

- the head loss in the exit side is calculated using equation 3.101:

$$h_2 = H\frac{\phi_2}{\phi_1 + \phi_2} = 3 \times \frac{2.3}{(1.58 + 2.3)} = 1.78\,m$$

- the flow rate is determined from 3.102:

$$q_l = k\frac{H}{\phi_1 + \phi_2} = k\frac{3}{(1.58 + 2.3)} = 0.773\,k \text{ per metre of perimeter}$$

- the *average* hydraulic gradient is estimated from equation 3.99:

$$i_{av} = \frac{h_2}{l} = \frac{1.78}{4.5} = 0.4$$

-on the right-hand side:

$$b = 2.5\,m, \quad d_1/T_1 = 0.555, \quad d_2/T_2 = 0.384, \quad T_2/b = 2.6$$

Whence the form factors: $\phi_1 = 1.1$, $\phi_2 = 1.45$.

- the head loss is: $h_2 = 3 \times \frac{1.45}{(1.1 + 1.45)} = 1.70\,m$

- the flow rate q is:

$$q_r = k\frac{3}{(1.1 + 1.45)} = 1.176\,k \text{ per metre of perimeter}$$

- the average hydraulic gradient: $i_{av} = \frac{1.7}{2.5} = 0.68$

The total flow rate per metre length of the trench is therefore the sum of the flow rates on each side:

$$Q = q_l + q_r = (0.773 + 1.176)\,k = 1.95\,k\ (m^3/s/m)$$

The different results summarised in the following, indicate the validity of both methods used above.

	flownet solution	*Davidenkoff and Franke method*
flow rate per metre length	$1.85\ k$	$1.95\ k$
exit gradient: left side	0.35	$i_{av} = 0.40$
exit gradient: right side	0.50	$i_{av} = 0.68$

Problems

3.1 Consider the sheet pile cut-off wall then use the corresponding flownet solution illustrated in figure *p*3.1 and calculate:

(*a*) the seepage forces per unit area at point *A, B, ..., G;*

(*b*) the effective stresses as well as the porewater pressures at the same points. Assume a saturated unit weight of sand of $\gamma_{sat} = 21\ kN/m^3$.

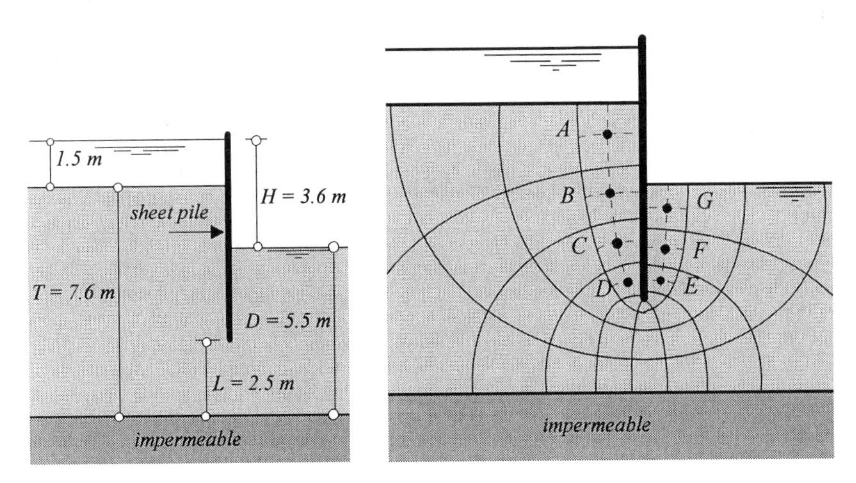

Figure p3.1

Ans: (*a*) *Seepage forces per unit area:*
$F_A = 1.89\,kN/m$, $F_B = 5.52\,kN/m$,
$F_C = 11.1\,kN/m$, $F_D = 18.8\,kN/m$, $F_E = -11.7\,kN/m$,
$F_F = -5.6\,kN/m$, $F_G = -1.8\,kN/m$.
(*b*) *Effective stresses:* $\sigma'_A = 11.8\,kN/m^2$, $\sigma'_B = 31.9\,kN/m^2$
$\sigma'_C = 51.8\,kN/m^2$, $\sigma'_D = 70.5\,kN/m^2$, $\sigma'_E = 16.9\,kN/m^2$
$\sigma'_F = 13.1\,kN/m^2$, $\sigma'_G = 4.8\,kN/m^2$.
(*c*) *Porewater pressures:* $u_A = 21.2\,kN/m^2$, $u_B = 32.6\,kN/m^2$
$u_C = 43\,kN/m^2$, $u_D = 48.4\,kN/m^2$, $u_E = 35\,kN/m^2$,
$u_F = 22.4\,kN/m^2$, $u_G = 7.8\,kN/m^2$.

3.2 A small filter tank 1.6 *m* long and 0.8 *m* wide is made up of five layers of sand of equal thickness $d = 0.2\,m$ (see figure *p*3.2), with the horizontal permeabilities of any two consecutive layers characterised by a constant ratio :
$k_{h2}/k_{h1} = k_{h3}/k_{h2} = k_{h4}/k_{h3} = k_{h5}/k_{h4} = 0.5$.

Calculate the permeability k_{h1} if the total flow quantity seeping through the sand layers is :
$Q = 8.33 \times 10^{-6}\,m^3/s$ (≈ 30 *litre per hour*).

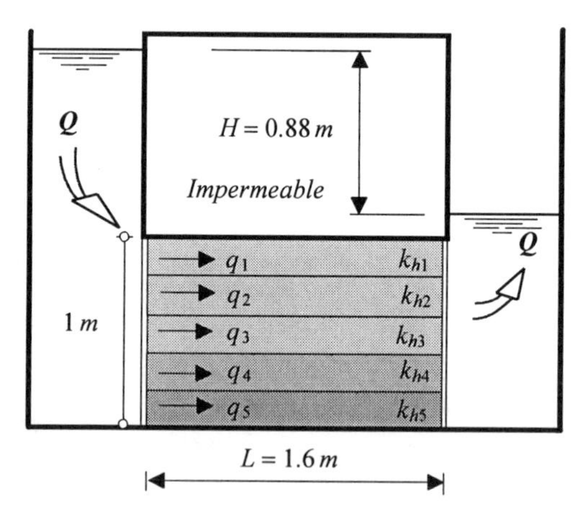

Figure p3.2

Ans: $k_{h1} = 3.91 \times 10^{-5}\,m/s$.

3.3 Calculate the vertical permeability k_{v1} in the case of the vertical flow illustrated in figure *p*3.3. Assume that the total flow seeping through each layer is $Q = 2.78 \times 10^{-6}\, m^3/s$ ($\approx 10\, l/h$), and that the ratio between consecutive permeabilities is identical to the previous case, *i.e.*
$k_{v2}/k_{v1} = k_{v3}/k_{v2} = k_{v4}/k_{v3} = k_{v5}/k_{v4} = 0.5$.

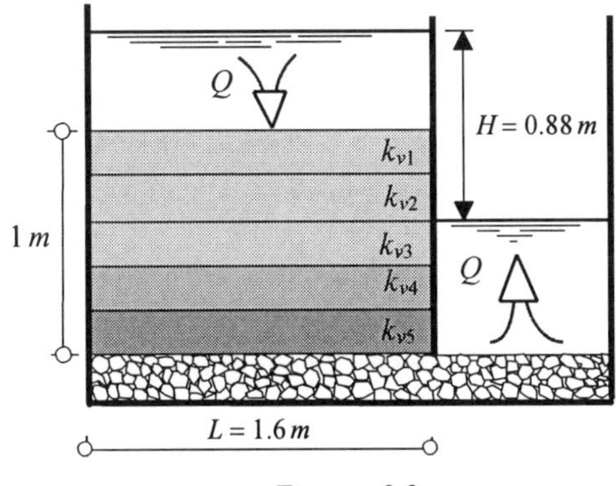

Figure p3.3

Ans: $k_{v1} = 1.53 \times 10^{-5}\, m/s$.

3.4 Assume now that the sand layers in figure *p*3.2 are characterised by the respective equivalent horizontal and vertical permeabilities: $k'_h = 1.51 \times 10^{-5}\, m/s$ and $k'_v = 2.47 \times 10^{-6}\, m/s$.

(*a*) Calculate the new dimensions of the tank were the five layers of sand to be replaced by a single 1 *m* thick isotropic layer with a permeability $k = \sqrt{k'_h k'_v}$.
(*b*) Estimate the corresponding quantity of vertical flow.

Ans: (*a*) $L' = 0.647\, m$, (*b*) $Q_v = 2.78 \times 10^{-6}\, m^3/s$.

3.5 Refer to the flownet illustrated in figure *p*3.1 and assume that it was sketched on a transformed section corresponding to the flow through an anisotropic soil having a ratio of anisotropy $k_h/k_v = 2$. Scale back the flownet in question to the natural section.

3.6 Consider the excavation in figure *p*3.6 in conjunction with the confined flow conditions of example 3.2. All things being equal, calculate the drawdown at points *A, B* and *C.*

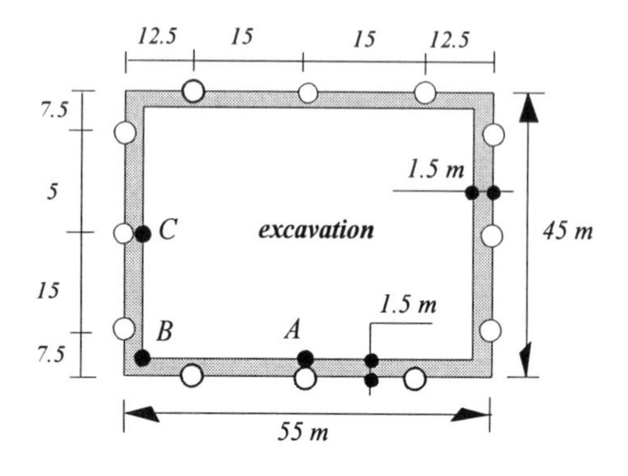

Figure p3.6

Ans: $d_A \approx 11.04\,m, \quad d_B \approx 10.21\,m, \quad d_C \approx 10.93\,m.$

3.7 Assume that the unconfined flow and soil conditions of example 3.3 apply, then estimate the drawdown achieved under points *D, E* and *F* in figure *p*3.7.

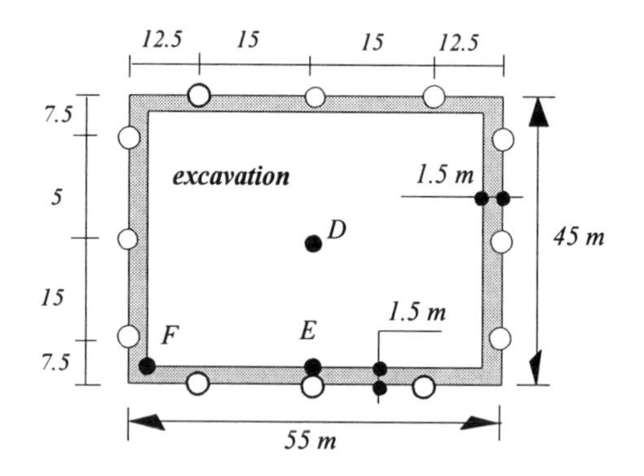

Figure p3.7

Ans: $d_D \approx 8.05\,m, \quad d_E \approx 8.38\,m, \quad d_F \approx 7.50\,m.$

3.8 Redraw the flownet corresponding to the unconfined flow through the earth dam of figure *p*3.8 in the absence of a drainage blanket, then estimate the porewater pressure at the centre of the dam (point *A* in figure *p*3.8),

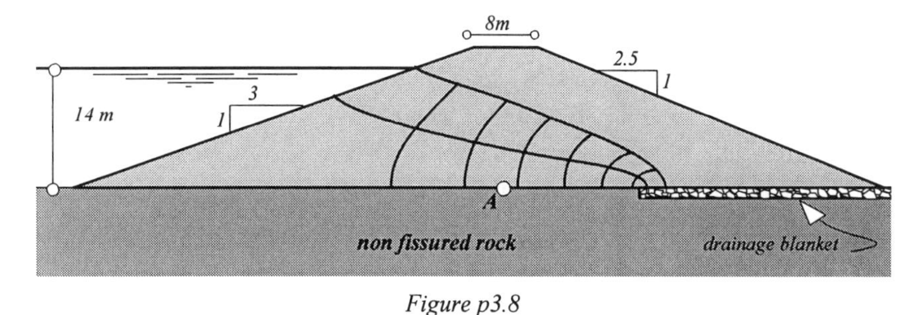

Figure p3.8

Ans: $u_A \approx 117\,kN/m^2$ *without drainage blanket, as opposed to* $u_A \approx 85\,kN/m^2$ *in figure p*3.8.

3.9 A circular cofferdam with the dimensions indicated in figure *p*3.9 is embedded 5.5 *m* into a clean sand having a saturated unit weight $\gamma_{sat} = 21\,kN/m^3$ and a permeability $k\,(m/s)$.
Sketch an accurate flownet, then estimate:

(*a*) the hydraulic gradient at the most critical point of flow within the cofferdam;
(*b*) the factor of safety against localised quicksand conditions (*i.e.* piping) at the most critical point of flow;
(*c*) the quantity of flow in $m^3/s/m$;
(*d*) the profile of vertical effective stresses along each side of the pile;
(*e*) the porewater pressure distribution along both sides of the pile.

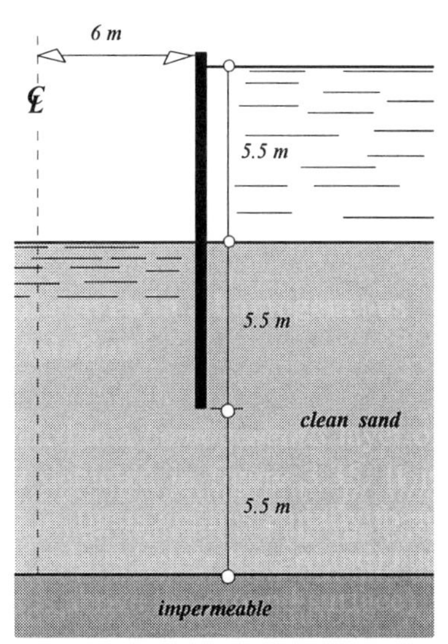

Figure p3.9

Ans: (*a*) $i_e \approx 0.41$, (*b*) $F = 2.4$,
(*c*) $q = 1.96k\,(m^3/s/m)$

3.10 Consider the circular cofferdam of problem $p3.9$ above, and use *Davidenkoff and Franke* charts, based on the method of fragments, to estimate:
(a) the exit hydraulic gradient,
(b) the factor of safety against piping,
(c) the quantity of flow in $m^3/s/m$.

How can you justify the difference between these results and those calculated previously from the flownet solution ?

Ans: (a) $i_e \approx 0.75$, (b) $F = 1.33$, (c) $q = 1.85k(m^3/s)$.

3.11 Use *Mandel* analysis in conjunction with figure $p3.11$ depicting a sheet pile cut-off wall embedded in a sand, then calculate the distribution along both sides of the pile of:
(a) the porewater pressure,
(b) the hydraulic gradient,
(c) the vertical effective stresses.

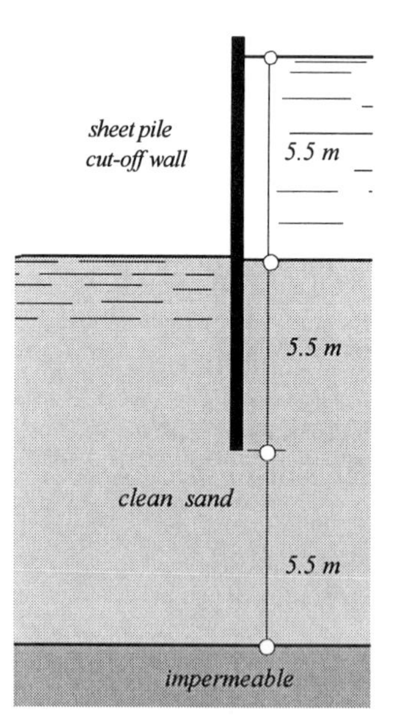

Figure p3.11

References

Bear, J. (1972) *Dynamics of Fluids in Porous Media.* Elsevier, New York.

Cedergren, H. (1990) *Seepage, Drainage and Flownets.* 3rd edn, John Wiley, New York.

Davidenkoff, R. N. and Franke, O. L. (1965) *Intersuchung der Räumlichen Sickerströmung in eine Umspundete Baugrube in Offenen Gewässern.* Die Bautechnik, 9, pp.298–307.

Davidenkoff, R. N. and Franke, O. L. (1966) *Räumliche Sickerströmung in eine Umspundete Baugrube im Grundwasser.* Die Bautechnik, 12, pp. 401–409.

Halek, V. and Svec, J. (1979) *Groundwater Hydraulics.* Elsevier, Amsterdam.

Harr, M. E. (1962) *Groundwater and Seepage.* McGraw-Hill, New York.

Hong Kong Geotechnical Engineering Office, (1993) *Review of Granular and Geotextile Filters.* Hong Kong.

Mandel, J. (1951) *Ecoulement de l'eau sous une ligne de palplanches: Abaque pour la condition de renard.* Travaux, 197, pp. 273–281.

Pavlovsky, N. N. (1933) *Motion of water under dams.* Transactions of the 1st Congress on Large Dams, Stockholm, Vol. 4.

Pavlovsky, N. N. (1956) *Collected Works.* Akad. Russia, Leningrad.

Polubarinova-Kochina, P. Ya. (1941) *Concerning Seepage in Heterogeneous (two layered) Media.* Inzhenernii Sbornik, Vol. 1 (2).

Polubarinova-Kochina, P. Ya. (1962) *Theory of Ground Water Movement.* (translated from Russian). Princeton University Press.

Somerville, S. H. (1986) *Control of Groundwater for Temporary Works.* CIRIA Report 113, Construction Industry Research and Information Association, London.

Tomlinson, M. J. (1995) *Foundation Design and Construction.* 6th edn, Longman, London.

CHAPTER 4

Transient flow: elastic and consolidation settlements

4.1 Introduction

The basic fundamental difference between (steady) seepage and transient flows relates to the change (or the lack of it) of the volume of soil through which flow occurs. Thus, when water flows into or out of a soil mass *without* causing the volume to change, the flow is known as *seepage* (refer to chapter 3). If, on the other hand, the flow of water within a soil mass induces a volume change, then the flow is referred to as *transient*. Obviously, exception should be made for critical seepage flow conditions whereby an erosion process might take place thus causing a decrease of soil volume.

The process of volume change triggered by a transient flow, known as *consolidation*, is related to the change in effective stresses within the soil matrix heralded by a surface loading (or unloading) or a variation in the water table. The *excess porewater pressure* (*i.e.* the load-induced porewater pressure) generated in both cases causes the water to be either squeezed out of the soil mass (case of a positive excess pressure) or sucked into the soil matrix (case of a negative excess pressure). This movement of water continues at a *changing rate* until all excess pressure has dissipated, and the equilibrium of stresses has been restored according to the effective stress principle.

Prior to dealing in detail with the consolidation process (*i.e.* the rate of volume change), which depends on the soil *permeability*, let us turn our attention to the final magnitude of the volume change which is related to the soil *compressibility*. In fact, the extent to which a soil deforms under the effect of an applied load depends on its stress history, that is the geological history of the soil through loading and (when applicable) unloading cycles. Thus, if at some stage during its geological history the soil has been subjected to unloading (due, for instance, to the disappearance of an ice cover or to a severe erosion), then the present pressure due to the soil weight and known as the *overburden pressure* is *smaller* than that which existed before the onset of the unloading process, and the soil is known as *overconsolidated.*

If, on the other hand, the soil has not been subjected to any unloading during its entire geological history, then the present overburden pressure constitutes the largest pressure that the soil has ever experienced, and the soil is in this case referred to as *normally consolidated.* These definitions can be made easier to understand at this stage with the illustration in figure 4.1 showing a soil covered during the ice age by a thick layer of ice which has since disappeared.

Obviously, at the peak of the Ice Age (*circa* 1.8 million years ago), the *maximum* vertical pressure at point A (σ_p in the figure) was due to the soil volume above A whose thickness is h_1 and to the volume of ice with a thickness h_2. However, the vertical stress σ_v has gradually decreased as the ice cover melted away, so that *at present,* σ_v corresponding to the same point is only due to the weight of soil above the point whose thickness h_1 has increased by an amount Δh due to the release of pressure.

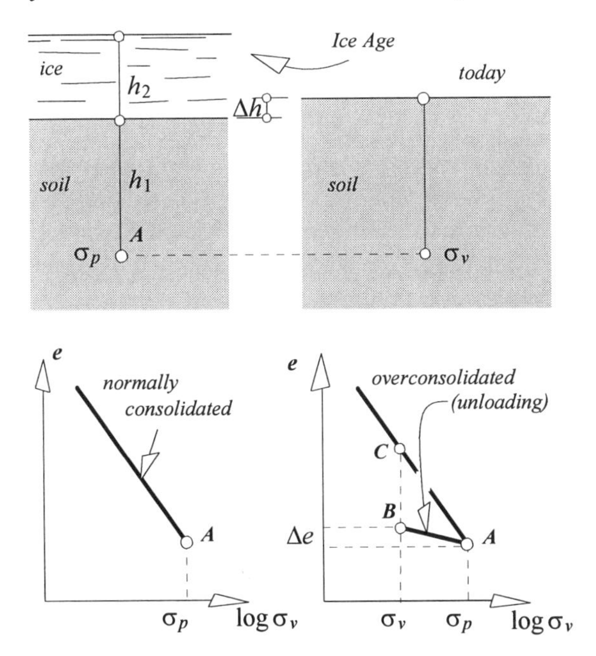

Figure 4.1: Effects of loading patterns on soil behaviour.

During the process of unloading, the soil, originally *normally consolidated* has become *overconsolidated.* The change in the void ratio Δe during overconsolidation is related to the deformation Δh (assumed to be elastic), the precise nature of the relationship being introduced in the following section. The figure shows that the change in the soil void ratio (*i.e.* the deformation) induced by the *same* pressure increment $\Delta\sigma = \sigma_p - \sigma_v$ depends on the state of soil: the deformation of the *normally consolidated soil* subjected to $\Delta\sigma$ from point C to point A is *larger* than that occurring under the same $\Delta\sigma$ between points B and A where the soil is *overconsolidated.* Though the mathematical aspects of this conclusion will be presented shortly, it is already clear that, from a practical perspective, an overconsolidated soil is less troublesome in terms of deformation and settlement than a normally consolidated one.

The consolidation characteristics of saturated clays can be measured in the laboratory using the one-dimensional consolidation test (also known as the oedometer test), a brief description of which follows.

4.2 Stress–strain relationships in one-dimensional consolidation

The stress–strain relationships that will be derived in what follows are based on different consolidation parameters that can be measured from the oedometer test. The corresponding apparatus used for the test and depicted in figure 4.2, consists mainly of a brass ring having an internal diameter of 75 *mm* and a height of 20 *mm*, two porous stones, a metal container and a loading hanger. The sharp end of the ring is used to cut through a bigger sample of clay, so that a clay specimen can be secured inside the ring with a minimum of disturbance. The specimen is thereafter sandwiched between the two porous stones, then placed inside the container. Note that in order to avoid any volume change of the clay sample *prior* to testing, the saturation process (*i.e.* the filling of the container with water as *per* figure 4.2) is undertaken *after* the first load increment is applied to the sample. The clay being *confined* within the ring, the *only* deformation induced by the load is *vertical* (one dimensional). The test, thus, allows for the vertical deformation of the sample to be measured as the experiment progresses under different effective stress increments $\Delta\sigma'_i$, applied successively in such a way that the soil is allowed to consolidate fully under each increment before the next increment is added.

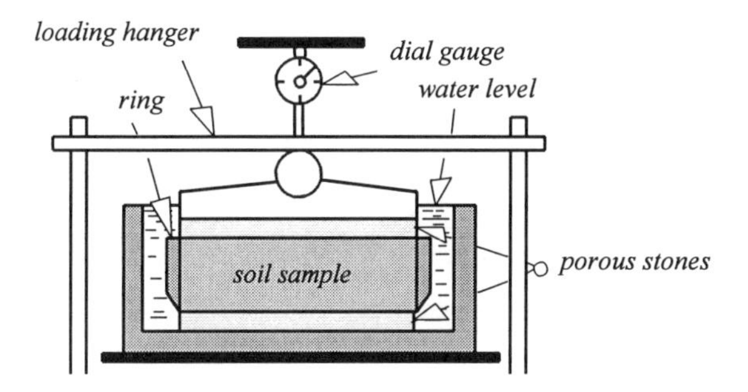

Figure 4.2: 1-D consolidation apparatus.

The variation in time of the sample deformation is typically as depicted in figure 4.3, where it can be seen that the higher the cumulative load, the smaller the voids between solid particles and the stiffer the soil.

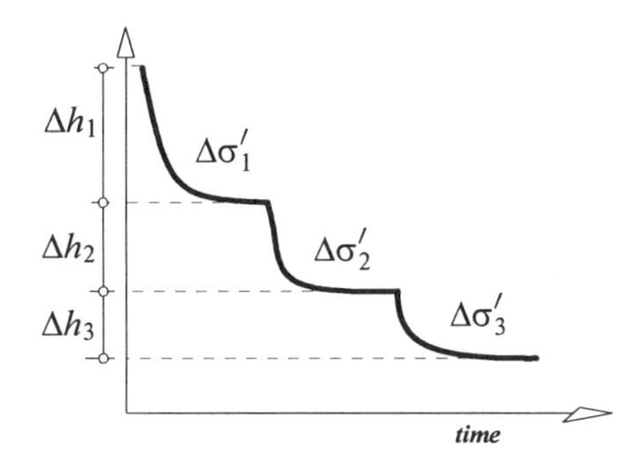

Figure 4.3: Typical results corresponding to 1-D consolidation test.

If the quantities $\Delta h_i/h$ (h being the original height of the sample) are plotted against the cumulative value of the effective vertical stress σ'_v, then a graph similar in shape to the one illustrated in figure 4.4 will ensue. The *changing* slope α of this graph is of particular interest and its use will be explained below.

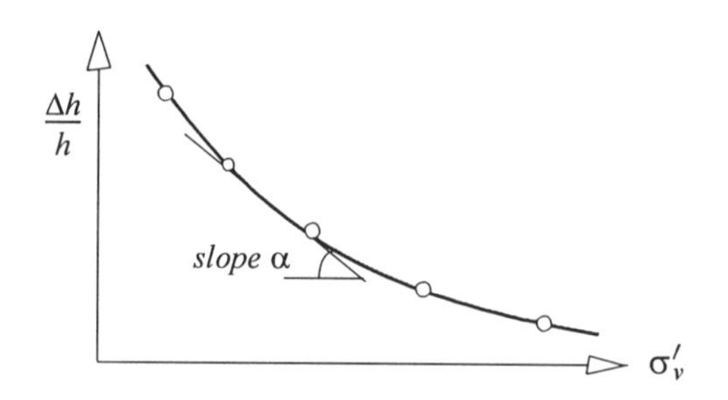

Figure 4.4: Typical 1-D consolidation test results.

In the meantime, the relationship between the variation of the void ratio and that of the sample deformation can now be established according to figure 4.5:

$$\frac{\Delta h}{h} = \frac{h_{wo} - h_{wf}}{h_s + h_{wo}} = \frac{(h_{wo}/h_s) - (h_{wf}/h_s)}{1 + (h_{wo}/h_s)} = \frac{e - e_f}{1 + e} = \frac{\Delta e}{1 + e} \qquad (4.1)$$

Giving the variation of the void ratio:

$$\Delta e = (1 + e)\frac{\Delta h}{h} \qquad (4.2)$$

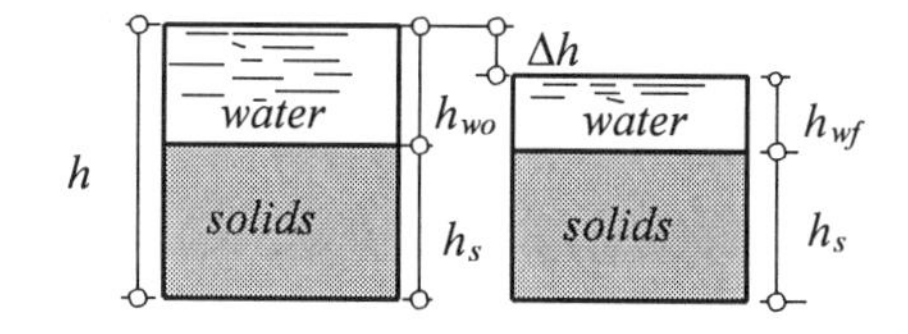

Figure 4.5: Initial and final stages during 1-D consolidation test.

Accordingly, the results of figure 4.4 can be analysed in terms of the variation of the void ratio e *versus* the effective vertical stress σ'_v. When plotted in $(e, \log \sigma'_v)$ space, the shape of the corresponding curve depends on the *state* of the soil.

Thus, for a *normally consolidated clay,* the curve corresponding to stress values greater than or equal to the *in situ* vertical effective stress σ'_{vo} is characterised by a slope C_c referred to as *the compression index*. For *overconsolidated clays*, the portion of the curve corresponding to vertical stresses smaller than σ'_{vo} has a different slope known as *the swelling index* C_s as depicted in figure 4.6. Note that in practice, C_s is measured from an unload–reload cycle to minimise the effects of clay disturbance.

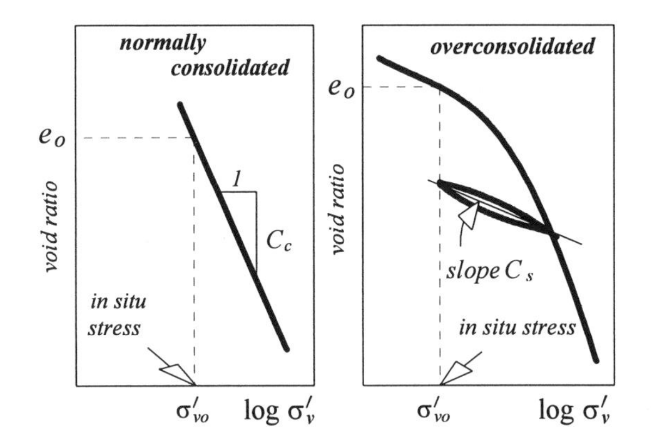

Figure 4.6: Normally consolidated and overconsolidated clays.

Meanwhile, if the *preconsolidation pressure* σ'_p is defined as being the maximum past effective overburden pressure to which the soil has been subjected during its entire geological history, then it is clear from both figures 4.1 and 4.6 that σ'_p in the case of a normally consolidated soil corresponds to the present *in situ* stress σ'_{vo}, whereas an overconsolidated soil is characterised by $\sigma'_p > \sigma'_{vo}$. In this respect, consider the layer of saturated soil with an effective unit weight $\gamma' = (\gamma_{sat} - \gamma_w)$ depicted in figure 4.7 where, through a natural process, the top layer of a thickness z_1 has since disappeared.

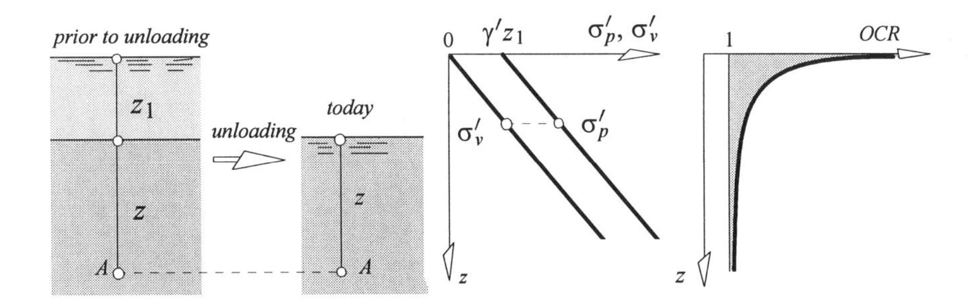

Figure 4.7: Variation with depth of the overconsolidation ratio.

Let us evaluate the stresses, past and present at point A situated at a depth z:

- the maximum effective vertical stress applied in the past: $\sigma'_p = \gamma'(z + z_1)$,

- the present effective vertical stress: $\sigma'_v = \gamma' z$.

Accordingly, while both preconsolidation pressure and effective vertical stress vary linearly with depth, the overconsolidation ratio *OCR,* defined as:

$$OCR = (\sigma'_p / \sigma'_v) = 1 + \frac{z_1}{z} \qquad (4.3)$$

is a hyperbolic function of z as indicated in figure 4.7. Clearly, as the depth increases, σ'_p increases and the *OCR* decreases towards its ultimate value of 1 corresponding to a normally consolidated state.

In practice, the magnitude of σ'_p can be determined using one of the two following graphical procedures.

(1) The procedure due to Casagrande (1936) is illustrated in figure 4.8, and consists of the following steps.

- Determine the point O on the $(e, \log \sigma'_v)$ graph that has the sharpest curvature (*i.e.* the smallest radius of curvature).
- Draw the horizontal line OA.
- Plot the tangent to the graph at O (*i.e.* OC in figure 4.8).
- Draw the line *OB* that bisects the angle *OAC*.
- Extend the straight line tail portion of the graph intersecting in the process the bisector *OB*.

The *estimated* preconsolidation pressure σ'_p of the soil corresponds to the abscissa of the point of intersection.

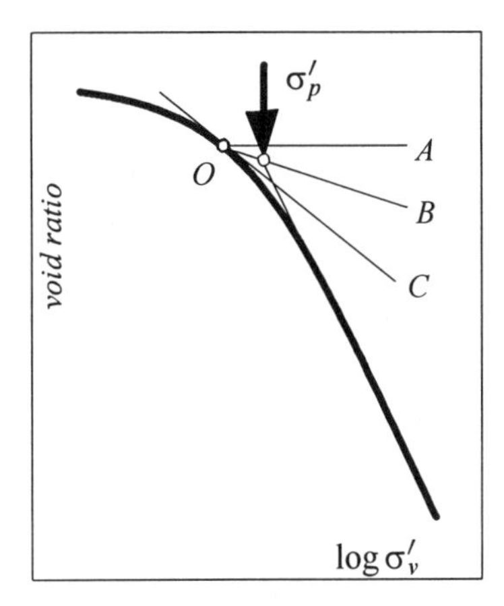

Figure 4.8: Casagrande procedure for the determination of σ'_p.

(2) The Schmertmann procedure (1953) was designed to minimise the effects of disturbance, represented by the shaded area in figure 4.9, due to *in situ* sampling and laboratory handling of the soil. Schmertmann suggested that the behaviour observed during a one-dimensional consolidation test does not reflect the actual soil behaviour *in situ* until the (cumulative) vertical effective stress applied to the sample reaches a magnitude corresponding to a void ratio of $0.42e_o$, where e_o is the initial void ratio corresponding to the *in situ* vertical stress σ'_{vo}. He then devised the following procedure to determine σ'_p.

(*a*) Apply incremental loading to the sample, including an unload–reload cycle, until point *A* corresponding to a void ratio of $0.42e_o$ is reached (refer to figure 4.9).
(*b*) Draw the line *BC* within the unload–reload loop.
(*c*) Plot point *D* with the co-ordinates $(e_o, \log \sigma'_{vo})$, from which the line *DE* is drawn parallel to *BC*.
(*d*) *Select* a point on *DE* (point *P*, for instance, in figure 4.9), thus fixing the presumed value of σ'_p.
(*e*) Plot the (presumed) virgin consolidation line *PA*.
(*f*) Measure the difference in void ratio Δe_p between the selected point and the actual experimental graph.
(*g*) Plot the point of co-ordinates $(\Delta e_p, \sigma'_p)$ in the space $(\Delta e, \log \sigma'_v)$ as shown in the figure.
(*h*) Repeat the procedure from step (*d*).

The *estimated* preconsolidation pressure of the clay, according to this method, is σ'_p which results in the most symmetrical (Δe, $\log \sigma'_v$) graph.

Notice that for stiff clays, the Schmertmann method necessitates the application of very high vertical pressures in order to achieve a void ratio of $0.42e_o$. These pressure levels, which can be in excess of $20\,MN/m^2$, require the use of specially adapted laboratory equipment.

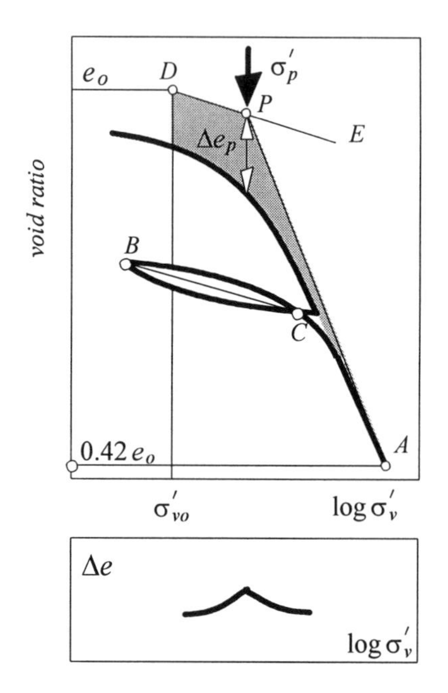

Figure 4.9: Schmertmann procedure.

4.3 One-dimensional consolidation theory

The one-dimensional consolidation test described previously simulates the behaviour of a thin layer of clay subjected to a wide surface load. Under such circumstances, the ensuing deformation, due to the expulsion of water from within the soil matrix, occurs mainly in the vertical direction, and therefore the drainage of water is predominantly vertical or one-dimensional.

In the case of a transient flow, water flow is caused by a gradual change in effective stresses that leads to a continuous rearrangement of solid particles, thus effecting a volume change. Accordingly, if a volume of soil is subjected at its surface to a constant total stress increase $\Delta\sigma$ then, in accordance with the effective stress principle of equation 2.4, the changes with time of both the effective stress increment $\Delta\sigma'$ and the excess porewater pressure Δu generated by $\Delta\sigma$ are such that:

$$\frac{\partial \sigma'_v}{\partial t} = -\frac{\partial u}{\partial t} \qquad (4.4)$$

Equation 4.4 indicates that the changes of both effective stress and porewater pressure occur at the same rate so that, for instance, the slopes at points A and B in figure 4.10 are equal and opposite.

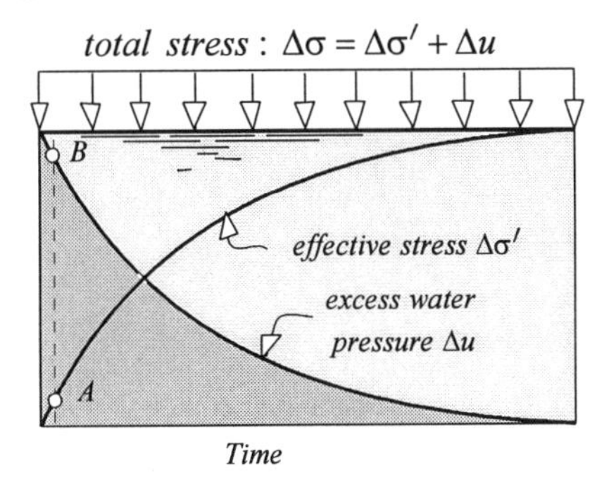

Figure 4.10: Excess water pressure dissipation with time.

On the other hand, it is clear that the behaviour exhibited in figure 4.4 can be replotted as the variation of the void ratio *versus* the effective vertical stress according to equation 4.2. Consequently, with reference to figure 4.11, an increment of stress $\delta\sigma'_v$, applied instantaneously over a wide area, would generate a change in void ratio δe at the following rate:

$$m_v = -\frac{1}{(1+e)}\frac{\delta e}{\delta \sigma'_v} \qquad (4.5)$$

m_v is known as the *coefficient of volume compressibility* of the soil, and is usually expressed in m^2/kN (notice that m_v depends on the stress increment level).

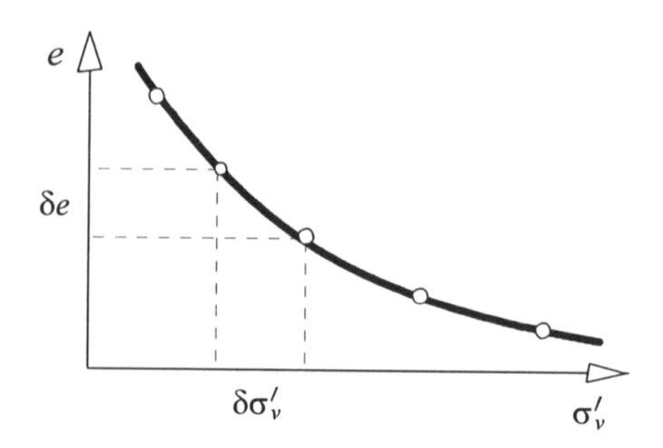

Figure 4.11: Relationship between the void ratio and the effective vertical stress.

Let us now try to establish the (one-dimensional) consolidation equation. To do so, consider, first, a volume V of a *saturated* clay such as the one depicted in figure 4.12. Assuming the water flows only in the z-direction, then it is clear that the variation (*i.e.* the decrease) in time of the volume V is related to the change in velocity in the following manner:

$$-\frac{\partial V}{\partial t} = \left[v_z + \frac{\partial v_z}{\partial z} dz \right] A - v_z A = \frac{\partial v_z}{\partial z} A\, dz$$

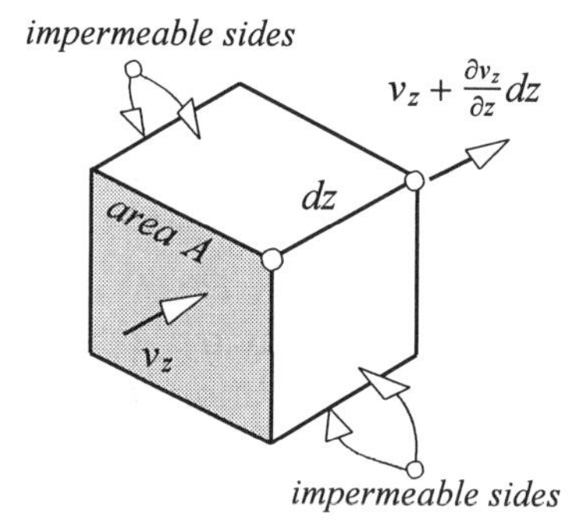

Figure 4.12: One-dimensional flow conditions within a saturated volume of soil.

Since $A\,dz = V$ (refer to figure 4.12), it follows that:

$$-\frac{\partial V}{\partial t} = \frac{\partial v_z}{\partial z} V \tag{4.6}$$

Moreover, a combination of equations 1.1 and 1.2 shows that:

$$V = (1+e)V_s$$

V_s being the (constant) volume of solid particles. Thus, substituting for V in equation 4.6, it is seen that:

$$-V_s \frac{\partial e}{\partial t} = V \frac{\partial v_z}{\partial z} \tag{4.7}$$

but $V_s/V = 1/(1+e)$ and, accordingly, equation 4.7 reduces to:

$$\frac{\partial v_z}{\partial z} = -\frac{1}{(1+e)} \frac{\partial e}{\partial t} \tag{4.8}$$

Introducing Darcy's law in the form of equation 3.6, $v_z = -k_z \dfrac{\partial h}{\partial z}$:

$$k_z \frac{\partial^2 h}{\partial z^2} = \frac{1}{(1+e)} \frac{\partial e}{\partial t} \tag{4.9}$$

where k_z represents the permeability of the soil in the vertical direction, and the total head h is the sum of the (constant) elevation head h_e and the pressure head $h_p = u/\gamma_w$ (u being the excess porewater pressure and γ_w the unit weight of water). Hence:

$$h = h_e + \frac{u}{\gamma_w} \quad \Rightarrow \quad \frac{\partial^2 h}{\partial z^2} = \frac{1}{\gamma_w} \frac{\partial^2 u}{\partial z^2}$$

Substituting into equation 4.9, it follows that:

$$k_z \frac{1}{\gamma_w} \frac{\partial^2 u}{\partial z^2} = \frac{1}{(1+e)} \frac{\partial e}{\partial t} \tag{4.10}$$

On the other hand, the variation in time of the void ratio can be expressed as follows:

$$\frac{\partial e}{\partial t} = \frac{\partial e}{\partial \sigma'_v} \frac{\partial \sigma'_v}{\partial t} \tag{4.11}$$

Therefore, a substitution for the quantities $\partial\sigma_v'/\partial t$ and $\partial e/\partial\sigma_v'$ from equations 4.4 and 4.5 respectively into equation 4.11 yields:

$$\frac{\partial e}{\partial t} = m_v (1+e) \frac{\partial u}{\partial t} \tag{4.12}$$

Finally, introducing this quantity into equation 4.10 and rearranging, it can be seen that:

$$k_z \frac{1}{\gamma_w m_v} \frac{\partial^2 u}{\partial z^2} = \frac{\partial u}{\partial t} \tag{4.13}$$

Equation 4.13 is the differential equation for *one-dimensional consolidation* derived by Terzaghi, and is often written as follows:

$$c_v \frac{\partial^2 u}{\partial z^2} = \frac{\partial u}{\partial t} \tag{4.14}$$

with the *coefficient of vertical consolidation*:

$$c_v = k_z \frac{1}{\gamma_w m_v} \tag{4.15}$$

where c_v is usually expressed in $m^2/year$.

An approximate solution to the parabolic equation 4.14 can be obtained using parabolic isochrones (see for instance Schofield and Wroth (1968) or Bolton (1991)). However, an exact solution can easily be established through the use of Fourier series.

Consider, for instance, a very wide area on which a uniform total stress increment $\Delta\sigma$ is applied as depicted in figure 4.13. Assuming the soil has a very low permeability then, initially, an excess porewater pressure equal to $\Delta u_i = \Delta\sigma$ is generated within the soil mass (*i.e.* $\Delta\sigma_i' = 0$).

It is accepted that under this type of uniform loading, an identical initial excess porewater pressure is generated throughout the entire depth of the soil mass, in a way that porewater pressure distribution is initially a rectangle (other types of loading generate different initial porewater pressure distributions as illustrated in figure 4.14).

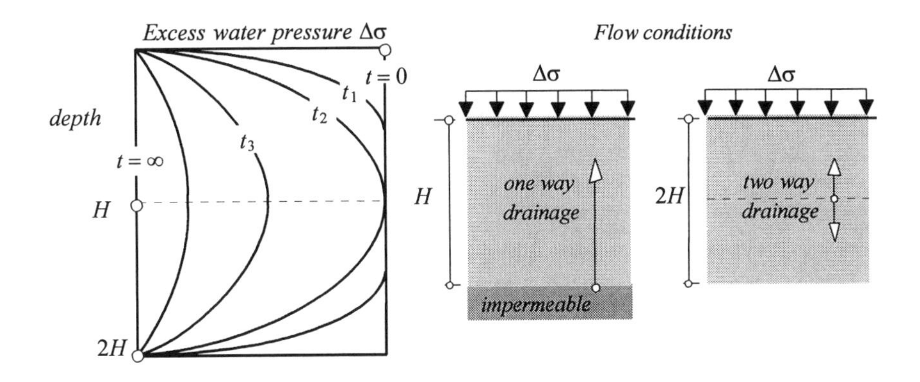

Figure 4.13: Transient flow conditions.

Under these circumstances, the boundary conditions corresponding to figure 4.13 in which H represents the longest drainage path are:

$$
\begin{aligned}
&t = 0, \quad 0 \le z \le H, \quad u = u_i = \Delta\sigma \\
&0 < t < \infty, \quad z = 0, \quad u = 0 \\
&0 < t < \infty, \quad z = H, \quad \frac{\partial u}{\partial z} = 0 \\
&t \to \infty, \quad 0 \le z \le H, \quad u = 0
\end{aligned}
\tag{4.16}
$$

Using Fourier series, together with the above boundary conditions, it can be shown (see Powrie (1997), for instance) that the exact solution to equation 4.14 in the case of a *rectangular* initial excess porewater pressure is:

$$\frac{u}{u_i} = \sum_{m=0}^{m=\infty} \frac{4}{\pi(2m+1)} \sin\left[\frac{\pi}{2}(2m+1)\frac{z}{H}\right] \exp\left[-\frac{\pi^2}{4}(2m+1)^2 T_v\right] \tag{4.17}$$

with

$$T_v = \frac{c_v t}{H^2} \tag{4.18}$$

representing a *time factor.* Notice that, depending on drainage conditions, the drainage path in figure 4.13 is either the total layer thickness if the soil is drained on one side only, or half the layer thickness were the soil to be drained on both sides. Also, in the case of one way drainage, only the upper half of the porewater pressure distribution with time applies.

Because the (dimensionless) time factor T_v is time dependent, equation 4.17 yields the precise dissipation with time of excess water pressure u at any depth z. Obviously, the distribution of u with depth at any given time such as t_3 in figure 4.13 is far from being uniform. However, in practice, the *average* distribution is usually used and is estimated as follows:

$$\left(\frac{u}{u_i}\right)_{av} = \frac{1}{H}\int_0^H \frac{u}{u_i}\,dz \tag{4.19}$$

Moreover, if the *average degree of vertical consolidation* is defined as:

$$U_v = \frac{u_i - u}{u_i} = 1 - \left(\frac{u}{u_i}\right)_{av} \tag{4.20}$$

then, using equations 4.17 and 4.19, it is easy to show that the *average* degree of vertical consolidation in the case of a *rectangular* initial excess water pressure distribution is as follows:

$$U_v = 1 - \sum_{m=0}^{m=\infty} \frac{8}{\pi^2(2m+1)^2} \exp\left[-\frac{\pi^2}{4}(2m+1)^2\, T_v\right] \tag{4.21}$$

Equation 4.21 corresponds to graph 1 in figure 4.15, which illustrates the variation of the average degree of vertical consolidation U_v *versus* the dimensionless time factor T_v.

Notice that with respect to the same figure, graph 1 (*i.e.* equation 4.21) is used in conjunction with any type of initial excess water pressure distribution as long as the soil is drained on both sides. In other words, the drainage path corresponds to half the layer thickness as indicated in figure 4.14 where the numbers in brackets correspond to the graph numbers in figure 4.15. In the case of one way drainage, the nature of dissipation of the excess pressure u_i within the soil depends on the type of loading, so that graph 1 in figure 4.15 is used if u_i were induced by a *uniform load.*

Graph 2 applies when the soil is subjected to a *concentrated load,* and graph 3 corresponds to *hydraulic fills* (such as a slurry mud) which, initially, do not carry any effective stresses.

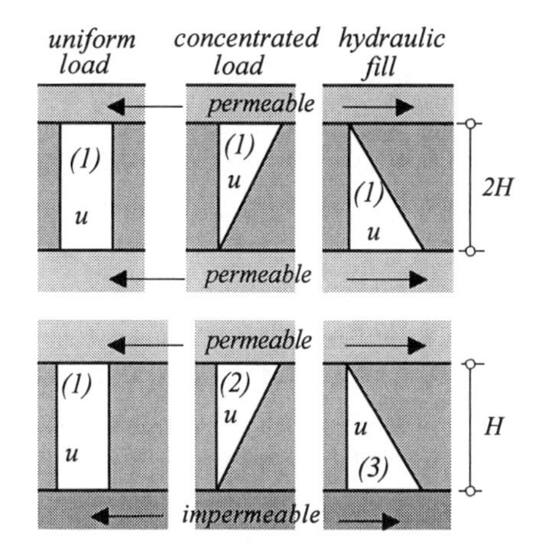

Figure 4.14: Initial porewater pressure distribution for different drainage conditions.

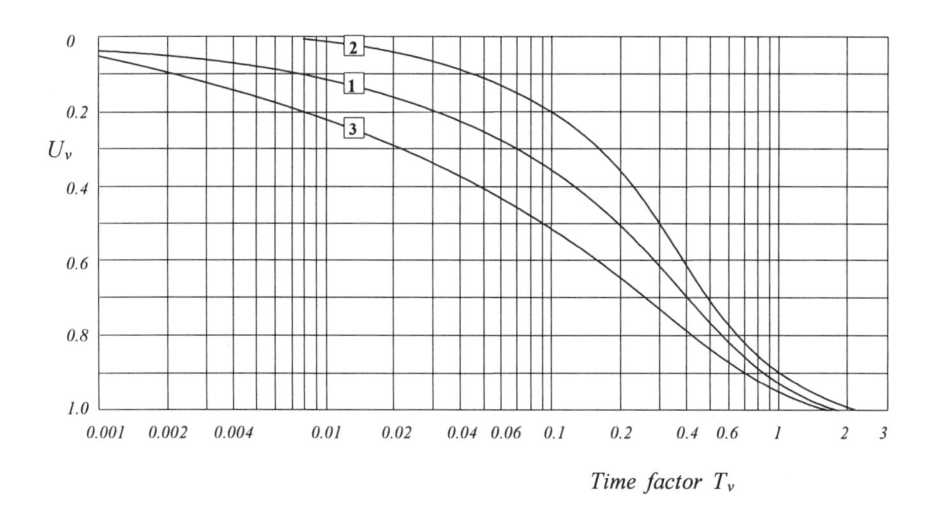

Figure 4.15: Average degree of consolidation corresponding to initial porewater pressure distributions of figure 4.14.

It is clear that the consolidation process, represented by the rate of dissipation of the excess porewater pressure is to all intents and purposes controlled by the permeability and compressibility of the soil. Thus a sand, for instance, with a high permeability will allow any excess porewater pressure to dissipate almost as fast as it was generated. A clay, in contrast, is characterised by a very low permeability relative to a sand and, as such, any consolidation process related to a clayey soil is most likely to be very slow. Because consolidation leads to a change in volume, it is obvious that such a process will induce time dependent deformation within a clay layer.

Accordingly, it is vital to remember that, while consolidation depends on the soil permeability, in other words on the ease (or the lack of it) with which water can flow out of or into the soil matrix, the magnitude of deformation generated at the end of consolidation depends on the compressibility of the soil and has very little to do with the consolidation process *per se*. Moreover, whenever a positive excess porewater pressure is generated within a clay deposit, a consolidation process is triggered during which the effective stresses increase gradually in the manner depicted in figure 4.10, thus increasing the clay density and therefore its resistance. In the meantime, the volume decreases at a rate which can be determined at any time using the graphs of figure 4.15. Ultimately, when all excess porewater pressure has dissipated, the final overall deformation better known as *settlement* can be estimated using appropriate compressibility characteristics of the clay as will be seen shortly. Consequently, both the *rate* of settlement and its final *magnitude* matter a great deal to a designer in order to ensure a safe environment on site and a sound structural design.

As already mentioned, the coefficient of consolidation c_v can be determined using the results of one-dimensional consolidation test, and in that respect, one has to go back to the one dimensional (vertical) transient flow equation 4.21. Replotting this equation as U_v *versus* $T_v^{1/2}$ in figure 4.16, it is seen that the corresponding graph is linear up to an approximate value $U_v \approx 0.5$. Taylor (1948) noticed that by extending the linear portion of the graph, then plotting a second line with a slope 15% less as shown in figure 4.16, the latter line intersects the graph at point A with the approximate co-ordinates ($U_v = 0.9$, $T_v^{1/2} = 0.92$). Taylor then suggested that, for a consolidation test undertaken under similar loading and drainage conditions, the graph corresponding to the variation of the (volumetric) strain $\varepsilon_v = \Delta H/H$ with the square root of time $t^{1/2}$ should be *similar* to that of figure 4.16, and therefore the time $t_{90}^{1/2}$ pertaining to 90% of consolidation can be determined in precisely the same graphical way. Now that the time factor $T_{v90} = (0.92)^2 = 0.848$ as well as the real time t_{90} are known, equation 4.18 can then be used to estimate the coefficient of vertical consolidation:

$$c_v = \frac{T_{v90}H^2}{t_{90}} = \frac{0.848H^2}{t_{90}} \tag{4.22}$$

As mentioned earlier, the coefficient c_v is usually expressed in $m^2/year$.

Conversely, for a clay deposit with a known c_v, the time needed to achieve 90% of consolidation can be estimated from equation 4.22:

$$t_{90} = 0.848\frac{H^2}{c_v} \tag{4.23}$$

where H in equations 4.22 and 4.23 represents the length of the drainage path.

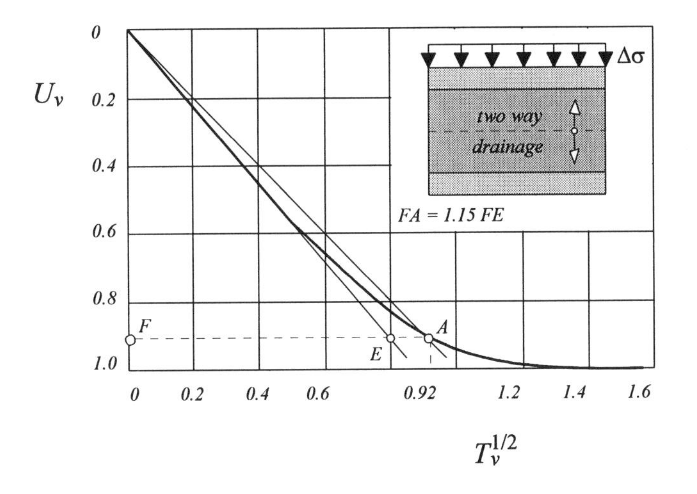

Figure 4.16: Consolidation graph (Taylor construction).

The coefficient of consolidation c_v can thus be determined from the results of a relatively simple one-dimensional consolidation test (as in figure 4.17) using equation 4.22.

However, one has to be aware of the theoretical as well as the practical limitations of the method which is known to yield smaller c_v values (in some cases by more than two orders of magnitude) than those estimated from field measurements.

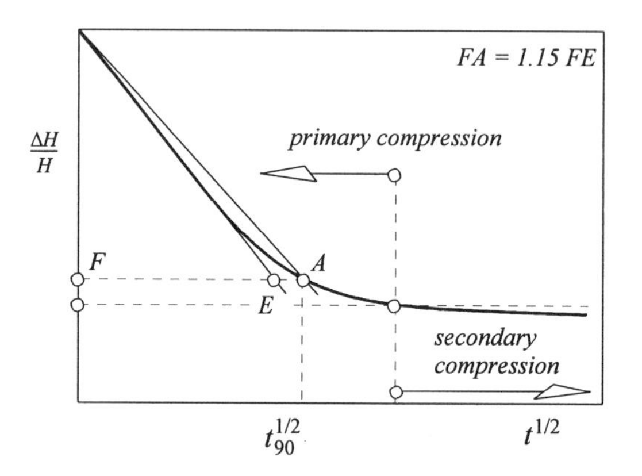

Figure 4.17: Interpretation of 1-D consolidation test results: primary and secondary consolidations.

Thus, the 1-D consolidation theory developed earlier *assumes* that c_v is constant in space and in time. According to equation 4.15, this assumption means that the ratio of vertical permeability to the coefficient of volume compressibility is constant with depth and with time. In other words, these two parameters are assumed to vary in the same proportion which may seem optimistic. However, there are instances for which the errors related to this assumption are insignificant provided that the ratio of the stress increment to the actual stress used during the test is such that $1 \leq \Delta\sigma'_v/\sigma'_v \leq 2$.

In this respect, figure 4.18 shows that c_v values which correspond to different levels of effective vertical stress increments compare very favourably. The variable effective stress increments applied to the *same* natural stiff clay during several 1-D consolidation tests (Azizi and Josseaume (1988)) are such that the ratio $\Delta\sigma'_v/\sigma'_v = 1.15$.

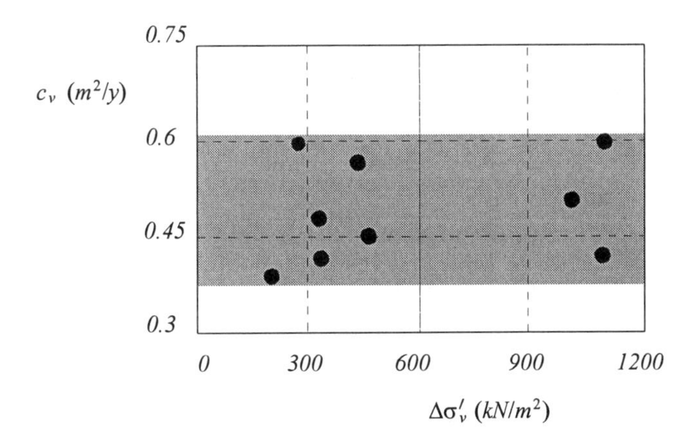

Figure 4.18: c_v values corresponding to a natural stiff clay.

On the practical side, the determination of c_v can be markedly affected by the testing procedure which requires some detailed consideration. In particular, care must be taken so as to minimise any soil disturbance during extraction from the site, transportation, storage and sample preparation.

More importantly (and perhaps notwithstanding the commercial pressures when applicable), the operator must have a clear procedure to follow before, during and after testing in order to ensure that high quality data are obtained. In particular, it is essential to realise that the outcome of a consolidation test can be adversely affected, to a large extent, during sample preparation. Not only are the cutting and trimming of the sample bound to disturb the soil fabric to some degree, but also, if not handled properly, the soil can be subjected to initial plastic deformation and/or a change in void ratio. The operator must therefore bear in mind that the quality of data depends almost exclusively on the stage of test that precedes the first loading, during which the very nature of the tested soil can be altered dramatically. It is well known that, once extracted from the ground, a saturated stiff clay, for instance, will be subjected to a negative porewater pressure due to the unloading process. Accordingly, once positioned inside the brass ring then trimmed, the clay sample will exhibit a tendency to expand in volume to offset the suction. However, a volume expansion can only occur if the clay has access to water and, therefore, it is fundamental to understand that *any* soil contact with water, *let alone saturation*, prior to loading must be prevented at all cost. The question that arises, then, is what load should be applied to the sample before saturation? The answer must be considered in conjunction with the entire testing procedure and, in this respect, the following logical steps are recommended:

(1) Once trimmed to the required dimensions, a dry filter paper is placed on each side of the sample as per figure 4.19.

(2) The ring containing the sample is thereafter placed on the pedestal, and a vertical stress slightly lower than the overburden pressure to which the soil was subjected *in situ* prior to its extraction is then gently applied, followed immediately by saturation of the sample and filter papers.

(3) The vertical pressure is then constantly adjusted so as to prevent any volume change (*i.e.* any axial deformation in this case) from occurring. This operation can last for up to one hour in the case of a stiff clay.

(4) Once equilibrium has been achieved, the actual incremental loading then begins. Each load increment is applied for at least 24 *h*, depending on the nature of the tested soil (for stiff clays, a duration of 48 *h* per increment might be needed). Subsequent increments must be such that the ratio of the stress increment to the current stress is within the range $1 \leq \Delta\sigma'/\sigma' \leq 2$ so as to minimise the effect of secondary compression (refer to figure 4.17).

(5) The loading is pursued until a void ratio $e \approx 0.4e_o$ is reached (where e_o is the initial void ratio as per figure 4.9), at which point an unload–reload cycle is then applied to the sample.

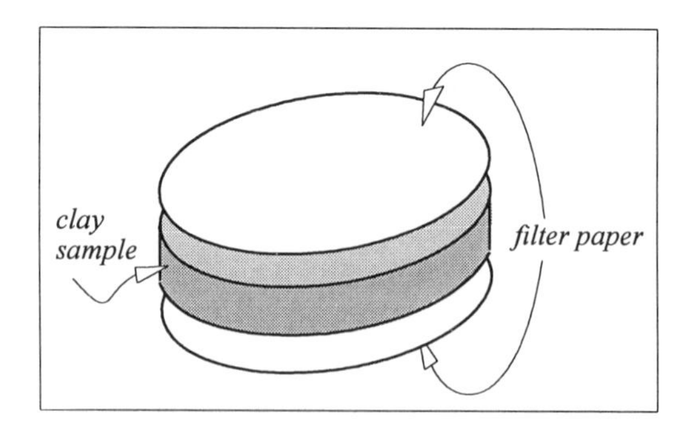

Figure 4.19: Sample preparation.

This procedure has the advantage of minimising any soil disturbance that would have otherwise occurred during testing, ensuring in the process the measurement of high quality data which can thence be used to estimate c_v values according to the Taylor procedure of figure 4.16, as well as the preconsolidation pressure σ'_p using the Schmertmann method of figure 4.9.

In so doing, the operator must be aware of the fact that, for tests involving high stress levels, the measured deformation might be due partly to the deformation of the apparatus itself. The 'artificial' component of the deformation can be as high as 1.5% in some instances as depicted in figure 4.20 where a standard oedometer equipment was calibrated using a steel sample, which was subjected to the same stress path followed during a 1-D consolidation test on a natural stiff clay (Azizi and Josseaume (1988)).

Clearly, any interpretation of data that does not take into account such fictitious sample deformation would be erroneous.

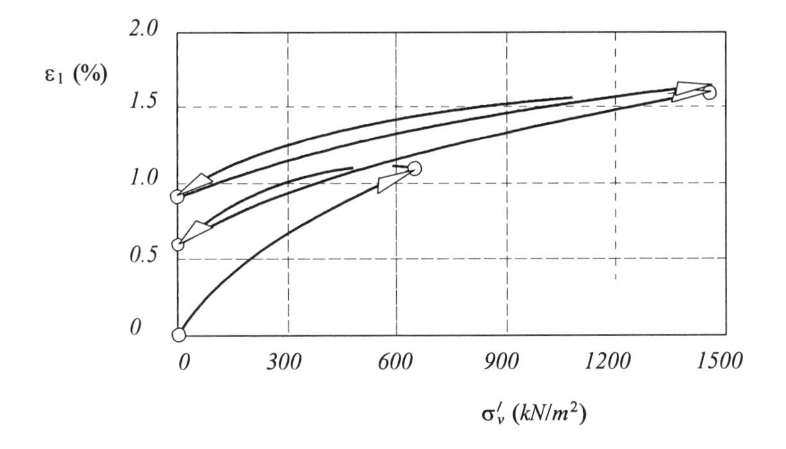

Figure 4.20: Deformation related to a standard 1-D consolidation apparatus.

Ample experimental evidence shows that standard 1-D consolidation tests yield c_v values which are *markedly smaller* than those estimated from *in situ* tests, due in part to the *size* of sample tested in the laboratory. To alleviate this problem, a Rowe cell (also known as a hydraulic compression cell (Rowe and Barden (1966)) can be used whereby samples 250 *mm* diameter and 100 *mm* thick can be tested. The cell having an impermeable base; the drainage occurs through the entire thickness of the sample within which the distribution of the porewater pressure generated by the loading is assumed to be parabolic as in figure 4.21. During the test, a back pressure can be applied to the sample (to ensure full saturation) which can then be loaded *continuously* in a variety of ways:

- constant stress rate whereby $d\sigma'/dt =$ constant,
- constant strain rate with $d\varepsilon_1/dt =$ constant,
- constant gradient so that the porewater pressure at the base of the sample remains constant.

The coefficient of (vertical) consolidation c_v can then be estimated as follows (Whitlow, 1995):

$$c_v = \frac{k}{m_v \gamma_w} \tag{4.24}$$

with the coefficient of permeability calculated from the strain–time curve:

$$k = \frac{H_o^2 \gamma_w}{2u_b} \frac{\partial \varepsilon}{\partial t} \tag{4.25}$$

and the coefficient of volume compressibility being estimated from the stress–strain curve:

$$m_v = \frac{1}{H_o} \frac{\partial H}{\partial \sigma'_v} \tag{4.26}$$

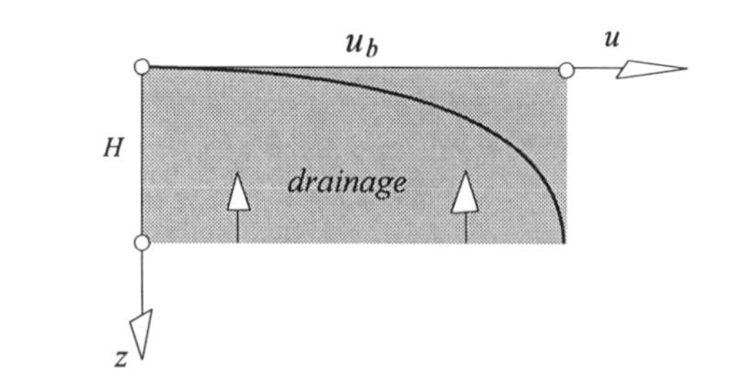

Figure 4.21: Parabolic porewater pressure distribution.

Alternatively, c_v can be estimated from the stress–time curve as follows:

$$c_v = \frac{H_o^2}{2u_b}\frac{\partial \sigma_v'}{\partial t} \qquad (4.27)$$

where ∂H represents the change in the sample thickness, H_o is the initial sample height, and u_b is the porewater pressure at the base of the sample as in figure 4.21.

The c_v values calculated from equations 4.24 and 4.27 still underestimate the *in situ* values, and are affected by the *secondary consolidation* depicted in figure 4.17. This component, which is due to *creep* rather than to the expulsion of water from within the soil matrix, can develop at a very early stage of testing. In particular, the magnitude of secondary consolidation in the case of a standard oedometer test is known to increase as the ratio of the vertical stress increment to the existing vertical stress $\Delta\sigma'/\sigma'$ decreases. These creep effects which can noticeably distort c_v values can be minimised by applying stress increments such that $1 \le \Delta\sigma'/\sigma' \le 2$, as suggested earlier.

Notwithstanding these shortcomings, carefully undertaken consolidation tests, using either standard equipment or Rowe cells, can yield reliable information that can be used in conjunction with field measurements. It is essential to understand that the time of consolidation calculated using a c_v value estimated from a one-dimensional consolidation test is bound to be considerably greater than the *actual* time of consolidation observed in the field. This is mainly due to the fact that clay deposits are usually anisotropic with a much larger horizontal permeability than the vertical one. In order to alleviate this problem, it is advisable to follow the suggestion made by Bishop and Al-Dhahir (1970) that consists of estimating c_v as follows:

$$c_v \approx \frac{k_{h\,(in\,situ)}}{m_{v\,(lab)}\gamma_w} \qquad (4.28)$$

where $k_{h\,(in\,situ)}$ corresponds to the horizontal permeability of the clay measured in the field from a pumping test, and $m_{v\,(lab)}$ is the coefficient of volume compressibility corresponding to the same vertical stress increment applied in the field and estimated from a laboratory consolidation test according to equation 4.5.

4.4 Practical aspects of vertical consolidation

In all the relationships that have been developed previously, the pressure generating the process of consolidation is assumed to be *constant* and to have been *applied instantaneously*. However, every practising engineer would realise that the construction of any type of project is realised in a time t_c throughout which the pressure would have increased gradually. A linear increase such as that depicted in figure 4.22 can be assumed, so that the final magnitude of the pressure p can be considered to have been applied instantaneously for a period of time $t_c/2$ (refer to figure 4.22). Under these conditions, the time factor corresponding to $t_c/2$, calculated from equation 4.18, is:

$$T_v = \frac{c_v}{H^2}\frac{t_c}{2} \tag{4.29}$$

Were the pressure to be assumed to have been applied in its entirety from the start of construction, the corresponding time factor T_v would double, resulting in an overestimate of the degree of consolidation achieved at the end of construction. This can be potentially unsafe as far as settlement calculations are concerned.

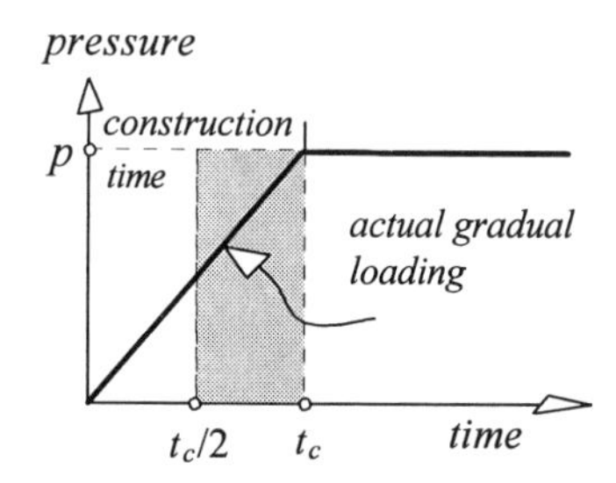

Figure 4.22: Actual and assumed pressure distribution with time.

Example 4.1

Consider the construction of a wide, 4 *m* high embankment on top of a 5 *m* thick layer of a clay deposit with a vertical coefficient of consolidation $c_v = 1.8\, m^2/y$. The clay is underlain by a thick layer of dense sand and, prior to placing the fill material, a thin drainage blanket of sand was applied on top of the clay deposit so as to ensure two-way drainage of the clay.

If it takes 6 months to build the embankment then, assuming a linear increase of the load during that time, equation 4.29 yields:

$$T_v = \frac{1.8}{2.5^2} \times \frac{0.5}{2} = 0.072$$

using graph 1 in figure 4.15, the corresponding average degree of vertical consolidation is $U_v \approx 30\%$.

If the embankment were (erroneously) assumed to have been built instantaneously then, according to equation 4.18:

$$T_v = \frac{1.8}{2.5^2} \times 0.5 = 0.144$$

thus yielding a degree of consolidation $U_v \approx 44\%$.

Due allowance should therefore be made for the *delaying* effect on consolidation of the gradual increase in loading during construction.

4.5 Three-dimensional consolidation theory

The consolidation depends on the permeability of the soil. Consequently if, for whatever reason a consolidation process is generated within a semi-infinite layer of a saturated homogeneous clay, having a very low vertical permeability, then the ensuing rate of vertical consolidation is bound to be very slow, thus reflecting the clay permeability. Knowing that most natural clays are *anisotropic* as they were usually deposited in horizontal (or nearly horizontal) layers, their permeability in the horizontal direction k_h is generally much higher than that in the vertical direction k_v. Therefore, if a horizontal flow can be triggered, the overall consolidation process will occur at a much faster rate.

To simulate the horizontal drainage, let us assume that perfectly cylindrical drains with radii r_w, consisting of a highly permeable material, are inserted in an arrangement such as that shown in figure 4.23. The governing two-dimensional transient flow equation can be established by considering the flow through a soil element situated at a height z and a distance r from the centre of the drain as indicated in the figure.

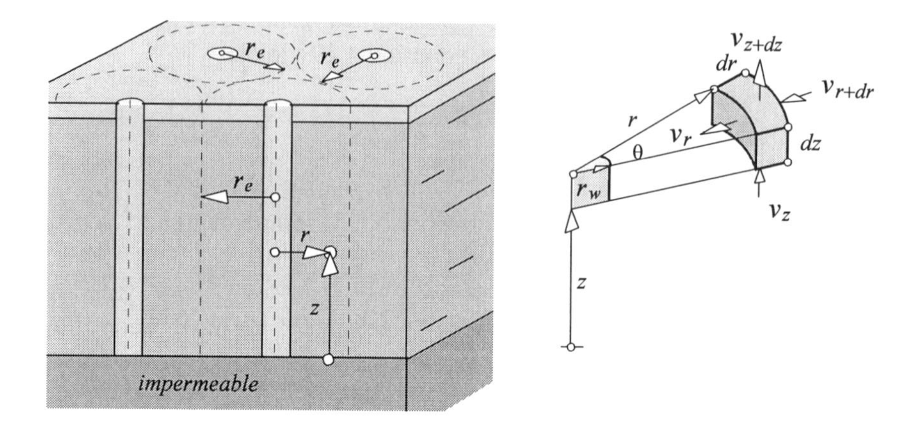

Figure 4.23: Practical arrangement for radial drainage.

Applying an approach similar to that used in the case of vertical consolidation, it can be shown that the ensuing three-dimensional consolidation equation is as follows:

$$\frac{\partial u}{\partial t} = c_v \frac{\partial^2 u}{\partial z^2} + c_h \left(\frac{\partial^2 u}{\partial r^2} + \frac{1}{r} \frac{\partial u}{\partial r} \right) \tag{4.30}$$

where u represents the excess porewater pressure, and c_h corresponds to the coefficient of *horizontal consolidation* calculated in a way similar to equation 4.15:

$$c_h = \frac{k_h}{\gamma_w m_v} \tag{4.31}$$

The average degree of *overall* consolidation U, corresponding to equation 4.30, reflects the average degree of consolidation in both vertical and radial directions. If the average degree of vertical consolidation U_v given by equation 4.21 represents the solution to the consolidation equation in the *vertical direction*, then Carrillo (1942) suggested that, provided the average degree of *radial consolidation* U_r corresponds to the solution of the following consolidation equation in the radial direction:

$$\frac{\partial u}{\partial t} = c_h \left(\frac{\partial^2 u}{\partial r^2} + \frac{1}{r} \frac{\partial u}{\partial r} \right) \tag{4.32}$$

The *overall* degree of consolidation U can then be estimated as follows:

$$(1 - U) = (1 - U_v)(1 - U_r) \qquad (4.33)$$

U_v having been established earlier (refer to equation 4.21), only U_r needs to be derived to resolve equation 4.33. The details of the analytical solution to equation 4.32 were given by Barron (1948) and are outside the scope of this text. However, the corresponding graphical solution is depicted in figure 4.24 with the average degree of *radial* consolidation U_r being related to the time factor:

$$T_r = \frac{c_h t}{r_e^2} \qquad (4.34)$$

where r_e represents the radius of influence of the drain (refer to figure 4.23). The family of curves represented in figure 4.24 corresponds to different values of the ratio n reflecting the spacing of two consecutive drains:

$$n = \frac{r_e}{r_w} \qquad (4.35)$$

r_w being the drain radius.

Notice that the spacing between drains depends on the way that they are arranged, as depicted on the top of figure 4.24.

Example 4.2

Consider the case of an 8 *m* thick layer of sand overlying a 6 *m* thick deposit of normally consolidated clay, resting on an impermeable shale. For technical reasons, the groundwater table, originally 1 *m* below the ground surface, is to be lowered permanently by 6 *m* over a wide area, causing an increase in effective vertical stresses.

To accelerate the drainage, it is proposed to install series of radial sand drains in sufficient numbers so that the time needed for 90% of overall consolidation to occur is cut to just two years. Prefabricated drains with 200 *mm* diameter are inserted throughout the clay layer in a triangular arrangement.

Calculate the spacing between two consecutive drains needed to achieve the stated objective. The coefficients of horizontal and vertical consolidation of the clay are respectively: $c_h = 3\,m^2/y,\ c_v = 2.2\,m^2/y$.

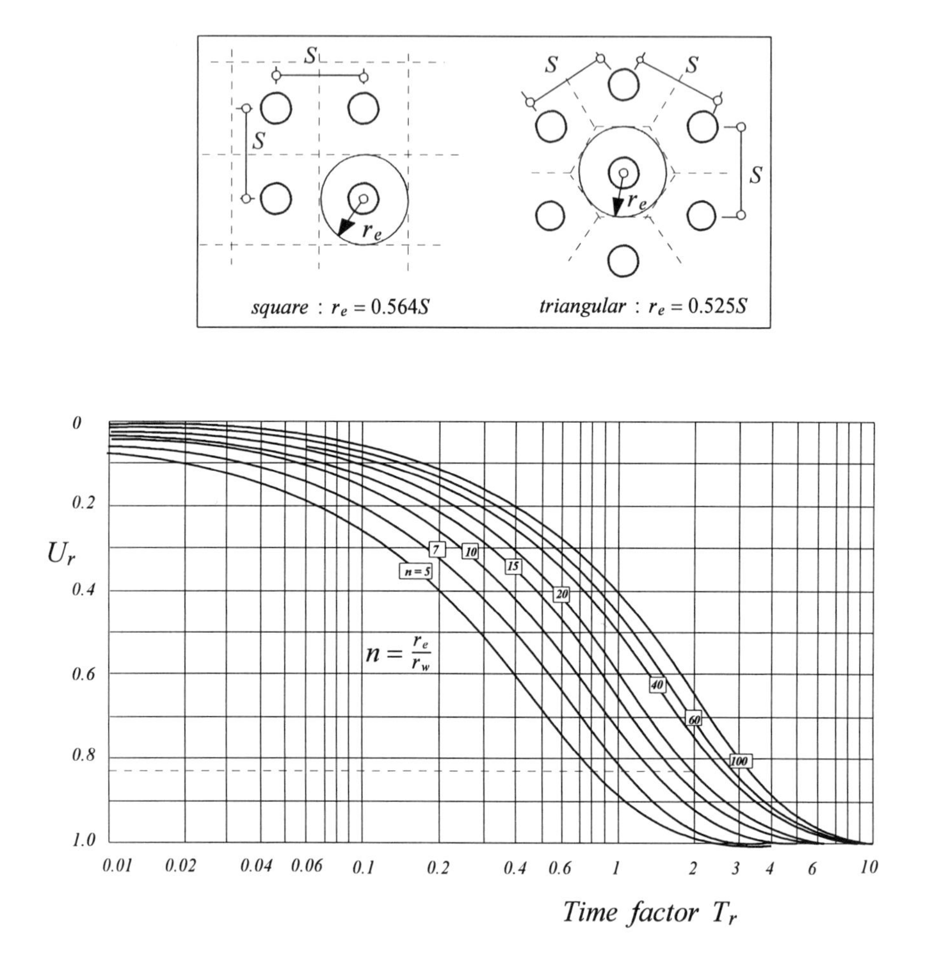

Figure 4.24: Graphical solution of equation 4.34 (Barron, 1948). (Reproduced by permission of the ASCE.)

First, the average degree of vertical consolidation U_v at $t = 2y$ can easily be found using equation 4.18 together with figure 4.15. Whence:

$$T_v = \frac{c_v t}{H^2} = \frac{2.2 \times 2}{36} = 0.122$$

Curve 1 in figure 4.15 then yields a value $U_v \approx 0.42$.

On the other hand, the degree of overall consolidation, calculated from equation 4.33, must be $U = 0.90$ at $t = 2y$. Substituting for U and U_v in equation 4.33, it follows that:

$$(1 - 0.9) = (1 - 0.42)(1 - U_r) \quad \Rightarrow \quad U_r \approx 0.83$$

Consequently, the spacing of drains must be such that $U_r = 0.83$ at $t = 2y$. The drains have the same radius $r_w = 100\,mm$, so that figure 4.24 can now be used to find the appropriate spacing. The procedure consists of choosing a value n_1, in other words selecting a curve on the figure from which the quantity T_r corresponding to $U_r = 0.83$ (refer to the broken line in figure 4.24) is read. Next, use the value of T_r and a combination of equations 4.34 and 4.35 to calculate n_2:

$$n_2 = \frac{1}{r_w}\left(\frac{c_h t}{T_r}\right)^{1/2} \qquad (4.36)$$

$$= \frac{1}{0.1}\left(\frac{3 \times 2}{T_r}\right)^{1/2}$$

If n_2 is different from n_1, then the same procedure is repeated using the newly calculated value of n_2 until convergence occurs. Usually, this iterative procedure converges very quickly. The results in this instance are summarised below:

n_1	T_r	n_2
20	2	17
17	1.85	18
18	1.9	18

It can be seen that, at convergence, $n = r_e/r_w = 18$, yielding a radius of influence $r_e = 1.8\,m$. Since the drains are arranged in a triangular way, their spacing S is calculated as follows (refer to figure 4.24):

$$S = \frac{1.8}{0.525} = 3.43\,m$$

Notice that the design of radial drains depends on the coefficient of horizontal consolidation c_h, whose value is usually higher than that of the coefficient of vertical consolidation. The value of c_h can be measured using a one-dimensional consolidation test during which the excess porewater pressure generated within the soil sample is dissipated *radially* through a central drain while preventing vertical drainage from occurring. The test results can then be interpreted in the same way as for vertical consolidation except that, in this case, ample experimental evidence shows that the value of $t_{90}^{1/2}$ corresponding to 90% of radial consolidation is obtained from the intersection of the graph (Δh, $t^{1/2}$) and the line with a slope 17% smaller than that of the initial linear portion of the graph (as opposed to 15% in the vertical consolidation case). The corresponding value of T_{r90}, which depends on n, is evaluated from figure 4.24, and c_h is then estimated as follows:

$$c_h = \frac{r_e^2\, T_{r90}}{t_{90}} \qquad (4.37)$$

4.6 Settlement analysis

4.6.1 Nature of settlement

Any type of loading that causes the effective stress within the soil mass to change induces a transient flow, in other words a consolidation process, the effect of which was illustrated in figure 4.17. From a design perspective, the final amount of vertical deformation, referred to as *settlement,* measured at the end of this process is most relevant. The *total settlement S* has three components as shown in figure 4.25:

$$S = S_i + S_c + S_s \qquad (4.38)$$

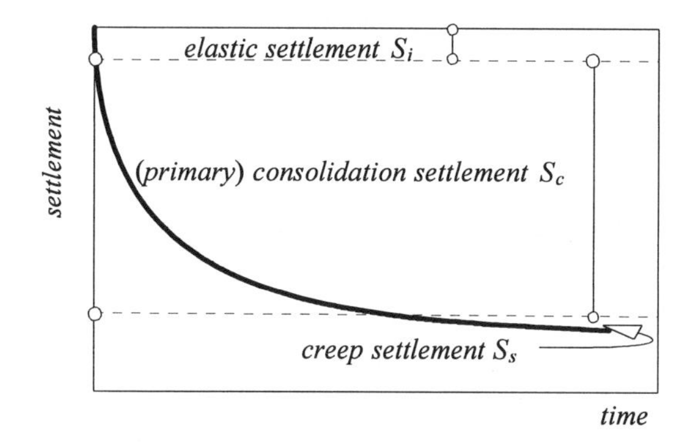

Figure 4.25: The three settlement components: S_i, S_c and S_s.

The (primary) *consolidation settlement* S_c represents by and large the major component. S_i is the (immediate) *elastic* constituent of settlement and S_s corresponds to the long term *creep settlement.*

Accordingly, whenever a process of consolidation is triggered, each of these three components must be calculated in order to estimate the ensuing final total settlement.

The oedometer test simulates the loading conditions applied through a *rigid foundation* in that, during the test, the increase in the vertical pressure across the sample top surface (that is the contact pressure) is almost uniform. For a *flexible foundation,* on the other hand, the contact pressure is always maximum under the foundation centre and minimum under its corners or edge.

It is therefore logical to expect the least favourable design conditions, *i.e.* the maximum total settlement, to occur under the centre of a flexible foundation (were it to exist) as shown in figure 4.26. Also, a difference in the magnitude of settlement across the area of a flexible foundation is bound to occur since points at the edge will undergo a minimum settlement compared with the centre of foundation.

The total settlement beneath a *rigid foundation* is assumed to be uniform and the approach that is generally adopted in practice to estimate its magnitude consists of calculating the total settlement which would result under the centre of an equivalent *flexible foundation*, then multiplying this value by a correction factor of 0.8 to account for the effects of rigidity of the foundation.

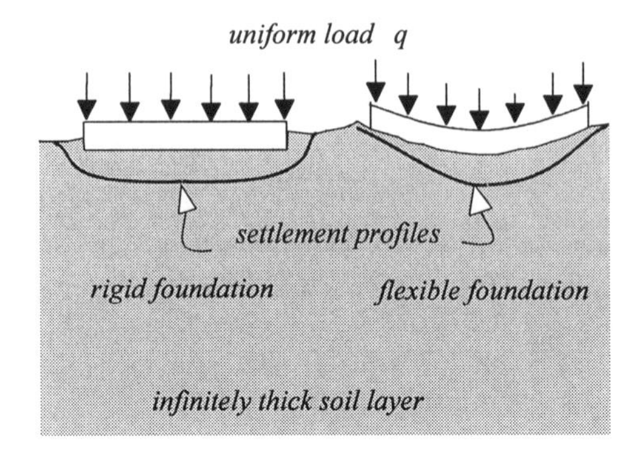

Figure 4.26: Settlement profile under rigid and flexible foundations.

4.6.2 Immediate settlement S_i

This elastic component of settlement is due to soil distortion and not to a volume change. It takes place during or immediately after construction of the structure, and can be calculated using the elasticity theory. Hence with reference to figure 4.26, assuming that the soil has a Poisson's ratio ν and an undrained elasticity modulus E_u, then the immediate settlement beneath a foundation of width B, transmitting a uniform pressure q can be estimated as follows:

- settlement at the *corner* of a *flexible* foundation:

$$S_i = \frac{Bq}{E_u}(1-\nu^2)\frac{I_s}{2} \tag{4.39}$$

- settlement at the *centre* of a *flexible* foundation:

$$S_i = \frac{Bq}{E_u}(1-\nu^2)I_s \tag{4.40}$$

- settlement at the *centre or corner* of a *rigid* foundation:

$$S_i = \frac{Bq}{E_u}(1 - \nu^2) I_r \tag{4.41}$$

The coefficients I_s and I_r depend on the ratio L/B (where L represents the foundation length and B its width), and can be read from figure 4.27 (produced from data by Giroud (1968) and Skempton (1951)).

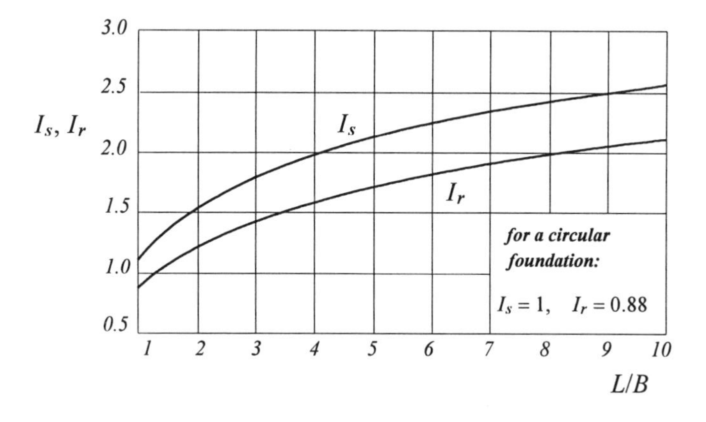

Figure 4.27: Settlement multipliers.

The derivation of the above relationships is based on the *assumption* that the soil layer beneath the foundation is infinitely thick. Consequently, the immediate settlement predictions should be interpreted cautiously. Although for a finite thickness, equations 4.39 to 4.41 yield an overestimate of S_i, leading thus to an error on the safe side.

If the soil beneath the foundation consists of a layer of saturated clay of a thickness H, then the immediate settlement occurs under *undrained* conditions, corresponding to a Poisson's ratio $\nu = 0.5$. Under these circumstances, the average S_i value can be estimated from the following equation:

$$S_i = \alpha_1 \alpha_2 \frac{Bq}{E_u} \tag{4.42}$$

The coefficients α_1 and α_2 are read from figure 4.28 (Christian and Carrier (1978)) in which L corresponds to the foundation length.

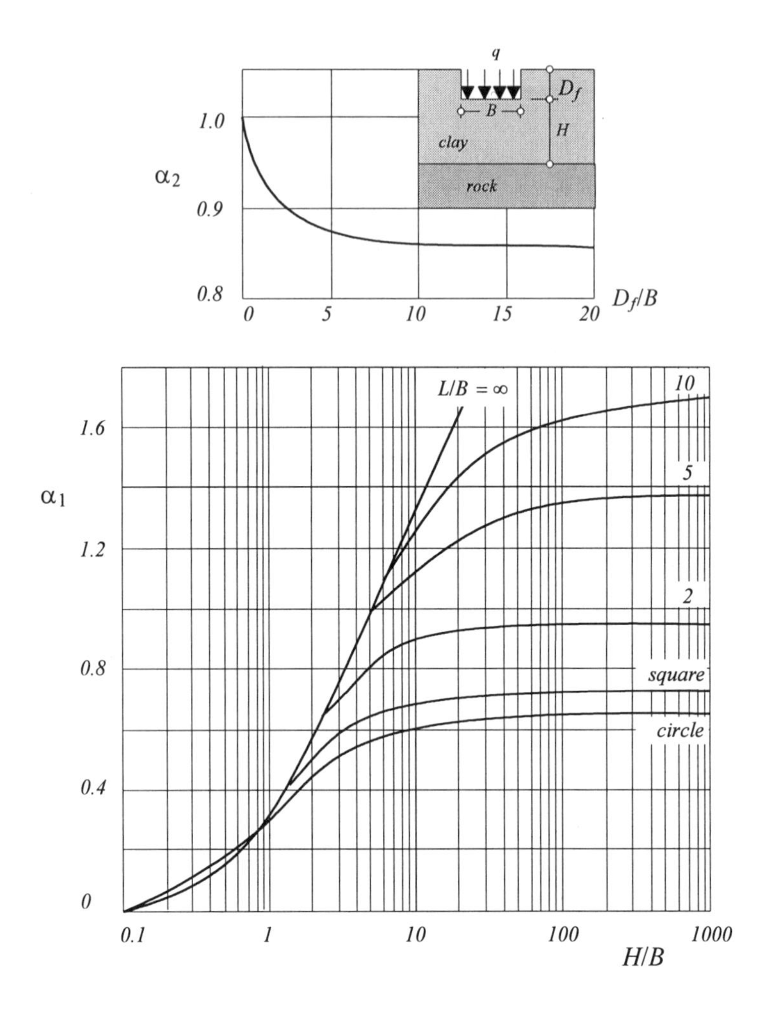

Figure 4.28: Settlement multipliers for undrained conditions. (Reproduced by permission of the National Research Council of Canada.)

Example 4.3

Consider the case of a square foundation founded at the surface of a stiff clay layer having a thickness H, a width B and subjected to a uniform pressure q.

Let us now use both previous methods to assess the average magnitude of the immediate settlement beneath the centre of foundation. Because this component of settlement is a short term phenomenon, it therefore occurs under undrained conditions for which the clay is characterised by a Poisson's ratio $\nu = 0.5$ (refer to section 6.6.1). Accordingly, equation 4.40 with a value $I_s \approx 0.9$ can be used to estimate the *average* immediate settlement at the foundation centre:

$$S_{i1} = 0.9\left(1 - 0.5^2\right)\frac{Bq}{E_u} \tag{4.43}$$

Similarly, S_{i2} can be calculated from equation 4.42 in which $\alpha_2 = 1$ since $D_f = 0$:

$$S_{i2} = \alpha_1 \frac{Bq}{E_u} \tag{4.44}$$

Obviously the *actual* settlement depends on the extent of the thickness H of the clay layer which is only taken into account in equation 4.44 *via* the coefficient α_1. The predictions of the two latter equations are plotted in figure 4.29 as the variation of the ratio S_{i1}/S_{i2} *versus* H/B. Clearly, in this case, both equations yield similar predictions for ratios $H/B > 3$.

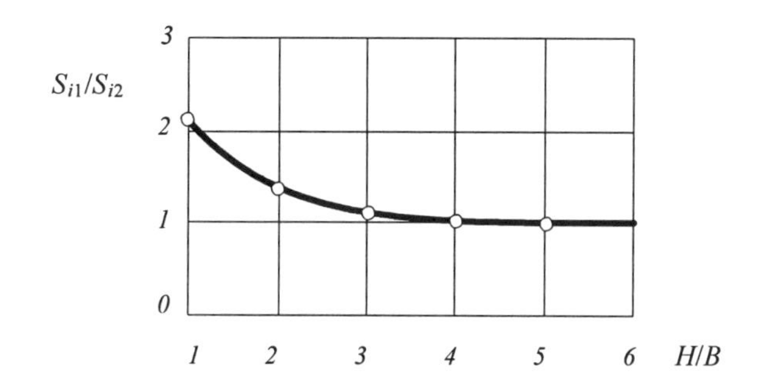

Figure 4.29: Comparison of settlement predictions for saturated clays.

4.6.3 Primary consolidation settlement

Primary consolidation (S_c in figure 4.25) represents the major component of settlement whose final magnitude is related to the *compressibility characteristics* of the soil (C_c and C_s in figure 4.6).

The magnitude of the consolidation settlement S_c can be estimated in a fairly accurate way provided that representative soil parameters are used in the calculations and, most importantly, a correction factor is applied to these calculations to take into account the effect of foundation size and soil type as suggested by Skempton and Bjerrum (1957).

Let us first tackle the calculations of primary consolidation using different relationships developed earlier in the case of the one-dimensional consolidation test. Consider the general case of a *saturated clay* which, after being unloaded during sampling, was subjected in the laboratory to a gradual increase in the vertical effective pressure, so that the variation of the void ratio covers phases of both overconsolidation and normal consolidation as shown in figure 4.30.

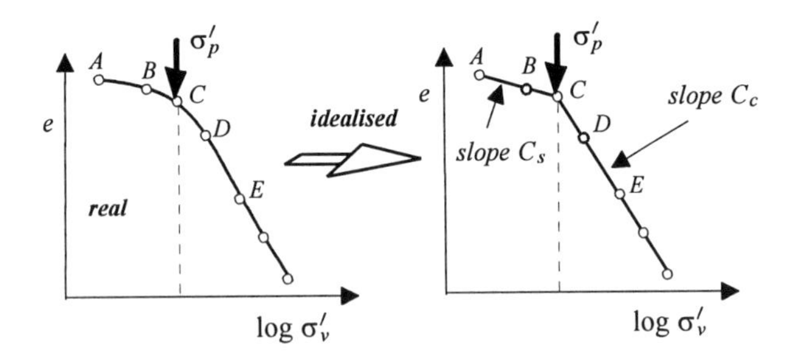

Figure 4.30: Actual and idealised behaviours of a saturated clay, subjected to 1-D consolidation.

The behaviour in figure 4.30 can be assimilated to two linear portions without any significant loss of accuracy, thus allowing for both compressibility coefficients C_c and C_s to be determined. Moreover if, at any stage during the test, the clay were subjected to a pressure increment $\Delta\sigma_v$, and allowed to consolidate, then the ensuing change in the void ratio would be related to the change in the clay thickness through equation 4.2 which can be rewritten as follows:

$$\frac{\Delta e}{1+e_o} = \frac{\Delta h}{h} \tag{4.45}$$

The quantity $\Delta h/h$ (*i.e.* the vertical strain which is identical to the volumetric strain ε_v in this case) can therefore be integrated throughout the thickness of the clay H_c, yielding the consolidation settlement:

$$S_c = \int_0^{H_c} \frac{\Delta e}{1+e_o}\, dz = \frac{\Delta e}{1+e_o} H_c \qquad (4.46)$$

Consequently, the consolidation settlement can be calculated at any stage of loading provided that the variation of the void ratio Δe is known. With reference to figure 4.30, three distinct cases of loading may occur, for which S_c is evaluated differently.

(*a*) *Overconsolidated clays with* $\sigma'_v + \Delta\sigma' < \sigma'_p$

When the applied vertical effective stress increment $\Delta\sigma'_v$ is such that the cumulative value $\sigma'_v + \Delta\sigma'_v$ is still smaller than the preconsolidation pressure σ'_p, then the quantity Δe is calculated using the swelling index C_s as depicted in figure 4.31:

$$e_f - e_o = C_s[\log(\sigma'_v + \Delta\sigma') - \log \sigma'_v]$$

or

$$\Delta e = C_s \log\left(\frac{\sigma'_v + \Delta\sigma'}{\sigma'_v}\right) \qquad (4.47)$$

now substituting for Δe into equation 4.46, it follows that :

$$S_c = \frac{C_s H_c}{1+e_o} \log\left(\frac{\sigma'_v + \Delta\sigma'}{\sigma'_v}\right) \qquad (4.48)$$

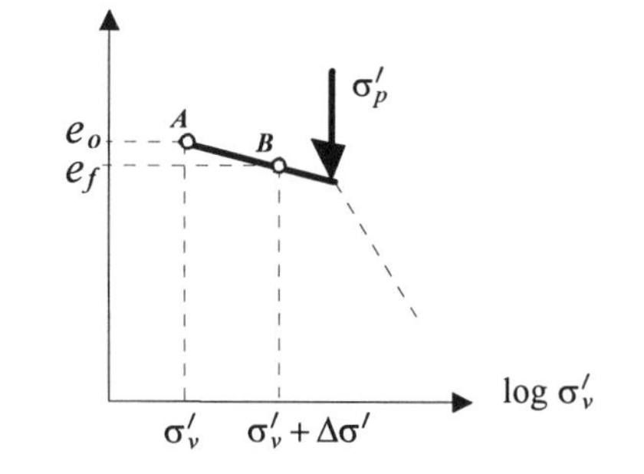

Figure 4.31: Settlement calculations: overconsolidated clays.

(*b*) *Overconsolidated clays with* $\sigma'_v < \sigma'_p < \sigma'_v + \Delta\sigma'$

This case corresponds to a load increment that causes the clay behaviour to change from an overconsolidated state to a normally consolidated one as illustrated in figure 4.32. It is straightforward to establish that:

$$\Delta e = C_s\left[\log \sigma'_p - \log \sigma'_v\right] + C_c\left[\log\,(\sigma'_v + \Delta\sigma') - \log \sigma'_p\right] \qquad (4.49)$$

Rearranging, then substituting for Δe into equation 4.46:

$$S_c = \frac{H_c}{1+e_o}\left[C_s \log\left(\frac{\sigma'_p}{\sigma'_v}\right) + C_c \log \frac{\sigma'_v + \Delta\sigma'}{\sigma'_p}\right] \qquad (4.50)$$

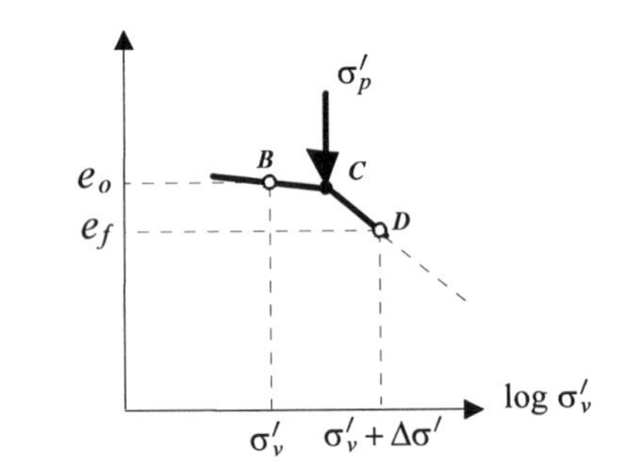

Figure 4.32: Settlement calculations: overconsolidated, then normally consolidated clays.

(c) *Normally consolidated clays*

The corresponding loading conditions are shown in figure 4.33 from which it can be seen that:

$$\Delta e = C_c \log\left(\frac{\sigma'_v + \Delta\sigma'}{\sigma'_v}\right) \qquad (4.51)$$

Substituting Δe in equation 4.46 yields:

$$S_c = \frac{C_c H_c}{1+e_o} \log\left(\frac{\sigma'_v + \Delta\sigma'}{\sigma'_v}\right) \tag{4.52}$$

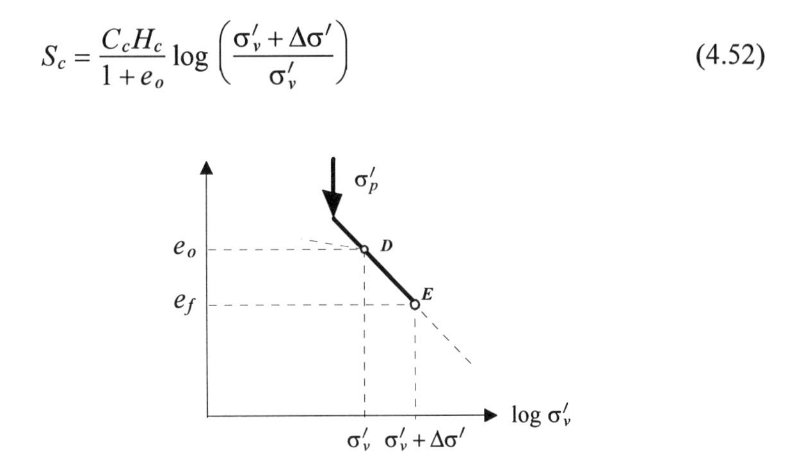

Figure 4.33: Settlement calculations: normally consolidated clays.

4.6.4 Effects of soil type and foundation size

The magnitude of consolidation settlement calculated from equations 4.48, 4.50 and 4.52 is only realistic if the dimensions of the loaded area are large in relation to the depth of the clay layer as in the case of an oedometer test (refer to figure 4.34*b*). However, in the majority of cases in the field, the size of the loaded area is too small compared with the thickness of the saturated clay layer as depicted in figure 4.34*a*.

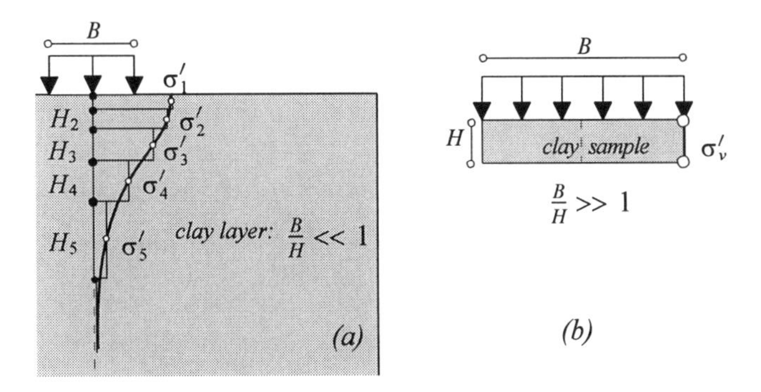

Figure 4.34: Effects of foundation size on settlement:
(a) field conditions and (b) laboratory conditions.

Under field conditions, the consolidation settlement is calculated by integrating the excess porewater pressure generated initially within the soil matrix by the load increment as suggested by Skempton and Bjerrum (1957), so that for *each* individual sublayer:

$$S_{ci} = \int_0^{H_i} m_{vi} \Delta u \, dz \tag{4.53}$$

where H_i represents the thickness of each sublayer (as in figure 4.34) having a coefficient of volume compressibility m_{vi}, and within which the initial excess porewater pressure Δu is assumed constant. Note that the limits used in conjunction with the integral sign in equation 4.53 represent the upper and lower limits of each individual sublayer.

Moreover, the initial increase in porewater pressure can be estimated from the Skempton relationship (Skempton, 1954):

$$\begin{aligned} \Delta u &= \Delta\sigma_3 + A(\Delta\sigma_1 - \Delta\sigma_3) \\ &= \Delta\sigma_1 \left[A + \frac{\Delta\sigma_3}{\Delta\sigma_1}(1 - A) \right] \end{aligned} \tag{4.54}$$

where $\Delta\sigma_1 \equiv \Delta\sigma_v$ and A represents a porewater pressure coefficient, whose values are discussed in detail in section 5.3.2.

Substituting for Δu in equation 4.53, it follows that:

$$S_{ci} = \int_0^{H_i} m_{vi} \Delta\sigma_1 \left[A + \frac{\Delta\sigma_3}{\Delta\sigma_1}(1 - A) \right] dz \tag{4.55}$$

On the other hand, equation 4.46 can be expressed as follows:

$$S_{c(oed)} = \int_0^{H_i} \frac{\Delta e}{1 + e_o} dz = \int_0^{H_i} m_{vi} \Delta\sigma_1 \, dz \tag{4.56}$$

Comparing equations 4.55 and 4.56, it emerges that:

$$S_{ci} = \mu S_{c(oed)} \tag{4.57}$$

with

$$\mu = A + (1 - A)\left(\frac{\int_0^H \Delta\sigma_3 dz}{\int_0^H \Delta\sigma_1 dz}\right) \tag{4.58}$$

It is therefore essential that when the loaded area is small in dimensions compared with the thickness of the clay on which it is founded, and more importantly, when the clay is overconsolidated (*i.e.* small A values), consolidation settlement calculated from equations 4.48, 4.50 and 4.52 is corrected to make due allowance for the effects of foundation size. The correction factor μ, which depends on the value of the *pore pressure parameter* A and the ratio H/B, can be read from figure 4.35.

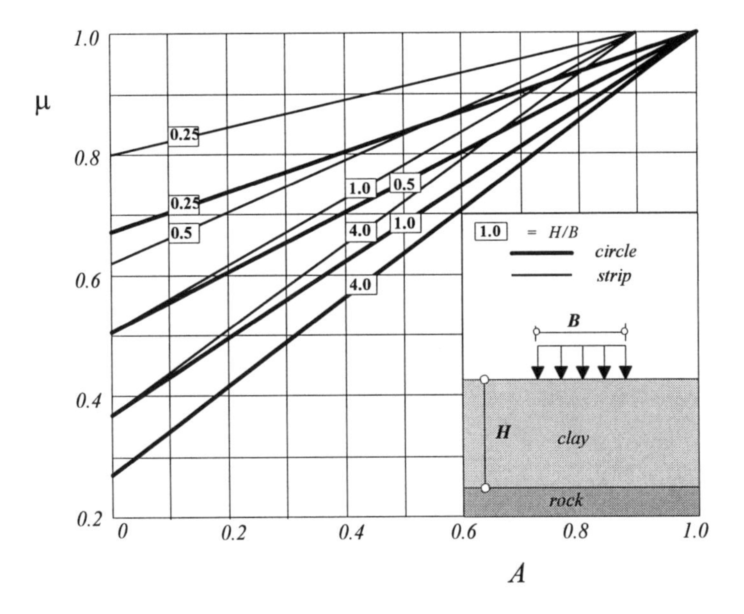

Figure 4.35: Correction factor for foundation size.(Skempton and Bjerrum, 1957). (Reproduced by permission of the Institution of Civil Engineers, London.)

The consolidation settlement of the entire clay layer corresponds to the sum of all individual contributions:

$$S_c = \sum_{i=1}^{n} S_{ci} \tag{4.59}$$

n being the number of sublayers as illustrated in figure 4.34. Logically, the number n is related to the type of pressure distribution generated by the loaded area. With reference to figures 2.37 and 2.38 in section 2.4, it is seen that for a circular or a rectangular uniformly loaded area, the vertical pressure due to the applied uniform load decays at a depth of about $3B$ (B being the foundation width), whereas for a long strip, the clay layer is unaffected by the pressure distribution generated by the strip below a depth of around $8B$. Accordingly, the calculations of consolidation settlement are restricted to within the above limits and sublayers of equal thickness are usually used. Sublayers with a thickness varying from $B/2$ for a circular or a rectangular foundation, to B in the case of a long strip are typical.

Example 4.4

A rigid $(2 \times 2)\,m^2$ square pad foundation is founded at the top of an 8 m thick layer of a lightly overconsolidated clay, sandwiched between a 2 m thick layer of dense sand at the top and a layer of soft rock at the bottom as shown in figure 4.36. The water table is situated at a depth of 1 m below the ground surface, and a unit weight $\gamma = 19\,kN/m^3$ is assumed to apply throughout the sand layer. The clay properties are as follows:

$\gamma_s = 20\,kN/m^3$, $C_c = 0.38$, $C_s = 0.1$,
$A = 0.55$, $E_u = 8000\,kN/m^2$.

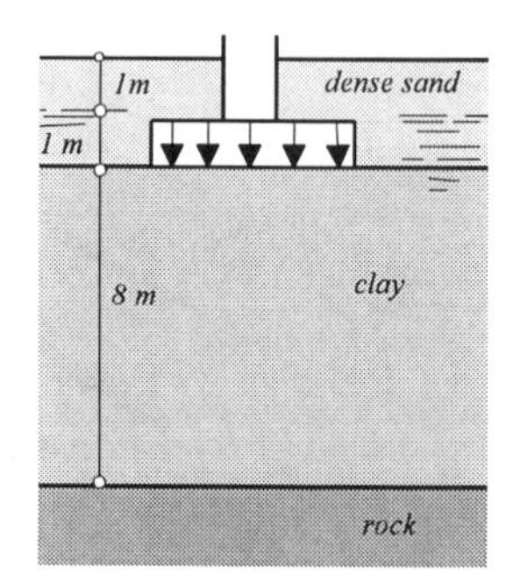

Figure 4.36: Soil conditions.

The preconsolidation pressure is assumed to vary linearly from $\sigma'_p = 100\,kN/m^2$ at the top of the clay layer to $\sigma'_p = 180\,kN/m^2$ at its bottom. The foundation will be transmitting a uniform net pressure $q = 300\,kN/m^2$, and the results of a laboratory one-dimensional consolidation test, undertaken on a clay sample extracted from a depth of 4 m below the ground surface, are as depicted in figure 4.37.

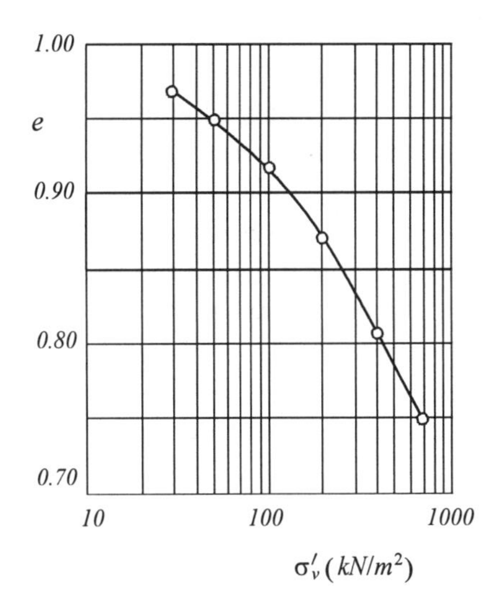

Figure 4.37: Consolidation properties of the clay.

The ensuing settlement includes an immediate component that can be estimated in a straightforward way using equation 4.42:

$$S_i = \alpha_1 \alpha_2 \frac{Bq}{E_u}$$

Since $H/B = 8/2 = 4$, $D_f/B = 2/2 = 1$, figure 4.28 then yields:

$$\alpha_1 = 0.62, \ \alpha_2 = 0.96$$

Whence:

$$S_i = 0.62 \times 0.96 \times \frac{2 \times 300}{8000} \approx 0.044 \ m$$

The component of settlement due to primary consolidation depends on the stress distribution generated by the uniform load q. Because of the square shape of the foundation, the effects of stress distribution will be limited to three times the foundation width, that is 6 *m*. Moreover, within these 6 *m* of clay, six sublayers of equal thickness $H_c = B/2 = 1\,m$ will be considered (refer to figure 4.38).

The increase in the vertical effective stress $\Delta\sigma'_v$ due to q, as well as the component σ'_{vo} due to self-weight are calculated at mid-depth, where the corresponding initial void ratio e_o is interpolated from figure 4.37. The results are summarised in the following table in which the stresses are expressed in kN/m^2, and where the depth z is taken with respect to the base of the foundation. Notice that the quantities $\Delta\sigma'_v$ are estimated from Fadum's influence chart of figure 2.23.

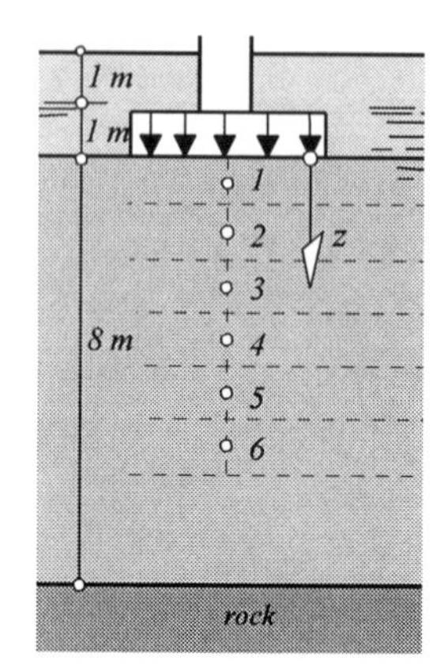

Figure 4.38: Selection of clay sublayers.

point	*depth (m)*	σ'_{vo}	e_o	$\Delta\sigma'_v$	$\sigma'_{vo}+\Delta\sigma'_v$	σ'_p
1	0.5	33.5	0.970	278.5	312	107.5
2	1.5	43.5	0.955	138	181.5	122.5
3	2.5	53.5	0.945	72	125.5	137.5
4	3.5	63.5	0.935	42	105.5	152.5
5	4.5	73.5	0.930	27.6	101.1	167.5
6	5.5	83.5	0.925	15.6	99.1	182.5

The clay being overconsolidated, the use of either equation 4.48 or equation 4.50 to calculate the oedometric consolidation settlement $\Delta S_{c(oed)}$ at the surface of each sublayer depends on the magnitude of the final stress level $\sigma'_{vo}+\Delta\sigma'_v$ with respect to σ'_p. Thus, with reference to the above table, it can be seen that equation 4.50 must be used in conjunction with the two top sublayers, whereas for the four remaining ones, equation 4.48 applies. It is easy to check that the calculations result in the following:

point	$\Delta S_{c\,oed)}\,(m)$
1	0.115
2	0.056
3	0.019
4	0.011
5	0.007
6	0.004

Now that the consolidation settlements for different sublayers are found, their sum must be corrected for soil type and foundation size effects. The correction factor μ is read from figure 4.32 in the knowledge that the clay has a porewater pressure parameter $A = 0.55$, and the ratio $H/B = 4$. Notice that the value of μ will be read from the graph corresponding to a circle with $H/B = 4$, yielding hence $\mu \approx 0.67$. Accordingly, the *corrected* consolidation settlement at the surface of the clay layer (*i.e.* at the base of foundation) is:

$$S_c = 0.67\,\Sigma\,\Delta S_{c(oed)} = 0.67 \times 0.212 = 0.142\,m$$

The immediate settlement must be added to the above value, then the sum corrected to take into account the rigidity of the foundation. Therefore, the foundation base will settle by an amount:

$$S = 0.8 \times (0.142 + 0.044) = 0.146\,m$$

4.6.5 Secondary compression settlement

At the end of the primary consolidation process when all excess porewater pressure will have dissipated, the clay continues to exhibit some deformation under constant effective stresses. This creep component, referred to as secondary consolidation settlement, occurs at a slow rate compared with primary consolidation as depicted in figure 4.39, and can be estimated during a given time increment $\Delta t = t_2 - t_1$ in the following way:

$$S_s = C_\alpha H_c \frac{1}{1 + e_o} \log \frac{t_2}{t_1} \tag{4.60}$$

where C_α is the (constant) slope at which secondary consolidation occurs. Given that creep tests on clays are notoriously time consuming, the following empirical relationship between the slope C_α of figure 4.39 and the compression index C_c of the clay (refer to figure 4.30) can be used cautiously:

$$C_\alpha \approx 0.05C_c \tag{4.61}$$

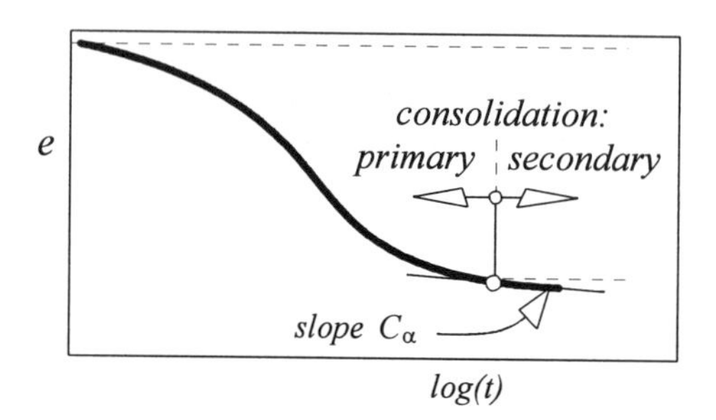

Figure 4.39: Creep settlement.

Accordingly, equation 4.60 can be used to evaluate the secondary consolidation settlement for each sublayer in the case of the previous example, between, say, $t_1 = 1\,y$ and $t_2 = 3\,y$. Substituting for $C_\alpha = 0.05 \times 0.38 = 0.019$ and for $H_c = 1\,m$ in equation 4.60, it follows that for each sublayer:

$$\Delta S_s = \frac{0.019}{1+e_o}\log 3 \approx \frac{9}{1+e_o} \times 10^{-3}\,m$$

The ensuing results are summarised below.

point	e_o	ΔS_s (m)
1	0.970	4.60×10^{-3}
2	0.955	4.60×10^{-3}
3	0.945	4.60×10^{-3}
4	0.935	4.65×10^{-3}
5	0.930	4.65×10^{-3}
6	0.925	4.70×10^{-3}

The overall secondary consolidation settlement of the entire clay layer corresponds to the sum of individual contributions and thus, over two years, the base of foundation will be subjected to a creep settlement with a magnitude:

$$S_s = \Sigma \Delta S_s \approx 0.028\, m$$

The magnitude of the creep settlement over two years is relatively small in comparison with the primary consolidation settlement calculated previously. This is due to the fact that creep deformation depends mainly on the compressibility of the clay and the logarithm of the time increment during which the deformation is estimated, whereas the extent of consolidation settlement is controlled by the stress increase due to surface loading, as well as the compressibility characteristics of the clay.

Problems

4.1 The results illustrated in figure *p*4.1 were measured on a sample of a stiff clay during an oedometer test. Apply the *Schmertmann* method and estimate the overconsolidation ratio of the clay.

Ans: $OCR \approx 4.5$

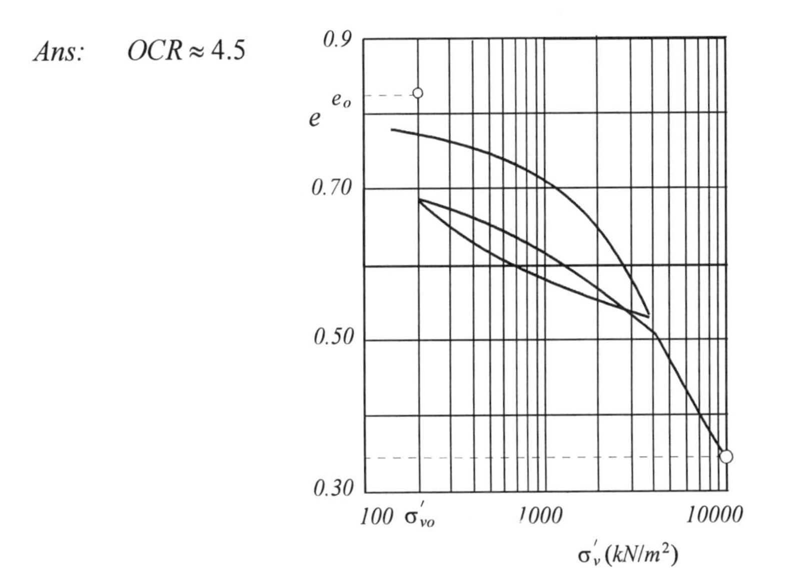

Figure p4.1

4.2 Estimate the same overconsolidation ratio using, this time, *Casagrande* method.

Ans: $OCR \approx 5.9$

4.3 The following results were measured from a one-dimensional consolidation test involving a 20 mm thick sample of a stiff clay, under a vertical stress $\Delta\sigma'_v = 960\, kN/m^2$ (the quantities dh represent the actual decrease in the sample thickness):

time	0.5 *min*	1 *min*	2 *min*	4 *min*	8 *min*	15 *min*	30 *min*	60 *min*	2 *h*	5 *h*	24 *h*
$dh\,(mm)$	0.61	0.67	0.72	0.79	0.89	0.99	1.14	1.25	1.31	1.33	1.35

(a) Plot these results according to the *Taylor* method, then estimate the coefficient of vertical consolidation c_v of the clay.

(b) If the test were undertaken to simulate the *in situ* behaviour of an 8 m thick layer of the same clay under similar loading conditions, estimate the time needed for 80% of vertical consolidation to be achieved on site. Assume that the clay layer is drained on both sides.

Ans: (a) $c_v \approx 0.55\, m^2/y$, (b) $t_f = 15.7\, y$.

4.4 A $20 \times 20\, m^2$ raft foundation for a tall building is to be founded at the top of an 8 m thick layer of firm clay, sandwiched between a 2 m thick layer of silty sand at the top and a thick layer of dense sand at the bottom. The total net pressure transmitted to the ground can be assumed to have been increased linearly from zero at the beginning to its full magnitude of $q = 186\, kN/m^2$ ten months later. The clay is characterised by the following coefficients of vertical and horizontal consolidation respectively:
$c_v = 2\, m^2/y$, $c_h = 3.1\, m^2/y$.

(a) Calculate the time needed for 90% of consolidation to take place if the clay is drained only vertically.

(b) It is suggested that 90% of the total settlement must take place by the end of the ten months construction period. For that to happen, prefabricated sand drains, 65 *mm* in diameter, need to be installed in a square arrangement. Estimate the number of drains needed, as well as their spacing.

Ans: (*a*) $t = 6.785\,y$, (*b*) $n = 23$, $S = 1.32\,m$.

4.5 A circular flexible foundation with a diameter $D = 10\,m$ is founded 1.5 *m* below the ground surface on top of a 6 *m* thick layer of firm overconsolidated clay as depicted in figure *p*4.5. The clay is characterised by an undrained modulus of elasticity $E_u = 12000\,kN/m^2$, and coefficients of vertical and horizontal consolidation: $c_v = 2.1\,m^2/y$, $c_h = 3.2\,m^2/y$.

(*a*) If the foundation is to transmit a net pressure to the top of the clay layer of a magnitude $q = 160\,kN/m^2$, estimate the initial average elastic settlement that will ensue.

(*b*) Calculate the time needed for the clay layer to reach a degree of consolidation $U_v = 90\%$.

(*c*) If the time needed to achieve an overall degree of consolidation $U = 90\%$ is to be cut to 18 months, calculate the required number of 150 *mm* diameter sand drains to be installed on site in a square arrangement, as well as their spacing.

Ans: (*a*) $S_i = 0.024m$, (*b*) $t = 14.51\,y$, (*c*) $n = 22$, $S = 2.92\,m$.

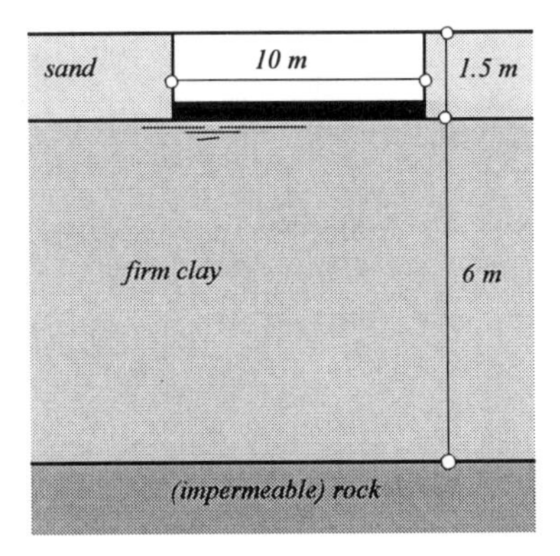

Figure p4.5

4.6 Consider the consolidation settlement of the layer of overconsolidated clay in problem *p*4.5, and assume that:
- the unit weight of sand is $\gamma = 19.5\, kN/m^3$,
- the clay properties are:
  $\gamma_{sat} = 20\, kN/m^3, \quad C_c = 0.3, \quad C_s = 0.12, \quad A = 0.25;$
- the preconsolidation pressure varies linearly from $\sigma'_p = 300\, kN/m^2$ at the top of the clay layer, to $\sigma'_p = 400\, kN/m^2$ at the bottom;
- the one-dimensional results on a clay sample corresponding to a depth of 4.5 *m* below the ground surface are as depicted in figure *p*4.6.

(*a*) Estimate the amount of consolidation settlement at the top of the clay layer.

(*b*) Apply any correction to the consolidation settlement that you deem necessary, then calculate the total settlement (including the elastic component) at the top of the clay layer.

(*c*) If the clay is characterised by a creep compression index $C_\alpha = 0.015$, estimate the creep deformation at the top of the clay layer between the times $t_1 = 2y$ and $t_2 = 7y$.

Ans: (*a*) $S_c = 0.156\, m$, (*b*) $S = 0.121 m$, (*c*) $S_s \approx 8.8 \times 10^{-3}\, m$.

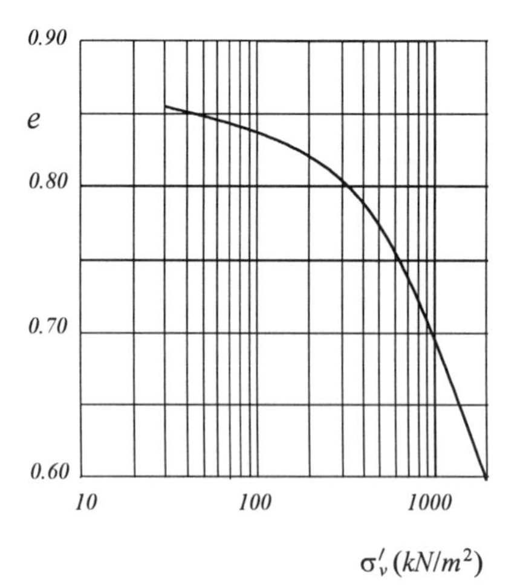

Figure p4.6

References

Azizi, F. and Josseaume, H. (1988) *Loi de comportement des sols raides: détermination de la courbe d'état limite de l'argile verte de Romainville.* Rapport des Laboratoires des Ponts et Chaussées. Série Géotechnique. GT-33.

Barron, R. A. (1948) *Consolidation of fine grained soils by drain wells.* Transactions of the *ASCE,* 113, pp. 718–742.

Bishop, A. W. and Al-Dhahir, Z. A. (1970) *Some comparisons between laboratory tests, in situ tests and full scale performance, with special reference to permeability and coefficient of consolidation.* Proceedings of the Conference on *in situ* Investigations in Soils and Rocks, ICE, London. pp. 251–264.

Bolton, M. D. (1991) *A Guide to Soil Mechanics.* M.D & K. Bolton, Cambridge.

Carrillo, N. (1942) *Simple two and three dimensional cases in the theory of consolidation of soils.* Journal of Mathematics and Physics, 21 (1), pp. 1–5.

Casagrande, A. (1936) *The determination of the preconsolidation load and its practical significance.* Proceedings of the 1st ICSM, Cambridge, Mass. Vol. 3, pp. 60–64.

Christian, J. T. and Carrier, W. D. (1978) *Janbu, Bjerrum, and Kjaernsli's Chart reinterpreted.* Canadian Geotechnical Journal, 15, pp. 124–128.

Giroud, J. P. (1968) *Settlement of a linearly loaded rectangular area.* Proceedings of the ASCE, 94, SM4, pp. 813–831

Powrie, W. (1997) *Soil Mechanics: Concepts and Applications.* E & FN Spon, London.

Rowe, P. W. and Barden, L. (1966) *A new consolidation cell.* Géotechnique, 16 (2), pp. 162–170.

Schmertmann, J. H. (1953) *Estimating the true consolidation behaviour of clay from laboratory test results.* Proceedings of the *ASCE,* 79. pp. 1–26.

Schofield, A. N. and Wroth, C. P. (1968) *Critical State Soil Mechanics.* McGraw-Hill, New York.

Skempton, A. W. (1951) *The bearing capacity of clays*. Proceedings of the Building Research Congress, London.

Skempton, A. W. (1954) *The pore pressure coefficients A and B.* Géotechnique, 4 (4). pp. 143–147.

Skempton, A. W. and Bjerrum, L. (1957) *A contribution to the settlement analysis of foundations on clay.* Géotechnique, 7 (4), pp. 168–178.

Taylor, D. W. (1948) *Fundamentals of Soil Mechanics.* John Wiley & Sons, New York.

Terzaghi, K. (1943) *Theoretical Soil Mechanics.* John Wiley & Sons, New York.

Terzaghi, K. and Peck, R. B. (1967) *Soil Mechanics in Engineering Practice.* Wiley, New York.

Whitlow, R. (1995) *Basic Soil Mechanics,* 3rd edn. Longman Scientific & Technical, England.

CHAPTER 5

Shear strength of soils

5.1 Introduction

The *shear strength* of a soil, that is the *maximum* resistance it can offer before the occurrence of shear failure along a specific failure plane, is intricately related to the soil *type* and *state*. Thus, the response of a granular soil to an applied load depends to a large extent on its density, whereas a cohesive overconsolidated soil exhibits markedly different behaviour to that of a normally consolidated soil as depicted in figure 5.1. In all cases, the shear strength is a quintessential design parameter on which depends the safety related to problems such as slope stability, bearing capacity and lateral thrust.

Figure 5.1(*a*) shows that the stress–strain behaviour of a stiff clay or a dense sand exhibits a *peak* beyond which the strength of the soil decreases towards a *critical* value as the strain increases. If the *same* soils were in a different *state*, that is if the clay were normally consolidated and the sand were loose, then their strength would show a logarithmic increase, tending towards the critical value as the soil is further strained. On the other hand, when a soil is subject to very large strains (as in the case of an old landslip for instance), then its post-peak strength decreases gradually until a *residual* value is reached as depicted in figure 5.1(*b*).

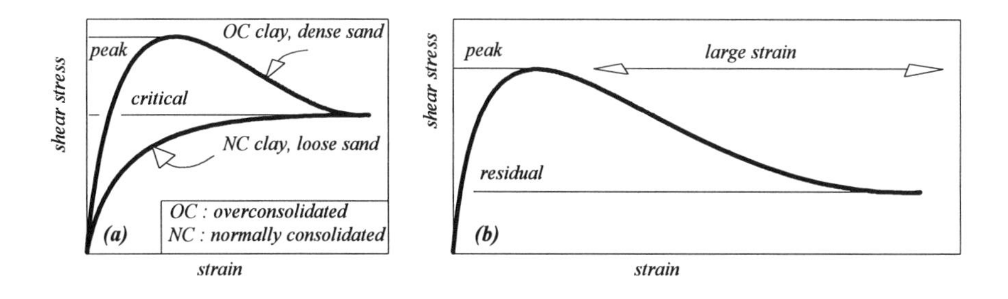

Figure 5.1: Stress–strain relationships as related to soil type and state.

From a practical perspective, the safety as well as the cost effectiveness of a design depend on how realistic are the mathematical relationships between stresses and strains. The relevance and shortcomings of an elastic analysis was highlighted in chapter 2 for the calculation of effective stress increases due to surface loading.

Figure 5.2 reinforces the idea that an elastic analysis can be deemed adequate provided that the effective stress level does not far exceed the *yield stress*. This applies principally to stiff overconsolidated clays whose initial behaviour is characterised by relatively large elastic strains. Normally consolidated clays on the other hand show an elastic–plastic behaviour from the onset of loading and, consequently, the assumption of elastic behaviour for such soils, even when used in conjunction with a large factor of safety, would be grossly in error and potentially unsafe. Rather, the *stiffness* of such soils, in other words the stress–strain ratio, depends on the stress level and can be estimated either as a *tangent modulus* (see equation 2.36, figure 2.41):

$$E = \frac{\delta\sigma'}{\delta\varepsilon} \tag{5.1}$$

or as a *secant modulus* (refer to figure 5.2):

$$E = \frac{\Delta\sigma'}{\Delta\varepsilon} \tag{5.2}$$

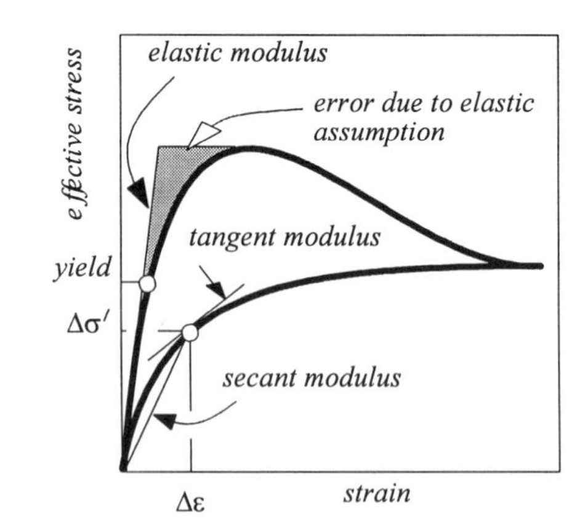

Figure 5.2: Selection of appropriate stiffness parameters.

The physical phenomena leading to soil failure under a known set of stresses are analysed in the following sections. In particular, the crucial effect of the porewater pressure on the strength of soils, with the implicit time effect, will be explored in order to dispel any confusion that may arise from the interpretation of experimental data.

5.2 Mohr's circle representation of stresses

To visualise the nature of plane stresses (*i.e.* two dimensional state of stresses), a simple method of analysis was established by Otto Mohr who suggested that the state of a sample subjected to a set of stresses can be represented graphically on a circle with a diameter corresponding to the difference between the major and minor stresses. Consider for instance the sample of cohesive soil subject to the set of principal stresses σ_1' (major stress) and σ_3' (minor stress) as depicted on the left-hand side of figure 5.3. Taking compressive stresses as well as anticlockwise shear stresses as *positive*, and considering that, under fully drained conditions, failure occurs along the plane *OD* at an angle α to the horizontal within the soil mass, then the same failure plane makes an angle 2α anticlockwise with respect to the horizontal in Mohr's representation. The mathematical proof of this fundamental property of Mohr's circle can be found in any textbook on stress analysis. Most importantly however, figure 5.3 can now be used to establish the expression of the shear stress τ_f and normal stress σ_n', acting on the failure plane.

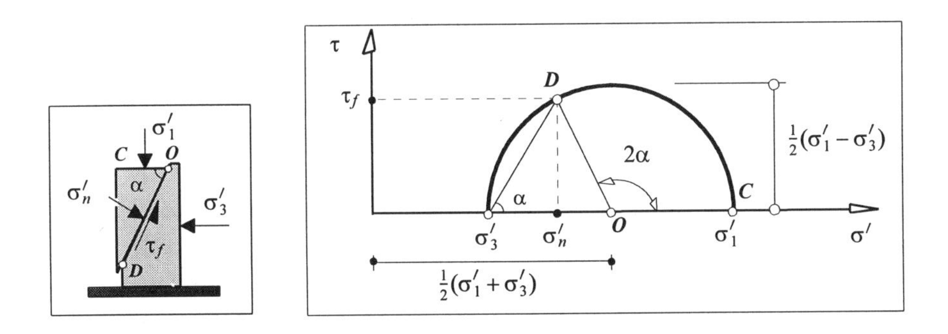

Figure 5.3: Mohr–Coulomb failure criterion for a fully drained cohesive soil.

Starting with the shear stress at failure, it is easy to see that:

$$\tau_f = OD \sin\beta = OD \sin(\pi - 2\alpha)$$
$$= \frac{1}{2}(\sigma_1' - \sigma_3') \sin 2\alpha \qquad (5.3)$$

As for the normal effective stress at failure σ_n':

$$\sigma_n' = \frac{1}{2}(\sigma_1' + \sigma_3') - \frac{1}{2}(\sigma_1' - \sigma_3') \cos(\pi - 2\alpha)$$

$$= \frac{1}{2}(\sigma_1' + \sigma_3') + \frac{1}{2}(\sigma_1' - \sigma_3') \cos 2\alpha \qquad (5.4)$$

Figure 5.3 also shows that the maximum shear stress occurs at an angle $\alpha = 45°$ for which the stresses corresponding to equations 5.3 and 5.4 reduce to:

$$t = \frac{1}{2}(\sigma_1' - \sigma_3') \qquad (5.5a)$$

$$s' = \frac{1}{2}(\sigma_1' + \sigma_3') \qquad (5.5b)$$

If, in addition, the two new stress variables, namely the *deviator stress q* and the *mean effective stress p'* are defined as follows:

$$q = (\sigma_1' - \sigma_3') \qquad (5.6a)$$

$$p' = \frac{1}{3}(\sigma_1' + 2\sigma_3') \qquad (5.6b)$$

then, by virtue of the effective stress principle $\sigma = \sigma' + u$, it is straightforward to establish that:

$$t = t', \quad q = q', \quad s = s' + u, \quad p = p' + u$$

meaning that the same equations 5.5a and 5.6a apply under effective or total stresses because water has no shear resistance. Accordingly, if Mohr's circle is represented in terms of total stresses, then the circle in figure 5.3 has simply to be shifted to the right-hand side by a distance corresponding to the porewater pressure u.

5.3 Stress–strain relationships at failure

5.3.1 Effective stress analysis

The behaviour exhibited in figure 5.1 is affected by the *type* and *state* of the soil. The failure stresses measured from different laboratory tests (whose details will be presented later) on the *same* natural clay are depicted in figure 5.4 (Josseaume and Azizi, 1991). It is very tempting *prima facie* to apply a linear regression analysis, so that the entire stress range at failure can be described by a unique linear equation. However, when the *state* of the clay is taken into account, a somewhat different picture emerges as shown in figure 5.5. All failure stresses corresponding to a *normal consolidation state* can be fitted with a line having a zero intercept and a slope β'_c.

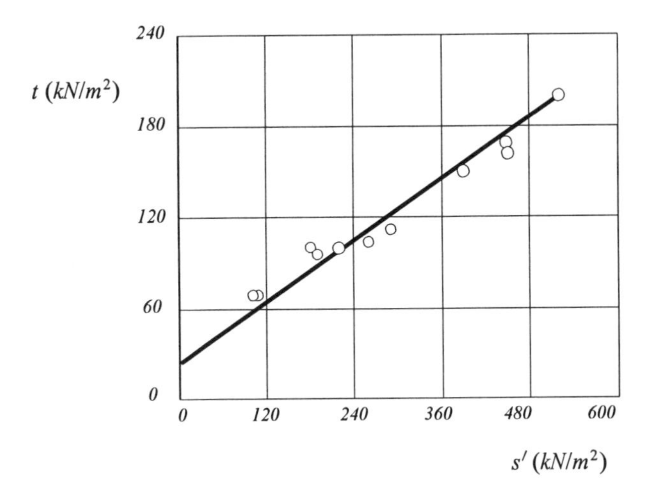

Figure 5.4: Failure stresses for a natural stiff clay.

As the clay becomes *overconsolidated* (due to the loading conditions of different tests), the relationship between stresses at failure becomes curved. For *heavily overconsolidated* samples, the behaviour at failure can be described by a linear equation with an intercept a' and a slope β'_p.

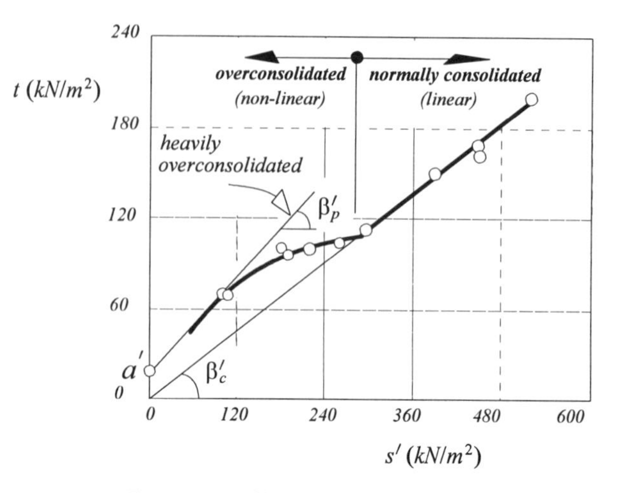

Figure 5.5: Effect of soil state on failure stresses.

Relating the behaviour depicted in figure 5.5 to that of figure 5.1, an overall picture of the shear strength associated with the state of natural soils can be drawn. Thus, for heavily overconsolidated clays, the peak shear strength can be estimated from a simple linear equation, established as early as 1776 by Charles Augustin Coulomb, who was the first to suggest that the shear strength, along a potential failure plane, of a soil subject to a normal stress σ'_n on the same plane (see figure 5.3) can be modelled with the following linear equation:

$$\tau_f = c' + \sigma'_n \tan\phi'_p \tag{5.7}$$

ϕ'_p represents the *peak angle of friction* and c', often referred to as *apparent cohesion,* was first thought to reflect the interparticle bonds otherwise known as cementation. It was not until Terzaghi established his effective stress equation (refer to section 2.1) some 150 years later that the nature of the intercept c' became clearer. In fact, experimental evidence suggests that, apart from some very sensitive soils, cementation in clays is insignificant (Bjerrum and Kenny, 1967). Rather, the apparent cohesion (that is the degree of shear strength exhibited by heavily overconsolidated clays under zero normal effective stress) is the result of curve fitting, and reflects the increase in effective stresses due to the negative porewater pressure and the suction that ensues within the clay mass. This process is tackled in detail in chapter 6.

The behaviour at failure of normally consolidated clays as well as granular soils does not exhibit any cohesion, and is therefore represented by the following simplified version of Coulomb equation 5.7:

$$\tau_f = \sigma'_n \tan\phi'_c \qquad (5.8)$$

ϕ'_c refers to the *critical angle of friction*. If the soil is subject to large deformations (refer to figure 5.1), then the behaviour at failure would be characterised by a *residual angle of friction* ϕ'_r as depicted in figure 5.6. Notice that in practice the angle ϕ'_r is usually used in conjunction with the design of slopes for which failure could be activated on an old landslip where the soil has experienced large displacements in the past; such problems will be analysed in chapter 7. Moreover, ϕ'_r can be significantly smaller than ϕ'_c depending on the nature of the clay and its mineralogy (Lupini *et al.*, 1981).

Figure 5.6 therefore represents the failure loci, better known as *failure envelopes,* for different types of soils at different states.

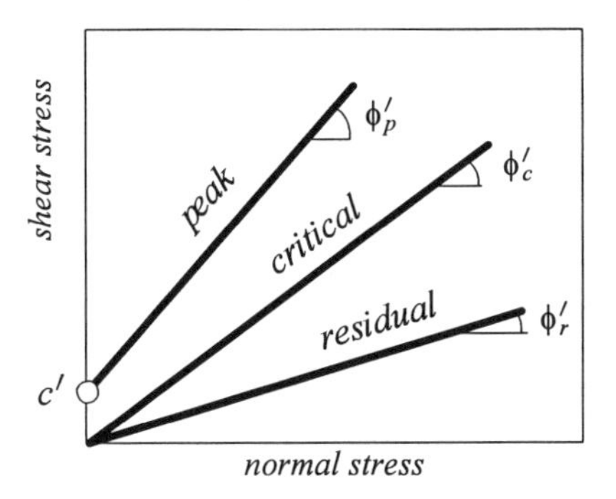

Figure 5.6: Failure envelopes.

Let us now use Mohr's circle of stresses to establish the relationships between different stresses at failure, as well as the link between the slopes β' in figure 5.5 and ϕ' in figure 5.6.

Thus with reference to figure 5.7 in the case of a heavily overconsolidated clay, it is seen that:

$$\alpha = \frac{\pi}{4} + \frac{\phi_p'}{2} \tag{5.9}$$

and therefore:

$$\sin 2\alpha = \sin\left(\frac{\pi}{2} + \phi_p'\right) = \cos\phi_p', \qquad \cos 2\alpha = -\sin\phi_p'$$

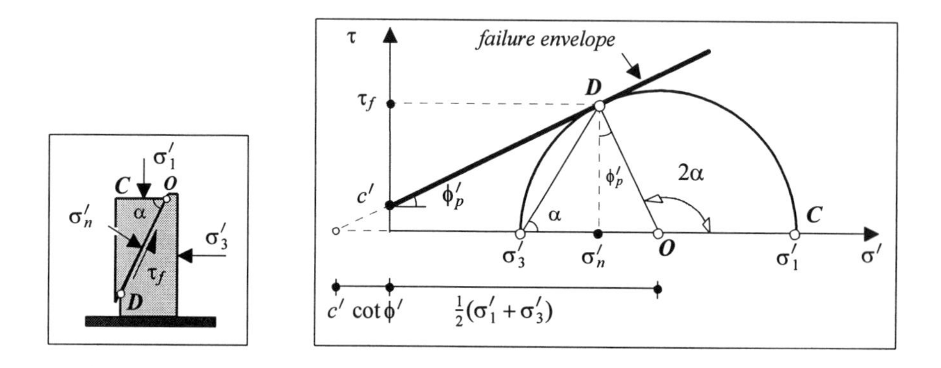

Figure 5.7: Mohr–Coulomb failure envelope for a heavily overconsolidated clay.

Substituting for α in equations 5.3 and 5.4, it follows that:

$$\tau_f = \frac{1}{2}(\sigma_1' - \sigma_3')\,\cos\phi_p' \tag{5.10}$$

$$\sigma_n' = \frac{1}{2}(\sigma_1' + \sigma_3') - \frac{1}{2}(\sigma_1' - \sigma_3')\,\sin\phi_p' \tag{5.11}$$

Inserting these quantities in equation 5.7 and rearranging :

$$(\sigma_1' - \sigma_3') = 2c'\,\cos\phi_p' + (\sigma_1' + \sigma_3')\,\sin\phi_p' \tag{5.12}$$

This equation can be further simplified, yielding thus the general relationship between effective minor and major stresses :

$$\sigma_1' = \sigma_3'\,\tan^2\left(\frac{\pi}{4} + \frac{\phi_p'}{2}\right) + 2c'\,\tan\left(\frac{\pi}{4} + \frac{\phi_p'}{2}\right) \tag{5.13}$$

or

$$\sigma'_3 = \sigma'_1 \tan^2\left(\frac{\pi}{4} - \frac{\phi'_p}{2}\right) - 2c' \tan\left(\frac{\pi}{4} - \frac{\phi'_p}{2}\right) \qquad (5.14)$$

The relationship between the angles β'_p and ϕ'_p, as well as the intercepts a' and c' in figures 5.5 and 5.6 respectively can now be established in a straightforward way by substituting for the quantities t and s' from equations 5.5 into equation 5.12. Whence:

$$t = c' \cos\phi'_p + s' \sin\phi'_p \qquad (5.15)$$

and therefore, according to figure 5.5, it is seen that:

$$c' = \frac{a'}{\cos\phi'_p} \qquad (5.16)$$

$$\phi'_p = \sin^{-1}(\tan\beta'_p) \qquad (5.17)$$

Also, using equations 5.6, it is easy to show that:

$$(\sigma'_1 + \sigma'_3) = \frac{2}{3}(\sigma'_1 + 2\sigma'_3) + \frac{1}{3}(\sigma'_1 - \sigma'_3) = 2p' + \frac{q}{3} \qquad (5.18)$$

Inserting this quantity into equation 5.12 then rearranging, it follows that:

$$q = \frac{6\cos\phi'}{3 - \sin\phi'}c' + \frac{6\sin\phi'}{3 - \sin\phi'}p' \qquad (5.19)$$

Equation 5.19 indicates that, when plotted in the space (p', q), the linear stress path *at failure* is characterised by a slope:

$$M = \frac{6\sin\phi'}{3 - \sin\phi'} \qquad (5.20)$$

Both equations 5.19 and 5.20 will be used to their full potential in chapter 6.

Example 5.1

The following failure stresses were measured on three samples of an overconsolidated clay:

σ_1' (kN/m^2)	250	40	645
σ_3' (kN/m^2)	110	180	305

These same results are plotted in terms of the stress variables (s',t) (refer to equations 5.5) in figure 5.8, thus yielding an intercept $a' \approx 10\,kN/m^2$ and a slope $\tan\beta_p' = 0.7/2 = 0.35$.

Making use of equations 5.16 and 5.17, it follows that:

$$\phi_p' = \sin^{-1}(0.35) = 20.5°$$

$$c' = \frac{a'}{\cos\phi_p'} = \frac{10}{\cos 20.5} = 10.7\,kN/m^2$$

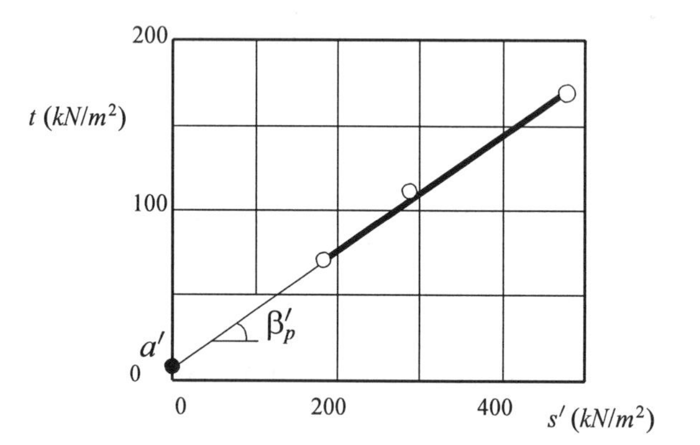

Figure 5.8: Measured results.

Granular soils, as well as normally consolidated clays, do not exhibit any cohesion intercept at failure, the corresponding Mohr's circle being as depicted in figure 5.9. Accordingly, equations 5.10 to 5.14 apply provided that all terms related to cohesion are dropped and that the critical angle of shearing resistance ϕ_c' is substituted for ϕ_p'.

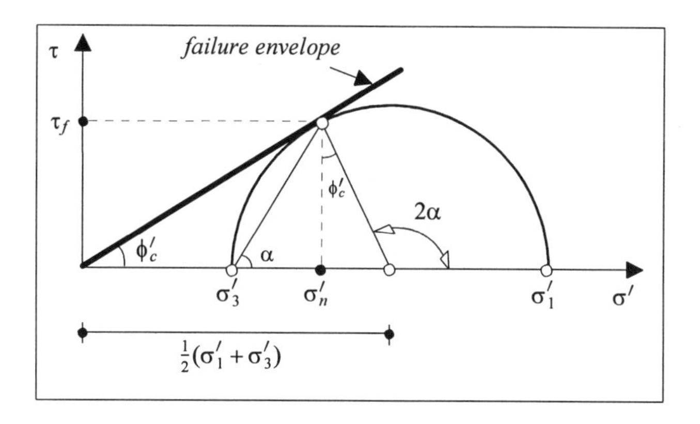

Figure 5.9: Mohr–Coulomb failure envelope for normally consolidated clays and granular soils.

5.3.2 Total stress analysis

Consider the time effect on the behaviour of a *saturated soil* subject at its surface to a *total stress increment* of magnitude $\Delta\sigma$. As soon as $\Delta\sigma$ is applied, one would expect the volume of soil to change (though it *is* possible for the stress increment to be such that it does not generate any volume change). For that to happen, water must be either expelled from within or absorbed by the soil matrix depending on the magnitude of $\Delta\sigma$ (a stress increase caused by a foundation loading, or a stress decrease generated by an excavation for instance); in other words, a *flow rate* must be established. If the soil in question is a *clay*, then the corresponding permeability is bound to be very small [bear in mind that a permeability $k = 10^{-8}\, m/s$ corresponds to water seeping at $31.5\, cm$ (roughly one foot) per year], and therefore in the *short term* (*i.e.* a few days or even a few weeks after the stress increment $\Delta\sigma$ is applied) any flow of water will be insignificant. As a result, the clay experiences virtually no volume change in the short term, meaning that the interlocking of solid particles within the clay matrix remains unchanged, and so does the effective stress.

In accordance with the effective stress principle $\sigma = \sigma' + u$, it follows that in the *short term*:

$$\sigma' = \text{constant} \quad \Rightarrow \quad \Delta\sigma' = 0 \quad \Rightarrow \quad \Delta\sigma = \Delta u$$

Thus in the *short term*, any stress increase $\Delta\sigma$ applied at the surface of a *saturated* clay is transmitted to the porewater, generating an *excess porewater pressure* $\Delta u = \Delta\sigma$. It is therefore essential to realise that, in the short term, shear failure of a saturated clay occurs under a *constant volume* prior to any dissipation of the porewater pressure taking place, that is before the occurrence of any *consolidation*. As such, the behaviour of the clay is referred to as *unconsolidated undrained*. Under these circumstances, the effective stress is zero, and equation 5.7 reduces to the following:

$$\tau_f = c_u \tag{5.21}$$

where c_u denotes the *undrained shear strength* of the clay.

Meanwhile, using the subscript u (for undrained), the shear strength of the clay can be estimated from equation 5.10:

$$c_u = \frac{1}{2}(\sigma_1' - \sigma_3') \cos\phi_u = t \cos\phi_u \tag{5.22}$$

Substituting for the stress variable t from equation 5.15 into the above equation, then rearranging:

$$c_u = c_u \cos^2\phi_u + s' \sin\phi_u \cos\phi_u \quad \Rightarrow \quad \phi_u = 0 \tag{5.23}$$

Therefore under *undrained conditions*, the *angle of shearing resistance* is reduced to zero (*i.e.* a horizontal failure envelope), and the shear resistance of the clay is provided by its undrained shear strength c_u as depicted in the Mohr's circle representation in figure 5.10. A thorough analysis of the undrained shear strength of clays follows in section 5.4.3.

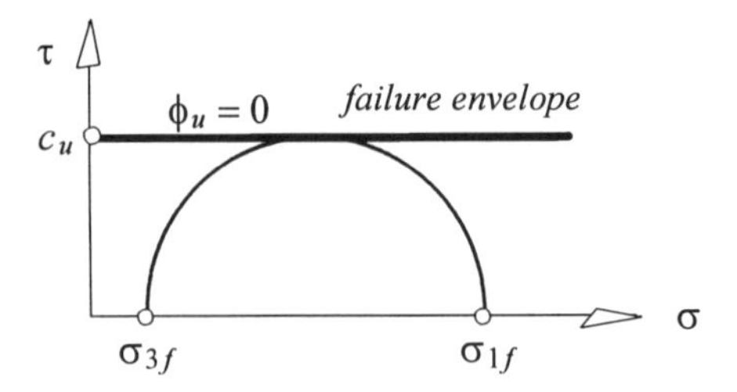

Figure 5.10: Mohr–Coulomb failure envelope for an undrained cohesive soil.

In practice, several techniques have been developed to estimate the shear strength parameters c', ϕ' and c_u, two of which are most widely used to test soils in the laboratory: the direct shear and the triaxial tests. Although both were developed to test a variety of soils, the shear box is used in the following analysis in conjunction with granular soils, whereas cohesive soils will be tested exclusively in a triaxial apparatus.

5.4 Shear strength of granular soils: direct shear test

Granular soils such as sands derive their shear strength from friction between solid particles, and the shear box test is well suited for such types of soils. The apparatus is depicted schematically in figure 5.11, and the test procedure consists basically of shearing a sand sample while subjecting it to a constant vertical pressure. The sample itself is confined in a metal box split horizontally so that the lower and upper halves can move relative to each other.

Because of the relatively high permeability of granular soils, the test is undertaken at a relatively high pace so that a horizontal speed of about 1 mm/min is usually used during testing. As the test progresses, the horizontal displacement Δl and the corresponding shear force S, as well as the change in sample thickness Δh (*i.e.* the vertical movement) are measured at regular time intervals.

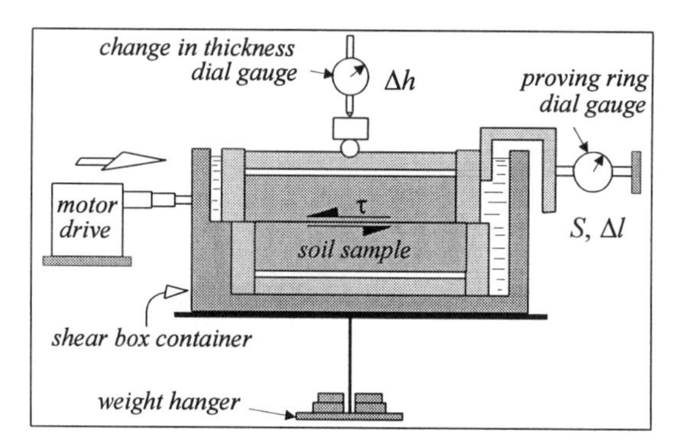

Figure 5.11: Shear box apparatus.

Knowing that the sample has a (square) cross-sectional area A, the vertical normal stress is calculated in a straightforward way:

$$\sigma = \frac{N}{A} \tag{5.24}$$

where N is the normal force, applied through a weight hanger (refer to figure 5.11), that takes into account any lever arm effect. In calculating the shear stress, however, care must be taken to divide the measured shear force by the corresponding (decreasing) cross-sectional area:

$$\tau = \frac{S}{A^*} \tag{5.25}$$

with

$$A^* = A\left(1 - \frac{\Delta l}{\sqrt{A}}\right) \tag{5.26}$$

The behaviour of a sand during a direct shear test depends on its state, in other words on its initial density. Thus, a sample of *dense sand* has a large degree of interlocking between solid particles so that, at the onset of a shear test, the sand has to go through a looser state prior to the occurrence of shear failure as depicted schematically in figure 5.12. This behaviour, known as *dilation*, implies therefore an increase in volume and hence a decrease in density. The shear stress reaches its peak at a relatively low strain on the corresponding stress–strain curve, then a strain softening ensues, leading eventually to a *critical shear strength* τ_c as shown in figure 5.13.

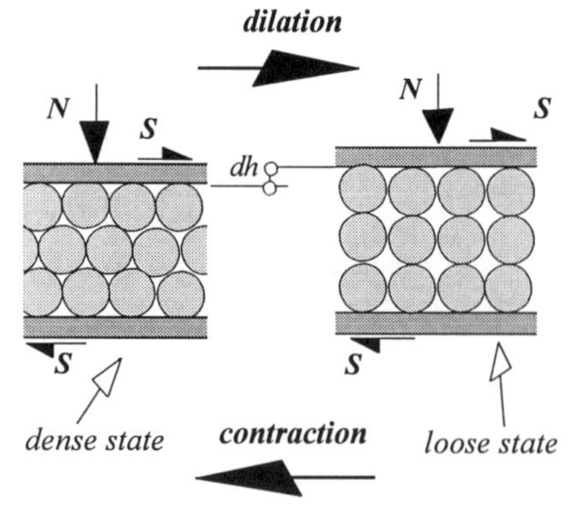

Figure 5.12: Loose and dense states for sands.

This *critical* value τ_c is linked to the rate of dilation represented by the angle υ in the figure. It is seen that, for a dense sand, dilation increases initially to reach a maximum υ_{max} under the peak shear stress τ_{max}, then decreases subsequently.

For a *loose sand* on the other hand, the normal stress applied during shear causes the sand to become denser at the beginning of test, thus heralding a decrease in volume. The corresponding *negative* angle of dilation eventually reaches a maximum value $\upsilon_{max} = 0$ when the shear stress on the stress–strain curve tends towards the maximum asymptotic value τ_c depicted in figure 5.13.

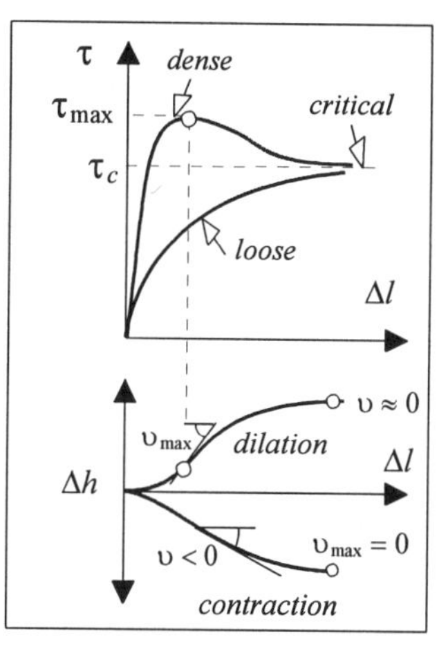

Figure 5.13: Contraction and dilation of sands.

Because sands are deprived of cohesion, the corresponding effective angle of friction ϕ' varies in proportion to the shear stress τ. Accordingly, when the shear stress reaches the critical value τ_c, equations 5.8, 5.15 and 5.19 apply:

$$\tau_c = \sigma'_n \tan \phi'_c \tag{5.27}$$

$$t = s' \sin \phi'_c \tag{5.28}$$

$$q = Mp' \tag{5.29}$$

M being the slope given by equation 5.20.

Consequently, if several shear tests were carried out on the same sand, each under a constant normal stress σ'_n, then the corresponding set of stresses at failure can easily be determined from graphs similar to those in figure 5.13, so that the measured values of τ_c can be plotted against the corresponding values of σ'_n; similarly, all τ_{max} values can be plotted in the same figure yielding thus a value ϕ'_{max} of the effective angle of friction as depicted in figure 5.14 from which it is seen that:

$$\phi'_{max} = \phi'_c + \upsilon_{max} \tag{5.30}$$

Figure 5.14 shows that, for loose sands, $\upsilon_{max} = 0$ and the maximum angle of friction ϕ'_{max} in this case corresponds to the critical angle ϕ'_c.

Accordingly, the maximum shear stress of a sand can be calculated as follows:

$$\tau_{max} = \sigma'_n \tan\phi'_{max} \qquad (5..31)$$

and the relationship between the *critical* and *maximum* shear stresses can then be established by combining equations 5.27, 5.30 and 5.31. Whence:

$$\tau_c = \frac{\tau_{max} - \sigma' \tan\upsilon_{max}}{\dfrac{\tau_{max}}{\sigma'}\tan\upsilon_{max} + 1} \tag{5.32}$$

The latter relationship indicates that for a *loose sand*, $\upsilon_{max} = 0$ and thus $\tau_c = \tau_{max}$.

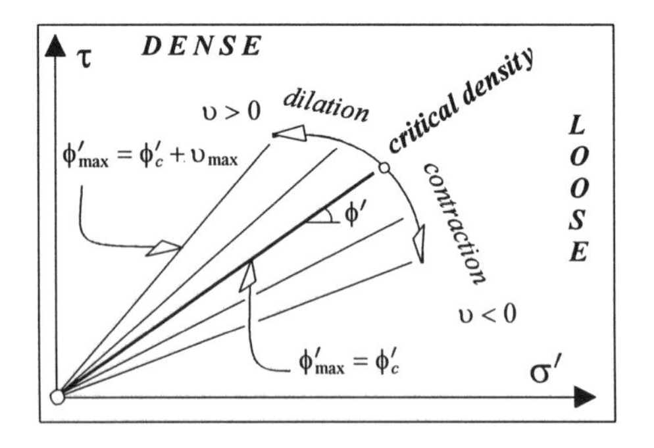

Figure 5.14: Critical density of a sand.

Typical values of the critical friction angle ϕ'_c, as well as the maximum dilation angle υ_{max} of some of the granular soils are listed below:

type of soil	ϕ'_c (°)	υ_{max} (°)
silt	28	3–4
silty sand	30	3–5
uniform sand	28	5–7
well graded sand	33	10–12
sandy gravel	35	12–15

5.5 Shear strength of saturated cohesive soils: the triaxial test

5.5.1 The triaxial apparatus

The triaxial test is a sophisticated experiment that needs to be undertaken with great care if the measured results are to have any physical meaning. Contrary to what the name might imply, the test is not truly *triaxial* since only two sets of stresses can be applied to a soil sample, namely a vertical stress $\sigma_v = \sigma_1$ and a horizontal stress $\sigma_h = \sigma_3$. However, it is easy to see that, were σ_1 and σ_3 to be *principal stresses* (*i.e.* assuming that no shear stress develops at each end of the sample), then the intermediate stress σ_2 becomes irrelevant as far as the *shear strength* of the soil is concerned.

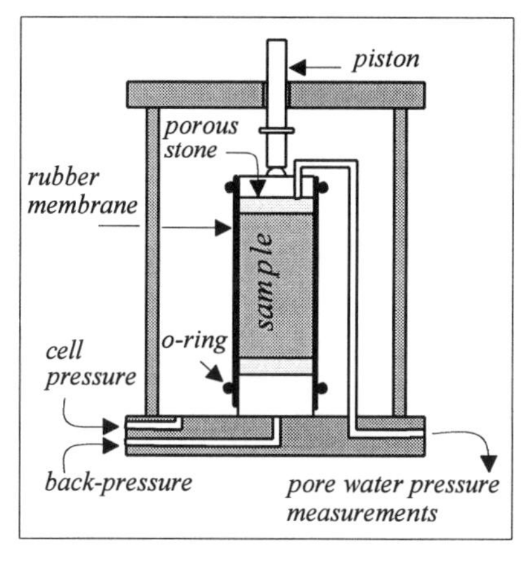

Figure 5.15: Triaxial apparatus.

The essential details of a triaxial apparatus are depicted in figure 5.15, where a cylindrical sample with a length to diameter ratio of 2, insulated by a rubber membrane, can be subjected to a chosen set of principal stresses $(\sigma_1, \sigma_2 = \sigma_3)$, under *controlled drainage conditions.* The all-round cell pressure σ_3 is applied hydraulically through a cell fluid, whereas the vertical stress σ_1 is applied mechanically through a piston. Because the contact between the piston and the top end of the sample is assumed to be reduced to a point contact, using the vertical load P applied through the piston and the known cross-sectional area of the sample A, the vertical stress σ_1 can be calculated in a straightforward way:

$$\sigma_1 = \sigma_3 + \frac{P}{A} \tag{5.33}$$

More importantly, the deviator stress, defined in equation 5.6*a*, can now be calculated at any stage of loading from equation 5.33:

$$q = (\sigma_1 - \sigma_3) = \frac{P}{A} \tag{5.34}$$

The quantity q can be expressed equally in terms of total or effective stresses. Also, in what follows, the stresses σ_v and σ_1 are used interchangeably, and so are σ_h and σ_3.

Tests corresponding to various *stress paths* (such as the one depicted in figure 5.5) can therefore be undertaken under *undrained* or *drained* conditions, simulating both *short term* and *long term* behaviours of the tested soil. Several of these stress paths will be presented in detail shortly, some of which will be subjected to a more rigorous theoretical analysis in chapter 6. However, before any scrutiny, it is essential to examine the difficulties related to soil sampling and sample preparation in the laboratory prior to testing. In this respect, it is useful to mention that, although it can accommodate different types of soils including sands, the triaxial apparatus will be used exclusively in what follows to test cohesive soils, mainly *saturated clays*.

5.5.2 State of a clay sample prior to testing

Consider the case of an element of a saturated clay located at a depth z as illustrated in figure 5.16. The corresponding state of *in situ* total stresses are such that:

$$\sigma_v = \sigma'_{vo} + u_o \tag{5.35}$$

$$\sigma_h = K_o\,\sigma'_{vo} + u_o \tag{5.36}$$

with

$$K_o = \frac{\sigma'_h}{\sigma'_v} \tag{5.37}$$

where σ'_{vo} represents the effective vertical stress *in situ*, u_o the porewater pressure at depth z, and K_o is known as the *coefficient of earth pressure at rest* and corresponds to the ratio of horizontal to vertical effective stresses *in situ.*

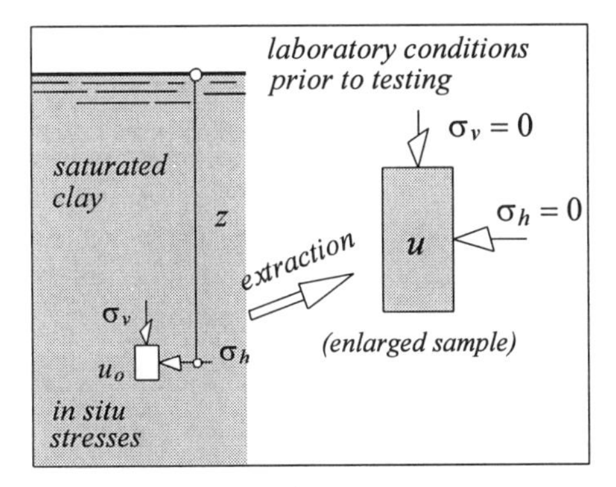

Figure 5.16: Stress release due to sampling.

As such, K_o depends on the type of soil, on its mode of deposition and especially on its stress history. Hence, for a *normally consolidated* clay, K_o can be estimated as follows (Jaky, 1944):

$$K_o = \left(1 + \frac{2}{3}\sin\phi'\right)\tan^2\left(\frac{\pi}{4} + \frac{\phi'}{2}\right) \approx 1 - \sin\phi' \tag{5.38a}$$

For overconsolidated clays, K_o is related to the overconsolidation ratio *OCR* of the soil and can be evaluated using the following (Mayne and Kulhawy, 1982):

$$K_o = (1 - \sin\phi')\,OCR^{\sin\phi'} \tag{5.38b}$$

A detailed examination of this coefficient is presented in section 10.2, yet already, it is clear from equations 5.38 that K_o varies from around 0.5 to

0.6 for normally consolidated clays, to well in excess of 1 in the case of heavily overconsolidated clays as indicated in figure 5.17. The clay becomes overconsolidated during the unloading process from point A to point B, and the corresponding K_o increases gradually with increasing OCR.

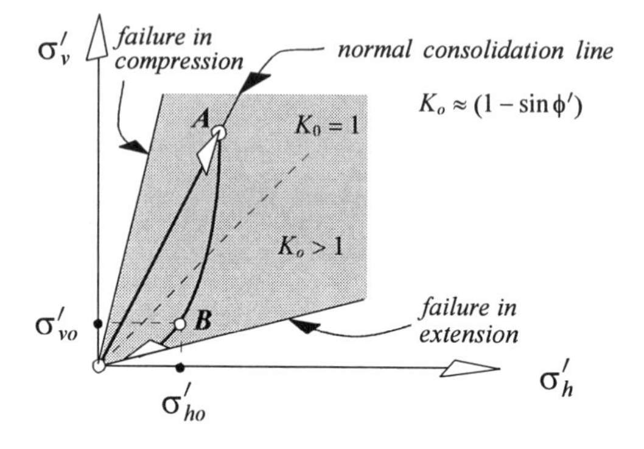

Figure 5.17: Variation of the coefficient K_o with the OCR of a soil.

Referring back to figure 5.16, it is seen that on extraction from the ground, the sample is relieved of the total stresses, so that the applied stress increments at the end of the extraction process are:

$$\Delta\sigma_v = -(\sigma'_{vo} + u_o) \tag{5.39}$$

$$\Delta\sigma_h = -(K_o\,\sigma'_{vo} + u_o) \tag{5.40}$$

At the end of this process of unloading, the clay sample *can* become overconsolidated, and consequently its volume can potentially increase (*i.e.* expand). However, any volume expansion can only occur physically if the clay has access to water. Failing that, the porewater pressure inside the sample becomes negative:

$$u = u_o + \Delta u < 0 \tag{5.41}$$

The porewater pressure increment Δu generated by the unloading can be estimated from the Skempton (1954) relationship:

$$\Delta u = B\left[\Delta\sigma_h + A\left(\Delta\sigma_v - \Delta\sigma_h\right)\right] \tag{5.42}$$

where A and B are porewater pressure parameters. In particular, B has a maximum value of one for *saturated* soils. A on the other hand, depends on the state of the clay, on the stress path, as well as on the magnitude of deformation. Figure 5.18 shows typical variations of parameter A with the axial deformation during an undrained triaxial test on a saturated sample of a lightly overconsolidated clay.

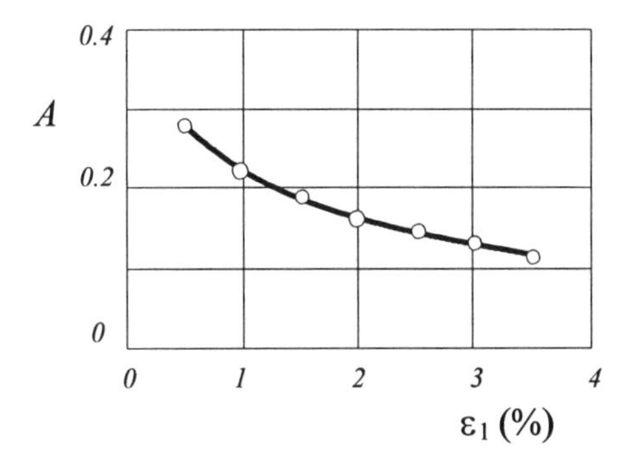

Figure 5.18: Typical variations of A with deformation.

There is ample experimental evidence to suggest that, at failure, the porewater pressure parameter A_f is related to the overconsolidation ratio of the clay. In general, $A_f \geq 0$ for normally consolidated to lightly overconsolidated clays, and $A_f < 0$ for heavily overconsolidated clays; the following limits being typical:

normally consolidated clays	$0.6 < A_f < 1.3$
lightly overconsolidated clays	$0 < A_f < 0.6$
heavily overconsolidated clays	$-0.5 < A_f < 0$

Thus, for a sample of a saturated clay (*i.e.* $B = 1$), a substitution for the quantities $\Delta\sigma_v$ and $\Delta\sigma_h$ from equations 5.39 and 5.40 into equation 5.42, then an insertion of the ensuing quantity Δu into equation 5.41 yields the expression of the (negative) porewater pressure within the clay mass after extraction from the ground:

$$u = -\sigma'_{vo}\,[K_o(1 - A_s) + A_s] \tag{5.43}$$

where A_s represents the porewater pressure parameter of the soil after extraction from the ground, whose value is *different* from that of A_f at failure.

Equation 5.43 indicates that, while the total stresses after extraction are zero, the corresponding effective stresses are somewhat different from those applied *in situ* prior to extraction, in that the new assumed isotropic set of effective stresses generated by the unloading process is:

$$\sigma'_v = \sigma'_h = -u = \sigma'_{vo}\,[K_o(1 - A_s) + A_s] \tag{5.44}$$

Equally important, equation 5.43 suggests that a *suction* of water develops within the clay sample, creating in the process a gradient of moisture content from the periphery of the sample towards its centre. In some instances, this gradient can cause the moisture content in the inner part of the sample to be higher than that in the outer limits by as much as *4 percentage points* (see for instance Bjerrum (1973)), mainly because the soil is more disturbed at the periphery of the sample. This fact presents the engineer with the following dilemma: prior to any testing, the clay sample should, ideally, have both stress and moisture conditions similar to that *in situ*. However, in practice, it is not physically possible to restore these two variables jointly to their field values since applying *in situ* stresses to the sample *will* cause the clay to have a moisture content different from the one in the field and *vice versa*. Nonetheless, given that the soil behaviour is very sensitive to the stress history, it is advisable in this case to restore the *in situ* stress conditions at the expense of having an initial moisture content slightly different from that in the field.

5.5.3 Undrained shear strength of saturated clays

If a clay layer is subjected to a total stress increase $\Delta\sigma$ at its surface, then, because of the low permeability of such type of soils, the immediate (short term) effect is manifested as a rise in porewater pressure of a magnitude $\Delta u = \Delta\sigma$; the effective stresses remaining unaltered. Under these circumstances, the clay resistance to the applied pressure is entirely provided by the *undrained shear strength* c_u. Accordingly, a simulation of the short term behaviour of a clay in a triaxial apparatus consists of preventing any drainage from taking place during testing, the corresponding triaxial test being known as *unconsolidated, undrained test.* The behaviour thus exhibited by a clay sample in the laboratory is representative of that of a saturated clay layer subject to a stress change lasting for a short period of time. For instance, the time needed to undertake an excavation or to build a foundation in a clay is relatively short, of the order of few weeks perhaps, during which the behaviour of the clay is virtually *undrained,* since the corresponding low permeability prevents any excess porewater pressure dissipation from occurring within this short time span.

The unconsolidated, undrained triaxial test consists of applying a confining pressure σ_3 around the clay sample, which is then sheared immediately under constant volume conditions; in other words, without allowing any drainage to take place. The state of stresses throughout the test can be summarised as follows.

- *Phase 1* - the sample is placed in the triaxial apparatus: both vertical and horizontal total stresses are nil, the values of the negative porewater pressure, as well as the (isotropic) effective stresses being given by equations 5.43 and 5.44 respectively:

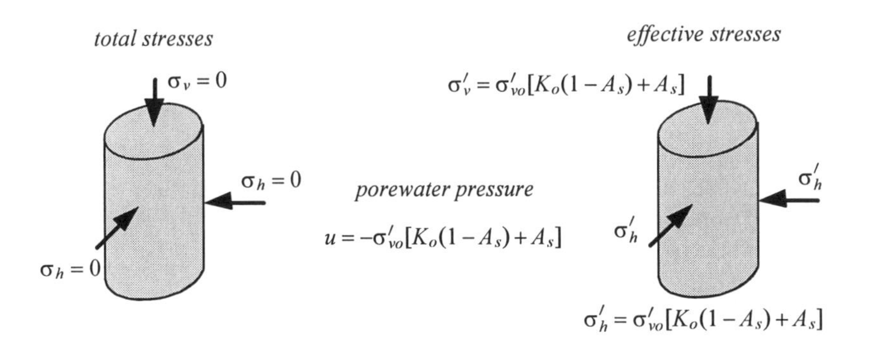

- *Phase 2* - a confining pressure of a magnitude σ_3 is applied around the sample, no drainage is allowed: the effective stresses therefore remain constant and the confining pressure is transferred entirely to the porewater pressure. Hence the ensuing stresses:

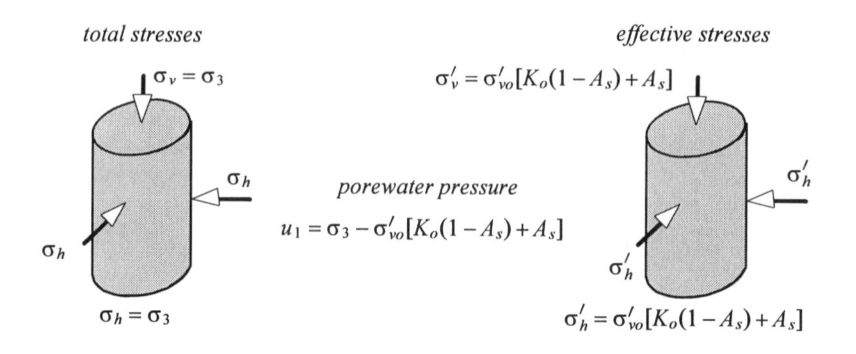

- *Phase 3* - the sample is immediately sheared under constant volume (*i.e.* no drainage allowed) until the occurrence of failure. The corresponding stresses *at failure* are σ_{vf}, σ_{hf} and u_f.

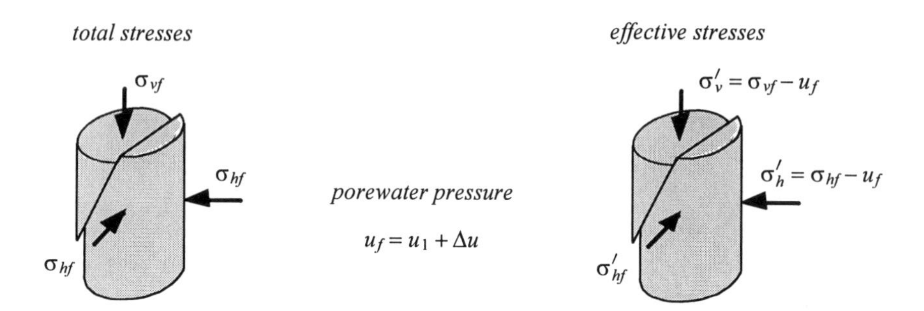

The porewater pressure increment Δu is calculated from equation 5.42 in which $B = 1$ (saturated clay) and $\Delta\sigma_3 = 0$ since the confining pressure σ_3 is maintained constant throughout the test. Accordingly, equation 5.42 reduces to the following:

$$\Delta u = A_f \, \Delta\sigma_1 \tag{5.45}$$

But the quantity $\Delta\sigma_1$ represents the difference between the final (*i.e.* at failure) and initial (*i.e.* at the onset of shear) values of vertical stress:

$$\Delta\sigma_1 = \sigma_{vf} - \sigma_h \tag{5.46}$$

Since the confining pressure is constant (*i.e.* $\sigma_3 = \sigma_h = \sigma_{hf}$), equation 5.45 can therefore be expressed as follows:

$$\Delta u = A_f(\sigma_v - \sigma_h)_f \qquad (5.47)$$

Hence the corresponding porewater pressure *at failure*:

$$u = u_1 + \Delta u$$

$$= \sigma_{hf} - \sigma'_{vo}[K_o(1 - A_f) + A_f] + A_f(\sigma_v - \sigma_h)_f \qquad (5.48)$$

The effective stresses *at failure* can now be established:

$$\sigma'_{vf} = \sigma_{vf} - u_f$$

$$= (\sigma_v - \sigma_h)_f + \sigma'_{vo}[K_o(1 - A_f) + A_f] - A_f(\sigma_v - \sigma_h)_f \qquad (5.49)$$

$$\sigma'_{hf} = \sigma_{hf} - u_f$$

$$= \sigma'_{vo}\,[K_o\,(1 - A_f) + A_f] + A_f\,(\sigma_v - \sigma_h)_f \qquad (5.50)$$

Moreover, the deviator stress *at failure* has already been established *via* equation 5.12:

$$(\sigma_v - \sigma_{hf}) = (\sigma'_v + \sigma'_h)\ \sin\phi' + 2c'\ \cos\phi' \qquad (5.51)$$

Thus, a combination of equations 5.49, 5.50 and 5.51 yields:

$$(\sigma_v - \sigma_h)_f = \frac{2\sigma'_{vo}\,[K_o(1 - A_f) + A_f]\ \sin\phi' + 2c'\ \cos\phi'}{1 + (2A_f - 1)\ \sin\phi'} \qquad (5.42)$$

At the onset of shear, the effective vertical stress, given by equation 5.44, is the *same* for all samples, regardless of the magnitude of the confining pressure applied to each one of them. Consequently, all samples tested under different confining pressures possess the *same* overconsolidation ratio, and therefore have identical A_f values, so that the right-hand side of equation 5.52 corresponds to a *constant.* This, in turn, indicates that the deviator stress at failure has a *constant* value, *irrespective* of the magnitude of the confining pressure. Figure 5.19 shows that the deviator stress at failure corresponds in fact to the diameter of Mohr's circle and, accordingly, the constancy of the deviator stress during *unconsolidated undrained* triaxial tests implies that all corresponding Mohr's circles at failure have the same diameter. Under such conditions, the angle of shearing resistance of the soil is reduced to $\phi_u = 0$, and equation 5.52 then becomes:

$$(\sigma_v - \sigma_h)_f = 2c_u \qquad (5.53)$$

c_u is the undrained shear strength of the clay as shown in figure 5.19, where the quantities σ_v and σ_1 (as well as σ_h and σ_3) are used interchangeably.

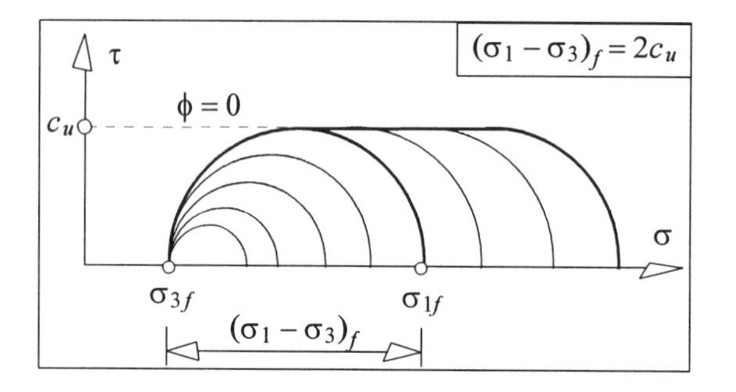

Figure 5.19: Undrained shear strength of clays.

The undrained shear strength c_u is a fundamental soil design parameter, in that it governs the clay behaviour in the short term. It will be shown (refer to section 6.5) that, for a *saturated normally consolidated clay* especially, the relationship between c_u and the moisture content w is similar to that depicted on the left-hand side of figure 5.20 where it is seen that a slight decrease in moisture content can lead to a significant increase in c_u. Accordingly, the undrained shear strength of a normally consolidated clay increases linearly with depth from a theoretical value of zero at the ground surface. However, in practice, any desiccation caused by whatever means (plant roots for instance) causes c_u to increase according to the graph shown on the-right hand side of figure 5.20.

Also, experimental evidence indicates that, for an *overconsolidated clay,* the variation of moisture content with depth is not as significant, and so the undrained shear strength in that case does not vary appreciably with depth.

It should be borne in mind, however, that the undrained shear strength of a (stiff) clay, measured in the laboratory on a sample of a small size (usually 76 mm long with a diameter of 38 mm), often represents an overestimate of the actual value in the field; the reason being that a small size sample is unlikely to contain slip planes or fissures that characterise a thick stiff clay layer in situ (see Simpson et al., 1979).

Skempton (1957) established the following empirical relationship between the ratio c_u/σ'_v of the undrained shear strength to the effective overburden pressure, and the plasticity index I_p, applicable exclusively to *normally consolidated clays:*

$$\frac{c_u}{\sigma'_v} \approx 0.11 + 3.7 \times 10^{-3} I_p \tag{5.54}$$

The *consistency* of a clay is related to its undrained shear strength c_u, and the following table can be used as a guide:

clay consistency	c_u *(kN/m²)*
very soft	< 20
soft	20–40
firm to medium	40–75
stiff	75–150
very stiff	> 50

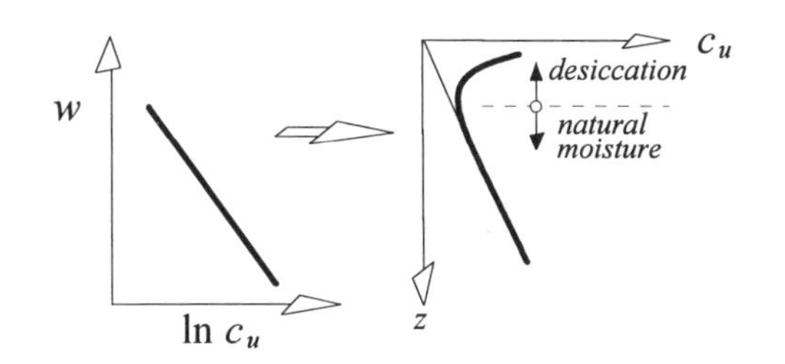

Figure 5.20: Variation of c_u with depth for a normally consolidated clay.

5.5.4 Drained shear strength parameters of saturated clays

The *unconsolidated undrained test* analysed previously can only yield the undrained shear strength of the clay because of the constancy of the deviator stress at failure, regardless of the magnitude of the confining pressure. Therefore a different type of test will have to be undertaken if the effective parameters (*i.e.* c' and ϕ') corresponding to the long term behaviour of the soil were to be measured. The *consolidated undrained test,* throughout which the variation of porewater pressure is measured, simulates the conditions of works extending over a period of time, long enough to assume that the excess porewater pressure induced by the loading has entirely dissipated, leading thus to a full consolidation of the clay. If, at that stage the soil is loaded rapidly, then its behaviour will be undrained, and provided that the porewater pressure is known at any stage of the undrained loading, then the effective stresses can easily be calculated, yielding in the process the drained shear strength parameters of the clay.

The *consolidated undrained triaxial test with porewater pressure measurement* consists therefore of *consolidating* the clay sample under a confining pressure σ_3 (read σ_h), then shearing it under *undrained conditions* while measuring the porewater pressure until the occurrence of failure. It is important at this stage to mention that for the measurement of porewater pressure to be of any significance, the sample must be sheared at *low speed* so that the excess porewater pressure generated continuously during this phase has time to *equalise* throughout the sample (see for instance Bishop and Henkel (1962)). In this respect, a shear speed lower than 0.001 *mm/min* (just under 1.5 *mm per day*) might be needed to fulfil this requirement in the case of a stiff clay.

Compared with the previous test, the state of stresses throughout a *consolidated undrained test* consists of the following phases.

- *Phase 1 is identical to that of the previous unconsolidated undrained test.*

- *Phase 2* - A confining pressure σ_3 is applied around the sample, which is then allowed to consolidate fully. Once all excess porewater pressure has dissipated, the corresponding stresses are:

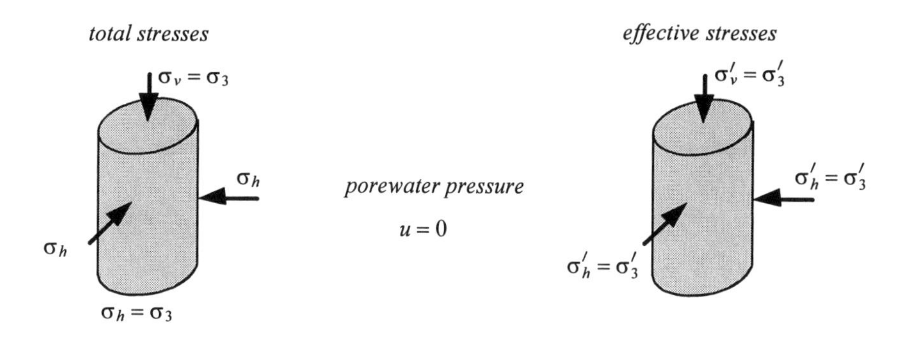

- *Phase 3* - The sample is sheared without allowing any drainage to take place (*i.e.* the volume of the sample remains constant throughout shear). The stresses at failure are therefore:

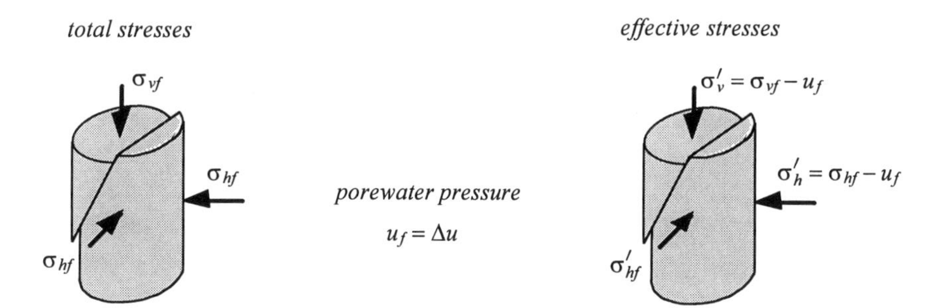

The porewater pressure at failure is identical to that given by equation 5.47:

$$u_f = \Delta u = A_f(\sigma_v - \sigma_h)_f \tag{5.55}$$

Hence the effective stresses at failure:

$$\sigma'_{vf} = \sigma_{vf} - A_f(\sigma_v - \sigma_h)_f \tag{5.56a}$$

$$\sigma'_{hf} = \sigma_{hf} - A_f(\sigma_v - \sigma_h)_f \tag{5.56b}$$

Finally, a combination of equations 5.51 and 5.56 yields the deviator stress at failure:

$$(\sigma_v - \sigma_h)_f = \frac{2\sigma_3 \sin\phi' + 2c' \cos\phi'}{1 + (2A_f - 1)\sin\phi'} \tag{5.57}$$

Significantly, equation 5.57 indicates that the deviator stress at failure (*i.e.* the diameter of Mohr's circle) depends on the magnitude of the confining pressure σ_3. This is illustrated in figure 5.21 where two samples of the same saturated clay have been consolidated under different pressures such that, at the end of consolidation, the volume of the sample subject to the lowest confining pressure has increased in a way that the clay has become *heavily overconsolidated.* The second sample, on the other hand, is in a *normally consolidated* state.

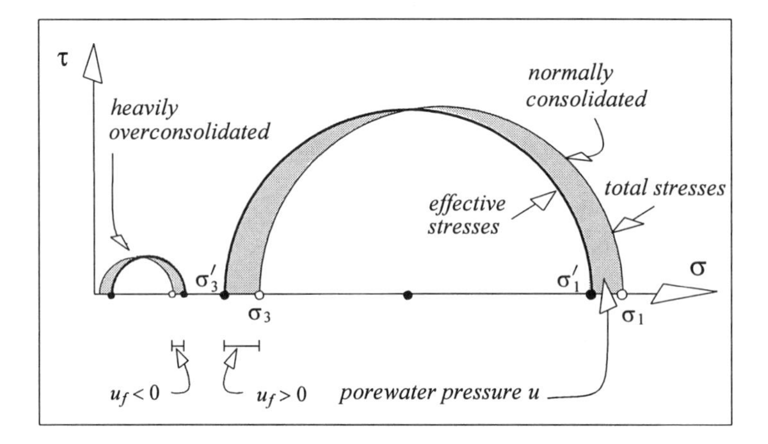

Figure 5.21: Effect of the state of a clay on its strength.

At the onset of undrained shear, the porewater pressure increases gradually within both samples and, eventually, at failure, a *positive* porewater pressure is measured in connection with the normally consolidated clay sample, whilst the heavily overconsolidated one exhibits a *negative* porewater pressure. The question of how the *effective shear strength* parameters ϕ' and c' can be determined then arises once the corresponding Mohr's circles are drawn. Evidently, the *total stress* circles are of little, if any, importance since ϕ' and c' are related to effective stresses. Yet, it is still unclear how figure 5.21 can possibly be of any use to measure these two parameters. More confusing perhaps is the fact that the behaviour depicted in this figure characterises the *same* saturated clay which, when subject to a confining pressure smaller than its preconsolidation pressure σ'_p (*i.e.* the maximum effective vertical stress ever applied to the soil) becomes overconsolidated. On the other hand, the state of clay becomes normally consolidated under consolidation pressures greater than σ'_p.

To clarify this somewhat confusing situation, let us concentrate for a moment on the *stress paths* during the undrained shear phase. In this respect, figure 5.22 depicts a typical behaviour where the variation of porewater pressure u clearly reflects the state of the soil: for a normally consolidated clay (*i.e.* $OCR = 1$), u increases continuously until the occurrence of failure, while in the case of an overconsolidated clay, the porewater pressure increases initially, only to decrease after having reached a peak, even becoming negative for heavily overconsolidated clays.

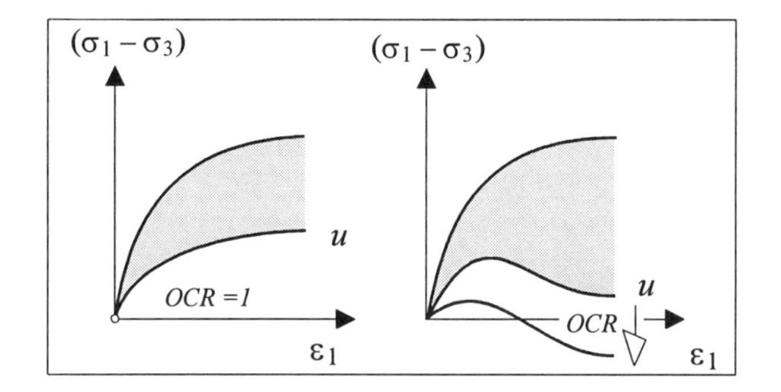

Figure 5.22: Effects of the state of a clay on the porewater pressure at failure.

These effects are better demonstrated in figure 5.23 using the stress variables *(q, p')* defined by equations 5.9 and 5.10. It will be shown in section 6.1 that the *total stress path* in the space *(q, p')* is linear with a slope of 3 as shown in the figure. The measurement of porewater pressure at the onset of shear allows *the effective stress path* to be found, and the figure further illustrates the effect of the state of clay on the shape of the effective stress path. Thus, the normally consolidated clay (consolidated under σ'_{31}) has a *positive* porewater pressure at failure. However, as the overconsolidation ratio increases, the porewater pressure at failure decreases (clay sample consolidated under σ'_{32}), only to become *negative* when the clay becomes heavily overconsolidated (sample consolidated under σ'_{33}).

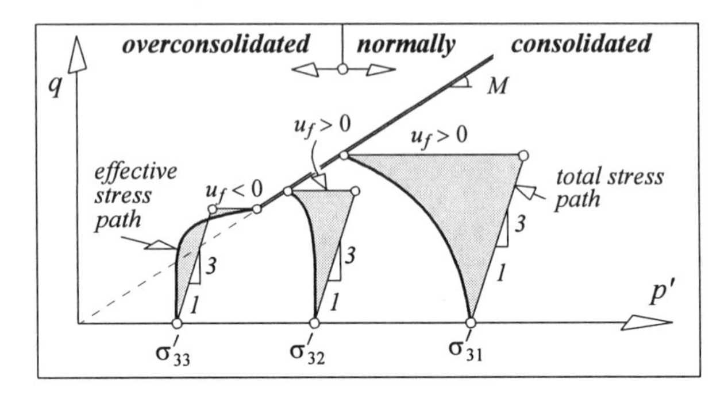

Figure 5.23: Effective and total stress paths for different states of a clay.

A clearer picture is now starting to emerge in relation to the *drained behaviour* of clays, such a picture being dominated by the *state* of the clay. Thus for a *soft normally consolidated clay*, the behaviour is characterised by an *ultimate* or a *critical* angle of shearing resistance ϕ'_c and no apparent cohesion ($c' = 0$) as shown in figure 5.24.

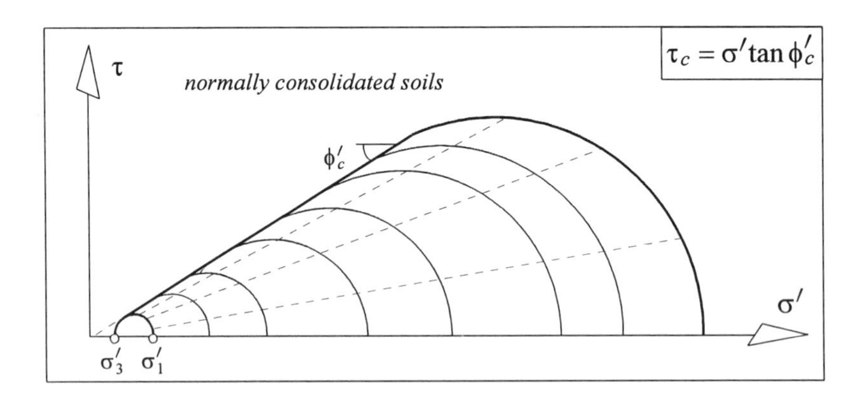

Figure 5.24: Behaviour of a normally consolidated clay.

In the absence of cohesion, equation 5.52 yields the following deviator stress at failure:

$$(\sigma_1' - \sigma_3')_f = \frac{2\sigma_3 \sin\phi_c'}{1+(2A_f - 1)\sin\phi_c'} \tag{5.58}$$

where σ_3' represents the (effective) consolidation pressure and A_f is the porewater pressure parameter at failure.

On the other hand, the behaviour of a *stiff overconsolidated clay* is more typical of that depicted in figure 5.25. The *critical* (or ultimate) angle of shearing resistance ϕ_c' of the clay is easily measured from the linear portion of the failure envelope where the clay is normally consolidated. As soon as the clay becomes overconsolidated, that is, when the confining pressure becomes smaller than the preconsolidation pressure, the failure envelope becomes markedly non-linear, and once more the question to be answered is what values should be assigned to both the effective cohesion c' and the angle of shearing resistance ϕ' so that the entire behaviour of the clay can be represented by a unique mathematical equation:

$$\tau_f = c' + \sigma' \tan\phi' \tag{5.59}$$

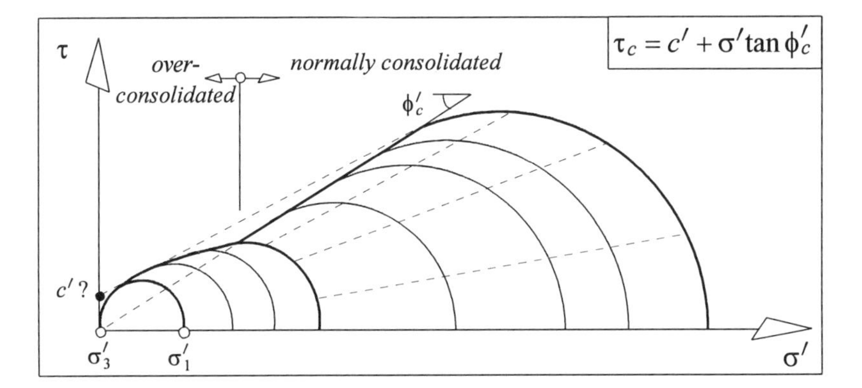

Figure 5.25: Typical behaviour of an overconsolidated clay.

The answer depends in part on the stress level to be applied in the field: if the magnitude of effective stresses is expected to cause the clay to become normally consolidated, then the ultimate critical conditions of $\phi' = \phi'_c$ and $c' = 0$ must be applied. If, however, after being loaded, the clay is still overconsolidated, then a (subjective!) engineering judgement must be made. In this respect, figure 5.25 makes it clear that neglecting the apparent cohesion might affect the shear strength of the soil, especially if the clay in question is heavily overconsolidated. Alternatively, assigning a high value to c' such as the one corresponding to the intercept of the tangent to the non-linear portion of the failure envelope (refer to figure 5.25) can be potentially dangerous, as this might increase dramatically (and artificially) the shear strength of the soil, knowing that such intercepts can be as high as $100\,kN/m^2$ in some instances.

Figure 5.26 depicts results measured from consolidated undrained triaxial tests with porewater pressure measurements undertaken on a natural stiff clay (Josseaume and Azizi, 1991). The slope M of the linear portion, calculated from equation 5.20, yields an *effective critical* angle of shearing resistance $\phi'_c = 22°$. Now, there is a need to consider carefully what value might be assigned to c', if the shear strength of the clay is to be modelled by an expression similar to equation 5.19.

$$q = \frac{6\cos\phi'}{3-\sin\phi'}\,c' + \frac{6\sin\phi'}{3-\sin\phi'}\,p'$$

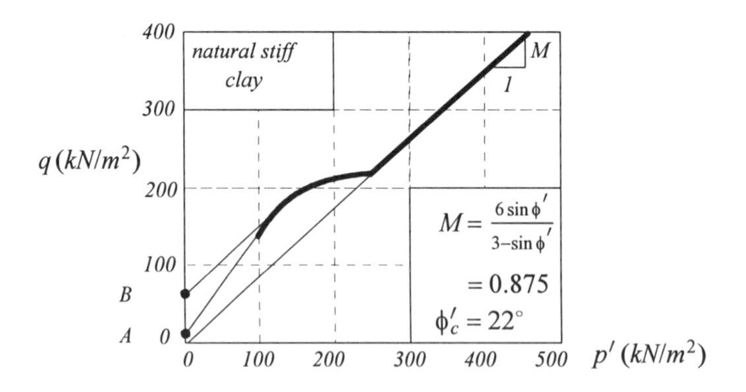

Figure 5.26: Behaviour of a natural stiff clay.

Manifestly, the somewhat contentious parameter c' is difficult to determine with sufficient accuracy because of its intricate link to the soil density and therefore to the void ratio, whose value varies with depth (*i.e.* with the effective overburden pressure) as illustrated in figure 5.27.

Referring back to figure 5.26, it is seen that, were the non-linear part of the failure envelope to be prolonged as illustrated, the intersection with the q-axis at point A then yields a deviator stress $q_A \approx 10\,kN/m^2$. Alternatively, if the tangent to the curved portion is used, the ensuing deviator stress at point B will be $q_B \approx 60\,kN/m^2$.

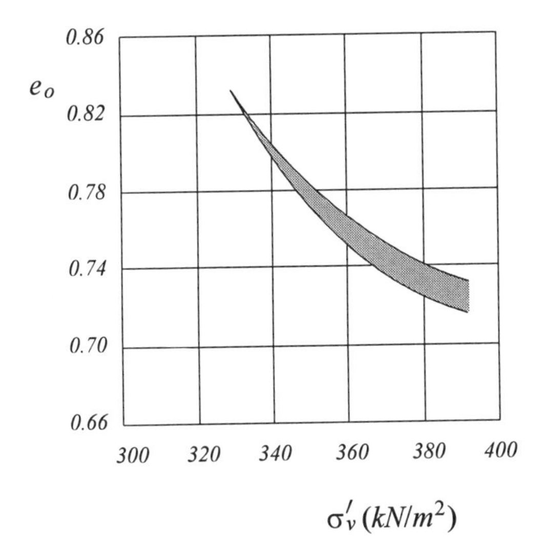

Figure 5.27: Variation of the initial void ratio with depth.

In both cases, c' is calculated from equation 5.19 in which p' is zero:

$$c' = q_{A,B}\left(\frac{3-\sin\phi'_c}{6\cos\phi'_c}\right) \tag{5.60}$$

Hence:

$$c'_A = 4.7\,kN/m^2, \qquad c'_B = 28.3\,kN/m^2$$

This simple example epitomises the practical difficulties that an engineer would face in selecting an appropriate value (if at all !) for c' and, in this particular instance, it is prudent to opt for the smaller of the two values.

Generally, it is advisable to limit c' to a maximum of $15\,kN/m^2$ for stiff heavily overconsolidated clays and to extend the value $c'=0$ from normally consolidated to lightly overconsolidated clays.

The drained shear strength parameters can also be determined from *consolidated drained triaxial tests* which consist of consolidating soil samples of the same clay under different confining pressures, then shearing them under *drained* conditions, thus allowing their volume to change. It is important to mention however that, as implied by its name, this type of test requires the sample to be sheared under drained conditions. In other words, the shear speed must be such that no build up of excess water pressure can occur during shear. To fulfil this requirement, shear speeds of as low as $5\times10^{-4}\,mm/min$ may be needed, implying an axial deformation of less than 1% *per day* in the case of a 76 *mm* long clay sample.

Taking into account the time needed for test preparation and consolidation, it is obvious that a carefully undertaken drained consolidated test necessitates weeks rather than days to be achieved. This can deter an engineer from opting for this type of test, especially when a consolidated *undrained* triaxial test with porewater pressure measurement can be achieved comparatively quickly, and can yield similar information about the *effective* shear strength parameters as detailed previously.

However, there is a need to undertake this type of *consolidated drained triaxial tests* to simulate some specific stress paths. The procedure of such tests is similar in nature to that applicable to a consolidated *undrained* test, with the exception of the shear phase which must be undertaken under drained conditions. Accordingly, *at failure*, the state of stresses is as shown below.

Evidently, the behaviour of the clay depends on the magnitude of the confining pressure, and therefore on the corresponding overconsolidation ratio.

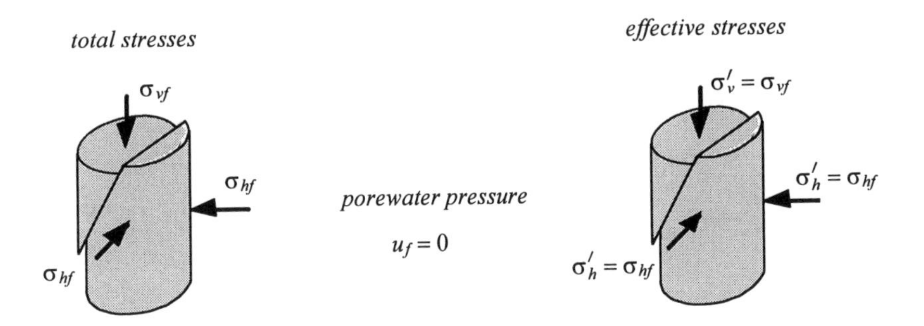

Figure 5.28 shows that a normally consolidated clay ($OCR = 1$) exhibits a behaviour characterised by a strain hardening, due to the (positive) compressive volumetric strains ε_v (*i.e. contraction* of volume). On the other hand, an overconsolidated clay initially displays a compressive volumetric strain, followed by a volume expansion that reaches a maximum when the deviator stress is at its peak.

The rate of expansion then starts to decrease, and the deviator stress tends towards its ultimate critical value q_c under which the expansion ceases. Figure 5.28 clearly indicates the link between an increasing expansion and an increasing *OCR*. In fact, it is well established that, as the *OCR* and the plasticity index increase, so does the difference between the peak shear stress q_{max} and the critical shear stress q_c. Moreover, the angle of shearing resistance associated with the critical shear stress of an overconsolidated clay is usually appreciably smaller than that corresponding to the peak shear strength.

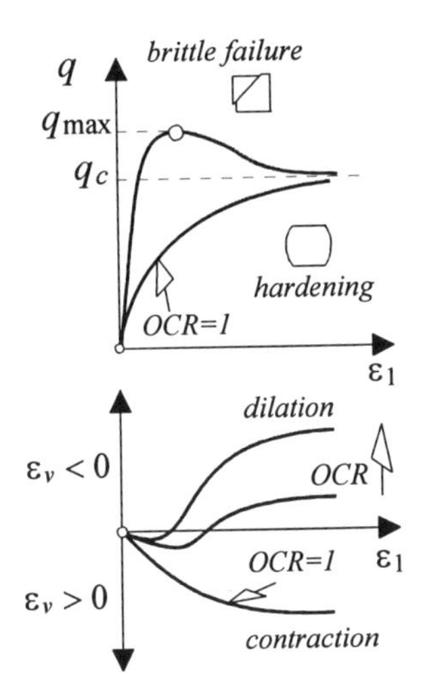

Figure 5.28: Effect of the OCR on the peak strength of a clay.

On the practical side, the *critical effective angle of friction* is an important parameter for the analysis of the stability of slopes. Although this topic is tackled in detail in chapter 8, it is important to mention in this respect that the use of peak shear parameters measured in the laboratory, on samples of small sizes, often leads to an overestimation of the factor of safety against shear failure (error on the unsafe side!). Several instances of slope failures reported in the literature are due precisely to this mechanical extrapolation of shear parameters measured in the laboratory to field conditions (see for example Skempton (1964), Skempton and La Rochelle (1965), Henkel (1957) and Palladino and Peck (1972)).

5.6 Quality assurance related to triaxial testing

The quality of the results measured during a triaxial test is inevitably linked to the testing procedure, especially that relating to sample preparation prior to testing. It has already been shown that sampling causes the porewater pressure in the clay to become negative, generating in the process a migration of moisture from the periphery of the sample where the soil fabric has been disturbed towards its centre. Ideally, prior to any testing taking place, each soil sample must be consolidated under stress field conditions ($\sigma'_v = \sigma'_{vo}$, $\sigma'_h = K_o \sigma'_{vo}$), where σ'_{vo} represents the effective overburden pressure and K_o is the coefficient of earth pressure at rest given by equation 5.38. In practice, however, the impediment to restoring these anisotropic stresses results from the fact that consolidation under these circumstances *must* occur under zero radial strain ($\varepsilon_r = 0$), known as K_o-condition, and refers to the fact that soils *in situ* are semi-infinite media, and as such, they do not deform radially when subjected to any type of loading. Though feasible, the stringent condition of zero lateral strain during consolidation is very difficult to maintain in the laboratory since there is a need to adjust the stress ratio, that is the coefficient K_o, almost continuously while consolidation progresses under $\varepsilon_r = 0$.

An alternative to the above procedure consists of assuming a uniform distribution of the negative porewater pressure throughout the sample, then consolidating the clay, prior to any testing, under an *isotropic* effective stress field, the magnitude of which is given by equation 5.43:

$$\sigma'_v = \sigma'_h = -u \qquad (5.61)$$

Under these conditions, applying an isotropic effective stress to the clay sample amounts to measuring its isotropic swelling pressure. This can easily and accurately be achieved using the set-up depicted in figure 5.29, whereby a *nil indicator* is connected through a valve to the drainage circuit on one side and to a container filled with de-aerated water on the other side. Prior to opening the valve, an isotropic fluid pressure, with a magnitude *estimated* from equation 5.43 is applied around the sample, both ends of the drainage circuit of the nil indicator being at atmospheric pressure, hence there is an initial horizontal level of mercury.

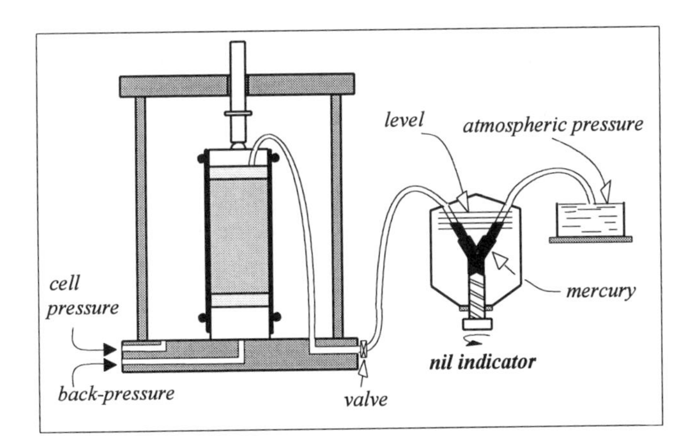

Figure 5.29: Use of a nil indicator to prevent any volume change of the sample prior to testing.

Once the valve is opened, one of three possibilities occurs:

(a) The applied pressure σ_h is such that it cancels out the negative porewater pressure, in which case the level of mercury in the nil indicator remains equalized.

(b) The porewater pressure is still negative, and consequently water is sucked into the sample making the mercury move upward in the left-hand side of the nil indicator. In this case, the pressure σ_h must be increased gradually until the mercury level is again equalized.

(*c*) The porewater pressure within the sample becomes positive causing water to be expelled through the drainage lead and mercury to move upward on the right-hand side of the nil indicator. Consequently, σ_h needs to be adjusted downward until a balanced mercury level is re-established.

Clearly, the first possibility above, though not impossible, is highly unlikely to occur, and the pressure σ_h almost inevitably will need to be adjusted until an equilibrium in the form of a horizontal mercury level (indicating zero porewater pressure inside the sample) is achieved. Because of the low permeability of clays, the adjustment can be a slow process, and can take several hours in the case of a stiff clay. The fluid pressure applied around the sample, under which an equilibrium is achieved is known as the *isotropic swelling pressure* σ_s of the clay. It corresponds to the limit confining (effective) pressure for which the porewater pressure within the clay is zero. Accordingly, any confining pressure smaller than σ_s will cause the porewater pressure to become negative, generating in the process a suction of water and causing the volume of the sample to expand.

The task of restoring the stress conditions to as near as practically possible to those applied *in situ* can become futile if the sample is subjected to substantial disturbances during the extraction process and the subsequent transportation, storage and preparation before testing. Assuming that an appropriate sampling technique is used to extract high quality samples, which are then adequately stored in the laboratory under controlled temperature and humidity conditions so that the moisture content of the clay remains virtually unchanged, the careful preparation of every sample should then become the focus of the experimenter in order to minimise the remoulding of the clay which can affect, in a major way, the quality of the measured results during testing. Of course, every step in sample preparation can potentially lead to the clay being markedly remoulded. These steps include cutting and trimming of the sample, applying a filter paper, then a rubber membrane around the clay, and saturating then connecting the drainage circuit.

Clearly, an accumulation of differing degrees of disturbance related to these steps can be detrimental to the quality of the subsequent measured results. For instance, cutting a 38 *mm* diameter, 76 *mm* long sample of a soft clay is a highly delicate operation which, if not handled carefully, can easily cause the sample to have an initial deformation before being loaded, thus altering the behaviour of the clay. Also, placing a *saturated* filter paper around the sample, then connecting a *saturated* drainage circuit (including saturated porous stones on each side of the sample) will cause a stiff clay to suck water through its periphery in order to balance the negative porewater pressure existing within the soil matrix, leading to an increase in volume prior to any stresses being applied or to any measurements being made. Similarly, applying a high back pressure in a single step to a sample of a stiff clay may cause the water to flow between the membrane and the periphery of the sample, leaving the clay unsaturated.

To minimise these 'side effects', the procedure detailed below, which has the advantage of not altering the volume of the sample prior to the application of a confining pressure, can be applied. Although it is more suited to stiff clays, the procedure can be used in conjunction with any type of soil, and consists of the following steps.

(*a*) Once cut to the required dimensions, a *dry* filter paper, shaped in the manner depicted in figure 5.30, is carefully placed around the sample. In so doing, care must be taken to ensure that only a very small amount of water is sprayed on two points to make the paper stick to the sample, which is then put on the pedestal of the triaxial cell on top of a *dry* porous stone overlain by a *dry* circular filter paper.

(*b*) A rubber membrane is placed around the sample on top of which a *dry* circular filter paper and a *dry* porous stone are then laid.

(*c*) The *empty* drainage circuit is then connected to the sample, and the triaxial cell filled with fluid under a pressure estimated from equation 5.43 (*i.e.* a cell pressure equivalent in magnitude to $-u$, where u represents the *negative* porewater pressure developed within the sample after extraction from the ground.

(*d*) The drainage circuit is closed at one end, then the air contained within it is pumped out at the other end. This step takes about two minutes to complete.

(*e*) The pump is carefully disconnected so as not to allow any air to re-enter the circuit, which is then immediately immersed in a bucket of de-aerated water. Because of the negative pressure inside the circuit created by pumping, water is automatically sucked in, saturating in the process the entire drainage circuit including the porous stones and filter papers. This step takes no longer than two to three minutes.

(*f*) Once saturated, the drainage circuit is thereafter connected to a nil indicator as depicted in figure 5.29, and the adjustment of the cell pressure is undertaken in accordance with the procedure described earlier.

Once equilibrium is reached (*i.e.* a confining pressure under which the porewater pressure inside the sample is reduced to zero), a back pressure is applied to the soil *in steps* in order to ensure a very high degree of uniformity of the back pressure throughout the sample. In this respect, it is strongly advisable to apply the same pressure *simultaneously* inside and outside the sample, so that the state of effective stresses is not altered. Accordingly, a back pressure Δu, which is *per se* used to saturate the soil, is applied inside the sample through the drainage circuit, while the cell pressure σ_3 is increased at the same time by the same increment Δu. In the absence of the possibility of increasing concurrently both the back pressure and the cell pressure by the same amount, it is preferable to increase the cell pressure first, then apply the back pressure, thus eliminating the possibility, albeit remote, of the back pressure causing the water to flow between the periphery of the sample and the rubber membrane.

Experimental evidence shows that in the case of stiff clays, back pressures as high as $1500\,kN/m^2$ are needed to achieve a high degree of saturation of 98% (Berre, 1981). However, these levels of back pressure can only be applied using specially adapted triaxial equipment. In fact, a standard piece of equipment (*i.e.* a Bishop triaxial cell) is designed to withstand a maximum cell pressure of $1200\,kN/m^2$ and, adopting the procedure described previously by which both back pressure and cell pressure are increased at the same time, it is clear that there is a limit to the maximum back pressure that can realistically be applied to the soil. For instance, if a stiff clay needs to be tested under confining pressures as high as 800 kN/m^2 using standard equipment, then the maximum back pressure to which the clay can be subjected will be limited to 400 kN/m^2.

On the practical side, applying a back pressure of, say, 400 kN/m^2 to a 38 mm diameter clay sample of low permeability (10^{-10} m/s for example) in one increment, may create preferential drainage paths, thus resulting in a differential consolidation of the soil. To alleviate this problem, the back pressure must be applied in gradually increasing increments, at time intervals long enough to allow the pressure throughout the sample to be as uniform as possible. An example related to the application of 400 kN/m^2 back pressure to a stiff clay with a permeability $k = 10^{-11}$ m/s is given in the following table. Notice the time related to each increment decreases as the back pressure increases, thus reflecting the gradual increase in saturation of the soil.

back pressure increment (kN/m^2)	*application time* (*days*)
50	3
50	2
100	1
200	1

The degree of saturation must be checked at the end of the last increment of back pressure. The corresponding procedure consists simply of closing the drainage circuit, then increasing the (isotropic) cell pressure by an increment $\Delta\sigma_3$ and measuring almost immediately, through a pressure transducer connected to the drainage circuit, the increment of porewater pressure Δu thus generated. Skempton equation 5.42 can then be used to calculate the coefficient B of saturation and, because the increment of deviatoric stress is zero ($\Delta\sigma_3 = \Delta\sigma_1$), the equation takes the simpler form:

$$\Delta u = B\,\Delta\sigma_3 \qquad (5.62)$$

Hence, the clay is fully saturated if $B = 1$, in other words when $\Delta u = \Delta\sigma_3$. In practice, one aims at achieving a degree of saturation corresponding to a minimum of $B = 0.98$.

The use of a lateral filter paper around the sample as depicted in figure 5.30 has the advantage of markedly reducing the time needed for the excess porewater pressure inside the sample either to dissipate (during a drained loading), or to become uniform (as in undrained shear, for example).

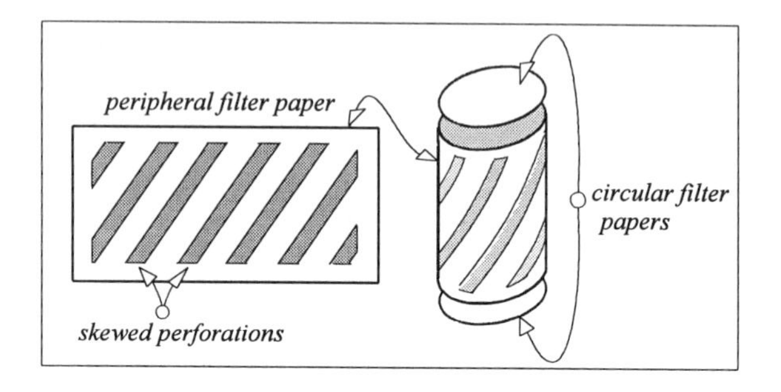

Figure 5.30: Use of peripheral filter paper during testing.

Gibson and Henkel (1954) established a relationship that yields the time t needed to achieve an average degree of consolidation U:

$$U = 1 - \frac{h^2}{\eta c_v t} \tag{5.63}$$

where h represents the length of the drainage path (half the sample length if the soil is drained on both sides), c_v corresponds to the coefficient of vertical consolidation given by equation 4.15 in section 4.3, and η is a coefficient that depends on drainage conditions, whose values are summarised in the following table.

Table 5.1: Coefficient η (from Bishop and Henkel, 1962)

drainage conditions	η
sample drained at one end	0.75
sample drained at both ends	3.0
sample drained radially	32.0
sample drained at both ends as well as radially	40.4

Equation 5.63 can be rearranged in the following way:

$$t = \frac{h^2}{(1-U)\eta c_v} \tag{5.64}$$

The latter equation indicates that a sample of clay for which drainage occurs radially through a lateral filter paper, as well as vertically at both ends ($\eta = 40.4$) consolidates 13.5 times faster than when the sample is drained at the ends only ($\eta = 3$). Also, equation 5.64 can be used to check the adequacy of the shear speed used during undrained shear, in the knowledge that any ensuing excess porewater pressure Δu cannot be representative of the undrained behaviour unless it is distributed uniformly throughout the sample.

Example 5.2

Consider the undrained shear of a $38\,mm$ diameter, $76\,mm$ long sample of a stiff clay, with a coefficient $c_v = 2 \times 10^{-8}\,m^2/s$ and with drainage conditions represented by a coefficient $\eta = 40.4$ in table 5.1. Considering that, in practice, a degree of uniformity of the excess porewater pressure $U = 95\%$ is deemed satisfactory, let us, first, use equation 5.64 to calculate the corresponding time t_{95}:

$$t_{95} = \frac{38^2 \times 10^{-6}}{(1-0.95) \times 40.4 \times 1.2 \times 10^{-6}} = 596\,min$$

Accordingly, the sample must be sheared undrained for about 10 *hours* before the excess porewater pressure reaches a degree of uniformity of 95%, which in turn indicates that any measurement of porewater pressure taken within 10 *hours* of the onset of shear is *not* representative of the clay behaviour. Thus, it is necessary to select a shear speed that yields a minimum axial deformation, say $\varepsilon_1 = 0.005$, in 10 *hours*, leading hence to a shear speed:

$$v = \varepsilon_1 \frac{l}{t_{95}} = \frac{0.005 \times 76}{600} = 6.3 \times 10^{-4}\,mm/min$$

Adopting this speed for the entire undrained shear phase, one can then easily calculate the degree of uniformity U of the excess porewater pressure throughout the test since, for any selected time t, U can be calculated from equation 5.63, while the axial strain is determined in a straightforward way:

$$\varepsilon_1 = \frac{tv}{l}$$

with $v = 6.3 \times 10^{-4}\ mm/min$ and $l = 76\ mm$ (the sample length).

The ensuing results, plotted in figure 5.31, show the extent to which the degree of uniformity U of the excess porewater pressure is below the required minimum value of 0.95. Clearly, any measurement of excess porewater pressure Δu for which U falls outside the shaded area in the figure are erroneous. In particular, Δu is not representative of the clay behaviour up to an axial deformation of 0.5% when the sample is drained radially as well as vertically (refer to the graph in bold in the figure).

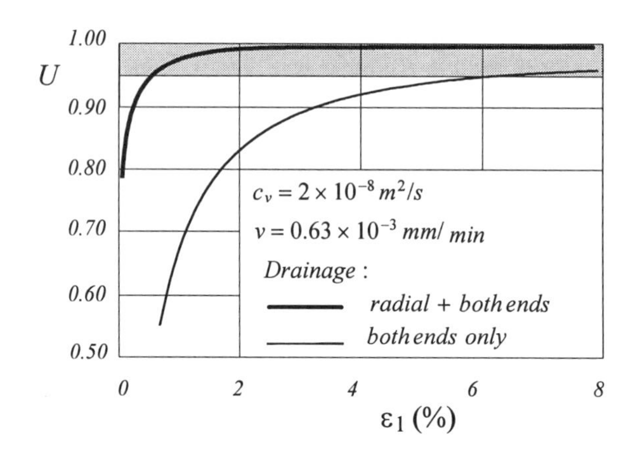

Figure 5.31: Effect of the shear speed on the degree of uniformity of excess porewater pressure.

However, were the sample to be drained only at the ends, the excess porewater pressure measured during shear will *not* be typical of the clay behaviour until the axial deformation reaches the significant value of 6.6%, which is most probably beyond failure considering the (stiff) nature of the clay in this case.

It is clear therefore that a good deal of care in sample handling, preparation, saturation and consolidation is not enough to ensure results of high quality with respect to (drained or undrained) shear. In fact it would be very unfortunate if, after going through all the trouble of meticulous preparations, the experimenter overlooked the details related to the selection of an adequate shear speed during an undrained triaxial test for example.

In conclusion, it is vitally important for the operator to be aware of the many pitfalls related to laboratory triaxial testing. A clear procedure that includes the means of checking the quality of measurements is therefore a necessity, without which the risk of making the wrong interpretations increases dramatically. Figure 5.32 (Azizi and Josseaume, 1988) shows results measured at the end of consolidation (*i.e.* prior to shear) for some undrained triaxial tests, carefully undertaken according to the procedure detailed previously that includes the use of a nil indicator. These results, together with those of figure 5.26 indicate that the stiff clay in question has a swelling pressure of around 250 kN/m^2.

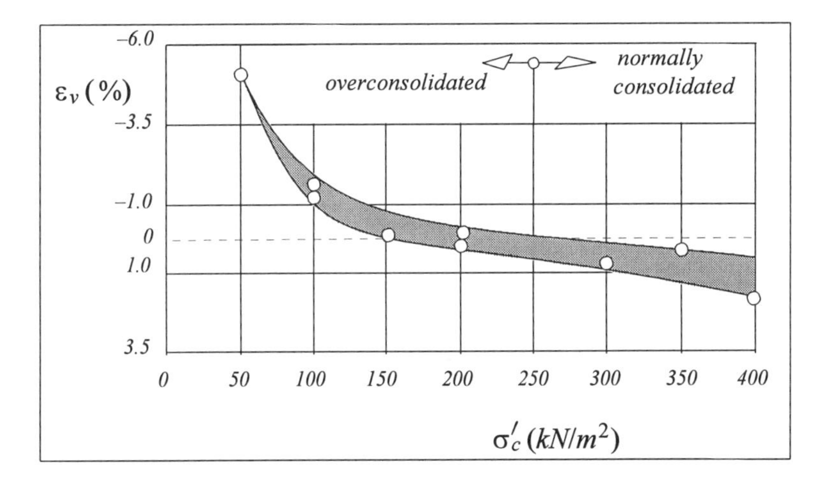

Figure 5.32: Natural stiff clay: volume change vs consolidation pressures. (Reproduced by permission of the Laboratoire Central des Ponts et Chaussées.)

5.7 The shear vane test

However careful an operator is, soils tested in the laboratory are almost inevitably remoulded to a certain extent. To minimise the disturbance, *in situ* tests, when properly carried out, can yield high quality results which complement those obtained under laboratory conditions. The *shear vane test* is a theory-based reliable *in situ* test, specifically developed for saturated clays with undrained shear strengths of up to $100\,kN/m^2$.

The equipment, depicted schematically in figure 5.33, consists of four thin stainless rectangular blades, welded to a steel rod. The length L of the vane and its width D are typically $100\,mm$ and $50\,mm$ respectively, though vanes with $L = 150\,mm$ are often used for clays with undrained shear strengths larger than $50\,kN/m^2$. Once pushed gently into the ground to the required depth, the vane is then rotated at a rate of about $0.1°$ *per second* until the occurrence of failure where a maximum torque T (*i.e.* moment of resistance) is recorded. Since the test is carried out quickly, the corresponding behaviour of the clay is therefore undrained.

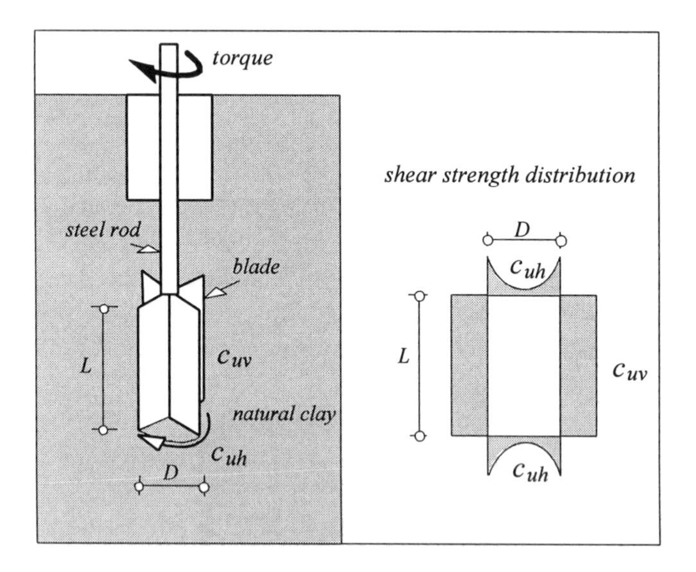

Figure 5.33: Vane shear apparatus.

If the clay is *anisotropic,* then its *vertical* undrained shear strength c_{uv} developed throughout the length of the vane, is distinct from its *horizontal* component c_{uh} occurring on the horizontal top and bottom sides of the vane as shown in figure 5.33. Accordingly, the maximum torque is the sum of a vertical and a horizontal components:

$$T = T_v + T_h \tag{5.65}$$

Referring to figure 5.33, the undrained shear strength c_{uv} is assumed to be fully mobilised throughout the length of the vane. Hence the corresponding torque is:

$$T_v = c_{uv}\pi DL\frac{D}{2} = c_{uv}\,\pi D^2\frac{L}{2} \tag{5.66}$$

On the other hand, experimental evidence (refer to Wroth (1984)) strongly suggests that the distribution of c_{uh} on the top and bottom sides of the vane is rather similar to that shown in figure 5.33. Under these circumstances, c_{uh} depends on the ratio r/R, where r represents the radial distance from the vane centre and R is the vane radius; so that in accordance with figure 5.34, the torque T_h is evaluated as follows:

$$T_h = 2\int_0^R c_{uh}\,2\pi r^2\,(r/R)^n\,dr \tag{5.67}$$

where the value of the exponent n can be as high as 5.

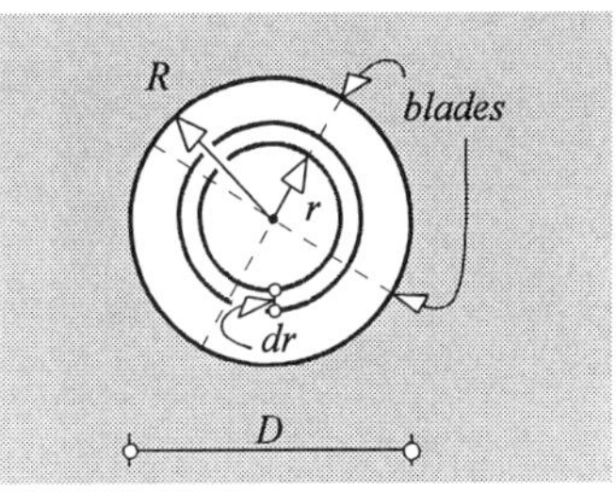

Figure 5.34: Boundary conditions related to the vane shear test.

A straightforward integration of equation 5.67 then yields:

$$T_h = c_{uh}\,\pi\frac{D^3}{2(n+3)} \tag{5.68}$$

Consequently, the total moment resistance $(T_h + T_v)$ is:

$$T = \pi D^2 \left[c_{uv} \frac{L}{2} + c_{uh} \frac{D}{2\,(n+3)} \right] \tag{5.69}$$

Equation 5.69 is the general expression relating the undrained shear strength of an *anisotropic clay* to the maximum torque recorded during an *in situ* shear vane test.

If the conditions of isotropy can be assumed to prevail, then $c_{uv} = c_{uh} = c_u$. Moreover, if the vane is characterised by a ratio $L/D = 2$, then equation 5.69 yields an *apparent undrained shear strength* c_u:

$$c_u = \frac{T}{\pi D^3} \frac{(n+3)}{(n+3.5)} \tag{5.70}$$

This equation is plotted in figure 5.35 as the variation of the dimensionless quantity $\pi D^3 c_u/T$ *versus* the coefficient *n*. The graph shows clearly the limited effect of *n* on the undrained shear strength of the clay and, for all intents and purposes, a value $n=2$ can be used in conjunction with equations 5.69 and 5.70.

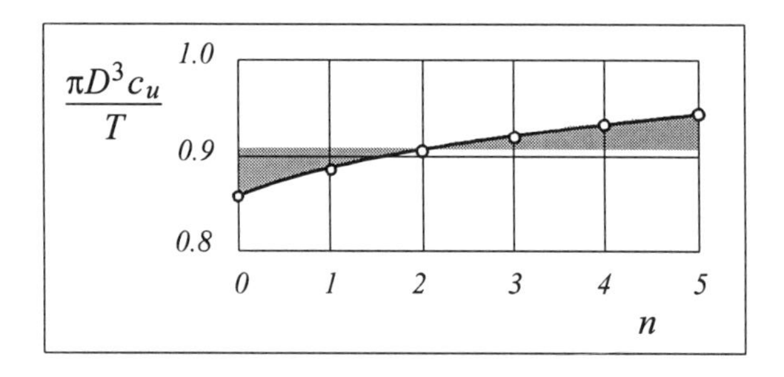

Figure 5.35: Effect of the exponent n on the undrained shear strength of a clay.

Defining the *strength ratio* as being:

$$\xi = \frac{c_{uh}}{c_{uv}} \tag{5.71}$$

then using $n = 2$, the quantity c_{uv} in equation 5.69 can be calculated as follows:

$$c_{uv} = \frac{10T}{\pi D^2(5L + \xi D)} \tag{5.72}$$

Obviously, when $L = 2D$, the above equation reduces to:

$$c_{uv} = \frac{10T}{\pi D^3(\xi + 10)} \tag{5.73}$$

so that when the clay is *isotropic*, $\xi = 1$ and equation 5.73 further reduces to the following:

$$c_u = \frac{10T}{11\pi D^3} \tag{5.74}$$

Bear in mind that both equations 5.73 (for anisotropic clays) and 5.74 (for isotropic clays) were derived using $n = 2$ and $L = 2D$, the quantity T being the total torque measured during the test.

Example 5.3

Consider the case of two shear vane tests, carried out on a clay. The first, made using a vane with a length $L = 150\,mm$ and a diameter $D = 50\,mm$, yielded, at failure, a torque $T_1 = 57\,Nm$, while the second, undertaken at the same depth, using this time a vane characterised by the dimensions $L = D = 50\,mm$, registered at failure a torque $T_2 = 23.9\,Nm$. Let us, in the first instance, explore the clay anisotropy by calculating the strength ratio ξ using equation 5.72 for both tests. Thus:

- from test 1:

$$c_{uv} = \frac{10 \times 57 \times 10^{-3}}{\pi \times 5^2 \times 10^{-4} \times (5 \times 0.15 + 0.05\xi)} = \frac{5700}{58.9 + 3.93\xi}\ kN/m^2$$

- a similar analysis of test 2 yields:

$$c_{uv} = \frac{2390}{19.63 + 3.93\xi}\ kN/m^2$$

Whence a strength ratio $\xi = 2.22$, leading to the following components of shear strength:

$$c_{uv} = 84.3\,kN/m^2, \qquad c_{uh} = 187.1\,kN/m^2$$

The shear vane test can also be used to determine the sensitivity of a clay. Consider for example the case of a third test carried out on a clay that can be described as isotropic, using a vane with a diameter $D = 50\,mm$ and a length $L = 100\,mm$. At failure, a torque $T = 37\,Nm$ was measured, following which, the vane was rotated rapidly several times to remould the clay, yielding in the process a maximum torque $T = 9\,N.m$. In order to estimate the sensitivity of the clay, both undrained shear strengths corresponding to undisturbed and remoulded clays must be determined using equation 5.74. Hence, for the undisturbed clay:

$$c_u = \frac{10 \times 37 \times 10^{-3}}{11 \times \pi \times 5^3 \times 10^{-6}} = 85.6\,kN/m^2$$

Similarly for the remoulded soil:

$$c_u = \frac{85.6 \times 9}{37} = 20.8\,kN/m^2$$

Whence a sensitivity:

$$S = \frac{85.6}{20.8} = 4.1$$

According to the classification table in section 1.5, the clay is described as being sensitive.

Problems

5.1 A series of *drained* triaxial tests were undertaken on samples of a *normally consolidated* clay. At failure, the following results were measured in terms of radial and deviator stresses :

$\sigma'_3\ (kN/m^2)$	50	100	150	200
$q\ (kN/m^2)$	64	129	193	257

Plot these results as the variation of deviator stress *versus* the mean effective stress at failure, then estimate the effective angle of shearing resistance of the clay.

Ans : $\phi' \approx 23°$

5.2 Four shear box tests were performed on a clean dense sand, under four different values of effective vertical stresses. The corresponding peak shear stresses are as follows.

σ_1' (kN/m^2)	30	60	90	120
τ_{max} (kN/m^2)	28	63	96	120

Estimate the effective angle of friction at the critical density, knowing that, on average, the sand has a maximum dilation angle of $\nu_{max} = 10°$.

Ans : $\phi_c' \approx 35°$

5.3 A drained triaxial test is to be performed on a *76 mm* long, *38 mm* diameter sample of firm clay with a coefficient of vertical consolidation $c_v = 2.7\, m^2/y$.

(*a*) Knowing that the minimum shear speed that can be generated by the testing equipment is $v_{min} = 6 \times 10^{-4} mm/min$, assume the sample will only be drained at both ends then check if the requirement of a minimum excess porewater pressure dissipation $U = 97\%$ beyond an initial axial deformation $\varepsilon_1 = 0.003$ can be fulfilled.
(*b*) Will the shear speed be adequate to fulfil the same requirements were the sample to be drained radially as well as vertically ?
(*c*) Plot the corresponding graphs of porewater pressure dissipation *versus* axial strain.

Ans : (*a*) $v = 7.3 \times 10^{-5}\, mm/min < 6 \times 10^{-4}\, mm/min$ (inadequate)
(*b*) $v = 9.83 \times 10^{-4} mm/min$ (adequate)

5.4 Two samples of a normally consolidated clay were consolidated in triaxial cells under different cell pressures σ_3, then sheared under undrained conditions by gradually increasing the deviator stress q while keeping σ_3 constant. The values at failure of the measured deviator stress q_f and the porewater pressure u_f were as follows.

$\sigma_3\,(kN/m^2)$	$q_f(kN/m^2)$	$u_f\,(kN/m^2)$
200	120	96
300	203	123

(*a*) Determine the shear strength parameters of the clay c' and ϕ'.
(*b*) Calculate the shear and effective normal stresses at failure for both tests.

Ans: (*a*) $c' = 0, \quad \phi' = 21.4^\circ$

(*b*)

$\sigma_3\,(kN/m^2)$	$\tau_f(kN/m^2)$	$\sigma'_f\,(kN/m^2)$
200	55.9	142.5
300	94.5	241.1

5.5 Two undrained triaxial tests were carefully undertaken on two clay samples *A* and *B*, which were consolidated then sheared under the respective (constant) cell pressures $\sigma_{3A} = 80\,kN/m^2$ and $\sigma_{3B} = 250\,kN/m^2$. The results were measured in terms of deviator stress and porewater pressure as follows.

sample *A*:

$q\,(kN/m^2)$	0	80	112	138	145
$u\,(kN/m^2)$	0	16	18	5	-12

sample *B*:

$q\,(kN/m^2)$	0	100	150	180	190	195
$u\,(kN/m^2)$	0	30	55	88	100	105

(*a*) Plot the effective and total stress paths in (q,p') and (q,p) spaces.

(*b*) Calculate the porewater pressure parameter at failure A_f and comment on the state of the clay.

Ans: (*a*) *Sample A:* $A_f = -0.08$
(*b*) *Sample B:* $A_f = 0.54$

References

Azizi, F. and Josseaume, H. (1988) *Loi de comportement des sols raides: détermination de la courbe d'état limite de l'argile verte de Romainville.* Rapport des Laboratoires des Ponts et Chaussées, Série Géotechnique, GT-33.

Berre, T. (1981) *Triaxial Testing at the Norwegian Geotechnical Institute.* N.G.I. Publication 134, Oslo.

Bishop, A. W. and Henkel, D. J. (1962) *The Measurement of Soil Properties in the Triaxial Test, 2nd edn.* Arnold, London. In Soils and Rocks. ICE, London, pp. 251–264.

Bjerrum, L. (1973) *Problems of soil mechanics and construction on soft clays and structurally unstable soils (collapsible, expansive and others).* Proceedings of the 8th I.C.S.M.F.E, Moscow, pp. 111–160.

Bjerrum, L. and Kenny, T. C. (1967) *Effect of structure on the shear behaviour of normally consolidated quick clays.* Proceedings of the Oslo Conference on Geotechnics, Vol. 2, pp. 19–27.

Gibson, R. E. and Henkel, D. J. (1954) *Influence of duration of tests at constant rate of strain on measured drained strength.* Géotechnique (4), pp. 6–15.

Henkel, D. J. (1957) *Investigation of two long term failures in London clay slopes at Wood Green and Northolt.* Proceedings of the 4th I.C.S.M.F.E, London, pp. 315–320.

Jaky, J. (1944) *The coefficient of earth pressure at rest.* Journal of the Society of Hungarian Architects and Engineers, 78 (22), pp. 355–358.

Josseaume, H. and Azizi, F. (1991) *Détermination expérimentale de la courbe d'état limite d'une argile raide très plastique: l'argile verte du Sannoisien.* Revue Française de Géotechnique, 54, pp. 13–25.

Lupini, J. F., Skinner, A. E. and Vaughan, P. R. (1981) *The drained residual strength of cohesive soils.* Géotechnique, 31 (2), pp. 181–213.

Mayne, P. W. and Kulhawy, F. H. (1982) *Relationships in soil.* ASCE Journal, 108, GT6, pp. 851–872.

Palladino, D. J. and Peck, R. B. (1972) *Slope failures in an overconsolidated clay in Seattle, Washington.* Géotechnique, 22 (4), pp. 563–595.

Simpson, B., Calabresi, G., Sommer, H. and Wallays, M. (1979) *Design parameters for stiff clays. General report.* Proceedings of the 7th European Conference on Soil Mechanics and Foundation Engineering, Brighton, Vol. 5, pp. 91–125.

Skempton, A. W. (1954) *The pore pressure coefficients A and B.* Géotechnique, 4 (4), pp. 143–147.

Skempton, A. W. (1957) *Discussion on the Planning and Design of the New Hong Kong Airport.* Proceedings of the I.C.E, Vol. 7, pp. 305–307.

Skempton, A. W. (1964) *Long term satbility of clay slopes.* Géotechnique, 14 (2), pp. 77–102.

Skempton, A. W. and Bjerrum, L. (1957) *A contribution to the settlement analysis of foundations on clay.* Géotechnique, 7 (4), pp. 168–178.
Skempton, A. W., and La Rochelle, P. (1965). *The Bradwell slip: a short term failure in London clay.* Géotechnique, 15 (3), pp. 221–242.
Wroth, C. P. (1984) *The interpretation of in situ tests.* 24th Rankine Lecture. Géotechnique, 34 (4), pp. 449–489.

CHAPTER 6

Modelling of soil behaviour: limit and critical states

6.1 Introduction

This chapter aims at linking different aspects of soil behaviour such as consolidation, shear strength, and elastic and plastic deformations through the use of a unique *predictive model.* The model combines elements of *critical state theory* that are particularly suited to normally consolidated or lightly overconsolidated clays, and Hvorslev *limit state* which applies to heavily overconsolidated clays. Such a combination yields a *state boundary surface*, the details of which are given in the following sections. It is however important to mention at this stage that the model is based on the assumptions that the soil is *isotropic* and *saturated,* and that its behaviour is represented through *global* (*i.e.* average) strain values induced by a set of *effective stresses*.

The *state* of a saturated soil, which depends entirely on the magnitude of the applied effective stresses and on the density of the soil, can be defined using the three following parameters:

the specific volume : $$v = 1 + e \qquad (6.1)$$

the deviator stress : $$q = \sigma_1' - \sigma_3' \qquad (6.2)$$

the mean effective stress : $$p' = \frac{1}{3}\left(\sigma_1' + 2\sigma_3'\right) \qquad (6.3)$$

Since water has no shear strength, the deviator stress can be expressed *equally* in terms of total or effective stresses. Also, both equations 6.2 and 6.3 correspond to the axisymmetric stress conditions of figure 6.1 (*i.e.* the conditions of a triaxial test whereby $\sigma_2' = \sigma_3'$).

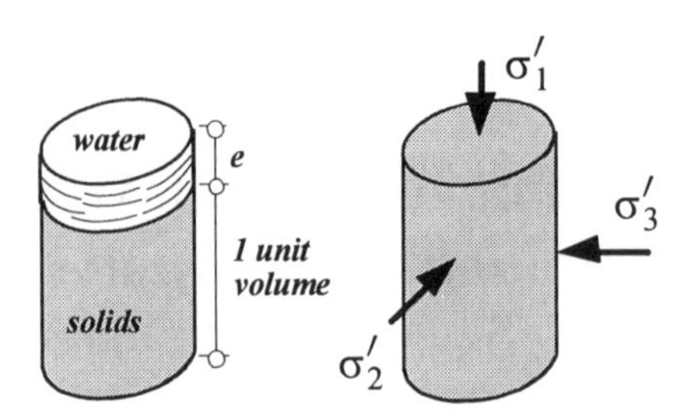

Figure 6.1: Triaxial stress field applied to a saturated soil.

Making use of the effective stress principle $\sigma' = \sigma - u$, and combining the two equations then rearranging yields:

$$p' + u = \sigma_3 + q/3 \qquad (6.4)$$

Accordingly, a soil sample consolidated under a set of effective stresses (σ'_1, $\sigma'_2 = \sigma'_3$) in a triaxial cell, then sheared under a constant cell pressure (σ_3 = *constant*) follows one of two possible stress paths depending on drainage conditions as follows.

(1) If shear occurs under *drained* conditions, then the excess porewater pressure is $u = 0$ and, according to equation 6.4, the *effective stress path* is linear with a slope of 1/3.
(2) If, on the other hand shear takes place under *undrained* conditions, then the excess porepressure u is no longer zero, and equation 6.4 indicates in this case that the *total stress path* is linear with a slope of 1/3 as depicted in figure 6.2.

Although the full implications of equation 6.4, as well as the detailed nature of different stress paths will be analysed in due course, it is important to specify a third stress path along which both principal effective stresses σ'_1 and σ'_3 vary at a *constant ratio:*

$$\sigma'_3/\sigma'_1 = K = constant \qquad (6.5)$$

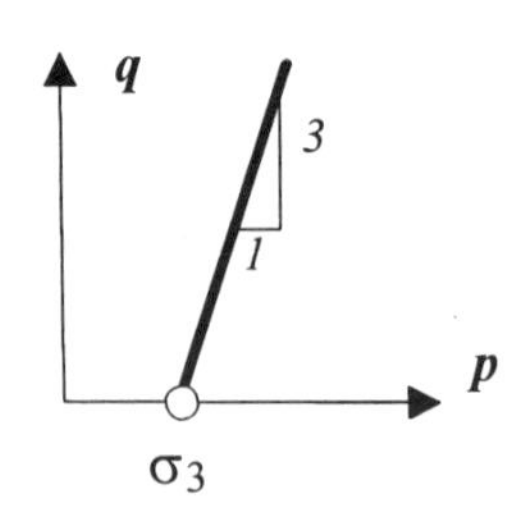

Figure 6.2: total stress path for equation 6.4.

In this case, it is easy to see that both equations 6.2 and 6.3 reduce to:

$$q = \sigma_1'(1-K), \qquad p' = \sigma_1' \frac{(1+2K)}{3}$$

thus yielding the following relationship:

$$\frac{q}{p'} = \eta = \frac{3(1-K)}{1+2K} \tag{6.6}$$

The corresponding effective stress path is shown in figure 6.3.

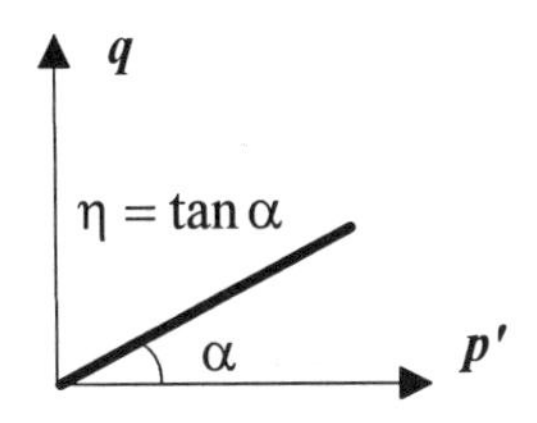

Figure 6.3: Effective stress path corresponding to equation 6.6.

6.2 Critical state theory: the modified Cam-clay model

The mathematical model known as *Cam-clay*, developed at the University of Cambridge (see for instance Schofield and Wroth (1968)), is based on the following assumptions.

- The soil is *isotropic.*
- The soil behaviour is *elasto-plastic.*
- The soil deforms as a *continuum.*
- The soil behaviour is *not affected by creep.*

The earlier version of *Cam-clay* was *modified* by Roscoe and Burland, (1968), to take into account the effect of the plastic component of the volumetric strain on the work dissipated per unit volume.

The critical state *concept* is based on the consideration that, when sheared, a soil sample will eventually reach a state at which (large) shear distortions

ε_s occur without any further changes in p', q or v. In other words, the onset of a critical state implies that:

$$\frac{\partial p'}{\partial \varepsilon_s} = \frac{\partial q}{\partial \varepsilon_s} = \frac{\partial v}{\partial \varepsilon_s} = 0$$

Such a *concept* has been validated experimentally, in that for a given soil, all critical states *do* form a unique line referred to as *the critical state line* (*CSL*) with the following equations in the space (p', q, v):

$$q = Mp' \qquad (6.7)$$

$$v = \Gamma - \lambda \ln p' \qquad (6.8)$$

where M, Γ and λ are soil constants. In particular, Γ corresponds to the specific volume when $p' = 1$ *unit*. Moreover, the plastic compression under isotropic stress conditions (*i.e.* $q = 0$) of a *normally consolidated soil* can be accurately represented by a unique line called *the normal consolidation line* (*NCL*), which has the following equations in (p', q, v) space :

$$q = 0 \qquad (6.9a)$$

$$v = N - \lambda \ln p' \qquad (6.9b)$$

with N corresponding to the specific volume when $p' = 1$ *unit pressure.*

If, at some point on the *NCL*, the soil were unloaded, then it will follow a path known as the *swelling line*, represented by the equations:

$$q = 0 \qquad (6.10a)$$

$$v = v_\kappa - \kappa \ln p' \qquad (6.10b)$$

Using these three lines and their respective equations, an early picture of the *state* of a soil emerges (see figure 6.4). In particular, notice that:

- a soil whose state lies on the *NCL* is *normally consolidated*,
- a soil with states along the swelling line is *overconsolidated*.

On the whole, a clay in a state within the shaded area in figure 6.4 (*i.e. between* the *CSL* and the *NCL*) is *lightly overconsolidated.* This shaded area is sometimes referred to as the *wet* side of the critical state, the reason being that a clay with a state within that area will generate positive excess porewater pressure when subjected to undrained loading, expelling water once the drainage is allowed to take place, thus appearing 'wet' in the process.

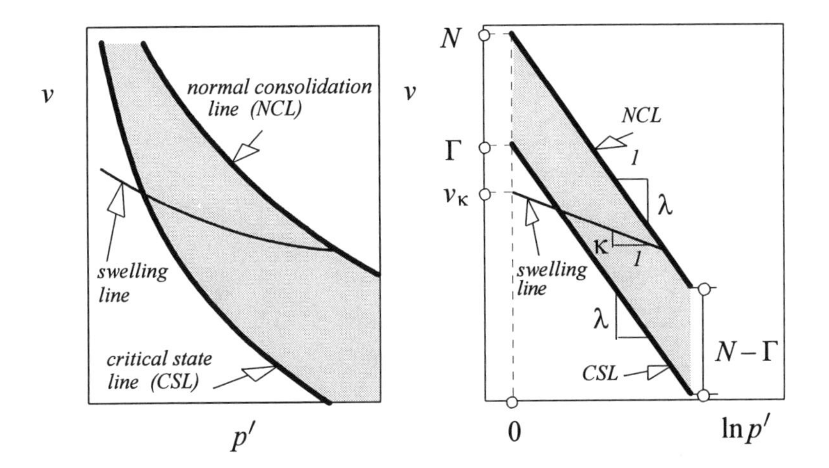

Figure 6.4: NCL, CSL and swelling lines.

Prior to developing the mathematical formalism, it is worth emphasising that the critical state theory is best suited to *normally consolidated* or *lightly overconsolidated clays* (refer to Roscoe and Burland (1968)). As such, the following formulation will be limited to the shaded area in figure 6.4.

Referring to the same figure, it is seen that both *CSL* and *NCL* plot parallel to each other in the space (v, $\ln p'$). Furthermore, it can be shown that, in the case of the modified Cam-clay model, the spacing between the two lines is (Wood, 1992):

$$N-\Gamma=(\lambda-\kappa)\ln 2 \tag{6.11}$$

substituting for N in equation 6.9*b*, the *NCL* equation then becomes:

$$v=\Gamma+(\lambda-\kappa)\ln 2-\lambda\ln p' \tag{6.12}$$

Although the *CSL* and *NCL* are defined in terms of the three (known) soil constants Γ, λ and κ, the nature of the stress path, however, remains partially undefined in the (q, p') space. Consider for instance the case of figure 6.5 in which an isotropic clay sample is sheared undrained from a normally consolidated state (point A on the *NCL*) to failure (point B on the *CSL*). Knowing the initial consolidation pressure p'_A, both specific volumes v_A and v_B can be calculated from equation 6.12 and 6.9*b* respectively. Similarly, the stresses at failure p'_B and q_B are found from equations 6.8 and 6.7, in that order. Nevertheless, it is seen in the figure that, while the specific volume during the *undrained* shear remains *constant*, both mean effective stress and deviator stress *vary* throughout the test, so that at point C, q_C and p'_C cannot be evaluated at this stage.

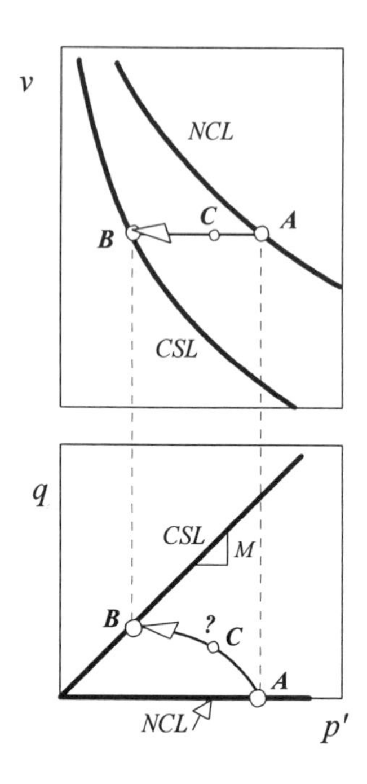

Figure 6.5: Stress path for undrained loading.

Because the clay sample is *normally consolidated*, all ensuing deformations will be *elastoplastic* since a *purely elastic* strain component occurs only in conjunction with *overconsolidated* samples that were unloaded along a swelling line prior to shearing. Consequently, the stress path lies entirely on the *boundary surface* which is yet to be defined.

To define the boundary surface, Roscoe and Burland (1968) adopted the following expression for the work dissipated per unit volume:

$$\delta W = p'\left[\left(\delta\varepsilon_v^p\right)^2 + \left(M\delta\varepsilon_s^p\right)^2\right]^{1/2} \tag{6.13}$$

Knowing that the dissipated work done by a load (q, p') is such that:

$$\delta W = p'\,\delta\varepsilon_v^p + q\,\delta\varepsilon_s^p \tag{6.14}$$

they then derived the *associated plastic flow* by combining these two equations:

$$\frac{\delta\varepsilon_v^p}{\delta\varepsilon_s^p} = \frac{M^2 - \eta^2}{2\eta} \tag{6.15}$$

the quantities $\delta\varepsilon_v^p$ and $\delta\varepsilon_s^p$ being the increments of plastic volumetric strain and plastic shear strain respectively, and $\eta = q/p'$. Both equations 6.13 and 6.15 will be discussed in detail in conjunction with the calculation of plastic strains that follow.

Next, Roscoe and Burland have combined the plastic flow equation 6.15 and the equations of plastic strains to derive the following expression of the *state boundary surface* (delimited by the *NCL* and the *CSL*):

$$\frac{p'}{p'_e} = \left[\frac{M^2}{M^2 + \eta^2}\right]^{\left(\frac{\lambda-\kappa}{\lambda}\right)} \tag{6.16}$$

p'_e being the Hvorslev *equivalent pressure* defined as the mean effective pressure on the *NCL* corresponding to the specific volume at failure. But for an undrained shear test, the specific volume remains constant (refer to figure 6.5), and thus the equivalent pressure can be found simply by rearranging equation 6.12:

$$p'_e = \exp\left[\frac{\Gamma + (\lambda - \kappa)\ln 2 - v}{\lambda}\right] \tag{6.17}$$

Substituting for p'_e into equation 6.16 and rearranging, yields the equation of the state boundary surface, also known as the *Roscoe surface*:

$$q = Mp'\left[2\exp\left(\frac{\Gamma - v - \lambda\ln p'}{\lambda - \kappa}\right) - 1\right]^{1/2} \tag{6.18a}$$

As will emerge, it is useful to express the *same* equation as follows:

$$v = \Gamma - \lambda\ln p' - (\lambda - \kappa)\ln\left[\left(\frac{q}{Mp'\sqrt{2}}\right)^2 + \frac{1}{2}\right] \tag{6.18b}$$

Equation 6.18 is that of the *Roscoe surface* (or the state boundary surface), depicted in figure 6.6, on which all elastoplastic deformations occur.

Now consider what happens if, instead of being sheared from a normally consolidated state, the clay is first allowed to become overconsolidated by unloading it along an elastic swelling line. In that case, the onset of shear will generate *elastic strains* while the state of the clay remains *inside* the Roscoe surface, irrespective of the stress path followed during shear.

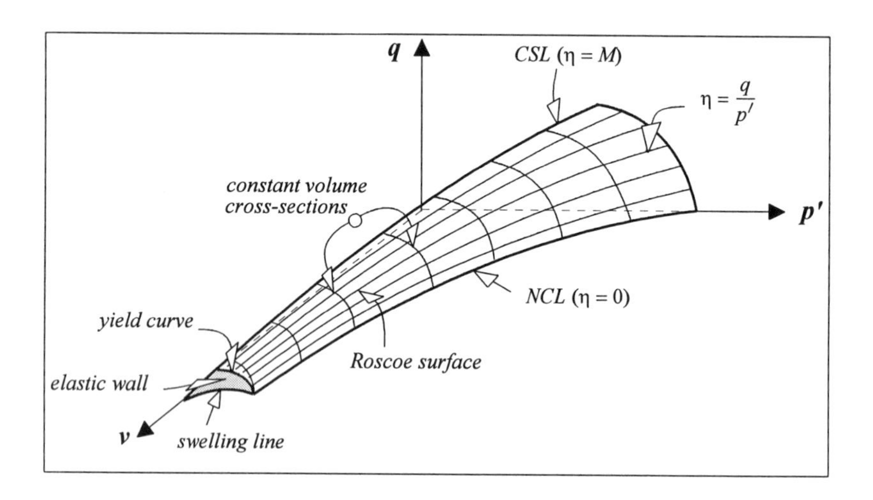

Figure 6.6: State boundary surface as per equation 6.18b.

As soon as the Roscoe surface is reached, the deformations become plastic. The *vertical surface* with a swelling line at its base and along which the deformations are purely elastic is known as the *elastic wall*. Because the base of this wall corresponds to a swelling line throughout which the specific volume is *not* constant, the curve resulting from the intersection of the elastic wall with the Roscoe surface, referred to as *the yield curve*, cannot be represented by equation 6.18. It is however easy to establish its appropriate equations knowing that the elastic wall is a vertical surface with, at its base, a swelling line whose equation has already been established (equation 6.10*b*):

$$v = v_\kappa - \kappa \ln p'$$

On the other hand, the *NCL* equation 6.12 is:

$$v = \Gamma + (\lambda - \kappa)\ \ln 2 - \lambda \ln p'$$

and according to figure 6.7, both equations become identical at point *A* where the mean effective pressure is $p' = p_o$.

Equating the two equations yields the intercept v_κ:

$$v_\kappa = \Gamma + (\lambda - \kappa)\ \ln 2 - (\lambda - \kappa)\ \ln p_o$$

Inserting this quantity into the swelling line equation and rearranging results in the expression of the yield curve equation in the (v, p') space:

$$v = \Gamma + (\lambda - \kappa)\ \ln\left(\frac{2}{p_o}\right) - \kappa \ln p' \tag{6.19}$$

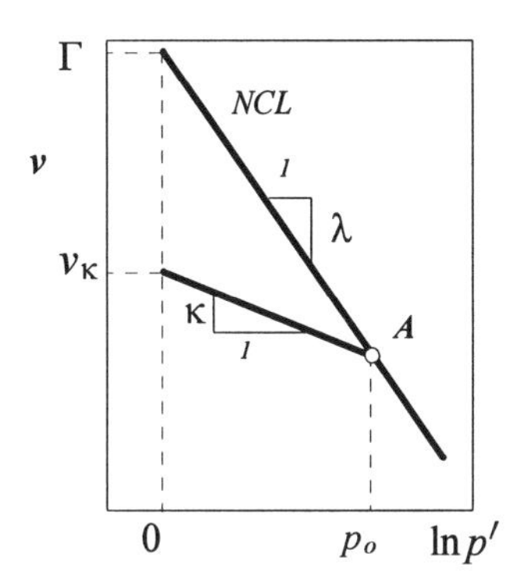

Figure 6.7: Normal consolidation and elastic lines.

In (q, p') space, the yield curve results from intersecting the elastic wall with the Roscoe surface. Thus equating equations 6.19 and 6.18*b* then rearranging gives:

$$\frac{p'}{p_o} = \frac{M^2}{M^2 + \eta^2} \tag{6.20}$$

$\eta = q/p'$. Hence, the yield curve is entirely defined by equations 6.19 and 6.20 in which the quantity p_o represents the normal consolidation pressure of the clay sample that determines the size of the yield curve. Note that p_o automatically becomes an overconsolidation pressure as soon as the clay is unloaded along a swelling line. Also, equation 6.20 is that of an *ellipse* with one axis coincident with the p' axis and, most importantly, the point where the yield curve intersects the *CSL* on which $\eta = M$ (point *C* in figure 6.9 for instance) has a mean effective stress $p' = p_o/2$.

6.3 Hvorslev limit state and the complete state boundary surface

The elegant critical state theory is best suited to 'wet' clays meaning normally consolidated to lightly overconsolidated clays. Heavily overconsolidated clays are characterised by a brittle behaviour and their states are on the dry side of the critical state line as depicted in figure 6.8. The loading of such clays can induce negative porewater pressures inside the sample which will then increase in volume were drainage to be allowed, appearing 'dry' in the process.

This type of behaviour is best represented by Hvorslev's shear law (Hvorslev, 1937):

$$\frac{q}{p'_e} = g + h\frac{p'}{p'_e} \tag{6.21}$$

where the equivalent pressure p'_e has been already defined through equation 6.17, and the slope h is a soil constant. The intercept g can easily be determined since, as shown in figure 6.9, point *C* is where both the Hvorslev equation and the *CSL* equation become identical. Equating equations 6.21 to 6.8 yields:

$$g = (M - h)\exp\left[\frac{(\kappa - \lambda)\ln 2}{\lambda}\right] \tag{6.22}$$

Substituting for g from equation 6.22 and for p'_e from equation 6.17 into equation 6.21, it follows that:

$$q = (M - h)\exp\left(\frac{\Gamma - v}{\lambda}\right) + hp' \tag{6.23a}$$

The same equation can be expressed differently:

$$v = \Gamma - \lambda \ln\left(\frac{q - hp'}{M - h}\right) \tag{6.23b}$$

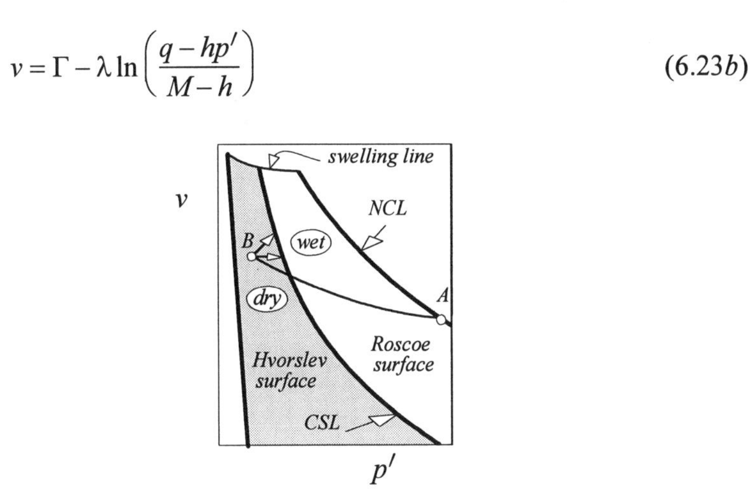

Figure 6.8: Dry and wet sides of the CSL.

Equations 6.23 define the *Hvorslev boundary surface* to the left of the *CSL* as sketched in figure 6.9. This surface is limited on its left by the *tension cut-off* since soils cannot withstand any tensile pressure, meaning that the minor stress σ_3' cannot fall below zero. Hence at the limit, when $\sigma_3' = 0$:

$$q = 3p' \tag{6.24}$$

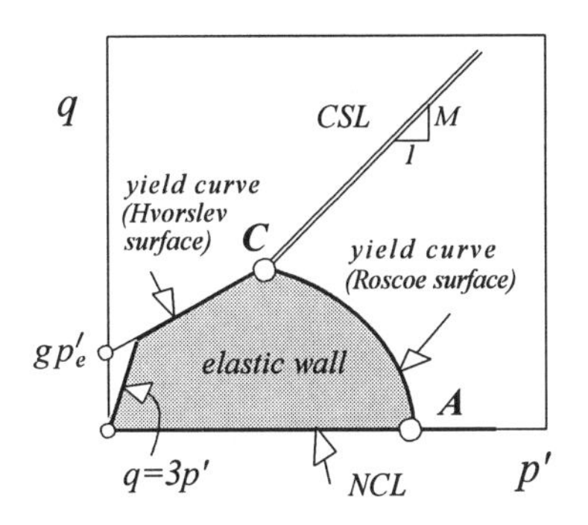

Figure 6.9: Roscoe and Hvorslev surfaces.

As for the yield curve to the left of the *CSL*, it can be established in a way similar to that used previously in conjunction with the Roscoe surface. In fact, the yield curve results from the intersection of the (vertical) elastic wall and the Hvorslev surface equation. Fortunately, the yield curve equation 6.19 established earlier is still valid on the left of the *CSL*. Thus, the second equation in (q, p') space can be obtained by equating the Hvorslev surface equation 6.23*b* to that of the elastic wall (equation 6.19), resulting in the following:

$$q = hp' + \frac{(M-h)}{2} p_o \left(\frac{2p'}{p_o}\right)^{\kappa/\lambda} \tag{6.25}$$

Both equations 6.19 and 6.25 represent the *yield curve* to the left of the *CSL*.

The *complete state boundary surface*, with its three components, namely the *Roscoe surface*, the *Hvorslev surface* and the *tension cut-off* can now be drawn, and figure 6.10 depicts a 3-D sketch of such a surface.

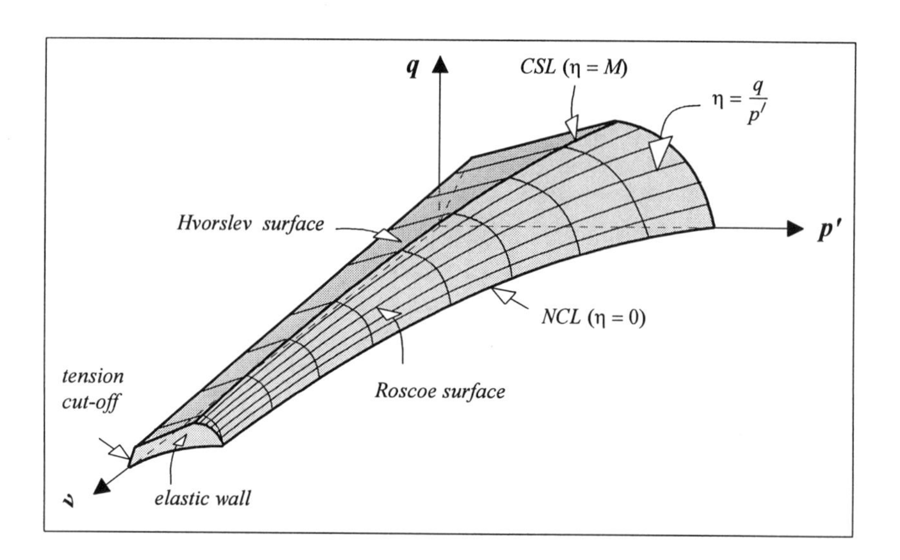

Figure 6.10: Complete state boundary surface.

It should be borne in mind that:
- the model applies to *isotropic saturated clays;*
- the Roscoe surface is used in conjunction with *normally consolidated to lightly overconsolidated clays*;
- the Hvorslev surface is used for *heavily overconsolidated clays*.

6.4 Stress paths within and on the state boundary surface during a triaxial test

Now that the mathematical formalism has been established, let us have a thorough examination of the complete state boundary surface (*SBS*) through the use of a numerical example.

Consider a sample of an isotropic saturated clay, consolidated in a triaxial cell under an isotropic effective mean pressure $p'_A = 400\, kN/m^2$. The clay is characterised by the following constants:

$$\lambda = 0.095, \quad \kappa = 0.035, \quad \Gamma = 2.0, \quad M = 0.9, \quad h = 0.75$$

Once consolidation is complete, it is proposed to determine the precise nature of different stress paths assuming that the sample is:

(*1*) sheared under undrained conditions until the occurrence of failure;
(*2*) sheared under drained conditions until failure;
(*3*) unloaded along a swelling line from $p'_A = 400\, kN/m^2$ to $p'_B = 320\, kN/m^2$, allowed to consolidate then:
- sheared under undrained conditions until failure;
- sheared under drained conditions until failure;

(*4*) unloaded along a swelling line from $p'_A = 400\, kN/m^2$ to $p'_B = 100\, kN/m^2$, allowed to consolidate then:
- subjected to undrained shear until failure;
- subjected to drained shear until failure.

6.4.1 Normally consolidated clay: undrained shear

In the first instance, the clay sample is *normally consolidated* under the initial isotropic pressure $p_o = p'_A = 400\, kN/m^2$, $q_A = 0$. Therefore, the state of the sample is *on* the *NCL* whose equation 6.12 can be used to calculate the corresponding specific volume at point A :

$$v_A = \Gamma + (\lambda - \kappa)\ln 2 - \lambda \ln p'_A$$

$$= 2 + (0.095 - 0.035)\ln 2 - 0.095 \ln 400 = 1.472$$

The sample is sheared under *undrained conditions*, so that failure occurs *on* the *CSL* at point *D* at a *constant specific volume* (since no drainage is allowed during shear). Whence:

$$v_D = v_A = 1.472$$

The stresses at failure can now be calculated from the *CSL* equations 6.8 and 6.7 respectively:

$$v_D = \Gamma - \lambda \ln p'_D \quad \Rightarrow \quad p'_D = \exp\left(\frac{\Gamma - v_D}{\lambda}\right) = \exp\left(\frac{2-1.472}{0.095}\right) = 259.3\, kN/m^2$$

$$q_D = Mp'_D = 0.9 \times 259.3 = 233.3\, kN/m^2$$

The entire stress path is sketched in figure 6.11 which indicates that the sample undergoes an undrained shear under constant volume from *A* (on the *NCL*) to *D* on the *CSL*.

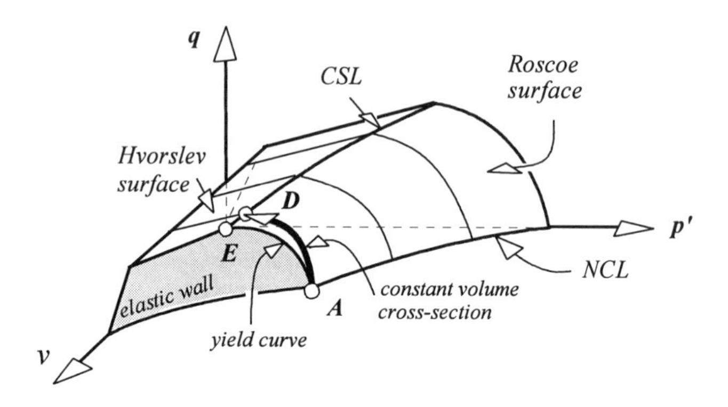

Figure 6.11: Normally consolidated clay; 3-D undrained stress path.

Also, the figure shows that the clay is on the *Roscoe surface* from the onset of shear and, therefore, the stress path *AD* in $(q,\ p')$ space can be determined using the Roscoe surface equation 6.18*a*. Because of the *undrained* shear, the volume remains constant, hence the projection in a straight line from *A* to *D* in $(v,\ p')$ space. If the sample were unloaded from *A* prior to shear, then it would follow a swelling line which forms the basis of the elastic wall. The curve *AE* represents the intersection between the

Roscoe surface and the elastic wall, and can be calculated using the yield curve equations 6.19 and 6.20. Thus, in (v, p') space, AE is calculated as follows:

$$v = \Gamma + (\lambda - \kappa)\ln(2/p_o) - \kappa \ln p'$$

$$= 2 + (0.095 - 0.035)\ln(2/400) - 0.035 \ln p'$$

$$= 1.628 - 0.035 \ln p'$$

Now, equation 6.20 is used to calculate the stress path AE in (q, p') space. Knowing that $\eta = q/p'$, the equation can easily be rearranged as follows:

$$\frac{p'}{p_o} = \frac{M^2}{M^2 + \eta^2} \quad \Rightarrow \quad q = Mp'\left(\frac{p_o}{p'} - 1\right)^{1/2}$$

The corresponding stress path is depicted in figure 6.12.

With regard to the *excess porewater pressure* generated during the undrained shear, the precise nature of its distribution can be calculated from equation 6.4 in which the quantity σ_3 is replaced by $p'_A = 400\, kN/m^2$. Rearranging this equation:

$$u = (400 - p') + \frac{1}{3}q$$

Accordingly, the porewater pressure can be calculated at any stage of the undrained shear, and in particular at points A and D:

$$u_A = (400 - p'_A) + \frac{1}{3}q_A = 0$$

$$u_D = (400 - p'_D) + \frac{q_D}{3} = (400 - 259.3) + \frac{233.3}{3}$$

$$= 218.5\, kN/m^2$$

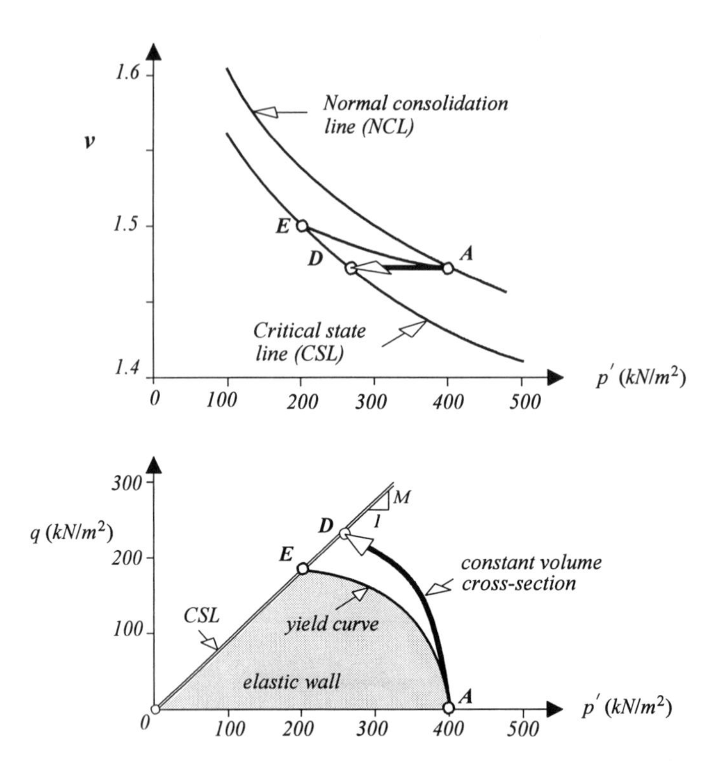

Figure 6.12: Undrained stress path for normally consolidated clay.

The variation of the porewater pressure during shear is illustrated by the shaded area in figure 6.13.

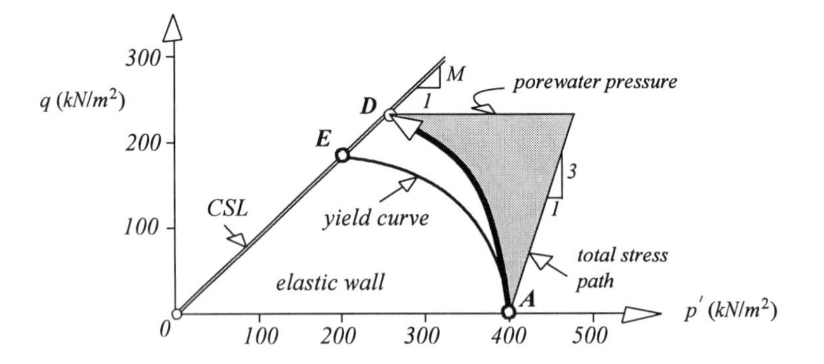

Figure 6.13: Variation of the porewater pressure during undrained loading.

6.4.2 Normally consolidated clay: drained shear

Now the clay is being sheared under *drained conditions* from point *A*, therefore, the initial *state* of the sample is identical to that calculated in the previous case: $v_A = 1.472$, $p'_A = 400\,kN/m^2$, $q_A = 0$.

When the clay fails at point *D*, the *CSL* equation 6.7 applies: $q_D = Mp'_D$. Moreover, because of the drained shear, there is no excess water pressure ($u = 0$) and equation 6.4 reduces to:

$$p'_D = p'_A + \frac{q_D}{3}$$

Combining this equation with that of the *CSL* yields the mean effective stress at failure:

$$p'_D = \frac{3p'_A}{3-M} = \frac{3\times 400}{3-0.9} = 571.4\ kN/m^2$$

and

$$q_D = Mp'_D = 0.9\times 571.4 = 514.3\,kN/m^2$$

the corresponding specific volume is calculated from the *CSL* equation 6.8:

$$v_D = \Gamma - \lambda \ln p'_D = 2 - 0.095\ln 571.4 = 1.397$$

The 3-D nature of the stress path for this drained loading is illustrated in figure 6.14 which shows that the entire stress path lies on the Roscoe surface from the onset of shear.

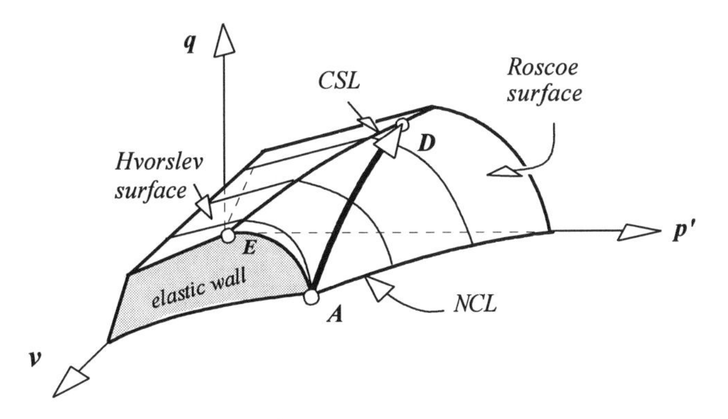

Figure 6.14: Normally consolidated clay; 3-D drained stress path.

Consequently, the Roscoe surface equation 6.18*b*:

$$v = \Gamma - \lambda \ln p' - \ln\left[\left(\frac{q}{Mp'\sqrt{2}}\right)^2 + \frac{1}{2}\right]^{(\lambda-\kappa)}$$

is used in conjunction with the effective stress path equation 6.4, which is rearranged as follows:

$$q = 3(p' - p'_A) = 3(p' - 400)$$

in order to calculate precisely the stress path *AD* represented in figure 6.15. Notice that since there is no generation of water pressure during the drained shear, both effective and total stress paths are *identical.*

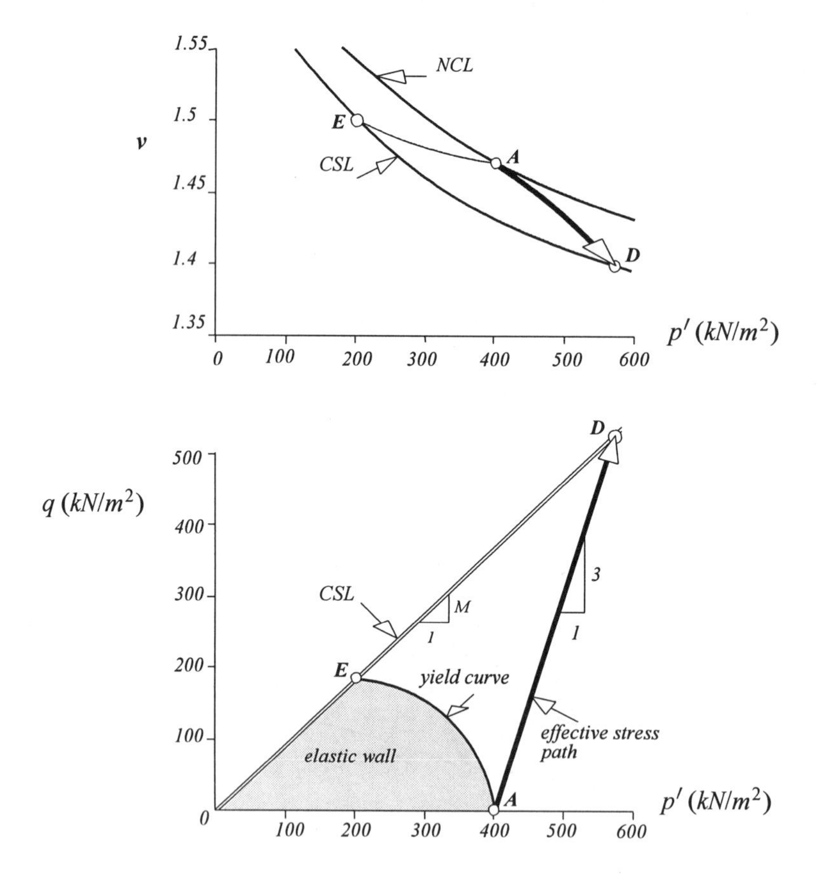

Figure 6.15: Drained stress path for a normally consolidated clay.

6.4.3 Lightly overconsolidated clay: undrained shear

In this case the clay sample is allowed to increase in volume, becoming in the process lightly overconsolidated; the overconsolidation ratio (*OCR*) being $OCR = 400/320 = 1.25$.

The initial state having already been determined previously, it follows that:

$$v_A = 1.472, \quad p'_A = 400\,kN/m^2, \quad q_A = 0$$

Once the sample is unloaded then consolidated under the mean effective pressure $p'_B = 320\,kN/m^2$, the specific volume is calculated from the swelling line equation 6.10*b*:

$$v_B = v_A + \kappa \ln \frac{p'_A}{p'_B} = 1.472 + 0.035 \ln \frac{400}{320} = 1.48$$

Next, the sample is sheared under *undrained conditions*, implying that the specific volume remains constant; therefore at failure $v_D = v_B = 1.48$.

The stresses at failure can thereafter be determined from the *CSL* equations 6.8 and 6.7 respectively:

$$v_D = \Gamma - \lambda \ln p'_D \quad \Rightarrow \quad p'_D = \exp\left(\frac{\Gamma - v_D}{\lambda}\right) = \exp\left(\frac{2-1.48}{0.095}\right) = 238.3\,kN/m^2$$

$$q_D = Mp'_D = 0.9 \times 238.3 = 214.5\,kN/m^2$$

The complete 3-D stress path is depicted in figure 6.16 which shows that the sample follows a path from *A* (on the *NCL*) to *B* (on a swelling line), then from *B* to *C* within the elastic wall, and finally from *C* to *D* on the Roscoe surface; point *D* where failure occurs being on the *CSL*.

While the precise positions of points *A, B* and *D* have been calculated, that of point *C* is yet to be determined. The shape of the stress path from *B* to *C* is of particular interest. Since it is within the elastic wall, the corresponding strains *must* be elastic. However, according to the elastic wall equation 6.19:

$$v = \Gamma + (\lambda - \kappa)\ln\left(\frac{2}{p'_o}\right) - \kappa \ln p'$$

hence:

$$dv = -\kappa\frac{dp'}{p'} \tag{6.26}$$

so that for *undrained* shear, the specific volume remains *constant*, in other words $dv = 0$. Thus, the stress path from B to C is such that $v_C = v_B = 1.48$, and

$$dv = -\kappa\frac{dp'}{p'} = 0 \quad \Rightarrow dp' = 0$$

This indicates that, under undrained conditions, the pressure p' remains constant inside the elastic wall (*i.e.* $p'_C = p'_B = 320\, kN/m^2$), whence the vertical shape of the stress path from B to C in figure 6.16.

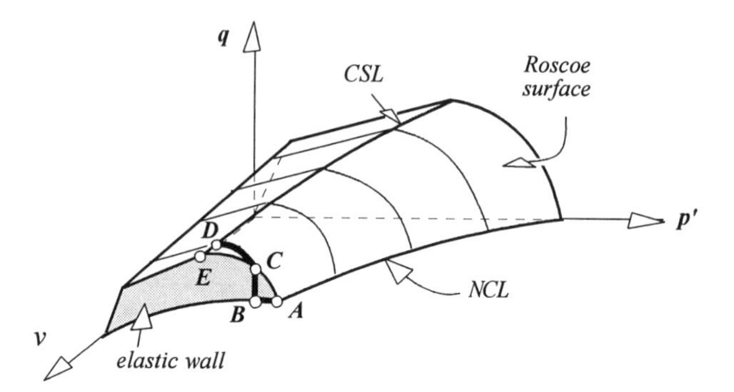

Figure 6.16 : Lightly overconsolidated clay; 3-D undrained stress path.

It is now clear from the figure that point C is where the stress path meets the yield curve, and because both the specific volume v_C and the mean effective stress p'_C are known, the value of the deviator stress q_C is obtained by rearranging the yield curve equation 6.20 :

$$q_C = Mp'_C\left(\frac{p'_A}{p'_C} - 1\right)^{1/2} = 0.9 \times 320\left(\frac{400}{320} - 1\right)^{1/2} = 144\, kN/m^2$$

Notice that because C is *also* on the Roscoe surface, it is easy to check that equation 6.18*b* yields a similar result.

Next, the stress path from C to D is calculated from the Roscoe surface equation 6.18*b* in which the volume v takes the constant value $v = 1.48$:

$$q = Mp'\left[2\exp\left(\frac{\Gamma - v - \lambda \ln p'}{\lambda - \kappa}\right) - 1\right]^{1/2}$$

$$= 0.9p'\left[2\exp\left(\frac{0.52 - 0.095\ln p'}{0.06}\right) - 1\right]^{1/2}$$

Finally, the yield curve between A and E is calculated from both equations 6.19 and 6.20 which, once rearranged, become:

$$v = \Gamma + (\lambda -)\ \ln\frac{2}{p'_A} - \kappa\ \ln p' \ = 1.682 - 0.035\ln p'$$

and

$$q = Mp'\left(\frac{p'_A}{p'} - 1\right)^{1/2} = 0.9p'\left(\frac{400}{p'} - 1\right)^{1/2}$$

As for the position of point E, it corresponds to the intersection of the CSL and the yield curve where:

$$p'_E = \frac{p'_A}{2} = \frac{400}{2} = 200\ kN/m^2$$

The CSL equations 6.7 and 6.8 can then be used to calculate q_E and v_E:

$$q_E = Mp'_E = 0.9 \times 200 = 180\ kN/m^2$$

$$v_E = \Gamma - \lambda\ln p'_E = 2 - 0.095\ln 200 = 1.497$$

The entire corresponding stress path is depicted in figure 6.17.

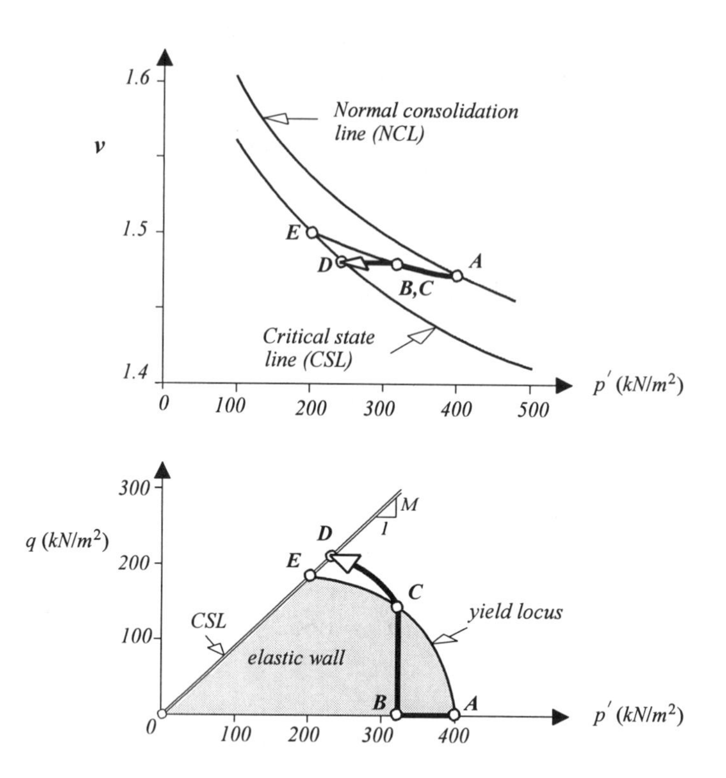

Figure 6.17: Undrained stress path for lightly overconsolidated clay.

The porewater pressure generated during the undrained shear can easily be calculated at any stage using equation 6.4 in which the quantity σ_3 is replaced by $p'_B = 320\,kN/m^2$. Whence the equation in a rearranged form is:

$$u = (320 - p') + q/3$$

so that the porewater pressure generated at point C, for instance, is:

$$u_C = \left(320 - p'_C\right) + \frac{q_C}{3} = \frac{144}{3} = 48\,kN/m^2$$

also, at point D, the porewater pressure at failure is:

$$u_D = (320 - 238.3) + \frac{214.5}{3} = 153.2\,kN/m^2$$

The variation of the porewater pressure throughout shear is depicted as the shaded area in figure 6.18.

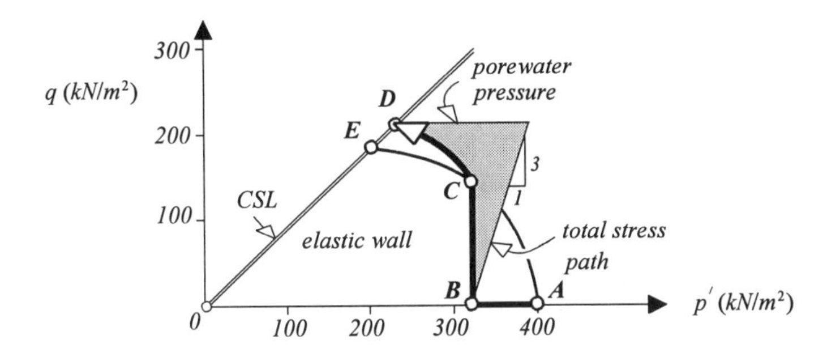

Figure 6.18: Porewater pressure generation during undrained loading.

6.4.4 Lightly overconsolidated clay: drained shear

This time, the (same) clay sample is being sheared from point B under *drained conditions*. The stresses and specific volumes for A and B are already known:

$$v_A = 1.47, \quad p'_A = 400\,kN/m^2, \quad q_A = 0$$

$$v_B = 1.48, \quad p'_B = 320\,kN/m^2, \quad q_B = 0$$

Also, the position of the yield curve (therefore that of point E) remains the same as under undrained conditions. However, according to equation 6.4, the *effective stress* path from the onset of *drained* shear (*i.e.* $u = 0$) is linear in (q, p') space with a slope of $1/3$:

$$p' = p'_B + q/3$$

so that, at point D, failure occurs when the effective stress path meets the CSL, whence:

$$q_D = Mp'_D \quad \text{and} \quad p'_D = p'_B + \frac{1}{3} q_D$$

and therefore:

$$p'_D = \frac{3}{(3 - M)} p'_B = \frac{3}{(3 - 0.9)} \times 320 = 457.1\ kN/m^2$$

$$q_D = Mp'_D = 0.9 \times 320 = 411.4\ kN/m^2$$

The specific volume at failure can now be calculated from the *CSL* equation 6.8:

$$v_D = \Gamma - \lambda \ln p'_D = 2 - 0.095 \ln 457.1 = 1.418$$

The complete stress path is sketched in figure 6.19 which shows that, as in the previous case of undrained shear, the position of point *C* corresponds to the intersection of the yield curve and the effective stress path.

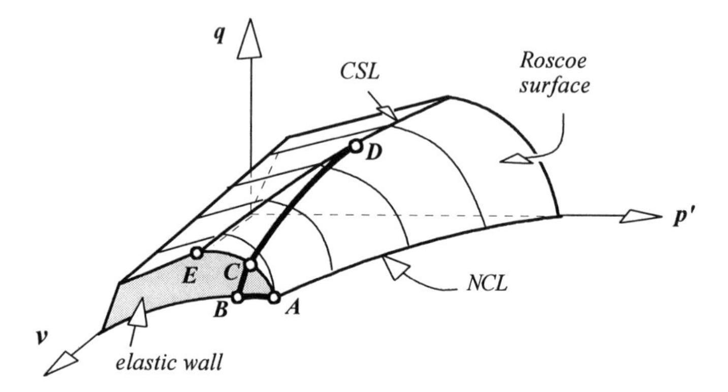

Figure 6.19: Lightly overconsolidated clay; 3-D drained stress path.

Accordingly, the yield curve equation 6.20 and the effective stress path equation at *C* are respectively:

$$q_C = Mp'_C \left(\frac{p'_A}{p'_C} - 1 \right)^{1/2}$$

$$p'_C = p'_B + q_C/3$$

Substituting for q_C in the effective stress path equation and rearranging:

$$p'^2_C\left(1+\frac{M^2}{9}\right) - p'_C\left(2p'_B + \frac{M^2}{9}p'_A\right) + p'^2_B = 0 \qquad \Rightarrow \quad p'_C = 357.1\ kN/m^2$$

Inserting this value into the yield curve equation yields the deviator stress:

$$q_C = 0.9 \times 357.1 \times \left(\frac{400}{357.1} - 1\right)^{1/2} = 111.4\ kN/m^2$$

The corresponding specific volume can then be calculated from the yield curve equation 6.19 in (v, p') space:

$$v_C = \Gamma + (\lambda - \kappa)\ \ln\frac{2}{p'_A} - \kappa \ln p'_C$$

$$= 2 + (0.095 - 0.035)\ \ln\frac{2}{400} - 0.035 \ln 357.1 = 1.476$$

The precise positions of the key points being determined, the stress path form B to C is calculated as follows.

- In (v, p') space, use the yield curve equation 6.19:

$$v = \Gamma + (\lambda - \kappa)\ \ln\frac{2}{p'_A} - \kappa \ln p' = 1.682 - 0.035 \ln p'$$

- In (q, p') space, use the effective stress path equation:

$$p' = p'_B + q/3$$

or $$q = 3(p' - 320) \qquad (a)$$

The stress path from C to D is determined in the following way.
- In (q, p') space, use equation (a) above.
- In (v, p') space, the stress path is on the state boundary surface, hence use both equation (a) and the Roscoe surface equation 6.18b:

$$q = 3(p' - 320)$$

$$v = \Gamma - \lambda \ln p' - (\lambda - \kappa)\ln\left[\left(\frac{q}{Mp'\sqrt{2}}\right)^2 + \frac{1}{2}\right]$$

The complete stress path is shown in figure 6.20. Notice the curvature of the stress path *CD* in figure 6.19, this is because of the curved nature of the Roscoe surface; the projection of the curve *CD* in (q, p') space results in a linear curve with a slope of 1/3 as depicted in figure 6.20. Also, because shear is undertaken under drained conditions, there is no generation of excess porewater pressure and hence the effective and total stress paths are *identical*.

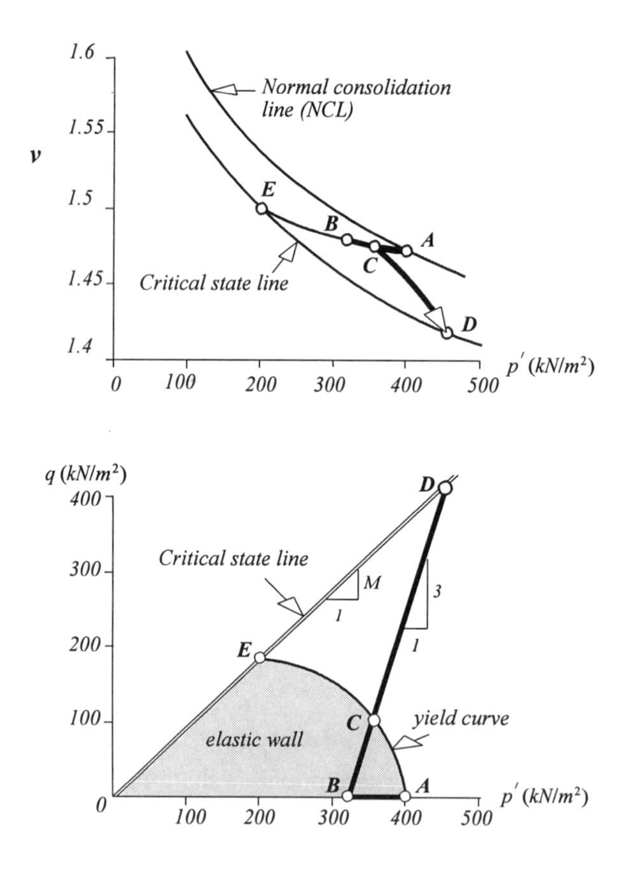

Figure 6.20: Drained stress path for lightly overconsolidated clay.

6.4.5 Heavily overconsolidated clay: undrained shear

In this instance, the sample being consolidated under an initial isotropic pressure $p_o = p'_A = 400\,kN/m^2$ is then unloaded along a swelling line to a pressure $p'_B = 100\,kN/m^2$, becoming in the process heavily overconsolidated with an overconsolidation ratio:

$$OCR = 400/100 = 4$$

Once consolidation under p'_B is complete, the sample is sheared under undrained conditions until failure.

The initial conditions are identical to those of the previous undrained case; thus at point *A*, the state of the clay is such that:

$$v_A = 1.472, \quad p'_A = 400\,kN/m^2, \quad q_A = 0$$

The unloading of the sample occurs along a swelling line, and therefore the specific volume at point *B* is found from the swelling line equation 6.10*b* which can be rearranged as follows:

$$v_B = v_A + \kappa \ln \tfrac{400}{100} = 1.52$$

From point *B*, the sample is sheared *undrained*, meaning that the specific volume remains constant throughout shear, so that at point *D* at failure:

$$v_D = v_B = 1.52$$

Since failure occurs on the *CSL*, both mean effective and deviatoric stresses at *D* can now be calculated from the *CSL* equations 6.8 and 6.7 respectively:

$$p'_D = \exp\left(\frac{\Gamma - v_D}{\lambda}\right) = \exp\left(\frac{2 - 1.52}{0.095}\right) = 155.6\,kN/m^2$$

$$q_D = Mp'_D = 0.9 \times 155.6 = 140\,kN/m^2$$

The entire stress path is as depicted in figure 6.21 in which point *C* corresponds to the intersection of the yield curve on the Hvorslev surface and the stress path *BC*.

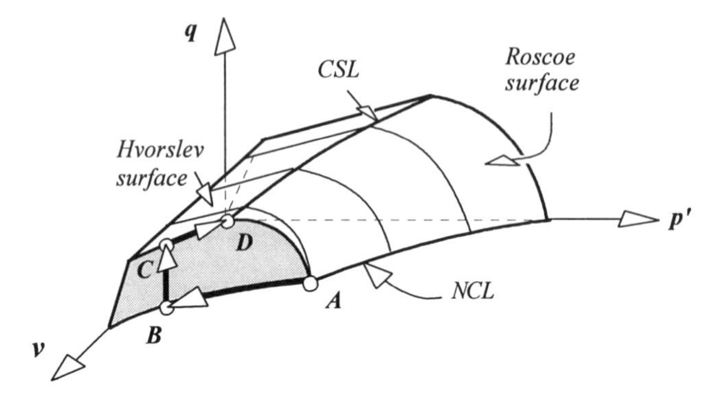

Figure 6.21: Heavily overconsolidated clay; 3-D undrained stress path.

As explained earlier, the constancy of volume between *B* and *C* implies that the corresponding stress path is vertical and hence:

$$p'_C = p'_B = 100\,kN/m^2$$

Inserting this quantity into the yield curve equation 6.25:

$$q_C = hp'_C + \left(\frac{M-h}{2}\right) p_o \left(\frac{2p'_C}{p_o}\right)^{\kappa/\lambda}$$

$$= 0.75 \times 100 + \left(\frac{0.9-0.75}{2}\right) \times 400 \times \left(\frac{2\times100}{400}\right)^{0.035/0.095} \qquad = 98.2\,kN/m^2$$

Notice that since point C is *also* on the Hvorslev surface, the same q_C value can be obtained using equation 6.23*a*.

The precise nature of the stress path followed throughout the test is shown in figure 6.22. Of particular interest, the position of point *E* in the $(v,\ p')$ space seems to be on the *CSL*. However, the $(q,\ p')$ space shows clearly that point E is in fact on the swelling line (which is an isotropic line through which $q = 0$), well below the *CSL*.

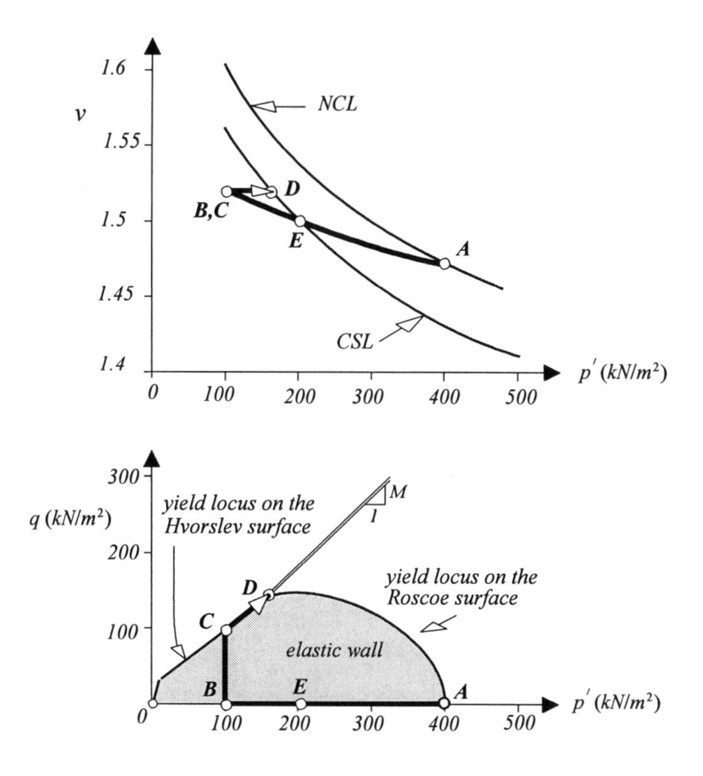

Figure 6.22: Undrained stress path for heavily overconsolidated clay.

The porewater pressure generated during shear can be calculated at any stage using equation 6.4 in which the quantity σ_3 is replaced by $p'_B = 100\,kN/m^2$:

$$u = (100 - p') + q/3$$

Thence, at C for instance, the excess porewater pressure is:

$$u_C = (100 - p'_C) + \frac{q_C}{3} = \frac{98.2}{3} = 32.7\,kN/m^2$$

whereas at D (*i.e.* at failure):

$$u_D = (100 - 155.6) + \frac{140}{3} = -8.9\, kN/m^2$$

This is precisely what happens in conjunction with clays on the dry side of the *CSL*: a negative excess porewater pressure is generated during *undrained shear*.

The nature of the variation of the excess porewater pressure generated throughout the undrained shear is represented by the shaded area in figure 6.23.

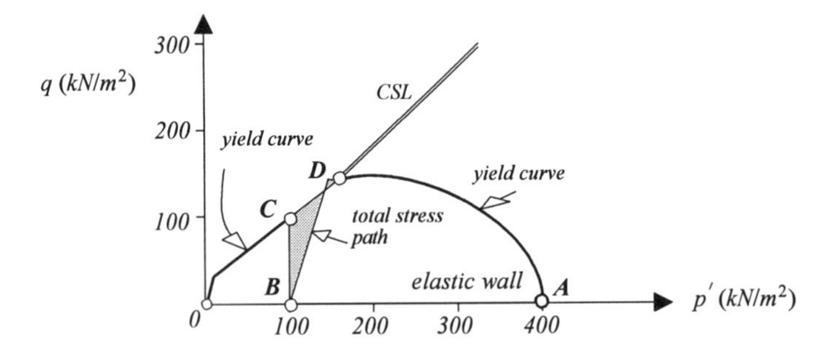

Figure 6.23: Porewater pressure generation during undrained loading.

6.4.6 Heavily overconsolidated clay: drained shear

The clay sample is now subjected to a *drained shear* after it has been unloaded from point A where:

$$v_A = 1.472, \quad p'_A = 400\, kN/m^2, \quad q_A = 0$$

to point B for which:

$$v_B = 1.52, \quad p'_B = 100\, kN/m^2, \quad q_B = 0$$

Because of the *drained* shear, no excess porewater pressure will be generated between point B (where shear starts) and point D (where failure occurs). Hence, according to equation 6.4, the *effective stress path* will be linear with a slope of 1/3, so that at D:

$$p'_D = p'_B + q_D/3$$

Also, D is on the *CSL*. Therefore, using equation 6.7:

$$q_D = Mp'_D$$

Combining the two latter equations:

$$p'_D = \frac{3}{(3-M)} p'_B = \frac{3}{3-0.9} \times 100 = 142.9\, kN/m^2$$

Thence the deviator stress at failure is:

$$q_D = Mp'_D = 128.6\, kN/m^2$$

and the specific volume at D (*i.e.* at failure), calculated from the *CSL* equation 6.8:

$$v_D = \Gamma - \lambda \ln p'_D = 2 - 0.095 \ln 142.9 = 1.53$$

The 3-D stress path corresponding to the entire test is sketched in figure 6.24.

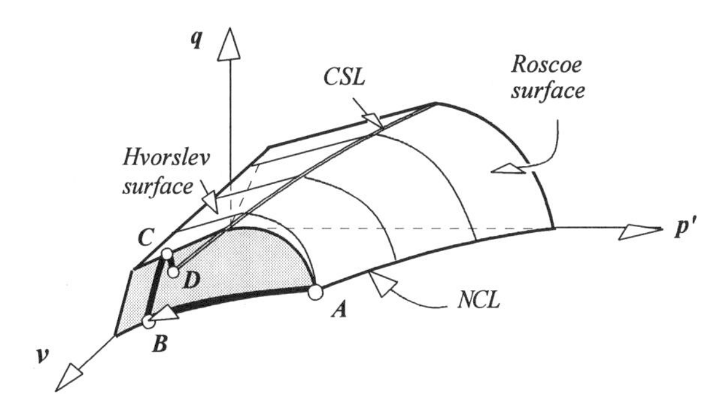

Figure 6.24: Heavily overconsolidated clay; 3-D drained stress path.

Once more, the position of point C in figure 6.24 is yet to be defined. It is however clear from the figure that point C results from the intersection of the yield curve and the effective stress path BC, whose equation is found by rearranging equation 6.4:

$$q_c = 3(p'_C - p'_B) \qquad (b)$$

Equating this equation to the equation 6.25 of the yield curve:

$$3(p'_C - p'_B) = hp'_C + \frac{M-h}{2} p_o \left(\frac{2p'_C}{p_o}\right)^{\kappa/\lambda}$$

A straightforward substitution for:

$$p'_B = 100\, kN/m^2, \quad p_o = p'_A = 400\, kN/m^2$$

into the above equation yields:

$$2.25p'_C - 4.26p_C'^{0.368} - 300 = 0$$

hence by trial and error:

$$p'_C = 145.5\, kN/m^2$$

The deviator stress at C is thereafter calculated from the stress path equation (b):

$$q_C = 3(145.5 - 100) = 136.5\, kN/m^2$$

and the specific volume at C is finally obtained from the elastic wall equation 6.19:

$$v_C = \Gamma + (\lambda - \kappa)\ \ln\left(\frac{2}{p_o}\right) - \kappa \ln p'_C$$

$$= 2 + 0.06 \ln \frac{2}{400} - 0.035 \ln 145.5 = 1.508$$

Notice that C is *also* on the Hvorslev surface and, as such, the Hvorslev surface equation 6.23*b* yields precisely the same value for v_C. The complete stress path followed during the test is shown in figure 6.25.

Interestingly, the negative porewater pressure generated towards the end of the *undrained shear* in the previous example is translated into an increase in volume now that the shear is occurring under *drained* conditions. This is clearly shown in both figures 6.24 and 6.25, whereby the specific volume at point D at failure ($v_D = 1.53$) is *larger* than that at point C ($v_C = 1.508$), this type of behaviour being typical of heavily overconsolidated clays. Accordingly, points C and D are *not* on the same elastic wall as depicted unambiguously in figure 6.24.

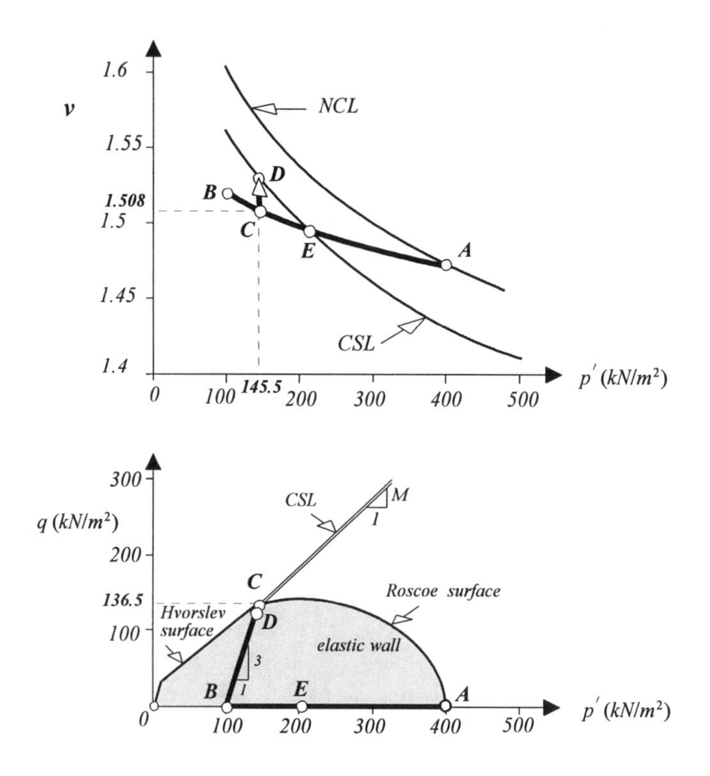

Figure 6.25: Drained stress path for a heavily overconsolidated clay.

6.5 Shear strength of clays related to the critical state concept

6.5.1 Undrained shear strength

It is widely accepted that the concept of critical state does apply to natural soils and in particular to clays. Mathematically, the critical state is represented by the *CSL* equations 6.7 and 6.8:

$$q = Mp'$$

$$v = \Gamma - \lambda \ln p'$$

rearranging the latter equation: $p' = \exp\left(\frac{\Gamma - v}{\lambda}\right)$, then substituting for p' into the first equation:

$$q = M \exp\left(\frac{\Gamma - v}{\lambda}\right) \tag{6.27}$$

If a clay sample were subject to *undrained shear* in a triaxial test, then the deviator stress at failure will correspond to the diameter of Mohr's circle as depicted in figure 6.26:

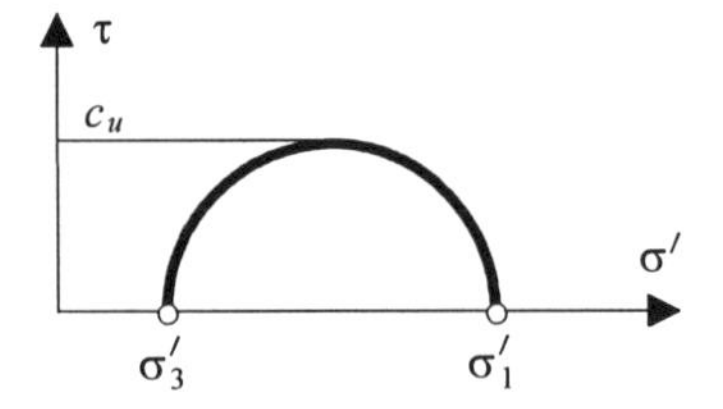

Figure 6.26: Stress field at failure during undrained shear.

$$q = (\sigma_1 - \sigma_3) = 2c_u$$

Accordingly, the undrained shear strength is:

$$c_u = \frac{M}{2} \exp\left(\frac{\Gamma - v}{\lambda}\right) \tag{6.28}$$

so that the specific volume at failure can be expressed as:

$$v = \left(\Gamma - \lambda \ln \frac{2}{M}\right) - \lambda \ln c_u \tag{6.29}$$

Knowing that for a saturated clay:

$$v = 1 + e = 1 + wG_s$$

it follows that the water content at failure is related to the undrained shear strength of the clay in the following way:

$$w = A - \frac{\lambda}{G_s}\ln c_u \tag{6.30}$$

where the *constant A* corresponds to:

$$A = \frac{1}{G_s}\left(\Gamma - 1 - \lambda \ln\frac{2}{M}\right)$$

and G_s represents the specific gravity of the clay.

Equation 6.30 is a linear relationship between the water content *at failure* and the logarithm of the undrained shear strength of the clay as can be seen from figure 6.27. The constant λ can hence be determined from a set of triaxial undrained shear tests.

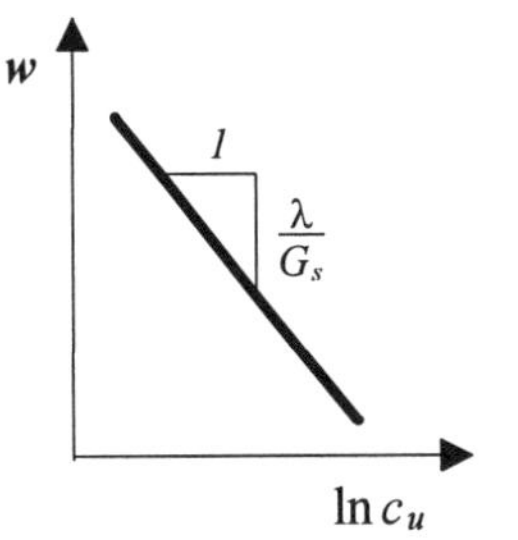

Figure 6.27: Determination of the critical state parameter λ.

6.5.2 Drained shear strength parameters

The compressive shear strength of a soil can be expressed in terms of Coulomb equation:

$$\tau_f = c' + \sigma'_f \tan\phi' \tag{6.31}$$

where the quantity c' represent the apparent cohesion of the soil. Referring to Mohr's circle depicted in figure 6.28, the co-ordinates of the centre of the circle are:

$$\tau = 0$$

$$\sigma' = \frac{1}{2}(\sigma'_1 + \sigma'_3) = \frac{1}{3}(\sigma'_1 + 2\sigma'_3) + \frac{1}{6}(\sigma'_1 - \sigma'_3) \ = p' + q/6$$

Thus, the expressions of the shear stress τ_f and the normal stress σ'_f at failure can now be established in a straightforward manner:

$$\tau_f = \tfrac{q}{2}\cos\phi'$$

$$\sigma'_f = p' + \tfrac{q}{6} - \tfrac{q}{2}\sin\phi' = p' + \tfrac{q}{6}(1 - 3\sin\phi')$$

Substituting for τ_f and σ'_f into equation 6.31 and rearranging:

$$q = \frac{6\cos\phi'}{3-\sin\phi'}c' + \frac{6\sin\phi'}{3-\sin\phi'}p' \tag{6.32}$$

but failure occurs on the *CSL* whose equation is:

$$q = Mp' \tag{6.33}$$

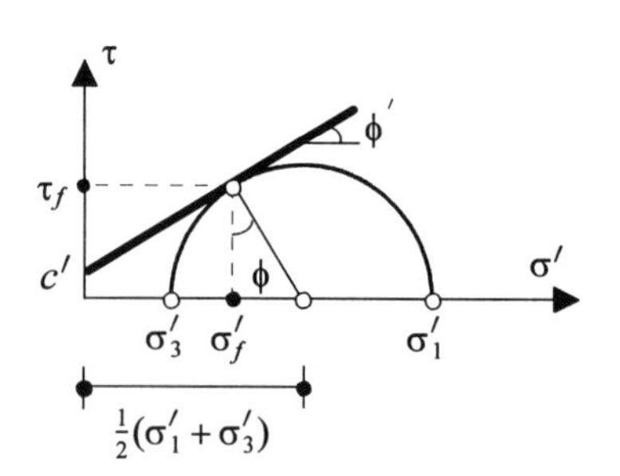

Figure 6.28: Drained behaviour of a cohesive soil at failure.

Comparing these two equations, it emerges that the critical state theory *assumes* that all soils are *frictional* (*i.e.* $c' = 0$), the critical state constant M and the angle of shearing resistance of the soil being related as follows:

$$M = \frac{6\sin\phi'}{3-\sin\phi'}$$

and

$$\phi' = \sin^{-1}\left(\frac{3M}{6+M}\right) \tag{6.34}$$

Although the assumption of zero apparent cohesion tends to be on the conservative side for heavily overconsolidated clays (remember $c' = 0$ for normally consolidated clays), in practice however, a representative value of c' for a given clay is usually very difficult to measure with sufficient accuracy.

6.6 Calculation of elasto-plastic strains

6.6.1 Elastic deformations prior to yielding

The ensuing formulation takes into consideration the axisymmetric stress conditions of a triaxial test during which the increments of radial stresses are equal: $\delta\sigma_2 = \delta\sigma_3$, and so are the increments of radial strains: $\delta\varepsilon_2 = \delta\varepsilon_3$. The formulation can easily be extended to apply to any set of stress increments.

The modified Cam-clay model assumes that, for a *saturated isotropic* soil sample subject to a set of stress increments ($\delta\sigma_1'$, $\delta\sigma_2'$, $\delta\sigma_3'$) so that its state remains *within* the elastic wall, the corresponding deformations are purely elastic, and as such, can be estimated from Hooke's generalised law of elasticity:

$$\begin{bmatrix} \delta\varepsilon_1 \\ \delta\varepsilon_2 \\ \delta\varepsilon_3 \end{bmatrix} = \frac{1}{E}\begin{bmatrix} 1 & -\nu & -\nu \\ & 1 & -\nu \\ sym & & 1 \end{bmatrix}\begin{bmatrix} \delta\sigma_1' \\ \delta\sigma_2' \\ \delta\sigma_3' \end{bmatrix} \tag{6.35}$$

E and ν being Young's modulus of elasticity and Poisson's ratio respectively.

Under axisymmetric stress conditions, the two strain parameters, namely the *volumetric strain* and the *shear strain*, are defined as follows:

- *volumetric strain:* $\delta\varepsilon_v = \delta\varepsilon_1 + 2\delta\varepsilon_3$ (6.36)

- *shear strain:* $\delta\varepsilon_s = \frac{2}{3}(\delta\varepsilon_1 - \delta\varepsilon_3) = \delta\varepsilon_1 - \frac{\delta\varepsilon_v}{3}$ (6.37)

Hence, according to the matrix form 6.35, the volumetric strain can be expressed as follows:

$$\delta\varepsilon_v = \frac{1}{E}(1-2\nu)\left(\delta\sigma_1' + 2\delta\sigma_3'\right) = \frac{3(1-2\nu)}{E}\delta p'$$

which can be written as:

$$\delta\varepsilon_v = \frac{\delta p'}{K} \tag{6.38}$$

where K represents the soil *bulk modulus*:

$$K = \frac{E}{3(1-2\nu)} \tag{6.39}$$

Similarly, a combination of equations 6.35 and 6.37 leads to:

$$\delta\varepsilon_1 - \delta\varepsilon_3 = \frac{3}{2}\delta\varepsilon_s = \frac{(1+\nu)}{E}\left(\delta\sigma_1' - \delta\sigma_3'\right) = \frac{(1+\nu)}{E}\delta q$$

or

$$\delta\varepsilon_s = \frac{2}{3}\frac{(1+\nu)}{E}\delta q = \frac{\delta q}{3G} \tag{6.40}$$

where G is the soil *shear modulus*:

$$G = \frac{E}{2(1+\nu)} \tag{6.41}$$

Relating the increments of the two stress parameters $(\delta p', \delta q)$ to the strain parameters $(\delta\varepsilon_v, \delta\varepsilon_s)$, the following elastic compliance matrix is then obtained:

$$\begin{bmatrix} \delta\varepsilon_v \\ \delta\varepsilon_s \end{bmatrix} = \begin{bmatrix} \frac{1}{K} & 0 \\ 0 & \frac{1}{3G} \end{bmatrix} \begin{bmatrix} \delta p' \\ \delta q \end{bmatrix} \tag{6.42}$$

The latter relationship indicates that, for *isotropic* soils, a change in volume can only be generated by a change in the mean effective stress, while the shear strain depends entirely on the change in deviator stress. Consequently, under undrained conditions, the soil volume remains constant (*i.e.* $\delta\varepsilon_v = 0$), in which case equation 6.38 is rewritten as:

$$\delta\varepsilon_v = \frac{\delta p'}{K_u} = 0$$

indicating that $K_u \to \infty$. Substituting for K from equation 6.39, it follows that:

$$K_u = \frac{E}{3\,(1-2\nu)} \rightarrow \infty \quad \Rightarrow \nu_u = 0.5$$

Thus, under undrained conditions (*i.e.* short term behaviour), Poisson's ratio takes the value $\nu_u = 0.5$. For drained behaviour of soils, Poisson's ratio is typically in the range: $0.2 \leq \nu \leq 0.4$.

6.6.2 Calculation of plastic strains

Once the stress path reaches the yield locus, elasto-plastic strains ensue. The onset of plasticity is governed by two sets of rules, the first being Druker *stability criterion* which postulates that, for the material to be stable, an increment of stress $\delta\sigma'$ that engenders yielding then hardening *must* do a positive or zero work during the increment of plastic strain $\delta\varepsilon^p$:

$$\delta\sigma' \delta\varepsilon^p \geq 0 \tag{6.43}$$

With reference to figure 6.29, it can be seen that the vectorial form of equation 6.43 is:

$$\delta p' \, \delta\varepsilon_v^p + \delta q \, \delta\varepsilon_s^p \geq 0 \tag{6.44}$$

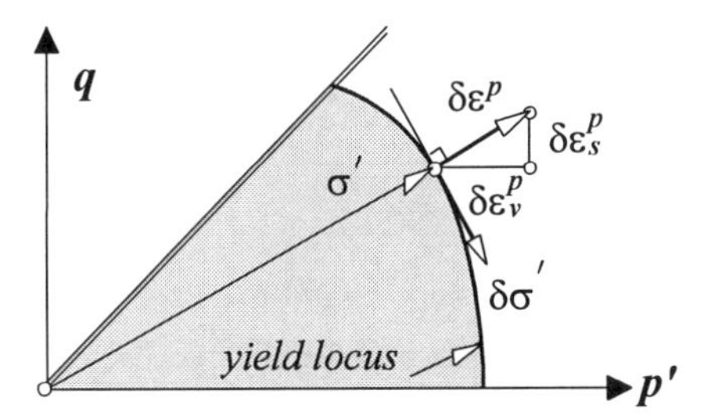

Figure 6.29: Associated flow rule.

The latter inequality means that $\delta\sigma'$ must have a component in the direction of $\delta\varepsilon^p$ and, consequently, the yield locus must have a *convex shape*. When the state of the soil is *on* the boundary surface, a strain hardening takes place through the expansion of the yield locus, and both $\delta\sigma'$ and $\delta\varepsilon^p$ become orthogonal as depicted in figure 6.29.

The second and perhaps most important criterion of plastic deformations is the *flow rule*. As mentioned earlier, the critical state theory assumes an *associated flow rule* defined by equation 6.15:

$$\frac{\delta\varepsilon_v^p}{\delta\varepsilon_s^p} = \frac{M^2 - \eta^2}{2\eta}$$

From a theoretical viewpoint (Hill, 1950), a plastic flow is said to be associated if the yield locus is also a *plastic potential*, that is if the plastic strains are normal to the yield locus as is the case in figure 6.29. Hence the associated flow rule is sometimes referred to as the *normality condition.*

Note that because equation 6.15 is related exclusively to the Roscoe surface, the following formulation of plastic strains will only apply in conjunction with the state boundary surface between the NCL and the CSL (refer to figure 6.10), heavily overconsolidated clays being characterised by a brittle behaviour on the Hvorslev surface.

Consider the same stress path depicted in figures 6.30 and 6.31 corresponding to a triaxial test on a normally consolidated clay that was first subjected to a drained loading (path *AB*), followed by an undrained shear until the occurrence of failure (path *BC*). It is clear that from the onset of shear, elasto-plastic strains are generated since point *A* is *on* the Roscoe surface.

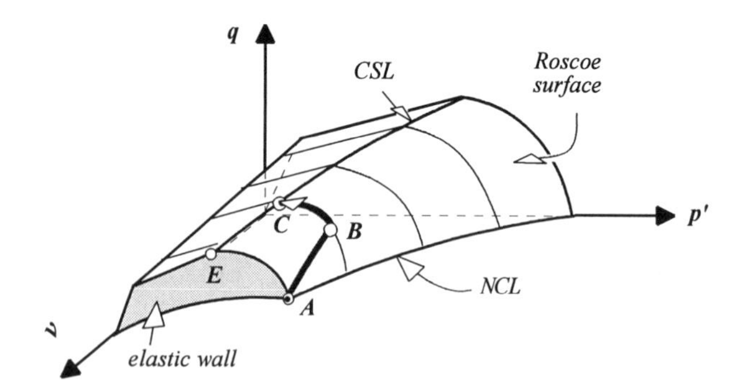

Figure 6.30: 3-D stress path on the state boundary surface.

On the other hand, figure 6.31 shows that the total volume change δv is the sum of an elastic component δv^e and a plastic component δv^p, so that:

$$\delta v^p = \delta v - \delta v^e \tag{6.45}$$

The quantity δv^e can easily be calculated from the swelling line equation 6.10*b*:

$$v^e = v_\kappa - \kappa \ln p'$$

whence:

$$\delta v^e = -\kappa \frac{\delta p'}{p'} \tag{6.46}$$

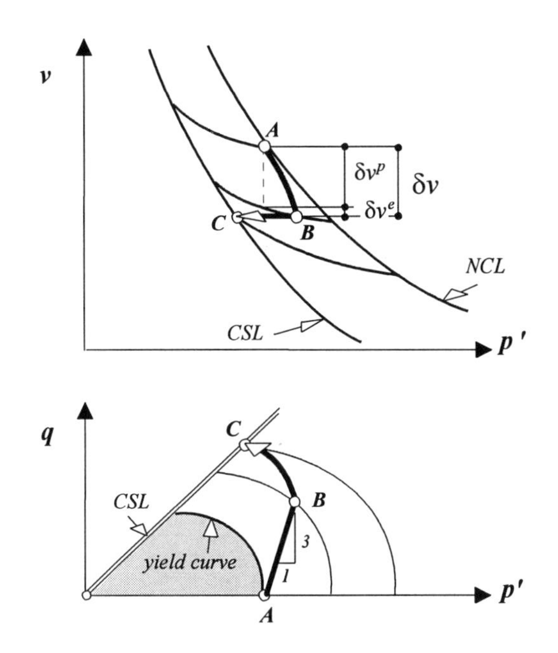

Figure 6.31: Elasto-plastic volume changes.

Moreover, the total volume change occurs on the Roscoe surface whose equation 6.18*b* has already been established:

$$v = \Gamma - \lambda \ln p' - (\lambda - \kappa)\ \ln\left[\left(\frac{q}{Mp'\sqrt{2}}\right)^2 + \frac{1}{2}\right]$$

This equation can now be differentiated:

$$\delta v = -\lambda \frac{\delta p'}{p'} - \frac{(\lambda - \kappa) q}{\left[\left(\frac{q}{Mp' \sqrt{2}} \right)^2 + \frac{1}{2} \right] (Mp')^2} \left(\delta q - q \frac{\delta p'}{p'} \right)$$

Inserting the quantity $\eta = q/p'$ into the latter equation and rearranging, it follows that:

$$\delta v = -\lambda \frac{\delta p'}{p'} - \frac{2\eta(\lambda - \kappa)}{p' \left(\eta^2 + M^2 \right)} (\delta q - \eta \delta p') \tag{6.47}$$

Substituting for δv^e and δv^p from equations 6.46 and 6.47 respectively into equation 6.45:

$$\delta v^p = \delta v - \delta v^e$$

$$= \frac{(\lambda - \kappa)}{p' \left(\eta^2 + M^2 \right)} \left(\eta^2 - M^2 \right) \delta p' - \frac{(\lambda - \kappa) 2\eta}{p' \left(\eta^2 + M^2 \right)} \delta q \tag{6.48}$$

Knowing that the plastic *strain increment* is related to the plastic component of the volume change in the following manner:

$$\delta \varepsilon_v^p = -\frac{\delta v^p}{v} \tag{6.49}$$

the final expression of the plastic volumetric strain is therefore:

$$\delta \varepsilon_v^p = \frac{(\lambda - \kappa)}{vp' \left(\eta^2 + M^2 \right)} \left[\left(M^2 - \eta^2 \right) \delta p' + 2\eta \, \delta q \right] \tag{6.50}$$

Also, the increment of the plastic shear strain is related to the increment of the plastic volumetric strain through the associated plastic flow equation 6.15 which can be rewritten as follows:

$$\delta\varepsilon_s^p = \frac{2\eta}{M^2 - \eta^2}\,\delta\varepsilon_v^p \tag{6.51}$$

Hence:

$$\delta\varepsilon_s^p = \frac{(\lambda - \kappa)}{vp'\left(\eta^2 + M^2\right)}\left(2\eta\,\delta p' + \frac{4\eta^2}{M^2 - \eta^2}\,\delta q\right) \tag{6.52}$$

Both equation 6.50 and 6.52 yield the plastic compliance matrix:

$$\begin{bmatrix} \delta\varepsilon_v^p \\ \delta\varepsilon_s^p \end{bmatrix} = \frac{(\lambda - \kappa)}{vp'\left(\eta^2 + M^2\right)}\begin{bmatrix} \left(M^2 - \eta^2\right) & 2\eta \\ 2\eta & \dfrac{4\eta^2}{\left(M^2 - \eta^2\right)} \end{bmatrix}\begin{bmatrix} \delta p' \\ \delta q \end{bmatrix} \tag{6.53}$$

Example 6.1

Consider, first, the case of the lightly overconsolidated clay sample subjected to an *undrained* loading as depicted in figure 6.16. Since the *state* of the sample has been defined in terms of stresses and specific volume, let us concentrate on the calculation of strains throughout the entire stress path, starting with the elastic strains within the elastic wall (stress path *ABC* in figure 6.16).

Along the path *AB*, the specific volume increases and the strains are elastic. The already established state of the sample at both points is summarised below:

- point *A*: $v_A = 1.472,\quad p'_A = 400\,kN/m^2,\quad q_A = 0$
- point *B*: $v_B = 1.48,\quad p'_B = 320\,kN/m^2,\quad q_A = 0$

The increment of elastic volumetric strain is defined as:

$$\delta\varepsilon_v^e = -\frac{\delta v^e}{v} \tag{6.54}$$

where the elastic volume change is determined from equation 6.46:

$$\delta v^e = -\kappa\frac{\delta p'}{p'}$$

Accordingly at point B:

$$\delta\varepsilon_v^e = \kappa \frac{\delta p'}{v p'} \tag{6.55}$$

with the quantities v and p' representing the *average values* of the specific volume and effective mean stress between the points A and B. Thus:

$$\delta\varepsilon_v^e = 0.035 \times \frac{(320 - 400)}{\left(\frac{1.472+1.48}{2}\right)\left(\frac{320+400}{2}\right)} = -5.27 \times 10^{-3}$$

Moreover, the clay *bulk modulus* is related to $\delta\varepsilon_v^e$ through equation 6.38, so that:

$$K = \frac{\delta p'}{\delta\varepsilon_v^e} = \frac{80}{5.27} \times 10^3 = 15,180\, kN/m^2$$

Assuming that the clay is characterised by a Poissons's ratio $\nu = 0.38$, its elastic modulus can then be calculated from equation 6.39:

$$E = 3K\,(1 - 2\nu) = 15180 \times 3 \times (1 - 0.76) = 10,930\, kN/m^2$$

Along the path BC, the volume remains constant because of the *undrained* conditions, and consequently the elastic volumetric strain is zero. Point C however marks the onset of *elastoplastic strains*. Considering that the loading along CD is undrained (*i.e.* no volume change), the increment of the overall volumetric strain $\delta\varepsilon_v$ must be zero. However, $\delta\varepsilon_v$ is the sum of two components:

$$\delta\varepsilon_v = \delta\varepsilon_v^e + \delta\varepsilon_v^p \tag{6.56}$$

consequently, the elastoplastic volumetric strains corresponding to an *undrained* loading are such that the elastic and the plastic components are equal and opposite to each other:

$$\delta\varepsilon_v = 0 \quad \Rightarrow \quad \delta\varepsilon_v^e = -\delta\varepsilon_v^p$$

The stresses and specific volumes at C and D have been established previously:

- point C: $v_C = 1.48$, $p'_C = 320\,kN/m^2$, $q_C = 144\,kN/m^2$
- point D: $v_D = 1.48$, $p'_D = 238.3\,kN/m^2$, $q_D = 214.5\,kN/m^2$

Also, the stress path CD is calculated from the Roscoe surface equation 6.18b:

$$q = Mp'\left[2\exp\left(\frac{\Gamma - v - \lambda \ln p'}{\lambda - \kappa}\right) - 1\right]^{1/2}$$

with $\Gamma = 2$, $v = 1.48$, $\lambda = 0.095$, $\kappa = 0.035$ and $M = 0.9$.

Using the above equation, the following results can be obtained.

$p'(kN/m^2)$		$q\,(kN/m^2)$		v		average		
	$\delta p'$		δq		δv	$p'(kN/m^2)$	$q\,(kN/m^2)$	$\eta = q/p'$
320		145		1.48				
	–20		23.5		0	310	156.8	0.51
300		168.5		1.48				
	–20		18.3		0	290	177.7	0.61
280		186.8		1.48				
	–20		14.8		0	270	194.2	0.72
260		201.6		1.48				
	–21.7		12.9		0	249.2	208	0.84
238.3		214.5		1.48				

Because of the undrained loading conditions from C to D, the total volumetric strain is zero and, according to equation 6.56, the plastic and elastic components of the volumetric strain have the same magnitude but opposite signs. Thence, using equation 6.55:

$$\delta\varepsilon_v^e = \kappa\frac{\delta p'}{vp'}$$

where $\delta p'$ and p' correspond to the pressure increment and the *average* mean pressure respectively between two points, so that with reference to the previous table, the first increment of pressure is $\delta p' = 300 - 320 = -20\, kN/m^2$, and the corresponding average pressure is:

$$p' = \frac{300+320}{2} = 310\, kN/m^2$$

The ensuing increment of elastic volumetric strain is thence:

$$\delta\varepsilon_v^e = -0.035 \times \frac{20}{1.48 \times 310} = -1.53 \times 10^{-3}$$

and the increment of plastic volumetric strain is:

$$\delta\varepsilon_v^p = -\delta\varepsilon_v^e = 1.53 \times 10^{-3}$$

A similar result for $\delta\varepsilon_v^p$ can be obtained from equation 6.50 in which the quantity η corresponds to the ratio of *average stresses* as indicated in the previous table. Whence:

$$\delta\varepsilon_v^p = \frac{(\lambda-\kappa)}{vp'\left(\eta^2+M^2\right)}\left[\left(M^2-\eta^2\right)\delta p' + 2\eta\delta q\right]$$

$$= \frac{0.06}{1.48 \times 310 \times \left(0.506^2+0.9^2\right)}\left[-\left(0.9^2-0.506^2\right) \times 20 + 2 \times 0.506 \times 23.5\right]$$

$$= 1.56 \times 10^{-3}$$

Finally, the corresponding increment of plastic shear strain is computed from equation 6.51:

$$\delta\varepsilon_s^p = \frac{2\eta}{M^2-\eta^2}\,\delta\varepsilon_v^p$$

$$= \frac{2 \times 0.506}{0.9^2-0.506^2} \times 1.56 \times 10^{-3} = 2.85 \times 10^{-3}$$

The results are summarised in the following table in terms of strain increments. Note that $\delta\varepsilon_v^p$ are calculated using equation 6.50.

$\delta p'$ (kN/m^2)	δq (kN/m^2)	$\delta\varepsilon_v^p$	$\delta\varepsilon_s^p$	$\delta\varepsilon_v^e$
–20	23.5	1.56×10^{-3}	2.85×10^{-3}	-1.56×10^{-3}
–20	18.3	1.62×10^{-3}	4.58×10^{-3}	-1.62×10^{-3}
–20	14.8	1.75×10^{-3}	8.59×10^{-3}	-1.75×10^{-3}
–21.7	12.9	2.06×10^{-3}	3.05×10^{-2}	-2.06×10^{-3}

The *cumulative* values of different strains throughout the test can now be calculated in a straightforward way as follows:

p' (kN/m^2)	q (kN/m^2)	ε_v^e	ε_v^p	$\delta\varepsilon_s^p$
400	0	0	0	0
320	0	-5.27×10^{-3}	0	0
320	145	-5.27×10^{-3}	0	0
300	168.5	-6.83×10^{-3}	1.56×10^{-3}	2.85×10^{-3}
280	186.6	-8.45×10^{-3}	3.18×10^{-3}	7.43×10^{-3}
260	201.6	-1.02×10^{-2}	4.93×10^{-3}	1.60×10^{-2}
238.3	214.5	-1.23×10^{-2}	6.99×10^{-3}	4.65×10^{-2}

Example 6.2

Let us now tackle the case of figures 6.19 and 6.20 where the lightly overconsolidated clay is subject to a *drained* loading from point *B*. The calculation of strains in this case will be undertaken in a way similar to that used in the previous example.

Since an identical clay sample with the same characteristics ($\Gamma = 2$, $\lambda = 0.095$, $\kappa = 0.035$, $M = 0.9$) is used, the loading from point *A* to point B will yield precisely the same elastic strain as in the previous example:

$$\delta\varepsilon_v^e = \kappa \frac{\delta p'}{v p'} = -5.27 \times 10^{-3}$$

From B to C, the stress path is within the elastic wall and therefore all strains will be elastic. The position of point C has already been established:

$$v_C = 1.476, \quad p'_C = 357.1\, kN/m^2, \quad q_C = 111.4\, kN/m^2$$

Also, the equations of the stress path BC have been defined thus (refer to section 6.4.4):

$$v = 1.682 - 0.035 \ln p' \tag{6.57}$$

$$q = 3\,(p' - 320) \tag{6.58}$$

These two equations yield the following values of the variables (p', q, v) within the elastic wall. Note that the calculations consist simply of selecting a value for p' then calculating the *corresponding values* of v and q from equations 6.57 and 6.58 respectively. The quantities $\delta\varepsilon_v^e$ are then calculated from equation 6.55: $\delta\varepsilon_v^e = \kappa \frac{\delta p'}{v p'}$ using the *average values* of p' and v.

For instance, during the first load increment in the following table, it can be seen that the pressure increased from $p' = 320\, kN/m^2$ to $330\, kN/m^2$; similarly, the corresponding specific volume changed from 1.48 to 1.479. Whence:

$$\delta p' = 330 - 320 = 10\, kN/m^2,$$

and the average values:

$$p' = \tfrac{1}{2}(330 + 320) = 325\, kN/m^2$$

and

$$v = \tfrac{1}{2}(1.48 + 1.479) = 1.4795$$

$p'(kN/m^2)$	$q(kN/m^2)$	$\delta p'$	v	average $p'(kN/m^2)$	average v	$\delta\varepsilon_v^e$
320	0		1.48			
		10		325	1.4795	7.28×10^{-4}
330	30		1.479			
		10		335	1.4785	7.06×10^{-4}
340	60		1.478			
		10		345	1.4775	6.87×10^{-4}
350	90		1.477			
		7.1		353.4	1.4765	4.76×10^{-4}
357.1	111.4		1.476			

The Roscoe surface is reached at point C where:

$$v_C = 1.476,\ p'_C = 357.1\, kN/m^2,\ q_C = 111.4\, kN/m^2$$

Also, point D where failure occurs is such that:

$$v_D = 1.418,\ p'_D = 457.1\, kN/m^2,\ q_D = 411.4\, kN/m^2$$

Along the path CD, the strains are elastoplastic; the corresponding stress path equations being as follows (refer to section 6.4.4):

$$q = 3(p' - 320) \qquad (6.59)$$

$$v = \Gamma - \lambda \ln p' - (\lambda - \kappa) \ln\left[\left(\frac{q}{Mp'\sqrt{2}}\right)^2 + \frac{1}{2}\right] \qquad (6.60)$$

Accordingly, for any selected p' value, both equation 6.59 and 6.60 yield the corresponding q and v values respectively. The results are summarised in the following table. Note that the stresses p', $\delta p'$, q and δq are expressed in kN/m^2.

p'		q		v		average			
	$\delta p'$		δq		δv	v	p'	q	$\eta = q/p'$
357.1		111.4		1.476					
	22.9		68.6		−0.013	1.469	368.6	145.7	0.39
380		180		1.463					
	20		60		−0.013	1.456	390	210	0.54
400		240		1.450					
	20		60		−0.012	1.444	410	270	0.66
420		300		1.438					
	20		60		−0.011	1.432	430	330	0.77
440		360		1.427					
	17.1		51.4		−0.009	1.422	448.6	385.6	0.86
457.1		411.4		1.418					

The three strain components are thereafter calculated in the following way:

- increment of elastic volumetric strain (equation 6.55):

$$\delta\varepsilon_v^e = \kappa \frac{\delta p'}{v p'}$$

- increment of plastic volumetric strain (equation 6.50):

$$\delta\varepsilon_v^p = \frac{(\lambda - \kappa)}{vp\left(\eta^2 + M^2\right)} \left[\left(M^2 - \eta^2\right) \delta p' + 2\eta\delta q\right]$$

- increment of plastic shear strain (equation 6.51):

$$\delta\varepsilon_s^p = \frac{2\eta}{M^2 - \eta^2} \delta\varepsilon_v^p$$

The results are tabulated below.

$\delta p'$ (kN/m^2)	δq (kN/m^2)	$\delta\varepsilon_v^e$	$\delta\varepsilon_v^p$	$\delta\varepsilon_s^p$
22.9	68.6	1.48×10^{-3}	7.93×10^{-3}	9.58×10^{-3}
20	60	1.23×10^{-3}	7.20×10^{-3}	1.49×10^{-2}
20	60	1.18×10^{-3}	7.05×10^{-3}	2.46×10^{-2}
20	60	1.14×10^{-3}	6.72×10^{-3}	4.65×10^{-2}
17.1	51.4	9.40×10^{-4}	5.44×10^{-3}	13.30×10^{-2}

Hence the *cumulative* strain values throughout the test are as follows:

p' (kN/m^2)	q (kN/m^2)	ε_v^e	ε_v^p	ε_s^p
400	0	0	0	0
320	0	-5.27×10^{-3}	0	0
330	30	-4.54×10^{-3}	0	0
340	60	-3.84×10^{-3}	0	0
350	90	-3.15×10^{-3}	0	0
357	111	-2.67×10^{-3}	0	0
380	180	-1.19×10^{-3}	7.93×10^{-3}	9.58×10^{-3}
400	240	0.04×10^{-3}	1.51×10^{-2}	2.45×10^{-2}
420	300	1.22×10^{-3}	2.22×10^{-2}	4.91×10^{-2}
440	360	2.36×10^{-3}	2.89×10^{-2}	9.56×10^{-2}
457	411	3.30×10^{-3}	3.43×10^{-2}	22.86×10^{-2}

The extent of the plastic shear strain at failure in the latter (drained) case is noticeable with a magnitude of just under 23%, compared with only 4.65% for the undrained case.

6.7 Shortcomings of Cam-clay theory and alternative models

The mathematical model developed earlier is based on the assumption that soils are isotropic. It is well known that natural soils, especially clays, are inherently *anisotropic* due to their mode of deposition. Accordingly, the properties of a natural clay differ in every direction depending on the degree (or the extent) of anisotropy.

There is ample experimental evidence to suggest that, while the critical state *concept* is valid, the predictions of the model fall short of the experimental behaviour of natural (anisotropic) clays. This is not surprising considering that Cam-clay models were validated using *reconstituted isotropic clays*.

The main behavioural difference between predictions and experimental measurements on natural clays relate to the *position* rather than the *shape* of the yield loci. In this respect, most natural clays exhibit a behaviour at failure similar to that shown in figures 6.32 in the case of a natural stiff clay (Josseaume and Azizi, 1991). The figure indicates clearly that while the yield locus is elliptical, the major axis of the ellipse is *not* horizontal as predicted by Cam-clay model.

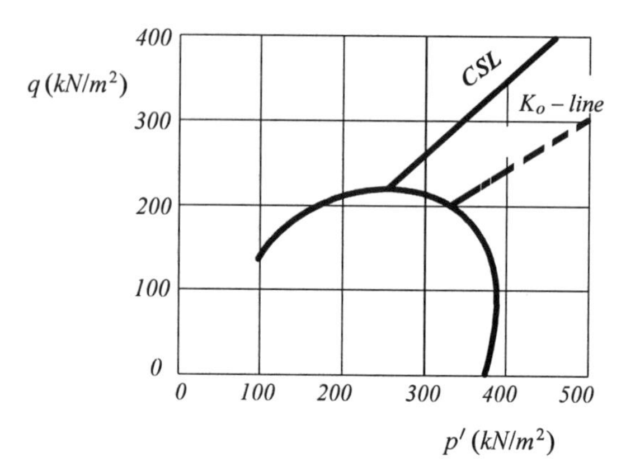

Figure 6.32: Yield locus for a natural stiff clay (Josseaume and Azizi, 1991).

In this case, the major axis of the yield locus seems to coincide with the K_o−line (K_o being the coefficient of earth pressure at rest as defined in equation 5.37), the departure from the prediction of a horizontal major axis being mainly due to anisotropy (see for instance Wood and Graham (1990)). This is illustrated schematically in figure 6.33 representing the three-dimensional state boundary surface for *natural clays*, as opposed to figure 6.34 which corresponds to the prediction of the *modified Cam-clay model* combined with the *Hvorslev surface* for *reconstituted isotropic clays*.

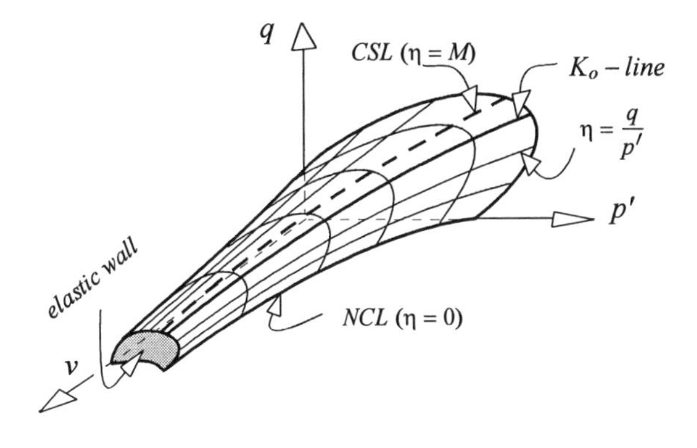

Figure 6.33: State boundary surface for anisotropic natural clays.

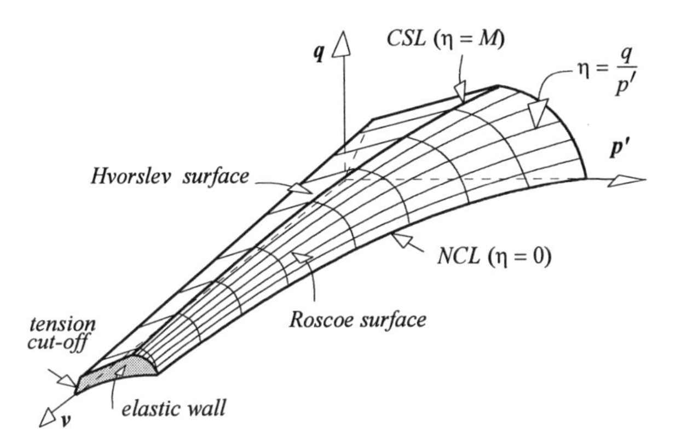

Figure 6.34: State boundary surface for reconstituted isotropic clays.

Predictive models such as the modified Cam-clay are attractive because of their simplicity. It is seen from equation 6.42 that for *isotropic soils*, only *two* independent parameters, namely the bulk and shear modulii, are required to calculate the strains generated by a set of stresses. Fully anisotropic soils, on the other hand, are characterised by no fewer than 21 independent parameters (see Graham and Houlsby, 1983) that need to be measured. However, the number can be reduced to five by assuming that soils are *transversely isotropic* or *cross-anisotropic*, in other words by assuming that their properties are identical in the two horizontal directions but different in the vertical direction.

Graham and Houlsby (1983) suggested using a relationship similar in nature to equation 6.42, in which *coupling effect* is taken into account. This coupling reflects the fact that for cross-anisotropic soils, shear stress *can* generate some volumetric strain, whereas a degree of shear strain is engendered by the mean effective stress:

$$\begin{bmatrix} \Delta p' \\ \Delta q \end{bmatrix} = \begin{bmatrix} K^* & J \\ J & G^* \end{bmatrix} \begin{bmatrix} \Delta \varepsilon_v \\ \Delta \varepsilon_s \end{bmatrix} \tag{6.61}$$

with only three independent parameters, that is a modified bulk modulus K^* (as opposed to the bulk modulus K in equation 6.42), a modified shear modulus G^* (in contrast to the shear modulus G in equation 6.42), and a cross-modulus J reflecting the degree of soil anisotropy:

- *isotropic soil:* $J = 0$
- *soil stiffer horizontally:* $J < 1$
- *soil stiffer vertically:* $J > 1$.

Experimental evidence shows that shear strains measured on *heavily overconsolidated clays* may differ substantially from values predicted by modified Cam-clay. This is essentially due to the fact that the behaviour within the elastic wall is by and large non-linear and, consequently, the shear modulus G does not have a constant value. Alternative models have been suggested to account for the non-linearity of the stress–strain behaviour within the elastic wall (Duncan and Chang, 1970, Jardine *et al.*, 1991). Other approaches suggest the inclusion, within the framework of Cam-clay, of kinematic surfaces inside the state boundary surface to delimit the elastic behaviour corresponding to very small strains (Al-Tabbaa and Wood, 1989, Atkinson, 1992).

As mentioned earlier, Cam-clay does not take into account the time effect on soil deformation. This effect, known as *creep* depends on the soil microstructure and its stress history, as well as on loading and drainage conditions (*i.e.* on the stress path).

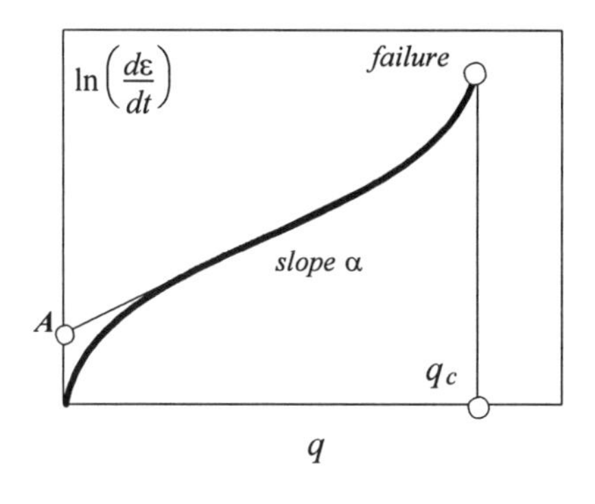

Figure 6.35: Creep curve.

Creep can be modelled using the following relationship between the strain rate and time (Mitchell, 1993):

$$\frac{d\varepsilon}{dt} = A\exp\left(\frac{\alpha q}{q_c}\right)\left(\frac{t_1}{t}\right)^m \tag{6.62}$$

with α: the slope of the creep curve as per figure 6.35,
t_1: a reference time,
q_c: deviator stress at failure,
m: a soil parameter ≤ 1, and
A: a fictitious strain rate corresponding to a deviator stress $q = 0$.

A straightforward integration yields the following creep strains:

- for $m = 1$:

$$\varepsilon_c = A\,t_1\exp\left(\frac{\alpha q}{q_c}\right)\ln\left(\frac{t}{t_1}\right) \tag{6.63a}$$

- for $m \neq 1$:

$$\varepsilon_c = \frac{A\,t_1}{1-m}\exp\left(\frac{\alpha q}{q_c}\right)\left(\frac{t}{t_1}\right)^{(1-m)} \tag{6.63b}$$

Experimental evidence suggests that creep effect is somewhat limited, and in this respect, figure 3.35 shows that the slope α remains moderate except when the deviator stress nears its magnitude q_c.

Alternatively, more sophisticated creep models based on hypoelasticity can be used (see for instance Yin *et al.*, 1989, 1990).

Finally, notwithstanding or perhaps because of its shortcomings, Cam-clay was and is still used as a framework for more sophisticated and more representative mathematical or numerical models. This work involves the modelling of increasingly complex soil behaviour as in the case of unsaturated soils (see Toll, 1990, Wheeler, 1991), or of thermally induced strains.

Problems

6.1 A clay sample was consolidated *isotropically* under a mean effective pressure $p'_o = 200\,kN/m^2$ and zero back pressure. Assume in what follows that the behaviour of the clay remains elastic throughout the test.

(*a*) Once consolidation under p'_o was achieved, the sample was then subjected to a stress increment $\delta p' = 50\,kN/m^2$ and $\delta q = 60\,kN/m^2$ under *drained conditions*. As a result, an increment of volumetric strain $\delta\varepsilon_v = 10^{-2}$ was measured. Knowing that the clay has a shear modulus $G = 10\,MN/m^2$, calculate its bulk modulus *K*, as well as the increments of axial and radial strains $\delta\varepsilon_1$ and $\delta\varepsilon_3$.

(*b*) At that state of stresses (*i.e.* $p' = 250\,kN/m^2$, $q = 60\,kN/m^2$), the drainage was closed and the deviator stress was increased by an additional increment $\delta q = 30\,kN/m^2$.

Calculate the ensuing value of $\delta\varepsilon_1$, as well as the porewater pressure that would have been generated.

(*c*) Were the drainage to be reopened at this stage, what would be the increment $\delta\varepsilon_v$ once all excess porewater pressure has dissipated ?

(*d*) Plot the effective stress path corresponding to the above stages of the test.

Ans: (*a*) $K = 5\,MN/m^2$, $\delta\varepsilon_1 = 5.3 \times 10^{-3}$, $\delta\varepsilon_3 = 2.35 \times 10^{-3}$
(*b*) $\delta\varepsilon_1 = 10^{-3}$, $\delta u = 10\,kN/m^2$
(*c*) $\delta\varepsilon_v = 2 \times 10^{-3}$

6.2 A normally consolidated clay has the following critical state parameters: $\Gamma = 2.4$, $\lambda = 0.15$, $\kappa = 0.05$, $M = 0.9$. A sample of the clay was first consolidated in a triaxial cell under isotropic stress conditions corresponding to $p'_o = 500\,kN/m^2$, then unloaded to $p'_1 = 300\,kN/m^2$, becoming in the process lightly overconsolidated. Once an equilibrium of volume was reached at the end of unloading, the sample was then sheared at a very low speed under *drained* conditions until a value $p'_2 = 350\,kN/m^2$ was reached.

(*a*) Show that at the end of shearing, the stress path is well within the elastic wall.

(*b*) Knowing that the (measured) elastic shear and volumetric strains corresponding to p'_2 are $\varepsilon_s = 4.2 \times 10^{-3}$ and $\varepsilon_v = 7 \times 10^{-3}$ respectively, estimate both bulk and shear modulii of the clay.

(*c*) If, at $p'_2 = 350\,kN/m^2$, the drainage was closed and the shearing of the sample was pursued under *undrained* conditions, calculate the specific volume, the mean effective stress, the deviator stress, as well as the porewater pressure generated at the point where the *yield curve* is attained.

(*d*) Calculate these same quantities at failure.

(*e*) Plot the entire stress path.

Ans: (*a*) *At yield,* $p'_y = 366.3\,kN/m^2 > 350\,kN/m^2 \Rightarrow$ *stress path is within the elastic wall.*
(*b*) $K = 7.14\,MN/m^2$, $G = 11.9\,MN/m^2$
(*c*) $v = 1.555$, $p' = 350\,kN/m^2$, $q = 206.2\,kN/m^2$, $u = 18.7\,kN/m^2$

(*d*) $v_f = 1.555$, $p'_f = 279.6\, kN/m^2$,
$q_f = 251.6\, kN/m^2$, $u_f = 104.3\, kN/m^2$

6.3 A clay has the following critical state parameters :
$\lambda = 0.1$, $\kappa = 0.05$, $\Gamma = 1.8$, $M = 0.9$.
A sample of the clay was consolidated *isotropically* under a cell pressure $\sigma_3 = 220\, kN/m^2$ and zero back pressure. At the end of consolidation, the drainage tap was closed and the deviator stress was increased gradually until a value $q = 74\, kN/m^2$ was reached, the cell pressure being kept constant.

(*a*) Calculate the specific volume v_1 and the mean effective pressure p'_1 corresponding to the value $q = 74\, kN/m^2$.

(*b*) Determine the value of the porewater pressure u generated by the same deviator stress $q = 74\, kN/m^2$.

At that stage of loading, the drainage was then opened for a period of time long enough for the excess porewater pressure generated during the undrained phase to dissipate.

(*c*) Calculate the values of the mean effective stress p'_2 and the deviator stress q_2, as well as the specific volume v_2 when all excess porewater pressure has dissipated.

(*d*) What is the increment of volumetric strain $\delta\varepsilon_v$ that would have been generated if the clay has a bulk modulus $K = 7\, MN/m^2$?

The test was thereafter resumed under *drained conditions* until the occurrence of failure on the critical state line.

(*e*) Calculate the values at failure of p'_f, q_f and v_f.

(*f*) Plot the entire stress path followed throughout the test.

Ans: (*a*) $v_1 = 1.295$, $p'_1 = 204.8\, kN/m^2$
(*b*) $u = 39.9\, kN/m^2$
(*c*) $p'_2 = 244.7\, kN/m^2$, $q_2 = 74\, kN/m^2$, $v_2 = 1.279$

(*d*) $\delta\varepsilon_v = 5.7 \times 10^{-3}$
(*e*) $p'_f = 314\,kN/m^2$, $q_f = 283\,kN/m^2$, $v_f = 1.225$

6.4 A clay sample identical to the one used in *6.2* was consolidated in a triaxial cell under an initial isotropic pressure $p'_o = 600\,kN/m^2$. At the end of consolidation, the clay was unloaded under isotropic conditions (*i.e.* $q = 0$) until an overconsolidation ratio $OCR = 1.5$ was achieved. At that point, the drainage tap was closed and the clay was subjected to an undrained shear phase until a mean effective stress $p'_2 = 360\,kN/m^2$ was reached.

(*a*) Calculate the specific volume v_1, the mean effective stress p'_1 and the deviator stress q_1 corresponding to the point where the Roscoe surface was met.

(*b*) Determine the specific volume v_2, the deviator stress q_2, as well as the excess porewater pressure corresponding to the mean effective stress p'_2.

(*c*) The drainage tap was then opened, thus allowing for the excess porewater pressure to dissipate fully. Once that happened, calculate the values of stresses p'_3 and q_3, and the corresponding specific volume of the sample v_3.

(*d*) The sample is then sheared under drained conditions at very low speed until failure occurred on the critical state line. Determine the values p'_f, q_f and v_f at failure.

(*e*) Plot the stress path corresponding to the entire test.

Ans: (*a*) $v_1 = 1.215$, $p'_1 = 400\,kN/m^2$, $q_1 = 254.3\,kN/m^2$
(*b*) $v_2 = v_1$, $q_2 \approx 300\,kN/m^2$, $u = 140\,kN/m^2$
(*c*) $p'_3 = 500\,kN/m^2$, $q_3 = q_2$, $v_3 = 1.195$
(*d*) $p'_f = 571.4\,kN/m^2$, $q_f = 514.3\,kN/m^2$, $v_f = 1.165$

6.5 Consider the clay sample used in conjunction with the previous problem *6.4*. Assume now that the clay was first consolidated under an isotropic pressure $p'_o = 600\,kN/m^2$, then unloaded isotropically (*i.e.* $q = 0$) to a value $p'_1 = 400\,kN/m^2$. A *drained* shear test then ensued until the mean effective pressure reached a value $p'_3 = 500\,kN/m^2$.

(*a*) Calculate the specific volume v_1 corresponding to p'_1.

(*b*) Determine the values q_2, p'_2 and v_2 corresponding to the point at which the yield curve was met.

(*c*) Calculate the quantities q_3 and v_3 occurring under the mean effective pressure p'_3.

(*d*) Once the stresses have reached the values (p'_3, q_3), the drainage was then closed and the clay was sheared under *undrained* conditions until the advent of failure. Calculate the quantities v_f, p'_f and q_f as well as the porewater pressure u_f at failure.

(*e*) Plot the corresponding stress path.

Ans: (*a*) $v_1 = 1.215$
(*b*) $q_2 = 220.1\,kN/m^2$, $p'_2 = 473.4\,kN/m^2$, $v_2 = 1.207$
(*c*) $q_3 = 300\,kN/m^2$, $v_3 = 1.195$
(*d*) $v_f = v_3$, $p'_f = 425\,kN/m^2$, $q_f = 382.5\,kN/m^2$, $u_f = 102.5\,kN/m^2$

6.6 Rework problem *6.5* above assuming that, at the point where the yield curve was met (*i.e.* point with the co-ordinates (p'_2, q_2, v_2)), the drainage tap was closed and the clay was sheared under undrained conditions up to the failure point. Estimate then the quantities p'_f, q_f and v_f at failure as well as the corresponding excess porewater pressure u_f generated. Also, plot the entire stress path.

Ans: $v_f = v_2 = 1.207$, $p'_f = 376.2\,kN/m^2$, $q_f = 338.5\,kN/m^2$, $u_f = 136.6\,kN/m^2$

6.7 A stiff clay is characterised by the parameters:
$\Gamma = 2$, $\lambda = 0.1$, $\kappa = 0.04$, $h = 0.75$ and $M = 0.95$. A carefully

prepared sample of the clay was initially consolidated under an isotropic pressure $p' = 550\,kN/m^2$ and zero back pressure. It was then unloaded isotropically (*i.e.* $q = 0$) to a value $p'_1 = 100\,kN/m^2$. The sample was thereafter subjected to an undrained shear at low speed so as not to generate any gradient of porewater pressure within the clay matrix.

(*a*) If the undrained shear was pursued until a deviator stress $q = 120\,kN/m^2$ was measured, show that yielding occurs on the Hvorslev surface. Then determine the quantities v_2, p'_2 and q_2 corresponding to the point at which this surface is met.

(*b*) Calculate the mean effective stress p'_3 and the specific volume v_3 when the deviator stress reaches the value $q = 120\,kN/m^2$.

(*c*) At that stage, the drainage was opened and the excess porewater pressure generated during the undrained loading was allowed to dissipate. Estimate the value of the mean effective stress p'_f and the corresponding specific volume v_f at failure.

(*e*) Plot the stress path corresponding to the entire test.

Ans: (*a*) $v_2 = 1.479$, $p'_2 = 100\,kN/m^2$, $q_2 = 111.6\,kN/m^2$
(*b*) $v_3 = v_2$, $p'_3 = 111.2\,kN/m^2$
(*c*) $p'_f = 126.3\,kN/m^2$, $v_4 = 1.516$

6.8 Consider problem *6.2* then:
(*a*) estimate the total elastoplastic strains (that is ε_v and ε_s) at failure;

(*b*) plot the graph (q, ε_s).

Ans: (*a*) $\varepsilon_v = 0$, $\varepsilon_s = 8.24 \times 10^{-2}$

6.9 Calculate the elastoplastic strains along the stress path relating to problem *6.3*. Plot the corresponding graphs (p', ε_v) and (q, ε_s).

6.10 Calculate the total (*i.e.* cumulative) strains at failure in the case of problem *6.4*.

Ans: $\varepsilon_v = 4.2 \times 10^{-2}$, $\varepsilon_s = 24.72 \times 10^{-2}$

6.11 Assuming that the clay has a bulk modulus $K = 8200\,kN/m^2$, estimate the elastic volumetric strain at the yield point corresponding to problem *6.5*. Calculate the variation of the elasto-plastic volumetric and shear strains along the stress path on the Roscoe surface, then plot the corresponding graphs (p', ε_v) and (q, ε_s).

Ans: *Initial elastic strain:* $\delta\varepsilon_v = 8.95 \times 10^{-3}$

References

Al-Tabbaa, A. and Wood, D. M. (1989) *An experimentally based 'bubble' model for clay.* Proceedings NUMOG III. Elsevier Applied Science, pp. 91–99.

Atkinson, J. H. (1992) *A note on modelling small strain stiffness in Cam-clay.* Predictive Soil Mechanics. Proceedings of the Wroth Memorial Symposium, Oxford. Thomas Telford, London, pp. 111–119.

Druker, D. C. (1954) *A definition of stable inelastic material.* Journal of Applied Mechanics. Trans. ASME 26, pp. 101–106.

Duncan, J. M. and Chang, C. Y. (1970) *Non-linear analysis of stress and strain in soils.* ASCE, Journal of Soil Mechanics and Foundation Engineering Division, 96 (SM5), pp. 1629–1653.

Graham, J. and Houlsby, G. T. (1983) *Elastic anisotropy of a natural clay.* Géotechnique, 33 (2), pp. 165–180.

Graham, J., Noonan, M. L. and Lew, K. V. (1983) *Yield states and stress–strain relationships in a natural plastic clay.* Canadian Geotechnical Journal, 20, pp. 502–516.

Hill, R. (1950) *The Mathematical Theory of Plasticity*. Oxford University Press. London.

Hvorslev, M. J. (1937) *Uber die fesigkeitseigenshaften gestorter bindinger boden.* Ingvidensk. Skr., A., No 45.

Jardine, R. J., Potts, D. M., St John, H. D. and Hight, D. W. (1991) *Some applications of a non-linear ground model.* Proceedings of the 10th European Conference on Soil Mechanics and Foundation Engineering. Florence. Vol. 1, pp. 223–228.

Josseaume, H. and Azizi, F. (1991) *Détermination expérimentale de la courbe d'état limite d'une argile raide très plastique*. Revue Française de Géotechnique, 54, pp. 13–25.

Mitchell, J. K. (1993) *Fundamentals of Soil Behaviour, 2nd edn.*

John Wiley & Sons, New York.

Roscoe, K. H. and Burland, J. B. (1968) *On the generalised stress-strain behaviour of "wet" clay.* Engineering Plasticity. Cambridge University Press. pp. 535–609.

Schofield, A. N. and Wroth, C. P. (1968) *Critical State Soil Mechanics.* McGraw-Hill, New York.

Toll, D. G. (1990) *A framework for unsaturated soil behaviour.* Géotechnique, 40 (1), pp. 31–44.

Wheeler, S. J. (1991) *An alternative framework for unsaturated soil behaviour.* Géotechnique, 41 (2), pp. 257–261.

Wood, D. M. (1992) *Soil Behaviour and Critical State Soil Mechanics.* Cambridge University Press.

Wood, D. M. and Graham, J. (1990) *Anisotropic elasticity and yielding of a natural plastic clay.* International Journal of Plasticity, 6, pp. 377–388.

Yin, J.-H., Graham, J., Saadat F. and Azizi, F. (1989) *Constitutive modelling of soil behaviour using three modulus hypoelasticity.* Proceedings of the 12th ICSMFE, Rio de Janeiro, Brazil, pp. 143–147.

Yin, J.-H., Saadat, F. and Graham, J. (1990) *Constitutive modelling of a compacted sand-bentonite mixture using three modulus hypoelasticity.* Canadian Geotechnical Journal, 27, pp. 365–372.

CHAPTER 7

The stability of slopes

7.1 Slope instability

Instability of natural slopes (generally the product of erosion), or artificial slopes (created by cuttings, excavations or the building of embankments for instance) results in most cases of a combination of gravitational forces and water pressures. Accordingly, when it occurs, failure involves a mass of soil within which the forces due to gravity (*i.e.* the weight of soil mass) are out of balance. The presence of water, if anything, increases the imbalance since it plays the role of a lubricant at the surface along which the soil mass is moving, gradually increasing its velocity and leading ultimately to total failure.

The mechanisms by which failure occurs differ depending on the nature and state of soil. Figure 7.1 depicts a translational slip whereby the failure surface is for all intents and purposes parallel to the slope, and the movement of the failing soil mass consists mainly of a translation.

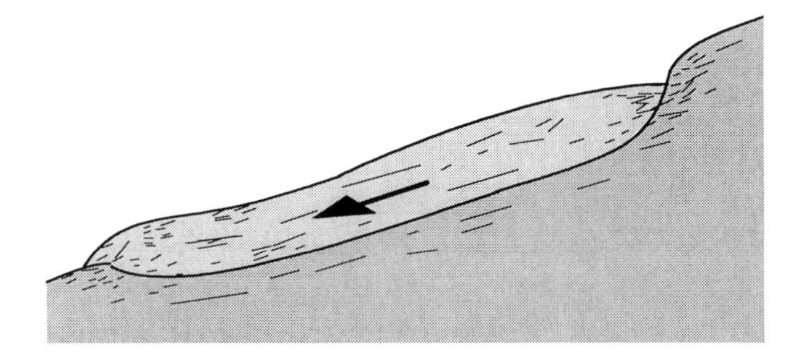

Figure 7.1: Translational slope failure.

On the other hand, figure 7.2 illustrates an almost circular slip surface where failure occurs mainly through rotation of the volume of the failing soil. As will be seen shortly, the position of the centre of rotation plays an important role in the stability analysis of such slopes.

In both cases, failure occurs (theoretically) once the driving shear forces due to the weight of the moving soil mass become equal to, if not larger than, the resisting intrinsic shear forces that the soil possesses.

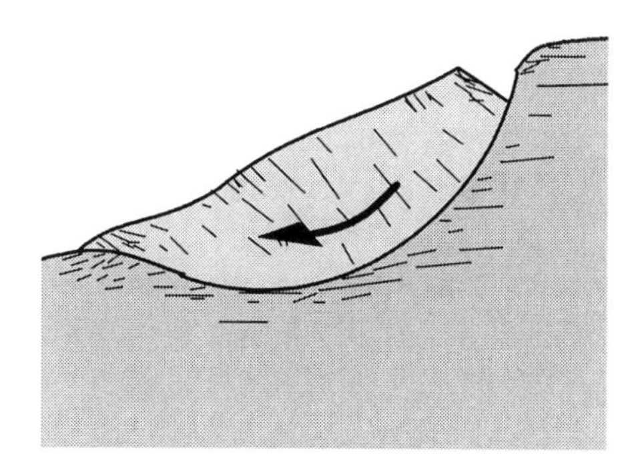

Figure 7.2: Rotational slope failure.

In stress terms, if the shear stress *mobilised* by the failing soil mass at its base is τ_{mob}, and if the shear *strength* of soil is referred to as τ_f, then the *factor of safety* against shear failure is simply defined as:

$$F = \frac{\tau_f}{\tau_{mob}} \qquad (7.1)$$

Accordingly, failure occurs (in theory) when $F \leq 1$. In practice though, failure is related more to the velocity with which the failing block of soil is moving; in this respect, experimental evidence shows that failures *do* occur for slopes with factors of safety of up to 1.25. Accordingly, it is tempting to assume that slopes with a factor of safety $F \geq 1.25$ are stable. However, such an assumption is *only realistic* as long as parameters *representative* of soil behaviour are used in the calculation. Several instances of slope failure occurring with a factor of safety of 1.5 or more were reported in the literature (see for instance Skempton (1964), Skempton and La Rochelle (1965), Palladino and Peck (1972)). In all cases, wrong shear strength values (usually overestimates) were used in the calculations. Potential problems related to the determination of soil parameters based on laboratory measurements were highlighted in section 5.6. In particular, one has to remember that the undrained shear strength c_u of a (stiff) clay, measured in the laboratory on a small size sample often represents an overestimate of the actual c_u value *in situ*, because it is unlikely that a small sample would contain fissures that characterise a thick stiff clay layer in the field.

It is therefore essential to bear in mind that the outcome of any design method depends *entirely* on the soil parameters used. This, in turn, indicates that not only the designer must master design techniques, but equally important, he (she) must be careful when selecting reliable values of relevant soil parameters.

Accordingly, the selection of appropriate and reliable soil strength parameters is not an option, but rather a precondition to a 'safe' design. This selection depends on the *state* of the soil (*i.e.* overconsolidated or normally consolidated, dense or loose), on *drainage conditions,* and on *failure criteria.* Thus, if failure is predicted to occur under drained conditions along an old slip surface, then the *residual* angle of shearing resistance will be most appropriate for the design, and the soil shear strength is calculated as follows:

$$\tau_f = \sigma_n' \tan \phi_r' \tag{7.2}$$

If, on the other hand, failure would occur within a soil mass that has never experienced failure in the past then, *à priori,* it would be appropriate to select the peak angle of shearing resistance in conjunction with overconsolidated or dense soils, or the critical angle if the soil is normally consolidated or loose (see figures 5.1 and 5.6). However, in practice, it is advisable to use the *critical* angle, so that for normally consolidated or lightly overconsolidated clays, as well as for (dense and loose) sands, the drained shear strength is estimated from equation 5.8:

$$\tau_f = \sigma_n' \tan \phi_c' \tag{7.3}$$

For heavily overconsolidated clays, the drained shear strength might be affected by the apparent cohesion, so that:

$$\tau_f = c' + \sigma_n' \tan \phi_c' \tag{7.4}$$

For slow draining soils (mainly clays), the excess porewater pressure generated at the onset of loading may create conditions whereby failure can take place in the short term (that is before the occurrence of any significant dissipation of excess porewater pressure). In which case, the undrained shear strength c_u must be used as in equation 5.21:

$$\tau_f = c_u \tag{7.5}$$

The detailed following slope stability analyses are based on equations 7.3 to 7.5.

7.2 Stability of infinite slopes

7.2.1 General case of flow

Slopes with a failure mechanism such as the one depicted in figure 7.1 are often assumed to be of infinite length because of the usually large ratio of slip plane length to its depth.

Consequently, the soil behaviour is modelled according to figure 7.4, whereby the potential slip plane, situated at a depth *z*, is assumed to be parallel to the ground surface, sloping at an angle β.

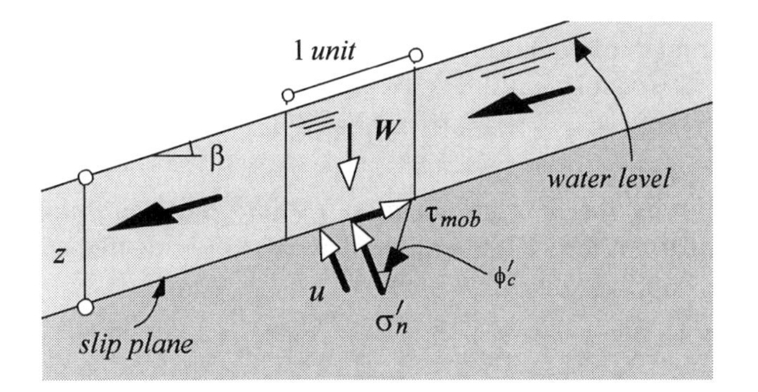

Figure 7.4: General case of an infinitely long slope and failure criteria.

Moreover, it is seen that (irrespective of the direction in which water is flowing), provided the pressure head h_p is constant along the slip plane, then the porewater pressure u can be estimated as follows:

$$u = h_p \gamma_w \tag{7.6}$$

γ_w being the unit weight of water.

In the case of a purely frictional material (*i.e.* $c' = 0$), the shear strength at failure is related to the effective normal stress according to equation 7.3. On the other hand, for an element of soil 1-*unit* long, 1-*unit* wide and z *units* deep as the one depicted in figure 7.4, the equilibrium requirements are such that:

$$\sigma'_n + u = W \cos\beta \tag{7.7}$$

$$\tau_{mob} = W \sin\beta \tag{7.8}$$

where τ_{mob} is the mobilised shear stress. The total weight W of the element is calculated as follows:

$$W = 1 \times 1 \times z\gamma \cos\beta \tag{7.9}$$

with γ being the total unit weight of soil (*i.e.* neglecting the difference between bulk and saturated unit weights).

The shear strength of equation 7.3 can now be expressed as follows:

$$\tau_f = \sigma'_n \tan\phi'_c = [(W\cos\beta) - u] \tan\phi'_c$$

Thus, substituting for W from equation 7.9 into the expression of τ_f and τ_{mob}, then rearranging, it is straightforward to show that equation 7.1 reduces to:

$$F = \frac{\tau_f}{\tau_m} = \frac{\tan\phi'_c}{\tan\beta}\left(1 - \frac{u}{z\gamma\cos^2\beta}\right) \tag{7.10}$$

now, introducing u from equation 7.6, it follows that:

$$F = \frac{\tan\phi'_c}{\tan\beta}\left(1 - \frac{\gamma_w h_p}{z\gamma\cos^2\beta}\right) \tag{7.11}$$

The factor of safety therefore depends on the slope angle β, the depth of the slip plane z and the pressure head h_p.

To establish the expression of h_p, consider the general case of flow sketched in figure 7.5, whereby the sliding soil mass of thickness z, through which water is flowing at an angle α to the horizontal, is assumed to be dry at the top, up to a depth z_w. If a standpipe is inserted to the depth of the slip plane, the water would rise to the depicted height h_p. Hence from the geometry of the figure:

$$h_p = (z - z_w)\,\frac{\cos\beta\cos\alpha}{\cos(\alpha-\beta)} \tag{7.12}$$

Substituting for h_p into equation 7.11, the general expression of the factor of safety in the case of a slope of infinite extent in a frictional soil affected by water seepage, is then established:

$$F = \frac{\tan\phi'_c}{\tan\beta}\left[1 - \frac{\gamma_w}{\gamma}\,\frac{\cos\alpha}{\cos\beta\cos(\alpha-\beta)}\left(1-\frac{z_w}{z}\right)\right] \tag{7.13}$$

Clearly, the factor of safety depends on the *direction of flow*.

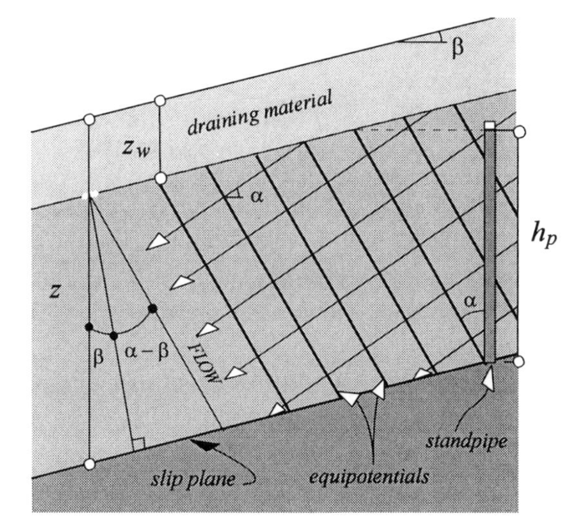

Figure 7.5: General water flow within a slope.

7.2.2 Flow parallel to the slope

In this case the angles α and β are identical, and equation 7.13 reduces to the following:

$$F = \frac{\tan\phi'_c}{\tan\beta}\left[1 - \frac{\gamma_w}{\gamma}\left(1 - \frac{z_w}{z}\right)\right] \tag{7.14}$$

so that for a dry slope ($z_w = z$), the factor of safety is:

$$F = \frac{\tan\phi'_c}{\tan\beta} \tag{7.15}$$

Similarly, for a waterlogged slope ($z_w = 0$):

$$F = \frac{\tan\phi'_c}{\tan\beta}\left(1 - \frac{\gamma_w}{\gamma}\right) \tag{7.16}$$

On the other hand, assuming a ratio $\frac{\gamma_w}{\gamma} \approx \frac{1}{2}$, then equation 7.14 can be expressed as follows:

$$F = \frac{1}{2}\frac{\tan\phi'_c}{\tan\beta}\left(1 + \frac{z_w}{z}\right) \tag{7.17}$$

The latter equation shows that, when the water is flowing parallel to the slope, the factor of safety is proportional to the ratio of the depth z_w of the water table (measured from the ground surface) to the thickness z of the sliding soil layer. Under such circumstances, the value of the factor of safety can be determined using the charts in figure 7.6 which are used in a straightforward way, in that for a soil having a critical angle of shearing resistance ϕ'_c and sloping at an angle β, the value F' is read on the appropriate (interpolated if need be) curve, and the factor of safety F is thereafter calculated as follows:

$$F = F'\left(1 + \frac{z_w}{z}\right) \tag{7.18}$$

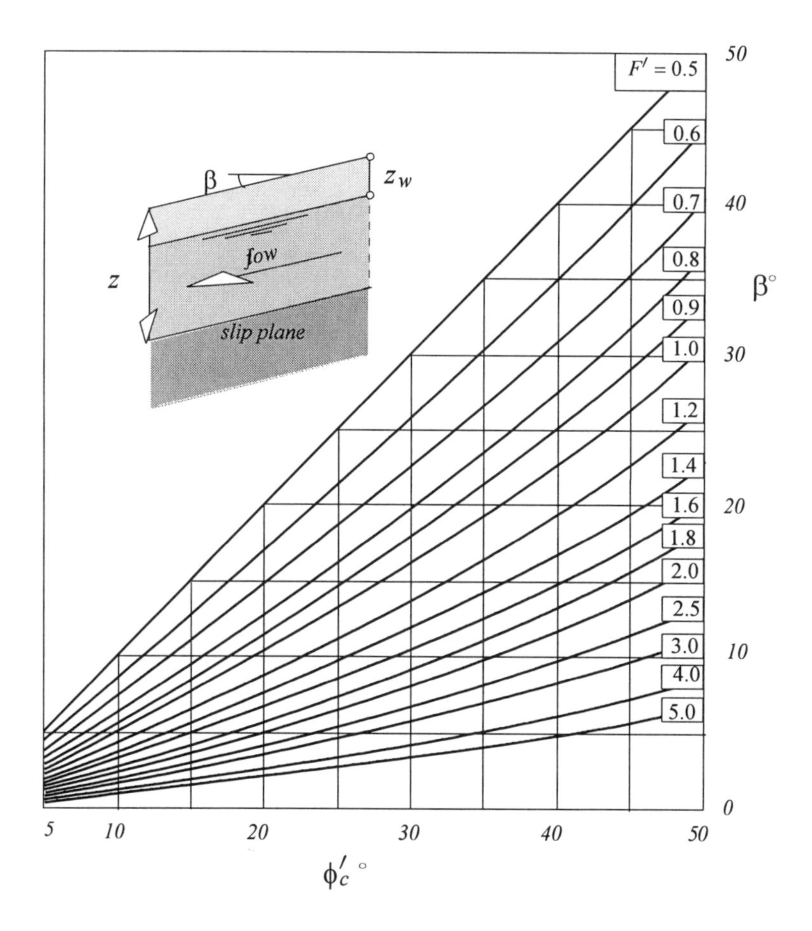

Figure 7.6: Charts for the calculation of the factor of safety for infinite slopes with water flowing parallel to the slope.

7.2.3 Horizontal flow

A horizontal flow implies an angle $\alpha = 0$ (refer to figure 7.5), in which case equation 7.13 yields:

$$F = \frac{\tan \phi_c'}{\tan \beta}\left[1 - \frac{\gamma_w}{\gamma}\frac{1}{\cos^2\beta}\left(1 - \frac{z_w}{z}\right)\right] \tag{7.19}$$

7.2.4 Vertical (downward) flow

The slope is, in this case, waterlogged meaning that $z_w = 0$ and, since the angle $\alpha = \pi/2$, equation 7.13 is then written as follows:

$$F = \tan\phi'_c/\tan\beta \tag{7.20}$$

Notice that equations 7.20 and 7.15 are identical, indicating that the stability of the slope is *not* affected by water seepage in the case of a vertical downward flow.

7.3 Effect of cohesion on the stability of infinite slopes

For heavily overconsolidated clays, the drained shear strength is affected by the apparent cohesion. However, the stability analysis is not altered a great deal; the reason being that, while the *mobilised* shear stress within the soil mass remains the same as defined by equation 7.8, the shear strength is now given by equation 7.4. Thus, substituting for τ_f from equation 7.4 into equation 7.10, it can readily be shown that the factor of safety is as follows:

$$F = \frac{c'}{z\gamma\sin\beta\cos\beta} + \frac{\tan\phi'_c}{\tan\beta}\left[1 - \frac{\gamma_w}{\gamma}\frac{\cos\alpha}{\cos\beta\cos(\alpha-\beta)}\left(1 - \frac{z_w}{z}\right)\right] \tag{7.21}$$

with z and z_w as per figure 7.5.

A comparison of equations 7.21 and 7.13 shows that, for cohesive soils, the factor of safety against sliding is simply increased by an amount $2c'/z\gamma\sin 2\beta$ as opposed to a granular soil.

For instance, a slope inclined at an angle $\beta = 15°$ to the horizontal, in a soil characterised by a cohesion $c' = 10\,kN/m^2$ and a total unit weight $\gamma = 20\,kN/m^3$ with a 6 m deep slip plane, will have its factor of safety against sliding increased by a respectable value of 0.33, irrespective of the flow direction or the water level within the slope.

Example 7.1

The frictional soil of a waterlogged slope, inclined at an angle $\beta = 15°$ to the horizontal, is characterised by a critical angle of shearing resistance $\phi'_c = 32°$ $(c' = 0)$. *In situ* tests, undertaken at different locations, indicated the presence of a layer of fissured hard rock at a depth $z = 5\,m$, running parallel to the surface. Also, piezometer readings indicated that water is flowing parallel to the slope. Required are:

(a) the factor of safety against sliding under the stated conditions;
(b) the depth z_w to which the water table must be lowered if a minimum factor of safety $F = 1.5$ were needed.

(a) First, estimate the constant F' from the charts of figure 7.6: for $\phi'_c = 32°$ and $\beta = 15°$, $F' \approx 1.2$. From the supplied information, the top of the rock layer represents an obvious potential slip plane, hence $z = 5\,m$ and $z_w = 0$ (waterlogged slope). From equation 7.18, the factor of safety is:

$$F = F'\left(1 + \frac{z_w}{z}\right) = 1.2$$

This value suggests that the slope is unstable (bear in mind that a minimum factor of safety of 1.25 is needed).

(b) This time, a factor of safety $F = 1.5$ is required. Hence, making use of equation 7.18, it follows that: $1.5 = 1.2 \times (1 + \frac{z_w}{5})$ or alternatively: $z_w = 5 \times (\frac{1.5}{1.25} - 1) = 1\,m$. Thus, the water table must be lowered by $1\,m$ for the factor of safety to meet the stated requirement.

Example 7.2

The clayey soil of a natural slope, inclined at an angle $\beta = 12°$ to the horizontal, is characterised by a residual angle of shearing resistance $\phi'_r = 14°$ $(c' = 0)$. A thorough site investigation revealed the presence of an old shallow slip plane, running roughly parallel to the slope, and situated at a depth $z = 3\,m$. The water table is at a depth $z_w = 1.5\,m$ below ground level, and the flow is assumed to be parallel to the slope.

From the charts of figure 7.6, it is seen that:

$$\phi_r' = 14°, \quad \beta = 12° \quad \Rightarrow \quad F' = 0.6.$$

Equation 7.18 then yields the following factor of safety:

$$F = F'\left(1 + \frac{z_w}{z}\right) = 0.6 \times \left(1 + \frac{1.5}{3}\right) = 0.9$$

indicating that the slope is actually failing.

7.4 Undrained analysis of the stability of infinite slopes

For slopes undertaken in clays, the excess porewater pressure generated during construction, or a short time after the end of construction (that is a few days or even few weeks), may trigger failure. Given the short time during which failure occurs, the excess porewater pressure remains virtually unchanged, and the soil resistance is provided by the undrained shear strength c_u as per equation 7.5.

Knowing that under undrained conditions, the angle of shearing resistance is $\phi_u = 0$, then substituting for c_u in equation 7.21 yields the following expression of the factor of safety of an infinite slope, with an angle β, under undrained conditions:

$$F = \frac{c_u}{2\gamma \cos\beta \sin\beta} = \frac{c_u}{\gamma \sin 2\beta} \qquad (7.22)$$

7.5 Stability of slopes with a circular failure surface

7.5.1 Total stress analysis

The shape of the failure surface such as the one illustrated in figure 7.2 is often assumed to be circular. The slip circle, as opposed to the sliding surface analysed earlier, involves a larger volume of soil and occurs, usually, in steeper slopes. The analysis at failure of such slopes consists mainly of finding the most unfavourable slip circle corresponding to the smallest value of the factor of safety. Moreover, because of the nature of the sliding surface, failure can occur in the short term under undrained conditions, or in the long term when any excess porewater pressure will have dissipated. Accordingly, separate analyses involving either total stresses (for the short term) or effective stresses (for the long term) must be undertaken.

Consider, for instance, the short term undrained behaviour of an embankment at the end of its (rapid) construction. Based on the previous analysis, it is seen that the shear strength of a fully saturated clay is represented by equation 7.5. Moreover, figure 7.7 shows that the instability of the slope is mainly due to the total weight W of the sliding soil mass (including any surcharge when applicable).

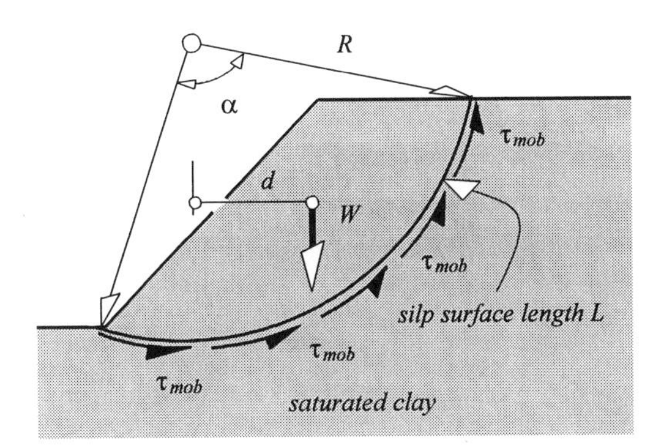

Figure 7.7: Circular slip surface.

The shear stress mobilised throughout the length L of the (circular) slip surface is related to the shear strength of the clay through equation 7.1:

$$\tau_{mob} = \frac{\tau_f}{F}$$

Now, considering the equilibrium of moments with respect to the centre of rotation in figure 7.7, it follows that:

$$Wd = \tau_{mob} LR \tag{7.23}$$

But the length of the slip surface is $L = R\alpha$. Hence, substituting for τ_{mob} and L into equation 7.23, then rearranging:

$$F = \frac{c_u R^2 \alpha}{Wd} \tag{7.24}$$

The factor of safety is related to the slip circle through the radius R and the angle α. Consequently, it is imperative to find the *smallest* value of F, locating in the meantime the likely slip surface. Although there may be cases whereby the location of the critical slip circle can roughly be guessed, the task of finding the minimum factor of safety is usually undertaken using a trial and error approach which can be tedious and might involve the use of literally dozens, sometimes in excess of a hundred, different potential slip circles (as opposed to the three depicted in figure 7.8, used to illustrate the method). Obviously, this type of analysis is generally undertaken numerically.

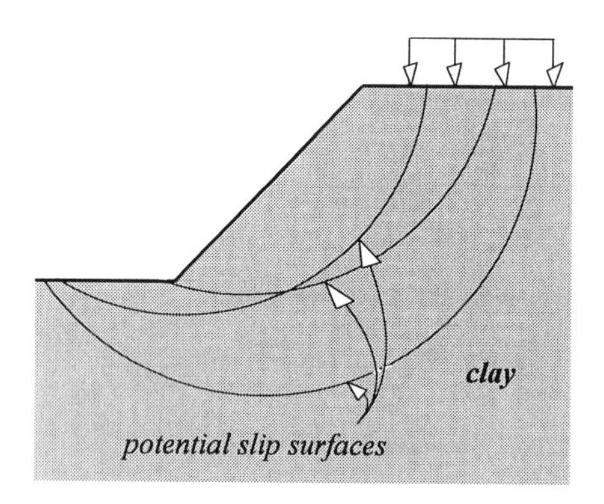

Figure 7.8: Determination of the critical slip circle.

For the specific case whereby the slope instability occurs within a layer of *homogeneous* soil, underlain by a layer of *hard* soil (such as rock), the charts produced by Taylor in 1948 and depicted in figure 7.9 can be useful. These charts were developed in terms of total stresses, and relate the *stability number N* to the *depth factor D* (refer to figure 7.9) and to the slope angle β. The *minimum* factor of safety is thereafter calculated as follows:

$$F_{\min} = \frac{c_u}{N\gamma H} \tag{7.25}$$

where γ represents the bulk unit weight of soil and H is the slope height as per figure 7.9.

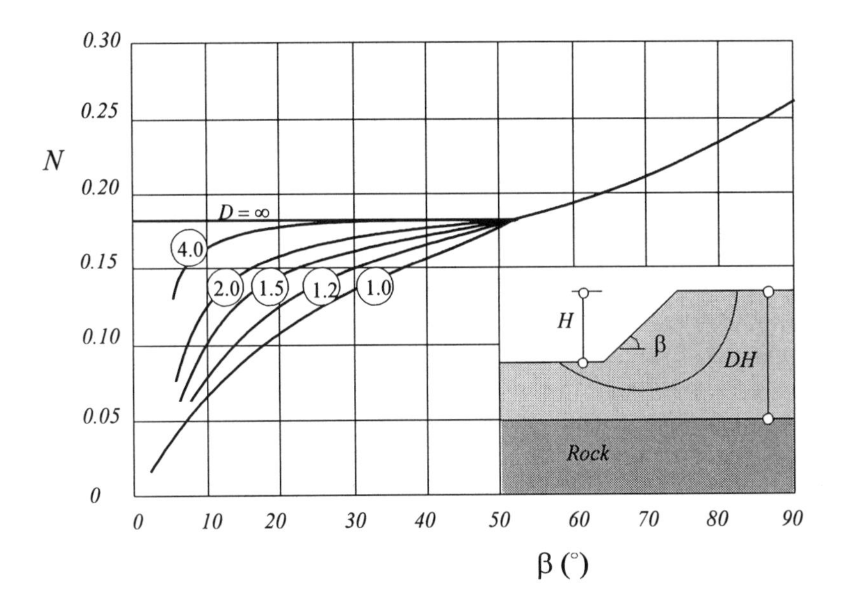

Figure 7.9: Taylor stability numbers for total stress analysis of slopes.

Consider for instance, the case of a slope with a height $H = 8\,m$, an angle $\beta = 35^\circ$, and a depth coefficient $D = 2$, made in a clay having an undrained shear strength $c_u = 50\,kN/m^2$ and a bulk unit weight $\gamma = 20\,kN/m^3$.

Figure 7.9 yields, in this case, a stability number $N \approx 0.178$. Thus, the *minimum* factor of safety for the slope is:

$$F_{\min} = \frac{50}{0.178 \times 20 \times 8} = 1.75$$

7.5.2 Effective stress analysis: Bishop procedure

In considering the long term stability of a slope, the *excess* porewater pressure is presumed to have all but dissipated and, accordingly, an effective stress analysis becomes a necessity. The aim being, as in the case of total stress analysis, to determine precisely the critical *circular* slip surface and through it the minimum factor of safety that must be used for design. This implies that the porewater pressure must be evaluated at any point so that the corresponding effective stress can be calculated.

Figure 7.10 depicts a slope with the water table as shown. If the water is assumed to be *static* (*i.e. no seepage*), then for any selected vertical element (or slice) within the failing soil mass, such as the one illustrated in the figure, the corresponding porewater pressure at base level can be estimated as follows:

$$u \approx z_w \gamma_w \tag{7.26}$$

where z_w represents the piezometric height of water and γ_w is the unit weight of water.

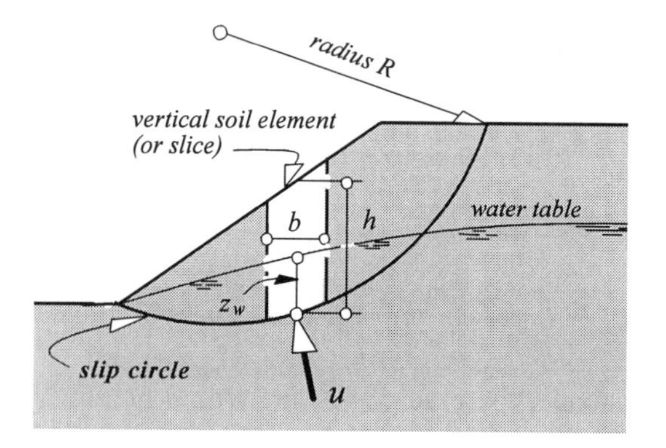

Figure 7.10: Evaluation of the porewater pressure at the base of a soil element in the case of a static water table.

The effective stress analysis can then begin in earnest, knowing that the failure criterion is represented in its general form by equation 7.4:

$$\tau_f = c' + \sigma'_n \tan\phi'_c$$

so that the mobilised shear stress is obtained by reducing τ_f by an adequate factor of safety according to equation 7.1:

$$\tau_{mob} = \frac{c'}{F} + \frac{\sigma'_n \tan\phi'}{F} \tag{7.27}$$

With reference to figure 7.11, let the entire soil mass above the slip circle be on the verge of failure, then consider in isolation, an element of a base length *l*, *one unit wide*, and having a total weight *w*. Since the element is on the verge of failure, equilibrium requires that vertical forces, as well as moments with respect to the centre of rotation O, must balance. Hence solving for moments for all elements into which the soil mass is divided, it follows that:

$$\Sigma \tau_{mob} l R = \Sigma w R \sin\alpha \tag{7.28}$$

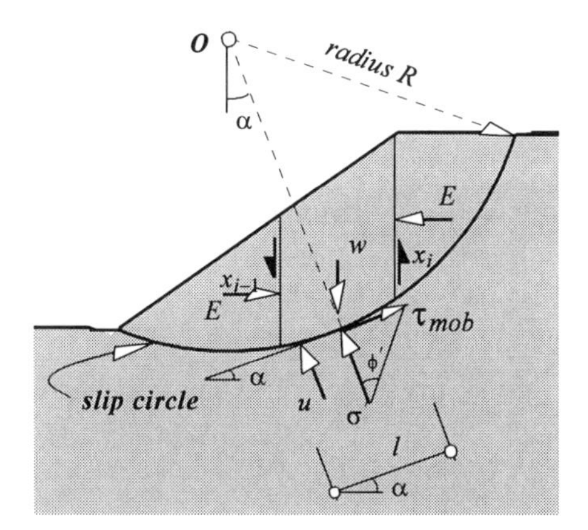

Figure 7.11: Stresses related to a slip circle failure mechanism.

Moreover, the equilibrium of vertical forces requires that:

$$w + (x_{i-1} - x_i) = (\sigma' + u)\, l \cos\alpha + \tau_{mob}\, l \sin\alpha \tag{7.29}$$

The tangential forces x_i are unknown, however the equilibrium would not be affected in any significant way were the quantity $(x_{i-1} - x_i)$ to be neglected (in fact an error estimated at less than 1% would be incurred by ignoring the latter quantity (see Bishop and Morgenstern (1960), for instance). Also, according to equation 7.27, the effective normal stress is such that:

$$\sigma'_n = \frac{\tau_{mob} F - c'}{\tan\phi'_c} \tag{7.30}$$

Thus, neglecting the quantity $(x_i - x_{i-1})$, then substituting for σ'_n from equation 7.30 into equation 7.29 yields:

$$\tau_{mob} = \frac{(w - u\,b) + \dfrac{c'b}{\tan\phi'}}{F\,l\left(\dfrac{\sin\alpha}{F} + \dfrac{\cos\alpha}{\tan\phi'}\right)} \tag{7.31}$$

with $b = l\cos\alpha$.

Inserting the quantity τ_{mob} from equation 7.31 into equation 7.28, then rearranging, it follows that:

$$F = \frac{1}{\Sigma\, w \sin\alpha} \sum \left\{ \left[c'b + w\left(1 - \frac{ub}{w}\right)\tan\phi'\right] \frac{1}{\cos\alpha + \dfrac{\tan\phi' \sin\alpha}{F}} \right\} \tag{7.32}$$

Now introducing the pore pressure ratio as defined by Bishop and Morgenstern (1960):

$$r_u = \frac{ub}{w} = \frac{u}{\gamma h} \tag{7.33}$$

where γ corresponds to the appropriate soil unit weight (*i.e.* bulk or saturated) and h is the height of the element as per figure 7.10. Equation 7.32 can then be rewritten as follows:

$$F = \frac{1}{\Sigma\, w \sin\alpha} \sum \left\{ [c'b + w(1 - r_u)\tan\phi'] \frac{1}{\cos\alpha + \dfrac{\sin\alpha \tan\phi'}{F}} \right\} \tag{7.34}$$

Equation 7.34 corresponds to the general expression of the factor of safety used in conjunction with Bishop's method of analysis of the stability of slopes (see also Bishop (1955)). The method is one of the most widely used in engineering practice because of its adaptability; and notwithstanding the fact that more sophisticated (and somewhat complex) methods have been developed (see for instance Morgenstern and Price (1965), Spencer (1967)), it remains one of the most reliable methods in use.

Note that, because the factor of safety F appears on both sides of equation 7.34, an iterative procedure is therefore required to solve it, and the details of the calculations (which are usually undertaken numerically) are presented in the following worked example.

Example 7.3

Consider the case of a slope made in a multi-layered soil consisting of a top layer of soft clay overlaying a thick layer of a moderately firm clay. Both clays have the same bulk unit weight $\gamma = 20\,kN/m^3$, and the slope geometry is as depicted in figure 7.12. Let us now apply a Bishop type effective stress analysis to assess (if only partly) the long term stability of the cut, assuming that the water table is well below the excavation base.

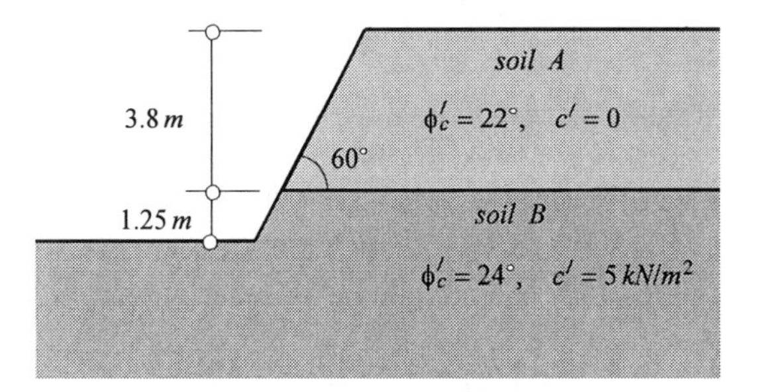

Figure 7.12: Slope geometry and soil characteristics.

In essence, the method aims at finding the most critical slip circle, and this can only be achieved through trial and error by selecting several (sometimes in excess of a hundred) different slip circles until the one yielding a minimum factor of safety is located. Two of these potential slip circles are depicted in figure 7.13 for purely illustrative purposes. Since the aim of this worked example is to present the working of the method, rather than to find the critical slip circle (which would be a tedious task to achieve using hand calculations anyway!), let us concentrate on one slip surface such as circle 1 in figure 7.13, and calculate the corresponding factor of safety.

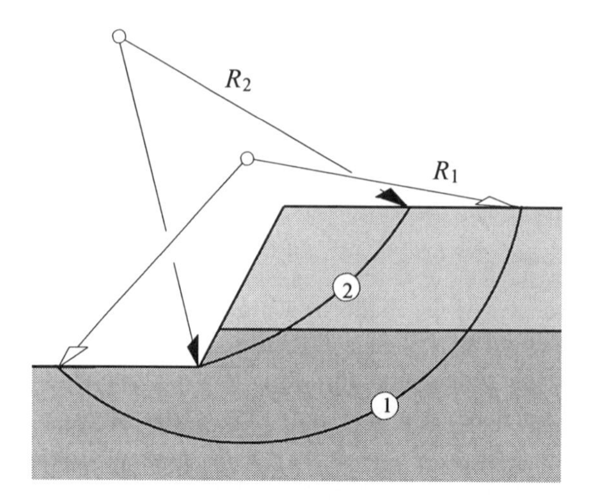

Figure 7.13: Selection of potential failure surfaces.

The procedure, as described earlier, consists of subdividing the (potentially) failing soil surface into vertical elements or slices. Because no restriction of any type is imposed on the selection of elements, and to make the hand calculations more palatable, five elements are selected as depicted in figure 7.14. Notice that the base of each element is assumed to be planar; this does not alter the accuracy of the outcome of the analysis in any way, if only because the circular shape of the slip surface 1 in figure 7.13 is roughly preserved.

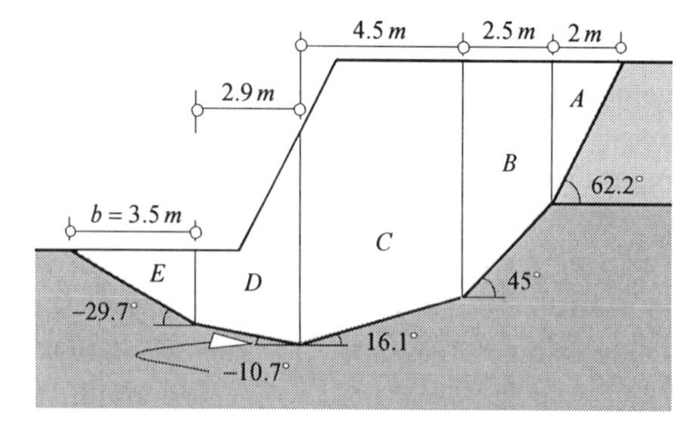

Figure 7.14: Selection of vertical slices.

Some of the most relevant dimensions are indicated in figure 7.14, so that the quantities needing to be used in conjunction with the iterative procedure related to equation 7.34 can be summarised as follows.

slice	$w\,(kN/m)$	$\phi'\,(^\circ)$	$c'\,(kN/m^2)$	$b\,(m)$	$\alpha\,(^\circ)$	$w\sin\alpha\,(kN/m)$
A	76	22	0	2.0	62.2	67.2
B	252.5	24	5	2.5	45.0	178.5
C	606.5	24	5	4.5	16.1	168.2
D	187.9	24	5	2.9	–10.7	–34.9
E	70	24	5	3.5	–29.7	–34.7

Whence the (constant) quantity: $\Sigma\, w\sin\alpha = 344.3\,kN/m$.

The iterative calculations can now be undertaken, and consist of:

- choosing an initial factor of safety (usually $F = 1$);
- calculating, for each slice, the quantity $m_\alpha = \cos\alpha + \dfrac{\tan\phi'_c\,\sin\alpha}{F}$;
- estimating, for each slice, the value $\xi = \dfrac{1}{m_\alpha}\left[c'b + w\tan\phi'_c\right]$, knowing that the porewater pressure is, in this case, $u = 0$.

Accordingly, it is easy to show that, when an initial factor of safety $F = 1$ is selected, the calculations yield the following results.

slice	$m_\alpha = \cos\alpha + \dfrac{\tan\phi'_c\,\sin\alpha}{1}$	$\xi\,(kN/m)$
A	0.82	37
B	1.02	122
C	1.08	271
D	0.90	109
E	0.65	75

Thus:

$$\sum \frac{1}{m_\alpha}\left[c'b + w\tan\phi'_c\right] = \sum \xi = 614\,kN/m$$

leading to a new factor of safety, calculated according to equation 7.34:

$$F = \frac{614}{344.3} = 1.78$$

It is therefore seen that the calculated factor of safety differs markedly from the selected (initial) value $F = 1$. Consequently, one or more iterations are needed for convergence to occur (that is for the selected and calculated factors of safety to be almost identical).

Let the new selected value of F be the last calculated one: $F = 1.78$. Under such circumstances, the calculations, undertaken in exactly the same way as above are such that:

slice	$m_\alpha = \cos\alpha + \dfrac{\tan\phi'_c \sin\alpha}{1.78}$	ξ (*kN/m*)
A	0.67	46
B	0.88	142
C	1.03	284
D	0.94	104
E	0.74	66

Whence the sum:

$$\Sigma\xi = 642\ kN/m$$

and the new factor of safety:

$$F = \frac{642}{344.3} = 1.86$$

This value is larger than the initial value of 1.78, and therefore another iteration needs to be undertaken using the latest calculated value $F = 1.86$ to start the calculations:

slice	$m_\alpha = \cos\alpha + \dfrac{\tan\phi'_c \sin\alpha}{1.86}$	ξ (*kN/m*)
A	0.66	47
B	0.88	142
C	1.03	284
D	0.94	104
E	0.75	65

Hence the quantity:

$$\Sigma\xi = 642\ kN/m$$

and the corresponding factor of safety:

$$F = \frac{642}{344.3} = 1.86$$

Both selected and calculated values being identical, the factor of safety corresponding to the slip circle 1 in figure 7.13 is therefore $F = 1.86$. This value, however, does *not necessarily* correspond to the minimum factor of safety since the slip surface 1 is unlikely to be the most critical. The way to determine this critical surface consists of selecting several more potential circular slip surfaces and undertaking the same iterative calculations for each one of them until a minimum factor of safety is obtained. Clearly, this task would be difficult to achieve by hand calculations. Rather, the entire procedure can be programmed fairly easily.

7.5.3 Effect of porewater pressure on the long term stability of slopes

In the previous example, the water table was assumed to be well below the slip surface so that the analysis was not affected in any way by the porewater pressure. Now let the water table rise to the level depicted in figure 7.15, in a way that it can still be considered static (*i.e.* neglecting any seepage forces).

Assuming that both bulk and saturated unit weights of the soil layers are identical $(\gamma = 20 kN/m^3)$, then the calculations of the factor of safety are undertaken in precisely the same way as previously except that, this time, the porewater pressure has to be evaluated for each element according to equation 7.26: $u = z_w \gamma_w$, where z_w is as per figure 7.15.

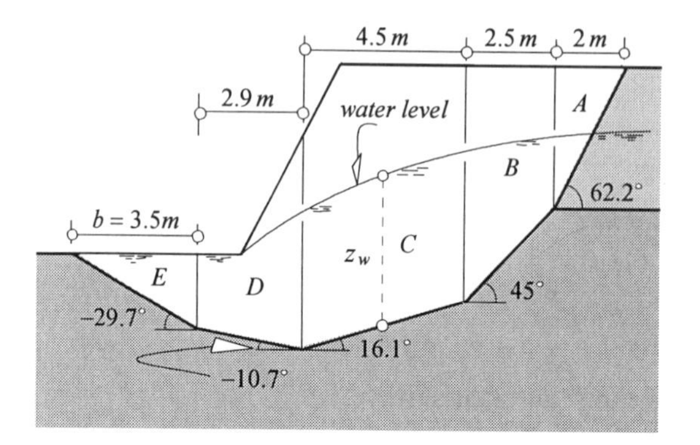

Figure 7.15: Effect of partial submersion of slopes on the overall stability.

In order to study the effect of porewater pressure on the stability, the same slip circle 1 in figure 7.13 has to be considered. Hence, with reference to figure 7.15, it follows that the total weight for each slice and the quantities $c'b$, as well as the sum $\Sigma w \sin\alpha$ have the same values as calculated previously for the dry slope. Accordingly, the quantities used in conjunction with equation 7.34 are as follows.

slice	$z_w\,(m)$	$u\,(kN/m^2)$	$ub\,(kN/m)$	$(w - ub)\tan\phi'_c\,(kN/m)$
A	0.7	7	14	25
B	2.5	25	62.5	83.6
C	3.7	37	166.5	193.5
D	2.2	22	63.8	54.6
E	1.0	10	35	15.4

The iterative calculations are identical to those corresponding to the dry case, except that the porewater pressure is now taken into account to estimate the factor of safety. Whence, starting with an initial factor of safety $F = 1$, the ensuing calculations yield the following:

slice	$m_\alpha = \cos\alpha + \dfrac{\tan\phi'_c \sin\alpha}{1}$	$\dfrac{1}{m_\alpha}\left[c'b + (w - ub)\tan\phi'_c\right]$ (kN/m)
A	0.82	30
B	1.02	94
C	1.08	200
D	0.90	77
E	0.65	51

leading to:

$$\sum \frac{1}{m_\alpha}\left[c'b + (w - ub)\tan\phi'_c\right] = 452\ kN/m$$

Hence the corresponding calculated factor of safety:

$$F = \frac{452}{344.3} = 1.31$$

The calculated value of F being different from the chosen value of 1, more iterations are therefore needed to achieve convergence. Now let the chosen value be $F = 1.31$, whence:

slice	$m_\alpha = \cos\alpha + \dfrac{\tan\phi'_c \sin\alpha}{1.31}$	$\dfrac{1}{m_\alpha}\left[c'b + (w - ub)\tan\phi'_c\right]$ (kN/m)
A	0.74	34
B	0.94	102
C	1.05	205
D	0.92	75
E	0.70	47

and:

$$\sum \frac{1}{m_\alpha}\left[c'b + (w - ub)\tan\phi'_c\right] = 463\ kN/m$$

yielding a factor of safety:

$$F = \frac{463}{344.3} = 1.34$$

It can be shown that, were a new iteration to be undertaken with an initial factor of safety $F = 1.34$, then the ensuing calculated factor of safety will be $F = 1.345$. Accordingly, it is seen that, when dry, the slip surface has a factor of safety $F = 1.85$, which decreases dramatically to $F = 1.34$ if the water table were to rise to the level depicted in figure 7.15, partially submerging the slope. One therefore expects the factor of safety to decrease even further were the slope to be waterlogged.

Consider the situation whereby the slope is totally submerged as illustrated in figure 7.16. Ignoring any seepage forces, and knowing that the total weight of individual slices, the quantities $c'b$, and the sum $\Sigma w \sin\alpha$ are all identical in value to the ones used in previous calculations, the basic quantities needed to start the iterative procedure related to equation 7.34 are then summarised as follows.

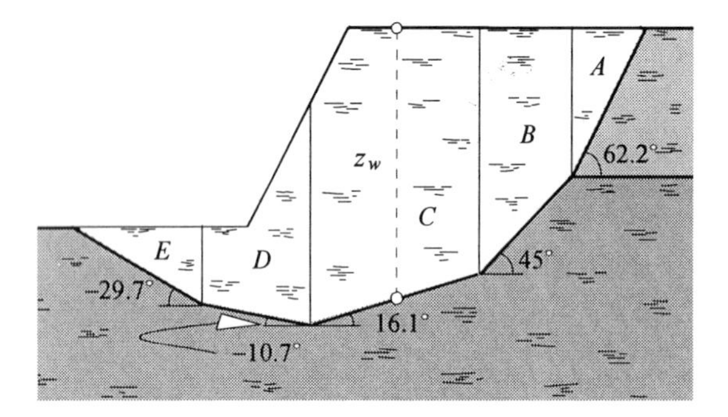

Figure 7.16: Waterlogged slope.

slice	$z_w\,(m)$	$u\,(kN/m^2)$	$ub\,(kN/m)$	$(w - ub)\tan\phi'_c\,(kN/m)$
A	1.9	19	38	15.3
B	5.05	50.5	126	55.5
C	6.73	67.3	303	133.5
D	3.25	32.5	94	41.2
E	1.00	10	35	15.4

Hence the following calculations using an initial factor of safety $F = 1$:

slice	$m_\alpha = \cos\alpha + \frac{\tan\phi'_c \sin\alpha}{1}$	$\frac{1}{m_\alpha}[c'b + (w - ub)\tan\phi'_c]$ (kN/m)
A	0.82	19
B	1.02	67
C	1.08	144
D	0.90	62
E	0.65	51

Therefore:

$$\sum \frac{1}{m_\alpha}\left[c'b + (w - ub)\tan\phi'_c\right] = 343\ kN/m$$

yielding a factor of safety:

$$F = \frac{343}{344.2} = 0.99$$

which is, for all practical purposes, identical to the initial chosen value $F = 1$, implying in the process that the slope would collapse. No wonder that slope failures are a common occurrence every time there is a prolonged period of sustained rainfall, leading to slopes being waterlogged.

These simple examples, though they may appear tedious in terms of calculations, are very useful to illustrate the practical effects of managing the water level within a slope. Failing that, stability problems may get out of hand.

7.5.4 Effect of seepage forces on the stability of slopes

In the presence of seepage forces, the analysis of the slope stability is still undertaken according to equation 7.34, except that this time the porewater pressure is estimated either from a flownet or through the use of an analytical procedure such as a conformal mapping technique as explained in chapter 3.

Such analyses are best illustrated by a practical example, and in this respect, let us consider once more the stability of the earth dam for which a flownet was constructed in section 3.4.3 (refer to figure 7.17). Clearly, the stability of both upstream and downstream sides of the dam is greatly affected by the slopes, which in turn depend on the fill material. Thus, for a compacted clay characterised by a (relatively) small angle of shearing resistance (say $\phi' \leq 20°$), a slope on the downstream side as low as 4 (horizontal) to 1 (vertical) might be needed to guarantee safety *vis à vis* slope failure. In general, and purely as a guideline, the design of an earth dam must be such that the slope on the *upstream* side is between 3 to 1 (for a small sized dam), decreasing to 3.5 to 1 for a medium sized dam. On the *downstream* side, slopes varying from 2.5 to 1, decreasing to 4 to 1 as the dam size increases, are usually suitable as far as stability is concerned. Obviously, these values are of a purely informative nature, and therefore *cannot* be used as a substitute for a full analysis such as the one to be undertaken shortly. Moreover, when the *downstream* slope of an earth dam is protected by a drainage blanket, as depicted in figure 7.17, then a minimum factor of safety $F_{\min} = 1.5$ is required when the dam is fully operational.

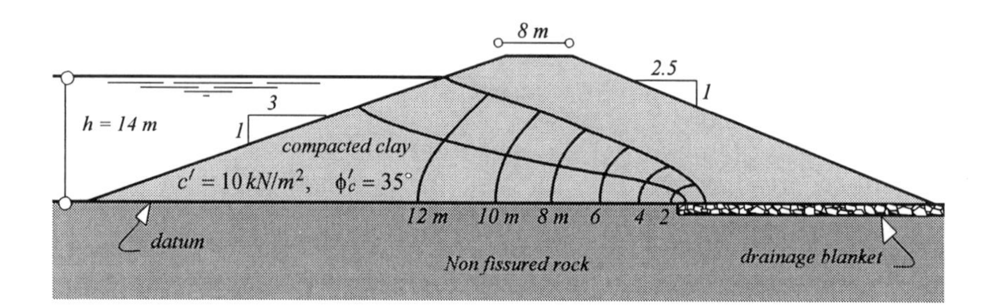

Figure 7.17: Stability of slopes subjected to seepage forces; case of an earth dam.

On the *upstream* side, however, the critical stability conditions occur when, for some reason, the water level drops very quickly. These conditions are usually referred to as *rapid drawdown*. Under such circumstances, the water pressure within the slope is hardly changed while the volume of water that used to be applied as an additional weight on top of the slope has disappeared (refer to figure 7.18). Accordingly, the stability of the *upstream* side must be analysed under this state of stresses, and a minimum factor of safety $F_{\min} = 1.25$, corresponding to *rapid drawdown* conditions is required.

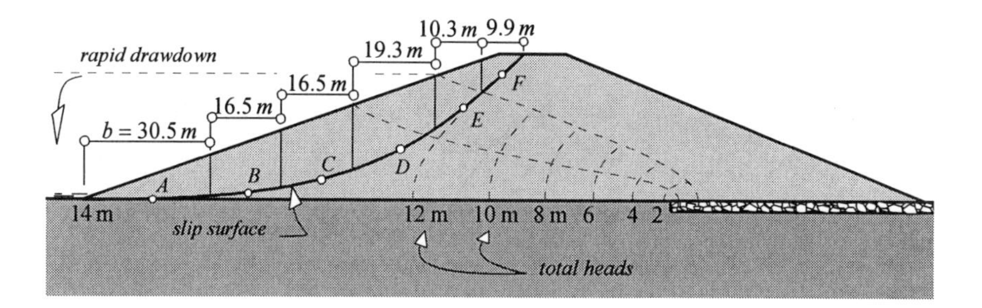

Figure 7.18: Simulation of rapid drawdown conditions.

Assuming, therefore, that rapid drawdown conditions have occurred, the analysis of the stability of the upstream slope can be undertaken in a manner similar to that used previously. Indeed, within the selected slip circle depicted in figure 7.18, six slices are chosen with the dimensions indicated in the figure. The total weight of each element can be calculated in a straightforward way assuming a saturated unit weight of clay $\gamma_{sat} = 21\ kN/m^3$. However, the porewater pressure, estimated at the middle of the base of each slice (*i.e.* at points *A, B, C, D, E* and *F* in figure 7.18), is calculated from the total head h (estimated from the flownet) and the elevation head h_e (scaled from the figure) of each point, so that for instance at *D*, the porewater pressure is:

$$u_D = (h - h_e)\,\gamma_w \approx (12.5 - 5.15) \times 10 = 73.5\ kN/m^2$$

Whence the following quantities are to be used in conjunction with equation 7.34 to estimate the factor of safety:

slice	$u\,(kN/m^2)$	$w\,(kN/m)$	$b\,(m)$	α (°)	$w \sin\alpha\,(kN/m)$	$(w - ub)\tan\phi'\,(kN/m)$
A	140	3,200	30.5	0	0	–749
B	130	3,995	16.5	6	418	1,292
C	114	5,065	16.5	13	1,139	2,235
D	74	5,605	19.3	27	2,545	2,929
E	21	2,138	10.3	38	1,316	1,346
F	0	856	9.9	42	573	599

and the sum: $\Sigma w \sin\alpha = 5991\ kN/m$.

The iterations, according to equation 7.34 can now be started, selecting initially a value $F = 1$ for the factor of safety; thence:

slice	$m_\alpha = \cos\alpha + \dfrac{\tan\phi'_c \sin\alpha}{1}$	$\dfrac{1}{m_\alpha}[c'b + (w - ub)\tan\phi'_c]\ (kN/m)$
A	1.00	–444
B	1.07	1,365
C	1.13	2,120
D	1.21	2,583
E	1.22	1,189
F	1.21	576

Accordingly:

$$\sum \frac{1}{m_\alpha}\left[c'b + (w - ub)\tan\phi'_c\right] = 7389\ kN/m$$

and the calculated value of the factor of safety is thus:

$$F = \frac{7389}{5991} = 1.23$$

This value is different from the initial value of one, and therefore another iteration is needed. Using the newly calculated value $F = 1.23$ as an initial value for iteration, it follows that:

slice	$m_\alpha = \cos\alpha + \dfrac{\tan\phi'_c \sin\alpha}{1.23}$	$\dfrac{1}{m_\alpha}[c'b + (w - ub)\tan\phi'_c]\ (kN/m)$
A	1.00	–444
B	1.05	1,383
C	1.10	2,177
D	1.15	2,716
E	1.14	1,273
F	1.12	621

yielding the quantity:

$$\sum \frac{1}{m_\alpha}\left[c'b + (w - ub)\tan\phi'_c\right] = 7726\, kN/m$$

and a factor of safety:

$$F = \frac{7726}{5991} = 1.29$$

It is easy to show that another iteration using an initial factor of safety $F = 1.29$ yields a calculated value $F = 1.31$, which is then assumed to be the actual factor of safety under the critical conditions of rapid drawdown for the selected slip surface on the downstream side.

In fact, in this particular case, it can be shown that the selected slip circle in figure 7.18 corresponds to the most critical slip surface, meaning that the value $F = 1.31$ is indeed the *minimum* factor of safety, and implying in the process that the dam is safe *vis à vis* rapid drawdown conditions.

Obviously, as mentioned previously, the process of calculating the *minimum* factor of safety F_{min} can be lengthy and tedious, and is usually undertaken numerically. Bishop and Morgenstern (1960) produced very useful charts, based on equation 7.34, which can be applied to calculate F_{min} (see also Chandler and Peiris (1989)). However, these charts cannot be used *ad lib,* and it is strongly advisable to read the papers in question thoroughly to be aware of the limitations related to their use then, when applicable, to use the charts cautiously.

7.6 Location of the critical failure surface

In practice, slope stability calculations are almost exclusively undertaken numerically, so that literally hundreds of failure surfaces can be used in conjunction with Bishop's method until the one with the lowest factor of safety is located. The computational process can therefore use a random selection of slip surfaces by varying the centre of rotation or the radius of the circle, for instance. However, the process can be optimised by concentrating the calculations on a prescribed area within which the critical slip surface is likely to be located. In this respect, some purely empirical methods can be used cautiously such as the one represented graphically in figure 7.19. The method, first suggested by Fellenius (1927), applies mainly to slopes in cohesive homogeneous soils under undrained conditions (*i.e.* $\phi = 0$). The initial position of the centre of the critical slip circle is estimated from both graphs of figure 7.19, depending on the slope angle β.

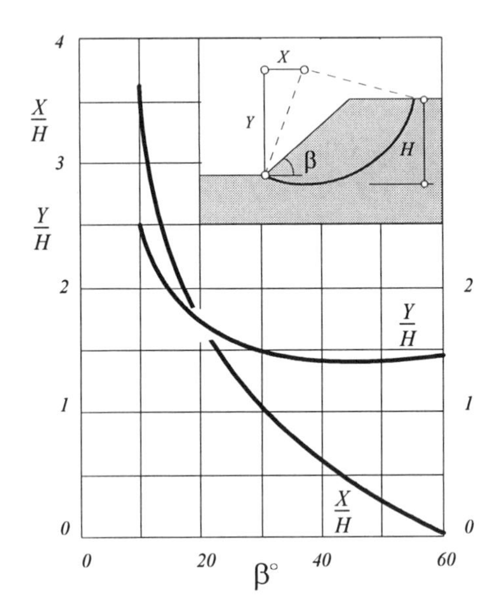

Figure 7.19: Initial location of the critical slip circle (from Whitlow (1995), reproduced by permission).

Using the initial centre of rotation, iterative calculations can then be made by varying the radius of the corresponding slip circle. Usually, a grid pattern is used whereby different centres of the slip circle estimated from figure 7.19 are tried until the critical circle yielding the minimum factor of safety is located.

Bear in mind that the method related to figure 7.19 is only used to help accelerate the search for the critical slip surface and, as such, it can be cautiously extended to homogeneous soils (*i.e.* soils consisting of one layer) under drained conditions, then used in conjunction with Bishop's routine presented earlier.

7.7 Non-circular failure surface

Apart from infinite slopes, the analysis developed so far assumes that the shape of the failure surface is circular. There are many instances in practice when this assumption is no longer valid, in that the actual failure surface is between a circle and an infinite slope (which is a circle with an infinite radius).

Several methods of analysis which can be applied to slope design independently of the failure surface shape are available (see Morgenstern and Price (1965), for instance). In particular, Janbu (1973) developed an analysis similar in nature to that related to Bishop's routine presented earlier, in which the expression of the factor of safety, which takes into account the resultant of the tangential forces $(x_{i-1} - x_i) = \Delta x$ applied on each slice (refer to figure 7.11), irrespective of the shape of the failure surface, is as follows:

$$F = \frac{1}{\Sigma(w + \Delta x)\sin\alpha} \sum \left\{ [c'b + w(1 + \Delta x - r_u)\tan\phi'] \frac{1}{\cos\alpha + \frac{\sin\alpha\tan\phi'}{F}} \right\} \qquad (7.35)$$

where all symbols are as defined in equation 7.34.

Janbu suggested calculating the factor of safety F from equation 7.35 for which the quantity Δx is assumed to be zero (in which case equation 7.35 becomes identical to equation 7.34), then applying an empirically based correction factor f_o to the computed value. The correction factor f_o, estimated from figure 7.20, depends on the ratio d/L reflecting the non-circular nature of the failure surface.

Although such a method is considered reliable in engineering practice, one has to bear in mind that the graphs in figure 7.20 were developed for a non-circular failure surface with a moderate slope and, accordingly, they should be used cautiously. Also, for fairly shallow inclines (with slopes smaller than 15°, say), the analysis related to infinite slopes (refer to sections 7.2 to 7.4) may be more appropriate.

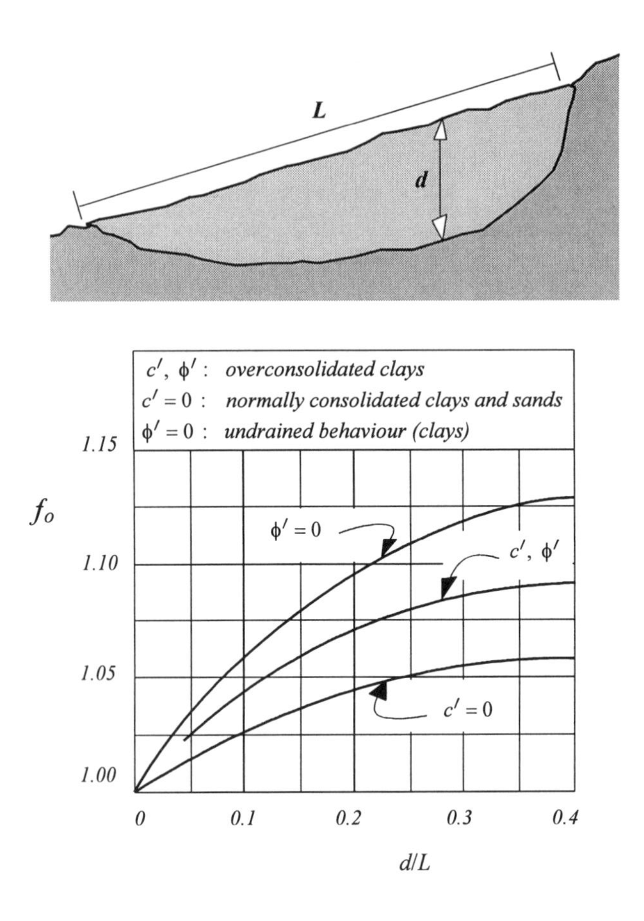

Figure 7.20: Correction factor for non-circular failure surface.

7.8 Improving the stability of unsafe slopes

7.8.1 Use of geotextile reinforcement

The use of *geotextiles* as a means of reinforcement of unsafe artificial slopes is, nowadays, a common engineering practice. Geotextiles are a by-product of petroleum, and are usually made from polyethylene or polypropylene.

To date, *geogrids* consisting of a polymer mesh geotextiles such as the one depicted in figure 7.21 are extensively used. The apertures in the geogrids ensure an interlocking with the surrounding soil, thus maximising the reinforcement, in other words, the tensile forces T_i shown in figure 7.22.

These tensile forces lead to a marked improvement of the factor of safety of the slope as shall be demonstrated below.

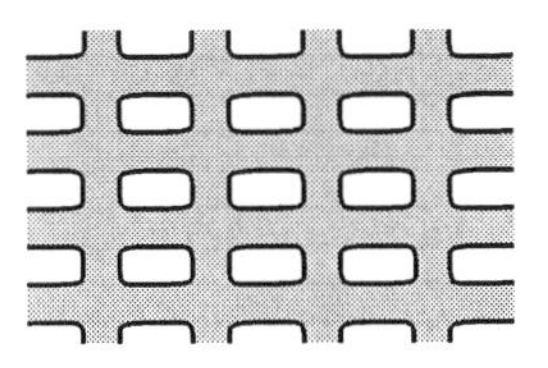

Figure 7.21: Geogrid.

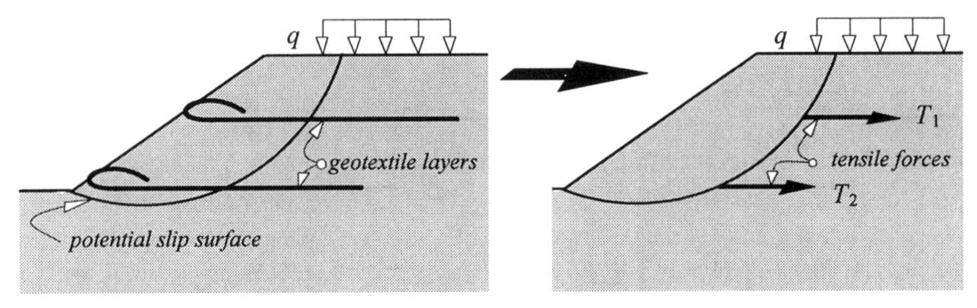

Figure 7.22: Equivalent tensile forces generated by geogrid layers.

Because of the nature of equation 7.34, a Bishop type analysis cannot easily be expanded to include the tensile forces in figure 7.22. Consequently, Fellenius' method (1927) will be applied to evaluate the effect on the overall stability of the slope of these tensile forces.

Consider an isolated element at the base of which a tensile force T, generated by a geogrid layer, and inclined at an angle β as depicted in figure 7.23, is applied. Fellenius' method considers both the equilibrium of forces perpendicular to the base of the element and the equilibrium of moments about the centre of rotation O. Thus, referring to figure 7.23, it is seen that:

$$w\cos\alpha = l\left(\sigma'_n + u\right) - T\cos\left(\frac{\pi}{2} - \beta\right) \tag{7.36}$$

Moreover, the equilibrium of moments about O is such that:

$$\Sigma R\tau_{mob}\, l + RT\cos\beta = \Sigma Rw\sin\alpha \tag{7.37}$$

but, according to equation 7.27:

$$l\tau_{mob} = \frac{lc'}{F} + \frac{\sigma'_n l}{F}\tan\phi'_c \tag{7.38}$$

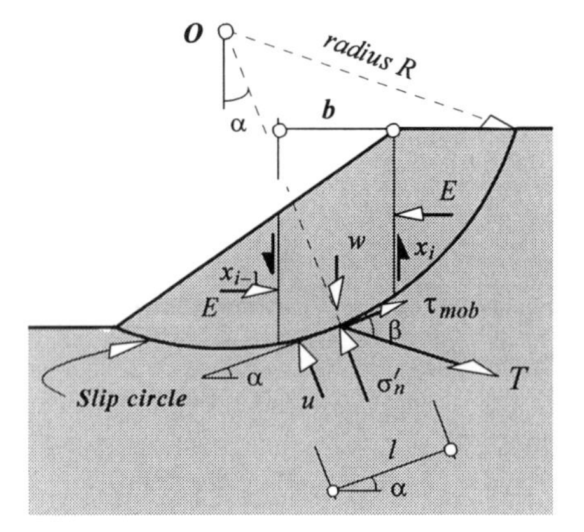

Figure 7.23: Boundary conditions related to Fellenius' method of slope analysis.

On the other hand, the quantity $\sigma'_n l$ can easily be found from equation 7.36:

$$\sigma'_n l = w\cos\alpha - ul + T\sin\beta \tag{7.39}$$

Therefore, substituting for $\sigma'_n l$ in equation 7.38, then for the resulting quantity $l\tau_{mob}$ in equation 7.37, and knowing that $l = b/\cos\alpha$ (refer to figure 7.23), the ensuing expression of the factor of safety can readily be established:

$$F = \frac{1}{\Sigma w\sin\alpha - T\cos\beta}\sum\left[\frac{c'b}{\cos\alpha} + \left(w\cos\alpha - \frac{ub}{\cos\alpha} + T\sin\beta\right)\tan\phi'_c\right] \tag{7.40}$$

As opposed to equation 7.34, equation 7.40 does not involve an iterative procedure for the calculation of the factor of safety.

Example 7.4

An 18.6 m high embankment, with a steep slope on each side as depicted in figure 7.24 is to be built using a dense silty sand characterised by a critical angle of shearing resistance $\phi_c' = 35°$ and a bulk unit weight $\gamma = 20\,kN/m^3$. Three geogrid layers, each providing a tensile force $T = 110\,kN/m$ are used to ensure the stability of both slopes. Under such circumstances, the factor of safety corresponding to the selected potential slip surface is evaluated using equation 7.40. The potentially failing soil mass is first subdivided into five elements (*a*, *b*,...., *e* in figure 7.24) with the following characteristics:

slice	$w\,(kN/m)$	$\alpha\,(°)$	$b\,(m)$	$T\,(kN/m)$
a	283	11	4	0
b	1,212	19	5.3	0
c	1,498	35	5.3	110
d	768	53	4	110
e	187	69	2.67	110

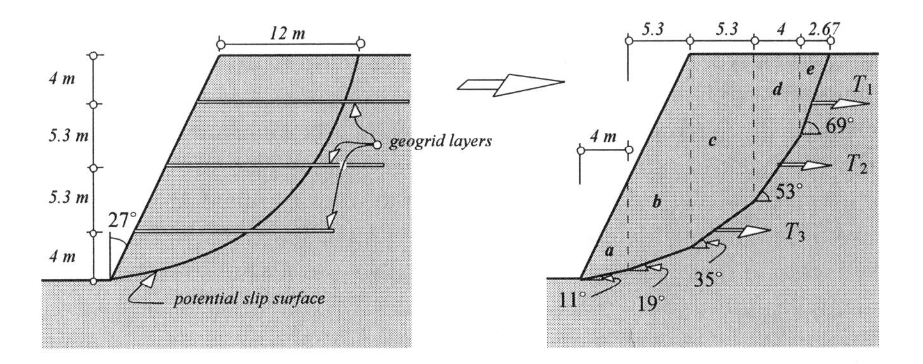

Figure 7.24: Embankment characteristics and potential slip surface.

It is easy to check that:

$$\Sigma w \sin\alpha = 2096\, kN/m\,,\ \ \Sigma w \cos\alpha = 3180\, kN/m,$$

$$\Sigma T \sin\beta = 0\,,\ \ \Sigma T \cos\beta = 330\, kN/m$$

(notice in this example, the angle β as per figure 7.23 is zero).

Therefore, assuming zero values for both cohesion and porewater pressure, equation 7.40 yields:

$$F = \frac{1}{(2096 - 330)} \times 3180 \times \tan 35 = 1.26$$

In the absence of any reinforcement, the same equation would have produced a factor of safety:

$$F = \frac{3180}{2096} \tan 35 = 1.06$$

meaning that the slope will not have stood on its own.

Notice that, in practice, Fellenius' method is found to be somewhat conservative in that it yields slight underestimates of the factor of safety.

7.8.2 Use of micropiles and soil nails

An alternative solution to the use of geotextiles consists of placing micropiles with a diameter usually not exceeding 250 *mm*, across the potential slip surface so that the forces resisting the slide are enhanced, and so is the factor of safety of the whole slope. Although the use of micropiles in conjunction with slopes is similar to the use of ground anchors in the case of retaining structures (see chapter 11), the fundamental behavioural difference is that anchors are essentially subject to tensile stresses whereas micropiles are usually placed either vertically or normal to the potential slip surface (refer to figure 7.25 for instance) and are, as such, subject principally to shear stresses. *Soil nails*, on the other hand, are essentially micropiles used exclusively for slope stability.

From a practical perspective, the designer must ensure that micropiles are designed in an effective way, thus avoiding the (embarrassing!) situation whereby they become ineffectual through the development of a potential slip surface deep beneath the piles as depicted in figure 7.25.

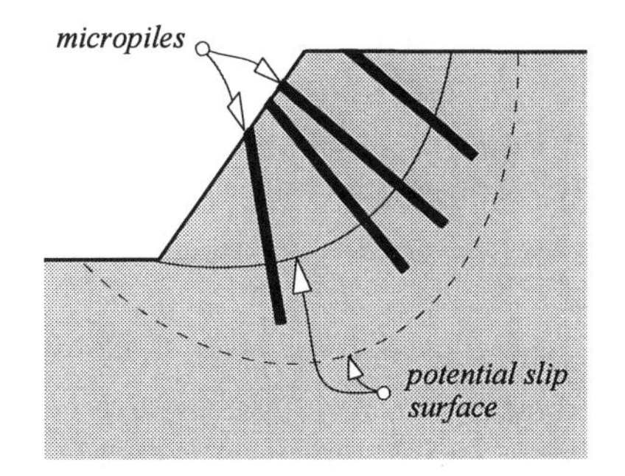

Figure 7.25: Potential problems related to the use of micropiles in conjunction with circular slip surfaces.

Figures 7.26, on the other hand, shows the potential pitfalls related to the inappropriate use of rigid piles to stabilise long slopes. Both figures indicate the need for a comprehensive analysis of the nature of the slope to be supported and the type of pile–soil interaction likely to be generated once piles are in place.

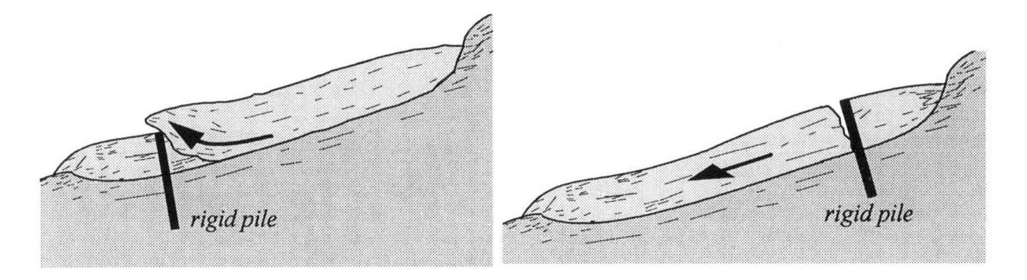

Figure 7.26: Practical aspects related to the use of rigid piles for slope stability.

However, the nature of soil–pile interaction can be very complex to analyse since, in this case, piles are subject mainly to passive lateral loading, in other words, the piles are loaded through the potentially failing soil mass and, as such, an appropriate design (in terms of pile dimensions and pile number) would require the prior knowledge of the profiles of bending moments, shear stresses and pressure distribution along the pile shaft. Moreover, piles are usually used in numbers, and therefore the design is bound to be affected by the group effect, which can be difficult to assess.

Nonetheless, this type of reinforced slopes can be designed reasonably well using, for instance, a Taylor type total stress analysis, based on the assumption that the pressure at which the soil in the vicinity of the pile fails (by literally flowing around the pile shaft) is equivalent to $10c_u$, where c_u represents the undrained shear strength of the soil (see for instance Randolph and Houlsby (1984), Poulos and Davis (1980)).

Consider the slope in a saturated clay depicted in figure 7.27. If n identical micropiles, each with a diameter a were inserted as per the figure, and if c_u is the undrained shear strength of the clay, then clearly the shear stress mobilised throughout the length of the slip plane $(L - na)$ is:

$$\tau_{mob} = \frac{c_u}{F} \tag{7.41}$$

In addition, the shear stress mobilised by the use of n piles is:

$$\tau = \frac{1}{F}(10nac_u) \tag{7.42}$$

F, in both equations, being the factor of safety.

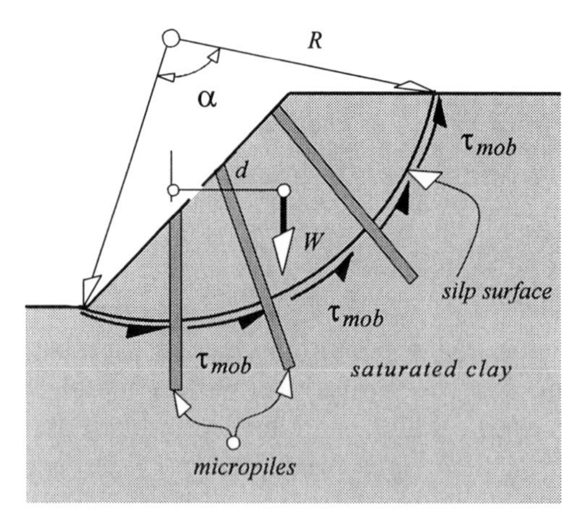

Figure 7.27: Use of micropiles in conjunction with circular slip planes.

Taking moments about the centre of rotation, it follows that:

$$Wd = \frac{1}{F}[(L - na)\, c_u R + 10nac_u R]$$

giving the expression of the factor of safety:

$$F = \frac{c_u R}{Wd}(L + 9na) \qquad (7.43)$$

Example 7.5

With reference to figure 7.27, assuming the weight of the failing soil mass is $W = 1000\, kN/m$, estimate the improvement in the factor of safety of the slope if 2 micropiles with a diameter $a = 180\, mm$ were used to stabilise the slope. The clay has an undrained shear strength $c_u = 40\, kN/m^2$, and the slip circle is characterised by a radius $R = 7.2\, m$, an angle $\alpha = 1.48\, rad$ and an eccentricity $d = 2.8\, m$.

First, the factor of safety prior to the installation of micropiles can be calculated from equation 7.22:

$$F_1 = \frac{40 \times 1.48 \times 7.2^2}{1000 \times 2.8} = 1.1$$

Obviously, this value indicates that the slope is unsafe; whence the use of micropiles. The new factor of safety is now estimated from equation 7.43:

$$F_2 = \frac{40 \times 7.2}{1000 \times 2.8} \times (7.2 \times 1.48 + 9 \times 2 \times 0.18) = 1.43$$

that is an increase in the factor of safety of 30%.

Notice that, if the magnitude and direction of the shear forces developed in the piles along the slip plane were known, then an effective stress analysis, similar to that used in conjunction with geotextiles, can be applied, and the factor of safety of the slope can be estimated from equation 7.40.

Problems

7.1 A 6 m thick layer of boulder clay has a natural angle of slope $\beta = 16^\circ$, and is underlain by a layer of rock inclined at roughly the same angle, constituting a potential slip surface. The saturated clay is characterised by an angle $\phi' = 24^\circ$, an apparent cohesion $c' = 7\,kN/m^2$, and a saturated unit weight $\gamma_{sat} = 21\,kN/m^3$, the water flow conditions being as illustrated in figure $p7.1$. Preliminary calculations indicate that the natural slope should be reduced to 12° for the requirement of a factor of safety against failure $F = 1.3$ to be fulfilled. Alternatively, the same factor of safety can be achieved by lowering the water table to a depth z_w while keeping the natural slope unaltered. Calculate z_w.

Ans: $z_w = 2.39\,m$

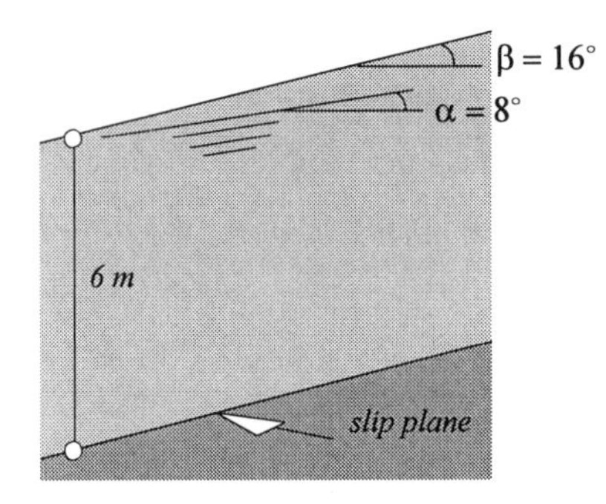

Figure p7.1

7.2 An analysis of the stability of a newly cut slope suggested that a slip circle failure mechanism, such as the one illustrated in figure $p7.2$ might develop in the short term. Knowing that the clay has an undrained shear strength $c_u = 50\,kN/m^2$ and a saturated unit weight $\gamma_{sat} = 20\,kN/m^3$, and assuming the failure surface is characterised by an area $A \approx 196\,m^2$, a radius $R = 12\,m$, an angle $\alpha \approx 90^\circ$, and an eccentricity $d = 2.5\,m$, estimate the number of $150\,mm$ diameter micropiles needed per metre to ensure a minimum factor of safety against failure $F = 1.6$.

Ans: $n = 6$

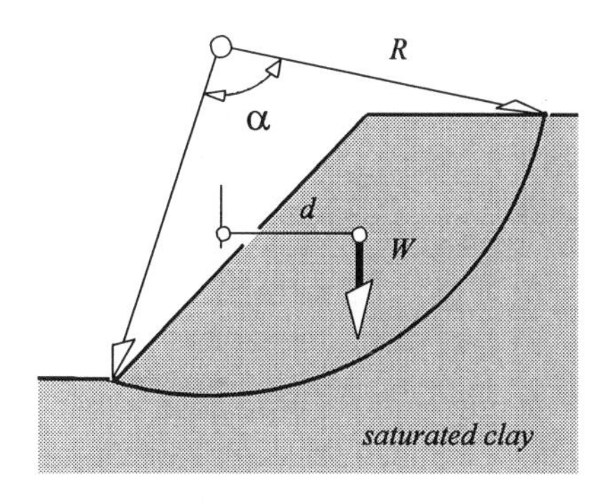

Figure p7.2

7.3 Refer to the failure surface related to the artificial slope of figure *p*7.3, then consider its short term stability by estimating the factor of safety against failure. Assume the soil has a bulk unit weight $\gamma = 20\, kN/m^3$ and an undrained shear strength $c_u = 65 kN/m^2$.

Ans: $F \approx 2$

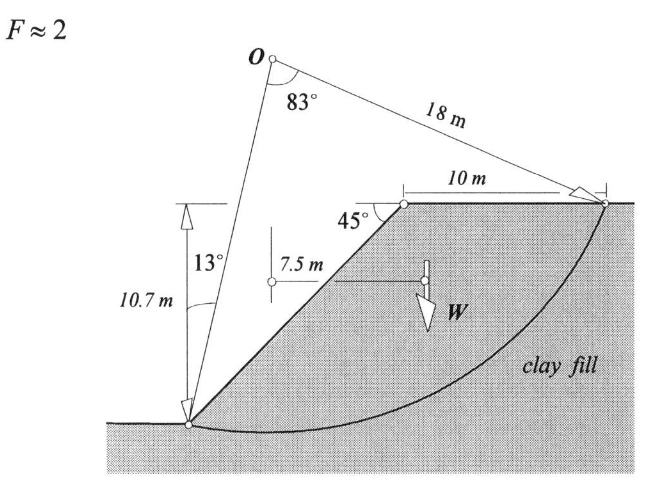

Figure p7.3

7.4 Consider the long term analysis of the stability of the slope depicted in figure *p*7.3. Assume now the long term behaviour of the clay fill is characterised by an angle $\phi' = 23°$ and an apparent cohesion $c' = 7\, kN/m^2$, then use a Bishop type analysis to estimate the factor of safety against failure.

Ans: $F \approx 1.34$

7.5 Calculate the long term factor of safety against shear failure if the slope in figure *p*7.3 were to be waterlogged.
Use $\gamma_{sat} = 21\ kN/m^3$.

Ans: $F \approx 0.7$

7.6 To improve the stability of the waterlogged slope of figure *p*7.3, layers of geogrids, each developing a horizontal tensile force of a magnitude $T = 150\ kN/m$, were placed at regular (vertical) intervals with respect to the toe of the slope. Estimate, using Fellenius' method, how many geogrid layers need be placed if a minimum factor of safety $F = 1.7$ were required when the slope becomes waterlogged.

Ans: $n = 4$

References

Bishop, A. W. and Morgenstern, N. R. (1960) *Stability coefficients for earth slopes.* Géotechnique, 10, pp. 129–150.

Chandler, R. J. and Peiris T. A. (1989) *Further extensions to the Bishop and Morgenstern slope stability charts.* Ground Engineering, May, pp. 33–38.

Fellenius, W. (1927) *Erdstatische Berechnungen mit Reibung und Kohasion (Adhasion) und unter Annahme Kreiszylindrischer Gleitflachen.* W. Ernst, Berlin.

Janbu, N. (1973) *Slope stability computations.* In Embankment Dam Engineering, Casagrande Memorial Volume (eds R. C. Hirschfield and S. J. Poulos). John Wiley, New York.

Morgenstern, N. R. and Price, V. E. (1965) *The analysis of the stability of general slip circles.* Géotechnique, 15 (1), pp. 79–93.

Palladino, D. J. and Peck, R. B. (1972) *Slope failures in an overconsolidated clay in Seattle, Washington.* Géotechnique, 22 (4), pp. 563–595.

Poulos, H. G. and Davis, E. H. (1980) *Pile Foundation Analysis and Design.* John Wiley & Sons, New York.

Randolph, M. F. and Houlsby, G. T. (1984) *The limiting pressure on a circular pile loaded laterally in cohesive soil.* Géotechnique, 34 (4), pp. 613–623.

Skempton, A. W. (1964) *Long term satbility of clay slopes.* Géotechnique, 14 (2), pp. 77–102.

Skempton, A. W. and La Rochelle, P. (1965) *The Bradwell slip: a short term failure in London clay.* Géotechnique, 15 (3), pp. 221–242.

Spencer, E. (1967) *A method of analysis of the stability of embankments using parallel interslice forces.* Géotechnique, 17 (1), pp. 11–26.

Taylor, D. W. (1948) *Fundamentals of Soil Mechanics.* John Wiley & Sons, New York.

CHAPTER 8

Limit analysis applied to the bearing capacity of shallow foundations

8.1 Introduction

Referring to section 6.6, it is seen that the calculation of strains in the entire elastic–plastic range requires the four following criteria to be fulfilled:

- the equilibrium of stresses;
- the compatibility of strains;
- the stress–strain relationship in the elastic range through Hooke's law;
- the normality condition.

A complete solution to any boundary value problem in geotechnics can be achieved if these criteria are satisfied *simultaneously*. However, it can be argued that for most engineering problems, a complete solution covering the entire elastic–plastic range can quickly become a very time consuming luxury that, given a viable alternative, most designers *will* do without. A possible alternative consists of focusing solely on the small range of soil behaviour during which collapse occurs, and to use the powerful bounds theorem of the plastic theory to calculate an upper and a lower limits to the *actual* collapse (or ultimate) load. These limits are often so close to each other and can in some cases be identical, corresponding thus to the *true* solution. The attractiveness of this alternative is obvious when the soil strength is primarily of interest to the designer, as in the case of bearing capacity problems, lateral earth pressure or slope stability.

In essence, the *upper* and *lower bounds methods*, to be detailed shortly, represent two different approaches that Calladine (1985), in his intelligent book refers to as the *geometry* and *equilibrium* approaches respectively. Both methods are based on the assumption that the soil is characterised by a *rigid perfectly plastic* behaviour with an *associated flow rule*, so that a lower bound solution to the collapse load can be achieved by satisfying only the *equilibrium equations* and *yield conditions*, whereas an upper bound solution is arrived at by considering the balance of both internal and external energies of the *mechanism* of failure. Consequently, while being always larger than or equal to a lower bound solution, the exact collapse load is always smaller than, if not equal to, an upper bound solution. As such, the lower bound represents a *safe* estimate of the soil strength.

8.2 The upper and lower bounds theorems

Upper bound theorem

Any chosen mechanism of deformation of a body, within which the rate of dissipation of energy is equated to the rate at which external forces do work, must give an estimate of the plastic collapse load that is higher than, if not equal to, the true collapse load.

Lower bound theorem

If there is a set of external forces for which the stress distribution is everywhere in internal equilibrium and nowhere exceeds the yield criteria, then those forces will be carried safely by the structure, and they therefore represent a lower bound to the true collapse load.

Leaving aside the mathematical proof of these theorems which is out of the scope of this book, and can be found in Calladine (1985), for instance, these two statements imply that an upper bound solution is achieved by choosing a *kinematically admissible* collapse mechanism that consists basically of rigid blocks sliding with respect to one another, then equating the energy dissipated through shear along the slip planes to the external work done by the set of applied loads. Similarly, a lower bound solution is obtained by selecting a *statically admissible* stress field, satisfying in the process the equilibrium equations as well as the yield criteria. However, as explicitly presented in chapter 5, the yield criteria of a soil depends on drainage conditions, in other words on whether or not the pore water pressure is allowed to build up or to dissipate. Let us therefore consider the bounds theorems separately in conjunction with both short term *undrained* and long term *drained* behaviours of cohesive soils, and determine the collapse load q_u in the case of a *surface* footing having a width B, and transmitting a stress q to a uniformly loaded soil as depicted in figure 8.1.

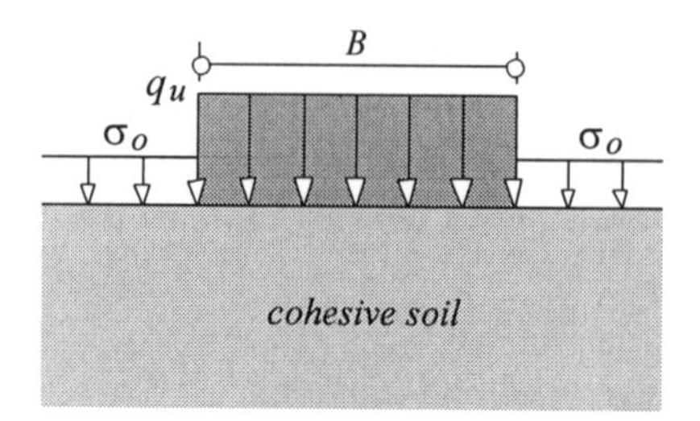

Figure 8.1: Loading conditions.

8.3 Kinematically admissible collapse mechanisms: undrained behaviour

(a) Slip circle mechanism

This mechanism, shown in figure 8.2, assumes that failure of the volume of soil beneath the foundation occurs through rotation with respect to point O. If the angle of rotation is $d\theta$, then the downward displacement of the external edge of foundation is $B d\theta$, and the work done by the external loads is then:

$$\delta E = \tfrac{1}{2}\, B d\theta (q_u - \sigma_o) B \tag{8.1}$$

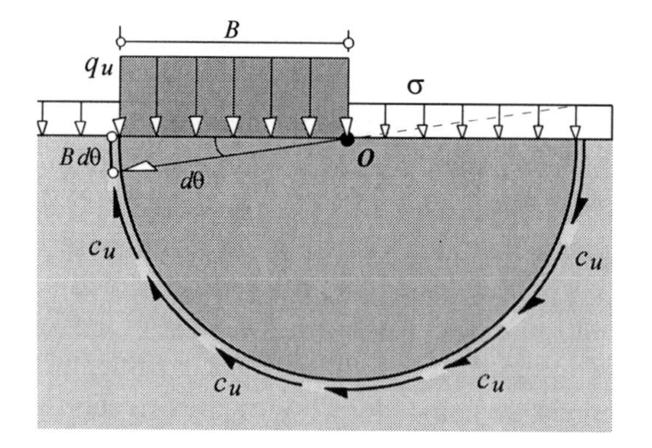

Figure 8.2: Slip circle collapse mechanism.

Similarly, the energy dissipated through shear along the slip circle is:

$$\delta W = \pi B\, c_u\, B\, d\theta \tag{8.2}$$

Equating both equations then rearranging, it follows that the upper bound solution of the collapse load for this mechanism is:

$$q_u = \sigma_o + 2\pi c_u \tag{8.3}$$

(b) Sliding rigid blocks

The corresponding collapse mechanism is shown in figure 8.3, where the rigid block beneath the foundation is pushed into the ground with a velocity V, while the middle block is pushed sideways, so making the last block move upward. All movements are resisted through friction on different slip planes with different velocities, which can easily be determined using the geometry of the blocks. Whence:

$$V_1 = V_3 = V\sqrt{2}\text{ , and } \quad V_2 = 2V$$

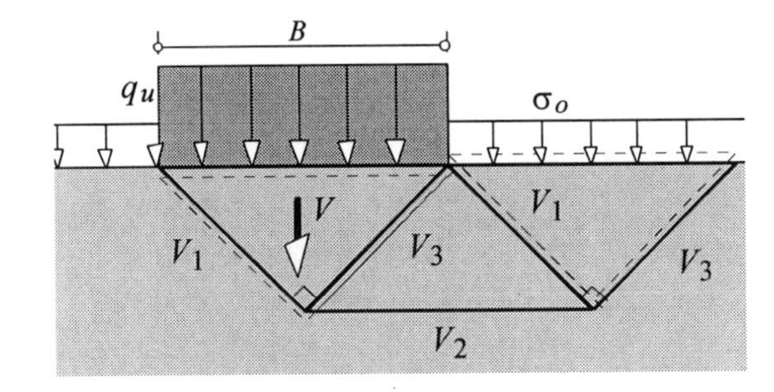

Figure 8.3: Rigid blocks collapse mechanism.

Accordingly, the work done by external forces is:

$$\delta E = (q_u - \sigma_o)\, B\, V \tag{8.4}$$

and the energy dissipated along all five slip planes is:

$$\delta W = 4\frac{B}{\sqrt{2}} c_u V\sqrt{2} + 2VBc_u = 6BVc_u \tag{8.5}$$

Thus, the upper bound solution related to this mechanism, obtained by equating the latter two equations:

$$q_u = \sigma_o + 6c_u \tag{8.6}$$

which represents an improvement on the value calculated from equation 8.3.

(c) Slip circle with a rotation centre above surface level

The corresponding collapse mechanism is depicted in figure 8.4, and as in the previous case, the work done by the external loads is calculated in the knowledge that the foundation width is $B = R\sin\alpha$. Whence:

$$\delta E = \frac{1}{2} R d\theta \sin\alpha \ \ (q_u - \sigma_o) R \sin\alpha \tag{8.7}$$

and the energy dissipated through shear is:

$$\delta W = 2\alpha R c_u R d\theta \tag{8.8}$$

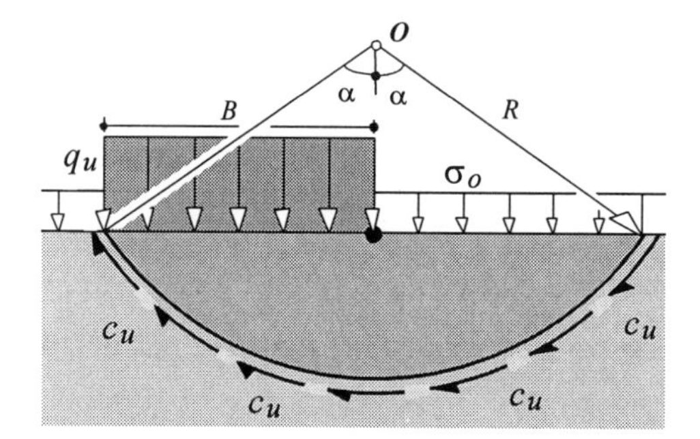

Figure 8.4: Collapse mechanism with a centre of rotation above surface level.

Equating these two quantities then yields:

$$(q_u - \sigma_o) = \frac{4\alpha}{\sin^2\alpha} c_u \tag{8.9}$$

The ultimate collapse load is thereafter found by optimising the angle α. From equation 8.9, it is easy to establish that:

$$\frac{\partial q_u}{\partial \alpha} = \frac{4c_u}{\sin^2\alpha}\left(1 - \frac{2\alpha}{\tan\alpha}\right) \tag{8.10}$$

The optimum q_u value is found by writing $\partial q_u/\partial\alpha = 0$ in equation 8.10, yielding the transcendental equation $\tan\alpha = 2\alpha$ which is satisfied when $\alpha = 0.37\,rad$ (or $67°$). Inserting this value into equation 8.9, then rearranging:

$$q_u = \sigma_o + 5.53c \tag{8.11}$$

which is a marked improvement on the previous upper bound load of equation 8.6.

(d) Sliding rigid blocks related by a shear fan

The mechanism depicted in figure 8.5 consists of two blocks sandwiching a shear fan constituted of an infinite number of sliding wedges. As the block beneath the foundation sinks, it pushes the first wedge of the fan which then rotates with respect to the centre *O* (refer to the figure) by a small angle $d\theta$ transmitting in the process the movement to the adjacent wedge. It is clear from the geometry of the figure that all slip planes, including the wedges inside the fan, have the same length $B/\sqrt{2}$. Also, the velocities along *ac, cO, Od* and *de* are identical with a magnitude of $V\sqrt{2}$. Consequently, the energy balance must be zero, and so the work due to external forces is:

$$\delta E = (q_u - \sigma_o)B\,V \tag{8.12}$$

the energy dissipated along the planes *ac, cd* and *de* is as follows:

$$\delta W_1 = \frac{B}{\sqrt{2}}c_u V\sqrt{2} + \frac{B}{\sqrt{2}}\frac{\pi}{2}c_u V\sqrt{2} + \frac{B}{\sqrt{2}}c_u V\sqrt{2} \tag{8.13a}$$

$$= B\,Vc_u\left(2 + \frac{\pi}{2}\right) \tag{8.13b}$$

The energy dissipated through the shear fan is easily integrated in the following manner:

$$\delta W_2 = \int_0^{\pi/2} c_u \frac{B}{\sqrt{2}} V\sqrt{2}\, d\theta = c_u B V \frac{\pi}{2} \tag{8.14}$$

Hence a total energy dissipated:

$$\delta W = c_u B V (2+\pi) \tag{8.15}$$

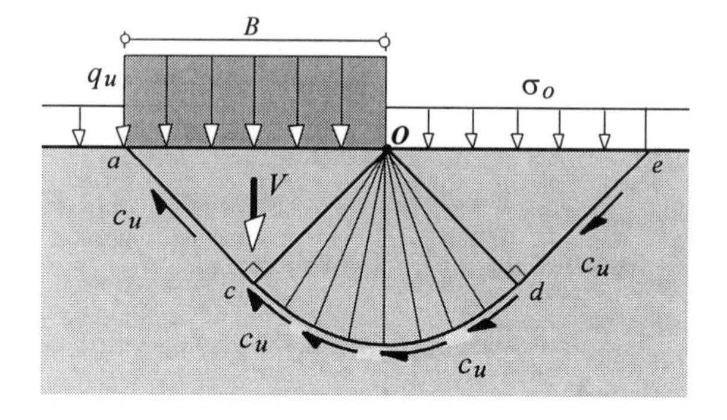

Figure 8.5: Sliding blocks related by a shear fan.

The upper bound solution corresponding to the mechanism is then found by equating equation 8.12 and 8.15; whence:

$$q_u = \sigma_o + (2+\pi)c_u \tag{8.16}$$

which is the *lowest* upper bound solution (often referred to as *Prandtl* solution) that can possibly be obtained.

8.4 Statically admissible stress fields: undrained behaviour

(a) Stress field with a single discontinuity

Consider the stress field depicted in figure 8.6 with a vertical discontinuity going through one edge of the foundation. In the absence of shear stresses, both vertical and horizontal stresses generated by the stress field are major stresses and, according to the corresponding Mohr's circles, the major vertical stress at point A with a magnitude $\sigma_A = \sigma_o$ rotates through an angle π (double that in figure 8.6) to reach point B, becoming in the process a major horizontal stress whose ultimate magnitude at failure is

$\sigma_B = 2c_u + \sigma_o$. When rotated a further 180°, σ_B then becomes a major vertical stress with an *ultimate lower bound value:*

$$q_u = 4c_u + \sigma_o \tag{8.17}$$

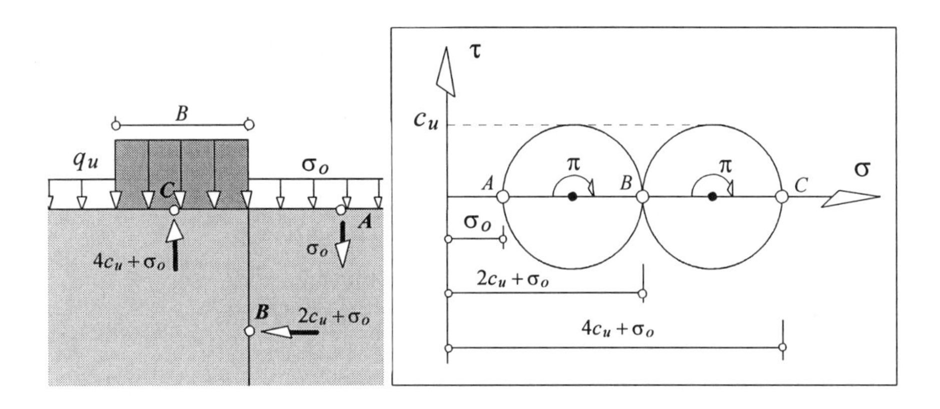

Figure 8.6: Stress field with a single discontinuity and its corresponding Mohr's circles.

(b) Stress field with three discontinuities

Figure 8.7 illustrates a stress field containing three discontinuities going, arbitrarily, through the right edge of the foundation. Assessing the stresses with the help of the corresponding Mohr's circles depicted in figure 8.8, it is seen that the vertical stress at point *A* rotates by an angle $\pi - 2\theta$ (double that in figure 8.7) to reach point *B*, where its magnitude increases to:

$$\sigma_B = \sigma_o + c_u + c_u \cos\beta \tag{8.18}$$

with

$$\beta = \frac{\pi}{2} - \theta \tag{8.19}$$

Therefore, the stress at *B* for an angle $\theta = \pi/6$ (refer to figure 8.7):

$$\sigma_B = \sigma_o + 1.5c_u \tag{8.20}$$

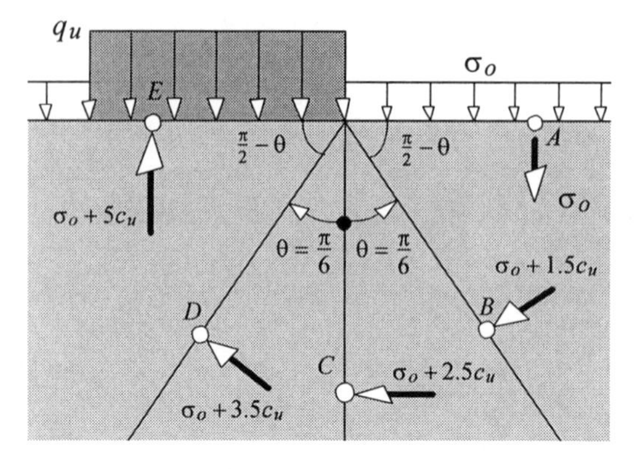

Figure 8.7: Stress field with three discontinuities.

Moreover, it is clear from figure 8.8 that the mean stresses p_1 and p_2 corresponding to any two consecutive circles are such that:

$$p_2 - p_1 = 2c_u \cos\beta \tag{8.21}$$

or

$$\Delta p = 2c_u \sin\theta \tag{8.22}$$

so that the ultimate lower bound stress at E is:

$$q_u = \sigma_o + 5c_u \tag{8.23}$$

This lower bound represents a marked improvement on the one obtained previously *via* equation 8.17.

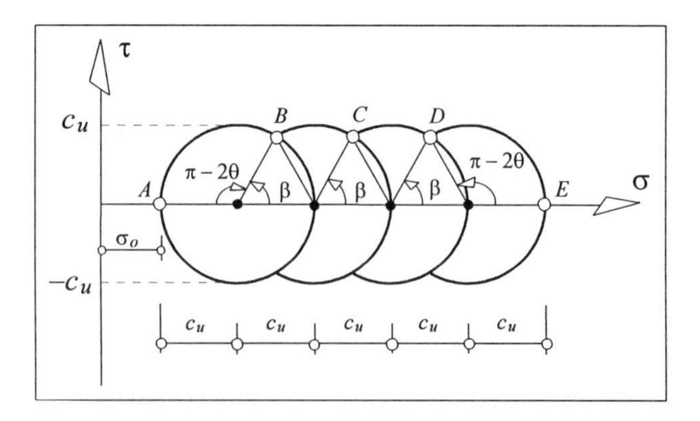

Figure 8.8: Mohr's circles corresponding to figure 8.7.

(c) Stress field containing a stress fan

The stress field depicted in figure 8.9 has a stress fan subtending an angle $\theta = \pi/2$ and consisting of a large number of discontinuities each subtending an angle $d\theta$. Referring to the corresponding Mohr's circles of figure 8.10, it is seen that the (normal) stress at point *A* rotates through 90° to reach point *B* where its magnitude will have increased to $\sigma_o + c_u$. Throughout the stress fan, the general relationship represented by equation 8.22 applies and, accordingly, when an elementary rotation $d\theta$ occurs within the fan, the change in the mean effective stress p is expressed as follows:

$$dp = 2c_u \sin d\theta \approx 2c_u \, d\theta \qquad (8.24)$$

Therefore, the change in the mean effective stress throughout the fan is:

$$\Delta p = \int_0^{\pi/2} 2c_u \, d\theta = \pi c_u \qquad (8.25)$$

Hence the ultimate lower bound stress at point *D*:

$$q_u = \sigma_o + (2+\pi)\, c_u \qquad (8.26)$$

which is the *highest* lower bound solution that can possibly be obtained.

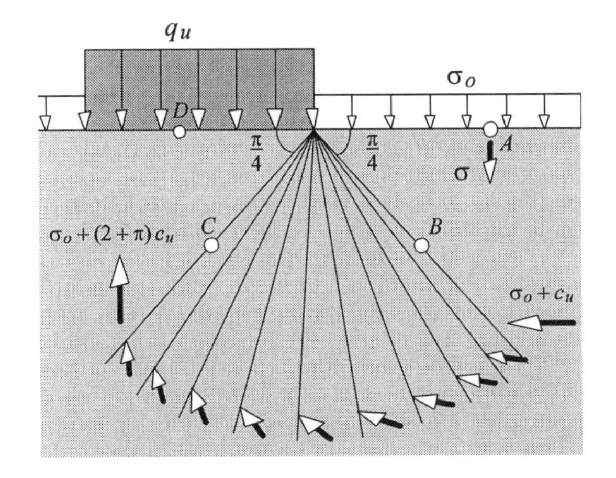

Figure 8.9: Stress field containing a stress fan.

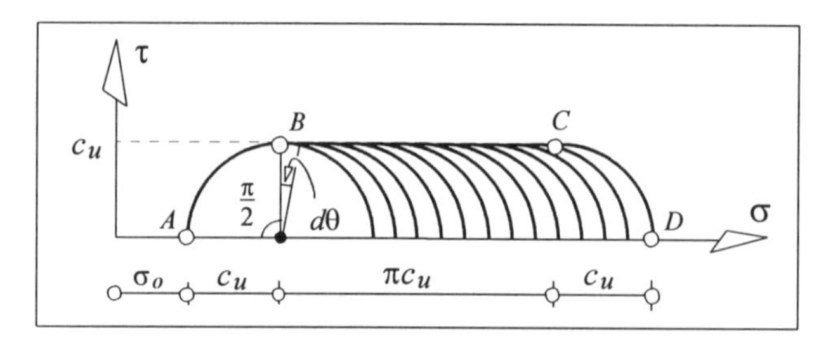

Figure 8.10: Mohr's circles corresponding to figure 8.9.

Because the highest lower bound of equation 8.26 is identical to the lowest upper bound of equation 8.16, an exact plastic solution to the ultimate bearing capacity of a surface footing is therefore achieved.

(d) Stress field containing a stress fan: case of an inclined load

Let us derive the ultimate lower bound solution to the bearing capacity of a surface footing transmitting a load q inclined to the ground as illustrated in figure 8.11. Manifestly, the inclination of the load q will generate a shear stress τ, thus causing the major principal stress to rotate by an angle α such that:

$$\alpha = \tan^{-1}\left(\frac{\tau}{q}\right) \tag{8.27}$$

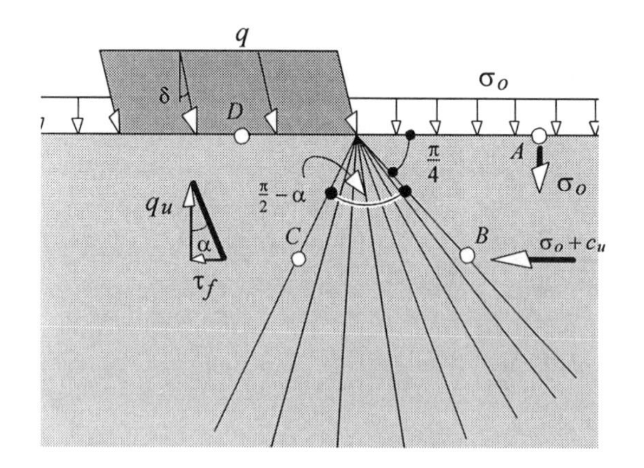

Figure 8.11: Case of inclined loading.

Moreover, any elementary rotation $d\theta$ occurring within the stress fan causes an increase in the mean stress dp according to equation 8.24, and consequently, the total change in the mean stress through the fan subtending an angle $(\frac{\pi}{2}-\alpha)$ is calculated as follows:

$$\Delta p = \int_0^{\frac{\pi}{2}-\alpha} 2c_u\, d\theta = c_u(\pi - 2\alpha) \qquad (8.28)$$

Referring to the corresponding Mohr circles in figure 8.12, the lower bound solution to the normal stress is written as:

$$q_u = \sigma_o + c_u(1 + \pi - 2\alpha + \cos 2\alpha) \qquad (8.29)$$

the corresponding shear stress being:

$$\tau = c_u \sin 2\alpha \qquad (8.30)$$

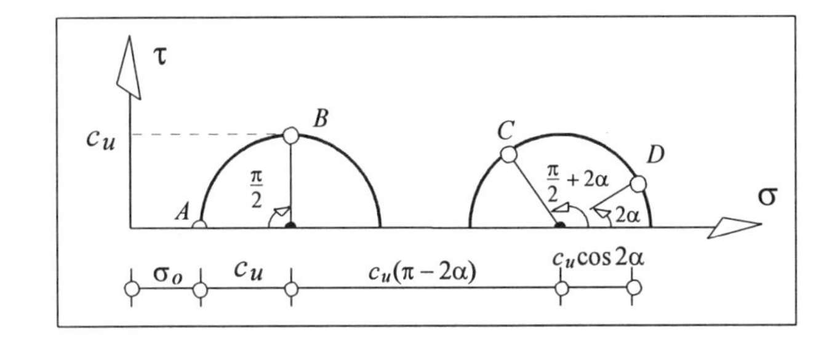

Figure 8.12: Mohr's circles corresponding to figure 8.11.

At the limit (*i.e.* at failure), the shear stress reaches its maximum value c_u when $\alpha = \pi/4$, yielding in the process the following ultimate bearing capacity:

$$q_u = \sigma_o + c_u\left(1 + \frac{\pi}{2}\right) \qquad (8.31)$$

Let both equations 8.29 and 8.30 be rearranged respectively in the following way:

$$\frac{q_u - \sigma_o}{c_u} = 1 + \pi - 2\alpha + \cos 2\alpha \qquad (8.32)$$

$$\frac{\tau}{c_u} = \sin 2\alpha \tag{8.33}$$

The graphs corresponding to these two equations, depicted in figure 8.13, indicate clearly that the ultimate bearing capacity corresponding to a surface footing subject to a load inclined at an angle $\alpha = \pi/4$, with a maximum shear stress, is precisely half that of a surface foundation transmitting a normal load with zero shear stress.

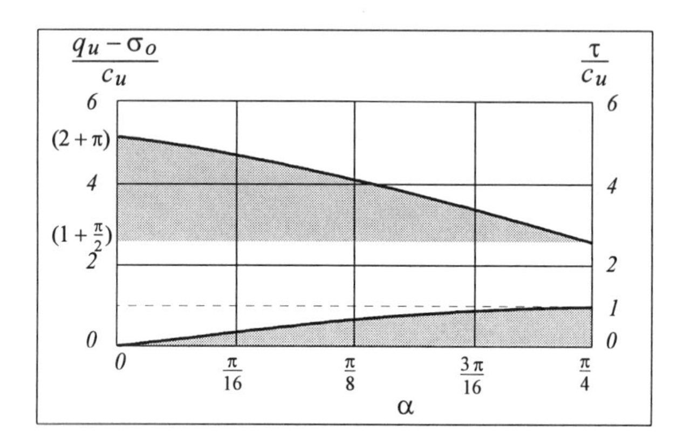

Figure 8.13: Effect of load inclination on the bearing capacity of a soil.

Also, figure 8.14, representing the variation of the normalised quantities τ/c_u and $(q_u - \sigma_o)/c_u$, shows that once it has reached its maximum value c_u, the shear stress at failure remains constant however small the corresponding normal stress may be.

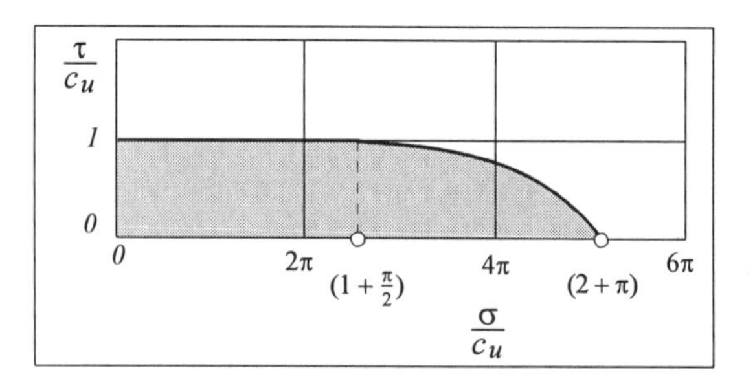

Figure 8.14: Relationship between normal and shear stresses at failure.

8.5 Kinematically admissible collapse mechanisms: drained behaviour

The following effective stress analysis of collapse mechanisms is based on the assumption that plastic deformations occur under an *associated flow rule*, meaning that the criterion of normality is fulfilled. Also, the soil is assumed to be *weightless*. Accordingly, the displacement within any chosen mechanism is characterised by a vector making an angle ϕ' (the effective angle of shearing resistance of the soil) with respect to any slip plane, thus reducing the internal work to zero, since any internal work done on a normal stress cancels out the work done on the corresponding shear stress. This therefore implies that an upper bound solution can be found by calculating the work done by the external loads, then equating it to zero.

Consider the collapse mechanism of figure 8.15 where the shape of the failing block is characterised by a logarithmic spiral with the following equation:

$$\frac{r_2}{r_1} = \exp\left(\theta \tan\phi'\right) \tag{8.34}$$

ϕ' being the angle of shearing resistance of the soil.

When the block rotates anticlockwise by an angle $d\theta$, the average displacement of foundation is $B/2d\theta$; also according to equation 8.34:

$$OA = B \exp(\pi \tan\phi')$$

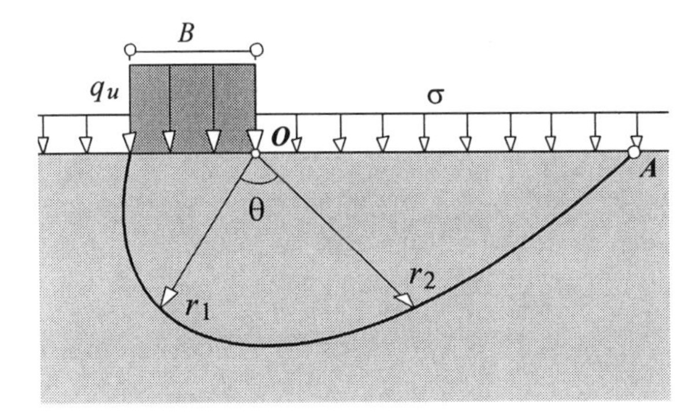

Figure 8.15: Logarithmic spiral collapse mechanism.

The same rotation $d\theta$ therefore causes the surface OA to have an upward displacement amounting to:

$$\frac{B}{2}d\theta \exp(\pi \tan\phi')$$

The total external work can now be determined:

$$\delta E = q_u \frac{B^2}{2} d\theta - q_o B \exp(\pi \tan\phi') \frac{B}{2} d\theta \exp(\pi \tan\phi')$$

and the upper bound solution to the bearing capacity of foundation is found by equating E to zero, hence:

$$q_u = q_o \exp(2\pi \tan\phi') \quad (8.35)$$

The latter equation yields a value $q_u = 18.7 q_o$ for an angle $\phi' = 25^\circ$.

The next mechanism to consider is depicted in figure 8.16 and consists of a shear fan, sandwiched between two rigid blocks, subtending an angle $\theta = \pi/2$. As indicated in the figure, the velocity V makes an angle ϕ' with respect to the slip planes, and within the slip fan bounded by a logarithmic spiral, the velocity increases in proportion to the radii ratio, thus:

$$V_e = V\frac{r_2}{r_1} = V \exp\left(\frac{\pi}{2}\tan\phi'\right) \quad (8.36)$$

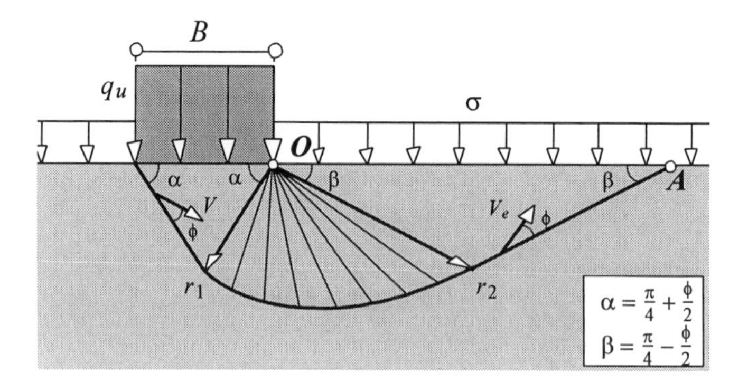

Figure 8.16: Rigid blocks related by a shear fan.

In addition, it easy to show that the distance OA is such that:

$$OA = B \tan\left(\frac{\pi}{4}+\frac{\phi'}{2}\right)\exp\left(\frac{\pi}{2}\tan\phi'\right) \tag{8.37}$$

Since only the vertical components of velocity are of interest (the soil is assumed to be weightless and the internal work within the slip fan is zero), it follows that:

- the (downward) vertical displacement of the block beneath the foundation is $V_1 = V\cos\left(\frac{\pi}{4}+\frac{\phi'}{2}\right)$
- the (upward) vertical displacement of the block to the right of the slip fan is $V_2 = V_e\cos\left(\frac{\pi}{4}-\frac{\phi'}{2}\right)$, or, substituting for V_e from equation 8.36: $V_2 = V\exp\left(\frac{\pi}{2}\tan\phi'\right)\sin\left(\frac{\pi}{4}+\frac{\phi'}{2}\right)$

The external work can now be evaluated:

$$\delta E = q_u BV\cos\left(\frac{\pi}{4}+\frac{\phi'}{2}\right) - q_o B\tan\left(\frac{\pi}{4}+\frac{\phi'}{2}\right)\exp\left(\frac{\pi}{2}\tan\phi'\right)$$

$$\times V\exp\left(\frac{\pi}{2}\tan\phi'\right)\sin\left(\frac{\pi}{4}+\frac{\phi'}{2}\right)$$

Finally, equating δE to zero, and rearranging yields the upper bound solution:

$$q_u = q_o\tan^2\left(\frac{\pi}{4}+\frac{\phi'}{2}\right)\exp(\pi\tan\phi') \tag{8.38}$$

which is the *lowest* obtainable upper bound.

For an angle $\phi' = 25^\circ$, equation 8.38 yields a value $q_u = 10.7q_o$, in marked contrast to the overestimate $q_u = 18.7q_o$ resulting from the previous mechanism of figure 8.15.

8.6 Statically admissible stress fields: drained behaviour

Prior to analysing a stress field, let us determine the effect that a rotation of a major stress has on the mean effective stress, that is the centre of the Mohr circle. Consider the case of a *frictional soil* (*i.e.* $c' = 0$) for which the failure criterion derived in section 5.3 (equation 5.13) is written as follows:

$$\sigma_1' = \sigma_3' \tan^2\left(\frac{\pi}{4} + \frac{\phi'}{2}\right) \tag{8.39}$$

Let the normal major stress rotate by a tiny angle $d\theta$, so that the corresponding Mohr circle centre moves by an amount dp' resulting in an angle $2d\theta$ as depicted in figure 8.17. Because $d\theta$ is very small, then $DA \approx DB$; also, it is clear from the figure that $\alpha = \frac{\pi}{2} - \phi' - d\theta$, and accordingly $\sin\alpha \approx \cos\phi'$.

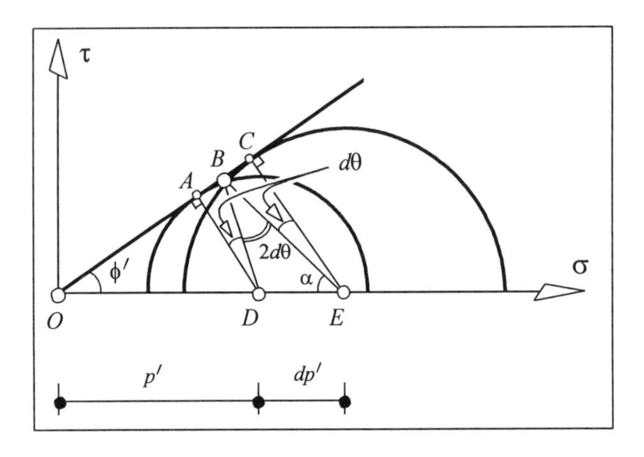

Figure 8.17: Effects of major stress rotation on p'.

On the other hand, the sine rule yields:

$$\frac{dp'}{\sin 2d\theta} = \frac{DB}{\sin\alpha} \tag{8.40}$$

Knowing that $DB \approx DA = p' \sin\phi'$ and $\sin 2d\theta \approx 2d\theta$, then substituting for both quantities DB and $\sin\alpha$ into equation 8.40 and rearranging, it follows that:

$$\frac{dp'}{p'} = 2\tan\phi'\, d\theta \tag{8.41}$$

Let us now consider the statically admissible stress field with one discontinuity, illustrated in figure 8.18. The corresponding Mohr circles of figure 8.19 show that the normal vertical stress σ_o at point A rotates through an angle of 180° to reach point B. Since the corresponding circle touches the failure envelope, the failure criterion of equation 8.39 is therefore applicable so that the stress at B is:

$$\sigma_B = \sigma_o \tan^2\left(\frac{\pi}{4} + \frac{\phi'}{2}\right) \tag{8.42}$$

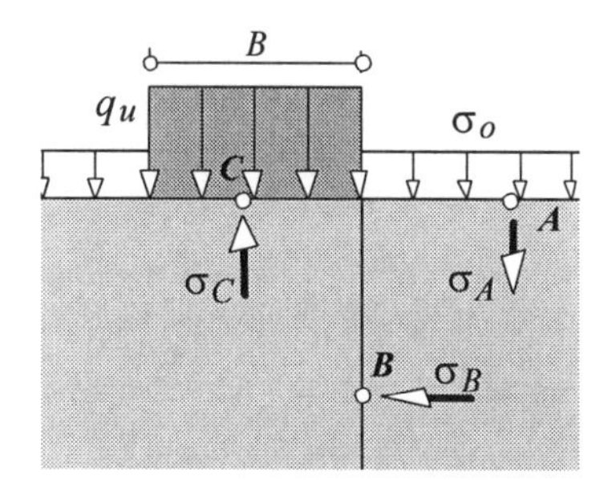

Figure 8.18: One discontinuity stress field.

As one moves from B to C, the stress rotates a further 180°; whence, applying the failure criterion as in the previous case, it follows that the ultimate lower bound to the bearing capacity at C is:

$$q_u = \sigma_o \tan^4\left(\frac{\pi}{4} + \frac{\phi'}{2}\right) \tag{8.43}$$

For $\phi' = 25^\circ$, equation 8.43 yields a value $q_u = 6.1\sigma_o$.

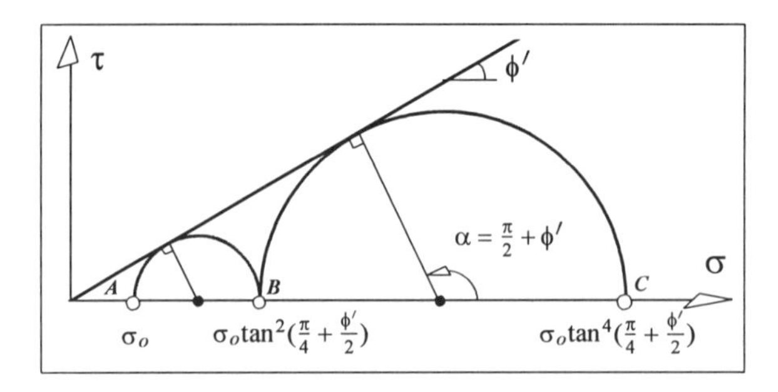

Figure 8.19: Mohr's circles corresponding to figure 8.18.

To improve on the value of the lower bound of equation 8.43, a stress field containing a stress fan can be considered. Referring to figure 8.20, the corresponding stress fan subtends an angle $\theta = \pi/2$ implying that the stress rotates by an angle π between points B and C on Mohr circles of figure 8.21. In addition, as has already been established, an elementary stress rotation $d\theta$ causes the mean effective stress p' (*i.e.* the centre of the Mohr circle) to move according to equation 8.41. Consequently, for an angle $\pi/2$ subtended by the fan, the displacement of the mean effective stress is easily integrated as follows:

$$\int_{p'_1}^{p'_2} \frac{dp'}{p'} = \int_0^{\pi/2} 2\tan\phi'\, d\theta$$

Hence:

$$p'_2 = p'_1 \exp(\pi \tan\phi') \tag{8.44}$$

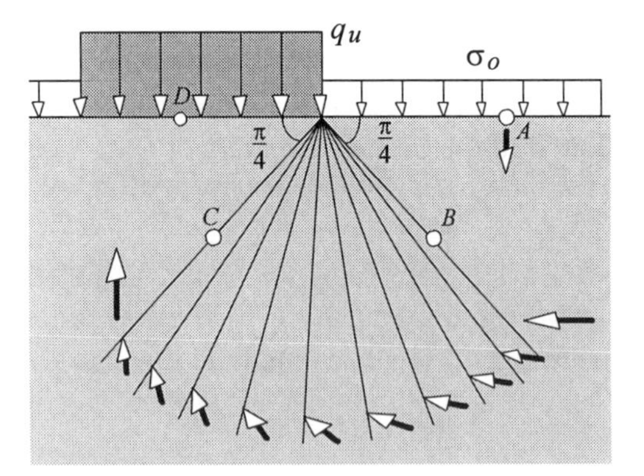

Figure 8.20: Stress field containing a stress fan.

Moreover, it is easy to show from the trigonometry of figure 8.21 that:

$$p'_1 = \frac{\sigma_o}{1-\sin\phi'} \qquad \text{and} \qquad p'_2 = q_u\left(\frac{1}{1+\sin\phi'}\right)$$

Substituting for both p'_1 and p'_2 in equation 8.44, then rearranging:

$$q_u = \sigma_o \tan^2\left(\frac{\pi}{4}+\frac{\phi'}{2}\right)\exp(\pi\tan\phi') \tag{8.45}$$

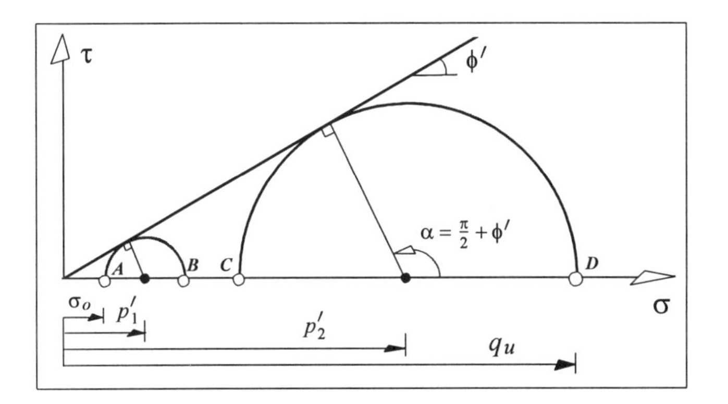

Figure 8.21: Mohr's circles corresponding to figure 8.20.

The latter equation thus yields the *highest lower bound* of the ultimate bearing capacity, which happens to be identical to the *lowest upper bound* established earlier *via* equation 8.38, whence the exact solution to the plastic collapse of a *drained weightless soil*, subject to a normal load transmitted by a surface footing.

The previous analysis undertaken can easily be expanded to include the general case of a surface footing subject to an inclined load depicted in figure 8.22. Clearly, equation 8.41 must be integrated within the limits of the angle subtended by the stress fan. Hence:

$$p'_2 = p'_1 \exp[(\pi - 2\alpha)\tan\phi']$$

but

$$p'_1 = \frac{\sigma_o}{1-\sin\phi'}$$

whence:

$$p_2' = \frac{\sigma_o}{1 - \sin\phi'} \exp\left[(\pi - 2\alpha)\tan\phi'\right] \tag{8.46}$$

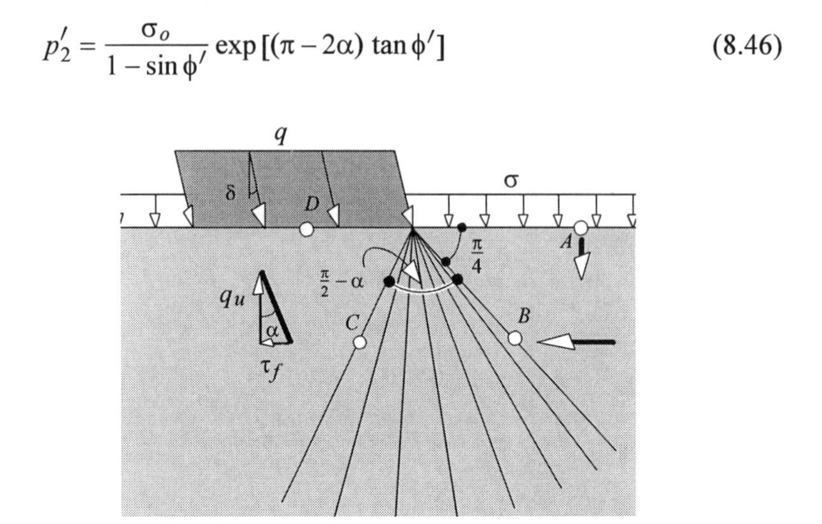

Figure 8.22: General case of an inclined loading.

Referring to figure 8.23, it is seen that both ultimate normal and shear stresses at point D are respectively:

$$q_u = p_2' + ED\cos 2\alpha \tag{8.47a}$$

$$\tau_f = ED\sin 2\alpha \tag{8.47b}$$

Making use of the trigonometry of the figure, it follows that:

$$\frac{p_2'}{\sin(2\alpha - \delta)} = \frac{ED}{\sin\delta} = \frac{p_2'\sin\phi'}{\sin\delta}$$

therefore:

$$\frac{\sin\delta}{\sin(2\alpha - \delta)} = \sin\phi' \quad \text{and} \quad ED = p_2'\sin\phi'$$

Substituting for ED and p_2' in equation 8.47 and rearranging yields:

$$q_u = \frac{\sigma_o}{1-\sin\phi'}(1+\sin\phi'\cos 2\alpha)\exp[(\pi-2\alpha)\tan\phi'] \qquad (8.48a)$$

$$\tau_f = \frac{\sigma_o}{1-\sin\phi'}\sin\phi'\sin 2\alpha\,\exp[(\pi-2\alpha)\tan\phi'] \qquad (8.48b)$$

with

$$\alpha = \frac{1}{2}\left[\delta + \sin^{-1}\left(\frac{\sin\delta}{\sin\phi'}\right)\right]$$

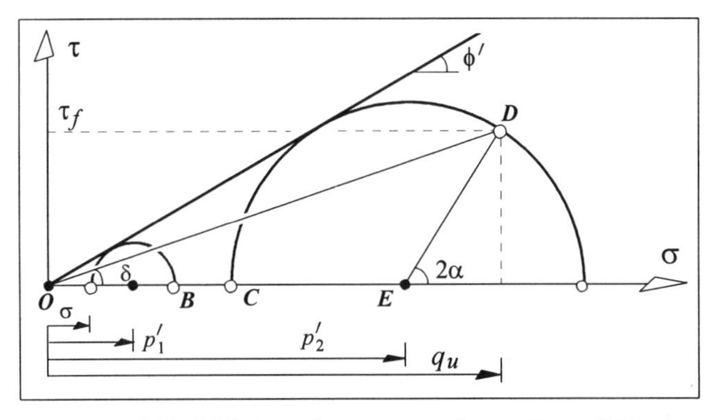

Figure 8.23: Mohr's circles corresponding to figure 8.22.

To illustrate the effects of a load inclination on the bearing capacity, the quantities q_u/σ_o and τ_f/σ_o are plotted in figure 8.24 for different angles δ in the case of a typical soil with an angle of shearing resistance $\phi' = 25^\circ$. It is seen that the ratio q_u/σ_o decreases from its maximum value for $\delta = 0$ to a value almost 4.5 times smaller for $\delta = \phi'$.

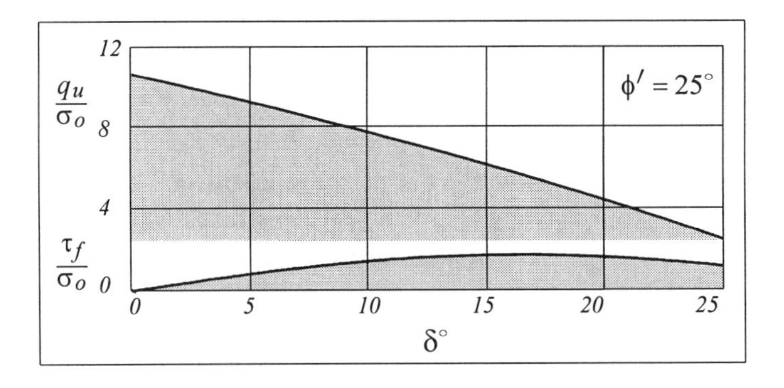

Figure 8.24: Effect of load inclination on the bearing capacity.

In fact, as suggested by Bolton (1991), there is a case to limit the load inclination to $\delta = 15^\circ$, regardless of the type of soil, in which case the bearing capacity is roughly half that corresponding to a normal load as illustrated in the figure.

8.7 Effects of soil weight and cohesion

The analysis undertaken so far was based on the assumption that the soil on which the foundation is built is weightless and, when drained, cohesionless. Let us, first, assess the contribution of the weight of soil to the ultimate bearing capacity q_u. In this respect, a quick glance at figure 8.25, for instance, makes you realise that neglecting the self weight of the block *OAB* can potentially lead to a markedly underestimated ultimate load (*i.e.* too safe!).

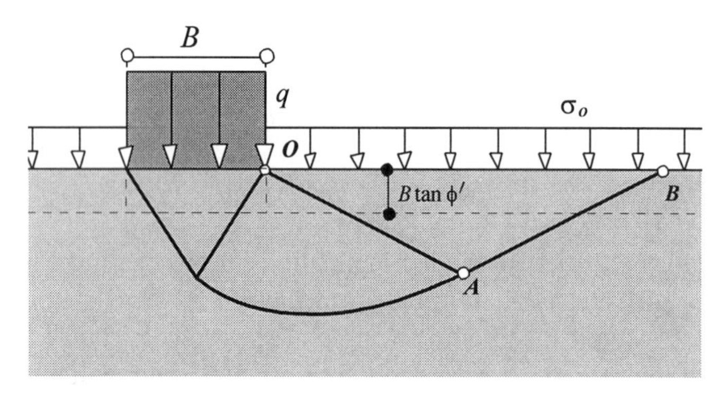

Figure 8.25: Effect of soil weight on its bearing capacity.

Several authors have suggested the use of different ways in which the weight of soil is included in the bearing capacity equation. However, a simple yet effective method based on the suggestion by Bolton can be used to the same effect. It consists of assuming that the foundation is built at a depth $B \tan\phi'$, then considering the pressure due to this extra soil weight as a load applied at foundation level. The net effect will be to increase all normal effective stresses by an amount $\sigma' = \gamma' B \tan\phi'$, with γ' referring to the effective unit weight of soil. Thus, for a normal load, equation 8.45 is altered as follows:

$$q_u + \gamma' B \tan\phi' = (\sigma_o + \gamma' B \tan\phi') \tan^2\left(\frac{\pi}{4} + \frac{\phi'}{2}\right) \exp(\pi \tan\phi')$$

Introducing the (dimensionless) factor $N_q = \tan^2\left(\frac{\pi}{4} + \frac{\phi'}{2}\right) \exp(\pi \tan\phi')$, then rearranging:

$$q_u = \sigma_o N_q + \gamma' B \tan\phi' (N_q - 1) \tag{8.49}$$

Moreover, if the foundation is built on a stiff heavily overconsolidated clay, then apparent cohesion can constitute a component of the long term drained shear strength of the soil, provided that its properly measured value is limited to a maximum of $15\,kN/m^2$ (refer to the discussion on cohesion in section 5.5.4). The effect of cohesion is to increase all normal stresses by an equal amount $c' \cot\phi'$; accordingly, equation 8.49 is modified:

$$q_u + c' \cot\phi' = (\sigma_o + c' \cot\phi') N_q + \gamma' B \tan\phi' (N_q - 1)$$

which can be rewritten as:

$$q_u = \sigma_o N_q + c' N_c + \frac{\gamma' B}{2} N_\gamma \tag{8.50}$$

with the bearing capacity factors:

$$N_q = \tan^2\left(\frac{\pi}{4} + \frac{\phi'}{2}\right) \exp(\pi \tan\phi') \tag{8.51a}$$

$$N_c = \cot\phi' (N_q - 1) \tag{8.51b}$$

$$N_\gamma = 2 \tan\phi' (N_q - 1) \tag{8.51c}$$

Equation 8.50, together with equations 8.51, represent the basic classical bearing capacity equation of a strip footing with a width B and an infinite length, transmitting a normal effective stress σ'_o to the soil at foundation level. While equations 8.51 can be used to calculate the bearing capacity factors N_c, N_q and N_γ, it is manifestly more straightforward to use the charts depicted in figure 8.26 to the same effect.

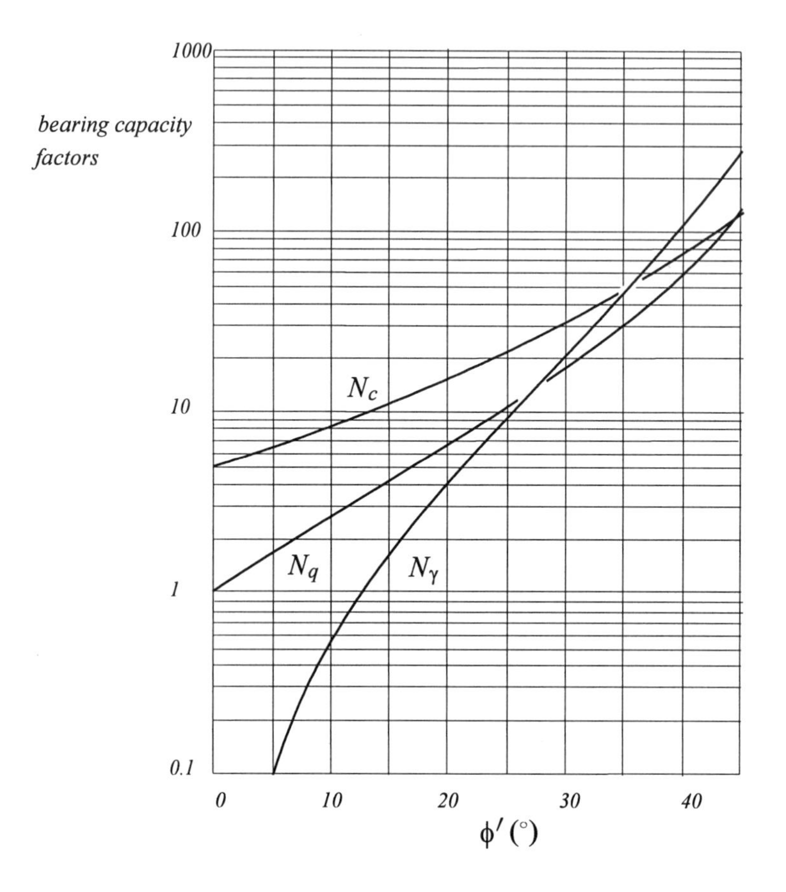

Figure 8.26: Bearing capacity factors for a strip footing of infinite length.

It is noticeable that, under undrained conditions (for which $\phi = 0$), equations 8.51 yield respectively $N_q = 1$, $N_c = (2 + \pi)$ and $N_\gamma = 0$. Accordingly, in the short term, the ultimate bearing capacity of a strip footing of infinite length, calculated from equation 8.50, is reduced to the following:

$$q_u = \sigma_o + c_u N_c \tag{8.52}$$

In 1951, Skempton showed that the bearing capacity factor N_c in equation 8.52 depends not only on the shape of foundation, but also on its depth of embedment. He then produced the charts depicted in figure 8.27 from which N_c can be estimated. It is seen that for a strip footing of infinite length, N_c value varies between $(2+\pi)$ at surface level and 8.5 for a depth to width ratio $D/B \geq 5$. Alternatively, a square or a circular foundation are characterised by a factor $N_c = 2\pi$ at surface level, increasing to $N_c = 9$ for a ratio $D/B \geq 5$.

For a rectangular foundation with an area $A = L \times B$, $(L \geq B)$, an interpolation is needed, and in this respect, the following relationship can be used cautiously:

$$N_{c(\text{rectangle})} = N_{c(\text{square})}(0.84 + 0.16B/L) \tag{8.53}$$

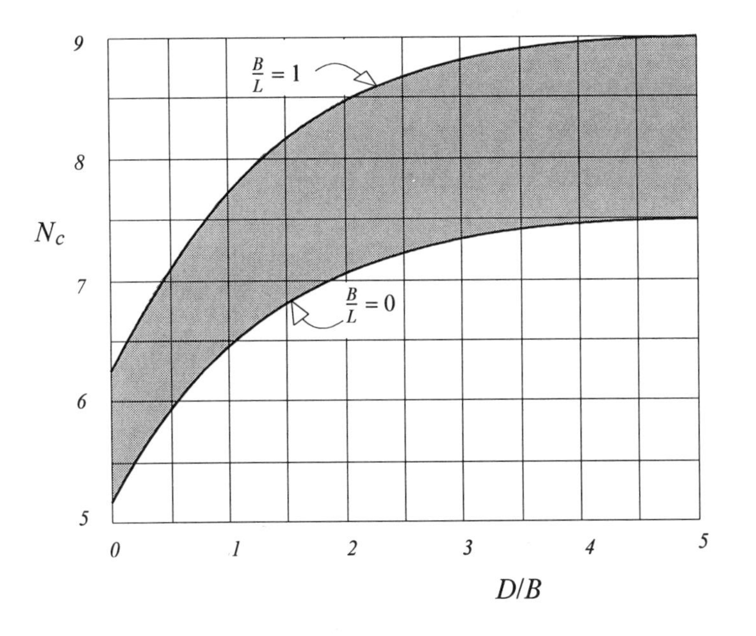

Figure 8.27: Bearing capacity factor N_c for undrained conditions. (Reproduced by permission of the Building Research Establishment.)

8.8 Effects of foundation shape and depth and load inclination

Figure 8.27 illustrates the effects that the depth of embedment as well as the shape of the foundation have on the soil ultimate bearing capacity. Although the use of this figure is restricted to undrained short term conditions, its implications are significant in that one surely expects similar effects to affect equation 8.50 which should then be expressed in the following general form:

$$q_u = \sigma_o N_q s_q d_q + c' N_c s_c d_c + \gamma' \frac{B}{2} N_\gamma s_\gamma d_\gamma \qquad (8.54)$$

where the (empirically derived) correction factors s and d reflect the effects of foundation shape and depth respectively. Although several methods were suggested in the literature by which these factors can be estimated, the following relationships (Hansen, 1970, De Beer, 1970) can be used with caution:

shape factors:

$$s_c = 1 + \frac{B}{L}\frac{N_q}{N_c} \qquad (8.55a)$$

$$s_q = 1 + \frac{B}{L} \tan\phi' \qquad (8.55b)$$

$$s_\gamma = 1 - 0.4\frac{B}{L} \qquad (8.55c)$$

depth factors:

$$d_c = 1 + 0.4\xi \qquad (8.55d)$$

$$d_q = 1 + \xi \tan\phi'(1 - \sin\phi') \qquad (8.55e)$$

$$d_\gamma = 1 \qquad (8.55f)$$

with $\xi = D/B$ if $D/B \leq 1$, and $\xi = \tan^{-1}(D/B)$ if $D/B > 1$.

Moreover, if the foundation is subject to a load inclined at an angle β with respect to the vertical as depicted in figure 8.28, then a fraction of the load will be transmitted horizontally and, accordingly, the three bearing capacity factors N_c, N_q and N_γ must be corrected to take account of the inclination effects. The respective correction factors can be evaluated in the following manner (Hanna and Meyerhof, 1981):

$$I_c = I_q = \left(1 - \frac{\beta^\circ}{90}\right)^2 \tag{8.56a}$$

$$I_\gamma = \left(1 - \frac{\beta^\circ}{\phi'}\right)^2 \tag{8.56b}$$

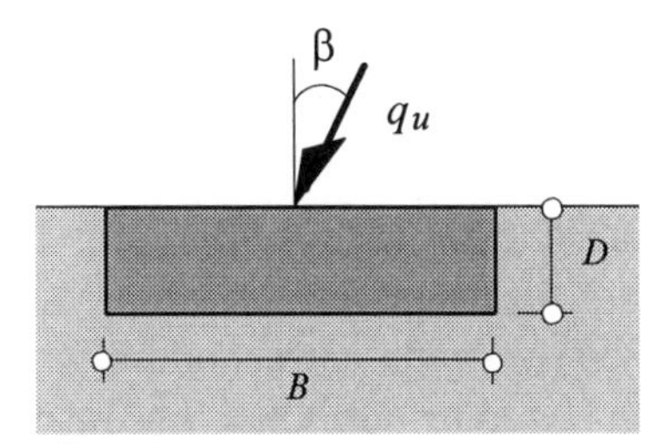

Figure 8.28: Effect on the bearing capacity of an inclined load.

It follows that, for a foundation of a *finite* shape, embedded at a depth D and subject to an inclined load, the general ultimate load capacity expression is:

$$q_u = \sigma'_o N_q s_q d_q I_q + c' N_c s_c d_c I_c + \gamma' \frac{B}{2} N_\gamma s_\gamma d_\gamma I_\gamma \tag{8.57}$$

Example 8.1

Assume that the foundation in figure 8.28 has a width $B = 2.5\,m$, a length $L = 4\,m$, and is embedded at a depth $D = 1.5\,m$ in a medium clay characterised in the short term by an undrained shear strength $c_u = 55\,kN/m^2$. Knowing that, under drained conditions, the clay has an angle of shearing resistance $\phi' = 21^\circ$ and an apparent cohesion $c' = 5\,kN/m^2$, estimate the *allowable load per linear metre* that can be safely applied to the foundation at an angle $\beta = 12^\circ$ with respect to the vertical, yielding a factor of safety $F = 3$ against shear failure. As far as the position of the water table is concerned, you may like to consider the three following scenarios : (*a*) the water is at the ground surface, (*b*) the water is at foundation level and (*c*) the water is several metres below foundation level. For the clay, it can be assumed that both saturated and bulk unit weights have the same value $\gamma = 20\,kN/m^3$.

Short term analysis

For the undrained short term behaviour, the bearing capacity equation 8.52 applies provided that appropriate correction factors are used. Thus, taking account of the linear aspect of the *ultimate* load, the (rectangular) shape of foundation and the load inclination, it follows that:

$$q_u = L(\sigma_o + c_u N_{cr}) I_c$$

- the total stress at foundation level is: $\sigma_o = \gamma D = 20 \times 1.5 = 30\, kN/m^2$;
- the bearing capacity factor for a square foundation with a ratio $D/B = 0.6$ is read from figure 8.27 on the graph corresponding to $B/L = 1$: $N_c \approx 7.2$. This value then has to be corrected (because of the rectangular shape of the foundation) through equation 8.53, whence:

$$N_{cr} = 7.2\left(0.84 + 0.16 \times \frac{2.5}{4}\right) = 6.8$$

- the correction factor I_c relating to the load inclination is calculated from equation 8.56*a*:

$$I_c = \left(1 - \frac{12}{90}\right)^2 = 0.75$$

So, the ultimate line load to which the foundation can be subjected in the short term is:

$$q_u = 4 \times (30 + 55 \times 6.8) \times 0.75 = 1212\, kN/m$$

Usually, the factor of safety against shear failure is defined as the ratio of the *net ultimate load* q_{un} to the *allowable net load* q_{an} applied to the foundation. The net ultimate and the net allowable loads are found by subtracting from the ultimate and allowable loads respectively the weight of soil above foundation level when applicable, thus:

$$q_{un} = q_u - L\gamma D = 1212 - 4 \times 20 \times 1.5 = 1092\, kN/m$$

Since a factor of safety of 3 is required in this instance, it follows that the allowable load per linear metre is:

$$\frac{q_{un}}{q_a - L\gamma D} = 3 \quad \Rightarrow \quad q_a = \frac{q_u - L\gamma D}{3} + L\gamma D$$

$$= \frac{1092}{3} + 4 \times 20 \times 1.5 = 484\, kN/m.$$

Long term analysis

For the long term analysis, effective stresses are used in conjunction with equation 8.57. Because the ultimate load is expressed per linear metre, it follows that:

$$q_u = L\left(\gamma' D N_q s_q d_q I_q + c' N_c s_c d_c I_c + \gamma' \frac{B}{2} N_\gamma s_\gamma d_\gamma I_\gamma\right)$$

γ' being the *effective* unit weight of soil. Prior to considering the position of the water table, let the different factors in the above equation be evaluated as follows:

- The bearing capacity factors are read from figure 8.26 for $\phi' = 21°$; thus:

$$N_q \approx 7, \ N_c \approx 17, \ N_\gamma \approx 5.$$

- The shape factors are calculated from equations 8.55(*a*)–(*c*); whence:

$$s_c = 1 + \tfrac{2.5}{4} \times \tfrac{7}{17} \approx 1.26, \quad s_q = 1 + \tfrac{2.5}{4} \tan 21 = 1.24,$$

$$s_\gamma = 1 - 0.4 \times \tfrac{2.5}{4} = 0.75$$

- The depth factors are computed from equations 8.55(*d*)–(*f*) in the knowledge that $\xi = 1.5/2.6 = 0.6$:

$$d_c = 1 + 0.4 \times 0.6 = 1.24, \quad d_q = 1 + 0.6 \tan 21(1 - \sin 21) = 1.15,$$
$$d_\gamma = 1$$

- The inclination factors are determined using equations 8.56:

$$I_c = I_q = \left(1 - \tfrac{12}{90}\right)^2 = 0.7 \ , \quad I_\gamma = \left(1 - \tfrac{12}{21}\right)^2 = 0.183$$

Inserting all these quantities in the bearing capacity equation, the result would then depend on the level of water, hence the following three distinct cases.

(1) *The water table is at the ground surface* in which case the effective unit weight of soil is $\gamma' = \gamma_{sat} - \gamma_w = 10\,kN/m^3$ and, accordingly:

$$
\begin{aligned}
q_u &= 4 \times (10 \times 1.5 \times 7 \times 1.24 \times 1.15 \times 0.75) \\
&\quad +4 \times (5 \times 17 \times 1.26 \times 1.24 \times 0.75) \\
&\quad +4 \times \left(\frac{10 \times 2.5}{2} \times 5 \times 0.75 \times 1 \times 0.183 \right) = 881.9\,kN/m
\end{aligned}
$$

Hence an allowable load per linear metre:

$$
q_a = \frac{q_u - L\gamma D}{3} + L\gamma D = \frac{881.9 - 120}{3} + 120 = 374\,kN/m
$$

(2) *The water table is at foundation level (i.e. at a depth of 1.5 m).* In this case, the effective unit weight $\gamma' = 10\,kN/m^3$ is used in conjunction with the factor N_γ, while the effective stress σ'_o at foundation level is calculated using a unit weight $\gamma = 20\,kN/m^3$. Thence:

$$
\begin{aligned}
q_u &= 4 \times (20 \times 1.5 \times 7 \times 1.24 \times 1.15 \times 0.75) \\
&\quad +4 \times (5 \times 17 \times 1.26 \times 1.24 \times 0.75) \\
&\quad +4 \times \left(\frac{10 \times 2.5}{2} \times 5 \times 0.75 \times 1 \times 0.183 \right) = 1331.1\,kN/m
\end{aligned}
$$

and the allowable load is in this case:

$$
q_a = \frac{1331.1 - 120}{3} + 120 = 523.7\,kN/m
$$

(3) *The water table is well below foundation level.* An effective unit weight $\gamma' = 20\,kN/m^3$ applies, and the calculation yields:

$$
\begin{aligned}
q_u &= 4 \times (20 \times 1.5 \times 7 \times 1.24 \times 1.15 \times 0.75) \\
&\quad +4 \times (5 \times 17 \times 1.26 \times 1.24 \times 0.75) \\
&\quad +4 \times \left(\frac{20 \times 2.5}{2} \times 5 \times 0.75 \times 1 \times 0.183 \right) = 1365.4\,kN/m
\end{aligned}
$$

leading to an allowable load:

$$q_a = \frac{1365.4 - 120}{3} + 120 = 535.3\ kN/m$$

The calculations are significant in many ways since it transpires that, in the long term, the water level has a detrimental effect on the bearing capacity of a foundation. It is seen that in this particular instance, were the clay to be totally submerged (*i.e.* the water level at the ground surface), the bearing capacity would be reduced by more than 35% compared with that corresponding to a water table well below foundation level. On the other hand, the bearing capacity in the short term depends to a large extent on the undrained shear strength of the clay and, once more, the calculations indicate that the short term bearing capacity, in this example, is more critical than the long term one unless the clay is waterlogged. This shows how crucial it is to manage the water level as far as the design of foundations is concerned.

8.9 Effects of load eccentricity

8.9.1 One way eccentricity

Consider the case of a shallow foundation subject to a moment $M_x = Qe$ as depicted in figure 8.29, where e represents the load eccentricity. Obviously, this type of loading is bound to affect the bearing capacity of foundation since the *effective* loaded area (*i.e.* the grey area in figure 8.29) is smaller in comparison with the actual foundation area $L \times B$. In fact, such foundations are designed in precisely the same way as described earlier, in other words both equations 8.52 and 8.57 still apply *provided that:*

- the width of foundation B in equations 8.52 (implicitly) and 8.53, as well as equation 8.57 is replaced by the effective width:

$$B' = B - 2e \tag{8.58}$$

- the effective width B' *must* be used in conjunction with equations 8.55(a)–(c) to evaluate the shape factors, the depth factors being unaffected.

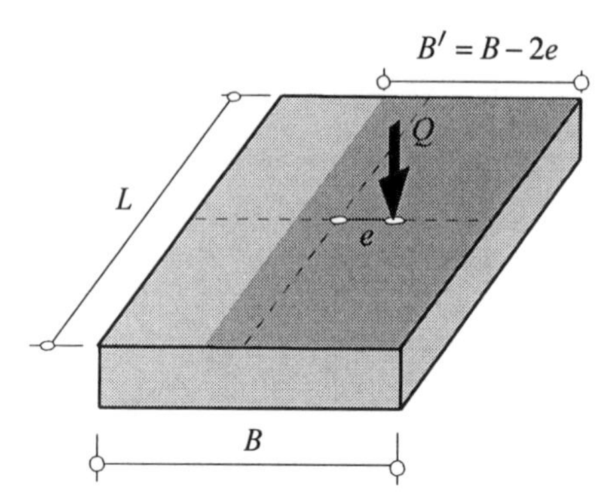

Figure 8.29: Effective width of an eccentrically loaded foundation.

A detailed examination of the calculations corresponding to this type of eccentricity (*i.e.* one way eccentricity) can be found in example 10.4, section 10.11.

In assessing the stresses beneath the foundation base, a distribution such as the one illustrated in figure 8.30 can be assumed, notwithstanding the fact that the *actual* distribution might be slightly different. Based on the stress diagram in figure 8.30, it can easily be shown that the optimum values at either extremity of foundation are as follows:

$$q_{min} = \frac{Q}{BL} - \frac{B}{2}\frac{M}{I} \tag{8.59a}$$

$$q_{max} = \frac{Q}{BL} + \frac{B}{2}\frac{M}{I} \tag{8.59b}$$

the moment of inertia being $I = LB^3/12$, and the moment is, with reference to figure 8.30, $M = Qe$.

Substituting for I and M in the above equations, it follows that:

$$q_{min} = \frac{Q}{BL}\left(1 - \frac{6e}{B}\right) \tag{8.60a}$$

$$q_{max} = \frac{Q}{BL}\left(1 + \frac{6e}{L}\right) \tag{8.60b}$$

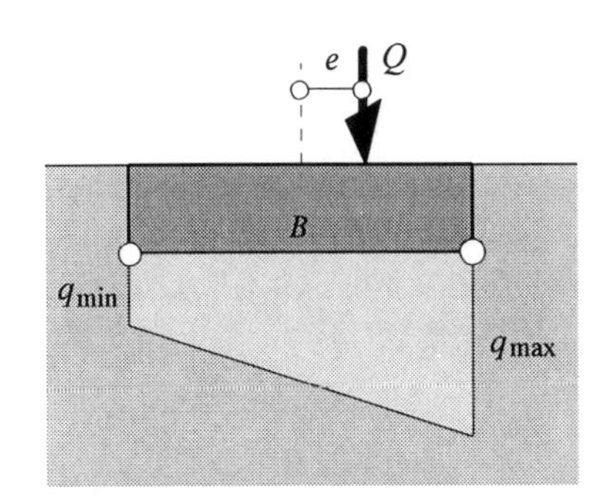

Figure 8.30: Assumed stress profile beneath an eccentrically loaded foundation.

Clearly, equation 8.60*a* implies that as soon as the eccentricity is larger than the value $B/6$, the quantity q_{min} becomes negative, meaning that the foundation starts to lift off the ground on one side. This can quickly give rise to a stress concentration that can cause a bearing capacity failure; hence, it is essential when designing a foundation with an eccentric loading, to ensure that the minimum stress beneath the foundation remains positive, in other words, the following criterion must be fulfilled:

$$e \leq \frac{B}{6} \tag{8.61}$$

8.9.2 Two-way eccentricity

When a foundation is subject to a two-way eccentricity, then its effective area, that is the area affected by the load, will be similar to the grey area depicted in figure 8.31 *provided* that the two following criteria relating to eccentricity are not violated:

$$e_L \leq \frac{L}{6} \tag{8.62a}$$

$$e_B \leq \frac{B}{6} \tag{8.62b}$$

In this case, it can easily be shown that the effective width of foundation B' is such that:

$$B' = \frac{(B+B_1)}{2} + \frac{L_1(B-B_1)}{2L} \tag{8.63}$$

L being the actual length of foundation and B_1 and L_1 (refer to figure 8.31) can be determined from the charts established by Higher and Anders (1985) and reproduced in figures 8.32 and 8.33.

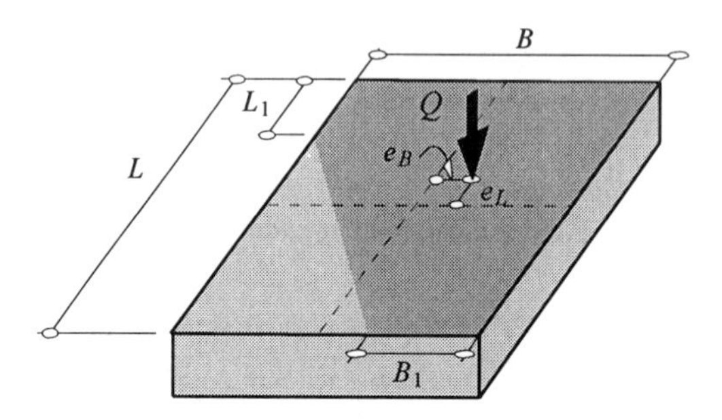

Figure 8.31: Effective area related to a two-way eccentricity load.

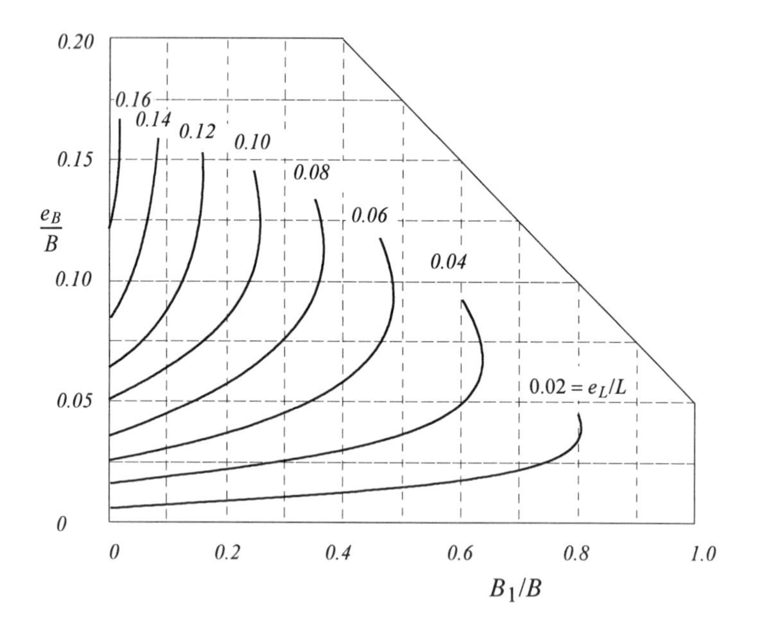

Figure 8.32: Higher and Anders charts for the determination of B_1. (Reproduced by permission of the ASCE.)

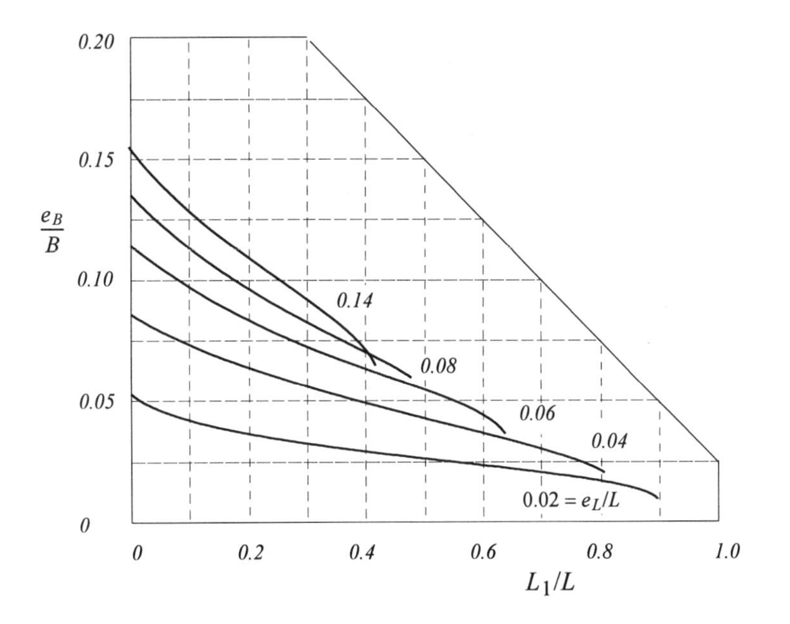

Figure 8.33 : Higher and Anders charts for the determination of L_1. (Reproduced by permission of the ASCE.)

Obviously, the ultimate load capacity of a foundation that is subject to such loading conditions will be affected through the use of the effective width B' in the bearing capacity equations. Once more, remember that only the shape factors of equations 8.55(*a*)–(*c*) are affected by B'; the depth factors being calculated in conjunction with the *actual* width B.

Thus, the ultimate total load that can be supported by a foundation is markedly reduced if the load in question is applied eccentrically as illustrated in figure 8.31, the reason being the decrease in the effective loaded area (refer to the grey area in figure 8.31). Since the total load is calculated from the ultimate pressure in the following way: $Q = A'q_u$, then any decrease in the area is bound to affect the loading capacity of the foundation.

Example 8.2

Consider the case of a rectangular foundation, similar to that depicted in figure 8.31, with a width $B = 3.5\,m$, a length $L = 5\,m$, embedded at a depth $D = 1.8\,m$ in a dense sand characterised by an effective angle of shearing resistance $\phi' = 40^\circ$ and a bulk unit weight $\gamma = 21\,kN/m^3$. The water level being well below the foundation base, calculate the maximum vertical load Q that can be applied eccentrically to the foundation. Assume the following eccentricities apply: $e_B = 0.5\,m$, $e_L = 0.7\,m$.

It is easy to check that the eccentricities do not violate the criteria set in equation 8.62.

Prior to calculating the effective width of foundation B' from equation 8.63, the quantities B_1 and L_1 need to be determined.

It is seen that:

$$\frac{e_L}{L} = \frac{0.7}{5} = 0.14 \quad \text{and} \quad \frac{e_B}{B} = \frac{0.5}{3.5} = 0.143$$

Accordingly, figures 8.32 and 8.33 yield respectively the values: $B_1/B \approx 0.075$, $L_1/L \approx 0.025$, from which the following values are then obtained:

$$B_1 = 0.26\,m, \quad L_1 = 0.125\,m$$

Equation 8.63 can now be used to calculate B':

$$B' = \frac{1}{2}(3.5 + 0.26) + \frac{0.125}{2 \times 5}(3.5 - 0.26) = 1.92m$$

The effective area is thus: $A' = L \times B' = 5 \times 1.92 = 9.6\,m^2$ (compared with the *actual* area $A = 3.5 \times 5 = 17.5\,m^2$).

Since the foundation is embedded in a cohesionless soil, and subject to a *vertical* load, albeit eccentrically, the bearing capacity equation 8.57 is then reduced to the following:

$$q_u = \gamma D N_q\, s_q\, d_q + \gamma \frac{B'}{2} N_\gamma\, s_\gamma\, d_\gamma$$

The bearing capacity factors corresponding to an angle $\phi' = 40^\circ$ are read from figure 8.26:

$$N_q = 60, \quad N_\gamma \approx 105$$

The shape factors are calculated using the *effective width B'*. Thus, using equations 8.55(*b*)–(*c*), it follows that:

$$s_q = 1 + \frac{B'}{L}\tan\phi' = 1 + \frac{1.92}{5}\tan 40 = 1.32$$

$$s_\gamma = 1 - 0.4\frac{B'}{L} = 1 - 0.4 \times \frac{1.92}{5} = 0.85$$

The depth factors are calculated in conjunction with the *actual width B* from equations 8.55(*e*)–(*f*) knowing that $\xi = \frac{D}{B} = \frac{1.8}{3.5} = 0.514 < 1$:

$$d_q = 1 + \xi\tan\phi'(1 - \sin\phi') = 1 + 0.514\tan 40(1 - \sin 40) = 1.154$$

$$d_\gamma = 1$$

Whence the ultimate pressure:

$$q_u = 21 \times 1.8 \times 60 \times 1.32 \times 1.154 + \frac{21 \times 192}{2} \times 105 \times 0.85 \times 1$$
$$= 5254\,kN/m^2$$

The corresponding ultimate vertical load is thence obtained as follows:

$$Q = q_u A' = 5254 \times 9.6 = 50.438\,MN$$

If a factor of safety of 2.5 were required against shear failure, then the *allowable load* can be calculated in a way similar to that used in the previous example:

$$q_a = \frac{q_u - L\gamma D}{2.5} + L\gamma D = \frac{5254 - 5 \times 21 \times 1.8}{2.5} + 5 \times 21 \times 1.8$$

$$= 2215\,kN/m^2$$

Whence an allowable vertical load:

$$Q_a = q_a A' = 2215 \times 9.6 = 21.264\,MN$$

Problems

8.1 An excavation of depth H is to be made in a (dry) soil with a bulk unit weight γ and an undrained shear strength c_u. The excavation sides are prevented from collapsing through the use of struts.

(*a*) Consider the short term collapse mechanism depicted in figure *p*8.1*a*, then use the upper bound theorem to establish the expression of the ultimate pressure q_u needed to counteract the bottom heave of the excavation.

(*b*) Use the same upper bound theorem in conjunction with the long term collapse mechanism of figure *p*8.1*b*, and establish the relationship between the ratios H/B and q_u/q_o.

Ans: (*a*) $q_u = q_o\left(1+\frac{H}{B}\right)+\gamma\frac{H^2}{2}\left(\frac{1}{B}+\frac{2}{H}\right)-2c_u\left(1+\frac{\pi}{2}+\frac{H}{B}\right)$

(*b*) $\frac{H}{B} = \exp\left(-\frac{\pi}{2}\tan\phi'\right)\left[\frac{q_u}{q_o}\tan^2\left(\frac{\pi}{4}+\frac{\phi'}{2}\right)\exp(\pi\tan\phi')-1\right]$

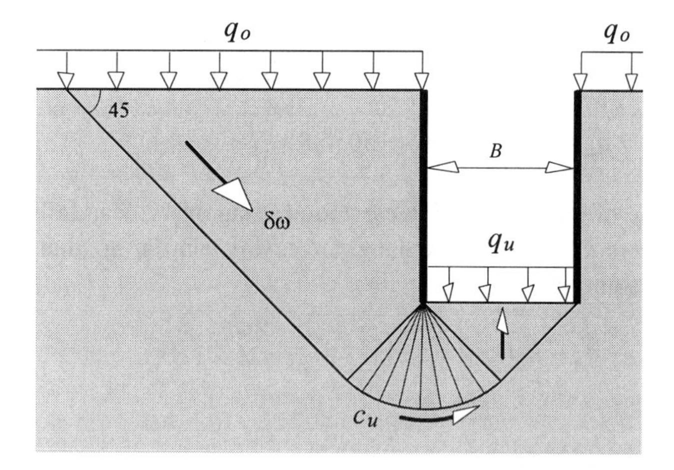

Figure p8.1a

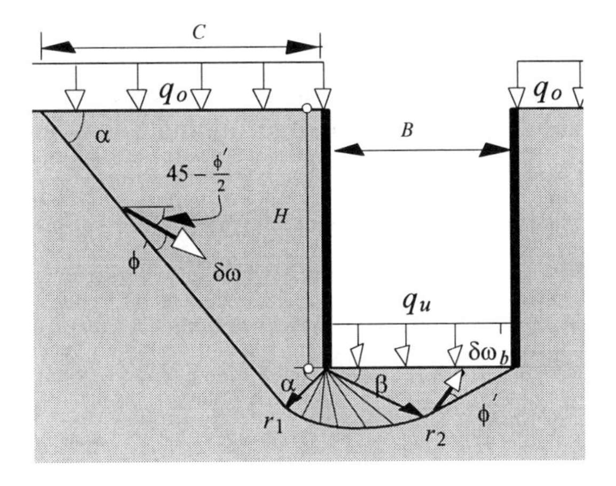

Figure p8.1b

8.2 Use the upper bound theorem and the collapse mechanism of figure *p*8.2 to establish the expression of the net ultimate pressure q_u that can be transmitted to the soil by the shallow foundation of width *B*.

Ans: $q_u = q_o \exp(\frac{\pi}{2}\tan\phi')\tan^2(\frac{\pi}{4}+\frac{\phi'}{2})\left[\exp(\frac{\pi}{2}\tan\phi') + \frac{H}{B}\right]$

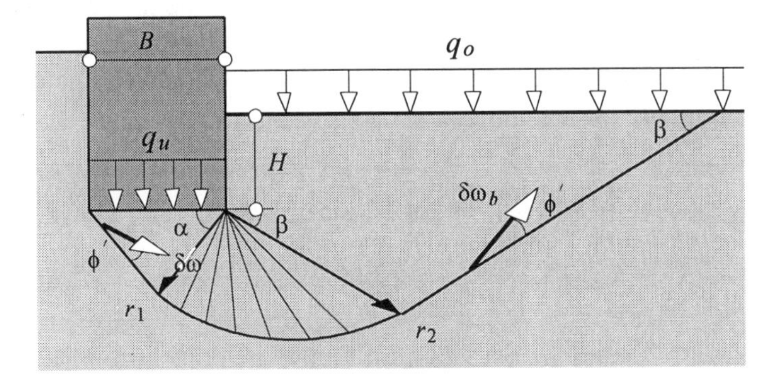

Figure p8.2

8.3 An excavation of depth H is carried out in a soil with a unit weight γ and an undrained shear strength c_u. Use the upper bound theorem in conjunction with the failure mechanism of figure *p*8.3, then establish the expression of the ultimate horizontal force P_1.

Ans: $$P_1 = \gamma\frac{H^2}{2}\left(1+\frac{2B}{H} - \tan\alpha\right) - 2c_u H\left[1+\frac{B}{H}(1+\frac{\pi}{2})\right]$$

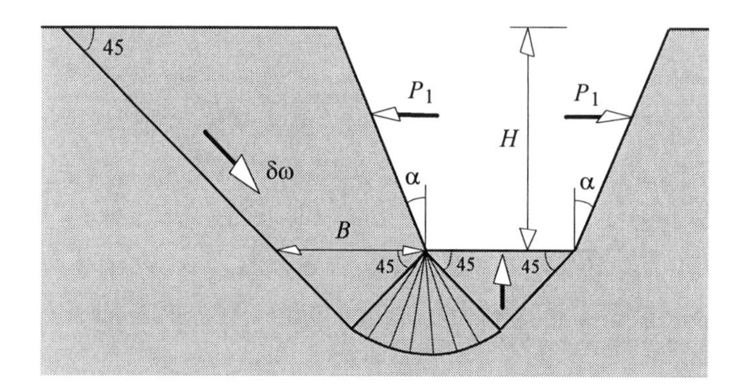

Figure p8.3

8.4 Consider the statically admissible stress field containing a stress fan, depicted in figure *p*8.4. Assuming that the excavation is undertaken in a (dry) clay with a bulk unit weight γ and an effective angle of shearing resistance ϕ', use the lower bound theorem to establish the expression for the ultimate heave resistance q_u at the bottom of the excavation, then estimate its magnitude for an angle $\phi' = 22°$.

Ans: $$q_u = \frac{(q_o + \gamma H)}{(1-\sin\phi')\tan^2(\frac{\pi}{4} - \frac{\phi'}{2})}\left[1+\sin\phi' + 0.845\exp\left(\frac{\pi}{2}\tan\phi'\right)\right]$$

$$\phi' = 22° \quad \Rightarrow \quad q_u = 10.43\,(q_o + \gamma H)$$

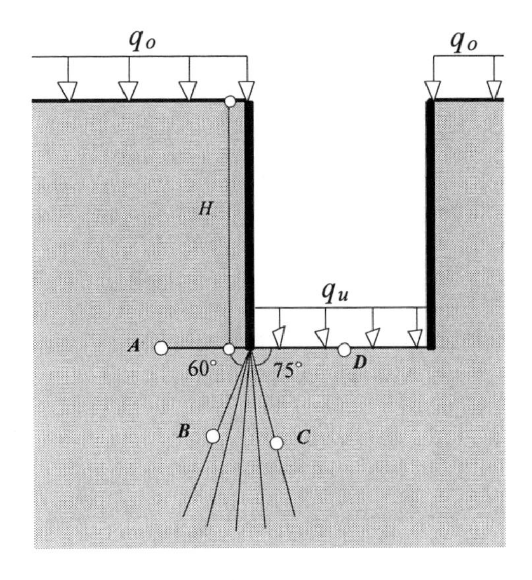

Figure p8.4

8.5 A long retaining wall with a 3 *m* wide base is founded at a depth of 1.2 *m* in a silty sand. The analysis of static forces applied to the wall yielded an eccentric resultant load $R = 400\,kN/m$, inclined at an angle $\alpha = 20°$ with respect to the vertical as depicted in figure *p*8.5. Above the ground water table situated at foundation level, the sand has a bulk unit weight $\gamma = 18\,kN/m^3$, and below the water, its saturated unit weight is $\gamma_{sat} = 19.8\,kN/m^3$.
The appropriate shear strength parameters of the sand are $c' = 0, \quad \phi' = 35°$. Calculate the eccentricity *e* if a factor of safety against shear failure $F = 3$ is required.

Ans: $e \approx 0.49\,m$

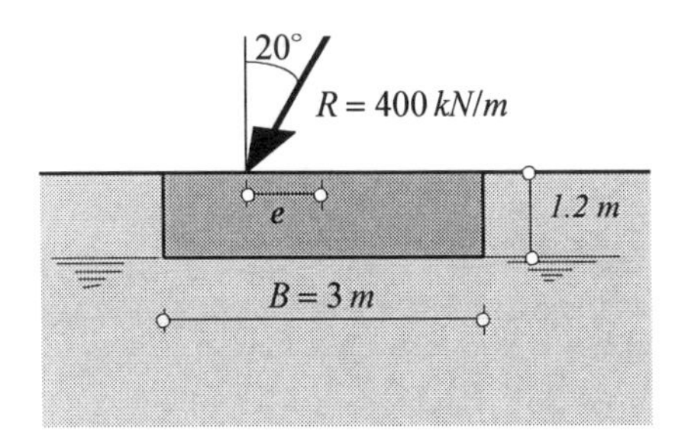

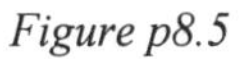

Figure p8.5

8.6 A rectangular foundation with a ratio $B/L = 0.8$, built at a depth $D = 2\,m$, is designed to support safely a net increase in vertical pressure of a magnitude $q_n = 150\,kN/m^2$, allowing for a factor of safety $F = 3$ against shear failure. The foundation is embedded in a thick layer of saturated firm clay having an undrained shear strength $c_u = 60\,kN/m^2$. Calculate the foundation size.

Ans: $B = 1.6\,m, \quad L = 2\,m$

8.7 The foundation of a prop is embedded at a depth $D = 2.5\,m$ in a layer of saturated dense sand with an effective angle of shearing resistance $\phi' = 39°$ and a saturated unit weight $\gamma_{sat} = 20\,kN/m^3$. The prop is assumed to be transmitting a concentrated load $R = 15,000\,kN$, inclined at an angle $\beta = 18°$ with respect to the vertical. The design of the foundation allows for a two-way eccentricity as depicted in figure *p*8.7, with $e_B = 0.3\,m$ and $e_L = 0.5\,m$.

(*a*) Estimate the ultimate bearing capacity of the foundation.

(*b*) Calculate the available factor of safety against shear failure.

Ans: (*a*) $q_u = 3364\,kN/m^2$, (*b*) $F = 2.91$

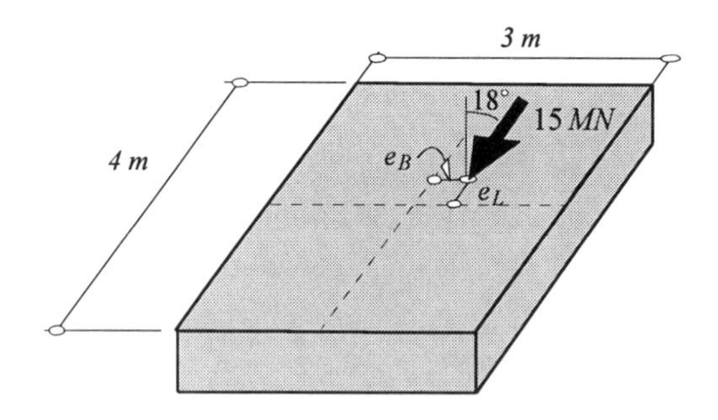

Figure p8.7

References

Bolton, M. D. (1991) *A Guide to Soil Mechanics.* M. D & K. Bolton, Cambridge.
Calladine, C. R. (1985) *Plasticity for Engineers.* Ellis Horwood Ltd, New York.
De Beer, E. E. (1970) *Experimental determination of the shape factors and bearing capacity factors of sands.* Géotechnique, 20 (4), pp. 387–411.
Hanna, A. M. and Meyerhof, G. G. (1981) *Experimental evaluation of bearing capacity of footings subjected to inclined loads.* Canadian Geotechnical Journal, 18 (4), pp. 599–603.
Hansen, J. B. (1970) *A Revised and Extended Formula for Bearing Capacity.* Danish Geotechnical Institute, Bulletin 28, Copenhagen.
Higher, W. H. and Anders J. C. (1985) *Dimensioning footings subjected to eccentric loads.* Journal of Geotechnical Engineering, ASCE, 111 (GT5), pp. 659–665.
Skempton, A. W. (1951) *The Bearing Capacity of Clays.* Proceedings of the Building Research Congress, London.

CHAPTER 9

Loading capacity of pile foundations

9.1 Type of piles

A *deep foundation* is defined as a foundation unit whose depth is at least five times larger than its width. Such a foundation unit provides support to an externally applied load through friction developing on its sides, better known as *skin* or *shaft friction*, and also through its *toe* or *base* as depicted in figure 9.1.

The proportions of both *skin friction* and *base resistance* depend on the soil in which the pile is embedded, and the graphs on each side of figure 9.1 give an indication of how the overall axial load applied at the pile head is transferred to the soil along the pile shaft. For a pile embedded in a thick layer of soft clay, most of the resistance develops along the shaft, and the pile is referred to as a *floating pile*. However, when the pile is driven through a layer of soft clay, to be embedded in a much stiffer soil, then toe resistance is predominant and the pile is known as an *end bearing pile*.

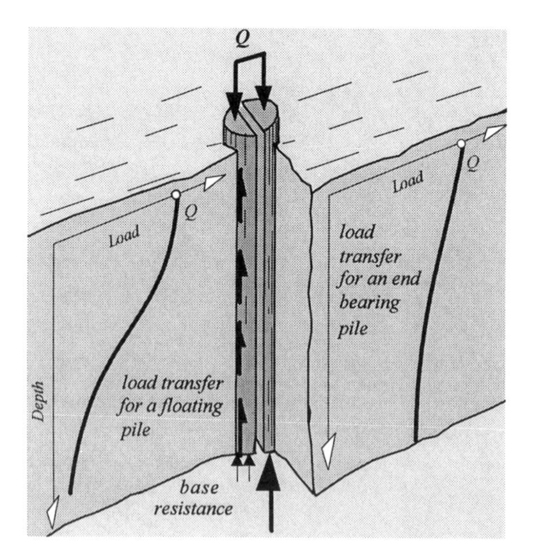

Figure 9.1: Axial load transfer through a piled foundation.

In both cases, both shaft friction and toe resistance must be calculated for the overall load capacity of the pile to be correctly evaluated, because the applied (axial) load is transferred from pile to soil through a combination of shaft friction and toe resistance.

There is evidence that deep foundations were used by the Romans when oak and olive wood piles were considered imperishable when submerged in water. Perhaps the most famous site where wooden piles were used extensively is Venice, where some buildings are thought to be supported by 1100 year old wooden piles. However, since the mid-twentieth century, piles are almost exclusively made of concrete and steel.

Piles are classified into two categories as follows:

(a) Displacement piles (or driven piles)

In the case of a *displacement pile*, the soil is moved laterally as the pile is driven into the ground and, obviously, the larger the cross-sectional area of the pile, the larger the volume of soil displaced to make room for the pile. Accordingly, a compaction process takes place in the vicinity of the pile, to an extent that a vertical soil movement can be observed at the ground surface were the pile to be driven into a dense granular material. The increased density during driving can be a mixed blessing, since it increases the shaft friction but, in the meantime, it can potentially impede the driving of the pile to the required depth. Also, *heave* may affect adjacent piles, particularly in dense sands.

Piles can be of different cross-sectional areas as shown in figure 9.2, and they can be made of precast concrete, prestressed concrete, steel tubes or steel boxes. The method of installation of displacement piles usually consists of using hydraulic diesel or compressed air hammers and, consequently, the pile head and toe should be protected so as to prevent any structural damage to the foundation. Obviously, the force needed to *drive* the pile to the desired depth *exceeds* the load the pile is designed to carry safely. Moreover, the driving generates tensile stresses within the pile due to wave reflections off its tip and, therefore, concrete piles must be designed to withstand these excess stresses. Tip reflections are a particular problem in clays where end resistance is small and reflections can be very large.

On the other hand, piles installed in sands can be driven using a vibration methods, in which case, care must be taken so as to minimise the structural damage to neighbouring buildings since vibration can cause quicksand conditions to occur (refer to section 3.2). In particular, vibration has important effects in looser sands where it causes a *volume decrease*. Also, piles with diameters of less than the nominal value of 250 *mm*, known as *micropiles*, can be jacked into the ground (usually hydraulically) against a fixed reaction.

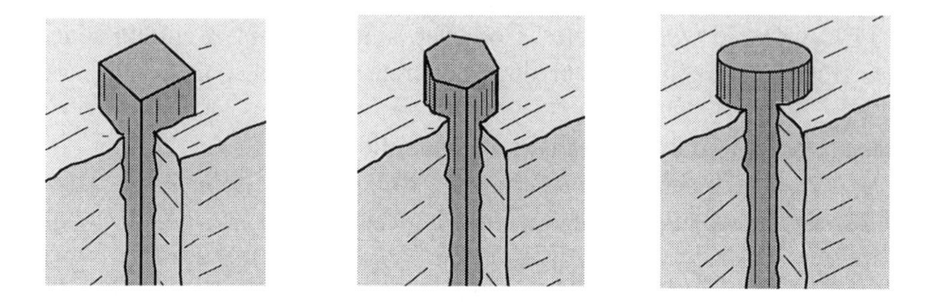

Figure 9.2: Some displacement piles.

Some of the most widely used types of displacement piles and their installation techniques are briefly introduced in the following.

Prefabricated concrete piles: these are constituted of reinforced concrete, and can be prestressed. These piles are installed either by driving or by vibration. Piles with diameters of between 200 *mm* and 600 *mm* can be prefabricated into up to 14 *m* long elements which are then driven *in situ* through jacking against a fixed reaction. During this process the different elements can be prestressed. For plain reinforced concrete piles, jointing systems allow installation of very long piles in some ground conditions.

Steel piles: these are made of steel with a minor copper content (usually less than 0.5%), and may be jacked into the ground against a fixed reaction. They may also be driven using conventional hammers.

Driven and cast-in-place piles (also known as *Franki piles*): these are plugged tubes, with a fabricated steel or a reinforced concrete shoe and a protected head, driven by a hammer to the required depth. The tube is then filled with concrete. A steel reinforcement cage is always used (reinforcement serves the purpose of resisting *heave forces* when adjacent piles are driven). In most cases, the tube is withdrawn during concreting.

Steel piles coated with concrete: steel tubes, steel H-sections or steel boxes, with steel plates larger than their cross-sectional areas welded at their bases are driven whilst concrete of high workability is poured to fill the void space including that created by the protruding base plate. Such a process requires that the external steel area is covered by at least a 40 *mm* thick concrete coat. Note that this type of piles is not common.

Screwed and cast-in-place piles: the process consists of obtaining a threaded hole by pushing a rotating drilling screw into the ground as depicted in figure 9.3. On reaching the required depth, the drilling equipment is then withdrawn through rotation in the opposite direction, while fresh concrete is poured into the hole. This process leads to a substantial increase in the carrying capacity of the pile (Imbo, 1984, Bustamente and Gianeselli, 1995). This type of *Franki* pile is especially suited for clays (excluding very soft and boulder clays).

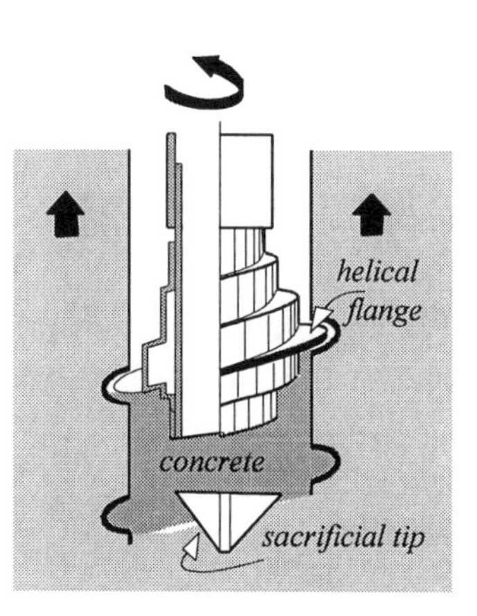

Figure 9.3: 'Atlas' screwed and cast in place pile.

Open base piles: these are of relatively small cross-sectional areas such as *steel H-sections*, *open end steel tubes* and *boxes* (see figure 9.4), and are mainly used *offshore*. In particular, *open end concrete shell piles* consist of elements of concrete shells, with a length of between 1.5 *m* and 3 *m*, an internal diameter of between 0.7 *m* and 0.9 *m* and a wall thickness of about 150 *mm*, driven with an open base to the required depth, then prestressed *in situ*. Sometimes, the upper layer of the soil plug inside the tube can be drilled using a rotating cutter, so as to facilitate the driving.

Regardless of their (cross-sectional) size, these piles, known as *low displacement piles*, offer a small cross-section to the soil and are easy to drive because, in this case, less compaction develops along the shaft during driving, thus lessening the friction between pile and soil. The reduced friction is an obvious advantage when the required depth of embedment is substantial.

Figure 9.4: Low displacement piles.

(b) Non-displacement piles (bored piles)

For a non-displacement pile, a borehole with a diameter corresponding to that of the pile is first excavated, then a concrete pile is cast *in situ*. For *small diameter piles* (up to 600 *mm*), boreholes are often executed using rigs of the type used for site investigations. Larger diameter boreholes are usually executed using rotary drilling methods (see Fleming *et al.* (1992) for instance).

The two main techniques used in conjunction with bored piles consist of *augering* and *grabbing*, the choice of which depends on soil conditions and pile dimensions. Hence, some *heavy duty rotary augers* can drill pile shafts with diameters of 4.5 *m* and depths of up to 70 m (Tomlinson, 1995). Depending on soil conditions, the borehole may have to be supported so as to prevent its collapse. A full casing can be used during drilling, then withdrawn during 'concreting' of the pile. Also, a *bentonite slurry* can provide lateral support to the walls of a borehole (refer to figure 9.5), due to its *high density* and *thixotropic properties* (*i.e.* it forms a gel at rest and becomes a fluid if rapidly agitated).

Bentonite consists largely of montmorillonite, a clay mineral known for its expansive properties as introduced in chapter 1. *Bentonite slurry* is made of a suspension of 5% bentonite (by weight) in water, thus forming a mud with a unit weight $\gamma_b \approx 11\, kN/m^3$, greater than that of water

($\gamma_w \approx 10\, kN/m^3$). This slurry is characterised by thixotropic properties, in that its high viscosity under low stresses decreases as the stresses are increased. For these properties, bentonite slurry is widely used in the construction industry to provide lateral support to the walls of an excavation while preventing any water ingress.

When the soil conditions are such that rotary augers cannot be used economically (drilling through boulder clays or very coarse gravel for example), then a *grabbing rig* is often preferred (see figure 9.5). This technique is used especially in conjunction with the construction of *diaphragm walls* (which are discussed in detail in chapter 11) and *barettes* which are piles with large cross-sectional areas with different shapes as illustrated in figure 9.6.

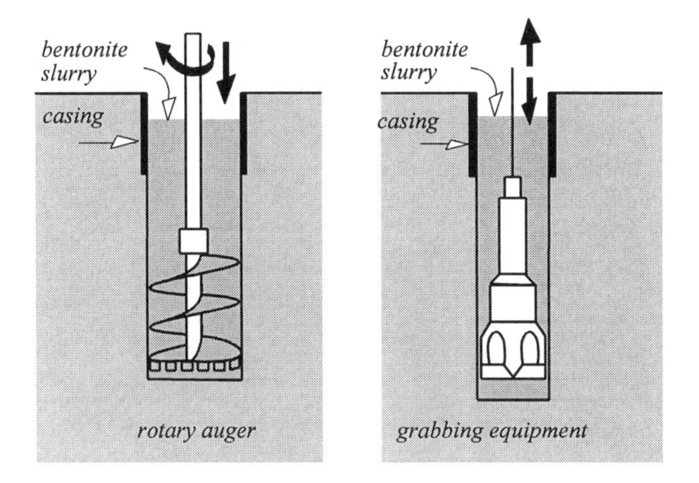

Figure 9.5: Augering and grabbing of pile shafts.

Generally, the cross-sectional dimensions of a barette are such that $2\,m \le e \le 3\,m$, and $0.5\,m \le b \le 1.5\,m$. Bentonite slurry is usually used as a support for the trench walls as depicted in figure 9.5. On achieving the required depth, a reinforcing steel cage is placed inside the borehole (or the trench in the case of a barette) then concrete is pumped through a *trémie* pipe (refer to figure 9.7).

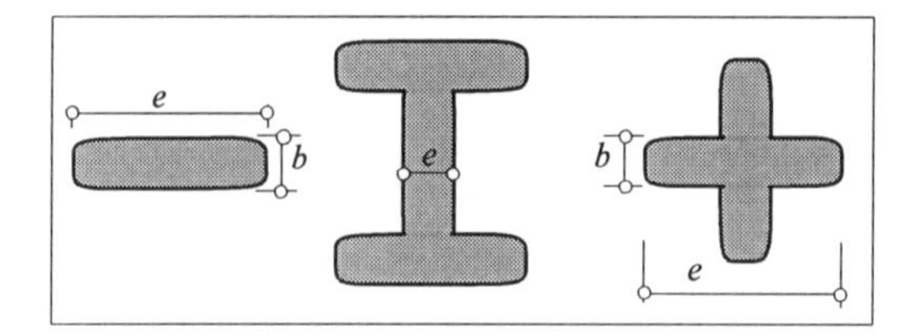

Figure 9.6: Barettes dimensions.

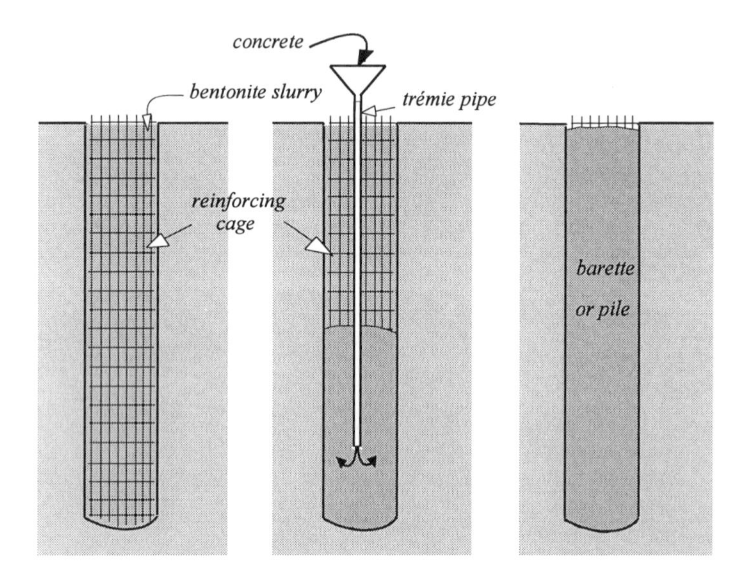

Figure 9.7: Steps of concreting bored piles and barettes.

Relatively small diameter bored piles (up to 1.5 *m* in diameter) can be executed using *short flight* or *continuous flight augers*. In particular, technologically advanced continuous flight augers are available, and figure 9.8 depicts the steps related to the execution of a bored pile using the *Starsol* rig developed and operated by *Solétanche*. On reaching the final depth, the auger is gradually withdrawn while concrete is injected under controlled pressure through the telescopic *trémie* pipe. Once concreting is finished, steel reinforcement is then pushed into the freshly poured concrete. Notice that the maximum size of concrete aggregates is limited to

20 *mm*; moreover, a plasticising agent is usually mixed with concrete so as to improve what Tomlinson (1995) refers to as 'pumpability' (*sic*).

Other systems are operated under full and continuous computer control, so that the construction process becomes almost automatic. This form of construction appears to lead to better and more consistent pile performance.

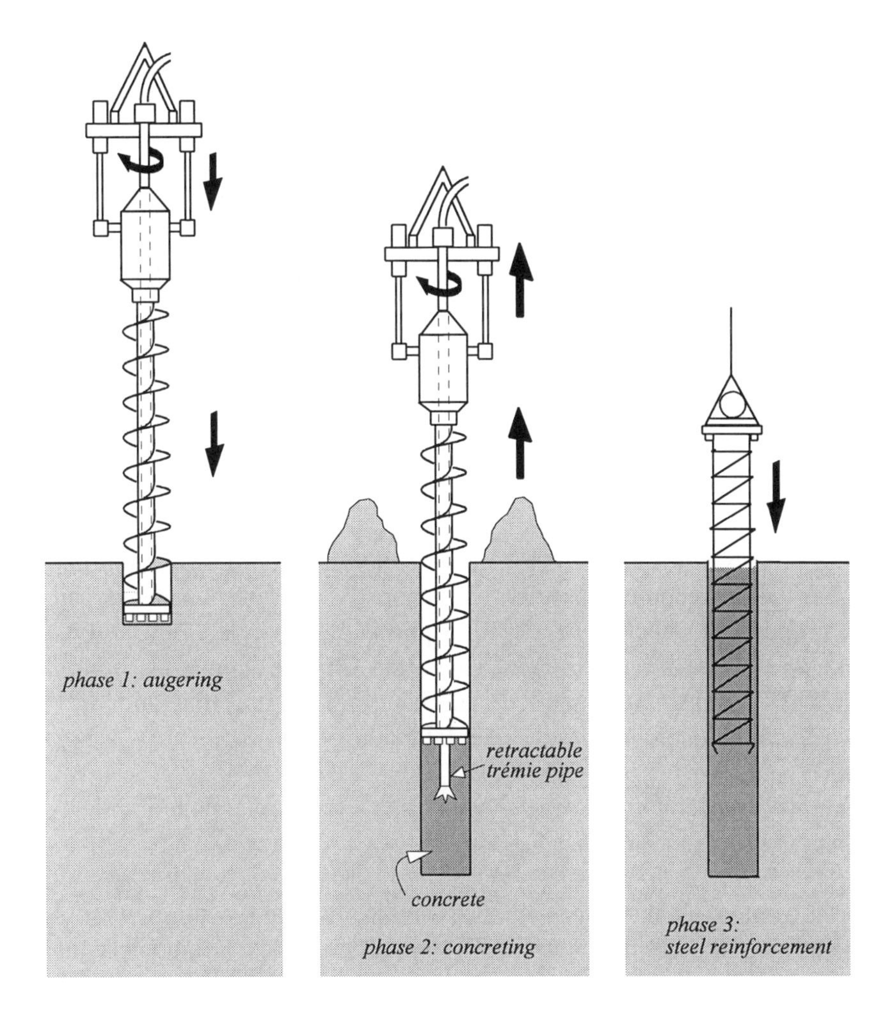

Figure 9.8: Starsol rig—Solétanche.
(Reproduced by permission of Solétanche.)

Other special cases of deep foundations include the following.

Hand excavated deep shafts are used where drilling equipment cannot be operated due to obstructed soil conditions or to a lack of access. For obvious safety reasons, the shaft is lined as the depth progresses. This method of drilling can be cost effective and offers an *ad hoc* solution to the inspection of the shaft base before concreting,

Open-well caissons are used in conjunction with water bearing strata ranging from soft clays to gravels. They are, however, unsuitable for boulder clays which are difficult to excavate by grabbing. The sinking of the caisson under its own weight occurs as the soil is excavated by grabbing from within the shaft. There are instances in which corrective measures may be required to ensure the verticality of the caisson as it is sunk. Once the required depth is reached, the bottom of the well is then sealed by pouring a thick layer of concrete. The well can thereafter be filled with water, sand or concrete. *Pneumatic caissons*, as opposed to open well caissons, are excavated by hand under air pressure and, accordingly, stricter safety regulations apply depending on the level of air pressure used, which should not exceed $350\,kN/m^2$. A detailed analysis of excavation under air pressure is presented in conjunction with tunnelling in cohesive soils in section 12.3.

There are no rules, as such, for the *selection* of a particular type of pile. Rather, some recommendations based mostly on common sense can be advocated. The selection of a type of pile is closely related to the type of soil, the type of structure to be supported (bridges, high rise buildings, quays *etc.*), and to the magnitude and nature of loading transmitted to the soil (*i.e.* axial loading and/or horizontal loading).

Generally, the contractor has a major say in the selection of pile type, depending on the following.

- The type of technology available (driving, augering, continuous flight augering or grabbing equipment). In particular, the continuous flight augering method accounts for about 40% of the bored pile market (1999 estimate).

- The site conditions [onshore or offshore, soil strata, and groundwater conditions are major factors affecting the selection of piles (steel or concrete piles, driven or bored, barettes, deep shafts, caissons *etc.*)].

- The cost of piling.

Piles with diameters exceeding 1 *m* are usually used for large projects (*i.e.* bridges, very tall buildings *etc.*). The length of a pile is dictated by the need to generate enough skin friction and/or base resistance to support the structural load. Bored and cast-in-place piles and barettes are reinforced wholly or partially. The depth of reinforcement depends on soil conditions, lateral and eccentric loading conditions, and the requirements of the construction process.

9.2 Pile testing

It is now becoming a common practice to check cast-in-place piles and barettes *selectively* for major defects that may arise, for instance, from the inclusion of soil during concreting or from lack of cover to steel reinforcement. This process involves the undertaking of *integrity tests* which are sometimes limited to a few preselected piles for reasons of cost effectiveness. Obviously, because of the limited number of tested piles, there is always a possibility that some defective piles may not be detected. The nature of these non-destructive tests can be acoustic, seismic, dynamic or radiometric. The principles involved consist of measuring pulses transmitted through the pile (case of acoustic and radiometric tests), or waves reflected from the pile base (case of seismic and dynamic tests). Signals thus measured can be interpreted to check, for example, major pile defects, the length of pile or the pile–soil contact at the pile base. Further technical details on how these tests are actually undertaken, and on the way they are interpreted can be found in Fleming *et al.* (1992).

Piles are also tested for their carrying capacity using *in situ* full scale *load tests*. These tests must be considered carefully because they can be expensive. They may involve the use of sophisticated instruments and complex testing set ups.

Preliminary load tests are undertaken *prior* to the construction of the actual deep foundation. They consist of applying incremental loads to a pile using the set up depicted schematically in figure 9.9. The pile settlement is monitored against the applied load until the occurrence of failure, or until a specified load has been reached. By definition, the *ultimate overall load at failure* is that at which the *full soil resistance* is mobilised. It is usual practice to apply one or more unload–reload cycles well before the (estimated) ultimate load is reached. However, the usefulness of several unload–reload cycles is debatable since they involve different stress paths to those of the normal incremental loads. Also, each load increment is maintained *constant* for about 60 *min* during loading and 5 *min* during unloading, before the next increment is applied. In this respect, it is important to bear in mind that the *constancy* of applied load is vital to permit useful analysis of test results.

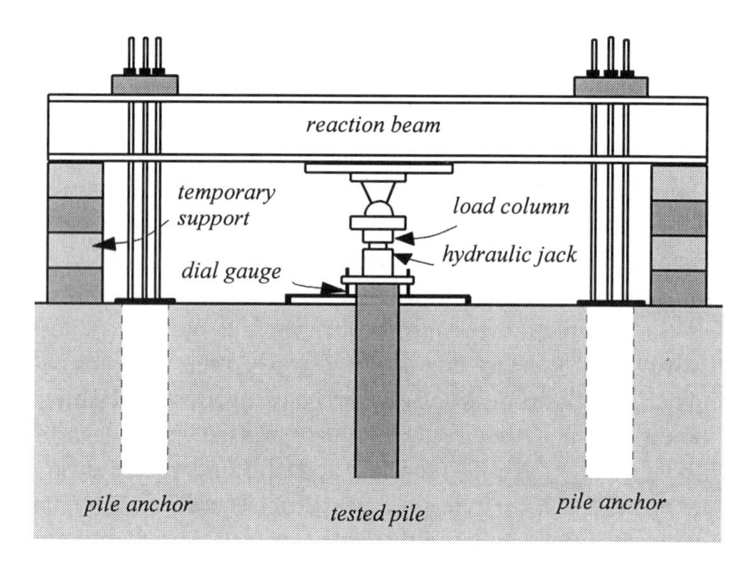

Figure 9.9: Set up used in conjunction with axial load testing of piles.

Because piles are sometimes load-tested until the occurrence of failure and to stress levels which may damage the pile structure, they cannot then be part of the actual foundation and, accordingly, preliminary load tests are undertaken prior to the construction of the foundation. On the other hand, unless the soil in which the pile is driven is highly permeable (sand or gravel), pile driving is bound to generate excess porewater pressures in the vicinity of the pile. If the pile is driven through a highly overconsolidated clay, then negative excess porewater pressures are generated (refer to section 5.5.4) meaning that, with time, the porewater pressure *increases* and the effective stress around the pile shaft (*i.e.* the skin friction) *decreases* until a large enough volume of water is absorbed by the soil matrix around the pile shaft so that the excess porewater pressure becomes zero. Alternatively, if the pile is driven through a normally consolidated clay, then positive excess porewater pressure is produced and, given time, the porewater pressure *decreases* and the effective stress (and hence the friction applied around the pile shaft) *increases*. As a result, load tests must allow for these adjustments to take place and, hence, a knowledge of dissipation rates may be necessary.

Depending on the type of pile instrumentation and the testing method, a load test can yield:

- the ultimate shaft resistance of the pile and the corresponding settlement (*i.e.* penetration);
- the penetration corresponding to the ultimate base carrying capacity of the pile;
- the overall load and settlement behaviour;
- the distribution of the unit skin friction along the pile shaft (which depends on the availability and accuracy of instrumentation).

Figure 9.10 depicts a typical load-penetration curve along the pile shaft and at the pile base in the case of a bored and cast-in-place or a driven pile. The graph shows, in particular, that the settlement S_s corresponding to the shaft friction at failure Q_s is noticeably smaller than the settlement S_f needed to mobilise the ultimate base resistance Q_b of the pile. Also noticeable is the difference in the carrying capacity of the pile along the shaft and at the base (the grey area in figure 9.10). Shaft capacity usually predominates in clay soils and base load does so in sands.

The pile must be designed so that not only is the structural load carried safely, but also the settlement is kept within an acceptable limit. In other words, the *safe load* or the *working load* must be a *fraction* of the ultimate carrying capacity Q_u in figure 9.10.

In any case, the overall settlement under the working load Q_a must be kept smaller than S_s in figure 9.10 by a reasonable *margin of safety*. In this respect, it is advised to use the lesser of the following two loads as the working load in the case of bored and cast-in-place piles:

$$Q_a = \frac{Q_u}{2.5} \tag{9.1}$$

or

$$Q_a = Q_s + \frac{Q_b}{3} \tag{9.2}$$

Notice that Burland *et al.* (1966) advocated the use of an overall factor of safety of 2 in equation 9.1 (instead of 2.5).

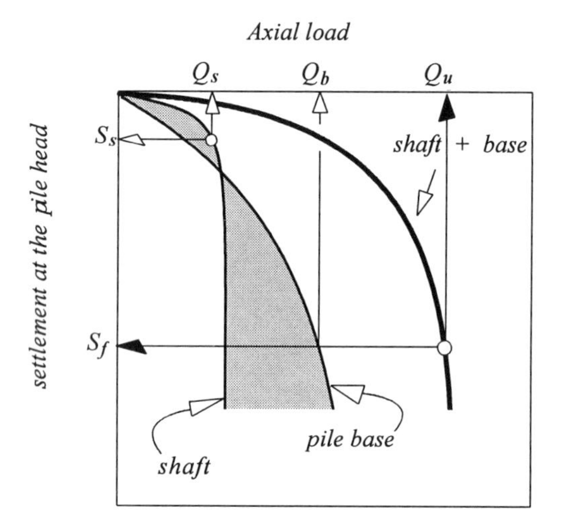

Figure 9.10: Typical load-penetration curves for a driven or bored and cast-in-place pile.

The use of such a working load in conjunction with bored and cast-in-place piles ensures that the overall settlement does not constitute any cause for concern since its magnitude S is kept well within the acceptable limits of shaft settlement as illustrated in figure 9.11.

At the end of a test, it is seen that, on removing the load, a *rebound* of the pile head is measured (soil recovery plus elastic component of settlement that consist of the axial displacement between points A and B in figure 9.11). The *residual settlement* under the applied load corresponds to the total settlement minus the rebound.

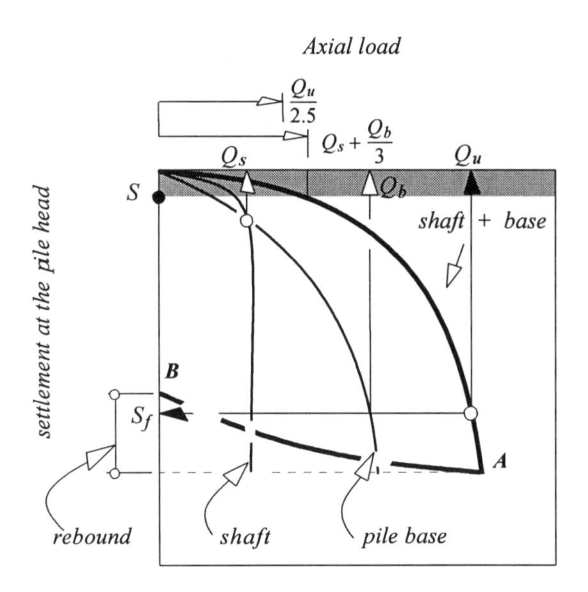

Figure 9.11: Typical overall settlement under working load (bored and cast-in-place piles).

However, equations 9.1 and 9.2 should be applied cautiously, and a degree of conservatism is even advised by using these equations in conjunction with driven piles (for which the skin friction and base resistance can be increased due to soil compaction in the vicinity of piles during driving). For bored and cast-in-place piles, Tomlinson recommends using a smaller working load Q_a (*i.e.* larger factors of safety), corresponding to the lesser of the two following values:

$$Q_a = \frac{Q_u}{2.5} \tag{9.3}$$

or

$$Q_a = \frac{Q_s}{1.5} + \frac{Q_b}{3.5} \tag{9.4}$$

Ample experimental evidence indicates that the settlement of small diameter piles (up to 600 *mm*) under working loads calculated from equation 9.3 rarely exceeds 10 *mm* (see Tomlinson (1995), for instance).

The pile can also be instrumented along the shaft with *extensometers*, which are reusable strain gauges utilised for measuring the local axial deformations $\Delta l/l$ of a pile under a constant load increment. These deformations are then used, in conjunction with appropriate calibration graphs, to determine the load distribution along the pile shaft and at its base, as well as the changes in skin friction along the shaft as the loading proceeds.

Results corresponding to a full scale load test, undertaken on a screwed and cast-in-place pile are depicted in figure 9.12 (Bustamente and Gianeselli, 1997). The pile, with a diameter of 360 *mm* and a length of 8.5 *m*, was installed in a site near *Lille* in the North of France characterised by a 5 *m* thick layer of sandy clay overlying a thick layer of dense clayey sand. The entire operation of pile execution, that is drilling, concreting and inclusion of an 8.5 *m* long steel reinforcement cage took less than 15 *min.*

In particular, figure 9.12*a* indicates that just over one third of the full load $Q = 1500\,kN$ is transferred to the pile base, whereas nearly $Q/6$ is resisted by skin friction along portion *C* of the pile (refer to shaded areas in figure 9.12*a*). Figure 9.12*b* on the other hand, shows that skin friction increases initially with settlement, then reaches limiting values as the settlement approaches the magnitude at which failure occurs.

More importantly, the same figure indicates that skin friction is not uniform along the pile shaft. It is seen that the friction developed along portion *D* is much higher than that generated along portions *A* and *B* due to the mechanism of load transfer from pile to soil (see figure 9.1). This remark is all the more important since the ultimate loads Q_s, Q_b and Q_u in figure 9.10 are not always determined from load tests. As mentioned earlier, such tests may be expensive and may only justifiably be used in conjunction with large projects. Alternatively, the ultimate loads can be estimated using the *static method* of design, provided that representative soil parameters are available, and a somewhat larger factor of safety is used to estimate the working (safe) load. Also, *back analysis* of normal pile load tests carried out to near failure constitutes another very useful way of design (Fleming, 1992, England and Fleming, 1994).

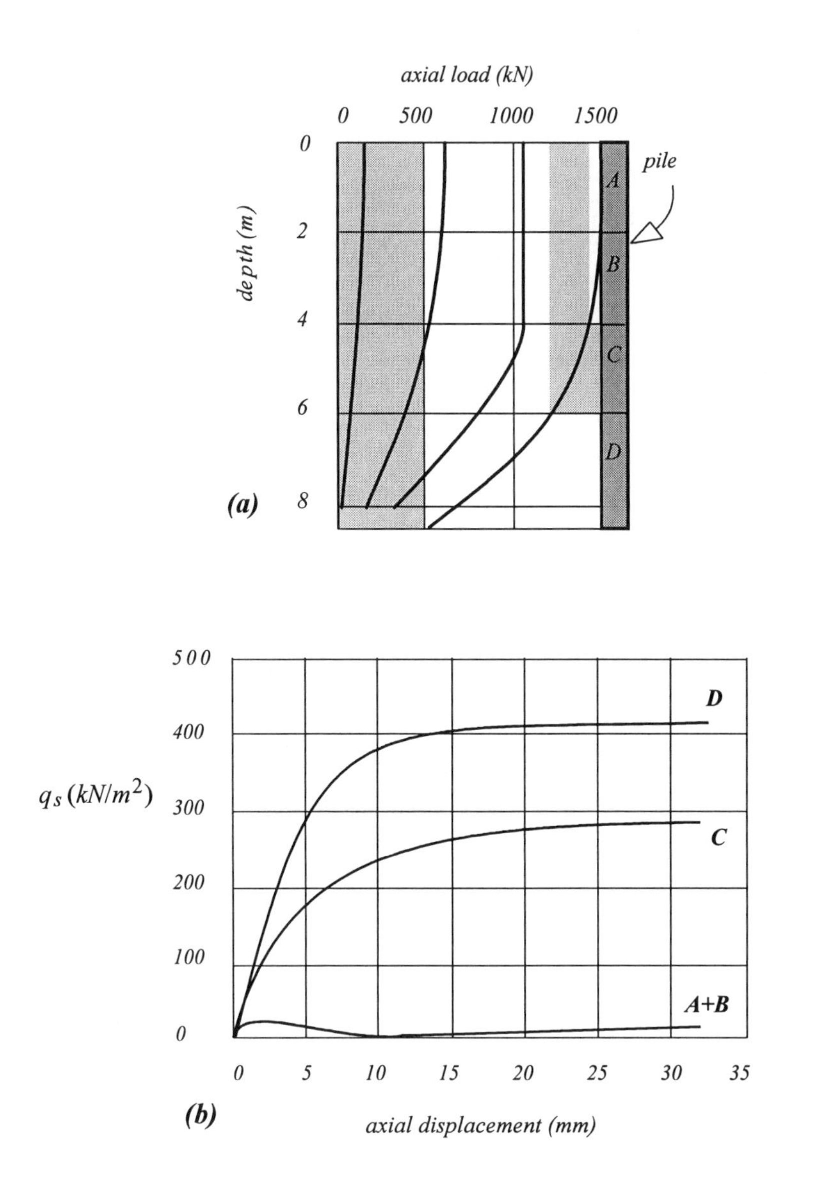

Figure 9.12: (a) Load transfer from pile to soil, and (b) variation of skin friction with axial displacement along the pile shaft (From Bustamente and Gianeselli (1997), by permission of the Laboratoire Central des Ponts et Chaussées.)

Prior to developing both methods, it is essential to realise that they do not constitute the only alternative way of pile design. In terms of design methods in general (including those applied to pile design), there are many regional variations which differ from country to country. The reader is therefore urged to appreciate the need to explore other methods of design and to make himself (herself) familiar with regional or even sometimes local variations. It is worth reiterating what was written in the preface to the extent that Geotechnics is not an exact science, rather, it can be described as an art in which the artist (*i.e.* the engineer) has to rely sometimes, if only partly, on his or her intuition (*i.e.* judgement). Consequently, the fact that different methods of design almost invariably yield different results for the same problem does not imply that all methods, other than method A with an outcome nearest to the measured results, are wrong. Rather, it shows that under the circumstances, method A is more suitable. Different conclusions may possibly be drawn were the circumstances to change.

There is often a need to utilise more than one method of design so that any aberrations can be discarded, bearing in mind that in the vast majority of cases in geotechnical design, the aim is limited to predicting a solution within x% of the behaviour observed in the field and to use adequate factors of safety (sometimes a solution within 50% either side of the exact measured solution can be deemed very acceptable). This explains the need to adjust the same method of design so that regional or even local variations of soil conditions are taken into account.

9.3 Ultimate loading capacity of axially loaded single piles using the static method

9.3.1 General considerations

In the static method of analysis, the *net ultimate loading capacity* of a single vertical pile, subject to an *axial load,* is evaluated as follows (refer to figure 9.13):

$$Q_u = Q_s + Q_b - W \qquad (9.5)$$

with

Q_u: ultimate loading capacity

Q_s : ultimate skin friction
Q_b : ultimate base resistance
W : weight of pile.

Note that, in many cases, W can safely be discarded since its effect on Q_u is relatively minor.

The (safe) working load related to the ultimate load of equation 9.5 can be estimated using, for example, a *minimum overall factor of safety* $F = 3$. Accordingly:

$$Q_a = \frac{Q_s + Q_b}{3} - W \qquad (9.6)$$

As mentioned earlier, the least favourable conditions for the ultimate carrying capacity may depend on the excess porewater pressure generated during pile execution. Thus, if a pile were driven through a thick layer of saturated sand, for instance, then no excess porewater pressure will be engendered because of the high permeability of such a soil. Consequently, provided that the water table remains relatively unaltered, the ultimate carrying capacity of the pile would be similar in the short term (that is a few days after pile installation) and in the long term (*i.e.*years after the pile execution). In other words, the state of effective stresses applied to the pile does not alter with time, and hence the analysis should be undertaken under *drained conditions* in terms of *effective stresses*.

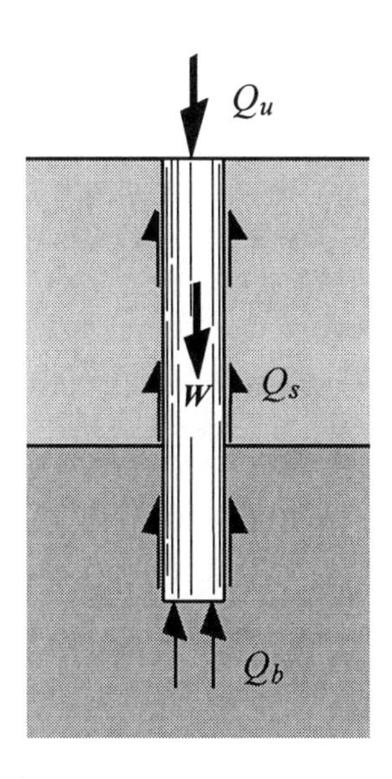

Figure 9.13: Load transfer from pile to soil.

On the other hand, if the pile were driven or bored and cast in place in a clayey soil, then its ultimate carrying capacity may initially depend on the nature of excess porewater pressure generated in the vicinity of the pile during installation. Because the pile must be designed to withstand the most unfavourable loading conditions when the soil develops the least resistance, the following possibilities arise.

(1) The clay is heavily overconsolidated, hence pile installation generates negative excess porewater pressure, thus causing the effective stresses in the vicinity of the pile (and therefore the friction on the pile shaft and the soil resistance at the pile base) to *increase.* The negative water pressure may take some time to dissipate (the time can be measured in days or even weeks in the case of a single pile, and can sometimes be expressed in years in the case of large pilegroups), during which, the porewater pressure gradually increases and the effective stresses *decrease* until zero excess water pressure is reached. Accordingly, the least favourable soil resistance (*i.e.* the smallest effective stresses) may occur in the *long term* when all excess negative porewater pressure has dissipated. Hence the critical carrying capacity of the pile must be calculated under *drained conditions* in terms of *effective stresses.*

(2) The pile is installed through a normally consolidated clay, in which case, positive excess porewater pressure is engendered, leading to a *decrease* in the effective stresses around the pile immediately after its installation. With time, as the excess water pressure slowly dissipates, the porewater pressure decreases and the effective stresses increase. Accordingly, the smallest effective stresses, or the least soil resistance to the load applied at the pile head, occur in the *short term* when the excess porewater pressure is hardly changed, and the critical carrying capacity of the pile must therefore be evaluated under *undrained conditions* in terms of *total stresses.*

(3) The clay in which the pile is driven or bored and cast in place is lightly overconsolidated, and pile installation generates relatively small excess porewater pressure. In this case, the critical carrying capacity of the pile may occur either in the long term or in the short term, and therefore both *drained* and *undrained conditions* ought to be checked, in that the design should be undertaken using both *total stresses* and *effective stresses*, so that the least favourable conditions are found.

(4) The pile is installed through multi-layered clay soil, whereby the top layer is still undergoing a process of consolidation. This implies that an excess porewater pressure already exists within the soil matrix of the top

layer and, consequently, any extra excess porewater pressure generated by pile installation is bound to induce a differential consolidation settlement between the pile and the top clay layer. This process generates a down force around the pile shaft in the consolidating clay layer known as *negative skin friction*. Because consolidation is a long term process (refer to chapter 4), the final settlement of the top layer occurs when all excess porewater pressure has dissipated and, hence, the critical carrying capacity of the pile occurs in the long term. Accordingly, whenever a negative skin friction is predicted, pile design must be undertaken under *drained conditions* in terms of *effective stresses*.

9.3.2 Undrained carrying capacity of single piles embedded in clays

Under undrained conditions, the net ultimate carrying capacity of an axially loaded single pile is calculated from equation 9.5. In the short term, the *ultimate skin friction* of a pile embedded in a homogeneous isotropic clay is evaluated as follows:

$$Q_s = A_s \alpha c_u \qquad (9.7)$$

with

$A_s = \pi dL$ shaft area (d and L are pile diameter and length respectively)

c_u : *average* undrained shear strength of the clay along the pile shaft

α : adhesion factor.

The *adhesion factor* corresponds to the ratio of the unit skin friction mobilised on the pile shaft to the undrained shear strength c_u of undisturbed clay. Because the process of pile driving or boring and casting in place is bound to disturb the soil, the skin friction developed in the short term around the pile shaft will only be a *fraction* of c_u, depending on the nature of the soil strata and pile dimensions. Naturally, one would expect the magnitude of skin friction developed around the shaft of a pile driven into a thick layer of a stiff overconsolidated clay to be different from that mobilised around the shaft of the same pile driven through a layer of soft clay and just reaching the top of an underlying layer of gravel for instance.

Reasonable estimates of the adhesion factor can be cautiously obtained from the useful graphs compiled by Tomlinson (1995) and Weltman and Healy (1978), and reproduced in figures 9.14 and 9.15.

In particular, the difference in behaviour exhibited in figures 9.14(*a*), (*b*) and (*c*) reflects the degree of clay disturbance generated during pile installation. It is seen that in figure 9.14(*a*) the pile is initially *driven* through a granular soil, and some granular frictional material is bound to 'adhere' to the pile shaft as it penetrates the clay layer. The skin friction mobilised around the shaft is thus maximised by the adhering thin layer of granular material, as long as the depth of embedment L within the clay layer remains modest ($L/d < 10$; d: pile diameter). This positive effect decreases gradually as the depth of embedment L increases (see graphs corresponding to $L/d = 20$, and $L/d > 40$ in figure 9.14(*a*)). Adhesion factors for intermediate values of L/d can be interpolated from the three graphs.

In figure 9.14(*b*), the pile is *driven* through a layer of soft soil overlying a layer of stiff soil. Under such circumstances, a process similar to that described previously develops with an opposite effect. In fact, while the pile is driven through the soft soil, a thin layer of soft material adheres to the shaft. As the pile embedment extends into the stiff soil, the soft material is gradually 'peeled', thus increasing the skin friction (*i.e.* the adhesion factor).

Figure 9.14(*c*) reflects the fact that, for piles *driven* into a thick layer of soil, a smaller degree of clay disturbance (hence a larger adhesion factor) is induced by a larger depth of embedment.

Bear in mind, however, that the empirical relationships in figure 9.14 are meant for driven piles and, as stated earlier, driving a pile through a stiff clay carries the risk of damaging the pile section due to wave reflection. Consequently, driving deep piles in firm or stiff clay soils is often not recommended. Moreover, the graphs in figures 9.14 and 9.15 should be used with caution since the magnitude of the adhesion factor is affected by the pile surface roughness. This emphasises the need for alternative methods of design (such as the use of an effective stress analysis) to be explored.

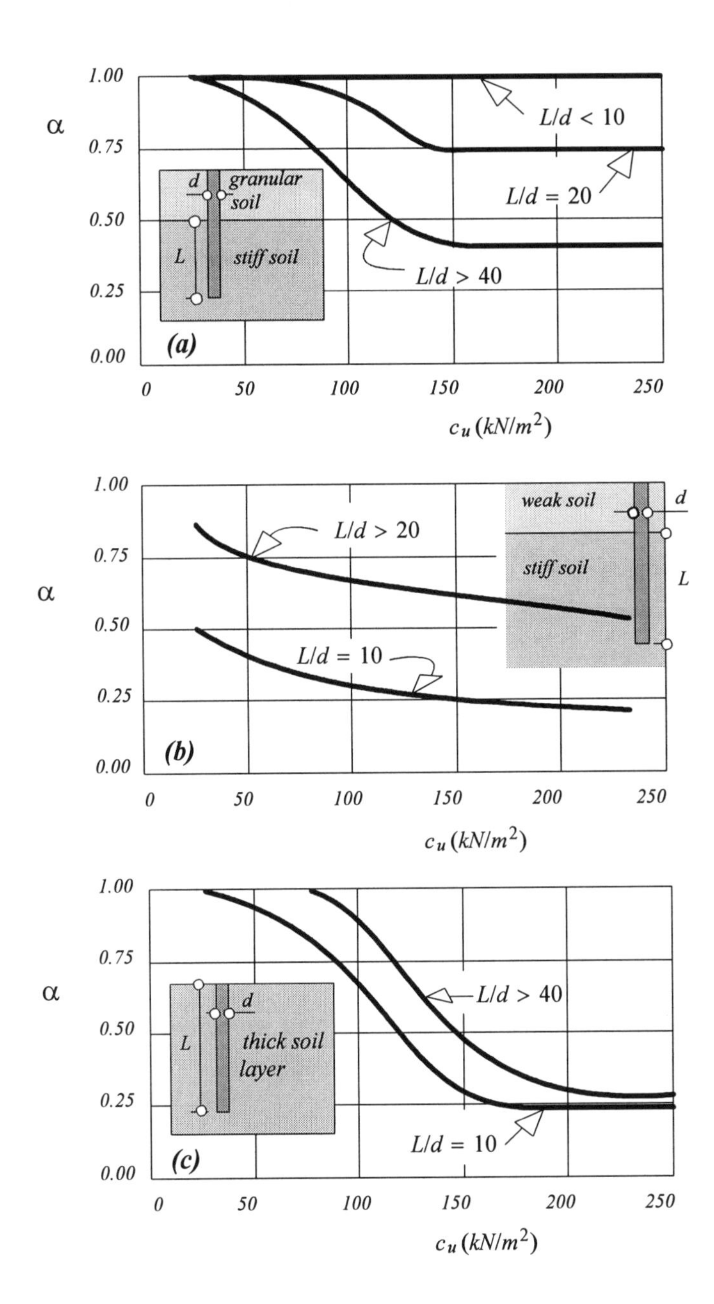

Figure 9.14: Adhesion factors applied to different soil strata: (a) granular soil overlying stiff soil; (b) weak soil underlain by stiff soil; (c) uniform layer of soil (from Tomlinson (1995), reproduced by permission).

Figure 9.15 on the other hand applies to piles bored and cast-in-place, driven or driven and cast-in-place in a boulder clay or a glacial till.

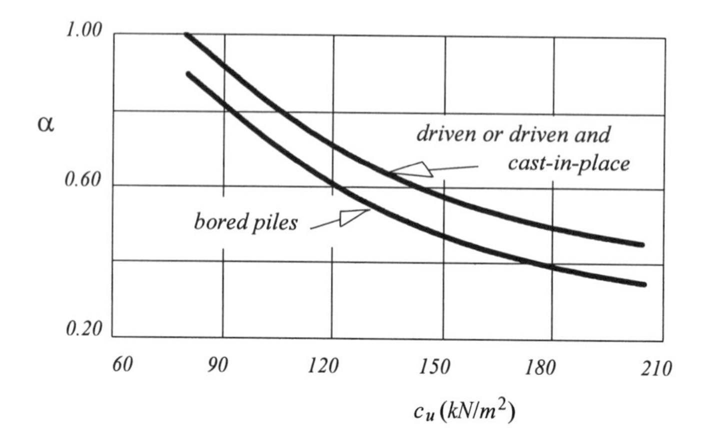

Figure 9.15: Adhesion factors for piles installed in boulder clays and glacial tills (from Weltman and Healy (1978), reproduced by permission).

If the soil properties vary with depth, then equation 9.7 would be expressed as follows:

$$Q_s = \int_0^L p\,\alpha\,c_u\,dz \tag{9.8}$$

with

L : pile length

p : pile perimeter (which can also vary with depth).

The *ultimate base resistance* of the pile is calculated from the bearing capacity equation 8.50 established earlier:

$$Q_b = A_b(c_u N_c + \gamma L N_q + \tfrac{1}{2}\gamma d N_\gamma) \tag{9.9}$$

with

A_b :cross-sectional area of the pile base, which includes the soil plug in the case of a steel tube pile or a H-section pile (see figure 9.16)

L : pile length

d : pile diameter (or equivalent).

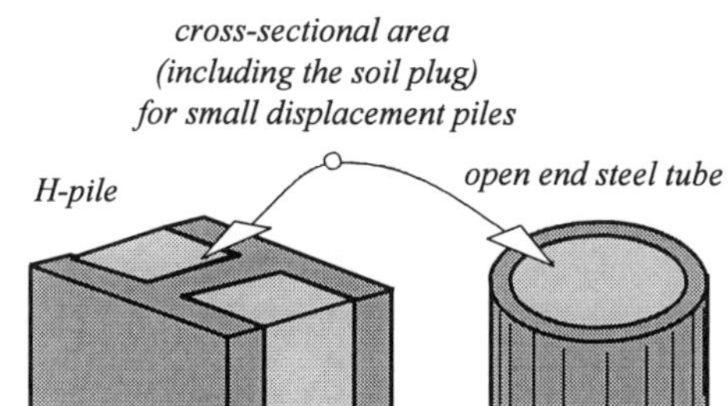

Figure 9.16: Cross-sectional areas used in conjunction with equation 9.8.

However, for undrained conditions, a clay is characterised by an angle of shearing resistance $\phi_u = 0$, in which case both equations 8.51*a* and *c* (refer to section 8.7) yield the bearing capacity factors $N_q = 1$ and $N_\gamma = 0$ respectively. Moreover, when the pile is driven through a layer of soft clay in a way that its base reaches a much higher bearing capacity stratum, Fleming *et al.* (1992) suggest that a linear interpolation should be made between a value $N_c = 6$ for the case of the pile base just reaching the stiff stratum, up to $N_c = 9$ where the pile base penetrates the stiff layer by three diameters or more as illustrated in figure 9.17.

On substitution for the values $N_q = 1$ and $N_\gamma = 0$ into equation 9.9, it follows that:

$$Q_b = A_b(N_c c_u + \gamma L) \tag{9.10}$$

Notice that the quantity $A_b \gamma L$ in equation 9.10 represents the total weight of soil at pile base level, and can be assumed, for all practical purposes, to be equal to the weight of pile W.

Consequently, equation 9.10 is rewritten as follows:

$$Q_b - W = N_c c_u A_b \tag{9.11}$$

and, according to equation 9.5, the short term *ultimate loading capacity* of the pile is in this case:

$$Q_u = Q_s + Q_b - W = c_u(\alpha A_s + N_c A_b) \tag{9.12}$$

where A_s is the area of the pile shaft, and A_b represents the cross-sectional area of the pile base.

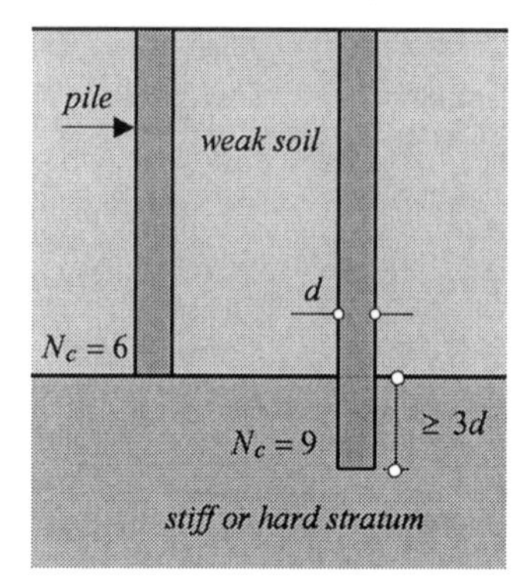

Figure 9.17: Short term effects of pile embedment on the bearing capacity factor N_c.

9.3.3 Drained carrying capacity of single piles embedded in clays

When a pile is embedded in a *stiff overconsolidated clay*, the long term *net ultimate loading capacity* calculated from equation 9.5 corresponds to the critical value for which the pile should be designed. Therefore, an effective stress analysis is required. Under drained conditions, the *ultimate skin friction* is calculated as follows:

$$Q_s = \int_0^L p\sigma'_v K \tan\delta \, dz \qquad (9.13)$$

with

σ'_v : vertical (overburden) effective stress *prior* to pile installation
p: pile perimeter
δ: friction angle of pile–soil interface which is dependent on the surface roughness of the pile, usually in the range $0.75\phi' \leq \delta \leq \phi'$ for steel or concrete piles, ϕ' being the effective angle of shearing resistance of the clay (Fleming *et al.*, 1992).

As for the coefficient of earth pressure K in equation 9.13, Poulos and Davis (1980) related it to the overconsolidation ratio OCR of the clay in the following manner:

$$K \approx (1 - \sin\phi')(OCR)^{1/2} \tag{9.14}$$

The *ultimate base resistance* of the pile is calculated from the bearing capacity equation 8.50, in which the drained cohesion of the clay is usually assumed to be $c' = 0$; whence:

$$Q_b = A_b(\sigma'_{vb}N_q + \frac{1}{2}\gamma' dN_\gamma) \approx A_b\sigma'_{vb}N_q \tag{9.15}$$

The quantity $\frac{1}{2}\gamma' dN_\gamma$ being negligible.

σ'_{vb} corresponds to the vertical effective stress at pile base level *prior* to pile installation, and the bearing capacity factor N_q varies with the soil angle of shearing resistance ϕ' as per figure 9.18 corresponding to average values compiled by Poulos and Davies (1980) and based on the work of Berezantzev *et al.* (1961).

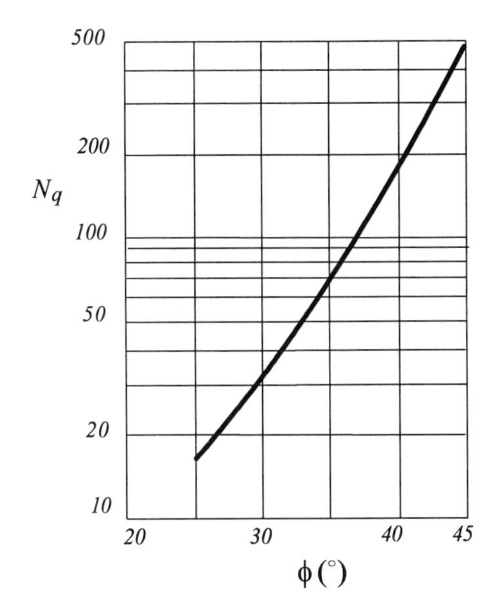

Figure 9.18 : Bearing capacity factor for drained analysis of piles in clays (Berezantzev et al. (1961), redrawn from Poulos and Davis, 1980).

Thence the *net ultimate loading capacity* under drained conditions, for piles embedded in clays characterised by $c' = 0$:

$$Q_u = Q_s + Q_b - W = A_b \sigma'_{vb} N_q - W + \int_0^L pK\sigma'_v \tan\delta\, dz \qquad (9.16)$$

in which the quantity W corresponds to the *effective weight* of the pile.

Example 9.1

A circular concrete pile with a diameter $d = 520\,mm$ is driven through a thick layer of soft saturated normally consolidated clay (see figure 9.19) having an undrained shear strength $c_u = 18\,kN/m^2$ at the ground surface, increasing with depth at a rate of $0.5\,kN/m^2$ per metre. The drained behaviour of the clay is characterised by an angle $\phi' = 20°$ and $c' = 0$. Estimate the length of pile required to carry safely a working vertical load $Q_a = 200\,kN$. The unit weight of concrete is $\gamma_c = 24\,kN/m^3$ and, for water, use $\gamma_w = 10\,kN/m^3$. Assume an overall factor of safety $F = 4$ applies.

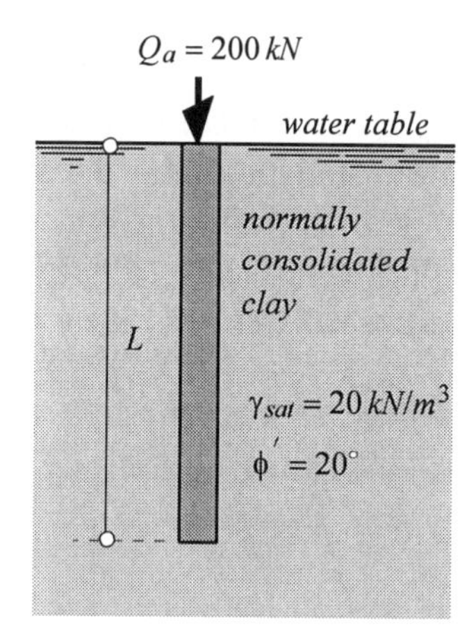

Figure 9.19: Soil characteristics.

This example is used to illustrate the fact that piles installed through a normally consolidated clay are characterised by a smaller carrying capacity in the short term. Accordingly, the prediction, or rather the expectation, is that a shorter pile is required in the long term (*i.e.* the least favourable carrying capacity conditions occur in the short term, for which the pile must be designed).

First, consider the *short term* conditions where the clay behaviour is undrained. The undrained shear strength varies from 18 kN/m^2 at the top of the clay layer to $(18 + \frac{L}{2})\,kN/m^2$ at the pile base, with an *average* value across the pile shaft of:

$$c_u = \left(18 + \frac{L}{4}\right) \, kN/m^2$$

The adhesion factor, estimated from figure 9.14*c*, is $\alpha \approx 1$. Because the bearing capacity factor $N_c = 9$ in this case, equation 9.12 then yields:

$$Q_u = \pi d L \alpha \left(18 + \frac{L}{4}\right) + \pi \frac{d^2}{4} N_c \left(18 + \frac{L}{2}\right)$$

Because an overall factor of safety $F = 4$ is assumed, the magnitude of the ultimate load is then:

$$Q_u = 4Q_a = 4 \times 200 = 800 \, kN$$

Substituting for $Q_u = 800 \, kN$ and $d = 0.52 \, m$ in the previous equation, the following quadratic can easily be established:

$$0.408L^2 + 30.36L - 789 = 0$$

leading to a pile length $L \approx 20.4 \, m$.

Now consider *the long term* behaviour when the excess porewater pressure will have dissipated. Under such circumstances, a drained analysis must be undertaken, and the net ultimate loading capacity is calculated from equation 9.16:

$$Q_u = A_b \sigma'_{vb} N_q - W + pK \tan\delta \int_0^L \sigma'_v \, dz$$

- the pile cross sectional area: $A_b = \pi \frac{d^2}{4} = 0.212 \, m^2$,

- the vertical *effective* stress at pile base level: $\sigma'_{vb} = (\gamma_{sat} - \gamma_w) L = 10L$,

- the bearing capacity factor corresponding to an angle $\phi' = 20^\circ$ is according to figure 9.18: $N_q \approx 10$,

- the coefficient of earth pressure is calculated from equation 9.14, with $OCR = 1$, and $\delta = 0.75\phi' = 15°$. Hence the *constant* quantity: $pK \tan\delta = \pi d(1 - \sin 20) \tan 15 = 0.288\ m$.

- the quantity corresponding to $\int_0^L \sigma'_v\, dz$ is in fact the shaded area in figure 9.20:

$A = \frac{1}{2}(L \times \gamma' L) = 5L^2$.

- the *effective* weight of the pile is:

$$W = \pi \frac{d^2}{4} L\gamma'_c = \pi \frac{d^2}{4} L(\gamma_c - \gamma_w) = 2.97L$$

γ_c being the unit weight of concrete.

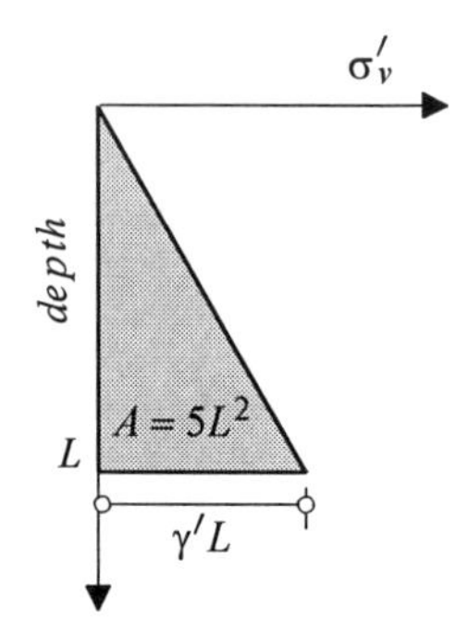

Figure 9.20: Integration of σ'_v along the pile shaft.

Finally, knowing that $Q_u = 800\, kN$, and substituting for all the above quantities into equation 9.16 then rearranging yields the following quadratic:

$$1.44L^2 + 18.23L - 800 = 0$$

whose solution is $L \approx 18.07\, m$.

A quick comparison of results suggests that, as expected, the short term conditions are *the least favourable.*

Example 9.2

A 20 *m* deep *H-section* steel pile is driven in an 18 *m* thick layer of stiff clay underlain by a thick layer of hard clay. Both clays are overconsolidated, and the *average* values of the undrained shear strength are:

- stiff clay: $c_u = 120\, kN/m^2$
- hard clay: $c_u = 400\, kN/m^2$.

The water table is 2 *m* beneath the ground surface and the soil properties are as indicated in figure 9.21. In estimating the ultimate loading capacity Q_u of the pile, assume that the steel cross-sectional area of the pile is $S = 22 \times 10^{-3}\ m^2$ and the steel unit weight is $\gamma = 77\ kN/m^3$. For water, use $\gamma_w = 10\ kN/m^3$.

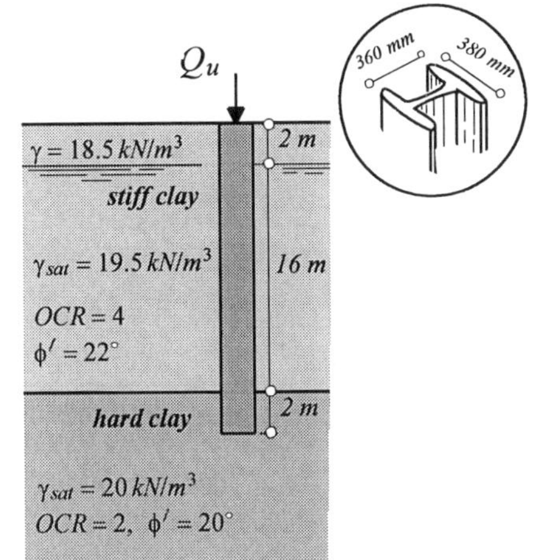

Figure 9.21: Soil conditions.

Remember that for this type of pile, the *total* cross-sectional area includes the soil plug as illustrated in figure 9.16.

Based on the analysis related to porewater pressure generation during pile driving, it is expected that, in this case, the least pile carrying capacity would occur in the long term.

For *undrained short term* conditions, the ultimate loading capacity is calculated from equation 9.5. Because the pile is driven through two different layers of clay, the *ultimate skin friction* has to be evaluated separately for both layers, hence:

- for the stiff clay (from the ground surface to a depth of 18 *m*), figure 9.14*b* yields $\alpha = 0.65$ for $c_u = 120\ kN/m^2$. Therefore:

$$Q_{s1} = A_s \alpha c_u = 2(0.36 + 0.38) \times 18 \times 0.65 \times 120 \approx 2078\ kN$$

- between 18 *m* and 20 *m* (hard clay), an interpolation of the graph in figure 9.14*b* yields $\alpha \approx 0.24$ for $c_u = 400\,kN/m^2$, and:

$$Q_{s2} = 2(0.36 + 0.38) \times 2 \times 0.24 \times 400 = 284\,kN$$

The *net ultimate base resistance* of the pile is calculated from equation 9.11:

$$Q_b - W = 9A_b c_u = 9(0.36 \times 0.38) \times 400 = 492\,kN$$

so that the short term *net ultimate loading capacity* of the pile is:

$$Q_u = Q_s + Q_b - W$$

$$= 2078 + 284 + 492 = 2854\,kN$$

The *long term* (drained) analysis necessitates the use of equation 9.16 to calculate the ultimate loading capacity:

$$Q_u = A_b \sigma'_{vb} N_q - W + pK \tan\delta \int_0^L \sigma'_v\, dz$$

- The cross-sectional area of the pile (including the soil plug) is:

$$A_b = 0.38 \times 0.36 = 0.1368\,m^2,$$

- the vertical *effective* stress at the pile base is:

$$\sigma'_{vb} = 2 \times 18.5 + 16 \times 9.5 + 2 \times 10 = 209\,kN/m^2,$$

- the bearing capacity factor corresponding to an angle $\phi' = 20°$ is, according to figure 9.18, $N_q \approx 10$,

- the coefficient of earth pressure K is calculated from equation 9.14:

- for the stiff clay: $K_1 = (1 - \sin 22)\sqrt{4} = 1.25$
$\delta_1 = 0.75 \times 22 = 16.5°$

- for the hard clay: $K_2 = (1 - \sin 20)\sqrt{2} = 0.93$
$\delta_2 = 0.75 \times 20 = 15°$

- the sum of the shaded areas in figure 9.22 corresponds to:

$$\int_0^L \sigma'_v \, dz = A_1 + A_2 + A_3$$

with

$$A_1 = \frac{1}{2}(18.5 \times 2 \times 2) = 37\, kN/m$$

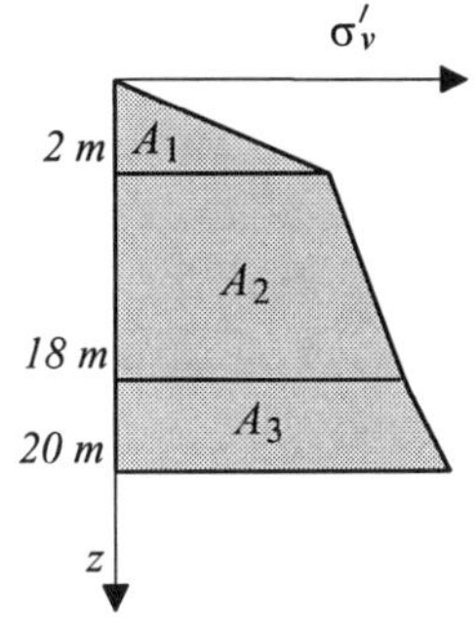

Figure 9.22: Integration of σ'_v along the pile shaft.

$$A_2 = (18.5 \times 2 \times 16) + \frac{1}{2}(9.5 \times 16 \times 16) = 1,808\, kN/m$$

$$A_3 = 2[(18.5 \times 2) + (9.5 \times 16)] + \frac{1}{2}(2 \times 2 \times 10) = 398\, kN/m$$

The total *skin friction* is therefore:

$$Q_s = p[K_1 \tan \delta_1 (A_1 + A_2) + K_2 \tan \delta_2 A_3]$$

$$= 2 \times (0.36 + 0.38)$$

$$\times [1.25 \times (37 + 1845) \times \tan 16.5 + 0.93 \times 398 \times \tan 15]$$

$$= 1178\, kN$$

and the *base resistance* of the pile is:

$$Q_b = A_b \sigma'_{vb} N_q = 0.1368 \times 209 \times 10 = 286\, kN$$

The *effective* weight of the pile is the combined effective weights of the steel and soil plug. Considering the depth of the water table indicated in figure 9.21:

- the effective weight of steel is:

$$W_1 = 22 \times 10^{-3}(2 \times 77 + 18 \times 67) = 30\, kN$$

- the effective weight of soil plug (assuming that the plug has a cross-sectional area $S_p = 0.36 \times 0.38\, m^2$) is:

$$W_2 = (0.36 \times 0.38)[(2 \times 18.5) + (16 \times 9.5) + (2 \times 10)] = 29\, kN$$

Whence the pile effective weight: $W = 30 + 29 = 59\, kN$

The long term *ultimate loading capacity* of the pile can now be calculated:

$$Q_u = 1178 + 286 - 59 \approx 1405\, kN$$

The *long term* ultimate carrying capacity is, as predicted, *smaller* than the short term one, and in this particular case it is seen that, with time, the pile loading capacity decreases by more than 50%, compared with the initial short term capacity.

9.3.4 Carrying capacity of single piles embedded in sand

The analysis of piles embedded in sand is undertaken in terms of effective stresses, where the long term *net ultimate loading capacity* is calculated from equation 9.5:

$$Q_u = Q_s + Q_b - W \tag{9.17}$$

The ultimate skin friction Q_s is evaluated according to the equation:

$$Q_s = \int_0^L p\sigma'_v K \tan\delta\, dz \tag{9.18}$$

with

p: pile perimeter
σ'_v: vertical effective stress (prior to pile installation)
δ: pile/soil friction angle
K: coefficient of earth pressure.

The pile/soil friction angle depends on the pile material, and can be estimated as follows (Kulhawy, 1984):

- smooth steel/sand $\delta = 0.5\phi'$ to $0.7\phi'$ (9.19*a*)
- rough steel/sand $\delta = 0.7\phi'$ to $0.9\phi'$ (9.19*b*)
- prescast concrete/sand $\delta = 0.8\phi'$ to ϕ' (9.19*c*)
- cast-in-place concrete/sand $\delta = \phi'$ (9.19*d*)

Moreover, the coefficient of earth pressure K is related to pile installation technique, and can be estimated from the coefficient of earth pressure at rest $K_o = (1 - \sin\phi')$ in the following way (Kulhawy, 1984):

- large displacement driven piles:

$$K = K_o \text{ to } 2K_o \tag{9.20a}$$

- small displacement driven piles:

$$K = 0.75K_o \text{ to } 1.75K_o \tag{9.20b}$$

- bored and cast-in-place piles:

$$K = 0.71K_o \text{ to } K_o \tag{9.20c}$$

- jacked piles:

$$K = 0.5K_o \text{ to } 0.7K_o \tag{9.20d}$$

It must be borne in mind that, according to equation 9.18, the ultimate skin friction developed around a pile shaft *embedded in a uniform sand* increases linearly with depth. However, suggestions made by different authors implying that the skin friction may reach an upper limit at penetration depths between 10*d* and 20*d* (*d* being the pile diameter) have constituted the basis of a common practice for some time (see for example Vesic, 1969,70; Meyerhof, 1976). Accordingly, some authors advocate the use of an upper limit to the unit skin friction $q_s = \bar{\sigma}'_v K \tan\delta$ (where $\bar{\sigma}'_v$ represents the average effective vertical pressure along the pile shaft) beyond a *critical depth* of about 20*d*.

Nevertheless, the notion of *critical depth* has been called into question lately by Fellenius and Altaee (1995), who argued that the critical depth is nothing but a 'fallacy' (*sic*) that derives from the misinterpretation of pile test measurements, and which can lead to unsafe designs. According to the same authors, the load distribution along a pile shaft during a full scale test *must* take into account the *residual loads* (*i.e.* loads generated by wave reflections or negative skin friction *prior* to pile testing). Failing that, the shape of the corresponding graph appears (erroneously) linear below what is commonly referred to as the critical depth. Similarly, neglecting the stress scale effects during the interpretation of model scale test results can equally be misleading.

Although this contentious issue is still a matter for debate, it is useful to remember that, in practice, driving a pile in a dense sand would induce a higher degree of compaction and therefore a higher density around the pile shaft, leading to early pile refusal (*i.e.* impossibility of further driving the pile into the sand). The refusal may develop for a depth of embedment into the dense sand of as little as $2d$ or $3d$, in which case, the debate related to the depth of embedment of a pile driven through a loose sand layer and into an underlying dense sand layer becomes irrelevant. In any case, caution must be exercised when using equation 9.18.

The *ultimate base resistance* Q_b, on the other hand, is calculated according to equation 9.15:

$$Q_b = A_b \sigma'_{vb} N_q$$

where A_b is the cross-sectional area of pile, σ'_{vb} the effective vertical stress (prior to pile installation) at pile base level, and N_q represents the bearing capacity factor that depends on the state of sand, and which can be cautiously estimated from figure 9.18 in which the angle ϕ corresponds to the following:

- for driven piles (Kishida, 1967):

$$\phi = \frac{\phi' + 40}{2} \tag{9.21a}$$

- for bored piles (Poulos and Davis, 1980):

$$\phi = \phi' - 3^\circ \tag{9.21b}$$

For driven piles, the *unit base resistance* q (*i.e.* the quantity $N_q \sigma'_{vb}$ developed at pile base level) is related to the type of sand and can reach a value of up to $25\ MN/m^2$ in the case of dense sands. Once more, it cannot be emphasised enough that the depth of embedment is controlled by pile refusal, and that under such circumstances, there is not much evidence for a critical depth beyond which q can be assigned a constant value.

Example 9.3

An open end rough steel tube pile with an external diameter $d = 400\ mm$ and a wall thickness of 5 *mm* is driven to a depth $L = 8\ m$, through a soil consisting of a 5 *m* thick layer of loose sand overlying a thick layer of medium dense sand. The water table is 1.5 *m* beneath the ground surface and the soil characteristics are as indicated in figure 9.23.

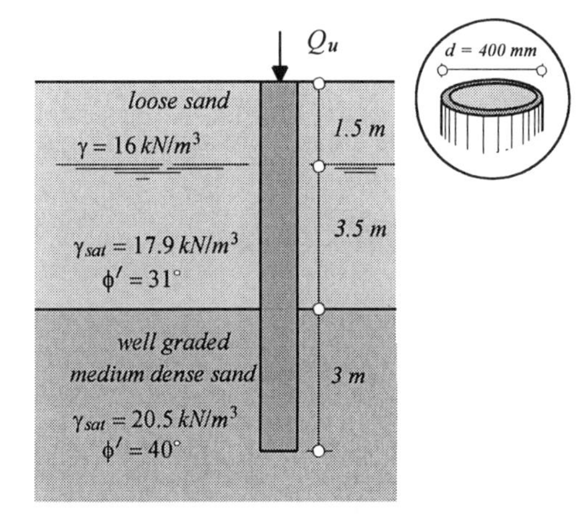

Figure 9.23: Soil characteristics.

In estimating the net ultimate loading capacity of the pile, assume a coefficient of earth pressure $K = 1.2K_o$, a soil/pile friction angle $\delta = 0.75\phi'$, a steel unit weight $\gamma = 77\ kN/m^3$ and a unit weight of water $\gamma_w \approx 10\ kN/m^3$.

The *net ultimate loading capacity* being:

$$Q_u = Q_s + Q_b - W$$

The ultimate skin friction:

$$Q_s = pK \tan \delta \int_0^L \sigma'_v\, dz$$

with

$p = 0.4 \times \pi$ (pile perimeter)

$K_1 = 1.2K_o = 1.2(1 - \sin 31)$ (loose sand)

$K_2 = 1.2(1 - \sin 40)$ (medium dense sand)

$\delta_1 = 0.75 \times 31 = 23°$

$\delta_2 = 0.75 \times 40 = 30°$.

The sum of the areas represented in figure 9.24 is:

$$\int_0^8 \sigma'_v \, dz = A_1 + A_2 + A_3 + A_4$$

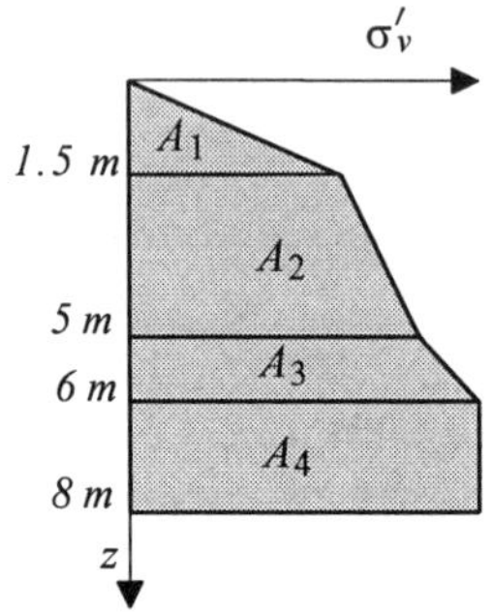

Figure 9.24: Integration of σ'_v along the pile shaft.

with

$$A_1 = \frac{1}{2}(1.5 \times 1.5 \times 16) = 18\,kN/m$$
$$A_2 = (1.5 \times 16 \times 3.5) + \frac{1}{2}(3.5 \times 3.5 \times 7.9) = 132\,kN/m$$
$$A_3 = (1.5 \times 16) + (3.5 \times 7.9) + \frac{1}{2}(1 \times 10.5) = 57\,kN/m$$
$$A_4 = 2[(1.5 \times 16) + (3.5 \times 7.9) + (1 \times 10.5)] = 124\,kN/m$$

Whence the *ultimate skin friction* of the pile:

$$Q_s = 0.4\pi[K_1 \tan\delta_1(A_1 + A_2) + K_2 \tan\delta_2(A_3 + A_4)]$$

$$= 0.4\pi[0.58 \times 0.42 \times (18 + 132) + 0.43 \times 0.58 \times (57 + 124)]$$

$$\approx 103\,kN$$

The effective weight W of pile is the sum of the effective weight of steel W_1 and that of the soil plug W_2.

- The cross-sectional area of steel:

$$A_s = \frac{\pi}{4}\left(0.4^2 - 0.39^2\right) = 6.2 \times 10^{-3}\, m^2$$

- The cross-sectional area of soil plug: $A_p = \pi \frac{0.39^2}{4} = 0.119\, m^2$

It follows that:

$$W = A_s(1.5 \times 77 + 6.5 \times 67) + A_p(1.5 \times 16 + 3.5 \times 7.9 + 3 \times 10.5) = 13\, kN$$

The *ultimate base resistance* of the pile is estimated from equations 9.15:

$$Q_b = A_b N_q \sigma'_v$$

- The cross-sectional area of the pile tip (including the soil plug) is:
$$A_b = \pi \frac{0.4^2}{4} = 0.125\, m^2$$

- At the pile base: $\sigma'_v = 1.5 \times 16 + 3.5 \times 7.9 + 3 \times 10.5 = 83.1\, kN/m^2$

- The bearing capacity factor for dense sand is read from figure 9.18:

medium dense sand: $\phi = \frac{40+40}{2} = 40^\circ \quad \Rightarrow \quad N_q \approx 180$

Accordingly, the *base resistance* is such that:

$$Q_b = 180 \times 83.1 \times 0.125 = 1870\, kN$$

Finally, the *ultimate loading capacity* of the pile is therefore:

$$Q_u = 103 + 1870 - 13 = 1960\, kN$$

9.4 Negative skin friction

The negative skin friction occurs when a pile (or a group of piles) is installed through a layer of clay undergoing a consolidation process. In some cases, the structural loads transmitted to the ground during construction can generate time dependent settlements that can trigger a negative skin friction in the upper part of a pile shaft. If the rate of settlement of the (loaded) pile is smaller than the rate of consolidation settlement of the clay, then a downward (negative) friction force (Q_n in figure 9.25) develops on the pile shaft reducing, markedly sometimes, the carrying capacity of the pile.

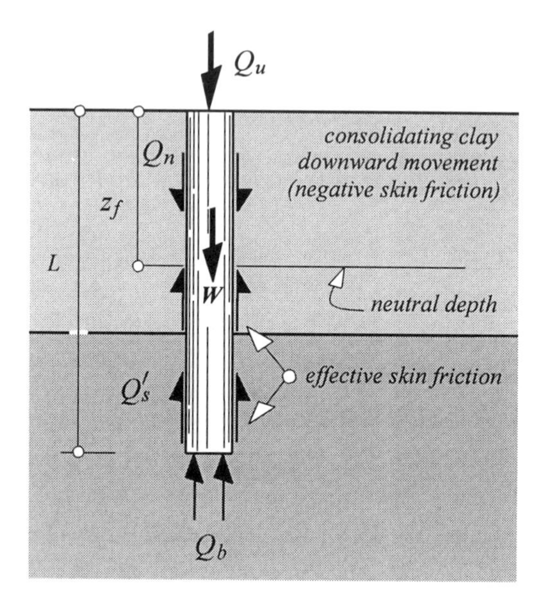

Figure 9.25: Negative skin friction due to a consolidating clay layer.

As consolidation develops with time, a drained analysis is therefore required in conjunction with negative skin friction. Figure 9.25 implies that the maximum displacement of the consolidating clay takes place at the top of the layer, and that the amount of settlement reduces with depth. The *neutral depth* z_f, below which no relative settlement occurs, can be estimated using the one-dimensional consolidation model developed in chapter 4. In practice, z_f can be taken as the depth at which a consolidation settlement $S_c = 0.01d$ is obtained (d being the pile diameter or equivalent).

Referring to figure 9.25, the ultimate loading capacity of the pile can be estimated as follows:

$$Q_u = Q_b + Q'_s - Q_n - W \tag{9.22}$$

The related quantities in equation 9.22 are such that:

- the base resistance:

$$Q_b = A_b \sigma'_{vb} N \qquad (9.23a)$$

- the *effective* skin friction:

$$Q'_s = pK \tan\delta \int_{z_f}^{L} \sigma'_v \, dz \qquad (9.23b)$$

- the *negative* skin friction:

$$Q_n = pK \tan\delta \int_{0}^{z_f} \sigma'_v \, dz \qquad (9.23c)$$

Note that with respect to equations 9.23, the bearing capacity factor N_q is calculated in the usual way using figure 9.18. The angle of soil/pile friction is such that $0.75\phi' \le \delta \le \phi'$, the same value being used to evaluate the *effective* and *negative* skin frictions. Finally the coefficient of earth pressure can be estimated from equation 9.14, adapted for normally consolidated clays (*i.e.* $OCR = 1$):

$$K = (1 - \sin\phi')$$

The (safe) *working load* that can be applied to a pile subject to a negative skin friction can now be calculated in the following manner using an overall factor of safety F:

$$Q_a = \frac{Q_b + Q'_s - 1.5Q_n}{F} - W \qquad (9.24)$$

A factor of 1.5 is used in conjunction with Q_n to cover any uncertainty related to the calculation of such a quantity. Also, notice that according to equation 9.22, the ultimate carrying capacity of the pile is affected, not only by the development of negative skin friction, but also by a reduction in the effective (or positive) shaft resistance.

The negative skin friction (or the downdrag) may be reduced by coating the pile with soft asphalt over the section within the consolidating layer (so reducing the value of δ). However, in practice, slip coats are expensive, difficult to apply and can get stripped off in upper frictional soils as in the case of fill materials. It is often more economical to provide extra loading capacity by making the pile longer.

Example 9.4

A square concrete pile, $0.4\,m \times 0.4\,m$ in cross-section is required to carry a *working* vertical load $Q_a = 800\,kN$ in the soil conditions illustrated in figure 9.26. It is predicted that consolidation settlement of the clay, triggered by the sand fill layer, becomes negligible at a depth $z_f = 4.5\,m$ below the ground surface.

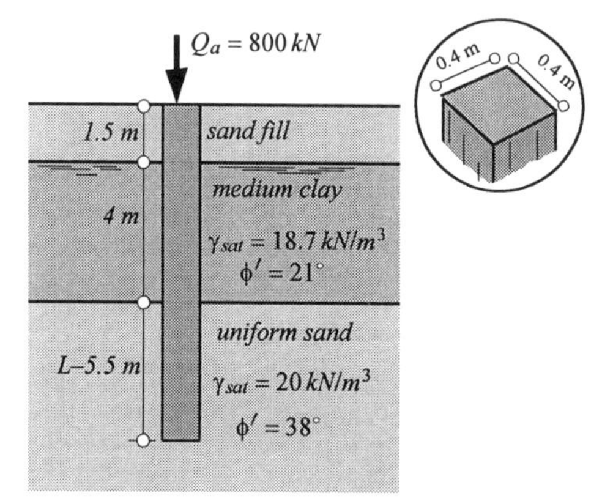

Figure 9.26: Soil conditions.

Determine the length L of the pile that satisfies the requirement of an overall factor of safety $F = 3$. The unit weight of concrete is $\gamma_c = 24\,kN/m^3$ and for water, $\gamma_w \approx 10\,kN/m^3$. Also, assume the following coefficients of earth pressure apply:

- sand fill and clay: $K = K_o = (1 - \sin\phi')$

- uniform sand: $K = 1.5K_o$

The sand fill has the following properties: $\gamma = 17\,kN/m^3$, $\phi' = 30°$.

The allowable and ultimate loading capacities are related through equation 9.24, accordingly:

$$Q_a = \frac{Q_b + Q_s' - 1.5Q_n}{3} - W = 800\,kN$$

On the other hand, the ultimate loading capacity is calculated from equation 9.22:

$$Q_u = Q_b + Q_s' - Q_n - W$$

The base resistance:

$$Q_b = A_b \sigma'_{vb} Nq$$

- The cross sectional area of the pile: $A_b = (0.4 \times 0.4) = 0.16\, m^2$

- The bearing capacity factor for an angle $\phi = \frac{38+40}{2} = 39^\circ$ is read from figure 9.18: $N_q \approx 155$.

- The *effective* vertical stress at the pile tip level is, according to figure 9.26: $\sigma'_{vb} = (1.5 \times 17) + (4 \times 8.7) + (L - 5.5) \times 10$

It follows that:

$$Q_b = (248L + 131.4)\, kN$$

The effective shaft resistance:

$$Q_s' = pK \tan\delta \int_{z_f}^{L} \sigma_v\, dz$$

where

$p = 2(0.4 + 0.4)\, m$ (pile perimeter)
$K_1 = (1 - \sin 30) = 0.5$ (sand fill)
$K_2 = (1 - \sin 21) = 0.64$ (clay)
$K_3 = 1.5(1 - \sin 38) = 0.58$ (uniform sand)
$\delta_1 = 0.75 \times 30 = 22.5^\circ$ (sand fill)
$\delta_2 = 0.75 \times 21 = 16^\circ$ (clay)
$\delta_3 = 0.75 \times 38 = 28.5^\circ$ (uniform sand).

Furthermore, it can be seen that according to figure 9.27:

$$\int_{z_f}^{L} \sigma'_v \, dz = A_3 + A_4$$

with

$$A_3 = 1 \times (17 \times 1.5 + 8.7 \times 3) + \frac{1}{2}(1 \times 8.7 \times 1) = 56\, kN/m$$

$$A_4 = (L - 5.5) \times (17 \times 1.5 + 8.7 \times 4) + \frac{1}{2}(L - 5.5) \times (L - 5.5) \times 10$$
$$= \left(5L^2 + 5.3L - 180.4\right) kN/m$$

Consequently:

$$Q'_s = p\,(K_2 \tan \delta_2 . A_3 + K_3 \tan \delta_3 . A_4)$$

$$= 2.52L^2 + 2.67L - 74.5$$

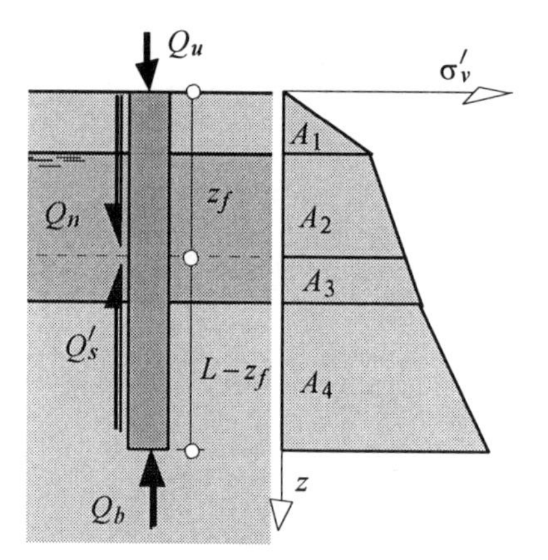

Figure 9.27: Integration of σ'_v along the pile shaft.

The negative shaft friction:

$$Q_n = pK \tan \delta \int_0^{z_f} \sigma'_v \, dz = p(K_1 \tan \delta_1 A_1 + K_2 \tan \delta_2 A_2)$$

where

$$A_1 = \tfrac{1}{2}(1.5 \times 1.5 \times 17) = 19\, kN/m$$
$$A_2 = (3 \times 1.5 \times 17) + \tfrac{1}{2}(3 \times 8.7 \times 3) = 116\, kN/m$$

Whence: $Q_n = 40.4\,kN$

The effective weight of pile:

$$W = A_b[1.5\gamma_c + (L-1.5)(\gamma_c - \gamma_w)]$$

$$= 0.4 \times 0.4 \times [1.5 \times 24 + (L-1.5) \times 14] = (2.24L + 2.4)\,kN$$

The *allowable loading capacity* of the pile is therefore:

$$Q_a = \frac{Q_b + Q_s' - 1.5Q_n}{3} - W = \left(0.84L^2 + 81.32L - 3.63\right)\,kN$$

Thus, substituting for the value $Q_a = 800\,kN$ into the above equation, then rearranging, it follows that:

$$L^2 + 96.8L - 956.7 = 0$$

leading to a length $L = 9.04\,m$, that is a depth of embedment into the dense sand of about $3.5\,m$.

9.5 Ultimate carrying capacity of pile groups

9.5.1 Group effect

Piles are usually used in groups related through a *pilecap* above, at, or below the ground surface (see figure 9.29) and as such, the behaviour of a pile within a group in terms of carrying capacity or settlement, for instance, is different from that of an isolated pile subject to the same loading conditions. The difference in behaviour is due to the *group effect* generated by:

- a higher degree of soil disturbance during the installation of a group of piles, as opposed to a relatively small degree of disturbance for an isolated pile;

- the loading and subsequent displacement of a pile within a group which is affected by the loading of adjacent piles. Because the size of the area of stressed soil beneath a single pile is relatively small compared with that developed at the base of a pile group (refer to figure 9.28), the ensuing settlement will be markedly different.

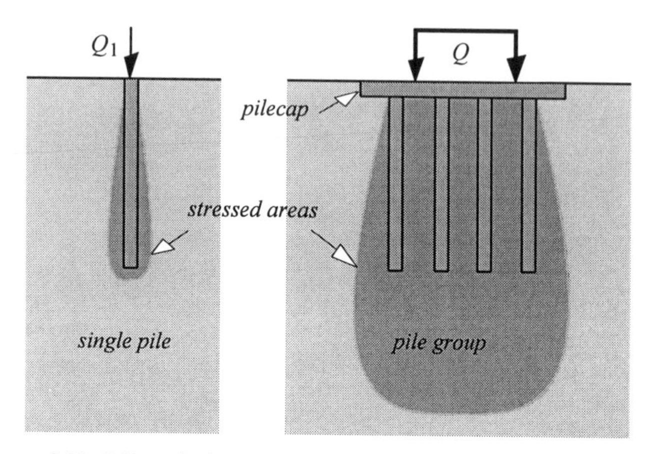

Figure 9.28: Effect of pile groups on the spreading of stressed soil areas.

Consequently, the carrying capacity of a pile group is *not necessarily* equal to the sum of the capacities of individual piles within the group taken in isolation. At this stage, one must distinguish between a *free standing group* where the pilecap is *above* the soil surface as depicted in figure 9.29*a*, and a *piled foundation* for which piles are joined by a pilecap *at* or *below* the ground surface (refer to figure 9.29*b*). Notice that a piled foundation is in fact a combined foundation since the pilecap plays the role of a shallow foundation supported by piles. *Often, however, the pilecap sits on fill or soft ground which may be unreliable in terms of bearing capacity.*

In both cases, the ratio of the ultimate loading capacity of the group to that of individual piles within the group, considered in isolation, is sometimes referred to as the *efficiency factor*:

$$\eta = \frac{Q_{ug}}{\sum_{i=1}^{n} Q_u} \qquad (9.25)$$

with

Q_{ug} : ultimate loading capacity of the group

Q_u : ultimate loading capacity of an isolated single pile within the group

n: total number of piles within the group.

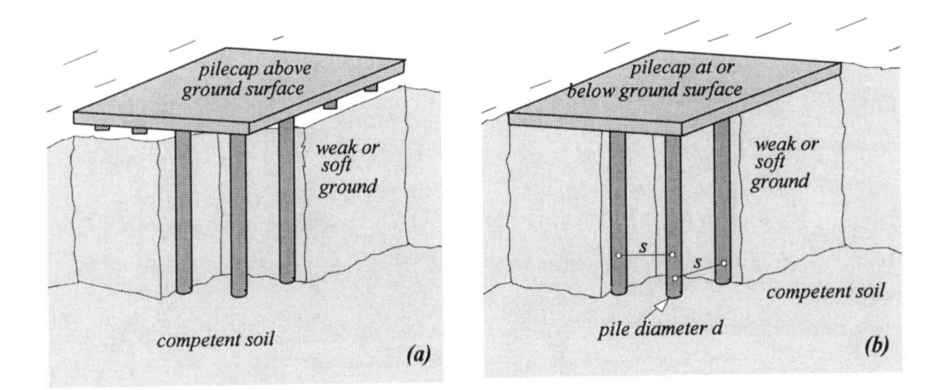

Figure 9.29: (a) Free standing group, (b) piled foundation.

The value of η depends on the type of foundation soil, the method of pile installation (displacement or non-displacement piles) and the ratio s/d where s represents the centre-to-centre spacing between two adjacent piles within the group as depicted in figure 9.29, and d is the pile diameter. Practical considerations related to pile installation dictate that the minimum value of s/d for driven piles is around 2.5 (about 2 for bored piles), and although no upper limit exists for this ratio, one has to bear in mind that the higher its value, the thicker the pilecap must be, and so in practice s/d rarely exceeds 8. Also, the spacing s indicated in figure 9.29*b* can have a different value in each direction.

From a physical view point, the efficiency factor η reflects the change in density of the material in which the piles are embedded. For example, if a group of piles is *driven* in a *loose sand* at a relatively close spacing, $s/d < 3$ say, the sand compaction in the vicinity of piles increases as the group expands. As the density of the sand progressively increases within the compound of the group, the loading capacity of each individual pile within the group becomes larger than that of an isolated pile. Hence, the efficiency factor related to a pile group driven in a loose sand is usually larger than unity and can be as high as 2.5 or 3 when $s/d \approx 2$. As the ratio s/d increases, the compaction effect becomes less significant. On the opposite side, when a group of piles is *driven* into a *dense sand*, dilation ensues causing the sand to loosen as a result of the sand particles being rearranged as pile driving proceeds. Once more, the loosening depends on the ratio s/d: the larger its value the smaller the loosening effect.

In practice, the depth of embedment to which a single pile can be driven into a dense sand is limited because of the increase in soil density during driving. The limitation in the depth of embedment is accentuated by the group effect in the case of a pile group, due to a larger increase in density as the stressed sand area spreads around the group (refer to figure 9.28).

However, there seems to be some confusion as to the use of the efficiency factor η, and in this respect, it is strongly advised to follow the suggestion made by Fleming *et al.* (1992), specifying that *the factor* η *should in no way be used to estimate the ultimate loading capacity of a pile group.* Instead, the ultimate capacities related to appropriate modes of failure must be assessed independently, and the *least favourable conditions* adopted with an adequate factor of safety.

Also, because of the large stressed area generated beneath the base of a pile group (see figure 9.28), provision must be made to check that the ensuing *settlement* under the working load is tolerable. It must be remembered that, while the ultimate loading capacity of a pile group is an important design parameter, its relevance depends on the amount of settlement needed to mobilise fully the base resistance of the group. This settlement can be very high and therefore unacceptable, in which case, it may be necessary to carry the entire load by skin friction, and totally overlook the end bearing of the group. Bear this in mind, the amount of settlement occurring before the end bearing is fully mobilised is usually much larger than that needed to develop the full skin friction (as illustrated in figure 9.10 in the case of single piles).

9.5.2 Carrying capacity of piled foundations embedded in clay: undrained analysis

A piled foundation is a combination of a shallow foundation represented by the pilecap and a deep foundation represented by the group of piles as illustrated in figure 9.29(*b*). The carrying capacity of such a foundation can be estimated by assessing the two following modes of failure.

- *Mode 1* whereby piles within the group may fail individually, thus causing a redistribution of the applied load as per figure 9.30(*a*).

- *Mode 2* in which the block of soil containing the piles fails once the shear stresses on the sides of the block exceed the ultimate strength of the soil as depicted in figure 9.30(*b*).

An analysis of the type developed by Poulos and Davis (1980) is adopted in what follows, in that for a piled foundation containing n piles, the ultimate loading capacity corresponds to the *smaller* of the following two quantities:

Q_1: the sum of the ultimate loading capacities of individual piles plus that of the pilecap

Q_2: the ultimate loading capacity of the block containing the piles plus that of the portion of the pilecap outside the perimeter of the block.

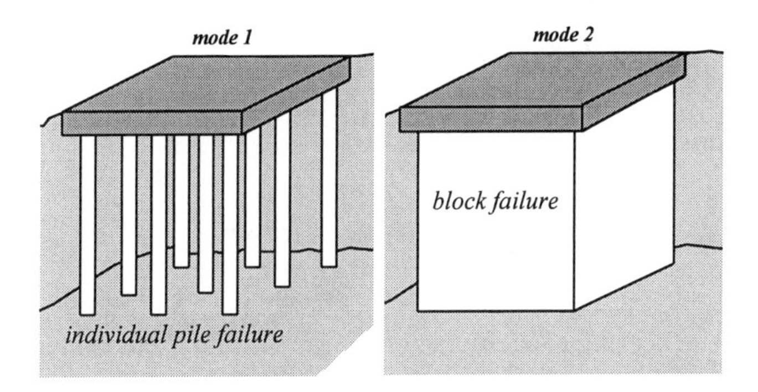

Figure 9.30: Modes of failure for a piled foundation.

Often, the contribution of the pilecap to the bearing capacity of the group is ignored because a large deflection of the cap is usually required before its bearing capacity is fully mobilised. Since sites are often disturbed during pile installation, it is advisable to be cautious and not to presume that the pilecap will always be effective. In the following analysis of piled foundations, the pilecap is assumed to be fully effective in order to cover the few cases where such an assumption can be justified. For the majority of cases, however, the contribution of the pilecap should be set to zero.

The ultimate loading capacity Q_1 can be estimated in the short term (*i.e.* under undrained conditions) as follows:

$$Q_1 = nQ_u + Q_c' - W_c \tag{9.26}$$

with

n: total number of piles within the group
Q_u: ultimate loading capacity of a single pile as per equation 9.12

$Q_c' = A_c'(N_{cc}c_u + \gamma D)$: ultimate loading capacity of the pilecap

$A_c' = A_c - nA_b$: effective area of pilecap as depicted in figure 9.31
A_c: pilecap area
A_b: cross-sectional area of a single pile
D: depth below ground surface of pilecap
W_c: total weight of pilecap.

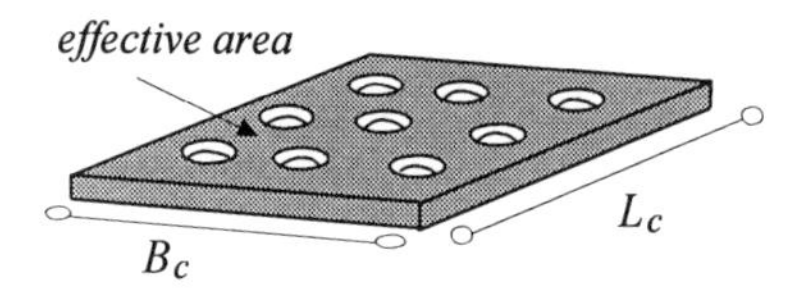

Figure 9.31: Effective area of pilecap.

The bearing capacity factor for the pilecap is calculated as follows (Skempton, 1951):

$$N_{cc} = (2+\pi)\left(1 + 0.2\frac{B_c}{L_c}\right) \tag{9.27}$$

The ultimate loading capacity Q_2 under undrained conditions is calculated in the following way:

$$Q_2 = Q_{sb} + Q_{bb} + Q_c'' - nW_p - W_c - W_s \tag{9.28}$$

with

$Q_{sb} = p_b L c_u$: ultimate side resistance of the block

$p_b = 2(b+e)$: block perimeter

$Q_{bb} = A_{bb}[c_u N_{cb} + \gamma(L+D)]$: ultimate base resistance of the block

$A_{bb} = b\,e$: cross-sectional area of the block (see figure 9.32)
L: block depth
D: depth of the pilecap below ground level
γ: total unit weight of soil in which the group is embedded.

The bearing capacity factor at the base of the block is estimated as follows (Skempton, 1951) [*L, b, e* as per figure 9.32 ($e > b$)]:

$$N_{cb} \approx 5\left(1 + 0.2\frac{b}{e}\right)\left(1 + \frac{L}{12b}\right) \qquad (9.29)$$

$Q_c'' = (A_c - A_{bb})(c_u N_{cc} + \gamma D)$: bearing capacity of the portion of the pilecap situated outside the block (refer to figure 9.32)

with

A_c : area of pilecap
N_{cc}: bearing capacity factor at pilecap level
$W_s = (A_{bb} - N A_b)\gamma L$: total weight of soil within the block.

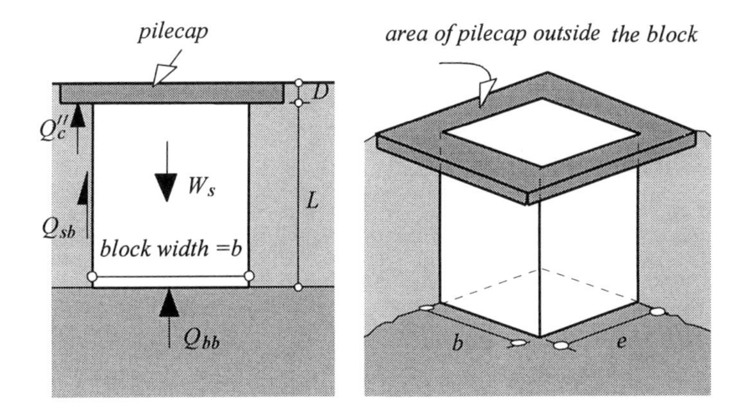

Figure 9.32: Dimensions used in conjunction with the design of piled foundations.

It is useful to reiterate that the ultimate load capacity of the group Q_{ug} adopted for design is the *lesser* of Q_1 and Q_2 and, in this respect, the ultimate load capacity Q_2 is usually smaller than Q_1 for ratios $s/d < 3$.

Example 9.5

A group of 8×10 square concrete piles is driven through a 12 *m* saturated soft clay layer, underlain by a thick slightly overconsolidated stiff clay as depicted in figure 9.33. Both layers have the following parameters:

- *soft clay*: $c_{uav} = 18\, kN/m^2$, $\gamma_{sat1} = 19\, kN/m^3$

-*stiff clay*: $c_{uav} = 90\, kN/m^2$, $\gamma_{ssat2} = 20\; kN/m^3$

(c_{uav} is the average undrained shear strength). Each pile is 18 *m* long and has a cross-sectional area of $(0.5 \times 0.5)\; m^2$. The group is characterised by a ratio $s/d = 3$, and the pilecap is 1.2 *m* thick and has an area of $(13 \times 16)\, m^2$.

Estimate the allowable loading capacity of the group if an overall factor of safety $F = 3$ is used. For concrete, use $\gamma_c = 24\; kN/m^3$.

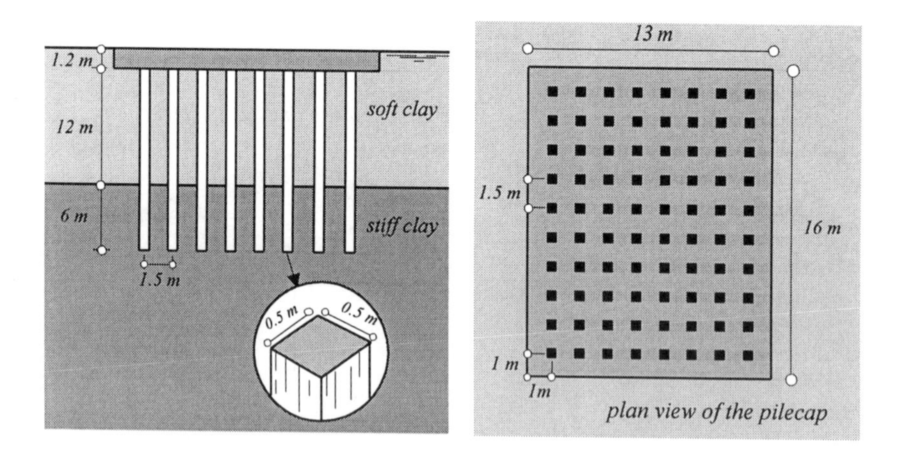

Figure 9.33: Soil characteristics and pile dimensions.

Considering the modes of failure described previously, the ultimate loading capacity of the group corresponds to the smaller of the two quantities Q_1 and Q_2.

Q_1 is calculated using equation 9.26:

$$Q_1 = nQ_u + Q_c' - W_c \quad (n = 80, \text{ total number of piles in the group}).$$

Taking into account the layered nature of soil, the *ultimate loading capacity* Q_u *of a single pile* is calculated from equation 9.12, in which the adhesion factors of both soft and stiff clay layers are read from figure 9.14(*b*), and have the respective values $\alpha_1 \approx 1$, $\alpha_2 \approx 0.7$. Referring to the dimensions in figure 9.33, it follows that (p being the pile perimeter):

$$Q_u = p(\alpha_1 c_{u1} L_1 + \alpha_2 c_{u2} L_2) + 9A_b c_{u2}$$

$$= 2 \times (0.5 + 0.5) \times (1 \times 18 \times 12 + 0.7 \times 90 \times 6)$$

$$+ 9 \times 0.5 \times 0.5 \times 90 \approx 1391 \, kN$$

The ultimate loading capacity of the pilecap:

$$Q_c' = (A_c - nA_b)(N_{cc} c_{u1} + \gamma_{sat1} D)$$

With the bearing capacity factor calculated from equation 9.27:

$$N_{cc} = (2 + \pi)\left(1 + 0.2 \times \frac{13}{16}\right) \approx 6$$

and the depth of pilecap below the ground surface $D = 1.2\,m$ (refer to figure 9.32), it is seen that:

$$Q_c' = (16 \times 13 - 80 \times 0.5 \times 0.5) \times (6 \times 18 + 19 \times 1.2)$$

$$= 24,590 \, kN$$

On the other hand, the total weight of pilecap is:

$$W_c = 16 \times 13 \times 1.2 \times 24 = 5990\, kN$$

Thus the ensuing first estimate of the ultimate loading capacity of the group:

$$Q_1 = 80 \times 1391 + 24590 - 5990 = 129,880\, kN \quad (129.88\, MN)$$

The ultimate loading capacity corresponding to the block failure mode is estimated from equation 9.28:

$$Q_2 = Q_{sb} + Q_{bb} + Q_c'' - nW_p - W_c - W_s$$

Notice that for the block failure, the block in question is 14 m long and 11 m wide as seen in figure 9.33. Therefore:

- the block perimeter: $p_b = 2 \times (11 + 14) = 50\, m$
- the block cross-sectional area: $A_{bb} = 11 \times 14 = 154\ m^2$

- the ultimate side resistance:

$$Q_{sb} = p_b (c_{u1} L_1 + c_{u2} L_2) = 50 \times (18 \times 12 + 90 \times 6)$$

$$= 37,800\, kN$$

- the ultimate base resistance:

$$Q_{bb} = A_{bb}[c_{u2} N_{cb} + \gamma_{sat1}(L_1 + D) + \gamma_{sat2} L_2]$$

with the bearing capacity factor calculated from equation 9.29:

$$N_{cb} = 5 \times (1 + 0.2 \times \frac{11}{14})(1 + \frac{18}{12 \times 11}) \approx 6.6$$

Whence:

$$Q_{bb} = 154 \times [90 \times 6.6 + 19 \times (12 + 1.2) + 20 \times 6] = 148,579\, kN$$

The bearing capacity of the portion of pilecap situated outside the block (see figure 9.32):

$$Q_c^{//} = (A_c - A_{bb})(c_{u1}N_{cc} + \gamma_{sat1}D)$$

$$= (16 \times 13 - 154) \times (18 \times 6 + 19 \times 1.2) = 7063\,kN$$

Moreover, the total weight of all piles within the group is:

$$nW_p = 80 \times 0.5 \times 0.5 \times 18 \times 24 = 8640\,kN$$

and the total weight of soil within the block:

$$W_s = (A_{bb} - NA_b)(\gamma_{sat1}L_1 + \gamma_{sat2}L_2)$$

$$= (154 - 80 \times 0.5 \times 0.5) \times (19 \times 12 + 20 \times 6) = 46,632 kN$$

Hence the second estimate of the ultimate carrying capacity of the group under undrained conditions:

$$Q_2 = 37800 + 148579 + 7063 - 8640 - 5990 - 46632 = 132,180\,kN$$

A quick comparison between the two ultimate loading capacities shows that the single pile failure mode is more critical than the block failure; hence the *ultimate loading capacity* of the pile group:

$$Q_{ug} = Q_1 = nQ_u + Q_c^{/} - W = 129,880\,kN$$

The *allowable loading capacity* is therefore:

$$Q_{ag} = \frac{nQ_u + Q_c^{/}}{F} - W = \frac{1391 \times 80 + 24590}{3} - 5990 = 39,300\,kN$$

Note that the loading capacity of pile groups may sometimes be restricted by settlement requirements rather than a simple factor of safety as in the case of the above example.

9.5.3 Design of free standing pile groups embedded in clay

For a free standing group, the pilecap is above the ground surface and does not as such contribute towards the loading capacity of the group except for its weight as illustrated in figure 9.34. The ultimate capacity of the group can then be estimated in a way similar to that used for piled foundations. That is to say, the ultimate loading capacity of the group is taken as the smaller of Q_1 and Q_2 (Terzaghi and Peck, 1967) whereby:

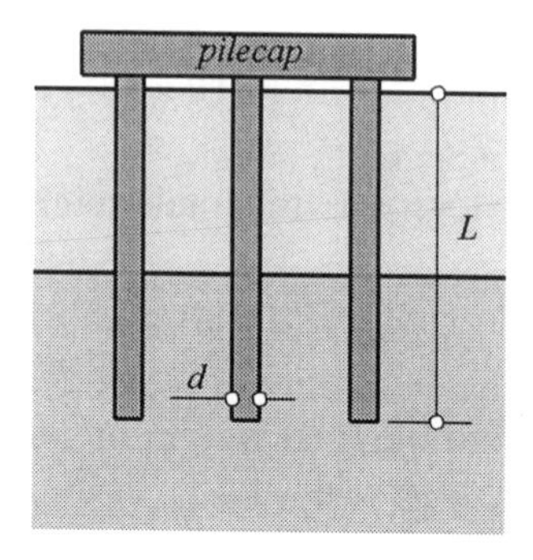

Figure 9.34: Free standing group.

$$Q_1 = nQ_u - W_c \tag{9.30}$$

$$Q_2 = Q_{sb} + Q_{bb} - nW_p - W_c - W_s \tag{9.31}$$

the different quantities having the same meaning (and are calculated in the same way) as in equations 9.26 and 9.28.

Example 9.6

Consider what would have been the ultimate loading capacity of the pile group in the previous example 9.5, had the pilecap been a small distance above the ground.

Manifestly, the calculations already undertaken in the previous example are valid and can be used in conjunction with both equations 9.30 and 9.31. Accordingly:

$$Q_1 = 80 \times 1391 - 5990 = 105,290\ kN$$

$$Q_2 = 37800 + 148570 - 8640 - 5990 - 46632 = 125,108\ kN$$

Thence the ultimate loading capacity of the free standing group:

$$Q_{ug} = Q_1 = nQ_u - W = 105,290\ kN$$

so that the allowable load, calculated using the same factor of safety $F=3$, is:

$$Q_{ag} = \frac{n\,Q_u}{3} - W = \frac{80 \times 1391}{3} - 5990 = 31,103\,kN$$

that is a decrease of around 21% with respect to the allowable load in the case of piled foundation.

9.5.4 Design of pile groups in sand: drained analysis

An effective stress analysis of a pile group embedded in non-cohesive soils can be undertaken in a way similar *in principle* to that used for clays under drained conditions.

Considering the two modes of failure adopted previously where piles can fail either individually or as a block. The ultimate loading capacity of a pile group driven through a non-cohesive soil can be taken as *the lesser* of the two quantities:

$$Q_1 = nQ_u + Q'_c - W \qquad (9.32)$$

$$Q_2 = Q_{bb} + Q_{sb} + Q''_c - nW_p - W_c - W_s \qquad (9.33)$$

with

n: total number of piles in the group
Q_u: ultimate loading capacity of a single pile as calculated from equation 9.16

$$Q'_c = (A_c - nA_b)\left(\gamma' D N'_q + 0.4\gamma' B_c N'_\gamma\right)$$

$$Q_{bb} = A_{bb}\left(\sigma'_{vb} N'_q + 0.4\gamma' b N'_\gamma\right) \quad \text{(refer to figure 9.35)}$$

$$Q_{sb} = p_b \int_0^L K_o \tan\phi' \sigma'_v \, dz: \text{ side friction of the block}$$

$$Q''_c = (A_c - A_{bb})\left(\gamma' D N'_q + 0.4\gamma' B_c N'_\gamma\right)$$

γ' : relevant *effective* unit weight of soil

W_p : effective weight of a single pile

W_c : effective weight of pilecap

W_s : effective weight of soil within the block

σ'_{vb} : effective overburden stress at the base of the block.

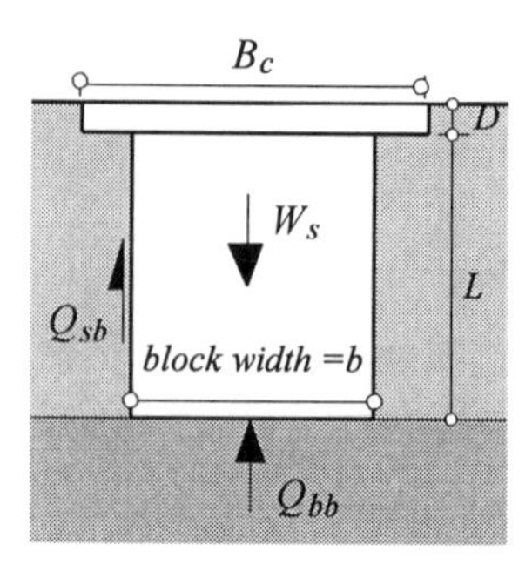

Figure 9.35: Dimensions used in conjunction with drained analysis.

The bearing capacity factors N'_q and N'_γ can be read from the graphs in figure 9.36 which take into account the localised shear failure (Terzaghi, 1943, Vesic, 1969). (They are *different* from those corresponding to figure 8.26, section 8.7.)

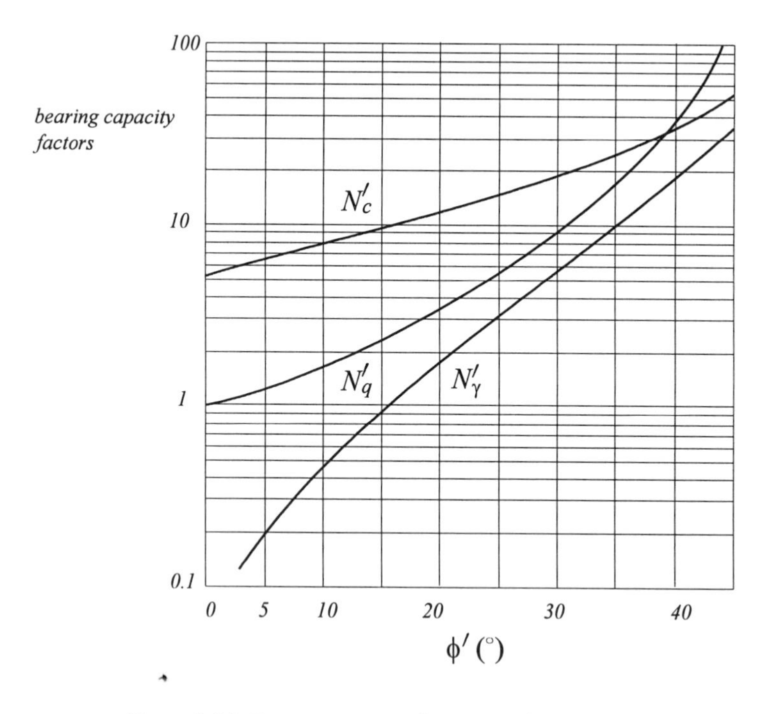

Figure 9.36: Bearing capacity factors used in conjunction with piles.

Example 9.7

A small group of 3×3 piles with a square pilecap is needed to support an ultimate load Q_{ug} (see figure 9.37). The dimensions of each individual tubular pile as well as the soil strata in which the group is driven are identical to those used in example 9.3. The loose sand is characterised by a bulk unit weight $\gamma = 16\,kN/m^3$, a saturated unit weight $\gamma_{sat} = 17.9\,kN/m^3$, and an effective angle of friction $\phi' = 31°$. The dense sand on the other hand has $\gamma_{sat} = 20.5\,kN/m^3$ and $\phi' = 40°$.

Estimate the allowable loading capacity of the group knowing that an overall factor of safety $F = 3$ is required.

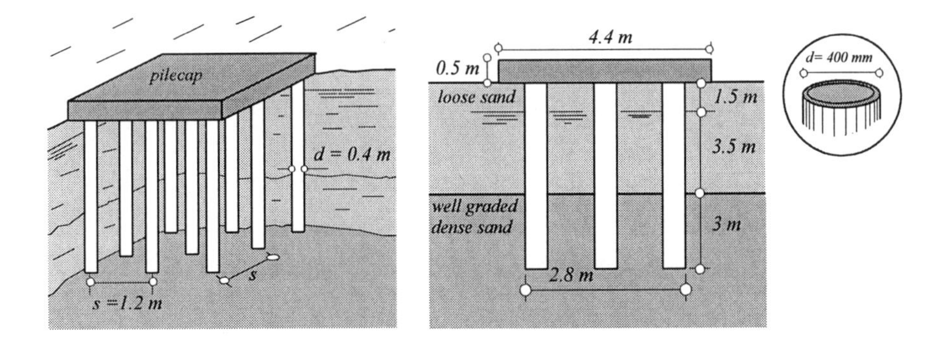

Figure 9.37: Group dimensions and soil conditions.

First, use equation 9.32 to assess the ultimate loading capacity Q_1:

$$Q_1 = nQ_u + Q'_c - W_c$$

$Q_u = 1566\,kN$: ultimate loading capacity of a single pile calculated in example 9.3

$$Q'_c = (A_c - nA_b)\left(\gamma' D N'_q + 0.4\gamma' B_c N'_\gamma\right)$$

$D = 0$ since the pilecap is on the ground surface (see figure 9.37), and for the pilecap:

$$A_c = (4.4 \times 4.4) = 19.36\,m^2 \text{ and } B_c = 4.4\ m$$

The cross-sectional area of a single pile is:

$$A_b = \pi\frac{0.4^2}{4} \approx 0.126\,m^2$$

Moreover, the bearing capacity factor, read from figure 9.36 for an angle $\phi' = 31°$ is $N'_\gamma \approx 6.2$

Whence:

$$Q'_c = (19.36 - 9 \times 0.126) \times (0.4 \times 16 \times 4.4 \times 6.2) = 3182\,kN$$

The effective weight of pilecap being:

$$W_c = 24 \times 4.4 \times 4.4 \times 0.5 = 232\,kN$$

and therefore the first estimate of the ultimate loading capacity of the group is as follows:

$$Q_1 = 9 \times 1566 + 3182 - 232 = 17{,}044\,kN$$

Next, the quantity Q_2 is calculated from equation 9.33:

$$Q_2 = Q_{bb} + Q_{sb} + Q''_c - nW_p - W_c - W_s$$

with:

$$Q_{bb} = A_{bb}\left(\sigma'_{vb}N'_q + 0.4\gamma' bN'_\gamma\right)$$

The block area being $A_{bb} = (2.8 \times 2.8) = 7.84\ m^2$, and its width is $b = 2.8\,m$. Furthermore, the effective overburden stress at the base of the block is as follows:

$$\sigma'_{vb} = (1.5 \times 16 + 3.5 \times 7.9 + 3 \times 10.5) = 83\,kN/m^2$$

The following bearing capacity factors are read from figure 9.36 for $\phi' = 40°$: $N'_q \approx 40$ and $N'_\gamma \approx 18$.

Whence:

$$Q_{bb} = 7.84 \times (83 \times 40 + 0.4 \times 10.5 \times 2.8 \times 18) = 27,688\,kN$$

Now, the side friction of the block is estimated as follows:

$$Q_{sb} = p_b \int_0^L K_o \tan\phi' \sigma_v' \,dz$$

with:

$$K_{o1} = (1 - \sin 31) = 0.48 \quad \text{for loose sand}$$

$$K_{o2} = (1 - \sin 40) = 0.36 \quad \text{for dense sand.}$$

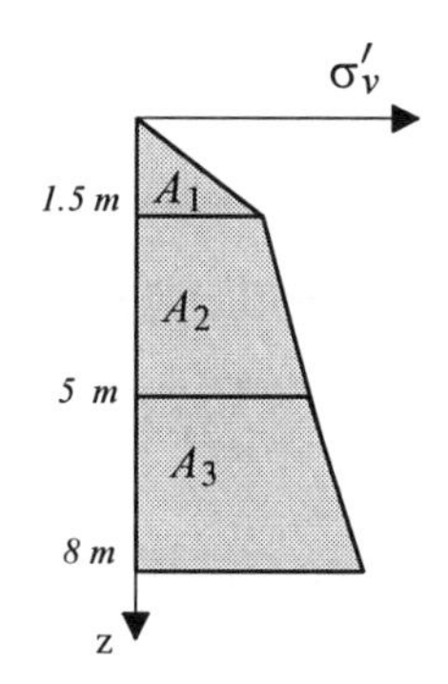

Figure 9.38: Integration of σ_v' along the block sides.

The sum of the shaded areas in figure 9.38 is:

$$\int_0^L \sigma_v' \,dz = A_1 + A_2 + A_3$$

where:

$$A_1 = \frac{1}{2}(1.5 \times 1.5 \times 16) = 18\,kN/m$$

$$A_2 = (1.5 \times 16 \times 3.5) + \frac{1}{2}(3.5 \times 3.5 \times 6) = 120.75\,kN/m$$

$$A_3 = 3 \times (1.5 \times 16 + 3.5 \times 6) + \frac{1}{2}(3 \times 3 \times 10.5) = 182.25\,kN/m$$

Thus:

$$Q_{sb} = 2(2.8 + 2.8)[0.48(18 + 120.75)\tan 31 + 0.36 \times 182.25 \tan 40]$$

$$= 1065\,kN$$

On the other hand:

$$Q_c'' = (A_c - A_{bb})\left(\gamma' D N_q' + 0.4\gamma' B_c N_\gamma'\right) \quad (note\ D = 0)$$

$$= (19.36 - 7.84)(0.4 \times 16 \times 4.4 \times 6.2) = 2011\,kN$$

The effective weight of a single pile was calculated earlier in example 9.3, whence the total effective weight of piles within the group is:

$$nW_p = 9 \times 13 = 117\,kN$$

The effective weight of soil within the block being:

$$W_s = (A_{bb} - nA_b)\gamma' L$$

$$= (7.84 - 9 \times 0.126) \times (16 \times 1.5 + 6 \times 3.5 + 3 \times 10.5) = 513\,kN$$

Hence the second estimate of the ultimate loading capacity of the group is:

$$Q_2 = 27688 + 1065 + 2011 - 117 - 232 - 513 = 29,902\,kN$$

Since Q_1 is smaller than Q_2, the ultimate loading capacity of the group is therefore:

$$Q_{ug} = Q_1 = nQ_u + Q_c' - W_c = 17,044\,kN$$

Using a factor of safety $F = 3$, the allowable loading capacity of the group is therefore:

$$Q_{ag} = \frac{nQ_u + Q_c'}{3} - W_c = \frac{9 \times 1566 + 3182}{3} - 232 \approx 5527\,kN$$

9.5.5 Negative skin friction on pile groups

As in the case of an isolated pile, a pile group may be subjected to a negative skin friction if it is embedded through a layer of consolidating soil. An effective stress analysis is dictated by the time dependent nature of consolidation. For a pile group, the group action arising from installing a cluster of piles has a limiting effect on the development of negative skin friction. In other words, the total negative skin friction force acting within the compound of a group of *n* piles is usually smaller than the sum of negative skin friction forces acting on the same number of isolated piles. However, it is difficult to determine accurately a negative skin force within a pile group. To overcome this difficulty, an upper limit can be imposed by stating that the total negative skin friction force cannot be larger than the force on the sides of the block of consolidating soil enclosed within the pile group, augmented by its the weight as illustrated in figure 9.39.

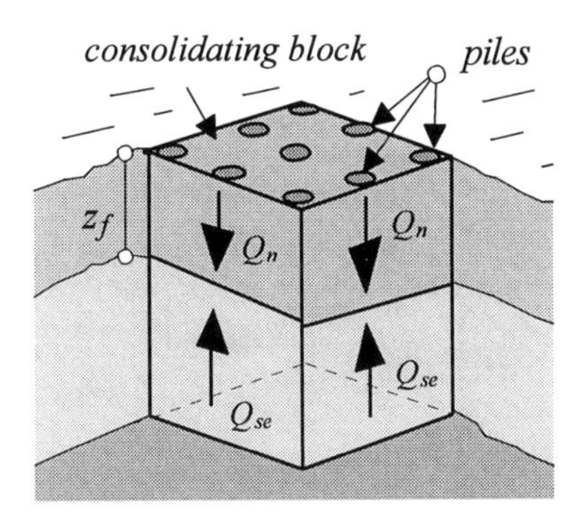

Figure 9.39: Development of negative skin friction along the block sides.

The negative skin friction force can therefore be taken as *the lesser* of the two following quantities:

$$Q_{n1} = np \int_0^{z_f} K \tan\delta\, \sigma'_v\, dz \tag{9.34}$$

$$Q_{n2} = (A_{bb} - nA_b)\gamma z_f + p_b \int_0^{z_f} K_o \tan\phi'\, \sigma'_v \tag{9.35}$$

with

n:	total number of piles within the group
p :	pile perimeter
p_b :	block perimeter
A_{bb} :	cross-sectional area of the block
A_b :	pile cross sectional area
z_f :	depth below which no negative skin friction occurs.

Example 9.8

A group of piles is required to support a structure transmitting an allowable load $Q_{ag} = 12.5\,MN$. A total of 31 circular concrete piles are driven in an arrangement characterised by a centre to centre spacing of 1.2 *m.* The 0.4 *m* diameter piles are related on the ground surface through a pilecap having a cross-sectional area $A_c = 57\,m^2$ and weighing 684 *kN.* The block of soil (represented by broken lines in figure 9.40) has a cross-sectional area $A_{bb} = 54.9\,m^2$ and a perimeter $p_b = 22.8\,m$.

The piles are installed through a 6 *m* thick layer of firm clay overlying a deep layer of stiff, lightly overconsolidated clay. Well before the pile installation and for reasons totally unrelated to the foundation, the water table, originally at the ground surface level, had to be permanently lowered by 2 *m*, causing the upper clay layer to undergo a consolidation process due to the increase in effective stresses.

It can be assumed that a resulting negative skin friction will develop around the shaft areas of piles down to a depth $z_f = 3.5\,m$ (neutral depth in figure 9.40). The soil properties are as follows:

- firm clay: $\gamma = 18.5\,kN/m^3$, $\gamma_{sat} = 19.5\,kN/m^3$, $\phi' = 24^\circ$, $c' = 0$

- stiff clay: $\gamma_{sat} = 20\,kN/m^3$, $\phi' = 22^\circ$, $c' = 0$

and the pile/soil friction angle can be taken as $\delta = 0.75\phi'$.

Estimate the length L of individual piles required to carry the aforementioned load allowing for an overall factor of safety $F = 3$.

Because of the consolidation process involved, a long term effective stress analysis must be undertaken.

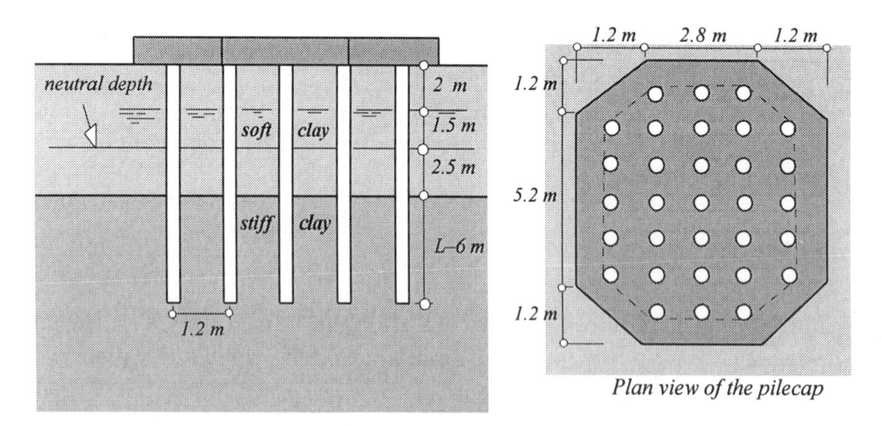

Figure 9.40: Group dimensions and soil conditions.

The negative skin friction exerted on the group down to a depth $z_f = 3.5\,m$ is assessed according to equations 9.34 and 9.35. Thus:

$$Q_{n1} = np \int_0^{z_f} K \tan\delta\, \sigma'_v\, dz$$

where

$n = 31$: total number of piles within the group
$p = 0.4\pi = 1.26\,m$: pile perimeter
$K = (1 - \sin 24) = 0.59$
$\delta = 0.75 \times 24 = 18^\circ$ and hence $\tan\delta = 0.32$

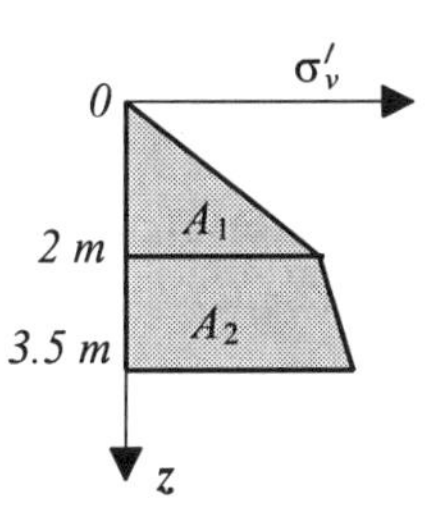

Figure 9.41: Integration of σ'_v along the block sides.

$$\int_0^{3.5} \sigma'_v\, dz = A_1 + A_2 \quad \text{(see figure 9.41).}$$

$$A_1 = 37\,kN/m, \quad A_2 = 66.2\,kN/m$$

Whence the quantity:

$$Q_{n1} = 31 \times 1.26 \times 0.59 \times 0.32 \times (37 + 66.2) = 761\,kN$$

Now, equation 9.35 is used, in turn, to estimate the negative skin friction:

$$Q_{n2} = (A_{bb} - nA_b)\gamma' z_f + p_b \int_0^{z_f} K_o \tan\phi'\, \sigma'_v\, dz$$

with

$A_{bb} = 54.9\,m^2$: cross-sectional area of the block
$A_b = \pi\frac{0.4^2}{4} = 0.1256\,m^2$: cross-sectional area of a pile
$p_b = 22.8\,m$: block perimeter
$K_o = (1 - \sin 24) = 0.59$
$\phi' = 24^\circ$.

Hence:

$$Q_{n2} = (54.9 - 31 \times 0.1256)(18.5 \times 2 + 9.5 \times 1.5)$$

$$+ (22.8 \times 0.59 \times 0.445) \times (37 + 66.2) = 3231\,kN$$

Since the smaller of the two values Q_{n1} and Q_{n2} applies, it follows that the negative skin friction is:

$$Q_n = 761\ kN$$

Next, adopting the two modes of failure of the previous example 9.7, the ultimate loading capacity of the group is therefore the *smaller* of the two values Q_1 (calculated using equation 9.32) and Q_2 (estimated from equation 9.33).

For the first mode of failure (*i.e.* individual pile failure within the group), the ultimate loading capacity of the group is:

$$Q_1 = nQ_u + Q_c' - Q_n - W_c$$

The ultimate loading capacity of an isolated pile:

$$Q_u = Q_b + Q_s - W_p.$$

- The base resistance of a single pile:

$$Q_b = A_b N_q \sigma'_{vb} = 0.1256 \times 36 \times [18.5 \times 2 + 4 \times 9.5 + 10(L - 6)]$$

$$= (67.8 + 45.2L)\ kN$$

N.B. at the base of the pile, $\phi' = 22^\circ$, and according to equation 9.21*a*, $\phi = (22 + 40)/2 = 31^\circ$ so that figure 9.18 yields a bearing capacity factor of $N_q \approx 36$.

- The shaft resistance of an isolated pile:

$$Q_s = p \int_{z_f}^{L} K \tan\delta\, \sigma'_v\, dz$$

where

$$K_1 = (1 - \sin 24) = 0.59, \quad \delta_1 = 0.75 \times 24 = 18^\circ \quad \text{(firm clay)}$$

$$K_2 = (1 - \sin 22) = 0.63, \quad \delta_2 = 0.75 \times 22 = 16.5^\circ \quad \text{(stiff clay)}$$

$$\int_{3.5}^{L} \sigma'_v \, dz = A_3 + A_4 \quad \text{(refer to figure 9.42)}$$

$$A_3 \approx 158 \, kN/m$$
$$A_4 = \left(5L^2 + 15L - 270\right) \, kN/m$$

Hence:

$$Q_s = \pi \times 0.4 \times (K_1 \times A_3 \times \tan\delta_1 + K_2 \times A_4 \times \tan\delta_2)$$
$$= \left(1.17L^2 + 3.51L - 25.2\right) \, kN$$

- The effective weight of a single pile:

$$W_p = 0.1256 \times (2 \times 24 + 14 \times [L - 2]) = (1.76L + 2.51) \, kN$$

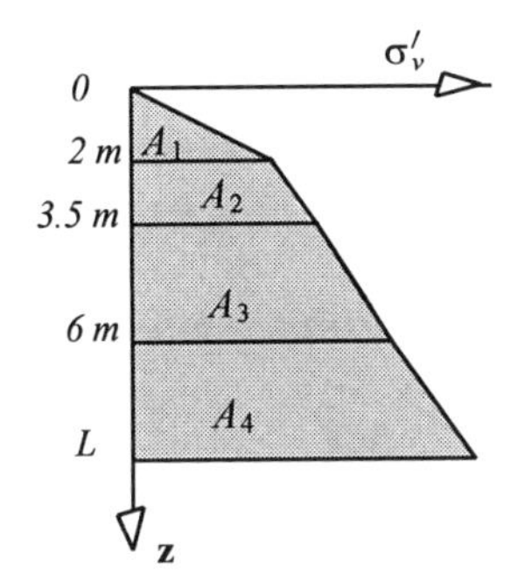

Figure 9.42: Integration of σ'_v along the block sides.

It follows that the ultimate loading capacity of an isolated pile is:

$$Q_u = Q_s + Q_b - W_p = \left(1.17L^2 + 47L + 40\right) \, kN$$

The ultimate loading capacity of the pilecap:

$$Q'_c = (A_c - nA_b) \times 0.4\gamma' B_c N'_\gamma$$

with

$A_c = 57\,m^2$: area of the pilecap
$B_c = 5.2\,m$: width of the pilecap
$N'_\gamma \approx 2.8$: (refer to figure 9.36, $\phi' = 24°$)

Accordingly:

$$Q'_c = (57 - 31 \times 0.1256) \times 0.4 \times 18.5 \times 5.2 \times 2.8 = 5722\,kN$$

Knowing that the weight of the pilecap is $W_c = 684\,kN$, the ultimate loading capacity of the group corresponding to the first mode of failure is therefore:

$$Q_1 = nQ_u + Q_c - Q_n - W_c = \left(36.3L^2 + 1457L + 5517\right)\,kN$$

The group is subjected to an *allowable load* $Q_{ag} = 12,500\,kN$, so that using a factor of safety of 3, Q_{ag} can then be expressed as follows:

$$Q_{ag} = nQ_a + \frac{Q'_c}{3} - Q_n - W_c \tag{9.36}$$

in which the allowable load on an *isolated* pile is:

$$Q_a = \frac{Q_s + Q_b}{3} - W_p$$

But the ultimate loading capacity of the pile group is:

$$Q_1 = nQ_u + Q'_c - Q_n - W_c$$

$$= n(Q_s + Q_b - W_p) + Q'_c - Q_n - W_c \tag{9.37}$$

Thus, subtracting equation 9.36 from equation 9.37 then rearranging, it is straightforward to establish that:

$$Q_{ag} = Q_1 - \frac{2}{3}\left[n(Q_s + Q_b) + Q'_c\right] \tag{9.38}$$

Substituting for all relevant quantities into equation 9.38, the following quadratic is then obtained:

$$L^2 + 37.1L - 963.5 = 0$$

leading to a pile length: $L_1 = 17.6\, m$.

For the second mode of failure (*i.e.* block failure), the ultimate load capacity of the group is:

$$Q_2 = Q_{bb} + Q_{sb} + Q_c'' - Q_n - nW_p - W_c - W_s \qquad (9.39)$$

The ultimate base resistance of the block:

$$Q_{bb} = A_{bb}\left(\sigma'_{vb} N'_q + 0.4\gamma' \, b N'_\gamma\right)$$

with

$A_{bb} = 54.9\, m^2$: cross-sectional area of the block

$N'_q \approx 4.6$, $N'_\gamma \approx 2.1$ (figure 9.36, $\phi' = 22°$)

$b = 5.2\, m$: width of the block.

Thus:

$$Q_{bb} = 54.9 \times (4.6 \times [18.5 \times 2 + 9.5 \times 4 + 10(L - 6)])$$

$$+ 54.9 \times (0.4 \times 10 \times 5.2 \times 2.1) = (2525.4L + 6186)\, kN$$

The ultimate side friction force of the block:

$$Q_{sb} = p_b \int_{z_f}^{L} K \tan\phi' \, \sigma'_v \, dz$$

$$= 22.8 \times (K_1 A_3 \tan 24 + K_2 A_4 \tan 22)$$

with

$K_1 = (1 - \sin 24) = 0.59, \quad K_2 = (1 - \sin 22) = 0.63$
A_3, A_4 as per figure 9.42.

so that:

$$Q_{sb} = \left(29L^2 + 87L - 621\right) kN$$

The ultimate loading capacity of the pilecap portion outside the block:

$$Q_c'' = (A_c - A_{bb}) \times 0.4\gamma' B_c N_\gamma'$$

$$= (57 - 54.9) \times 0.4 \times 18.5 \times 5.2 \times 2.8 = 226.3\, kN$$

The effective weight of soil within the block:

$$W_s = (A_{bb} - n.A_b)\gamma' L$$

$$= (54.9 - 31 \times 0.1256) \times (18.5 \times 2 + 9.5 \times 4 + 10 \times [L - 6])$$

$$= (510L + 765)\, kN$$

Substituting for the different quantities in equation 9.39, the ensuing expression of the ultimate loading capacity Q_2 can easily be established:

$$Q_2 = \left(29L^2 + 2048L + 3504\right) kN$$

The *allowable loading capacity* of the group is now calculated as follows:

$$Q_{ag} = \frac{1}{3}\left(Q_{bb} + Q_{sb} + Q_c''\right) - Q_n - nW_p - W_c - W_s \qquad (9.40)$$

Hence, if equation 9.40 were subtracted from equation 9.39, it follows that:

$$Q_{ag} = Q_2 - \frac{2}{3}\left(Q_{bb} + Q_{sb} + Q_c''\right) \qquad (9.41)$$

so that a substitution for the different quantities into equation 9.41 results in the quadratic:

$$L^2 + 31.6L - 1325.5 = 0$$

the solution of which is $L_2 \approx 23.9\,m$.

A comparison between the two modes of failure indicates that the block failure is more critical than individual pile failure and, accordingly, the required pile's length is $L \approx 23.9\,m$.

9.6 Settlement of single piles and pile groups

The settlement of a deep foundation generated by a working load is usually difficult to predict because of the many variables upon which settlement depends. In fact, the magnitude of settlement of such a foundation is related to:

- the magnitude of applied load;
- the pile–soil interaction in terms of relative stiffness;
- the pile shape (straight sides, enlarged base) and dimensions (ratio L/d);
- the soil strata and its behaviour in terms of stiffness, density, compressibility, non-linearity;
- the group dimensions and the interaction between piles within the group (group depth and spacing ratio s/d).

Empirical, *semi-empirical*, *analytical,* and *numerical* methods have been advocated by different authors to estimate the settlement of deep foundations. As mentioned earlier (refer to section 9.2), the settlement of a single pile subject to axial loading is usually very small *provided* that an adequate factor of safety is used to reduce the ultimate load. In this respect, results relating to a comprehensive full scale load testing programme, undertaken over a period of years in different sites on different types of piles with a shaft length of between 6 *m* and 45 *m,* and diameters varying from 0.3 *m* to 1.5 *m* (LCPC-SETRA, 1985) indicate that, in all but a very few cases, the settlement under the working load measured at the pile head, rarely exceeds 10 *mm*.

A *crude estimate* of settlement at the head of a single pile with a diameter d, subject to a working Q_a can be obtained from the following *empirical relationships* (Frank, 1995):

$$\text{- bored piles:} \quad 0.003d \leq S \leq 0.01d \tag{9.42}$$

$$\text{- driven piles:} \quad 0.008d \leq S \leq 0.012d \tag{9.43}$$

Other methods of analysis of single pile settlement include the following:

The elastic methods (Poulos and Davis, 1980): this is an analytical method based on the assumption that the *isotropic* soil in which the pile is embedded is characterised by a *linear elastic* behaviour. Such assumptions may quickly become unrealistic since the behaviour of most soils is non-linear and, accordingly, caution is advocated when using such a method.

The load transfer method, better known as the (t, z) method (Meyer *et al.*, 1975, Vijayergiya, 1977, Frank and Zhao, 1982): this can easily accommodate non-linear soil behaviour and different soil conditions. However, this somewhat complex method requires that graphs corresponding to the variation of unit skin friction with settlement are known along the pile shaft at different depths. Representative graphs are often difficult to determine, and settlement predictions should therefore be considered carefully.

Numerical techniques such as finite difference and finite element methods (refer to chapter 14) can always be used to estimate settlement. One has to bear in mind, however, that the outcome of a numerical analysis depends entirely on the quality of the input in terms of soil parameters.

Alternatively, settlement of single piles can be predicted from the ***modified hyperbolic method*** (Fleming, 1992). The method, originally developed by Chin (1970, 1972, 1983), is based on the idea that the load–settlement graphs for a pile shaft and its base can be closely represented by hyperbolic functions. Both the *ultimate shaft load* Q_{su} *and the ultimate base resistance* Q_{bu} correspond to the *asymptotic values* depicted in figure

9.43(*a*) (Terzaghi, 1943). Under these circumstances, both graphs can then be linearised by plotting the variation of the quantity S/Q *versus* the settlement S as shown in figure 9.43(*b*).

The method was further refined by Fleming (1992) who showed that the total settlement predictions of the original Chin method were distorted by the *elastic shortening* of the pile body. Fleming then suggested a simplified technique by which the elastic pile deformation can be determined with sufficient accuracy, subtracting it thereafter from the settlement prior to plotting the graphs of figure 9.43. Based on the dimensions shown in figure 9.44, the elastic shortening of the pile depends on the magnitude of the *applied load* Q, compared with the *ultimate shaft friction* Q_{su}.

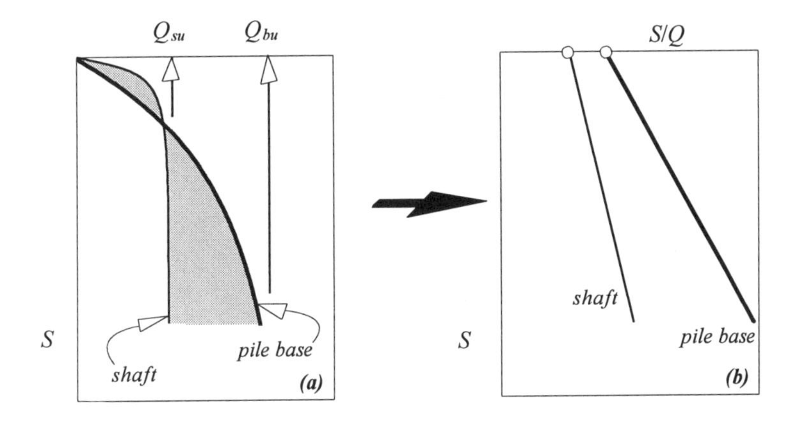

Figure 9.43: (a) Hyperbolic and (b) linearised settlement graphs.

Thus, if $Q \leq Q_{su}$, the elastic deformation of the pile body corresponds to the sum of the pile shortening along the *free or low friction depth* L_o, and that developed along the depth L_a of *the active shaft*, leading to the following expression:

$$S_e = \frac{4}{\pi}\frac{Q(L_o + K_e L_a)}{d_s^2 E_c} \tag{9.44}$$

If, on the other hand, the applied load Q exceeds the ultimate shaft friction, an additional shortening of the active length L_a occurs, which must then be added to the elastic deformation of equation 9.44. Accordingly, for $Q > Q_{su}$:

$$S_e = \frac{4}{\pi} \frac{1}{d_s^2 E_c} [Q(L_o + L_a) - L_a Q_{su}(1 - K_e)] \tag{9.45}$$

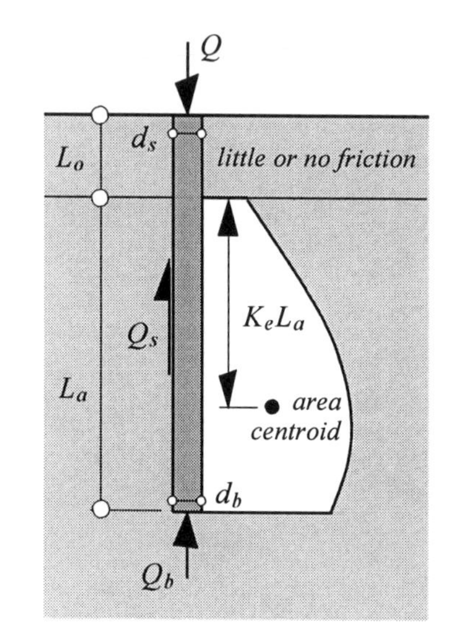

Figure 9.44: Pile dimensions related to the calculation of pile elastic shortening.

The different quantities in both equations are such that:

d_s: pile shaft diameter

E_c: stiffness modulus of the pile material whose value can be linearly interpolated between the values $E_c = 26 \times 10^6\ kN/m^2$ for a concrete specified strength of $20\ N/mm^2$, and $E_c = 40 \times 10^6\ kN/m^2$ for a concrete strength of $40\ N/mm^2$

$L_o,\ L_s$ as per figure 9.44

K_e: ratio of the equivalent shaft length to the active shaft length L_a. Its value can be taken as $K_e = 0.5$ when a uniform friction develops along

L_a or when the pile is installed in a sand or a gravel. For piles in a clay characterised by a strength increasing with depth, $K_e = 0.45$ can be adopted.

Now the settlement of the *rigid pile* can be calculated knowing that the sum of the shaft friction and base resistance corresponds to the total load applied at the pile head:

$$Q = Q_s + Q_b \tag{9.46}$$

Because the pile is rigid, the total settlement at the pile head is identical to the settlement occurring along the pile shaft, as well as that measured at the pile base:

$$S_t = S_s = S_b \tag{9.47}$$

Moreover, it can easily be shown that, by virtue of the linear graphs in figure 9.43(*b*), the settlement along the pile shaft can be calculated as follows:

$$S_s = \frac{M_s d_s Q_s}{(Q_{su} - Q_s)} \tag{9.48}$$

with

M_s: a dimensionless soil/shaft flexibility factor
d_s : pile shaft diameter
Q_s: shaft friction
Q_{su} : long term (*i.e. drained*) *ultimate* shaft friction determined from the static method (refer to sections 9.3.3 and 9.3.4).

The equation of settlement at the pile base derived by Fleming is based on the settlement formula of a circular footing, in which the *secant soil modulus* corresponds to the load $Q_{bu}/4$:

$$S_b = \frac{0.6 Q_{bu} Q_b}{d_b E_b (Q_{bu} - Q_b)} \tag{9.49}$$

with

d_b: diameter of the pile base
Q_b : base resistance
Q_{bu} : ultimate base resistance
E_b : soil secant modulus corresponding to $Q_{bu}/4$.

Hence, taking equations 9.49 and 9.48 to be equal, and using the total load from equation 9.46, it can readily be shown that the *total settlement of the rigid pile* is as follows:

$$S_t = \frac{-g \pm \sqrt{\left(g^2 - 4fh\right)}}{2f} \tag{9.50}$$

with only the positive value considered. The variables in equation 9.50 are such that:

$$f = \eta(Q - \alpha) - \beta \tag{9.51a}$$

$$g = Q(\delta + \lambda\eta) - \alpha\delta - \beta\lambda \tag{9.51b}$$

$$h = \lambda\delta Q \tag{9.51c}$$

where $\alpha = Q_{su}$, $\beta = d_b E_b Q_{bu}$, $\lambda = M_s d_s$, $\delta = 0.6Q_{bu}$, and $\eta = d_b E_b$.

The overall settlement of the pile consists therefore of the component of total settlement calculated from equation 9.50 and the elastic shortening of the pile determined from equations 9.44 or 9.45, whichever is appropriate.

The soil modulus below the base E_b is related to the soil characteristics and is highly sensitive to pile construction techniques. Fleming (1992) advocates that it is highly desirable for this key parameter to be determined from full scale tests whereby piles are loaded to the point where a substantial amount of base resistance is mobilised. Failing that, E_b can be *cautiously* selected from the following range of values related to the type of soil and pile installation technique. In all cases of bored piles, the pile base is assumed to be *well cleaned.*

- bored piles in clays

clay consistency	E_b (kN/m^2)
very soft	<3000
soft	3000 to 6000
firm	6000 to 15,000
stiff	15,000 to 25,000
very stiff	25,000 to 40,000
hard	>40,000

- bored piles in sands and gravel

soil type	E_b (kN/m^2)
very loose	<15,000
loose	15,000 to 30,000
medium dense	30,000 to 100,000
dense	100,000 to 200,000
very dense	>200,000

- bored piles in marls, shales and mudstones

soil type	E_b (kN/m^2)
unweathered	150,000 to 250,000
relatively weathered	80,000 to 150,000
weathered	50,000 to 80,000
highly weathered	10,000 to 50,000

- bored piles in chalk

chalk type	E_b (kN/m^2)
unstructured	<100,000
chalk with apertures > 3 mm	100,000 to 200,000
apertures < 3 mm	150,000 to 250,000
closed discontinuities	>250,000

For piles driven without an enlarged base, the above relevant values apply. When the pile base is *enlarged*, then the above values must be multipied by a factor of 1.5.

As regards the shaft/soil flexibility factor M_s, a comprehensive database amassed by *Cementation Piling and Foundations Limited* indicates that the variation of this *dimensionless factor* is remarkably small, and that its values are invariably between 0.001 and 0.0015. Moreover, these values seem to be unaffected by the type of pile or the nature of soil.

Example 9.9

Consider the pile of example 9.3, for which the different quantities have been calculated, neglecting the weight of pile:

$$Q_{s_u} \approx 100\,kN, \quad Q_{bu} = 1870\,kN, \quad Q_u = 1970\,kN.$$

Assuming that the pile is made of concrete, and that the depth along which little or no friction develops is $L_o = 1.5\,m$, so that $L_a = 6.5\,m$, calculate the elastic shortening and the settlement of the rigid pile knowing that:

$$K_e = 0.5, \quad M_s = 0.001, \quad E_c = 3 \times 10^7\,kN/m^2,$$
$$E_b = 2 \times 10^5\,kN/m^2, \quad d_s = d_b = 0.4\,m$$

The following calculations are undertaken for an applied load $Q = 600\,kN$. Because $Q > Q_{su} = 100\,kN$, the elastic shortening of the pile is calculated from equation 9.45:

$$S_e = \frac{4}{\pi} \times \frac{1}{0.4^2 \times 3 \times 10^7}[600 \times 8 - 6.5 \times 100 \times (1 - 0.5)] \approx 1.2mm$$

Next, the rigid pile settlement is evaluated using equation 9.50. The intermediate variables are as follows:

$$\alpha = Q_{su} = 100\,kN$$

$$\beta = d_b E_b Q_{bu} = 0.4 \times 2 \times 10^5 \times 1870 = 149.6 \times 10^6\,kN^2/m$$

$$\lambda = M_s d_s = 0.001 \times 0.4 = 4 \times 10^{-4}\,m$$

$$\delta = 0.6Q_{bu} = 0.6 \times 1870 = 1122\,kN$$

$$\eta = d_b E_b = 0.4 \times 2 \times 10^5 = 8 \times 10^4\,kN/m$$

Whence:

$$g = Q(\delta + \lambda\eta) - \alpha\delta - \beta\lambda$$

$$= 600 \times \left(1122 + 8 \times 10^4 \times 4 \times 10^{-4}\right) - 100 \times 1122$$
$$-4 \times 10^{-4} \times 149.6 \times 10^6 = 520.36 \times 10^3\,kN^2$$

$$f = \eta(Q - \alpha) - \beta$$

$$= 8 \times 10^4 \times (600 - 100) - 149.6 \times 10^6 = -109.6 \times 10^6\,kN^2/m$$

$$h = \lambda\delta Q = 4 \times 10^{-4} \times 1122 \times 600 = 269.3\,kN^2 m$$

Accordingly, the settlement of the rigid pile corresponding to the positive quantity calculated from equation 9.50 is:

$$S_t = \frac{520.36 \times 10^3 \pm 10^3\sqrt{(520.36)^2 - 4 \times 109.6 \times 269.3}}{2 \times 109.6 \times 10^6}$$

$$= 5.22\,mm$$

If similar calculations are undertaken for different loads, the results can then be plotted in terms of load–settlement and are depicted in figure 9.45. The graph can be used to select a safe working load instead of the usual factor of safety related to the ultimate load as in the static method of design. Figure 9.45 indicates that loads of up to 800 kN can safely be carried out by the pile.

The above calculations may appear to be tedious. However, the formulae can easily be programmed, and commercial software packages based on this method are available.

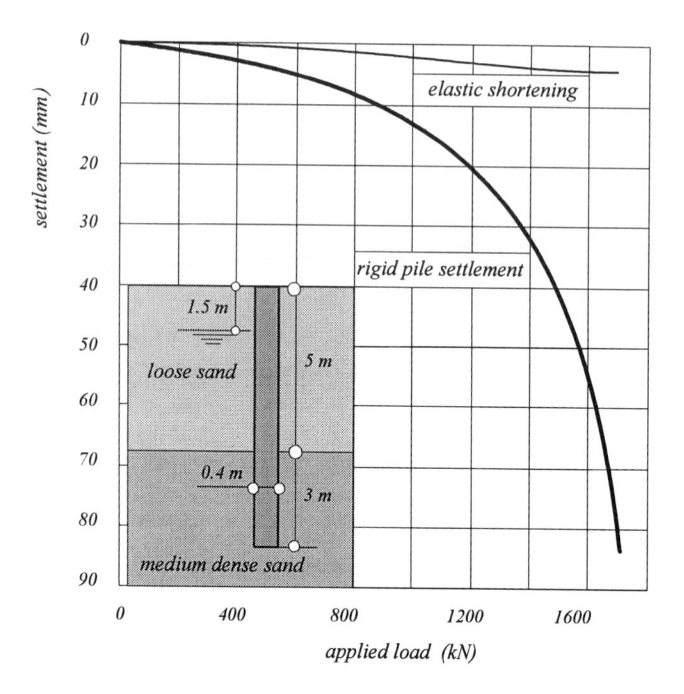

Figure 9.45: Load–settlement analysis using the hyperbolic method.

Example 9.10

Assuming the *driven* pile in example 9.2 is made of concrete with a circular cross-section corresponding to a diameter $d = 0.38\,m$, all other things being equal.

The different pile characteristics are as follows:

$$E_c = 3 \times 10^7\,kN/m^2, \quad E_b = 6 \times 10^4\,kN/m^2, \quad M_s = 0.001,$$
$$K_e = 0.45, \quad L_o = 2\,m, \quad L_a = 18\,m$$

Neglecting the self weight of the pile, and considering the ultimate loads calculated in example 9.2:

$$Q_{su} = 1178\,kN, \quad Q_{bu} = 286\,kN, \quad Q_u = 1464\,kN$$

The calculation of both the pile elastic shortening and its settlement can then be undertaken in the same way using equations 9.44 or 9.45 (whichever is appropriate) and 9.50. The ensuing results, plotted in figure 9.46, indicate that the pile can safely support a load of up to $800\,kN$ for which the elastic shortening ($2.4\,mm$) is 3.2 times larger than the actual total settlement of the pile ($\approx 0.75\,mm$).

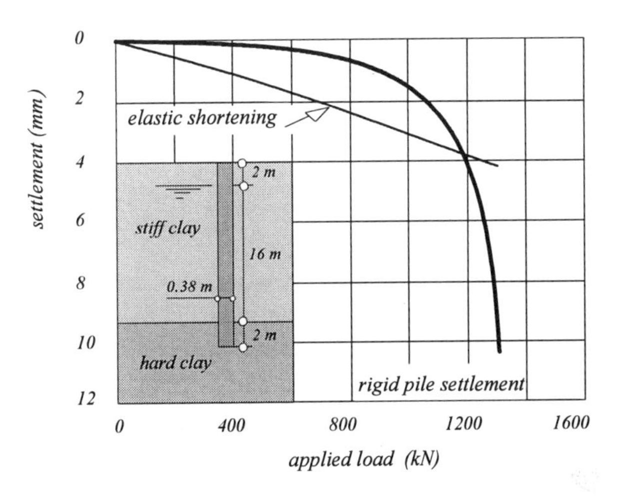

Figure 9.46: Total settlement of a single pile driven in clay.

Interestingly, both examples 9.9 and 9.10 reflect the effects of soil type on the overall settlement of single piles. The ultimate base resistance in example 9.9 ($Q_{bu} = 1870\,kN$) was predominant compared with the ultimate shaft friction ($Q_{su} = 100\,kN$), hence the relatively large settement calculated before the end bearing is mobilised. In contrast, a much smaller settlement is induced by the full mobilisation of the shaft resistance in example 9.10 ($Q_{su} = 1178\,kN$), which is much larger than the end bearing ($Q_{bu} = 286\,kN$).

The hyperbolic method can be used in the same way to determine the load–settlement behaviour of pile groups and shallow foundations. Nevertheless, it ought to be emphasised that, in practice, problems related to pile group design arise from the *differential settlement* of individual piles within the group rather than from the uniform settlement of the group *en masse*. It has been mentioned earlier that the contribution of the pilecap to the loading capacity of a pile group is usually discarded since large displacements are required to mobilise such a contribution in full. This implies that individual piles within a group are subjected to *different* loading through the pilecap, thus inducing different settlements depending on the position of each pile within the group compound. The potentially dangerous excessive differential settlement between individual piles may result in one or more piles failing, increasing in the process (markedly and suddenly) the load on adjacent piles. The hyperbolic method is well suited to predict reliably the amount of differential settlement, so that any potential problems can be remedied prior to the group being fully operational.

Problems

9.1 An open end steel tube pile (figure *p*9.1) having an external diameter $B = 0.85\, m$ and a wall thickness of $7.5\, mm$ is embedded in a layer of stiff overconsolidated clay with an $OCR = 2$, an undrained shear strength varying with depth : $c_u = (120 + 2z)\, kN/m^2$, a bulk unit weight $\gamma = 18.7\, kN/m^3$, and a saturated unit weight $\gamma_{sat} = 19.8\, kN/m^3$.

Estimate the pile length required to carry an ultimate load $Q_u = 3.8\, MN$. Assume that an adhesion factor $\alpha = 0.35$ applies. In the long term, the clay behaviour is characterised by $c' = 0$, $\phi' = 22°$. For steel, use $\gamma = 77\, kN/m^3$.

Ans : $L \approx 24.1\, m$

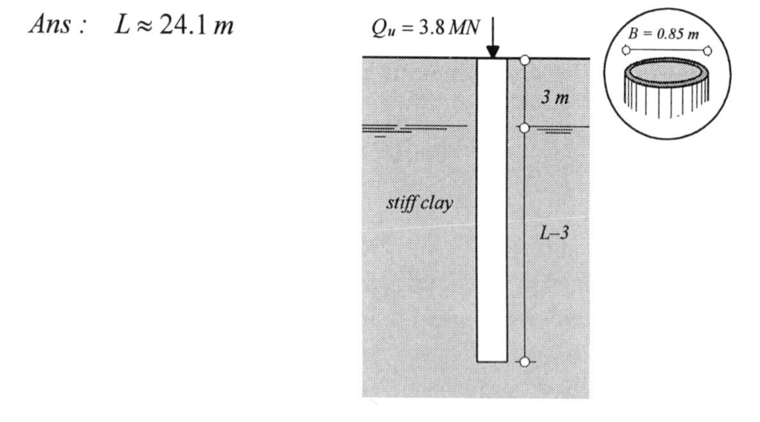

Figure p9.1

9.2 A 15 m long H-section steel pile, with the dimensions indicated in figure *p*9.2, has a steel cross-sectional area $S = 22 \times 10^{-3}\,m^2$. The pile is embedded in a normally consolidated clay characterised by an *average* shear strength $c_u = 30\,kN/m^2$ along the pile shaft, increasing to $c_u = 40\,kN/m^2$ at the pile tip. A unique unit weight $\gamma = 19\,kN/m^3$ can be assumed to apply throughout the clay layer, which is characterised in the long term by $c' = 0,\ \phi' = 23^\circ$. Estimate the ultimate loading capacity of the pile.

Ans: $Q_u \approx 615\,kN$

Figure p9.2

9.3 A circular concrete pile with a diameter $B = 0.5\,m$ is embedded in the soil conditions depicted in figure *p*9.3. The soft clay is characterised by a bulk unit weight $\gamma = 18.6\,kN/m^3$, assumed to apply throughout the layer, and an effective angle of friction $\phi' = 24^\circ$. The stiff clay has a saturated unit weight $\gamma_{sat} = 19.8\,kN/m^3$ and an angle $\phi' = 22^\circ$. Assume that for both layers, the coefficient of earth pressure is $K = 1 - \sin\phi'$, and for concrete, use $\gamma_c = 24\,kN/m^3$.

The upper layer of soft clay is undergoing a process of consolidation due to the load q applied at its surface, and the neutral depth is estimated at 5 m below the ground surface. Calculate the length of pile required to carry an allowable (safe) load $Q_a = 500\,kN$ corresponding to an overall factor of safety $F = 3$.

Ans: $L = 27.3\,m$

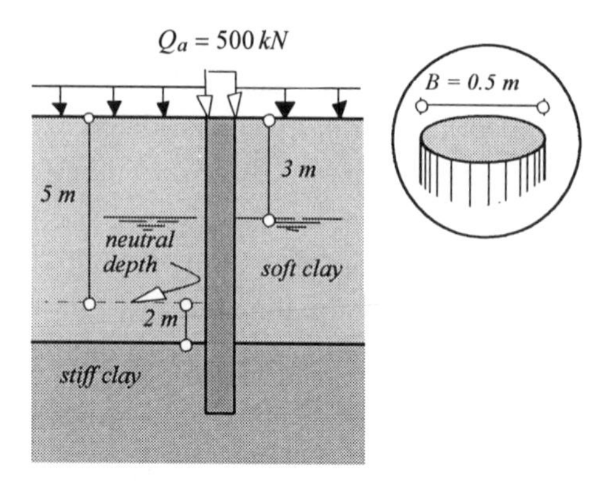

Figure p9.3

9.4 A circular concrete pile, driven in the soil conditions illustrated in figure *p*9.4, has a diameter $B = 0.52\,m$. It is assumed that a negative skin friction will develop along the pile shaft down to a depth $z_f = 15\,m$ below the ground surface. The parameters related to different layers are as follows:

- clay fill: $\gamma = 19.6\,kN/m^3$, $\phi' = 21^\circ$, $K = (1 - \sin\phi')$
- soft clay: $\gamma = 18.5\,kN/m^3$, $\gamma_{sat} = 19.5\,kN/m^3$, $\phi' = 22^\circ$, $K = (1 - \sin\phi')$
- dense sand: $\gamma_{sat} = 19.5\,kN/m^3$, $\phi' = 38^\circ$, $K = 1.5(1 - \sin\phi')$

For concrete, use $\gamma_c = 24kN/m^3$. Estimate the allowable loading capacity Q_a that can be carried safely by the pile with an overall factor of safety $F = 3.5$.

Ans : $Q_a \approx 2077\,kN$

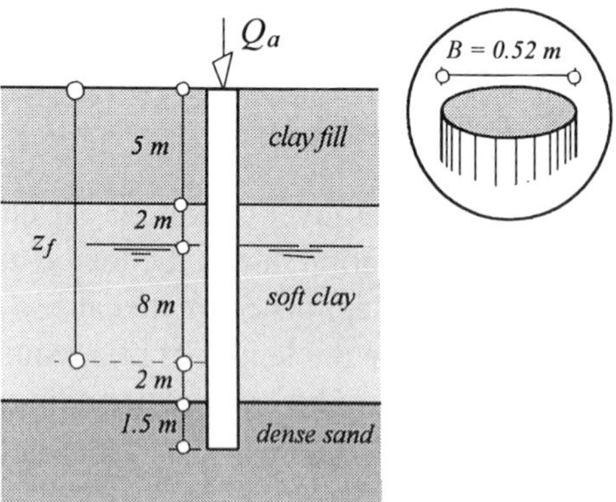

Figure p9.4

9.5 A concrete pile with a diameter $B = 0.45\,m$ is driven through two sand layers as shown in figure *p9-5*. The upper loose sand layer has a bulk unit weight $\gamma = 16.5 kN/m^3$ and an effective angle of friction $\phi' = 32°$. The saturated dense sand layer is characterised by $\gamma_{sat} = 21\,kN/m^3$ and $\phi' = 41°$. Estimate the length of the pile needed to carry an ultimate load $Q_u = 1.5\,MN$. Assume for both layers a coefficient of earth pressure $K = 1.6\,(1 - \sin\phi')$.

Ans : $L = 5.64\,m$

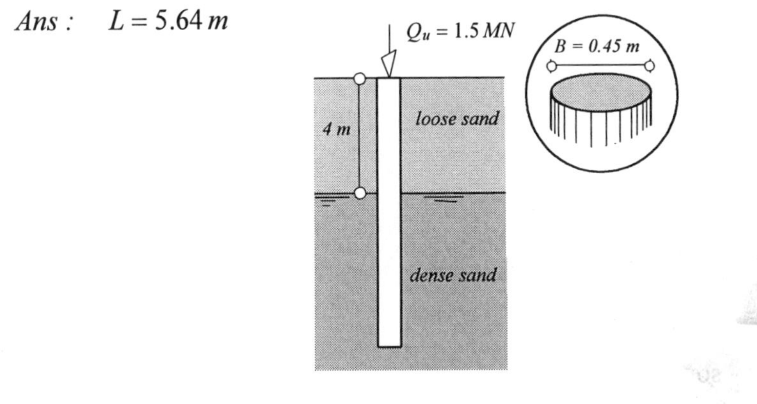

Figure p9.5

9.6 The foundation of a very tall building consists of 12 caissons, each with a diameter $B = 4\,m$, extending 45 m into a boulder clay having a saturated unit weight $\gamma_s = 22\,kN/m^3$ and an average undrained shear strength $c_u = 400\,kN/m^2$. The saturated clay layer is overlain by a *5m* thick layer of sand fill with a bulk unit weight $\gamma = 20\,kN/m^3$.

The caissons are related by a 5 m thick pile cap as depicted in figure *p*9.6. The average skin friction along the caissons shafts is estimated at $f_s = 175\,kN/m^2$, and the bearing capacity factor $N_c = 6$ can be assumed to apply. For concrete, use $\gamma_c = 24\,kN/m^3$.

(*a*) Because large (unacceptable) settlements have to occur before the caissons end bearing can be mobilised, it is suggested that the working load $Q_a = 500\,MN$ should be carried entirely through skin friction.

Calculate in this case the factor of safety on skin friction.

(*b*) Discarding the contribution of the pilecap to the loading capacity of the group, and assuming that at a depth of 45 *m*, the undrained shear strength of the clay is $c_u = 1000\,kN/m^2$ and N_c value is still 6, determine the factor of safety if the end bearing of the caissons is taken into account.

Ans : (*a*) $F = 1.9$, (*b*) $F = 3.7$

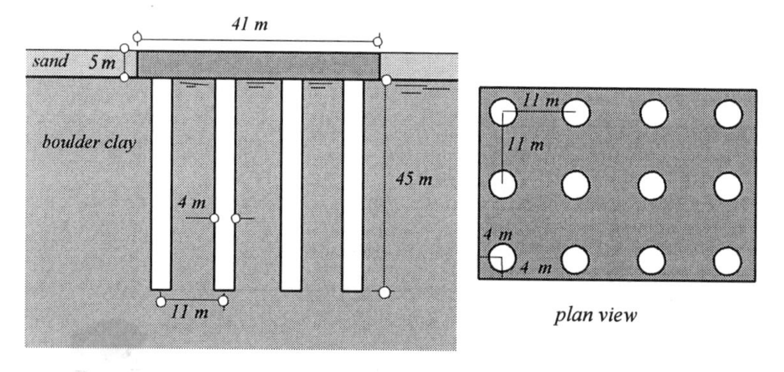

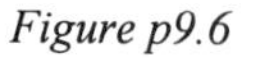
Figure p9.6

9.7 A group of 4×4 circular concrete piles are related through a square pilecap (refer to figure *p*9.7) having a total weight $W_c = 720\,kN$. The dimensions of each individual pile (*i.e.* length $L = 5.64\,m$, diameter $B = 0.45\,m$), as well as the soil conditions in which the piles are driven, are identical to those used in problem *9.5*

- loose sand: $\gamma = 16.5\,kN/m^3$, $\phi' = 32°$
- dense sand: $\gamma_{sat} = 21\,kN/m^3$, $\phi' = 41°$
- for both sand layers: $K = 1.6(1 - \sin\phi')$

Assuming that only a proportion of the loading capacity of the pilecap is mobilised through the use of a factor $N_\gamma = 3$ at ground level, estimate the allowable loading capacity of the group corresponding to an overall factor of safety $F = 3$.

Ans: $Q_{ag} = 8021\,kN$

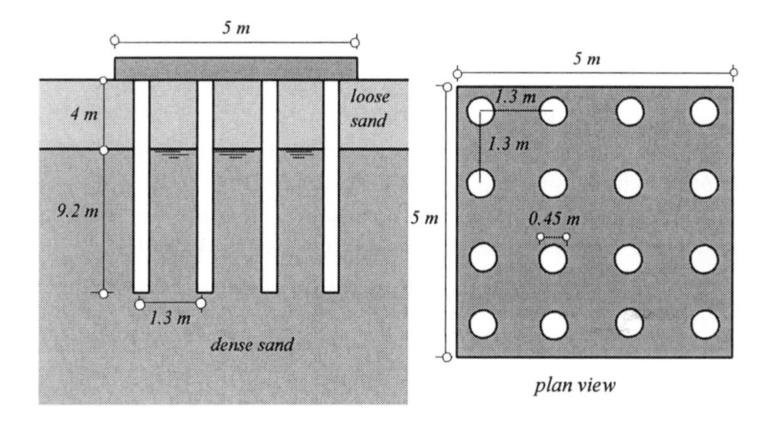

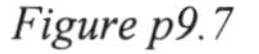
Figure p9.7

9.8 Assume that the pilecap in problem 9.7 is above the ground surface, all other things being equal. Estimate under the same circumstances (*i.e.* $F = 3$) the allowable loading of the free standing group.

Ans: $Q_{ag} = 7280\,kN$

References

Berezantzev, V. C., Kristoforov, V. and Golubkov, V. (1961) *Load Bearing Capacity and Deformation of Piled Foundations.* Proceedings of the 5th International Conference on Soil Mechanics and Foundation Engineering. Vol. 2, pp. 11–15.

Burland, J. B., Butler, F. G. and Dunican, P. (1966) *The behaviour and design of large diameter bored piles in stiff clay.* Proceedings of the Symposium on Large Bored Piles. Institution of Civil Engineers, Reinforced Concrete Association, London, pp. 51–71.

Bustamente, M. and Gianeselli, L. (1995) *Portance d'un pieu vissé moulé dans une marne infragypseuse.* Bulletin des Laboratoires des Ponts et Chaussées, 199, pp. 19–25.

Bustamente, M. and Gianeselli, L. (1997) *Portance d'un pieu de Waal vissé moulé dans un sable sous nappe.* Bulletin des Laboratoires des Ponts et Chaussées, 208, pp. 107–115.

Chin, F. K. (1970) *Estimation of the ultimate load of piles from tests not carried to failure.* Proceedings of the 2nd S.E. Asian Conference on Soil

Engineering, Singapore, pp. 81–92.

Chin, F. K. (1972) *The inverse slope as a prediction of ultimate bearing capacity of piles.* Proceedings of the 3rd S.E. Asian Conference on Soil Engineering, Hong Kong, pp. 83–91.

Chin, F. K. (1983) *Bilateral plate bearing tests.* Proceedings of International Symposium on In Situ Testing, Paris, pp. 29–33.

England, M. and Fleming, W. G. K. (1994) *Review of foundation testing methods and procedures.* Proceedings of the Institution of Civil Engineers, Geotechnical Engineering. 107, pp. 135-142.

Fellenius, B. H., Altaee, A. A. (1995). *Critical depth: how it came into being and why it does not exist.* Proceedings of the Institution of Civil Engineers, Geotechnical Engineering, 113, pp. 107–111.

Fleming, W. G. K (1992) *A new method for single pile settlement prediction and analysis.* Géotechnique, 42 (3), pp. 411–425.

Fleming, W. G. K., Weltman, A. J., Randolf, M. F. and Elson, W. K. (1992) *Piling Engineering,* 2nd edn. Wiley, New York.

Frank, R. (1995) *Fondations profondes.* Collection Technique de l'Ingénieur, C248.

Frank, R. and Zhao, S. R. (1982) *Estimation par les paramètres pressiométriques de l'enfoncement sous charge axiale de pieux forés dans des sols fins.* Bulletin des Laboratoires des Ponts et Chaussées, 119, pp. 17–24.

Imbo, R. P (1984) *The Atlas screw pile : an improved foundation technique for the vibration free execution of piles with larger bearing capacity.* Proceedings of the 6th C. S.M.F.E, Vol. 5, Budapest, pp. 363–372.

Kishida, H. (1967)*Ultimate bearing capacity of piles driven into loose sand.* Soil and Foundations, 7 (3), pp. 20–29.

Kulhawy, F. H. (1984) *Limiting tip and side resistance, fact or fallacy.* Symposium on Analysis and Design of Pile Foundations. A.S.C.E, San Francisco, pp. 80–98.

LCPC-SETRA (1985) *Règles Provisoires de Justification des Fondations sur Pieux à Partir des Résultats des Essais Pressiométriques.*

Meyer, P. L, Holmquist, D. V. and Matlock, H. (1975) *Computer predictions for axially-loaded piles with non-linear supports.* Proceedings of the 7th Offshore Technology Conference, Houston, Texas.

Meyerhof, G. G. (1976) *Bearing capacity and settlement of pile foundations.* Journal of the Geotechnical Engineering Division. A.S.C.E, 102, pp. 197–228.

Poulos, H. G. and Davis, E. H. (1980) *Pile Foundation Analysis and Design.* John Wiley & Sons, New York.

Skempton, A. W. (1951) *The bearing capacity of clays.* Proceedings of the Building Research Congress, London.

Terzaghi, K. (1943) *Theoretical Soil Mechanics.* John Wiley & Sons, New York.

Terzaghi, K. and Peck, R. B. (1967) *Soil Mechanics in Engineering Practice*. Wiley, New York.

Tomlinson, M. J. (1995) *Foundation Design and Construction.* 6th edn. Longman, London.

Vesic, A. S. (1969) *Experiments with Instrumented Pile Groups in Sand.* American Society for Testing and Materials, Special Technical Publication 444, pp. 177–222.

Vesic, A. S. (1970) *Tests on Instrumented piles, Ogeechee River site.* Proceedings of the A.S.C.E, 96 (SM-2), pp. 561–584.

Vijayvergiya, V. N. (1977) *Load-movement characteristics of piles.* Proceedings of the Ports'77 Conference, Long Beach, California.

Weltman, A. J. and Healy, P. R. (1978) *Piling in boulder clay and other glacial tills.* Construction Industry Research and Information Association, Report PG5.

CHAPTER 10

Lateral earth pressure exerted on retaining structures

10.1 Coefficient of earth pressure

The analysis of the stability of a retaining structure, such as the one depicted in figure 10.1, indicates that apart from its self weight, the structure is subjected to lateral thrusts whose intensity and direction depend on the movement (or the lack of it) of the structure itself. These thrusts are best examined using the *coefficient of earth pressure* defined as:

$$K = \frac{\sigma'_h}{\sigma'_v} \qquad (10.1)$$

where σ'_h and σ'_v are respectively the effective horizontal and vertical stresses at any given depth below the ground surface. The ratio K in equation 10.1 depends on the wall movement, its value being characterised by the following three quantities.

(1) If the wall, subjected to lateral pressures at its back, does not move *at all*, then K is referred to as *the coefficient of earth pressure at rest* K_o, and at rest stress conditions prevail as indicated in figure 10.1(*a*).

(2) If the wall is pushed into the soil then, at the advent of failure, the coefficient K reaches its maximum value known as *the coefficient of passive earth pressure* K_p, hence the passive mode of failure of the block of soil behind the wall (figure 10.1(*b*)).

(3) If the wall is moved away from the soil it supports, then at failure (figure 10.1(*c*)) the ratio K reaches its minimum value, referred to as *the coefficient of active earth pressure* K_a.

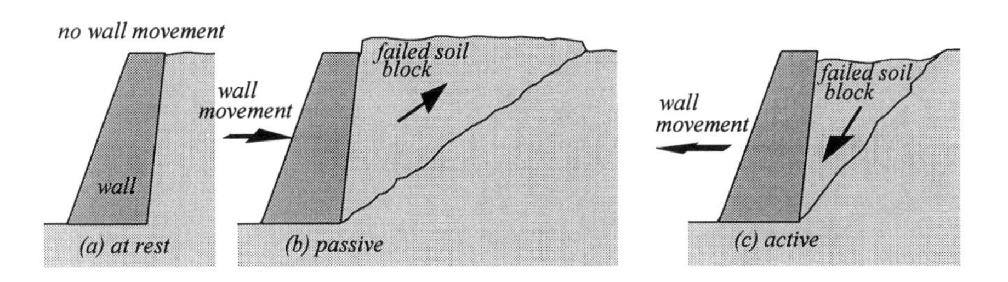

Figure 10.1: Coefficients of earth pressure.

The coefficients of active and passive earth pressure represent respectively the lower and upper limits of the coefficient of earth pressure at rest:

$$K_a < K_o < K_p \tag{10.2}$$

Moreover, the actual wall displacement that causes active or passive failure conditions to develop depends on the type of retained soil and on the mode of failure (*i.e.* active or passive). For soils without cohesion, figure 10.2 shows that the displacement prior to failure in the passive mode (y_p) is much larger than that corresponding to the active mode (y_a).

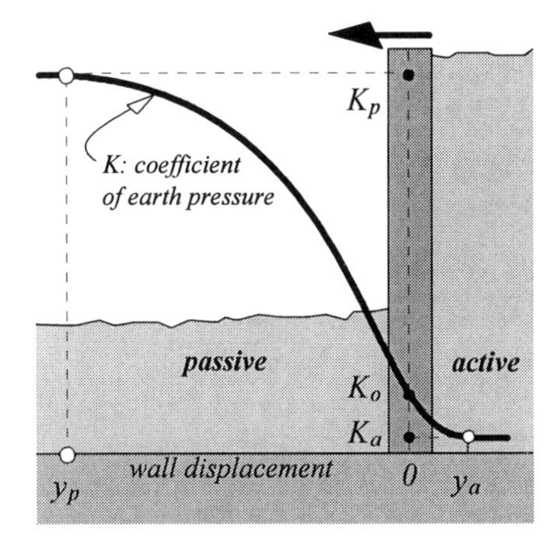

Figure 10.2: Displacement at failure related to active and passive stress conditions for cohesionless soils.

On the other hand, if the retained soil were a clay, experimental evidence shows that the displacements y_p and y_a are comparatively similar in magnitude, and the following values can be considered as *typical* (H represents the height of the wall):

type of soil	y_a	y_p
dense sand	0.001H	0.005H
loose sand	0.005H	0.01H
stiff clay	0.01H	0.01H
soft clay	0.05H	0.05H

10.2 At rest stress conditions: coefficient K_o

At its natural state, an element of soil located at a depth z beneath the ground surface, such as the one depicted in figure 10.3, is subject to the *in situ principal effective stresses*:

$$\sigma'_{vo} = \gamma z_w + (\gamma_{sat} - \gamma_w)(z - z_w) = \gamma z_w + \gamma' (z - z_w) \qquad (10.3a)$$

$$\sigma'_{ho} = K_o \sigma'_{vo} \qquad (10.3b)$$

where γ and γ_{sat} represent the bulk and saturated unit weights of soil respectively, and γ_w is the unit weight of water. For any given soil, the coefficient K_o in equation 10.3*b* is one of the most difficult parameters to measure, its value being linked to the *type* and *state* of soil.

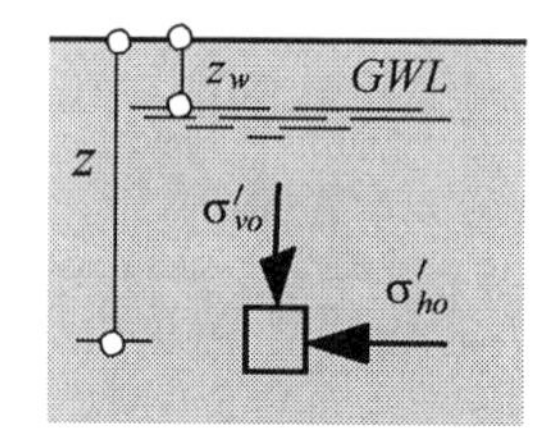

Figure 10.3: At rest stress conditions.

For *loose sands* as well as for *normally consolidated clays,* K_o can be thought of as an intrinsic soil parameter, and its value can be estimated from the relationship derived by Jaky (1944):

$$K_o = \left(1 + \frac{2}{3}\sin\phi'\right)\tan^2\left(\frac{\pi}{4} + \frac{\phi'}{2}\right)$$

$$\approx 1 - \sin\phi' \qquad (10.4)$$

For *dense sands*, experimental evidence indicates that K_o is intricately linked to the density. Sherif *et al.* (1984) suggested the following empirical relationship, which can be used with caution:

$$K_o = (1 - \sin\phi') + 5.5\left(\frac{\gamma_d}{\gamma_{d\min}} - 1\right) \tag{10.5}$$

where γ_d is the *in situ* dry unit weight of sand, and $\gamma_{d\min}$ represents the minimum dry unit weight of sand corresponding to its loosest state.

As far as *overconsolidated clays and sands* are concerned, the coefficient K_o is no longer an intrinsic soil characteristic, but is rather closely linked to the stress history of the soil. Many investigators have examined the link between the overconsolidation ratio (*OCR*) of a clay and the coefficient K_o. In particular, Brooker and Ireland (1965) produced very useful charts, reproduced in figure 10.4, relating K_o to the plasticity index I_P of a clay for different values of *OCR*. Notice that the points corresponding to $I_P = 0$ on the six curves are data corresponding to overconsolidated sands, obtained by Hendren (1963).

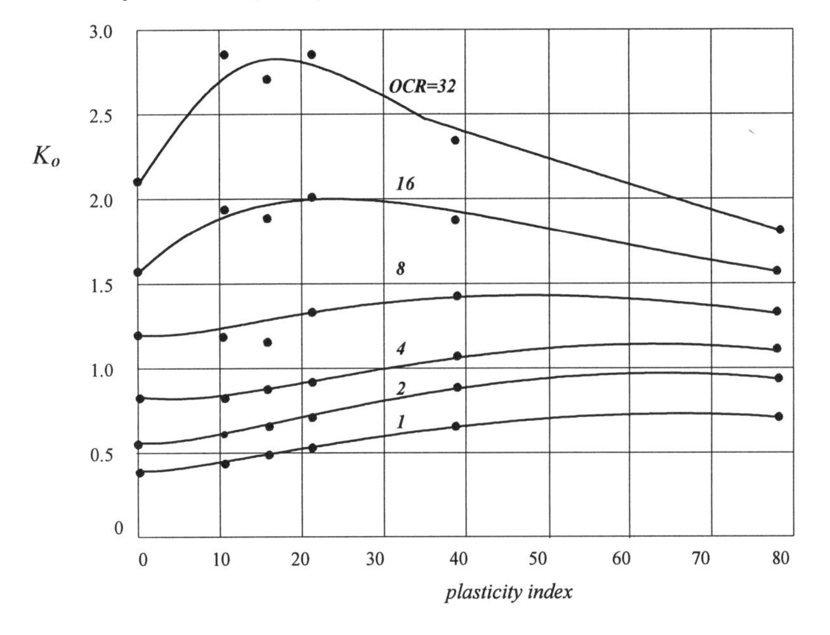

Figure 10.4 : Relationship between plasticity index, overconsolidation ratio (OCR) and coefficient of earth pressure at rest. (Reproduced by permission of the National Research Council of Canada.)

The coefficient K_o for an *overconsolidated clay* can also be estimated from the empirical relationship established by Mayne and Kulhawy (1982):

$$K_o = (1 - \sin\phi')(OCR)^{\sin\phi'} \tag{10.6}$$

ϕ' being the effective angle of shearing resistance of the clay.

10.3 Active and passive stress conditions: Rankine theory

10.3.1 Case of a smooth wall with a vertical back, retaining a horizontal cohesive backfill: drained analysis

Rankine theory, established as early as 1857, considers the limit equilibrium of an isotropic and homogeneous soil mass, subject at its *horizontal* surface to a uniform pressure q. The retaining wall with a *vertical back* is assumed to be *smooth* (*i.e.* there is no friction at the soil–wall interface).

For stiff, highly plastic and heavily overconsolidated clays, the corresponding failure envelope is characterised by an angle of shearing resistance ϕ' and a cohesion intercept c' as illustrated in figure 10.5.

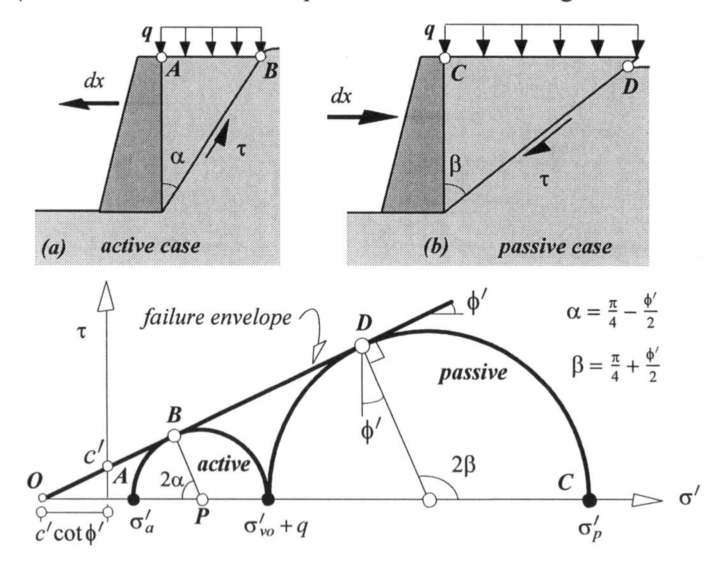

Figure 10.5: Active and passive stress conditions in relation to a smooth wall.

Let us assume that the smooth wall with a vertical back is *moved away* from the soil it retains in the active case (figure 10.5*a*) and *pushed into* the soil in the passive case (figure 10.5*b*) a distance *dx,* large enough to cause the soil to be in a state of plastic equilibrium. Consider, under these circumstances, the *limit equilibrium* of the two failing soil blocks behind the wall. Starting with the *active* case, it is seen from the corresponding Mohr's circle that:

$$\sin\phi' = \frac{PB}{OP} = \frac{(\sigma'_{vo}+q)-\sigma'_a}{(\sigma'_{vo}+q)+\sigma'_a+2c'\cot\phi'} \tag{10.7}$$

This equation can be reorganised thus:

$$(\sigma'_{vo}+q)(1-\sin\phi') = \sigma'_a(1+\sin\phi') + 2c'\cos\phi' \tag{10.8}$$

But

$$\cos\phi' = \sqrt{(1+\sin\phi')(1-\sin\phi')} \quad \text{and} \quad \frac{1-\sin\phi'}{1+\sin\phi'} = \tan^2\left(\frac{\pi}{4}-\frac{\phi'}{2}\right)$$

Substituting for these quantities into equation 10.8, then rearranging, yields the *effective active pressure*:

$$\sigma'_a = (\sigma'_{vo}+q)\tan^2\left(\frac{\pi}{4}-\frac{\phi'}{2}\right) - 2c'\tan\left(\frac{\pi}{4}-\frac{\phi'}{2}\right) \tag{10.9}$$

It is straightforward to show that a similar analysis undertaken using Mohr's circle corresponding to the passive case leads to the following *effective passive pressure*:

$$\sigma'_p = (\sigma'_{vo}+q)\tan^2\left(\frac{\pi}{4}+\frac{\phi'}{2}\right) + 2c'\tan\left(\frac{\pi}{4}+\frac{\phi'}{2}\right) \tag{10.10}$$

Equations 10.9 and 10.10 are usually presented in the following way:

$$\sigma'_a = (\sigma'_{vo}+q)K_a - 2c'\sqrt{K_a} \tag{10.11}$$

$$\sigma'_p = (\sigma'_{vo}+q)K_p + 2c'\sqrt{K_p} \tag{10.12}$$

with the *coefficient of active pressure*:

$$K_a = \tan^2\left(\frac{\pi}{4} - \frac{\phi'}{2}\right) \tag{10.13}$$

and the *coefficient of passive pressure*:

$$K_p = \frac{1}{K_a} = \tan^2\left(\frac{\pi}{4} + \frac{\phi'}{2}\right) \tag{10.14}$$

Knowing that the overburden effective stress at a given depth z corresponds to $\sigma'_{vo} = \gamma'$, with γ' being the appropriate effective unit weight of soil, it is seen from equation 10.11 that the active pressure σ'_a has a *negative value* (*i.e.* becomes tensile) down to a depth:

$$z = \frac{1}{\gamma'}\left(\frac{2c'}{\sqrt{K_a}} - q\right)$$

Accordingly, for a 5 m high wall retaining an overconsolidated stiff clay characterised by $c' = 10\, kN/m^2$, $\gamma = 20\, kN/m^3$ and $\phi' = 24°$, assuming the clay is totally submerged as a result of a prolonged period of heavy rainfall and considering a uniform pressure $q = 10\, kN/m^2$ applied on the horizontal surface behind the wall:

$$z = \frac{1}{(20-10)}\left(\frac{2 \times 10}{\tan(45-12)} - 10\right) = 2.08\, m$$

meaning, practically, a tensile active effective stress along the top 40% of the height of wall. This simple example implies that caution must be exercised when choosing an appropriate value for c' which is, alas, quite difficult to evaluate (refer to the discussion on this parameter in chapter 5).

In all cases, it seems logical to avoid using clay fills in conjunction with retaining walls. In the absence of an alternative, it is advisable to adopt a conservative design by assuming $c' = 0$.

10.3.2 Case of a smooth wall with a vertical back, retaining a horizontal frictional backfill

For sands as well as for normally consolidated clays, the failure envelope is characterised by a zero intercept, as illustrated in figure 10.6, and the strength of such soils is independent of cohesion.

In terms of effective stresses, the same analysis used earlier for cohesive soils applies except that, this time, no cohesion is involved. Accordingly, the *active* and *passive pressures* exerted at the *vertical back* of the *smooth* wall are determined from equations 10.11 and 10.12 in which $c' = 0$:

$$\sigma'_a = (\sigma'_{vo} + q) K_a \tag{10.15}$$

$$\sigma'_p = (\sigma'_{vo} + q) K_p \tag{10.16}$$

where the coefficients of active and passive earth pressure, K_a and K_p respectively, are calculated from equations 10.13 and 10.14.

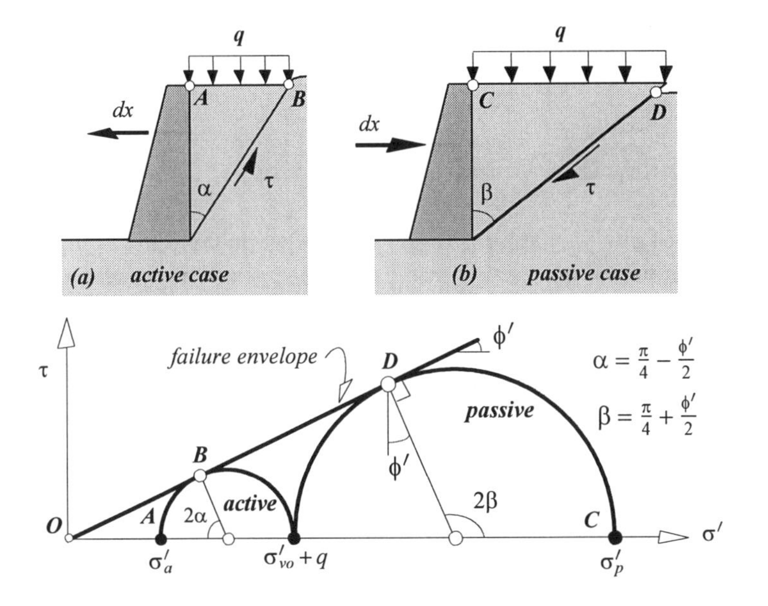

Figure 10.6: Active and passive pressures related to a smooth wall retaining a cohesionless soil.

Figure 10.7 indicates that, in the long term, a state of equilibrium of stresses is achieved as long as the horizontal effective stress σ'_h is within the range represented by the shaded area and corresponding to the condition:

$$\sigma'_a < \sigma'_h < \sigma'_p$$

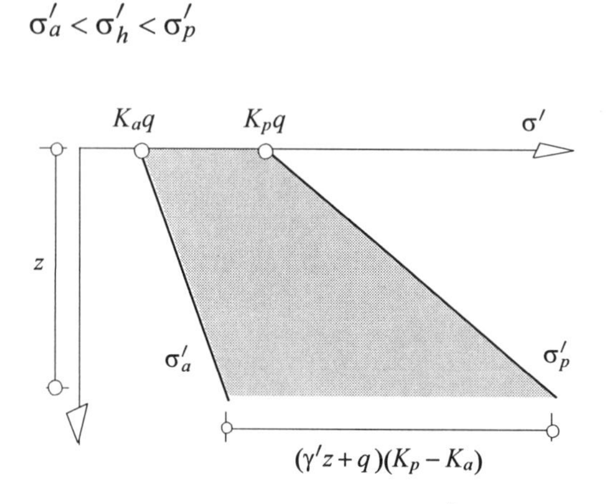

Figure 10.7: Variation with depth of σ'_h at equilibrium.

The effects due to changes in the horizontal stress level in the case of a frictional soil ($c' = 0$), without any surcharge ($q = 0$), are illustrated in the 3-D graphics of figure 10.10.

10.3.3 Rankine theory: undrained analysis

This type of analysis applies to the short term behaviour of clays where the stresses induced by an externally applied load are transmitted to the porewater, leaving the effective stresses practically unchanged. Prior to any significant consolidation (*i.e.* dissipation of excess porewater pressure) taking place, the clay's behaviour is characterised by an angle of shearing resistance $\phi = 0$ and an undrained shear strength c_u as shown in figure 10.8.

Under these circumstances, both coefficients of active and passive pressure of equations 10.13 and 10.14 reduce to:

$$K_a = K_p = \tan^2 45 = 1 \qquad (10.17)$$

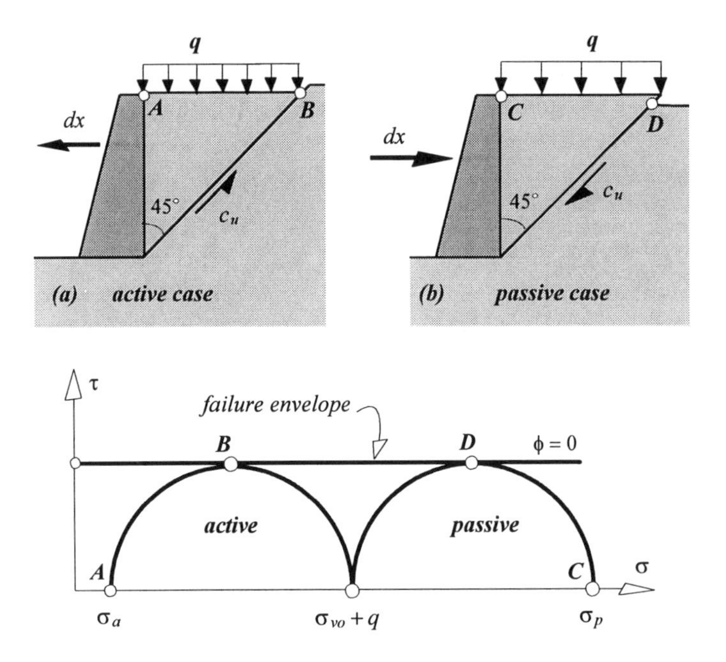

Figure 10.8: Active and passive stresses under undrained (short term) behaviour.

Substituting for $c' = c_u$ into equations 10.11 and 10.12 yields the expressions of active and passive *total* stresses:

- *active total stress*:

$$\sigma_a = \sigma_{vo} + q - 2c_u \tag{10.18}$$

- *passive total stress*:

$$\sigma_p = \sigma_{vo} + q + 2c_u \tag{10.19}$$

It follows that, in the short term, a state of equilibrium of stresses is achieved as long as the total horizontal stress σ_h is within the range:

$$\sigma_a < \sigma_h < \sigma_p$$

Notice that equation 10.18 indicates that the active pressure has a negative value down to a depth $z_c = (2c_u - q)/\gamma$. However, because a soil cannot withstand tensile stresses, *tension cracks* are very likely to appear down to the depth z_c. This, in turn, will have a restricting effect on the stress equilibrium, in that the range of stresses established earlier is now limited to:

$$\sigma_a = 0, \qquad z \leq (2c_u - q)/\gamma$$

$$\sigma_a = \gamma z + q - 2c_u, \quad z > (2c_u - q)/\gamma \qquad (10.20)$$

$$\sigma_a < \sigma_h < \sigma$$

These conditions are represented by the shaded area in figure 10.9.

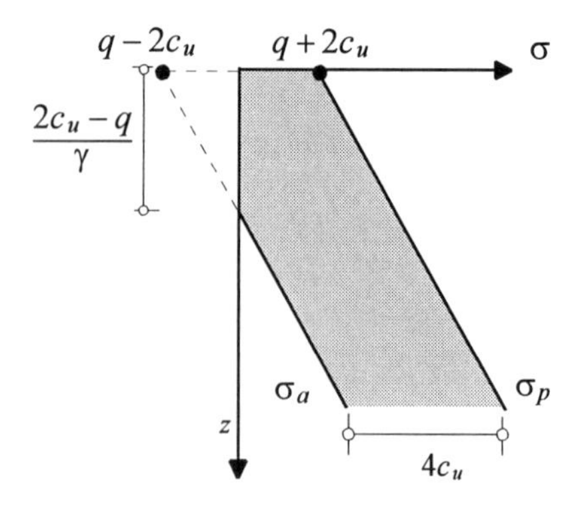

Figure 10.9: Effect of short term tensile stresses on σ_a.

The long and short term variations of active and passive pressures as represented by equations 10.15, 10.16, 10.18 and 10.19 are shown in figures 10.10 and 10.11. The third dimension representing σ'_h (σ_h in the short term) is used to illustrate the effects relative to any change in the horizontal stress (increase or decrease) with respect to the *in situ* vertical stress σ'_{vo} (σ_v in the short term). Notice that both figures correspond to a surcharge $q = 0$, and that the value of the horizontal *in situ* stress σ'_{ho} (σ_{ho} in figure 10.11) represented in figure 10.10 corresponds to a normally consolidated or a lightly overconsolidated clay.

$$K_a = \frac{1-\sin\phi'}{1+\sin\phi'} = \tan^2\left(\frac{\pi}{4} - \frac{\phi'}{2}\right), \qquad K_p = \frac{1}{K_a} = \tan^2\left(\frac{\pi}{4} + \frac{\phi'}{2}\right)$$

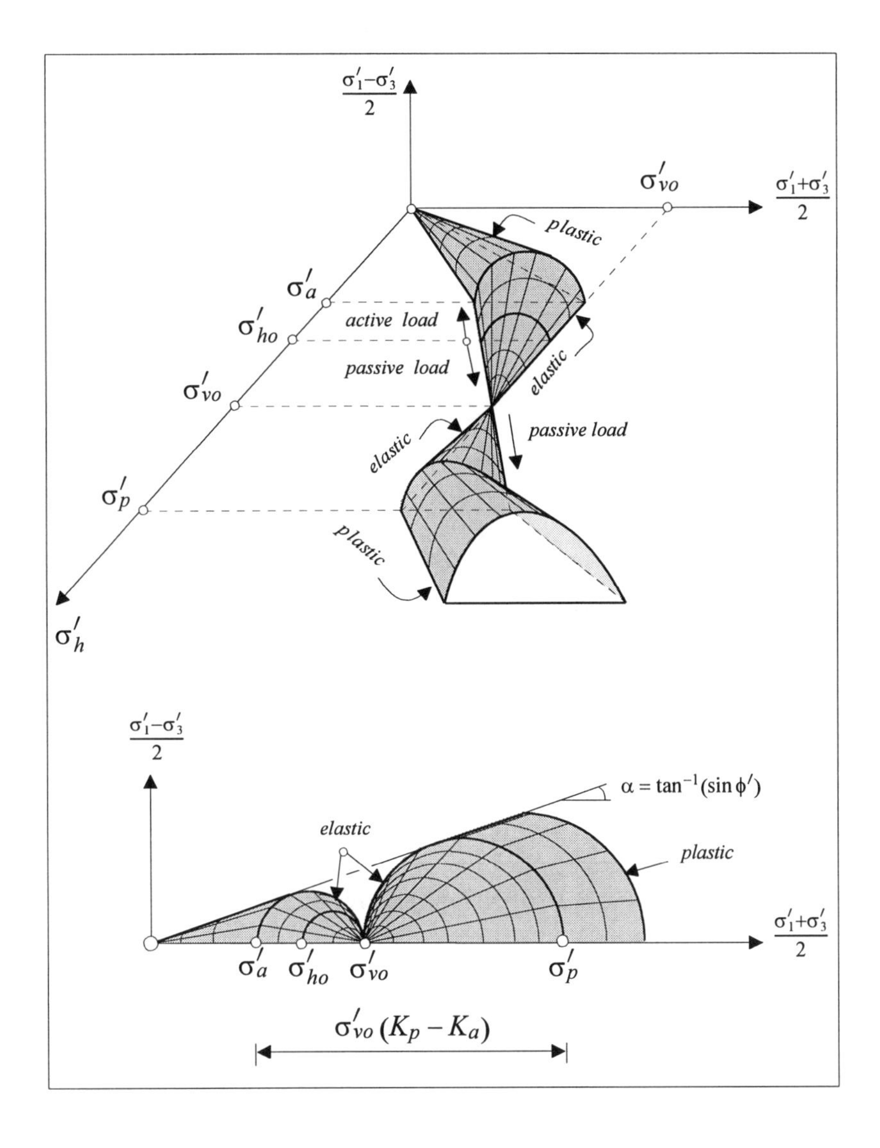

Figure 10.10: 3-D effects due to the variation of the horizontal effective stress in the case of a cohesionless drained soil.

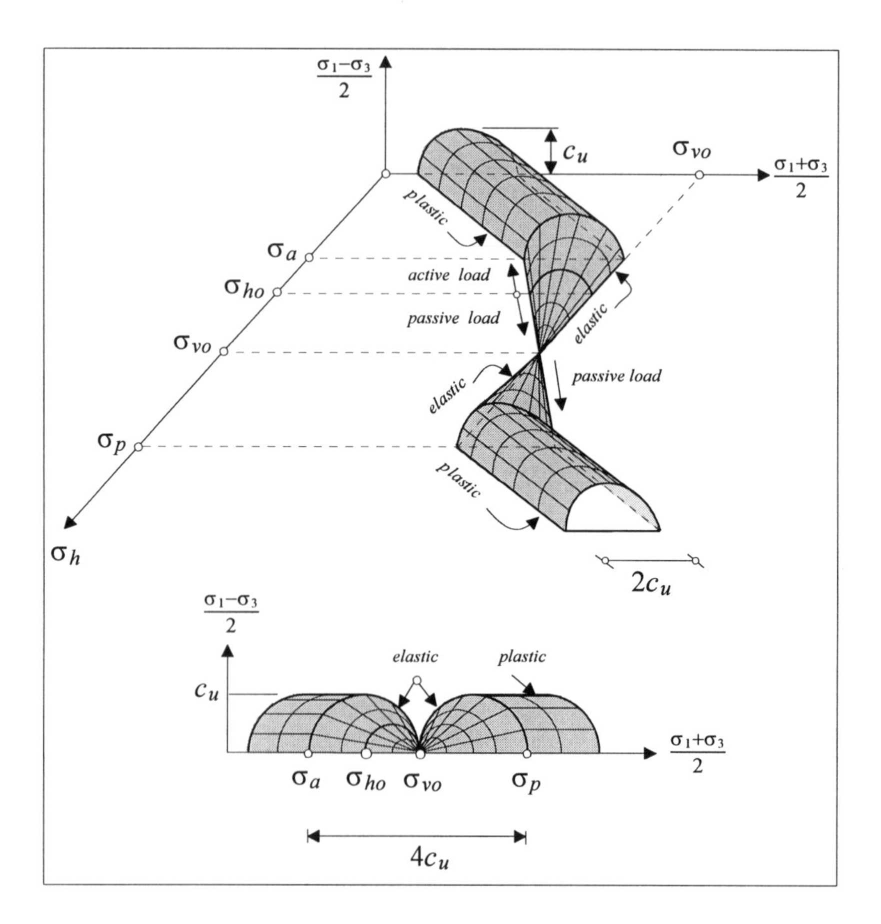

Figure 10.11: 3-D effects due to the variation of the horizontal stress in the case of an undrained clay.

10.3.4 Rankine theory: case of a sloping frictional backfill

The analysis undertaken previously related to a *smooth wall*, with a *vertical back*, retaining a *horizontal* backfill, so that no shear stresses are developed at the soil–wall interface. Although such ideal structures do not exist, this analysis can sometimes be justified, especially when little or no soil movement occurs relative to the structure. Apart from these cases, *wall friction* can significantly affect both the magnitude and the direction of the lateral thrust.

Rankine analysed these effects by considering the limit equilibrium of an element of an *isotropic homogeneous* soil in the case of an *inclined ground surface* as illustrated in figure 10.12. In order to simulate *friction* at the *vertical* back of wall, Rankine considered the forces acting on the sides of the element to be parallel to the ground surface and, in so doing, he assigned the same value of the angle of inclination of the ground surface to the angle of wall friction. Under these circumstances, the coefficient of active pressure can be written as:

$$K_a = \frac{X_a}{\sigma_v} = \frac{X_a}{\gamma' z} \tag{10.21}$$

where the *active stress* is $X_a = \left(\sigma_h^2 + \tau_a^2\right)^{1/2}$

Referring to figure 10.12, the *normal stress* at the base of the element is:

$$\sigma = W \cos\beta = \gamma' z \cos^2\beta \tag{10.22}$$

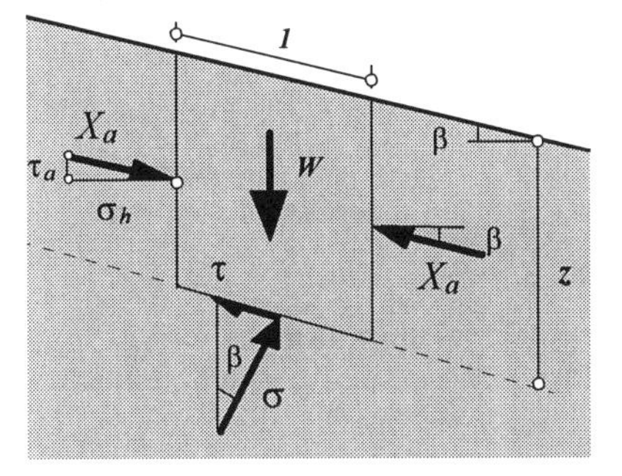

Figure 10.12: Limit equilibrium of an element of soil inclined at the surface.

The different quantities in equation 10.21 can easily be determined using Mohr's circle representation of stresses illustrated in figure 10.13. It is seen that:

$$DA = DB = OD \sin\phi'$$

$$DC = OD \sin\beta$$

hence:

$$AC = \left(AD^2 - DC^2\right)^{1/2} = OD\left(\sin^2\phi' - \sin^2\beta\right)^{1/2}$$

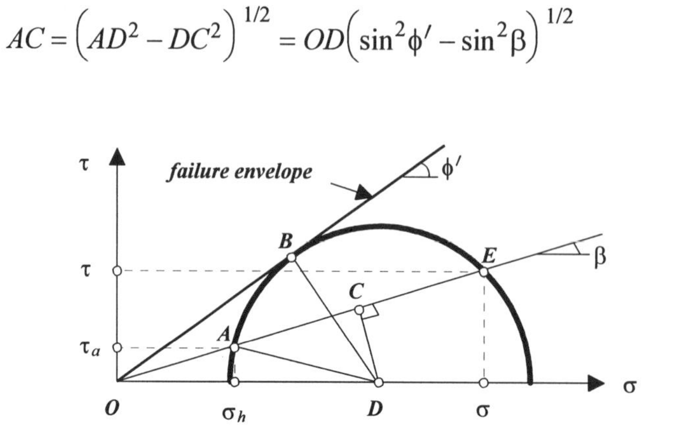

Figure 10.13: Mohr's circles related to figure 10.12.

Also, according to the same figure:

$$OC = OD\cos\beta$$

$$OA = OC - AC = OD\left[\cos\beta - \left(\sin^2\phi' - \sin^2\beta\right)^{1/2}\right]$$

$$OE = OC + CE = OC + AC = OD\left[\cos\beta + \left(\sin^2\phi' - \sin^2\beta\right)^{1/2}\right]$$

From the geometry of the figure, the following quantities are established in a straightforward manner:

$$X_a = OA \qquad \text{and} \qquad \frac{\sigma}{\cos\beta} = OE$$

Now, substituting for σ from equation 10.22 and rearranging:

$$\gamma' z = \frac{OD}{\cos\beta}\left[\cos\beta + \left(\sin^2\phi' - \sin^2\beta\right)^{1/2}\right]$$

finally, substituting for the quantities X_a and $\gamma' z$ in equation 10.21 yields the expression of the *coefficient of active pressure*:

$$K_a = \cos\beta\,\frac{\cos\beta - \sqrt{\cos^2\beta - \cos^2\phi'}}{\cos\beta + \sqrt{\cos^2\beta - \cos^2\phi'}} \tag{10.23}$$

A similar analysis using the passive stress on the element sides yields the coefficient of *passive earth pressure*:

$$K_p = \cos\beta \frac{\cos\beta + \sqrt{\cos^2\beta - \cos^2\phi'}}{\cos\beta - \sqrt{\cos^2\beta - \cos^2\phi'}} \tag{10.24}$$

Hence the respective resultant *active* and *passive thrusts*, acting on a *vertical wall* of height *H*, *parallel to the slope*:

$$P_a = \tfrac{1}{2}\gamma H^2 K_a, \quad P_p = \tfrac{1}{2}\gamma H^2 K_p$$

Note that for $\beta = 0$ (horizontal soil surface), equations 10.23 and 10.24 reduce to equations 10.13 and 10.14, because Rankine theory does not take into account any *wall friction* in conjunction with a *vertical wall* retaining a *horizontal ground surface.*

10.4 Coulomb theory

The pioneering work of Coulomb presented to the *Académie Royale des Sciences* in 1773, and published three years later marked a turning point in the analysis and evaluation of soil behaviour in general and of the lateral earth pressure exerted on a retaining structure in particular. Coulomb developed the theoretical basis for the analysis of the stability of retaining walls through the use of slip plane collapse mechanisms such as the two illustrated in figure 10.14. He showed that by assuming a planar failure surface, originating at the heel of the wall, the algebraic expressions of both active and passive thrusts can be established, and the ensuing *critical failure surface* can be determined by differentiating these expressions with respect to the angle α in figure 10.14.

It is essential to bear in mind that Coulomb's method of analysis, in contrast with Rankine's, yields an upper bound solution (i.e. it overestimates the passive thrust while underestimating the active one). As such, it can potentially be unsafe to use, and therefore its limitations, which are explained below, must be well understood.

With reference to figure 10.14, the analysis of forces is relatively straightforward in relation to a smooth wall. In either case (active or passive), the weight W of the failing block is known, and so are the directions of the reaction force R and the lateral thrust P. Hence for a *cohesionless soil* (*i.e.* $c' = 0$) having a unit weight γ and an angle of shearing resistance ϕ':

$$W = \frac{1}{2}\gamma H^2 \tan\alpha \tag{10.25}$$

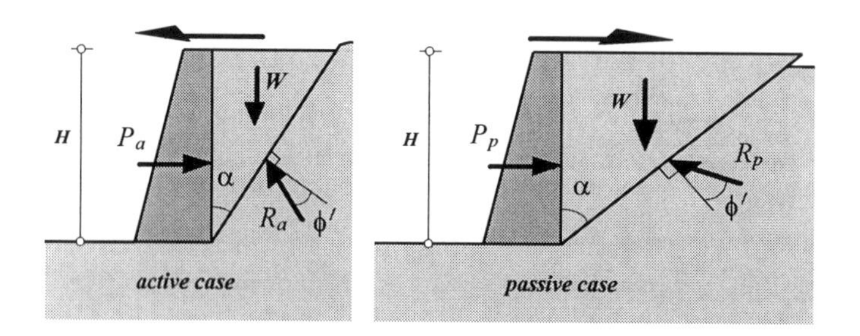

Figure 10.14: Active and passive pressures, case of a smooth wall.

Also, from the geometry of the figure, it can be seen that:

- on the active side:

$$P_a \sin(\alpha + \phi') = W \cos(\alpha + \phi')$$

- on the passive side:

$$P_p \sin(\alpha - \phi') = W \cos(\alpha - \phi')$$

Alternatively, using equation 10.25, the active and passive thrusts are respectively:

$$P_a = \frac{1}{2}\gamma H^2 \frac{\tan\alpha}{\tan(\alpha + \phi')} \tag{10.26a}$$

$$P_p = \frac{1}{2}\gamma H^2 \frac{\tan\alpha}{\tan(\alpha - \phi')} \tag{10.26b}$$

To find the *maximum active thrust*, the two following conditions must be fulfilled:

$$\frac{\partial P_a}{\partial \alpha} = 0 \qquad \text{and} \qquad \frac{\partial^2 P_a}{\partial \alpha^2} < 0$$

$$\frac{\partial P_a}{\partial \alpha} = \frac{1}{2}\gamma H^2 \frac{1}{\tan^2(\alpha + \phi')}\left[\frac{\tan(\alpha + \phi')}{\cos^2\alpha} - \frac{\tan\alpha}{\cos^2(\alpha + \phi')}\right]$$

writing $\partial P_a / \partial \alpha = 0$ yields:

$$\cos^2(\alpha + \phi')\tan(\alpha + \phi') = \cos^2\alpha \tan\alpha$$

or

$$\sin 2(\alpha + \phi') = \sin 2\alpha$$

leading to the solution: $\alpha = \left(\frac{\pi}{4} - \frac{\phi'}{2}\right)$

Using this value, it can be shown that $\partial^2 P_a / \partial \alpha^2 < 0$.

A back-substitution for α into equation 10.26*a* then yields the expression of *active thrust*:

$$P_a = \frac{1}{2}\gamma H^2 \tan\left(\frac{\pi}{4} - \frac{\phi'}{2}\right)\cot\left(\frac{\pi}{4} + \frac{\phi'}{2}\right)$$

$$= \frac{1}{2}\gamma H^2 \tan^2\left(\frac{\pi}{4} - \frac{\phi'}{2}\right) \qquad (10.27)$$

For the passive thrust, the aim is to find the *minimum* force and this can be achieved if the following conditions are met:

$$\partial P_p / \partial \alpha = 0 \qquad \text{and} \qquad \partial^2 P_p / \partial \alpha^2 > 0$$

Applying the same procedure used in the active case, it is easy to establish that:

$$\alpha = \frac{\pi}{4} + \frac{\phi'}{2}$$

Once substituted into equation 10.26*b*, the *passive thrust* expression is thus established:

$$P_p = \frac{1}{2}\gamma H^2 \tan^2\left(\frac{\pi}{4} + \frac{\phi'}{2}\right) \tag{10.28}$$

Several investigators used Coulomb theory to explore more complex problems pertaining to lateral pressures such as accounting for wall friction. Figure 10.15 depicts the case of a vertical *rough* wall with a *given* angle of friction δ retaining a *cohesionless* horizontal backfill.

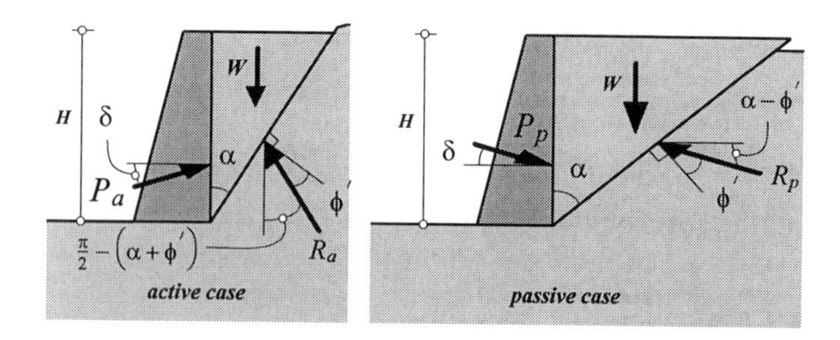

Figure 10.15: Active and passive pressures, case of a rough wall.

Using Coulomb theory, and solving perpendicular to the soil reactions *R*, it follows that:

- on the active side:

$$P_a \sin(\delta + \alpha + \phi') = W \cos(\alpha + \phi')$$

- on the passive side:

$$P_p \sin(\alpha - \phi' - \delta) = W \cos(\alpha - \phi')$$

So that when equation 10.25 is used, the active and passive thrusts become respectively:

$$P_a = \frac{1}{2}\gamma H^2 \frac{\tan\alpha \cos(\alpha + \phi')}{\sin(\delta + \alpha - \phi')}$$

$$P_p = \frac{1}{2}\gamma H^2 \frac{\tan\alpha \cos(\alpha - \phi')}{\sin(\alpha - \phi' - \delta)}$$

These two expressions can then be optimised to find the *maximum active* and the *minimum passive* thrusts; this is achieved by writing:

$$\partial P_a/\partial\alpha = 0, \quad \partial^2 P_a/\partial\alpha^2 < 0 \quad \text{and} \quad \partial P_p/\partial\alpha = 0, \quad \partial^2 P_p/\partial\alpha^2 > 0.$$

Notwithstanding the cumbersome nature of the derivations, the solution established by Mayniel in 1808 leads to:

- the active thrust:

$$P_a = \frac{1}{2}\gamma H^2 \frac{\cos^2\phi'}{\cos\delta\left[1 + \sqrt{\dfrac{\sin(\delta+\phi')\sin\phi'}{\cos\delta}}\right]^2} \tag{10.29}$$

- the passive thrust:

$$P_p = \frac{1}{2}\gamma H^2 \frac{\cos^2\phi'}{\cos\delta\left[1 - \sqrt{\dfrac{\sin(\delta+\phi')\sin\phi'}{\cos\delta}}\right]^2} \tag{10.30}$$

This solution was further refined in 1906 by Müller-Breslau who investigated the general case of a rough inclined wall retaining an inclined *cohesionless* backfill as depicted in figure 10.16. By adopting Coulomb analysis, the following expressions of active and passive thrusts were established:

- the active thrust:

$$P_a = \frac{1}{2}\gamma H^2 \frac{\sin^2(\eta + \phi')}{\sin^2\eta \sin(\eta - \delta)\left[1 + \sqrt{\dfrac{\sin(\phi'+\delta)\sin(\phi'-\beta)}{\sin(\eta-\delta)\sin(\beta+\eta)}}\right]^2} \tag{10.31}$$

- the passive thrust:

$$P_P = \frac{1}{2}\gamma H^2 \frac{\sin^2(\eta - \phi')}{\sin^2\eta \sin(\eta + \delta)\left[1 - \sqrt{\frac{\sin(\phi'+\delta)\sin(\phi'+\beta)}{\sin(\eta+\delta)\sin(\beta+\eta)}}\right]^2} \qquad (10.32)$$

Clearly, both equations 10.31 and 10.32 encompass all previous cases derived from Coulomb theory and represent, as such, the general relationship for a *dry granular backfill.* For instance when $\eta = \pi/2$ (*i.e.* vertical wall) and $\beta = 0$ (horizontal backfill), equations 10.31 and 10.32 reduce to equations 10.29 and 10.30 (Mayniel solution). If, in addition, the wall were frictionless ($\delta = 0$), then the equations are further reduced to equations 10.27 and 10.28 (Coulomb solution). Notice that for a homogeneous dry backfill, the active thrust is assumed to be acting at the third of the wall height, measured from the base of the wall. The same applies to the passive thrust where the height refers to the embedment depth of the wall.

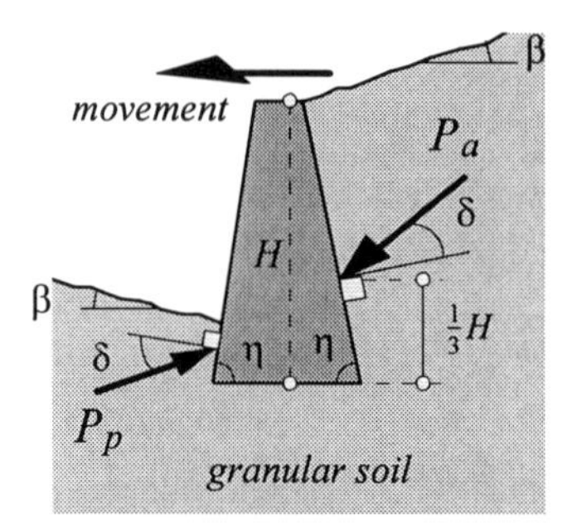

Figure 10.16: General case of a rough inclined wall retaining an inclined soil.

Experimental evidence shows that Coulomb theory yields reasonable estimates of lateral thrusts on the active side of wall. However, the passive thrusts calculated from equations 10.28, 10.30 and 10.32 are well in excess of the actual passive thrusts. This is due to the nature of slip surface assumed by Coulomb theory on the passive side, which is markedly different from the actual slip surface as depicted in figure 10.17. Accordingly, Coulomb theory is inherently (and dangerously) unsafe when used in conjunction with the calculation of passive thrusts.

10.5 Boussinesq theory

10.5.1 Introduction

Experimental evidence shows that, while the assumed shape of the failure surface for both Rankine and Coulomb theories corresponds by and large with the actual failure surface on the active side, the shape of the block failing in the passive mode differs markedly from that observed in practice as illustrated in figure 10.17. Moreover, Rankine theory leads to a *lower bound solution* whereas Coulomb analysis yields an *upper bound solution*, hence the substantial difference between Rankine and Coulomb lateral thrusts on the passive side.

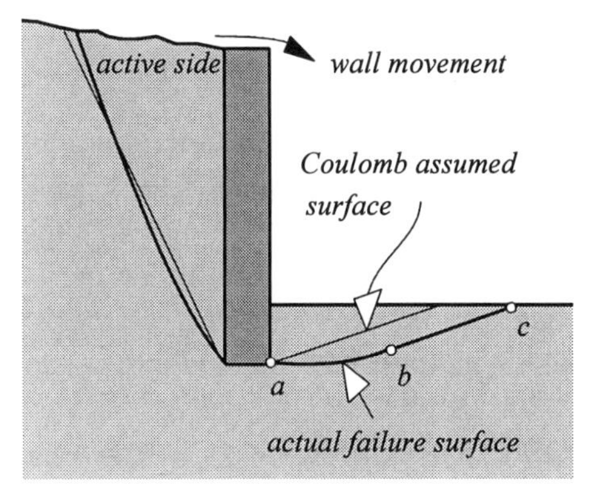

Figure 10.17: Actual and theoretical failure surfaces.

In 1882, Boussinesq correctly argued that by considering a linear failure surface, Coulomb analysis is being restrictive, whereas Rankine theory is even more so since it only takes into account a wall friction angle corresponding to the slope of the retained material. In order to offset these shortcomings, Boussinesq developed his theory by considering that the soil wedge *OAB* shown in figure 10.18 is in a state of *limit equilibrium*. The analysis hinges on the fact that, while Rankine equilibrium applies to the zone immediately to the right-hand side of the slip surface *OB,* it can no longer be used in the zone *OAB* since it violates the boundary conditions along the back of the wall where the friction developed depends only on the relative soil–wall movement.

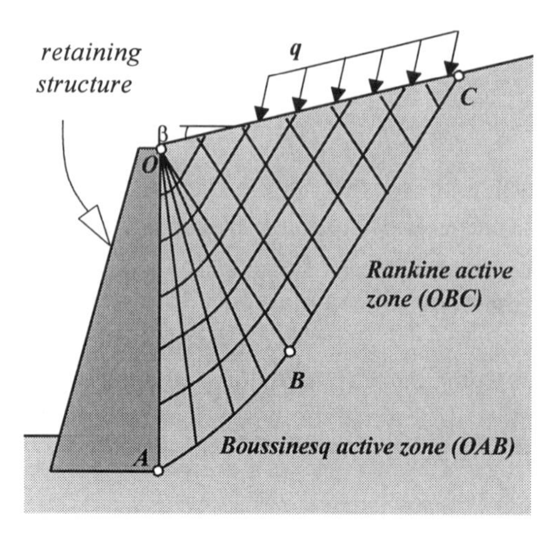

Figure 10.18: Boussinesq failure surface.

Boussinesq based his theory on the assumption that, for the general case of a structure retaining an *isotropic homogeneous* soil with a unit weight γ, a cohesion c' and a friction angle ϕ', sloping at an angle β and subject to a surcharge q as depicted in figure 10.18, the active (or passive) pressure exerted on the wall can be considered as a *combination* of the three following components, that can be evaluated independently:

(*a*) the pressure due to the weight of the soil, deprived of its cohesion (*i.e.* $c' = 0,\ \ \gamma \neq 0,\ \ \phi' \neq 0$);

(*b*) the pressure due to the surcharge q applied to a weightless, cohesionless soil (*i.e.* $\gamma = 0,\ \ c' = 0,\ \ \phi' \neq 0$);

(*c*) the pressure due to a weightless cohesive soil (*i.e.* $\gamma = 0,\ \ c' \neq 0,\ \ \phi' \neq 0$).

10.5.2 Pressure due to the weight of a frictional cohesionless soil

Consider the general case of a wall retaining a dry *isotropic homogeneous* frictional material with a unit weight γ as depicted in figure 10.19. Let us isolate the radial element *OCD*, whose base is at a distance r from point *O*. The element has a weight:

$$W = \frac{r^2}{2}\gamma \, d\lambda$$

Along OD, both normal and shear stresses vary linearly with depth from zero at O to the respective values of σ and τ at the base of the element; hence the triangular distribution shown in figure 10.19. Similarly, the stresses along OC vary from zero at O to $\sigma + d\sigma$ and $\tau + d\tau$ at the base of the element.

The equivalent forces per unit length exerted on OD and OC, as well as their lines of action can now be established as shown on the right-hand side of figure 10.19 in the case of normal forces. Obviously, a similar analysis applies to the shear forces, whence the complete set of normal and shear forces to which the element is subjected, represented in the figure.

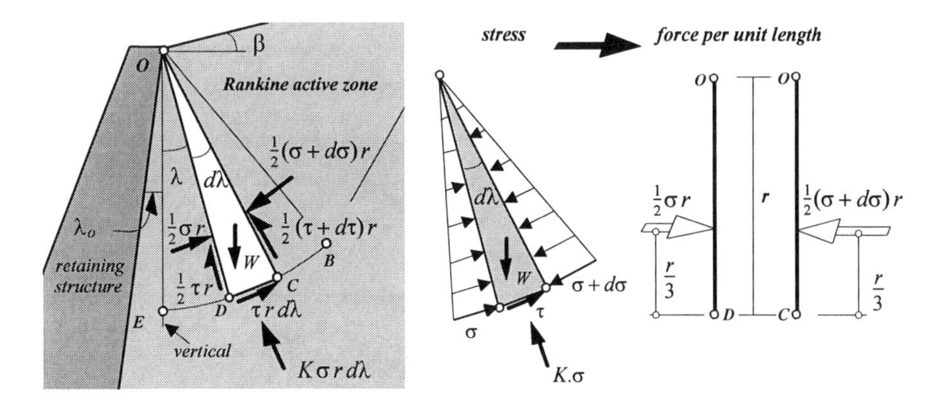

Figure 10.19: Boundary conditions related to Boussinesq theory.

Notice that for the normal force at the base of the element, the coefficient K is dependent on the ratio τ/σ. With reference to figure 10.20, it is seen that under *active* stress conditions, the mean stress at the centre of Mohr's circle (point G) is as follows:

$$p = \tfrac{1}{2}(\sigma + K\sigma) \tag{10.33}$$

Also, the radius of the circle is: $FG = p \sin\phi'$.

Thence the normal stresses:

$$K\sigma = p + p\sin\phi'\cos(\alpha-\delta) \tag{10.34a}$$

$$\sigma = p - p\sin\phi'\cos(\alpha-\delta) \tag{10.34b}$$

It follows that for the *active* case:

$$K = \frac{1+\sin\phi'\cos(\alpha-\delta)}{1-\sin\phi'\cos(\alpha-\delta)} \tag{10.35a}$$

It is straightforward to establish that under *passive* stress conditions:

$$K = \frac{1-\sin\phi'\cos(\alpha-\delta)}{1+\sin\phi'\cos(\alpha-\delta)} \tag{10.35b}$$

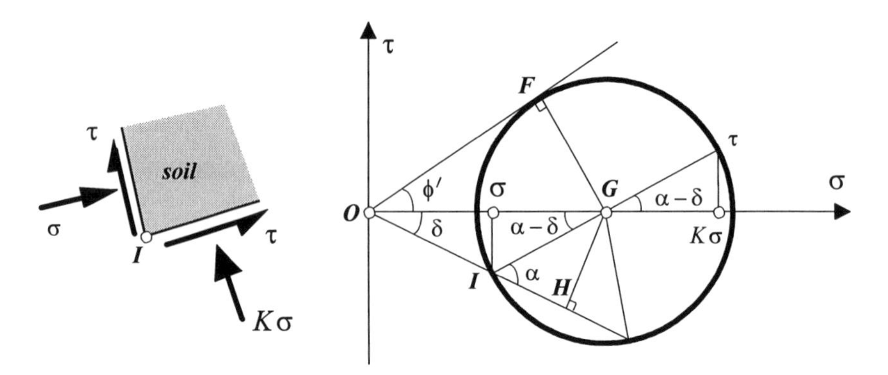

Figure 10.20: Coefficient of earth pressure; active conditions.

The angle α in equations 10.35 is such that $\sin\alpha = GH/GI$, and according to figure 10.20: $GH = p\sin\delta$, and $GI = FG = p\sin\phi'$. Whence:

$$\alpha = \sin^{-1}\left(\frac{\sin\delta}{\sin\phi'}\right) \tag{10.36}$$

the angle δ in equation 10.36 being the *known* value of the wall friction angle.

Solving for the moments, with respect to O in figure 10.19, it follows that:

$$\sigma \frac{r}{2} \times \frac{2r}{3} - (\sigma + d\sigma)\frac{r}{2} \times \frac{2r}{3} + \tau r^2 d\lambda - \frac{2}{3}\gamma \frac{r^2}{2} d\lambda \sin\lambda = 0$$

or

$$\frac{d\sigma}{d\lambda} = 3\tau - r\gamma \sin\lambda \qquad (10.37)$$

Now, resolving parallel to the surface OD:

$$\frac{(\sigma + d\sigma) r}{2} \sin d\lambda - K\sigma r\, d\lambda \cos\frac{d\lambda}{2} - \tau\frac{r}{2}$$

$$+ \frac{(\tau + d\tau) r}{2} \frac{1}{\cos d\lambda} + \frac{\gamma}{2} r^2\, d\lambda \cos\lambda = 0$$

but

$$\sin d\lambda \approx d\lambda, \quad \cos d\lambda \approx \cos\frac{d\lambda}{2} \approx 1, \quad d\lambda\, d\sigma \approx 0$$

hence:

$$\frac{d\tau}{d\lambda} = (2K - 1)\sigma - \gamma r \cos\lambda \qquad (10.38)$$

The ensuing set of first order differential equations, together with the boundary conditions are thus:

$$\frac{d\sigma}{d\lambda} = 3\tau - \gamma r \sin$$

$$\frac{d\tau}{d\lambda} = \sigma(2K - 1) - \gamma r \cos\lambda$$

with

$$K = \frac{1 + \sin\phi' \cos(\alpha - \delta)}{1 - \sin\phi' \cos(\alpha - \delta)} \qquad \text{(active case)}$$

$$K = \frac{1 - \sin\phi' \cos(\alpha - \delta)}{1 + \sin\phi' \cos(\alpha - \delta)} \qquad \text{(passive case)}$$

$$\alpha = \sin^{-1}\left(\frac{\sin\delta}{\sin\phi'}\right)$$

where δ is the *known* value of the wall friction angle.

The sign conventions are represented in figure 10.21. With reference to both figures 10.19 and 10.21, the boundary conditions for the above set of differential equations are such that:

$$\lambda = \lambda_o, \qquad \delta = \delta_a \ (or\ \delta_p)$$

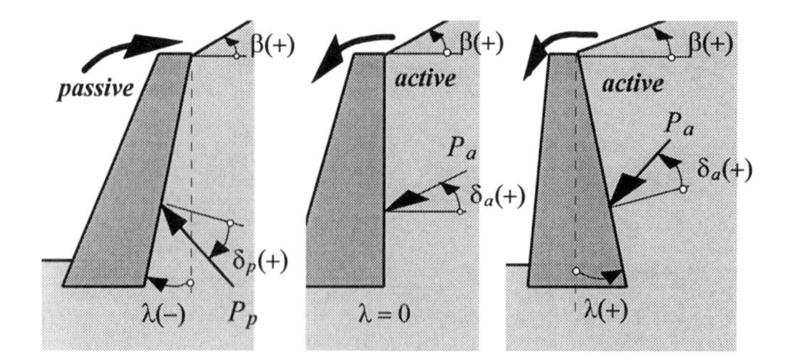

Figure 10.21: Sign conventions used in conjunction with the boundary conditions related to figure 10.20.

A closed form solution to the above system of equations is not available. However, in 1948, Caquot and Kerisel published the numerical solution to this system in the form of tables giving both active and passive pressure coefficients for different combinations of the wall inclination λ, the slope β of the retained soil, its angle of shearing resistance ϕ' and the wall friction angle δ. Very few of these results, corresponding to the case of a vertical back of wall (*i.e.* $\lambda = 0$), are reproduced in the following table and reference should be made to the complete set of results in the useful book of Kerisel and Absi (1990). Notice that the table contains the values of the coefficients K_a and K_p, *and not their horizontal components* $K_a \cos\delta$ and $K_p \cos\delta$.

Table 10.1: Coefficients of active and passive pressure corresponding to $\delta = \frac{2}{3}\phi'$. *(Reproduced by permission from Kerisel and Absi (1990).)*

Coefficients of active and passive earth pressure K_a and K_p = 0, $\delta = 2\phi'/3$									
ϕ (°)		10	15	20	25	30	35	40	45
$\frac{\beta}{\phi} = 1$	K_a	0.990	0.964	0.927	0.879	0.822	0.756	0.683	0.603
	K_p	1.910	2.800	4.300	7.000	12.50	25.00	58.00	163.0
0.8	K_a	0.806	0.715	0.628	0.546	0.469	0.397	0.330	0.269
	K_p	1.860	2.650	4.000	6.300	10.90	20.50	45.00	115.0
0.6	K_a	0.748	0.644	0.551	0.468	0.395	0.329	0.271	0.219
	K_p	1.790	2.500	3.700	5.700	9.300	16.60	34.00	78.00
0.4	K_a	0.710	0.598	0.503	0.422	0.352	0.291	0.239	0.193
	K_p	1.730	2.350	3.400	5.000	7.800	13.00	24.50	52.00
0.2	K_a	0.680	0.564	0.469	0.389	0.322	0.266	0.218	0.176
	K_p	1.660	2.200	3.100	4.300	6.500	10.00	17.50	33.00
0.0	K_a	0.656	0.537	0.442	0.364	0.300	0.247	0.202	0.163
	K_p	1.590	2.050	2.750	3.700	5.300	8.000	12.00	20.00
–0.2	K_a	0.636	0.515	0.420	0.343	0.282	0.231	0.189	0.153
	K_p	1.520	1.900	2.400	3.100	4.200	5.700	8.200	12.50
–0.4	K_a	0.619	0.496	0.401	0.326	0.266	0.218	0.177	0.144
	K_p	1.430	1.720	2.100	2.550	3.200	4.100	5.400	7.100
–0.6	K_a	0.603	0.479	0.384	0.311	0.253	0.206	0.167	0.135
	K_p	1.330	1.520	1.760	2.050	2.400	2.800	3.400	4.000
–0.8	K_a	0.590	0.464	0.370	0.298	0.241	0.195	0.158	0.127
	K_p	1.200	1.300	1.400	1.530	1.580	1.650	1.800	1.700

N.B. According to the sign conventions used in Kerisel and Absi's book, the angle of wall friction δ *is positive in the active case and negative in relation to the passive case. Therefore, all the values* K_p *listed in table 10.1 are easily found in the book by assigning a negative sign to the angle* δ *shown at the top of the table.*

For instance, if $\lambda = 0$, $\delta = 2\phi/3$, $\beta/\phi = 0.0$ and $\phi = 35°$, *table 10.1 yields the coefficients* $K_a = 0.247$ and $K_p = 8.0$. *Using Kerisel and Absi's book, the same* K_a *value is read from the table on p.78 corresponding to:* $\beta/\phi = 0$, $\delta/\phi = 0.66$, $\lambda = 0$, and $\phi = 35°$. *As for* K_p, *its value is read from the table on p.14 in conjunction with:* $\beta/\phi = 0$, $\delta/\phi = -0.66$, $\lambda = 0$ and $\phi = 35°$.

10.5.3 Pressure due to a uniform surcharge on a weightless cohesionless soil

Boussinesq's theory makes due allowance for the effects of an inclined uniform load of infinite extent applied at the surface of a weightless cohesionless soil (*i.e.* $c' = 0,\ \gamma = 0,\ \phi' \neq 0$). With reference to the sign convention used in Kerisel and Absi's book depicted in figure 10.22, it can be seen that:

$$\Omega = \frac{\pi}{2} + \beta - \lambda \tag{10.39}$$

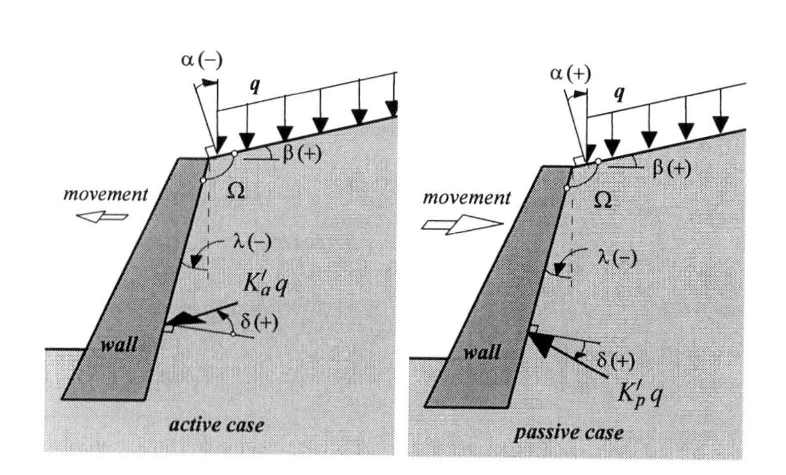

Figure 10.22: Sign convention used by Kerisel and Absi (1990).

The resulting lateral pressure exerted on the wall in the active case is:

$$\sigma_a = K'_a\, q \tag{10.40}$$

On the passive side, the pressure can be calculated in a similar way:

$$\sigma_p = K'_p\, q \tag{10.41}$$

Notice that, for the passive case, the coefficient K'_p is obtained in a very unconventional way from Kerisel and Absi's tables, in that the values of the angles α and δ (figure 10.22) are interchanged, thus yielding K'_a from which K'_p is then calculated as follows:

$$K_p' = \frac{1}{K_a'} \tag{10.42}$$

For example, consider the case of a wall retaining a silty sand with an angle $\phi' = 30^\circ$ as depicted in figure 10.23.

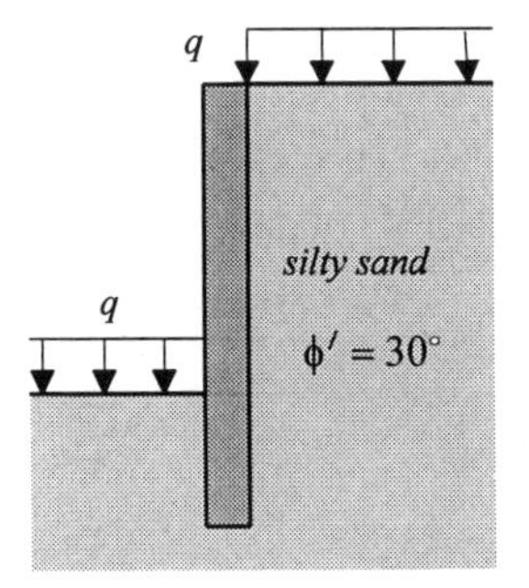

Figure 10.23: Example related to the use of Kerisel and Absi's tables.

Assuming an angle of wall friction $\delta = 2\phi'/3$, both coefficients of active and passive pressure, related to the uniform load q applied on either side of the wall, are found from Kerisel and Absi's book as follows.

- With reference to figure 10.22, it is seen that, on the active side, figure 10.23 corresponds to $\alpha = 0,\ \Omega = 90^\circ$. Thus, the coefficient K_a' is read from the table on p.132 for $\phi' = 30^\circ,\ \Omega = 90^\circ,\ \alpha = 0$ and $\delta = 20^\circ \quad \Rightarrow \quad K_a' = 0.304$.

- On the passive side, the values of α and δ are now interchanged, so that $\alpha = 2\phi'/3 = 20^\circ$ and $\delta = 0$. The corresponding K_a' value is read from the same table for $\phi' = 30^\circ$, $\Omega = 90^\circ$, $\alpha = 20^\circ$ and $\delta = 0$: $K_a' = 0.203$, and the coefficient of passive pressure is thereafter calculated from equation 10.42:

$$K_p' = \frac{1}{0.203} = 4.926$$

Very few of the K_p' values have been *calculated* from Kerisel and Absi's book according to the procedure described above. These values, together with the coefficients K_a' corresponding to the active case, are presented in table 10.2.

Table 10.2: Coefficients of active and passive pressure applicable to a uniform surcharge. (Reproduced by permission from Kerisel and Absi (1990).)

Coefficients K_a' and K_p' : $\alpha = 0$, $\delta = 0.66$								
ϕ (°)		15	20	25	30	35	40	45
$\Omega = 90°$	K_a'	0.542	0.447	0.369	0.304	0.249	0.206	0.165
	K_p'	2.024	2.645	3.554	4.926	7.075	10.64	17.24
95	K_a'	0.517	0.420	0.340	0.274	0.221	0.177	0.139
	K_p'	2.123	2.822	3.861	5.464	8.000	12.40	20.41
100	K_a'	0.494	0.394	0.314	0.248	0.196	0.153	0.116
	K_p'	2.222	3.006	4.178	6.024	9.063	14.28	24.39
105	K_a'	0.471	0.370	0.289	0.224	0.173	0.132	0.098
	K_p'	2.331	3.202	4.545	6.667	10.20	16.57	29.41
110	K_a'	0.450	0.347	0.266	0.203	0.153	0.114	0.082
	K_p'	2.445	3.417	4.918	7.353	11.54	19.23	34.48
115	K_a'	0.429	0.326	0.246	0.183	0.135	0.098	0.069
	K_p'	2.557	3.636	5.346	8.130	12.99	22.22	41.67
120	K_a'	0.409	0.305	0.226	0.166	0.120	0.085	0.058
	K_p'	2.681	3.876	5.803	9.010	14.78	25.42	50.00
125	K_a'	0.391	0.286	0.209	0.150	0.106	0.073	0.049
	K_p'	2.809	4.126	6.263	10.00	16.57	29.70	58.82
130	K_a'	0.373	0.269	0.192	0.136	0.094	0.063	0.041
	K_p'	2.941	4.405	6.802	10.99	18.87	34.48	71.43
135	K_a'	0.356	0.252	0.177	0.123	0.083	0.054	0.034
	K_p'	3.086	4.687	7.389	12.19	21.28	40.00	83.33

10.5.4 Pressure due to the cohesion of a weightless soil

For cohesive soils, Boussinesq considered the behaviour of such materials to be similar to that of purely frictional soils subject throughout their surface areas to an extra normal stress of a magnitude $c'\cot\phi'$. This effect can easily be established from the failure criterion of a cohesive soil, written as follows:

$$\tau = c' + \sigma' \tan\phi'$$

so that after rearranging:

$$\tau = (c'\cot\phi' + \sigma')\tan\phi' = \sigma^{*\prime}\tan\phi' \qquad (10.43)$$

Equation 10.43 is that of a frictional soil in which part of the normal stress $\sigma^{*\prime}$ is due to cohesion. Consequently, the cohesion effect on the lateral thrust induced by a cohesive weightless soil can be evaluated by considering that the soil in question is frictional ($c' = 0$), but subject throughout to an external normal stress $c'\cot\phi'$ as depicted in figure 10.24.

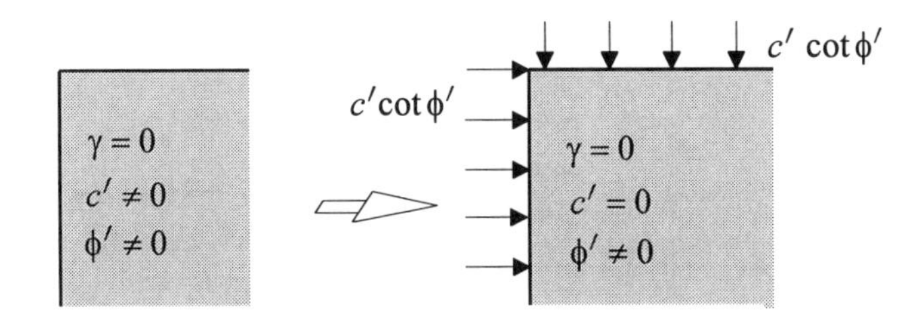

Figure 10.24: Effect of cohesion in Boussinesq theory.

In the general case of a *rough wall* with an inclined back, retaining a sloping cohesive soil, the active pressure due to cohesion is as depicted in figure 10.25. Whence the ensuing value of the component of the active pressure, *normal to the wall*:

$$\sigma_{an} = -c'\cot\phi'\left(1 - K_a'\cos\delta\right) \qquad (10.44)$$

the corresponding shear stress along the back of the wall being:

$$\tau = c'\cot\phi' K_a' \sin\delta \qquad (10.45)$$

It is straightforward to show that, for the passive case, the component of passive pressure due to cohesion, *normal to the wall* is given by:

$$\sigma_{pn} = c'\cot\phi'\left(K_p'\cos\delta - 1\right) \qquad (10.46)$$

the corresponding shear stress due to wall friction being:

$$\tau = c'\cot\phi' K_p' \sin\delta \qquad (10.47)$$

The angle δ in equations 10.44 to 10.47 is the *angle of wall friction,* the coefficient of active pressure K'_a being the same as in the case of a uniformly loaded weightless cohesionless soil. The coefficient K'_p is related to K'_a through equation 10.42.

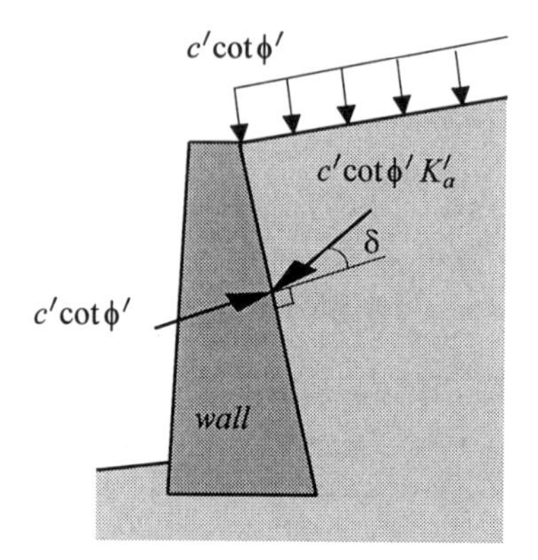

Figure 10.25: Coefficient of active pressure due to cohesion in the case of a sloping rough wall.

As can be seen from equations 10.44 and 10.46, the cohesion causes the overall active pressure at the back of a retaining wall to *decrease* (notice the negative sign of the quantity σ_{an}, since $K'_a \cos\delta < 1$), and the overall passive pressure in front of the wall to *increase*. However, these variations depend to a large extent on the value of cohesion and, as mentioned earlier, caution must be exercised when choosing a value for c'; it is even advisable to ignore the cohesion effects in the absence of representative reliable measurements of this parameter.

10.6 Lateral thrust due to different types of ground loading

10.6.1 Uniform load of infinite extent

As seen in figure 10.22, the analysis of the general case of an inclined uniform ground pressure q of infinite extent led to equations 10.40 and 10.41 of the active and passive effective pressure. These stresses propagate constantly throughout the appropriate height of wall, so that the induced lateral thrusts are found by simple integration of both equations. For instance, the lateral active thrust in the case of figure 10.26, is calculated as follows:

$$P'_a = \int_0^H q\,K'_a\,dz = qK'_aH \tag{10.48}$$

where K'_a refers to the coefficient of active earth pressure. The lateral passive thrust is calculated in a similar way if a surcharge is applied in front of the wall.

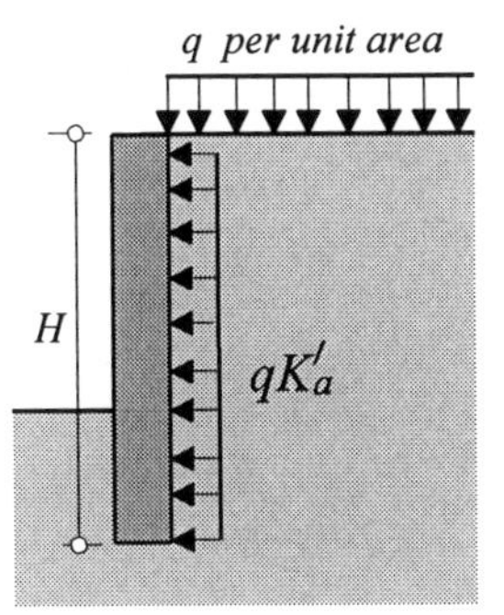

Figure 10.26: Lateral thrust due to a uniform load of infinite extent.

10.6.2 Strip load

The increase in horizontal pressure σ'_h due to a strip load can be calculated using Boussinesq's elastic solution. However, the value of σ'_h is doubled to make allowance for the effects of rigidity of the wall. Hence, with reference to figure 10.27:

$$\sigma'_h = 2\frac{q}{\pi}\,[\beta - \sin\beta\,\cos(\beta + 2\alpha)] \tag{10.49}$$

where β and α are expressed in radians.

The lateral thrust due to such a stress distribution as well as the location of its line of action can be found numerically.

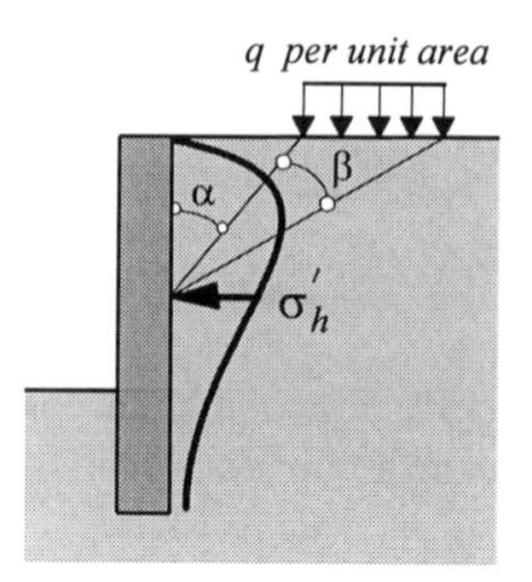

Figure 10.27: Lateral thrust due to a strip loading.

10.6.3 Line load

The increase in horizontal pressure is calculated, in this case, according to the following relationships suggested by Terzaghi (1954) (refer to figure 10.28 for m and n):

- for $m \leq 0.4$: $$\sigma'_h = q\frac{n}{5H\left(0.16 + n^2\right)^2} \qquad (10.50)$$

- for $m > 0.4$: $$\sigma'_h = q\frac{4m^2 n}{\pi H\left(m^2 + n^2\right)^2} \qquad (10.51)$$

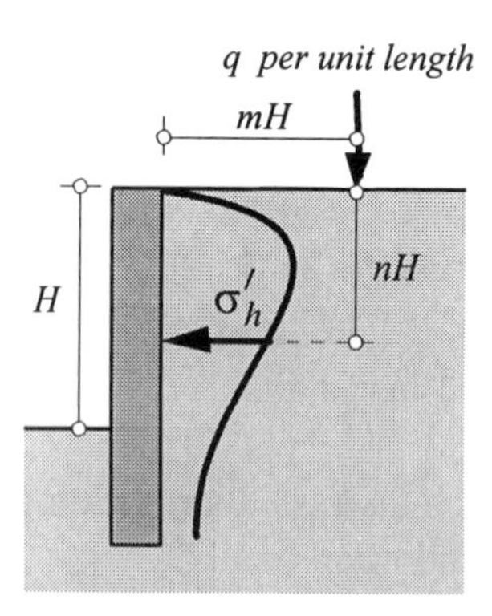

Figure 10.28: Lateral thrust due to a line load.

10.6.4 Point load

Terzaghi suggested using the following relationships for the calculation of the lateral pressure caused by a point load as depicted in figure 10.29:

- for $m \leq 0.4$:

$$\sigma'_h = 0.28 \frac{q}{H^2} \frac{n^2}{\left(0.16 + n^2\right)^3} \tag{10.52}$$

- for $m > 0.4$:

$$\sigma'_h = 1.77 \frac{q}{H^2} \frac{m^2 n^2}{\left(m^2 + n^2\right)^3} \tag{10.53}$$

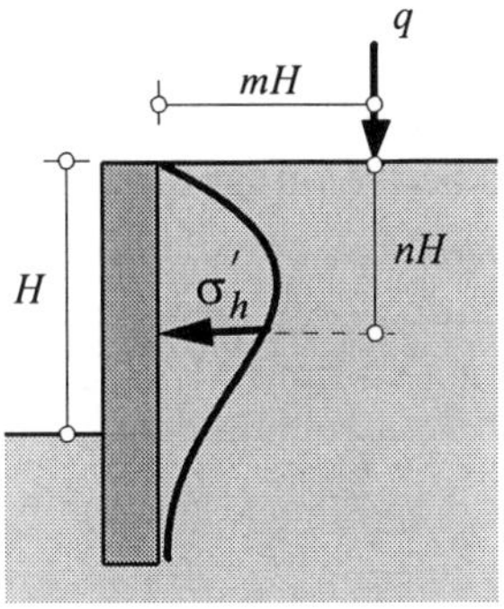

Figure 10.29: Lateral thrust due to a point load.

10.7 Effect of water pressure on the lateral thrust exerted on retaining structures

The *total* lateral thrust per metre length exerted on a retaining structure is the sum of the *effective* lateral thrust *augmented* by the hydrostatic force due to *water pressure* when applicable. It can easily be calculated from the integration of lateral pressures (including water pressure) throughout the relevant height of the structure.

Consider the wall with the dimensions shown in figure 10.30, retaining a totally submerged *isotropic homogeneous* soil characterised by a saturated unit weight γ_{sat}, a cohesion c' and an angle of shearing resistance ϕ'. The back of the wall is inclined at an angle λ with respect to the vertical (positive anticlockwise as in figure 10.21) and the backfill is horizontal.

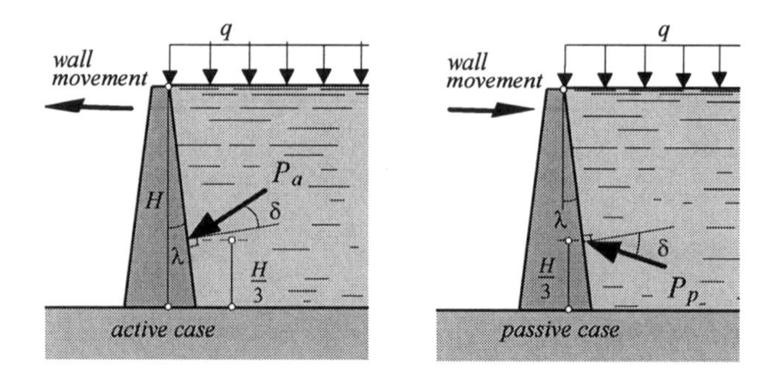

Figure 10.30: Effect of water pressure on lateral thrust.

Assuming that there is little or no seepage around the wall so that there is a *static* water table, then for a given wall friction angle δ and a uniform pressure q, the *total* lateral active thrust P_{an}, *normal* to the wall is the sum of the *effective normal thrust* P'_{an} calculated using Boussinesq theory, for instance, and the hydrostatic force per linear metre P_w due to water pressure. Thus, with reference to figure 10.31:

$$P_{an} = P'_{an} + P_w \qquad (10.54)$$

$$P'_{an} = A_1 + \int_0^{H/\cos\lambda} \left[K'_a q \cos\delta - \frac{c'}{\tan\phi}\left(1 - K'_a \cos\delta\right) \right] dz \qquad (10.55)$$

where A_1 represents the *effective thrust, normal to the wall* due to the effective weight of soil, and the integral quantity corresponds to the contribution to the *effective normal thrust* of both the soil cohesion c' and the uniform load q. Accordingly, the quantity A_1 is obtained by multiplying the *average* effective active pressure (*i.e.* one half of the effective active pressure at point O in figure 10.31) by the length $oa = H/\cos\lambda$:

$$A_1 = \frac{H^2}{2} K_a (\gamma_{sat} - \gamma_w) \frac{\cos\delta}{\cos\lambda}$$

γ_w being the unit weight of water.

Substituting for A_1 in equation 10.55 yields the expression for the *effective normal active thrust*:

$$P'_{an} = \frac{H^2}{2} K_a (\gamma_{sat} - \gamma_w) \frac{\cos\delta}{\cos\lambda}$$

$$+ \frac{H}{\cos\lambda}\left[K'_a q \cos\delta - \frac{c'}{\tan\phi}\left(1 - K'_a \cos\delta\right)\right]$$

The lateral thrust due to the water pressure is calculated in a way similar to that used to calculate A_1:

$$P_w = A_2 = \frac{\gamma_w}{2\cos\lambda} H^2$$

so that the *total normal active thrust* exerted on the wall is:

$$P_{an} = \frac{H^2}{2\cos\lambda}[\gamma_w + K_a \cos\delta\,(\gamma_{sat} - \gamma_w)]$$

$$+ \frac{H}{\cos\lambda}\left[K'_a q \cos\delta - \frac{c'}{\tan\phi}\left(1 - K'_a \cos\delta\right)\right] \quad (10.56a)$$

For the passive case, the *total normal passive thrust* is calculated in precisely the same way using the passive pressures. Consequently:

$$P_{pn} = \frac{H^2}{2\cos\lambda}[\gamma_w + K_p \cos\delta\,(\gamma_{sat} - \gamma_w)]$$

$$+ \frac{H}{\cos\lambda}\left[K'_p q \cos\delta + \frac{c'}{\tan\phi}\left(K'_p \cos\delta - 1\right)\right] \quad (10.56b)$$

Notice that, with respect to equations 10.56, K_a, K'_a, K_p and K'_p represent the coefficients of earth pressure obtained from the Boussinesq theory (Kerisel and Absi's tables).

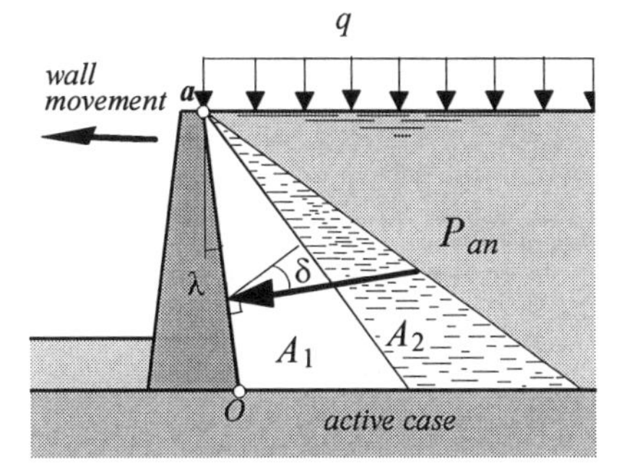

Figure 10.31: Evaluation of the effect of water pressure on the total lateral thrust.

Let us now appraise the effect of water pressure on the magnitude of the *total normal lateral thrust* by considering that the soil behind the wall is *totally dry*, in which case the thrust due to water becomes zero. Assuming the dry soil has a bulk unit weight γ, it is easy to see that both equations 10.56 reduce to:

$$P_{an} = K_a \gamma H^2 \frac{\cos\delta}{2\cos\lambda} + \frac{H}{\cos\lambda}\left[K'_a q\cos\delta - \frac{c'}{\tan\phi}\left(1 - K'_a\cos\delta\right)\right] \qquad (10.57a)$$

$$P_{pn} = K_p \gamma H^2 \frac{\cos\delta}{2\cos\lambda} + \frac{H}{\cos\lambda}\left[K'_p q\cos\delta + \frac{c'}{\tan\phi}\left(K'_p\cos\delta - 1\right)\right] \qquad (10.57b)$$

It is always useful to have a feel for these effects by using simple examples. Take the wall depicted in figure 10.30: for a height $H = 7\,m$ and a vertical back (*i.e.* $\lambda = 0$), consider first the case of a clayey backfill with the following characteristics: $c' = 0$, $\phi' = 20°$, $\gamma = 19\,kN/m^3$ and $\gamma_{sat} = 20\,kN/m^3$. Assume a wall friction angle $\delta = 2\phi'/3$, a uniform pressure $q = 20\,kN/m^2$, and the unit weight of water $\gamma_w = 10\,kN/m^3$.

Prior to any calculation of lateral thrusts, the values of the coefficients of active and passive earth pressure must be found from different tables. Hence:

- from table 10.1: ($\lambda = 0$, $\delta = 2\phi/3$, $\beta/\phi = 0$, $\phi = 20°$):

$$K_a = 0.442, \qquad K_p = 2.75$$

- using table 10.2: ($\alpha = 0$, $\delta = 2\phi'/3$, $\Omega = 90°$, $\phi = 20°$):

$$K'_a = 0.447, \qquad K'_p = 2.645$$

For a *totally submerged* backfill, the *total lateral thrust* is calculated using equation 10.56. Thus, using the superscript *w* for wet:

- the active normal thrust:

$$P_{an}^{w} = \frac{7^2}{2}[10 + 0.442 \times (20 - 10) \cos 13.3]$$

$$+ 7 \times 0.447 \times 20 \cos 13.3 = 411\, kN/m$$

- the passive normal thrust:

$$P_{pn}^{w} = \frac{7^2}{2}[10 + 2.75 \times (20 - 10) \cos 13.3]$$

$$+ 7 \times 2.645 \times 20 \cos 13.3 = 1261\, kN/m$$

For a *dry* backfill, equation 10.57 applies for the calculation of *total normal* active and passive thrusts. Using the superscript *d* for dry:

- the active normal thrust:

$$P_{an}^{d} = 19 \times \frac{7^2}{2} \times 0.442 \cos 13.3$$

$$+ 7 \times 0.447 \times 20 \cos 13.3 = 261\, kN/m$$

- the passive normal thrust:

$$P_{pn}^{d} = 19 \times \frac{7^2}{2} \times 2.75. \cos 13.3$$

$$+ 7 \times 2.645 \times 20 \cos 13.3 = 1606\, kN/m$$

Comparing these values, it follows that:

$$P_{an}^{w} \approx 1.57\, P_{an}^{d} \qquad \text{and} \qquad P_{pw}^{w} \approx 0.78\, P_{pn}^{d}$$

These simple findings indicate that, in this particular instance, if the backfill were to be totally submerged, the active thrust, normal to the wall will *increase* by more than 50%, whereas the passive thrust will *decrease* by more than 20%.

Now what about the effect of the nature of backfill? Consider a granular backfill in conjunction with the same wall with a vertical back, and assume that the soil has the following parameters:

$$\gamma = 18\,kN/m^3,\ \gamma_{sat} = 20\,kN/m^3,\ c' = 0,\ \phi' = 35^\circ.$$

Also, assume a uniform load and a wall friction angle identical to those used previously (*i.e.* $q = 20\,kN/m^2,\ \delta = 2\phi'/3$).

Prior to calculating the active and passive lateral thrusts for the limiting cases of totally dry and totally submerged backfill, the coefficients of earth pressure need to be determined:

- using table 10.1: $(\lambda = 0,\ \delta = 2\phi'/3,\ \beta/\phi' = 0,\ \phi' = 35^\circ)$:

$$K_a = 0.247, \quad K_p = 8.00$$

- from table 10.2: $(\alpha = 0,\ \alpha = 2\phi'/3,\ \Omega = 90^\circ,\ \phi' = 35^\circ)$:

$$K'_a = 0.249, \quad K'_p = 7.08$$

Starting with a *totally submerged* backfill, equations 10.56 yield the following:

- active normal thrust:

$$P^w_{an} = \frac{7^2}{2}[10 + 0.247 \times (20 - 10)\cos 23.3]$$

$$+ 7 \times 0.249 \times 20 \cos 23.3 = 333\,kN/m$$

- passive normal thrust:

$$P^w_{pn} = \frac{7^2}{2}[10 + 8 \times (20 - 10)\cos 23.3]$$

$$+ 7 \times 7.08 \times 20 \cos 23.3 = 2955\,kN/m$$

For a *dry* backfill, use equations 10.57:

- active normal thrust:

$$P_{an}^{d} = \frac{7^2}{2} \times 18 \times 0.247 \cos 23.3$$

$$+ 7 \times 0.249 \times 20 \cos 23.3 = 132\, kN/m$$

- passive normal thrust:

$$P_{pn}^{d} = \frac{7^2}{2} \times 18 \times 8 \cos 23.3$$

$$+ 7 \times 7.08 \times 20 \cos 23.3 = 4151\, kN/m$$

Whence:

$$P_{an}^{w} \approx 2.52\, P_{an}^{d} \qquad \text{and} \qquad P_{pn}^{w} \approx 0.71\, P_{pn}^{d}$$

It is clear that the water pressure has a more marked effect in the case of a granular backfill as opposed to a cohesive (clayey) backfill. It is seen that in conjunction with a granular backfill, in this instance, if the soil conditions were to change from totally dry to fully saturated, the *total active normal thrust* would more than double, while the *total passive normal thrust* (*i.e.* the resistance offered by the soil) would reduce by about one third. These simple examples show how crucial it is for the drainage behind the wall to take place, as the structural stability might be compromised due to the build-up of excessive water pressure.

10.8 Relevance of analytical solutions

The analyses undertaken thus far in relation to active and passive stresses point to the fact that, *when applicable*, the Boussinesq theory yields the most realistic predictions for either active or passive lateral thrusts exerted on retaining structures. The Rankine and Coulomb theories, on the other hand, lead to lower and upper bound solutions respectively. Moreover, the restrictive nature of Rankine's theory makes it the least advisable for use in practice. Specifically, the assumption of a wall friction angle equal to the slope of the ground surface for vertical walls (refer to equations 10.23 and 10.24) is at best arguable and at worst unrealistic (see for example Terzaghi (1936)).

There is, however, a particular case in which Rankine analysis might be useful, namely the design of cantilever walls as depicted in figure 10.32. In this case, failure of the backfill occurs along the shear planes *ad* and *de* and it can be seen from the Mohr's circle in figure 10.33 that, for a wall friction δ (equal to the slope of the ground surface in figure 10.32), the shear surface *FI* is inclined at an angle η with respect to the horizontal. The relationships relating different angles can be established in a straightforward manner. Thus:

$$\xi = \frac{\pi}{2} + \phi' - \alpha + \delta \quad \text{and} \quad \xi + 2(\eta + \alpha - \delta) = \pi$$

hence:

$$\eta = \frac{\pi}{4} - \frac{\phi'}{2} + \frac{1}{2}(\delta - \alpha) \tag{10.58}$$

where the angle α is given by equation 10.36:

$$\alpha = \sin^{-1}(\sin\delta / \sin\phi')$$

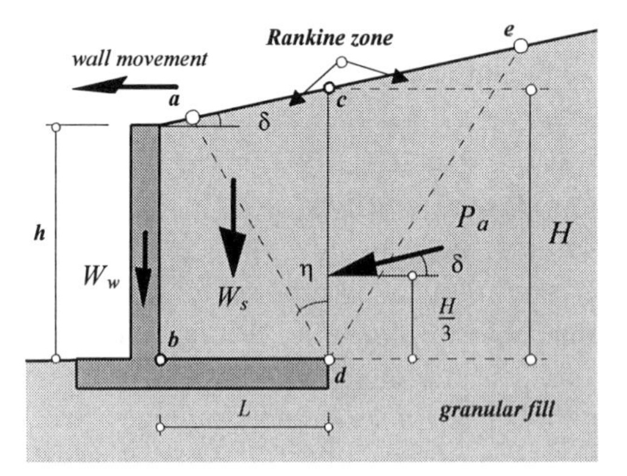

Figure 10.32: Lateral thrust exerted on cantilever walls.

Checking on equation 10.58, it can be seen that for no wall friction (*i.e.* $\delta = \alpha = 0$), the shear surface corresponds to *FC* (figure 10.33) and the angle η becomes:

$$\eta = \frac{\pi}{4} - \frac{\phi'}{2}$$

On the opposite side when $\delta = \phi'$, then $\alpha = \pi/2$ and $\eta = 0$. Consequently, with reference to figure 10.32, provided that:

$$\tan^{-1}\left(\frac{L}{h}\right) \geq \eta \qquad (10.59)$$

then the magnitude of the lateral thrust P_a is almost unaffected by wall friction along *ab,* and Rankine analysis can be applied. Notice that, in this particular instance, the lateral thrust is assumed to be acting on the *virtual* back of wall *cd*, so that the weight W_s of the volume of soil above the base of the wall *abdc* is added to the actual weight of the wall W_w for stability calculations. Also, the active pressure is assumed to be increasing linearly with depth and, consequently, the line of action of the lateral thrust is situated at a height *H*/3 as depicted in figure 10.32.

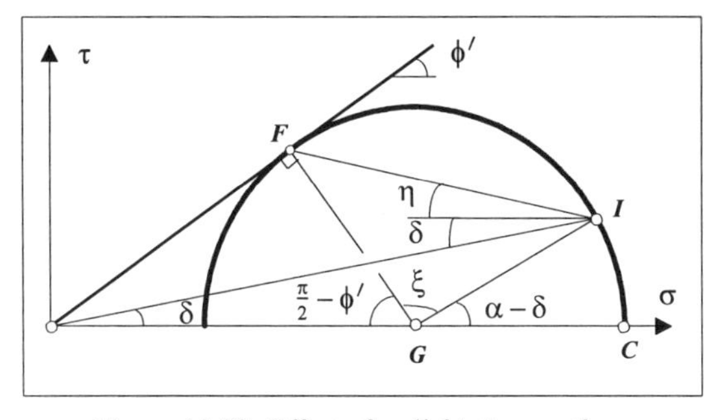

Figure 10.33: Effect of wall friction on the magnitude of total lateral thrust.

By considering the assumptions used in each of the three theories presented earlier, it quickly emerges that they can only be applied under very specific conditions relating to the type of soil (*i.e.* isotropic and homogenous) and to the geometry of the problem (*i.e.* backfill either horizontal or sloping at a constant gradient). In particular, the long term analytical solutions derived from both Rankine and Coulomb theories do not allow for any cohesion effect to be considered.

When the retained soil is *cohesive*, *partially* or *totally submerged* and *non-homogeneous* (*i.e.* multi-layered), with *uneven sloping surface* and subject to a given loading as depicted in figure 10.34, the limitations of the previous theories become apparent. Faced with such a problem, the designer has to resort to other techniques and, in this regard, finite element modelling can be a very useful tool for calculating the distribution of the effective horizontal stresses along the wall (the working of such a method is presented in chapter 14). Bear in mind that the method requires the input of several soil parameters such as elastic modulii, Poisson's ratios, elastic–plastic stress–strain relationships, and water pressures. The implications are that the output of a finite element analysis is as accurate (or inaccurate!) as the soil parameters are representative of the soil behaviour, and it is therefore advisable to assess the input and appraise the output with a critical eye.

A simpler and sufficiently accurate alternative to finite element modelling consists of using either an approximate analytical solution based on Boussinesq theory or a graphical method of solution utilising the principles of Coulomb theory.

Notwithstanding the approximate nature of either of these methods, they are deemed to be sufficiently accurate in engineering practice.

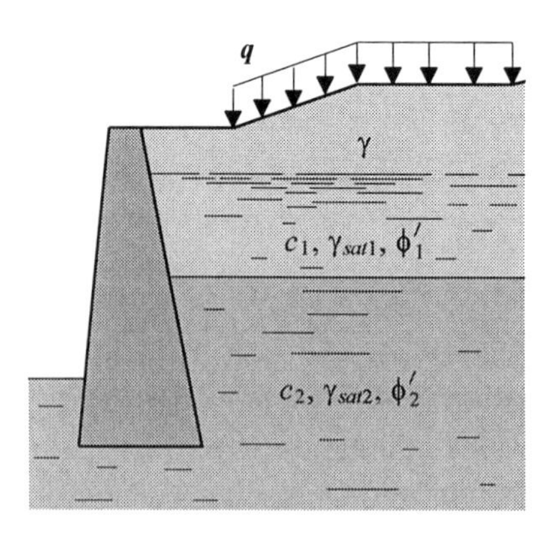

Figure 10.34: Case of non-homogeneous soils.

10.9 Boussinesq solution adapted to isotropic homogeneous backfill with uneven surface

Let us consider the two retaining walls depicted in figure 10.35, where the retained *dry, isotropic and homogeneous* soil has an uneven profile. Starting with figure 10.35(*a*), the magnitude of the lateral (active in this case) thrust P_a due to the weight of the soil corresponds to the shaded area $(A_1 + A_2 + A_3)$. This area is determined by *assuming* that, at the top of the wall, the lateral stress σ'_a is mainly due to a sloping backfill, hence the linear distribution of σ'_a from the depth *b* (corresponding to the crest of the wall) at a slope $\gamma K_a(\beta)$, where γ represents the unit weight of soil and $K_a(\beta)$ is the coefficient of active pressure corresponding to the angle β and read from Kerisel and Absi's tables. At the depth *z*, the slope then becomes $\gamma K_a(0)$ where $K_a(0)$ represents the coefficient of active pressure corresponding to an angle $\beta = 0$, and reflects the fact that below the depth *z*, the lateral stress is mainly due to a *horizontal* backfill.

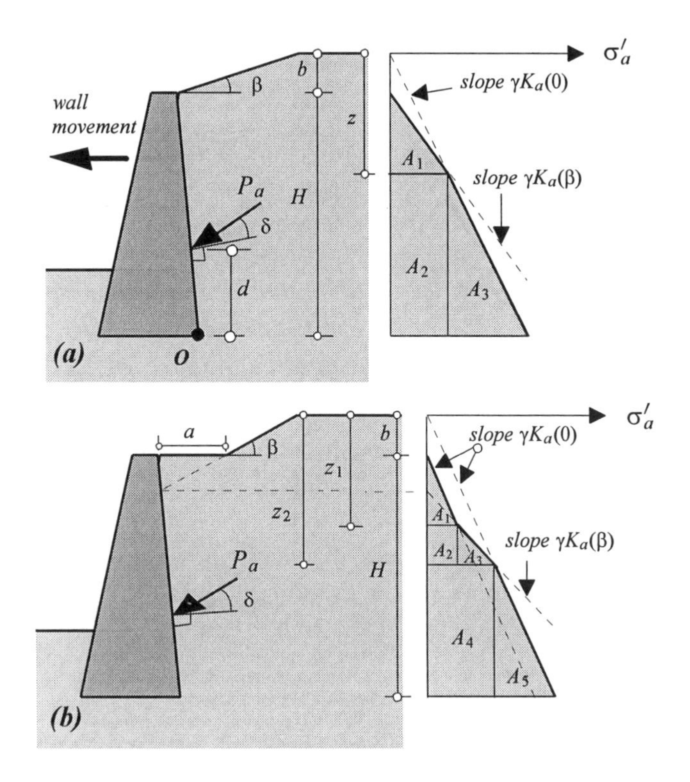

Figure 10.35: (a) Initially sloping backfill and (b) initially horizontal backfill.

Referring to figure 10.35(*a*), the lateral thrust as well as its line of action can be determined in a straightforward way if the depth z were known. But z is where the two slopes intersect, thence:

$$K_a(\beta)\gamma(z-b) = K_a(0)\gamma z$$

or

$$z = \frac{bK_a(\beta)}{K_a(\beta) - K_a(0)} \tag{10.60}$$

Using equation 10.60, the lateral thrust is thereafter calculated as follows :

$$P_a = A_1 + A_2 + A_3 \tag{10.61}$$

with

$$A_1 = (z-b)^2 \frac{\gamma}{2} K_a(\beta) = \frac{\gamma}{2} b^2 K_a(\beta) \left[\frac{K_a(0)}{K_a(\beta) - K_a(0)} \right]^2$$

$$A_2 = \gamma(H+b-z)(z-b)K_a(\beta)$$

$$A_3 = \frac{\gamma}{2}(H+b-z)^2 K_a(0)$$

The line of action of the lateral thrust is such that the distance d (refer to figure 10.35(*a*)) is found by taking moments about point O:

$$d = \frac{A_1 d_1 + A_2 d_2 + A_3 d_3}{A_1 + A_2 + A_3} \tag{10.62}$$

with $d_1 = H + \frac{2}{3}(b-z)$, $d_2 = \frac{1}{2}(H+b-z)$ and $d_3 = \frac{1}{3}(H+b-z)$.

With respect to the backfill profile in figure 10.35(*b*), the shaded area corresponding to the lateral pressure distribution has a slope $\gamma K_a(0)$ between a depth b and a depth z_1 (*i.e.* the coefficient of active pressure corresponds to a horizontal backfill). The slope then changes to $\gamma K_a(\beta)$ from z_1 to z_2, only to rejoin the initial slope $\gamma K_a(0)$ below the depth z_2. Both z_1 and z_2 can easily be calculated by writing:

$$\gamma K_a(0)(z_1 - b) = \gamma K_a(\beta)(z_1 - b - a\tan\beta)$$

and

$$z_2 \gamma K_a(0) = \gamma K_a(\beta)(z_2 - b - a\tan\beta)$$

Hence the ensuing values of z_1 and z_2:

$$z_1 = \frac{K_a(\beta)(b+a\tan\beta) - bK_a(0)}{K_a(\beta) - K_a(0)} \tag{10.63}$$

$$z_2 = \frac{K_a(\beta)(b+a\tan\beta)}{K_a(\beta) - K_a(0)} \tag{10.64}$$

The lateral thrust is in this case:

$$P_a = \sum_{k=1}^{5} A_k \tag{10.65}$$

with

$$A_1 = \gamma\frac{(z_1 - b)^2}{2}K_a(0)$$

$$A_2 = \gamma(z_1 - b)(z_2 - z_1)K_a(0)$$

$$A_3 = \gamma\frac{(z_2 - z_1)^2}{2}K_a(\beta)$$

$$A_4 = \gamma(H + b - z_2)[(z_1 - b)K_a(0) + (z_2 - z_1)K_a(\beta)]$$

$$A_5 = \gamma\frac{(H + b - z_2)}{2}K_a(0)$$

The line of action of the lateral thrust is then calculated in the same way used in conjunction with figure 10.35(*a*).

If the soil were cohesive, then the lateral thrust *normal* to the wall and due to cohesion is evaluated as detailed in the Boussinesq theory, except that the appropriate coefficients of active pressure are taken as follows:

- with respect to figure 10.35(*a*), use $K_a'(\beta)$ from the crest of the wall to the depth z, below which $K_a'(0)$ must be used.

- in the case of a backfill profile such as in figure 10.35(*b*), first use $K_a'(0)$ between the top of the wall and the depth z_1, then use $K_a'(\beta)$ between z_1 and z_2, and finally revert to $K_a'(0)$ for any depth below z_2.

Example 10.1

Consider the case of the wall depicted in figure 10.36, retaining a dry clayey isotropic and homogeneous backfill with the following characteristics: $\gamma = 18\,kN/m^3$, $\phi' = 20°$, $c' = 5\,kN/m^2$. The wall friction angle is $\delta = 2\phi'/3$ and the dimensions of the uneven backfill surface are as depicted in the figure.

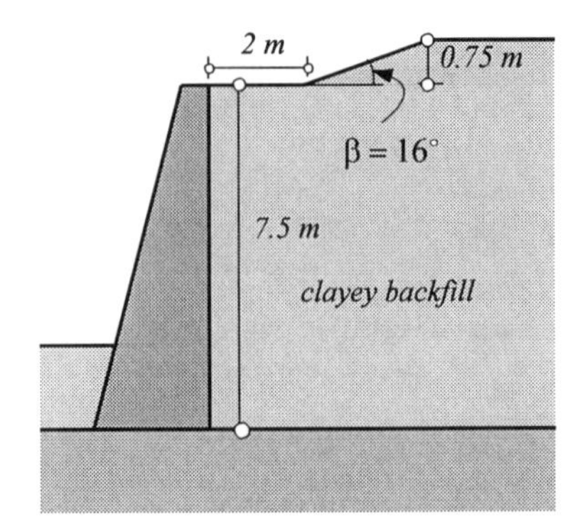

Figure 10.36: Uneven surface of backfill.

The procedure developed in conjunction with figure 10.35(*b*) applies to this problem and, accordingly, the component of the lateral thrust due to the self weight of backfill is calculated from equation 10.65 for which the following quantities need to be evaluated:

- *The coefficients of active earth pressure are read from table 10.1*:

$$\lambda = 0,\ \delta = 2\phi'/3,\ \phi' = 20°,\ \beta/\phi' = 0 \quad \Rightarrow \quad K_a(0) = 0.442$$

$$\lambda = 0,\ \delta = 2\phi'/3,\ \phi' = 20°,\ \beta/\phi' = \frac{16}{20} = 0.8 \quad \Rightarrow \quad K_a(16) = 0.628$$

- *The depths z_1 and z_2 are determined from equations 10.63 and 10.64*:

$$z_1 = \frac{0.628 \times (0.75 + 2\tan 16) - 0.75 \times 0.442}{0.628 - 0.442} = 2.69\,m$$

$$z_2 = \frac{0.628 \times (0.75 + 2\tan 16)}{0.628 - 0.442} = 4.47\,m$$

whence:

$$A_1 = \frac{18}{2}(2.69 - 0.75)^2 \times 0.442 = 15\,kN/m$$

$$A_2 = 18\,(2.69 - 0.75)(4.47 - 2.69) \times 0.442 = 27.5\,kN/m$$

$$A_3 = \frac{18}{2}(4.47 - 2.69)^2 \times 0.628 = 17.9\,kN/m$$

$$A_4 = 18 \times (7.5 + 0.75 - 4.47)$$

$$\times [(2.69 - 0.75) \times 0.442 + (4.47 - 2.69) \times 0.628] = 134.4\,kN/m$$

$$A_5 = \frac{18}{2}(7.5 + 0.75 - 4.47) \times 0.442 = 15\,kN/m$$

The lateral thrust, due to the weight of backfill is therefore:

$$P_{a1} = \sum_{k=1}^{5} A_k = (15 + 27.5 + 17.9 + 134.4 + 15) = 209.8\,kN/m$$

This component is applied at a distance d_1 (from the base of the wall) which can easily be found by taking the moments about the point O (see figure 10.37). It is straightforward to show that in this case $d_1 \approx 2.73\,m$.

The second component of the lateral thrust is due to cohesion. In this regard, prior to any calculations, the appropriate coefficients of active pressure need to be found. Referring to the table on p.111 in Kerisel and Absi's book, it is seen that:

$$\alpha = \frac{2\phi'}{3} = 13.3^\circ,\ \delta = 0,\ \phi' = 20^\circ,\ \Omega = 90^\circ \quad \Rightarrow \quad K'_a(0) = 0.378$$

$$\alpha = \frac{2\phi'}{3},\ \delta = 0,\ \Omega = 90 + 16 = 106^\circ \quad \Rightarrow \quad K'_a(16) \approx 0.312$$

Hence the active pressure due to cohesion and *normal* to the back of the wall, calculated according to equation 10.44:

$$\sigma_{an} = -\frac{c'}{\tan\phi}\left(1 - K'_a \cos\delta\right)$$

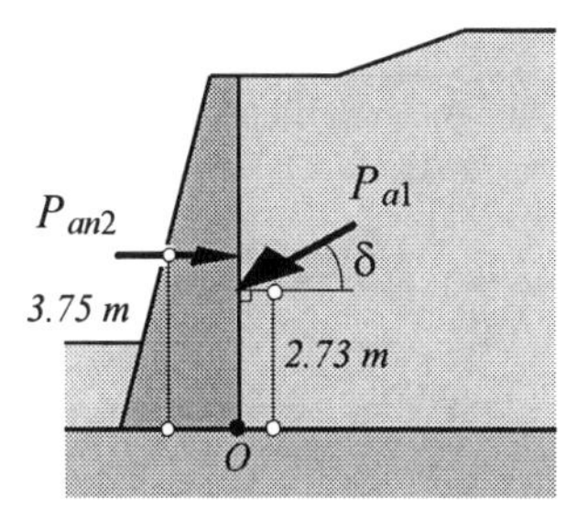

Figure 10.37: components of the lateral thrust applied to the wall.

The corresponding lateral thrust P_{an2} *normal to the wall* is found by integrating the expression of the normal active pressure throughout the height of wall. Care must be taken when choosing the appropriate value for the coefficient K'_a, and with reference to figure 10.35(*b*), P_{an2} is calculated as follows:

$$P_{an2} = \int_0^{z_1-b}\left[c'\cot\phi\left(\cos\delta\, K'_a(0) - 1\right)\right]dz$$

$$+ \int_{z_1}^{z_2}\left[c\cot\phi\left(\cos\delta\, K'_a(16) - 1\right)\right]dz$$

$$+ \int_{z_2-b}^{H}\left[c\cot\phi\left(\cos\delta\, K'_a(0) - 1\right)\right]dz$$

$$= c'\cot\phi\,(z_1 - b)\left(\cos\delta\, K'_a(0) - 1\right)$$
$$+ c'\cot\phi\,(z_2 - z_1)\left(\cos\delta\, K'_a(16) - 1\right)$$
$$+ c'\cot\phi\,(H - z_2 + b)\left(\cos\delta . K'_a(0) - 1\right)$$

knowing that: $c' = 5\,kN/m^2$, $\phi = 20°$, $\delta = 2\phi/3$, $H = 7.5\,m$, $b = 0.75\,m$, $z_1 = 2.69\,m$, $z_2 = 4.47\,m$, $K'_a(0) = 0.378$, $K'_a(16) = 0.312$, the above expression yields a value:

$$P_{an2} = -66.7\,kN/m$$

The line of action of this *normal thrust* is situated at a distance $d_2 = \frac{H}{2} = 3.75\,m$ from the base of the wall.

The *overall lateral thrust normal to the back of wall* is thus:

$$P_{an} = P_{an2} + P_{a1}\cos\delta = -66.7 + 209.8\cos\left(\frac{40}{3}\right) \approx 137.4\,kN/m$$

and the line of action of this total normal thrust is situated at a distance *d* such that:

$$d = \frac{1}{P_{an}}(d_1 P_{a1}\cos\delta + d_2 P_{an2})$$

$$= \frac{1}{137.4} \times (2.73 \times 209.8 \times \cos 13.3 - 3.75 \times 66.7) = 2.24\,m$$

10.10 Coulomb solution adapted to multi-layered soils with uneven surface

The solutions to lateral thrust problems based on any of the three theories (Boussinesq, Coulomb or Rankine) are *only* valid if the retained soil is *isotropic and homogeneous*. In practice however, there can be cases whereby a structure is needed to retain a multi-layered soil (*i.e.* non homogeneous), subjected at its surface to a complex loading as depicted in figure 10.34. Such cases *can* be solved satisfactorily using a graphical technique, based on the principles of Coulomb theory. At this stage, it must be emphasised that this technique is *only realistic* when used in conjunction with active pressure, since the actual failure surface on the passive side may differ markedly from the linear surface assumed in Coulomb theory (refer to figure 10.17).

The graphical solution to the problem corresponding to figure 10.34 is introduced in steps. First, consider a wall having a back inclined at an angle λ and retaining an *isotropic homogeneous dry cohesionless soil* ($c' = 0$) with a unit weight γ, an angle of shearing resistance ϕ' and an even sloping surface as shown in figure 10.38.

The method consists first of choosing a number of slip planes such as oa_1, oa_2, a_3,... on which the soil reactions R_1, R_2, R_3,... are drawn. At this stage, only the *directions* of the reactions are known since each one is inclined at an angle ϕ' with respect to the normal to each slip plane. Similarly, the wall reaction to the active thrust P_a, whose magnitude is not yet known, is drawn inclined at an angle δ with respect to the normal to the back of the wall, δ being the angle of wall friction. The next step consists of calculating the weight per linear metre w_1 of the block nearest to the wall oa_1a_2, then using the polygon of forces to find graphically the magnitude of the soil reaction R_1 as well as the active thrust P_{a1} in the knowledge that w_1 is *known in magnitude and in direction*, and the *directions* of both R_1 and P_{a1} are as depicted in figure 10.38.

Any appropriate scale can be used to plot the corresponding polygon of forces from which the magnitude of both R_1 and P_{a1} can be scaled. Physically, P_{a1} represents the magnitude of the active thrust that would be mobilised were the wedge oa_1a_2 to be the *actual* block that fails.

Notice that, with reference to figure 10.38, the weight per linear metre w_n of any block oa_1a_n is calculated as follows:

$$w_n = \frac{dH\gamma}{2\cos\lambda}\cos(\beta-\lambda) \tag{10.66}$$

where d represents the distance a_1a_n.

If the same procedure were repeated for the remaining blocks (*i.e.* oa_1a_3, oa_1a_4, oa_1a_5), then the magnitude of each corresponding active thrust can be determined. These values can thereafter be plotted at any convenient scale above the retained soil in a way that the slip planes are projected vertically as depicted in figure 10.38. A curve can then be drawn, leading to the *maximum* value of the active thrust as well as the location of the corresponding failure surface.

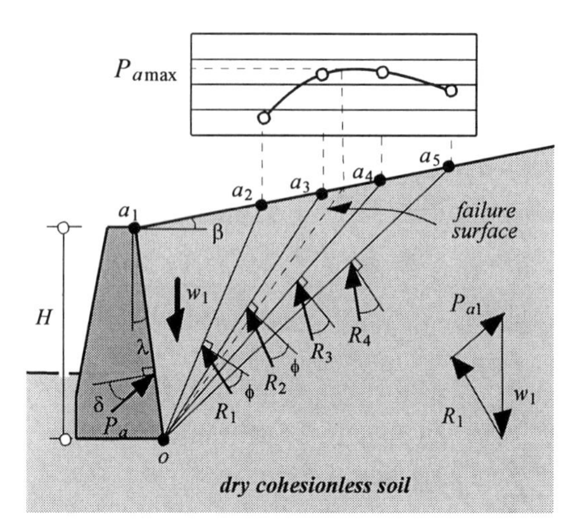

Figure 10.38: Coulomb graphical solution adapted to a backfill with uneven surface.

Let us now tackle the case of a wall retaining a *dry, homogeneous, isotropic cohesive soil* ($c' \neq 0$). The graphical solution, drawn in figure 10.39, is determined in precisely the same way as the one depicted in figure 10.38 *except* that, in this case, the polygon of forces includes the *shear forces per linear metre* T_1 and T_2 developed along the shear planes

oa_1 and oa_2 respectively. If d_1 represents the distance oa_1, then the shear force per linear metre developed along the back of wall is evaluated as follows:

$$T_1 = c_w d_1 \tag{10.67}$$

with c_w representing the cohesion at the soil–wall interface. In practice, its value can be taken as:

$$c_w \approx \frac{1}{2}c' \tag{10.68}$$

Similarly, the force T_2 is calculated using the following:

$$T_2 = c' d_2 \tag{10.69}$$

where d_2 represents the length of the shear plane (*i.e.* the distance oa_2).

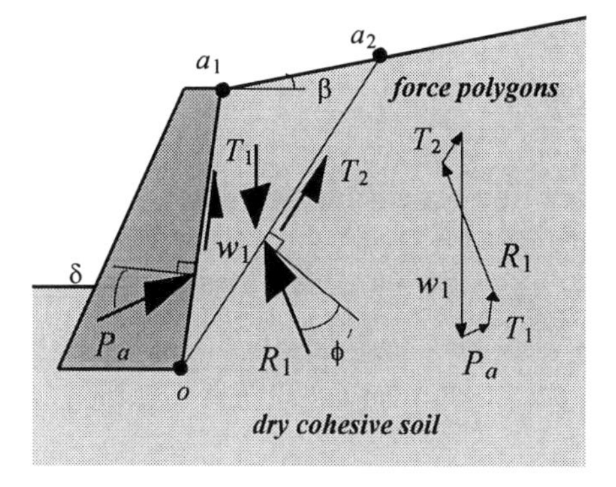

Figure 10.39: Case of a sloping cohesive backfill.

The next instance corresponds to the general case of a wall retaining a *totally submerged cohesive soil* (figure 10.40). The groundwater level is assumed to be *static* and the soil is *isotropic* and *homogeneous*. Once more, the same principles as in the graphical method used previously apply. However, care must be taken as regards the following.

- The *total* weight w_1 is calculated using the soil saturated unit weight. Any surface loading q applied between a_1 and a_2 must be *added* to w_1. Notice that q has a unit of a *force per linear metre.*

- The *average* water pressure is the same along oa_1 and oa_2 and has a value $u = h_w\gamma_w/2$, with γ_w being the unit weight of water.

Accordingly, the thrust due to the water pressure along these planes (see figure 10.40) are such that:

$$U_1 = ud_1 = \frac{1}{2} h_w d_1 \gamma_w \tag{10.70}$$

$$U_2 = ud_2 = \frac{1}{2} h_w d_2 \gamma_w \tag{10.71}$$

where d_1 and d_2 represent the distances oa_1 and oa_2 respectively.

- The cohesion force per linear metre T_2 is calculated according to equation 10.69, however, when calculating the quantity T_1, equation 10.68 is in all probability no longer realistic because the *lubricating* effect of water would result in a substantial reduction in the value of c_w. Since the estimation of the *actual* value of soil cohesion c' is often fraught with difficulties, it is advisable in this case to discard the thrust T_1.

- The thrust P'_a and the force R'_1 determined from the force polygons are *effective forces per linear metre.*

- The porewater pressures being *identical* along the planes oa_1 and oa_2, the *horizontal components* of the thrusts U_1 and U_2 are equal and opposite.

- The *total active thrust* applied to the wall is the (vectorial) sum of the quantities P'_a and U_1.

Note that all previous remarks and calculations regarding the thrust due to water pressure are no longer applicable under steady state seepage conditions.

In such a case, the porewater pressure profile can be evaluated either analytically (using Mandel theory presented in section 3.5 for instance) or graphically (from a flownet), then integrated along the slip planes thus leading to the appropriate thrusts. The details of such calculations will be presented in chapter 11.

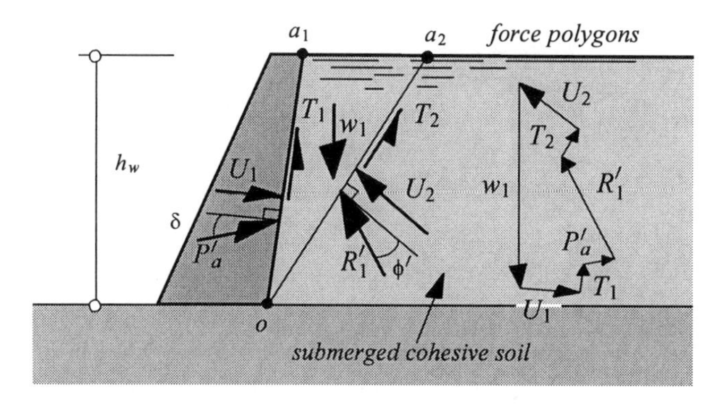

Figure 10.40: Case of a totally submerged cohesive backfill.

Let us now make use of the knowledge accumulated from the previous cases to attempt a graphical solution to the more complex problem of figure 10.34. Since we know how to take into account both cohesion and water pressure (refer to figure 10.40), the graphical solution sought can be made clearer if we assumed that both c_1 and c_2 are zero, and that both soil layers are dry (no water pressure) as illustrated in figure 10.41. The wall is characterised by the friction angles δ_1 across the top layer and δ_2 throughout the bottom layer. For the sake of clarity, the procedure will be introduced in steps as follows.

Step 1

The first step of the graphical method is to consider the top layer *on its own* and find the maximum thrust P_{a1} using precisely the same technique as that developed in conjunction with figure 10.34, bearing in mind that the portion of the external load q applied at the surface of the block (*i.e.* between a_2 and a_3) in figure 10.41 should be added to the weight w_1 as a force per linear metre.

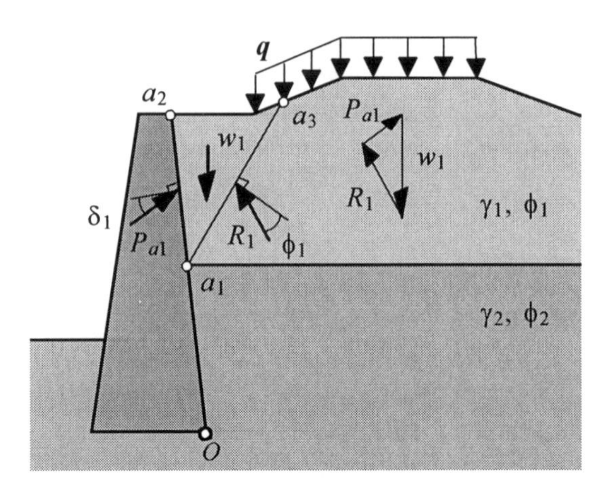

Figure 10.41: Lateral thrust due to the top layer.

Step 2

A failure plane is assumed across the bottom layer, originating at the toe of the wall (point O) as illustrated in figure 10.42. This plane meets the top layer at point a_4 and is then projected vertically, intersecting the surface at point a_5. At this stage, the block $a_4a_5a_6$ is considered on its own and the same technique as that used in step 1 is applied to determine the *maximum* thrust F_a that would develop along a_4a_5 if the block were to fail on its own. Once more, remember to include the external load as a force per linear meter when calculating the weight w_2 of the block.

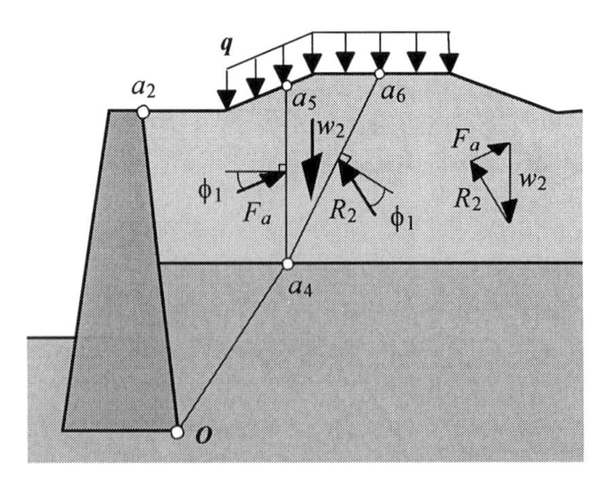

Figure 10.42: Selection of a failure plane across the bottom layer.

Step 3

Now that the magnitude, as well as the direction, of both thrusts P_{a1} and F_a are known, the next phase consists of finding the contribution to the active thrust exerted on the wall by the bottom layer of soil. In this respect, the force polygons depicted in figure 10.43 yield the component P_{a2}. Notice that the *total* weight W used in conjunction with the force polygons corresponds to the weight of the block $oa_2a_5a_4$, augmented by the appropriate external load applied at the surface of the block. Naturally, the different unit weights γ_1 and γ_2 of the two layers must be used accordingly.

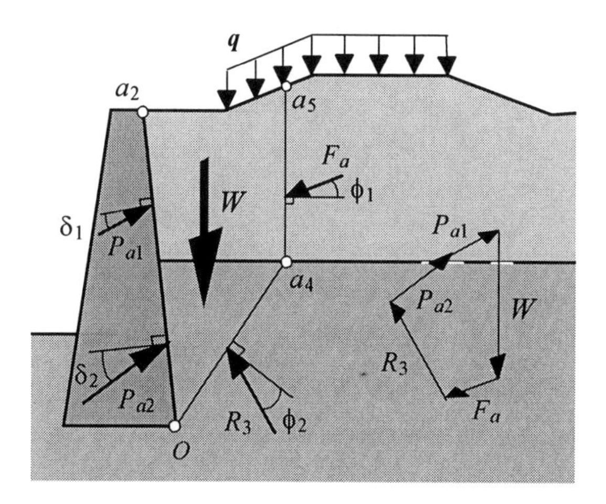

Figure 10.43: Contribution of the bottom layer to the total lateral thrust.

Step 4

The procedure is then repeated from step 2 by assuming different failure planes oa_6, oa_8, oa_{10} across the bottom layer as depicted in figure 10.44, and calculating the magnitude of the respective thrusts P_{a2} as detailed in step 3. These values are thereafter plotted at any convenient scale above the retained soil as illustrated in the figure, and a curve is plotted across the different points leading to the *maximum* value of the thrust $P_{a2\,max}$ as well as to the location of the critical failure plane in the bottom layer. The point of intersection of this failure plane with the top layer is projected vertically and step 2 is thenceforth used to determine the location of the critical failure plane in the top layer.

The *total active thrust* P_a exerted on the wall can now be calculated and corresponds to the vectorial sum of P_{a1} and $P_{a2\,max}$, so that the total thrust *normal to the wall* is:

$$P_{an} = P_{a1}\cos\delta_1 + P_{a2\,max}\cos\delta_2 \tag{10.72}$$

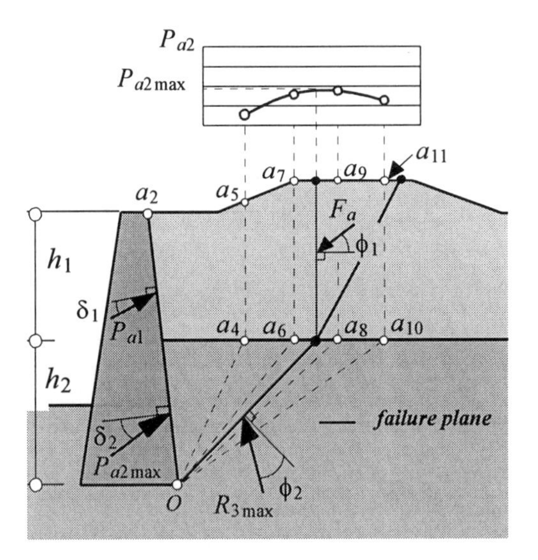

Figure 10.44: Determination of the optimum failure plane across both layers.

The line of action of the total thrust is difficult to determine with certainty. However, its position can be *approximated* by assuming that the thrusts P_{a1} and $P_{a2\,max}$ act at a distance $h_2 + \frac{h_1}{3}$ and $\frac{h_2}{3}$ respectively from the base of the wall. Whence, the total thrust acts at a distance d from the base of the wall such that:

$$d = \frac{(h_2 + \frac{h_1}{3})P_{a1}\cos\delta_1 + \frac{h_2}{3}P_{a2\,max}}{P_{an}} \tag{10.73}$$

Note that, with regard to figure 10.44, the external load q was omitted for the sake of clarity. In addition, the thrusts P_{a1} and $P_{a2\max}$ represent the reaction of the wall and therefore the *actual* thrusts are equal and opposite to P_{a1} and $P_{a2\max}$ respectively.

The graphical solution in relation to multi-layered soils appears to be arduous and time consuming, especially when cohesion and water pressures are to be included. However the method can be programmed reasonably well.

10.11 Critical appraisal of engineering practice

There seems to be a certain degree of confusion (lack of critical analysis perhaps?) when it comes to choosing the appropriate method of calculations related to lateral pressures. Most of the examples presented so far were tackled with the view to simulating the real problems an engineer would most likely face in practice. Yet, any practising geotechnical engineer will recognise that making a decision with regard to the level of the water table and the nature of flow, for example, is far from being a straightforward matter. Similarly, the nature of the retained soil with respect to isotropy can be difficult to determine with sufficient accuracy.

However, assuming that these points of detail (important though they may be!) are solved, there is a problem specific to soil homogeneity, which appears to be ignored (even trivialised). This problem is depicted in figure 10.45 which shows a wall retaining a partially submerged layer of an otherwise *isotropic homogeneous* soil. The temptation to apply an analysis of the Boussinesq type is somewhat irresistible. However, the error engendered by a mechanical decision in terms of the evaluation of the *actual* lateral thrust can be substantial. The reason for this is that Boussinesq analysis can only be applied in conjunction with *isotropic homogenous* soils, and a partially submerged soil is *no longer homogeneous* since the effective lateral pressures are calculated using a bulk unit weight γ above the water table and an effective unit weight $(\gamma_{sat} - \gamma_w)$ below the water table. This amounts *precisely* to having two different layers of soil, thus making the problem become one of a muti-layer nature that can be solved using, for instance, the graphical technique developed earlier.

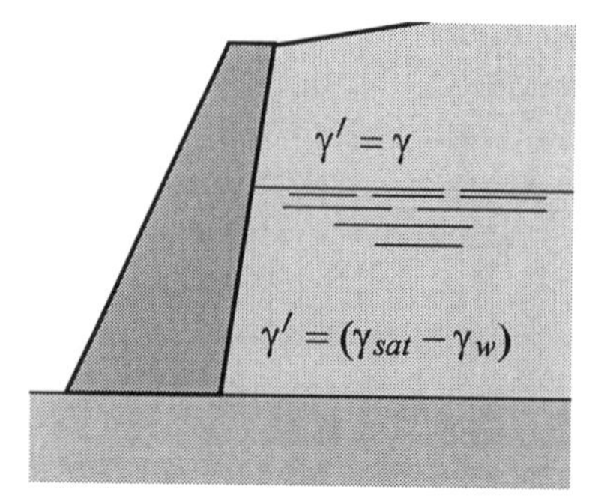

Figure 10.45: Effect of non-homogeneity on the method of analysis of lateral thrusts.

Consider the wall depicted in figure 10.46 with a rough back inclined at an angle λ, retaining an *à priori homogeneous isotropic* granular fill having an angle of shearing resistance ϕ', a bulk unit weight γ (which applies above water level) and a saturated unit weight γ_{sat}. The fill material is sloping at an angle β and the water table (assumed to be static) is situated at a depth z_w from the crest of the wall whose height is H.

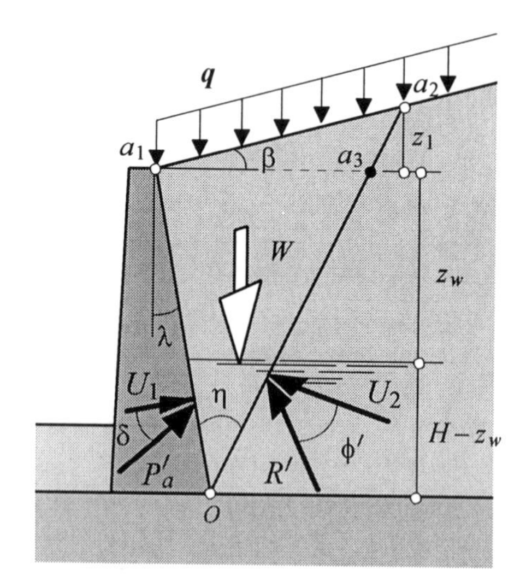

Figure 10.46: Effect of water pressure on the total lateral thrust in the case of a rough inclined wall.

Let us now examine the effects that a *partially submerged* backfill has on the procedure of analysis, starting with a Coulomb type method. Since Coulomb analysis can be applied either analytically or graphically, let us first attempt to establish the analytical solution.

The analytical procedure described in section 10.4 yields the *maximum* active lateral effective thrust $P'_{a\,max}$ exerted on the back of the wall (the angle of soil–wall friction being δ). This is achieved analytically by, first, establishing the expression of the active lateral effective thrust P'_a with respect to the angle η (refer to figure 10.46), then optimising it by writing $\partial P'_a/\partial\eta = 0$.

The corresponding polygon of forces, and the angles between the different components are as shown in figure 10.47.

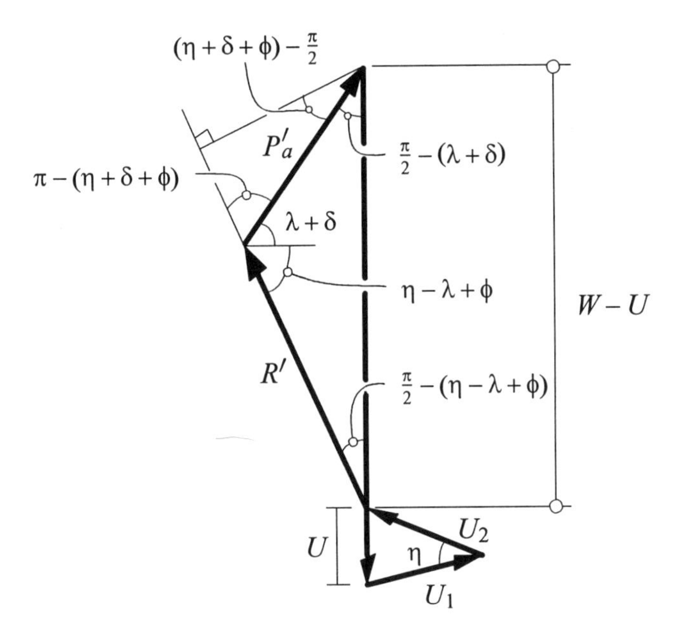

Figure 10.47: Polygon of forces corresponding to figure 10.46.

Thus, solving perpendicular to the soil reaction R', it follows that:

$$P'_a \cos\left((\eta + \phi + \delta) - \frac{\pi}{2}\right) = (W - U) \sin\left[\frac{\pi}{2} - (\eta - \lambda + \phi)\right]$$

or

$$P'_a \sin(\eta + \delta + \phi) = (W - U) \cos(\eta - \lambda + \phi)$$

Whence the effective active thrust:

$$P'_a = (W - U)\frac{\cos(\eta - \lambda + \phi)}{\sin(\eta + \delta + \phi)} \tag{10.74}$$

the quantity W in equation 10.74 corresponds to the *total weight* of the block oa_1a_2 (figure 10.46), and is calculated as follows:

$$W = q\,\frac{(H + z_1)\tan(\eta - \lambda) + H\tan\lambda}{\cos\beta} \tag{10.75}$$

$$+ \frac{\tan(\eta - \lambda) + \tan\lambda}{2}\left[\gamma\, z_w (2H - z_w) + \gamma H z_1 + \gamma_{sat}(H - z_w)^2\right]$$

As for the quantity U, it represents the vertical component of the resultant of the water thrust as can be seen from figure 10.47, and is calculated from the following expression:

$$U = \left(U_1^2 + U_2^2 - 2U_1U_2\cos\eta\right)^{1/2} \tag{10.76}$$

Once these two quantities are substituted into equation 10.74, the ensuing equation can then be optimised by writing $\partial P'_a/\partial\eta = 0$, from which the value of the angle η corresponding to the *maximum* active lateral effective thrust $P'_{a\,max}$ can be determined. This value is then substituted into equation 10.74, yielding thus the quantity $P'_{a\,max}$.

Obviously, establishing the first derivative of equation 10.74 with respect to η by hand, after having substituted for W and U from equation 10.75 and 10.76 is a tedious task in this case. In practice, such derivations can be undertaken numerically.

However, if the back of the wall were vertical (*i.e.* $\lambda = 0$) as illustrated in figure 10.48, then the problem can readily be handled analytically. The corresponding polygon of forces is, in this case, identical to the one represented in figure 10.47 except for two minor changes: (*a*) the angle λ is now *zero*, and (*b*) the water thrust U_1 becomes *horizontal*.

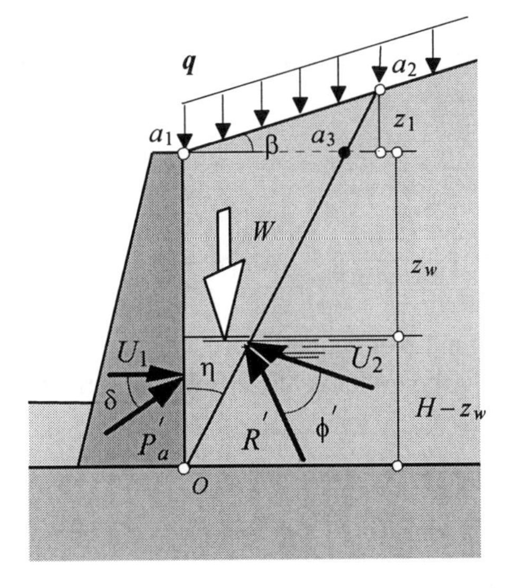

Figure 10.48: Effect of water pressure; case of a vertical wall.

Under these circumstances, equations 10.75 and 10.76 reduce to:

$$W = \tan\eta \left[q\frac{(H+z_1)}{\cos\beta} + \gamma\frac{z_w(2H-z_w) + Hz_1}{2} + \gamma_{sat}\frac{(H-z_w)^2}{2} \right] \quad (10.77)$$

$$U = U_1 \tan\eta = \gamma_w \frac{(H-z_w)^2}{2} \tan\eta \quad (10.78)$$

Substituting for these two quantities in equation 10.74, then rearranging, it follows that:

$$P'_a = C\tan\eta \frac{\cos(\eta+\phi)}{\sin(\eta+\phi+\delta)} \quad (10.79)$$

where the quantity C is expressed as follows:

$$C = \frac{q}{\cos\beta}(H+z_1) + \frac{\gamma}{2}[z_1 H + z_w(2H - z_w)]$$

$$+\frac{1}{2}(\gamma_{sat} - \gamma_w)(H - z_w)^2 \qquad (10.80)$$

At this stage, a straightforward derivation of equation 10.79 yields the condition related to the optimum angle at failure η, and it can readily be shown that:

$$\partial P'_a / \partial \eta = 0 \quad \Rightarrow$$

$$\tan(\eta+\delta+\phi)\left[\frac{1}{\cos\eta\,\sin\eta} - \tan(\eta+\phi)\right] = 1 \qquad (10.81)$$

Once the angle η is found from equation 10.81, its value is substituted into equation 10.79 which then yields the *maximum* effective lateral thrust exerted on the wall.

N.B. According to figure 10.48, the quantity z_1 *can be expressed as follows*:

$$z_1 = H\frac{\tan\beta\,\tan\eta}{(1 - \tan\beta\,\tan\eta)}$$

However, because the effects of any variation of the angle η *on* z_1 *are minimal, and to avoid any unnecessary complication of equation 10.79,* z_1 *is assumed to be independent of* η. *Accordingly, equation 10.79 yields an effective active thrust estimated to be less than 2% in excess of the actual thrust (i.e. the error is on the safe side). Therefore, for all practical purposes, the assumption made as regards* z_1 *does not, in any way, affect the accuracy of the results obtained from equations 10.77 to 10.81.*

Example 10.2

Consider the wall depicted in figure 10.49, in conjunction with the following characteristics: a height $H = 6\,m$, a granular backfill sloping at an angle $\beta = 15°$ and having $\phi' = 35°$, $\gamma = 19\,kN/m^3$, $\gamma_{sat} = 21\,kN/m^3$, a uniform pressure $q = 15\,kN/m^2$ applied at the surface, and an angle of wall friction $\delta = 24°$. Assuming that the drainage behind the wall is not functional up to a height of 3 m above the base of the wall $(i.e.\ z_w = 3\,m)$, find the potential plane of failure and the corresponding maximum total active thrust applied to the wall.

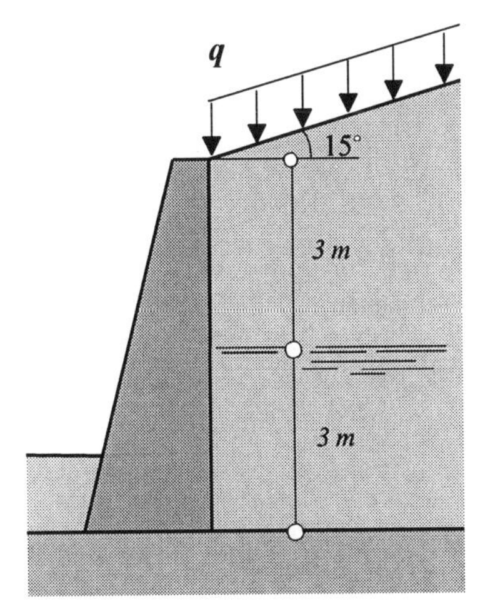

Figure 10.49: Wall dimensions and soil conditions.

First, let us apply the graphical technique to find the maximum effective active thrust as well as the plan of failure (*i.e.* the angle η).

The graphical procedure is illustrated in figure 10.50 and is of a straightforward nature. To apply the method, first *choose* different slip planes such as oa_3, $oa_4, \ldots$, then, starting from the slip plane nearest to the wall (oa_3 in this case), proceed as detailed in section 10.10.

(a) Calculate the *total weight* w_1 of the wedge oa_1a_3 using equation 10.77.

(b) Compute the magnitude of the thrust U_1 due to the water pressure from equation 10.70.

(c) Because the horizontal components of the thrusts U_1 and U_2 are equal and opposite, the magnitude of U_2 can now be determined using the relationship $U_2 = \dfrac{U_1}{\cos\eta}$.

(d) Use the polygon of forces to determine the magnitude of the soil reaction R' and the lateral effective active thrust P'_a, their direction being as depicted in figure 10.50.

(e) Select the next slip plane oa_4, then repeat the procedure from step (a).

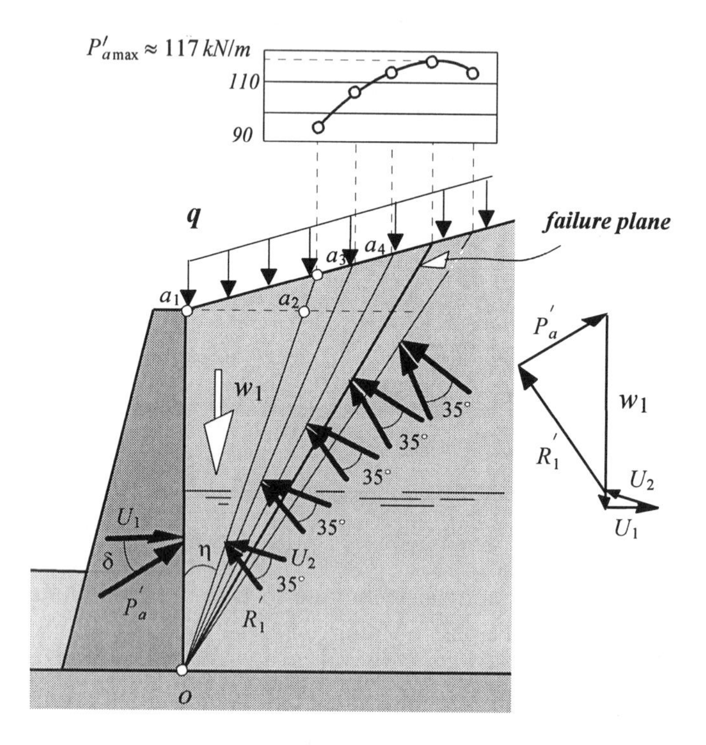

Figure 10.50: Graphical solution to example 10.2.

The computed values of P'_a are then plotted as illustrated in figure 10.50, and a curve is constructed, the peak of which represents the *maximum effective active thrust*, which in this case has a value $P'_{a\max} \approx 117\,kN/m$.

The *maximum total active thrust, normal to the wall* corresponds to the (vectorial) sum of $P'_{a\max}$ and U_1, whence:

$$P_{a\max} = U_1 + P'_{a\max}\cos\delta = 45 + 117 \times \cos 24 = 151.9\,kN/m$$

the corresponding *failure plane* being at angle $\eta \approx 30^\circ$ with respect to the vertical.

If the analytical procedure were applied instead of the graphical method, then the solution will consist of:

- finding the optimum angle of failure η from equation 10.81:

$$\tan(\eta + 24 + 35)\left[\frac{1}{\cos\eta\ \sin\eta} - \tan(\eta + 35)\right] = 1 \Rightarrow \eta \approx 31.147^\circ$$

N.B. The solution to the above equation is very sensitive to small changes in η.

- estimating the height z_1 from the following relationship (refer to figure 10.48):

$$z_1 = H\frac{\tan\beta\ \tan\eta}{1 - \tan\beta\ \tan\eta} = 6 \times \frac{\tan 15\ \tan 31.1}{1 - \tan 15\ \tan 31.1} \approx 1.157\,m$$

- calculating the constant C from equation 10.80:

$$C = \frac{15}{\cos 15}(6 + 1.157) + \frac{19}{2}[1.157 \times 6 + 3(12 - 3)]$$

$$+\frac{1}{2}(21 - 10)(6 - 3)^2 = 483.1\,kN/m$$

- computing the *maximum effective active thrust* exerted on the wall from equation 10.79:

$$P'_a = 483.1 \times \tan 31.1 \times \frac{\cos(31.1 + 35)}{\sin(31.1 + 35 + 24)} \approx 118\,kN/m$$

- finally, calculating the *total thrust normal to the wall* as follows:

$$P_{an} = U_1 + P'_a \cos\delta = 45 + 118\cos 24 = 152.8\,kN/m$$

Obviously the solutions, in terms of the active thrust P_{an} and the angle of failure η are, for all practical purposes, very similar to the ones determined from the graphical method.

Let us now apply a Boussinesq type analysis to the problem of figure 10.49, and calculate both *effective* and *total thrusts* applied to the wall.

Prior to any calculations, the relevant coefficients of active pressure are determined as follows:

- from table 10.1:

$$\lambda = 0,\ \phi = 35^\circ,\ \delta = 24^\circ \approx \frac{2}{3}\phi,\ \beta = 15^\circ \approx 0.43\phi \qquad \Rightarrow \qquad K_a \approx 0.297$$

- from Kerisel and Absi (1990), p.152:

$$\alpha = -15^\circ,\ \phi = 35^\circ,\ \delta = \frac{2}{3}\phi,\ \Omega = 105^\circ \quad \Rightarrow \quad K'_a \approx 0.308$$

The effective active thrust applied to the wall can now be estimated:

$$P'_a = 15 \times 6 \times 0.308 + \frac{1}{2} \times 3 \times 3 \times 19 \times 0.297$$

$$+\frac{3}{2}(11 \times 3 + 11 \times 6) \times 0.297 = 97.2\, kN/m$$

so that the *total thrust, normal to the wall* is in this case:

$$P_{an} = U_1 + P'_a \cos\delta$$

$$= 45 + 97.2 \cos 24 = 133.8\, kN/m$$

Compared with the value of total thrust determined previously using Coulomb's procedure, it is seen that Boussinesq analysis underestimates the total active thrust by about 14% in this instance.

Now consider the passive case depicted in figure 10.51 where a rough wall, with a back inclined at an angle λ, is retaining a *homogeneous isotropic granular fill* having an angle of shearing resistance ϕ', a bulk unit weight γ (which applies above water level) and a saturated unit weight γ_{sat}. The fill material is horizontal and the water table (*assumed to be static*) is situated at a depth z_w from the crest of the wall whose height is H.

The corresponding polygon of forces, and the angles between the different components are as shown in figure 10.52. Thus, solving perpendicular to the soil reaction R' then rearranging, it follows that:

$$P'_p \sin(\eta - \delta - \phi') = (W - U)\cos(\eta - \lambda - \phi')$$

Whence the *effective passive thrust*:

$$P'_p = (W - U)\frac{\cos(\eta - \lambda - \phi)}{\sin(\eta - \delta - \phi)} \qquad (10.82)$$

the quantity W in equation 10.82 corresponds to the *total weight* of the block oa_1a_2 (figure 10.51), and is calculated as follows:

$$W = [\tan(\eta - \lambda) + \tan\lambda]\left[qH + \tfrac{\gamma}{2} z_w(2H - z_w) + \tfrac{\gamma_{sat}}{2}(H - z_w)^2\right] \qquad (10.83)$$

As in the previous case, the quantity U represents the vertical component of the resultant of the water thrust, and is calculated from equation 10.76.

Once these two quantities are substituted into equation 10.74, the ensuing expression can then be optimised by writing $\partial P'_p/\partial\eta = 0$, from which the value of the angle η corresponding to the *maximum* passive lateral effective thrust $P'_{p\,\max}$ can be determined. This value is thereafter substituted into equation 10.74, yielding the quantity $P'_{p\,\max}$.

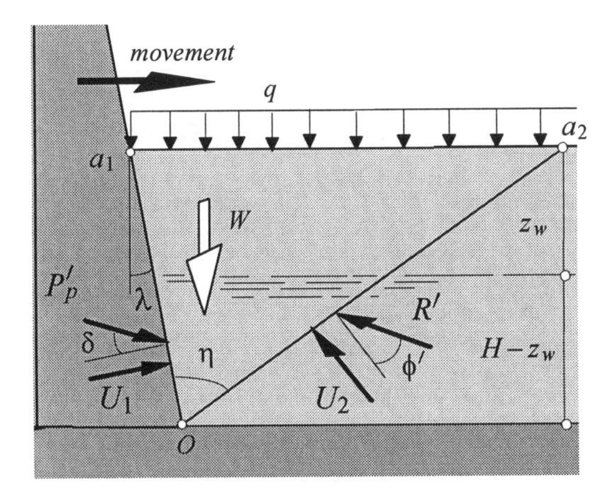

Figure 10.51: Effect of water pressure; passive case.

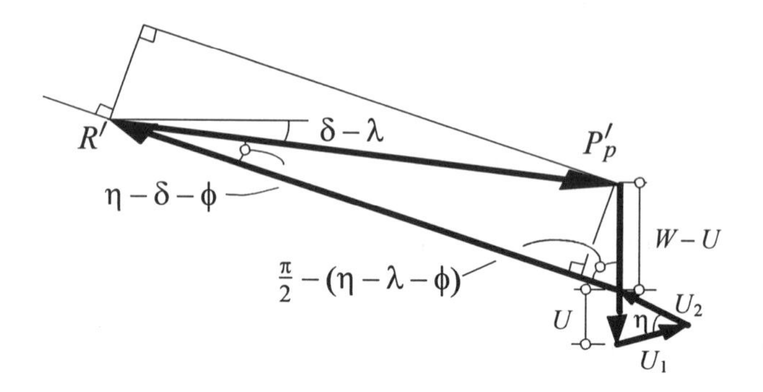

Figure 10.52: Polygon of forces corresponding to figure 10.51.

If the back of wall were vertical (*i.e.* $\lambda = 0$) as illustrated in figure 10.53, then the water thrust U_1 in figure 10.52 becomes *horizontal*.

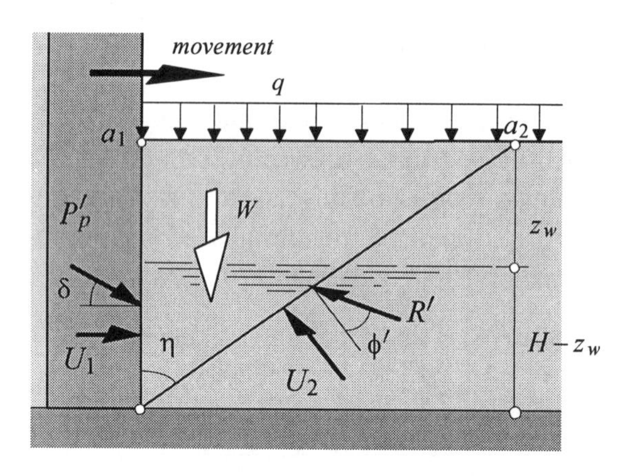

Figure 10.53: Case of a vertical wall.

Under these circumstances, both equations 10.83 and 10.76 reduce to:

$$W = \tan\eta\left[qH + \frac{\gamma}{2}\, z_w(2H - z_w) + \frac{\gamma_{sat}}{2}(H - z_w)^2\right] \tag{10.84}$$

$$U = U_1 \tan\eta = \frac{\gamma_w}{2}(H - z_w)^2 \tan\eta \tag{10.85}$$

Substituting for these two quantities into equation 10.82, then rearranging, it follows that:

$$P'_p = C \tan\eta \frac{\cos(\eta-\lambda-\phi)}{\sin(\eta-\delta-\phi)} \tag{10.86}$$

where the *constant* C represents the following quantity:

$$C = qH + \frac{\gamma}{2} z_w (2H - z_w) + \frac{1}{2}(\gamma_{sat} - \gamma_w)(H - z_w)^2 \tag{10.87}$$

At this stage, a straightforward derivation of equation 10.86 yields the condition related to the optimum angle of failure η, and it can readily be shown that:

$$\partial P'_p / \partial\eta = 0 \quad \Rightarrow$$

$$\tan(\eta - \delta - \phi)\left[\frac{1}{\cos\eta \, \sin\eta} - \tan(\eta - \phi)\right] = 1 \tag{10.88}$$

Once the angle η is found from equation 10.88, its value is substituted into equation 10.86 which then yields the *maximum effective passive thrust* exerted on the wall.

Example 10.3

Consider the wall depicted in figure 10.54, retaining a partially submerged sand similar to the one used in the previous example 10.2, that is: $\phi' = 35°$, $\gamma = 19\,kN/m^3$, $\gamma_{sat} = 21\,kN/m^3$. The angle of wall friction is $\delta = 24°$, and a uniform pressure $q = 15\,kN/m^2$ is applied at the ground surface.

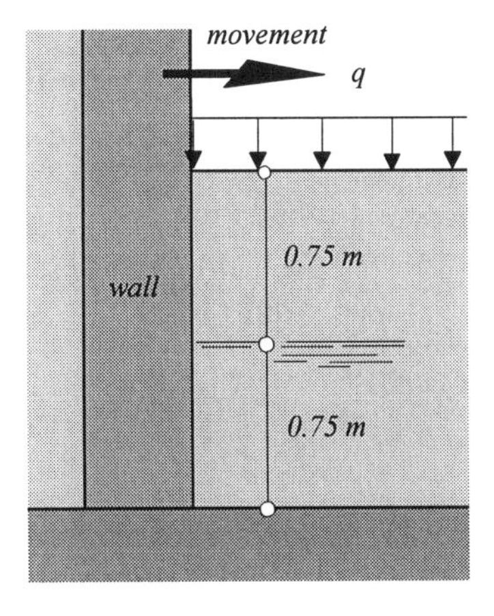

Figure 10.54: Wall dimensions and soil conditions.

Let us apply both Coulomb and Boussinesq analyses to estimate the *passive thrust* resisting the wall movement.

First, the analytical solution based on Coulomb theory consists of solving the transcendental equation 10.88 to find the optimum angle η, whence:

$$\tan(\eta - 35 - 24)\left[\frac{1}{\sin\eta\cos\eta} - \tan(\eta - 35)\right] = 1 \quad \Rightarrow \quad \eta \approx 76^\circ$$

Next, the constant C is estimated from equation 10.87:

$$C = 15 \times 1.5 + \frac{19}{2} \times 0.75 \times (3 - 0.75)$$

$$+\frac{1}{2} \times (21 - 10) \times (1.5 - 0.75)^2 = 41.6\, kN/m$$

Equation 10.86 can now be used to estimate the *effective passive thrust* developed by the soil:

$$P_p' = 41.6 \times \tan 76 \times \frac{\cos(76 - 35)}{\sin(76 - 35 - 24)} = 431\, kN/m$$

so that the *total passive thrust normal to the wall* is calculated as follows:

$$P_{pn} = 45 + 431 \cos 24 = 439\, kN/m$$

An alternative solution to the above analytical method consists of applying the graphical technique used in conjunction with the previous active case example depicted in figure 10.50. The outcome in terms of the magnitude of the total passive thrust would be similar.

As regards Boussinesq analysis, the relevant coefficients of passive pressure related to our problem are as follows:

- using table 10.1:

$$\lambda = 0,\ \phi' = 35^\circ,\ \delta = 24^\circ \approx \frac{2}{3}\phi',\ \beta = 0 \quad \Rightarrow \quad K_p = 8$$

- from table 10.2:

$$\alpha = 0,\ \phi' = 35^\circ,\ \delta = \frac{2}{3}\phi',\ \Omega = 90^\circ \quad \Rightarrow \quad K_p' = 7.09.$$

Whence the *effective passive thrust*:

$$P'_p = 15 \times 1.5 \times 7.09 + \frac{1}{2} \times (0.75 \times 0.75 \times 19 \times 8)$$

$$+ \frac{0.75}{2} \times 8 \times (11 \times 0.75 + 11 \times 1.5) = 276.5\ kN/m$$

and the *total passive thrust normal to the wall*:

$$P_{pn} = 45 + 276.5 \cos 24 = 297.6\ kN/m$$

A quick comparison of these results reinforces the point made earlier; that is Coulomb analysis yields an unrealistic and *unsafe passive thrust.*

It is therefore essential to bear in mind that, while Coulomb analysis is deemed acceptable on the active side, its use in conjunction with passive thrusts can be dangerously optimistic and therefore ineffectual. This is illustrated in figure 10.55 where the coefficient of passive pressure is calculated from both Coulomb and Boussinesq analyses for an angle of wall friction $\delta = 2\phi'/3$.

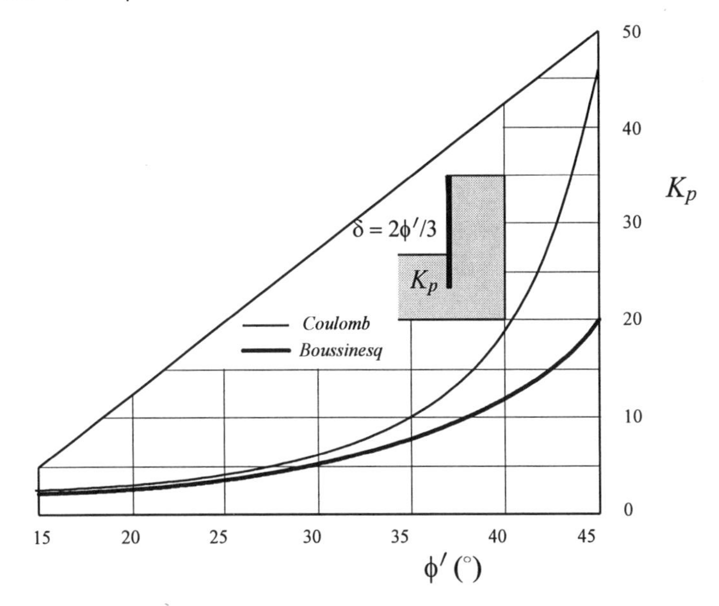

Figure 10.55: Comparison between Coulomb and Boussinesq theories for passive stress conditions.

It is clear that, as the angle ϕ' increases, both analyses yield increasingly different K_p values. Accordingly, it is strongly advisable to use a Boussinesq type analysis on the passive side in conjunction with any type of retaining structures. On the active side, either Boussinesq or Coulomb analyses can be applied. In particular, the Coulomb graphical solution can be very handy in the case of multi-layered soils subjected to complex loading conditions applied on uneven surfaces. As far as Rankine theory is concerned, its use should be restricted to the case of cantilever retaining walls for which the effects of wall friction are very limited.

10.12 Practical aspects of the design of retaining walls

Considering that, in essence, an earth retaining structure is used to provide support for steep, often vertical, cuts in different types of soils, the nature and shape of the structure can vary to suit the site conditions. Figure 10.56 depicts *some* of the wall types used in practice. Their height can vary up to a maximum of about 8 *m*, and they include:

- gravity (or semi-gravity) walls;
- cantilever walls;
- counterfort (or buttress) walls.

Besides the gravitational forces due to its self weight and the weight of soil immediately above its base, a retaining wall is invariably subjected to *lateral thrusts* caused by the already detailed *active and passive pressures*. In practice, the first steps in the design process of a retaining wall consist of choosing some of the dimensions according to the *empirical relationships* shown in figure 10.57.

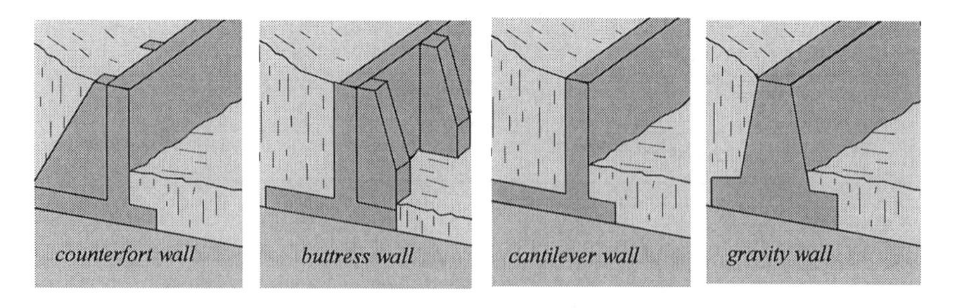

Figure 10.56: Some types of retaining walls.

This procedure, referred to as *proportioning*, allows for an iterative design process to take place whereby the dimensions are adjusted at the end of calculations if need be.

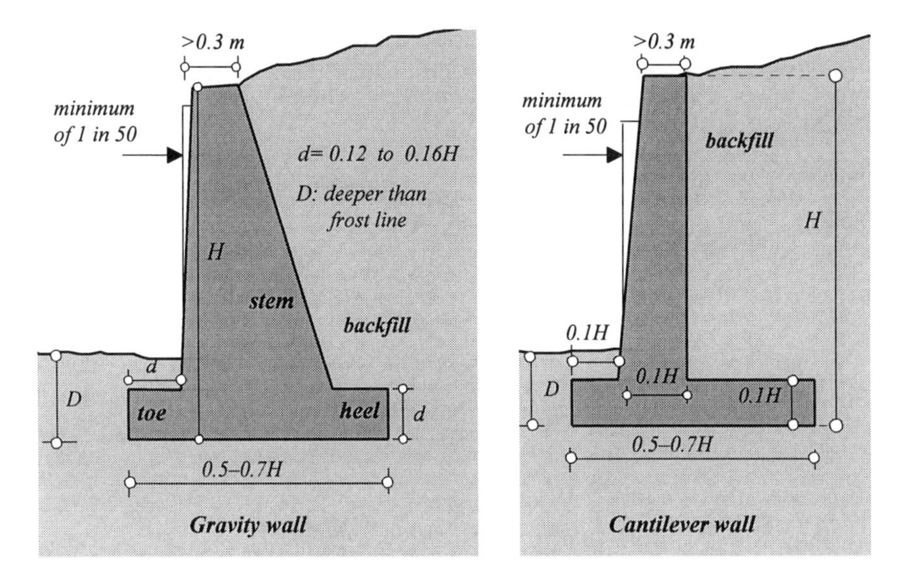

Figure 10.57: Empirical relationships related to the design of walls.

Notice that, in conjunction with figure 10.56, the counterfort or buttress slabs in the case of a cantilever wall are usually placed at a spacing of around 0.5 to 0.7*H*.

10.13 Stability criteria relating to the design of retaining walls

The designer must ensure that a retaining wall is stable *vis à vis* the following failure criteria.

(*a*) Forward sliding

The combination of *horizontal forces* may cause the wall to slide along its base the moment the sum of horizontal *sliding forces* ΣF_s (due entirely to the horizontal component of the active thrust) equals the sum of horizontal *resisting forces* ΣF_r (due to the horizontal shear force developed at the base of the wall). In practice, any component of the resisting force due to the passive thrust is usually discarded when calculating the factor of safety against sliding, the reason being that the full mobilisation of passive thrust requires a large translational movement as can be seen from figure 10.58. Accordingly, the quantity ΣF_r is reduced to the shear force F_b mobilised at the base of the wall, which is the product of the mobilised shear stress, times the area of the base.

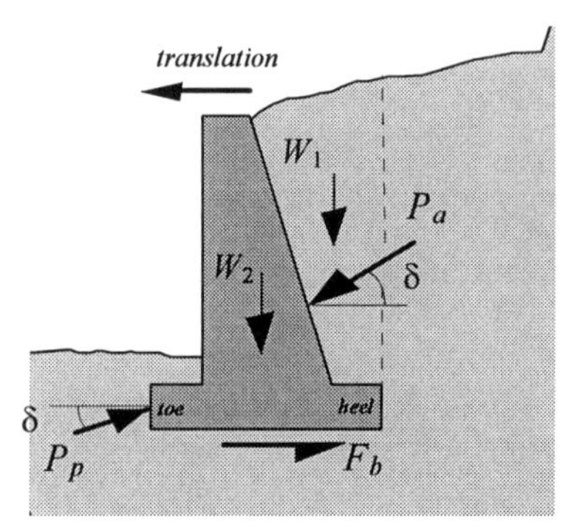

Figure 10.58: Forces resisting wall sliding.

Thus, when expressed *per unit length*, F_b is written as follows:

$$F_b = B \times 1 \times \tau_{mob} \tag{10.89}$$

where B represents the width of the wall base, and the mobilised shear stress at foundation level is:

$$\tau_{mob} = c'_{mob} + \sigma \tan \phi'_{mob} \tag{10.90}$$

But σ represents the stress, *normal to the base,* generated by the *sum of vertical forces* $\sigma = \Sigma F_v / B$.

Moreover, the mobilised angle of shearing resistance can be taken as $\phi'_{mob} \approx 2\phi'/3$; the same applies to cohesion (if any): $c'_{mob} \approx 2c'/3$. Whence the force F_b:

$$F_b = \frac{2}{3} B c' + \Sigma F_v \tan \frac{2}{3} \phi' \tag{10.91}$$

As can be seen from figure 10.58:

- the sum of vertical forces:

$$\Sigma F_v = W_1 + W_2 + P_a \sin\delta \tag{10.92}$$

- the sliding force:

$$F_s = P_a \cos\delta \tag{10.93}$$

Whence the factor of safety against sliding:

$$F = \frac{F_r}{F_s} = \frac{\frac{2}{3}Bc' + (W_1 + W_2 + P_a \sin\delta)\tan\frac{2}{3}\phi'}{P_a \cos\delta} \tag{10.94}$$

For all practical purposes, a minimum factor of safety of 1.5 is considered to be adequate. However, the designer must always bear in mind that, were a large factor of safety against sliding to be considered, the consequence might be to prevent the wall from moving forward at all, increasing, inadvertently perhaps, the possibility of generating large pressures behind the wall during compaction of backfill, which can result in structural damage. No wonder that an upper limit to the factor of safety in this case must be imposed, and in this respect, a maximum value of 2.5 is advisable. Thence for sliding:

$$1.5 \leq F = \frac{\Sigma F_r}{\Sigma F_s} \leq 2.5 \tag{10.95}$$

(*b*) **Overturning**

Another mode of failure consists of the overturning of the wall about its toe (point O in figure 10.59). Failure occurs once the *overturning moment* M_o (entirely due to the active thrust) exceeds the *resisting moments* ΣM_r developed by the self weight of the wall and any soil above its base, as well as the passive thrust. However, this mode of failure is only credible when the wall is founded on a hard clay or on rocks. If this is not the case, then not only is the centre of rotation no longer at point O, but more importantly a bearing capacity failure would in all probability occur before any significant rotation takes place because of the high concentration of stresses around the toe of the wall.

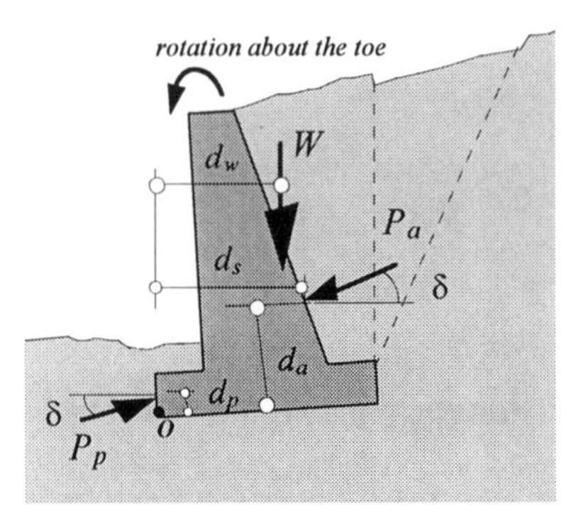

Figure 10.59: Resisting and overturning forces.

When applicable, the resisting moments can be written, according to figure 10.59, as follows:

$$\Sigma M_r = d_w W + d_p P_p \cos\delta + d_s P_a \sin\delta \tag{10.96}$$

and the overturning moment is:

$$M_o = d_a P_a \cos\delta \tag{10.97}$$

A minimum factor of safety of 2 is deemed to be adequate when dealing with this type of failure, hence:

$$F = \frac{\Sigma M_r}{M_o} \geq 2.0 \tag{10.98}$$

(*c*) **Bearing capacity failure**

The third and perhaps most serious mode of failure relates to the bearing capacity of the soil on which the wall is built as sketched in figure 10.60. This type of failure occurs when the *mobilised shear stresses* τ_{mob} due to the vertical components of the active pressure as well as the weight of the wall and any soil above its base exceed the *shear strength* $\tau_{\max}$ of the soil on which the wall is founded. Consequently, one must ensure that at no time is the bearing capacity of the soil beneath the foundation exceeded.

This can be achieved by adopting an adequate factor of safety in the design.

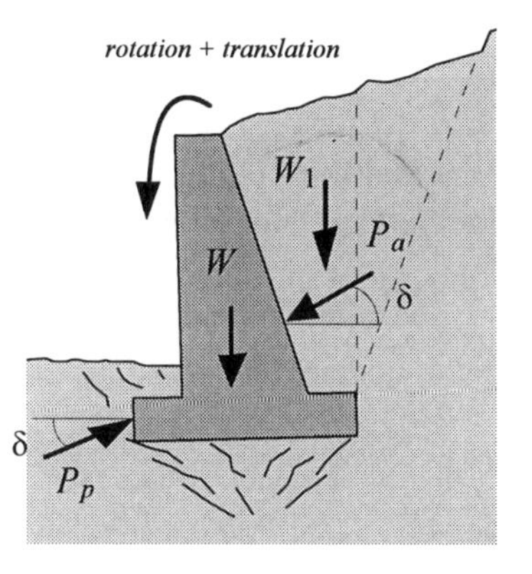

Figure 10.60: Bearing capacity failure.

Consider the retaining wall prior to the occurrence of a bearing capacity failure. The system of forces established earlier are:

- the sum of *all* vertical forces ΣF_v given by equation (10.92);
- the horizontal component of the active thrust: $P_a \cos\delta$.

The resultant eccentric force R will be inclined at an angle α with respect to the vertical as shown in figure 10.61. It is straightforward to establish that the *net moment* with respect to the toe of the wall (point O), generated by the resultant force R is:

$$M_{net} = \Sigma M_r - M_o \tag{10.99}$$

where ΣM_r and M_o are given by equations 10.96 and 10.97 respectively. Using equation 10.99, the eccentricity e can then be derived:

$$e = \frac{B}{2} - \frac{M_{net}}{\Sigma F_v} \tag{10.100}$$

Now that the eccentricity is known, the pressure distribution at the wall base can be established:

$$q = \frac{\Sigma F_v}{B \times 1} \pm \frac{My}{I} = \frac{\Sigma F_v}{B \times 1} \pm \frac{ye\Sigma F_v}{I} \tag{10.101}$$

with I being the moment of inertia *per unit length* of the base of the wall:

$$I = \frac{LB^3}{12} = \frac{B^3}{12}$$

Substituting for $y = B/2$ in equation 10.101 and rearranging, the *maximum* (at the toe) and *minimum* (at the heel) *pressures* can be calculated as follows:

$$q_{max} = \frac{\Sigma F_v}{B \times 1}\left(1 + \frac{6e}{B}\right)$$
$$q_{min} = \frac{\Sigma F_v}{B \times 1}\left(1 - \frac{6e}{B}\right) \qquad (10.102)$$

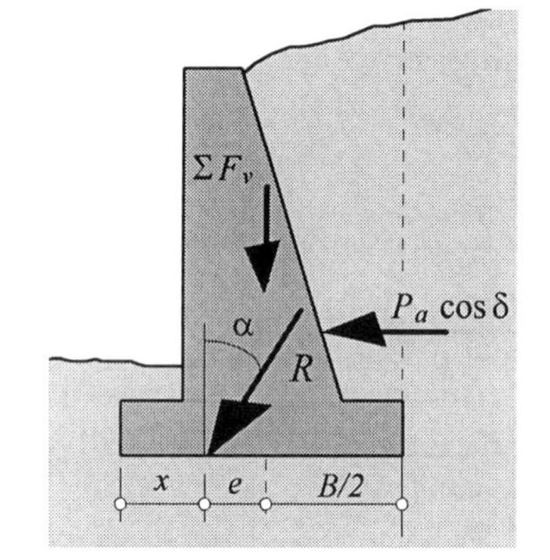

Figure 10.61: Load eccentricity at the base of wall.

Equations 10.102 indicate that the pressure at the heel becomes *negative* as soon as the eccentricity e exceeds the value $B/6$. Therefore, the well-known engineering rule stating that the line of application of the resultant force must be within the middle third of the base of the wall. This criterion can be expressed as:

$$e \le \frac{B}{6} \qquad (10.103)$$

At this stage, it is important to remember that the pressure distribution represented by equation 10.101 is only an *approximation* of the *actual* pressure distribution observed in the field.

Once the maximum pressure q_{max} is calculated, the ultimate bearing capacity of the soil q_u can be determined by applying the procedure used in conjunction with the design of shallow foundations as per chapter 8. It is a usual practice to consider a minimum factor of safety of 3 against shear failure, and accordingly:

$$F = \frac{q_u}{q_{max}} \geq 3.0 \qquad (10.104)$$

Notice, however, that the bearing capacity criterion is affected by settlement considerations. In fact, a structure such as a retaining wall with a large base is most likely to settle by a large amount before the occurrence of a bearing capacity failure and the settlement needed to herald such a failure will be well in excess of the tolerable limit. Consequently, a much larger factor of safety might be needed to keep the settlement within an acceptable range.

(*d*) Deep slip failure

This type of failure, depicted in figure 10.62, was dealt with in detail in chapter 7.

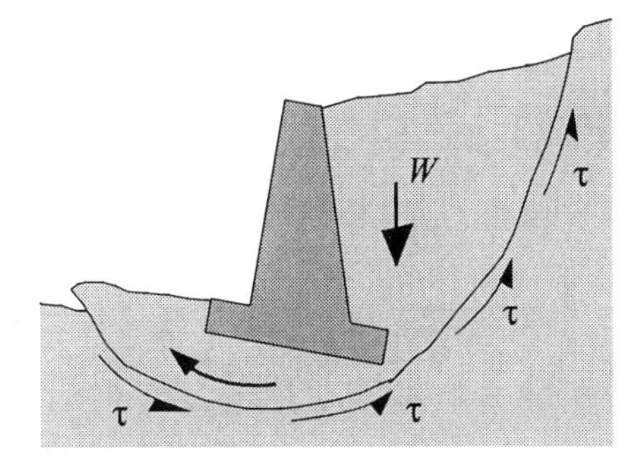

Figure 10.62: Slip failure beneath a wall.

Example 10.4

Consider once more example 10.3 and use the results obtained using the Boussinesq method to check the stability of the wall *vis à vis* sliding, overturning and bearing capacity failures. The 10 m long wall is founded on a stiff clay having an effective cohesion $c' = 10\,kN/m^2$, a bulk unit weight $\gamma = 21\,kN/m^3$ and an effective angle of shearing resistance $\phi' = 22°$, the water level being well below the base of wall. Assume the unit weight of concrete is $\gamma_c = 24\ kN/m^3$.

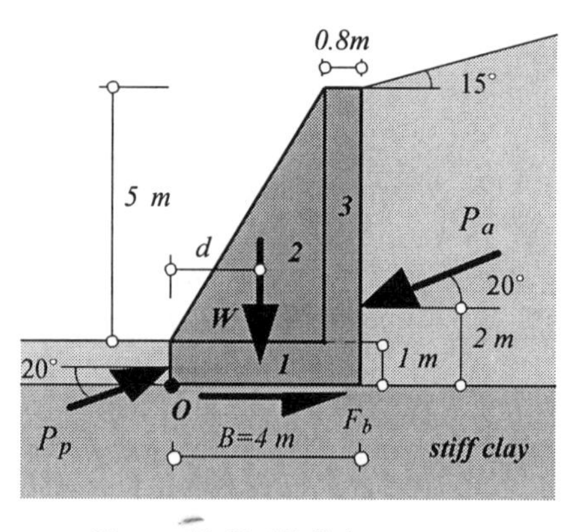

Figure 10.63: Wall dimensions.

Since the magnitude, direction and line of action of both active and passive thrusts have been already established, only the calculation of the weight *per linear metre* as well as the shear force at the base of the wall are required. The total weight W of the wall and its line of action are easily determined using the three elementary areas *1, 2* and *3* in figure 10.63:

- *area 1*: $w_1 = 1 \times 4 \times 24 = 96\ kN/m$

  moment arm from O: $d_1 = \frac{4}{2} = 2\,m$

- *area 2*: $w_2 = \frac{1}{2} \times 5 \times 3.2 \times 24 = 192\,kN/m$

  moment arm from O: $d_2 = \frac{2}{3} \times 3.2 = 2.13\,m$

- *area 3*: $w_3 = 0.8 \times 5 \times 24 = 96\,kN/m$

  moment arm from O: $d_3 = 3.2 + 0.4 = 3.6\,m$

The total weight is therefore:

$$W = w_1 + w_2 + w_3 = 384\ kN/m$$

and the corresponding moment arm from point O is:

$$d = \frac{96 \times 2 + 192 \times 2.13 + 96 \times 3.6}{384} = 1.824\,m$$

(a) ***Factor of safety against forward sliding***

The sum of vertical forces is calculated from equation 10.92:

$$\Sigma F_v = W + P_a \sin 20 = 384 + 113.4 \sin 20 = 422.8\ kN/m$$

Using equation 10.91, it follows that:

$$F_b = \frac{2}{3} \times 30 \times 4 + 422.8 \tan \frac{2 \times 22}{3} = 190.7\,kN/m$$

Since the passive thrust is ignored in the calculation of the factor of safety against translational movements, the force resisting the sliding is:

$$F_r = F_b = 190.7\,kN/m$$

The sliding force, on the other hand, is calculated from equation 10.93:

$$F_s = P_a \cos 20 = 113.4 \cos 20 = 106.5\,kN/m$$

Hence the factor of safety against sliding:

$$F = \frac{F_r}{F_s} = \frac{190.7}{106.5} = 1.79$$

this value being within the limits set by the inequality 10.95.

(b) ***Factor of safety against overturning***

The resisting as well as the overturning moments with respect to O are required to evaluate the factor of safety according to equation 10.98. As can be seen from figure 10.59, the resisting moments are exerted by the weight of the wall, the *vertical* component of the active thrust and the *horizontal* component of the passive thrust (the moment arms being respectively $d = 1.82\,m$ for the weight, $d = 4\,m$ for the vertical active thrust and $d = 0.33\,m$ for the horizontal passive thrust). Hence:

$$\Sigma M_r = 1.82W + 4P_a \sin 20 + 0.33P_p \cos 20$$

$$= 1.82 \times 384 + 4 \times 120.8 \sin 20 + 0.33 \times 47.7 \cos 20$$

$$= 878.9\,kN\,m/m$$

The overturning moment is entirely due to the *horizontal* component of the active thrust whose moment arm is $d = 2\,m$:

$$M_o = 2P_a \cos 20 = 2 \times 120.8 \cos 20 = 227\,kN\,m/m$$

Equation 10.98 then yields a factor of safety against overturning:

$$F = \frac{\Sigma M_r}{M_o} = \frac{878.9}{227} = 3.87$$

this value does not violate the criterion set by equation 10.98.

(*c*) ***Factor of safety against bearing capacity failure***

In order to calculate a factor of safety, we need to determine the shear stress distribution beneath the base of the wall caused by an inclined, eccentric resultant force. The eccentricity is estimated from equation 10.100:

$$e = \frac{B}{2} - \frac{\Sigma M_r - \Sigma M_o}{\Sigma F_v} = \frac{4}{2} - \frac{878.9 - 227}{422.8} = 0.46\,m$$

the criterion of equation 10.103 is not violated since:

$$e = 0.46\,m < \frac{B}{6} = 0.67\,m$$

The shear stress can now be calculated using equation 10.101:

$$q_{\max,\min} = \frac{\Sigma F_v}{B}\left(1 \pm \frac{6.e}{B}\right)$$

hence:

$$q_{\max} = \frac{422.8}{4}\left(1 + \frac{6 \times 0.46}{4}\right) = 178.6\ kN/m^2$$

$$q_{\min} = \frac{422.8}{4}\left(1 - \frac{6 \times 0.46}{4}\right) = 32.8\ kN/m^2$$

The angle of inclination is found from the ratio of horizontal to vertical forces, with the horizontal forces being:

$$\Sigma F_h = (P_a - P_p)\cos 20 = (120.8 - 47.7)\cos 20 = 68.7\ kN/m$$

therefore:

$$\alpha = \tan^{-1}\left(\frac{\Sigma F_h}{\Sigma F_v}\right) = \tan^{-1}\left(\frac{68.7}{422.8}\right) \approx 9^\circ$$

The ultimate bearing capacity is then calculated from equation 8.57 in which $\sigma'_o = \gamma' D$ (D being the depth of embedment of the wall). Because of the load eccentricity, the foundation width will have to be reduced to:

$$B' = B - 2e = 4 - 2 \times 0.46 = 3.08\, m$$

Thus:

$$q_u = \left(\gamma' D N_q s_q d_q I_q + c' N_c s_c d_c I_c + \gamma' \frac{B'}{2} N_\gamma s_\gamma d_\gamma I_\gamma\right)$$

The bearing capacity as well as the correction factors that will have to be applied in conjunction with the above equation are as follows.

- The bearing capacity factors are estimated from figure 8.26. Hence for $\phi' = 22^\circ$, it is seen that:

$$N_c = 18.5, \quad N_q = 8, \quad N_\gamma = 6$$

- The shape factors are calculated from equations 8.55*a–c* using the effective width B' (section 7.8):

$$s_c = 1 + \frac{3.08}{10}\frac{8}{18.5} = 1.13$$

$$s_q = 1 + \frac{3.08}{10}\tan 22 = 1.12$$

$$s_\gamma = 1 - 0.4 \times \frac{3.08}{10} = 0.88$$

- The depth factors are determined using equations 8.55*d–f*, knowing that $\xi = 1/4 = 0.25$:

$$d_c = 1 + 0.4 \times 0.25 = 1.1, \quad d_q = 1 + 0.25\tan 22\,(1 - \sin 22) = 1.05$$

$$d_\gamma = 1$$

- The inclination factors calculated from equations 8.56 are as follows:

$$I_c = I_q = \left[1 - \frac{9}{90}\right]^2 = 0.81$$

$$I_\gamma = \left[1 - \frac{9}{22}\right]^2 \approx 0.35$$

The ensuing ultimate bearing capacity is thence:

$$q_u = (21 \times 1 \times 8 \times 1.12 \times 1.05 \times 0.81)$$

$$+ (10 \times 18.5 \times 1.13 \times 1.1 \times 0.81)$$

$$+ \left(\frac{21 \times 3.08}{2} \times 6 \times 0.88 \times 1 \times 0.35\right) = 406\ kN/m^2$$

The factor of safety against shear failure is the ratio of the ultimate capacity of the clay to the maximum shear stress caused by the loading from the wall. Hence, according to equation 10.104:

$$F = \frac{\tau_{max}}{\tau_{mob}} = \frac{q_u}{q_{max}} = \frac{406}{178.6} = 2.27$$

It is seen that the criterion set in equation 10.104 is not fulfilled, and therefore in order to improve the factor of safety against shear failure, one might either increase the base (*i.e.* the width B) of the wall, or increase its embedment.

Problems

10.1 A 15 *m* long cantilever rough wall, with the dimensions indicated in figure *p*10.1, is retaining a dense silty sand sloping at an angle $\delta = 16^\circ$. The sand has an effective angle of friction $\phi' = 35^\circ$ and a bulk unit weight $\gamma = 20\, kN/m^3$.

(a) Check that the magnitude of the active lateral thrust P'_a is unaffected by the wall friction, then apply a Rankine type analysis to estimate P'_a and its line of action.

(b) If the wall were founded on a clean dense sand having an angle $\phi' = 38^\circ$, calculate the factors of safety against sliding (F_s) and bearing capacity failure (F_c) discarding the lateral passive thrust.

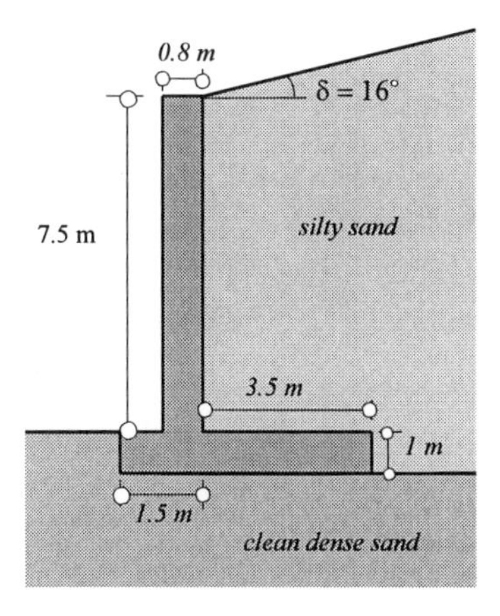

Figure p10.1

Ans: (a) $P'_a = 216.8\ kN/m$ *applied at* $d = 3.83\ m$ *from the base of the wall.*
(b) $F_s = 2$, $F_c \approx 7.6$

10.2 Estimate the active thrust applied to the wall of figure *p*10.1 using Coulomb analysis.

Ans: $P'_a \approx 139\ kN/m$

10.3 Assuming that the rough wall in figure *p*10.1 has an angle of friction $\delta = 2\phi'/3$, with $\phi' = 35°$, all other things being equal, use Boussinesq analysis to calculate the ensuing active lateral thrust.

Ans: $P'_a \approx 170\, kN/m$

10.4 A rough diaphragm wall is retaining a (dry) firm clay with an effective angle of friction $\phi' = 22.5°$ $(c' = 0)$ and a unit weight $\gamma = 20\, kN/m^3$. The angle of wall friction is $\delta = 15°$, and on the active side, the sloping ground surface is subjected to a vertical uniform pressure $q_a = 15\, kN/m^2$, the passive side having a surcharge $q_p = 25\, kN/m^2$ (refer to figure *p*10.4).

Use Boussinesq analysis and estimate the components of active and passive lateral thrusts applied *normal* to the wall.

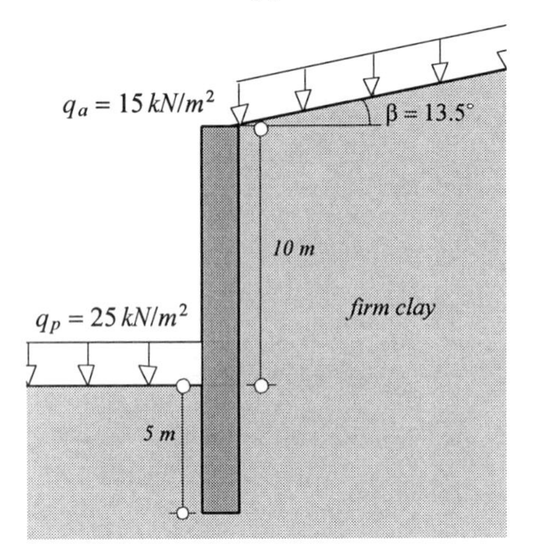

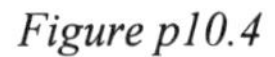
Figure p10.4

Ans: $P_{an} \approx 1182\, kN/m$, $\quad P_{pn} \approx 1593\, kN/m$

10.5 A 7 *m* high concrete wall is retaining a cohesive fill material having an effective friction angle $\phi' = 25°$, an apparent cohesion $c' = 5\, kN/m^2$, a bulk unit weight $\gamma = 19 kN/m^3$, and a saturated unit weight $\gamma_{sat} = 20.5\, kN/m^3$. The wall friction angle is $\delta = 2\phi'/3$ and, on the active side, the ground surface sloping at an angle $\beta = 13°$ as depicted in figure *p*10.5, is subjected to a uniform pressure

$q = 40\,kN/m^2$.

(*a*) The fill behind the wall being fully drained, use Boussinesq analysis to calculate the *normal* component of the total active thrust applied to the wall, as well as its line of action.

(*b*) Assume now, because of faulty drainage, the soil behind the wall is waterlogged. Estimate the new value of the *normal* component of the total active thrust, as well as the location of its point of application.

Ans: (*a*) $P_a \approx 272\,kN/m$ *applied at a distance* $d_1 = 2.65\,m$ *above the wall base.*
(*b*) $P_{an} \approx 418\,kN/m$ *applied at* $d_2 = 2.54\,m$ *above the wall base.*

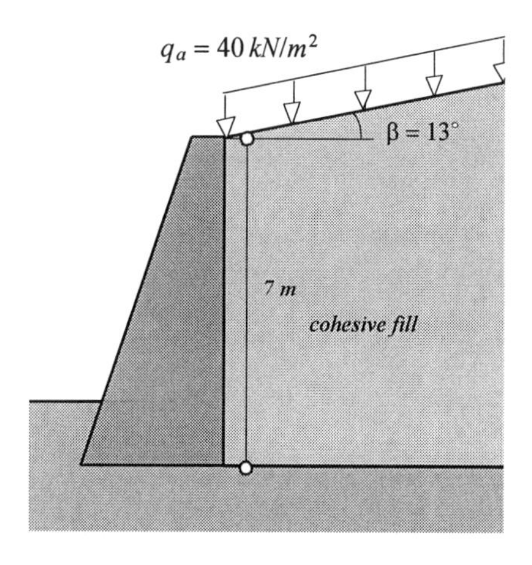

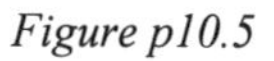
Figure p10.5

10.6 Use the Boussinesq adapted analytical solution to calculate the *normal* component of the total active thrust, as well as its point of application, in conjunction with the wall of figure *p*10.6 retaining a cohesive fill with the following properties:
$\phi' = 21°,\ c' = 5\ kN/m^2,\ \gamma = 20\ kN/m^3$.
Assume the wall friction angle is $\delta = 2\phi'/3$.

Ans: $P_{an} \approx 133.5\,kN/m$ *applied at a distance* $d = 1.68\,m$ *above the base of the wall.*

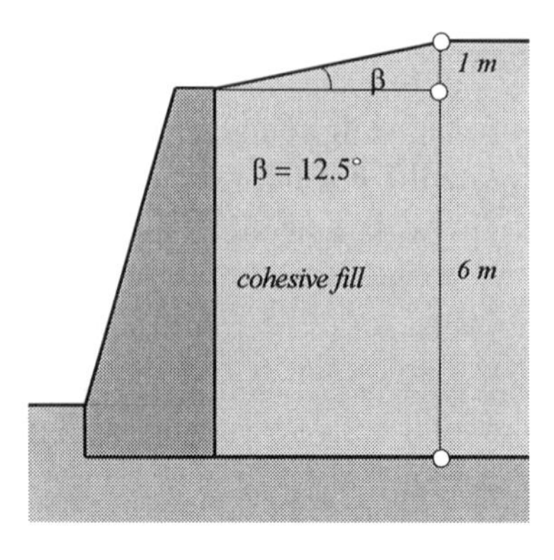

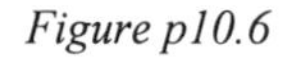
Figure p10.6

10.7 Consider the wall depicted in figure *p*10.7, retaining a normally consolidated clay having an effective angle of friction $\phi' = 22°$, a bulk unit weight $\gamma = 19\,kN/m^3$, and a saturated unit weight $\gamma_{sat} = 20.5\,kN/m^3$. The retained ground, subjected to a uniform pressure $q = 30\,kN/m^2$, is sloping at an angle $\beta = 14°$.

A faulty drain has caused the clay behind the wall to be partially submerged. Assuming the wall friction angle is $\delta = 2\phi'/3$, use the Coulomb analytical method to locate the potential failure surface, then calculate the maximum total active thrust *normal* to the wall.

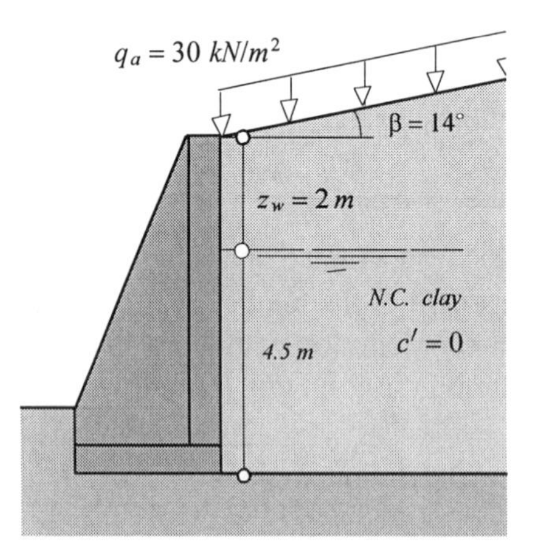

Figure p10.7

Ans: *The failure plane is situated at an angle* $\eta = 38.8^\circ$ *from the heel of the wall.*

$P_{an} \approx 363\ kN/m$

References

Brooker, E. and Ireland, H. (1965) *Earth pressure at rest related to stress history.* Canadian Geotechnical Journal, 2 (1), pp. 1–15.

Caquot, A. and Kerisel, J. (1948) *Tables for the Calculation of Passive Pressure, Active Pressure and Bearing Capacity of Foundations.* Gauthier-Villars, Paris.

Caquot, A. and Kerisel, J. (1966) *Traité de Mécanique des Sols,* 4th edn. Gauthier-Villars, Paris.

Hendren, A. J. (1963) *The behaviour of sand in one-dimensional compression.* PhD thesis, Department of Civil Engineering, University of Illinois.

Jaky, J. (1944) *The coefficient of earth pressure at rest.* Journal of the Society of Hungarian Architects and Engineers, 78 (22), pp. 355–358.

Kerisel, J. and Absi, E. (1990) *Active and Passive Earth Pressure Tables,* 3rd edn. A.A. Balkema, Rotterdam.

Mayne, P. W. and Kulhawy, F. H. (1982) $K_o - OCR$ *relationships in soil.* ASCE Journal, 108 (GT6), pp. 851–872.

Sherif, M. A., Fang, Y. S. and Sherif, R. I. (1984) K_a *and* K_o *behind rotating and non-yielding walls.* Journal of Geotechnical Engineering, ASCM, 110 (GT1), pp. 41–56.

Terzaghi, K. (1936) *The shearing resistance of saturated soils.* Proceedings of the 1st International Conference on Soil Mechanics, Harvard, pp. 54–56.

Terzaghi, K. (1954) *Anchored bulkheads.* Trans. ASCE, 119.

CHAPTER 11

Design of sheet-pile and diaphragm walls

11.1 Introduction

The retaining structures dealt with in the previous chapter related to walls of *modest height*, founded at *shallow depths* and retaining a vertical or a sloping backfill. In contrast, the execution of a deep excavation necessitates the use of a different type of retaining structure to provide support for the ensuing *high lateral thrusts* that are dependent on the depth of excavation and the type of soil in which it is executed.

Steel sheet-piles are widely used in practice as a means of support for the sides of deep excavations. Apart from their flexibility, such structures present the advantage of being widely available in different shapes, easy to drive through a wide range of soils, and simple to weld on site if the need arises for a deeper wall. In practice, a sheet-pile can be *cantilevered*, *encastré*, or *anchored*, depending on the height of the retained soil. *Cantilever sheet-piles* are of modest height and quickly become uneconomical once the height above the dredge level exceeds the nominal value of about 5 *m* (refer to figure 11.1).

If the height of the retained soil exceeds 5 *m*, then piles can be anchored in the manner depicted in the figure whereby the use of a single anchor (or a line of anchors) results in a controlled deflection and bending moments, thus allowing for a greater height of soil retention. From a design view point, a combination of the depth of embedment D and anchor size can provide support for heights up to 20 *m* as indicated in figure 11.1.

There are three ways in which sheet pile walls are installed: (*1*) piles are first driven then backfill is placed, (*2*) the soil is first dredged then the piles are driven, followed by the placement of backfill and (*3*) the soil is excavated after the piles are driven to the required depth of embedment. In all cases, it is *assumed* that active stress conditions are developed behind the wall.

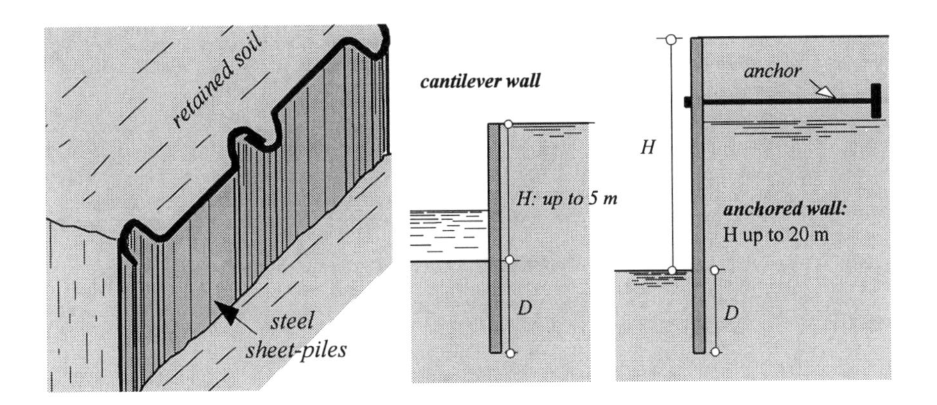

Figure 11.1: Types of sheet-pile walls.

When the use of steel sheet-piles becomes either uneconomical or impractical, an alternative solution consists of using reinforced concrete *diaphragm walls*, as well as *contiguous* or *secant pile* walls. The technical progress achieved in this area is such that reinforced concrete diaphragm walls of depths in excess of 50 m can be constructed, though in practice, the depth usually varies up to 30 m. Although there are instances in which they can be anchored or propped, most diaphragm walls are designed as cantilevers, and depending on the type of soil through which they are constructed, such walls can support up to 10 m high of retained soil as depicted in figure 11.2.

Bored pile walls, which can be constructed in any soil conditions, are of two types depending on the nature of the soil to be supported, and also on the need to control water seepage. *Contiguous pile walls* are constructed with a spacing between consecutive piles of 75 mm to 10 mm and are depicted schematically in figure 11.3. This type of wall is ideally suited for clayey soils with little or no water seepage, although, as pointed out by Fleming *et al.* (1992), there are instances in which contiguous piles can be used to support granular materials, so long as soil collapse between consecutive piles is prevented from occurring. If the structure has to be watertight, then *secant pile walls* may be used whereby the interlocking of piles can prevent water from seeping through the wall as indicated in the figure.

The height of soil retained depends primarily on the size of the piles and in this regard, a cantilever contiguous pile wall, made of piles of about 500 *mm* diameter, is capable of retaining up to 5 *m* of soil.

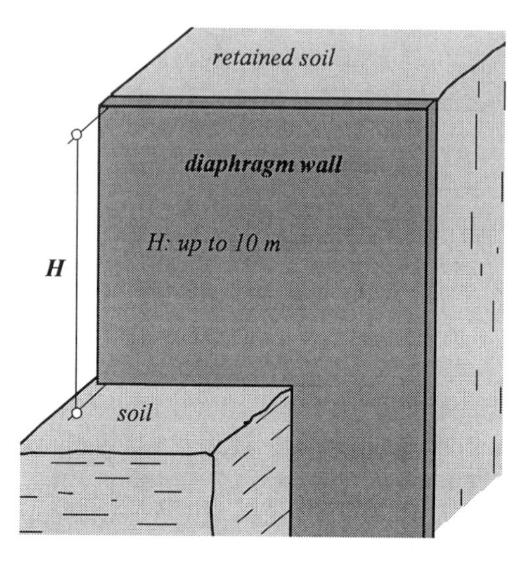

Figure 11.2: Typical diaphragm wall.

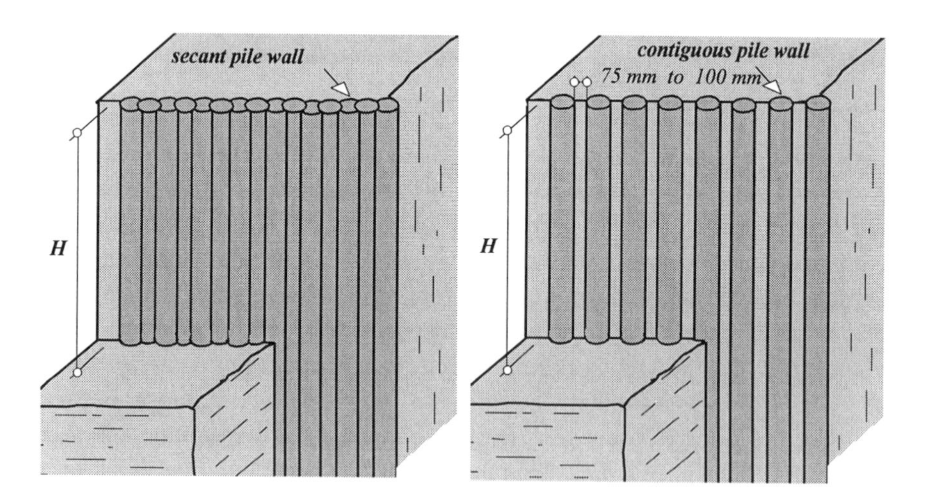

Figure 11.3: Contiguous and secant pile walls.

11.2 Methods of design

Several methods can be used to design *anchored* as well as *cantilever sheet piles*, each necessitating a specific input, yielding thus a wide range of results, some of which can be most conservative. These methods can be numerical (*i.e.* mainly based on finite element modelling), semi-empirical (such as the one based on subgrade reaction modulii), or analytical. The first two categories are outside the scope of this text and, consequently, only the two analytical methods of design, namely *the modified free earth support* and *the fixed earth support* will be presented in details in what follows.

Both methods assume that active stress conditions are fully developed behind the wall, so that the actual active stress distribution at the back of the wall can be calculated fairly accurately using Coulomb or Boussinesq theories. For this to happen, the wall is supposed to have yielded sufficiently, thus leading to the active stress distribution depicted diagrammatically in figure 11.4 (Terzaghi, 1954). Only under these circumstances can a Coulomb or a Boussinesq type analysis be applied.

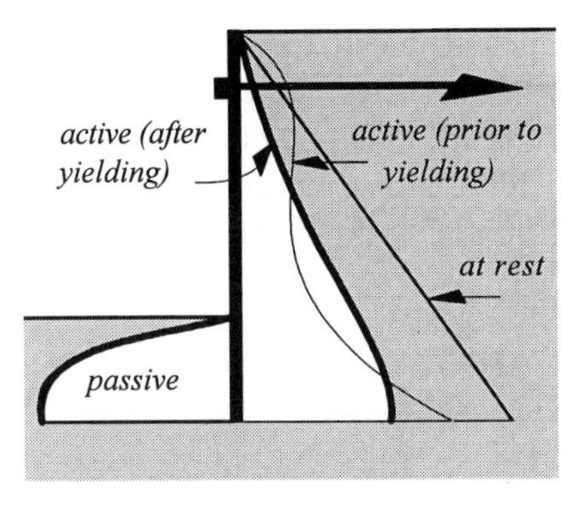

Figure 11.4: Active and passive stress profiles.

11.3 Design of anchored sheet-piles using the modified free earth support method

Anchored sheet piles are usually used when the height of the retained soil is in excess of 5 *m*. With increasing height, a series of anchors, as opposed to a single anchor, might be necessary, so as to limit the deflection and control the bending moments as depicted diagrammatically in figure 11.5.

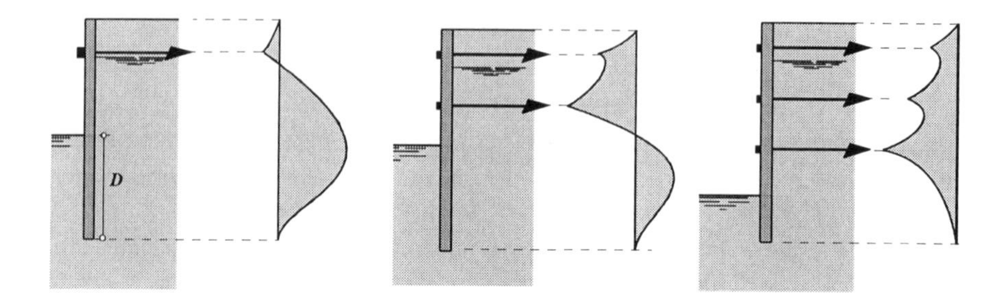

Figure 11.5: Bending moment diagrams related to different anchorage systems.

Notwithstanding (and perhaps because of) its somewhat conservative nature, the classical *free earth support method* of design is still widely used in practice. The method assumes (rather arguably) that the shear strength of the soil is mobilised throughout the depth of embedment D (refer to figure 11.5). In other words, the entire volume of soil below dredge level is presumed to be in limit equilibrium. Moreover, the *net lateral pressure* (that is the difference between active and passive *total* pressures) is *assumed* to cause the *rigid anchored pile* to rotate about the anchorage point O, yielding a distribution of displacements, shear forces, and bending moments, similar to those depicted in figure 11.6.

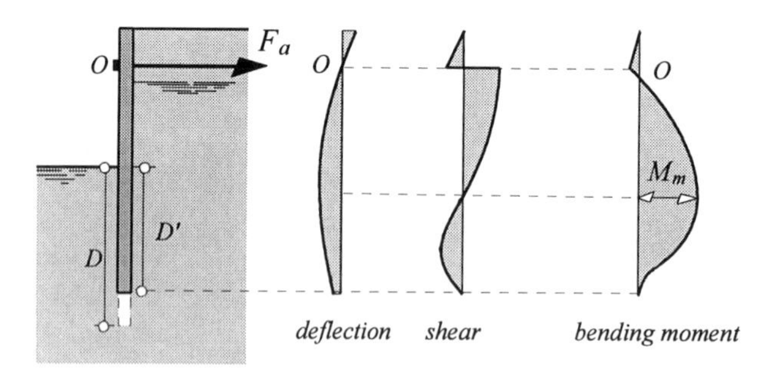

Figure 11.6: Deflection, shear force and bending moment diagrams corresponding to the free earth support method.

The *net pressure diagram* is therefore a prerequisite for the establishment of the equilibrium equations. In this respect, *either* of the two following procedures may be applied to produce the pressure diagram which is then used to determine the depth of embedment D, the anchorage force F_a, and the maximum bending moment M_m in the pile (remember that the active pressure is *assumed* to be fully mobilised at the back of the wall).

(1) The distribution of the effective active and passive thrusts along both sides of the pile is calculated using a factor of safety on the coefficient of passive earth pressure. It is a usual practice to reduce the actual coefficient of passive pressure K_p by a factor of safety $F = 2$. The net pressure diagram is thereafter determined by adding the total active and passive thrusts (*i.e.* the effective active and passive thrusts augmented by the net thrust due to water when applicable) on both sides of the pile. At that stage, the moment equilibrium equation can then be established with respect to the anchorage point O; the depth of embedment D being the only unknown contained in the equation. The anchor force F_a is then found from the equilibrium equation of horizontal forces, and the diagram of shear forces along the pile is determined. This diagram is then used to locate the depth corresponding to zero shear force, at which the maximum bending moment M_m is calculated. Apart from the case of stiff structures such as thick diaphragm walls, this procedure usually overestimates both the maximum bending moment and the anchorage force.

(2) The distribution of the effective active and passive thrusts along both sides of the pile is calculated using the limiting conditions (*i.e.* a factor of safety $F = 1$ on the coefficient of passive earth pressure). The net pressure diagram is then determined, taking care to add any net thrust due to water when applicable, and the depth of embedment D' (refer to figure 11.6) as well as the anchorage force F_a are calculated from the moment equilibrium equation and the equilibrium equation of horizontal forces respectively. The actual depth of embedment D is then obtained by multiplying the calculated depth D' by a factor of between $\sqrt{2}$ and $\sqrt{3}$ depending on the soil type [Tschebotarioff (1973) reported that a depth $D = \sqrt{2}\,D'$ is equivalent to using a factor of

safety $F \approx 1.7$ on the coefficient of passive pressure in conjunction with procedure (1) above]. The diagram of shear forces can now be found, from which the maximum bending moment M_m (occurring at the depth of zero shear force) can then be calculated.

Experimental evidence tends to confirm the *conservative* nature of the design of anchored sheet piles using the *free earth support method,* and this may result, in some instances, in overdesigned uneconomical structures. These experimental findings seem to suggest that a *pressure redistribution* takes place around the pile once yielding has occurred. In 1952, Peter Rowe embarked on a series of experimental and theoretical analyses and his findings (see also Rowe (1955, 1957)), related to model laboratory tests, led to the publication of his *moment reduction factors* in relation to sheet piles stiffness, the details of which will be explained shortly.

However, there seems to be a degree of confusion surrounding the use of the otherwise excellent analysis of Rowe, since its application is sometimes advocated in a somewhat perfunctory way in conjunction with the free earth support method . What is clear though, is that the graphs in figure 11.7 were generated from *model laboratory tests,* and that a redistribution of lateral earth pressure around a yielding wall under these circumstances cannot be extrapolated mechanically to *field conditions* if only because of the *scale effects.* Accordingly, because the reduction in the maximum bending moment of a sheet pile, due to its flexibility under field conditions, may not be as high as the one implied by Rowe's method (see for example Terzaghi (1954)), the graphs in figure 11.7 (in the case of sands) should be used with great caution. In this respect, it is *advisable* to disregard any reduction in bending moments if the sheet pile is embedded in clays as suggested by Skempton (1951).

If, on the other hand, the pile is embedded in a sand or a silty sand, then the graphs in figure 11.7 are *only useful in the case of flexible piles (such as steel sheet piles) for which the net pressure diagram is calculated according to procedure* (1) (*i.e.* using a factor of safety $F = 2$ in conjunction with the coefficient of passive pressure). In which case, only a fraction of the soil passive resistance is mobilised and, for that to happen, some displacement of the toe of the pile, relative to the anchorage point, has to occur as depicted in figure 11.6, leading forcibly to a stress redistribution behind the wall. This movement at the toe is bound to cause a reduction in the value of the maximum bending moment M_{max}, which should be, therefore, corrected to allow for a *degree* of pile flexibility. Nevertheless, it is important to bear in mind that the justification of a reduction in the value of M_{max} is entirely based on the movement at the toe of the sheet pile *after* it has been driven to the required depth. However, most of the severe stress conditions are applied during driving, and therefore the question that arises is, when applicable, of by how much M_{max} should be reduced. Too large a reduction could result in a sheet pile with a steel section that is too thin for it to be driven to the required depth. For these reasons, it is advisable to be cautious and to limit any reduction to the following.

- *For sheet piles embedded in clean sands*:

$$M_c = M_m - \frac{1}{2}(M_m - M_r) \tag{11.1}$$

- *For sheet piles embedded in dense or medium silty sands*:

$$M_c = M_m - \frac{1}{4}(M_m - M_r) \tag{11.2}$$

with

M_m : maximum bending moment determined from the *net pressure diagram*

M_r : reduced bending moment obtained from Rowe's method

M_c : corrected bending moment used to select an appropriate sheet pile section.

The reduced bending moment M_r is obtained using the value M_m and the graphs in figure 11.7 according to the following *modus operandi*.

(1) Calculate the ratio α (refer to figure 11.7) from the pile geometry, then select the appropriate curve in the figure.

(2) Use the selected curve to read the values of $\rho = H^4/EI$ corresponding to different ratios M/M_{max}.

(3) Multiply the different ratios M/M_{max} by the maximum bending moment M_m, then plot the curve $(M, \log\rho)$.

(4) Use tables to select different sheet pile sections and calculate their stiffness and maximum bending moments:

$$\rho_t = \frac{H^4}{EI} \qquad \text{and} \qquad M_t = \sigma_a \frac{I}{y}$$

σ_a being the working (*i.e.* allowable) stress for steel (for permanent work, $\sigma_a = 180\,N/mm^2$ for mild steel (S270GP, British Steel), and $\sigma_a = 230\,N/mm^2$ for high yield steel (S355GP). For temporary work, $\sigma_a = 200\,N/mm^2$ for mild steel and $\sigma_a = 260\,N/mm^2$ for high yield steel). The steel modulus of elasticity can be taken as $E = 2.1 \times 10^5\,N/mm^2$, and y represents the distance from the neutral axis to the edge of section. In Steel sheet piling manuals, the section modulii I/y are usually provided.

(5) Plot the corresponding points (M_t, $\log \rho_t$) on the same curve produced in step (3) and select the sheet pile section nearest to the curve from above. Sections relating to all points situated below the curve are inadequate (*i.e.* unsafe).

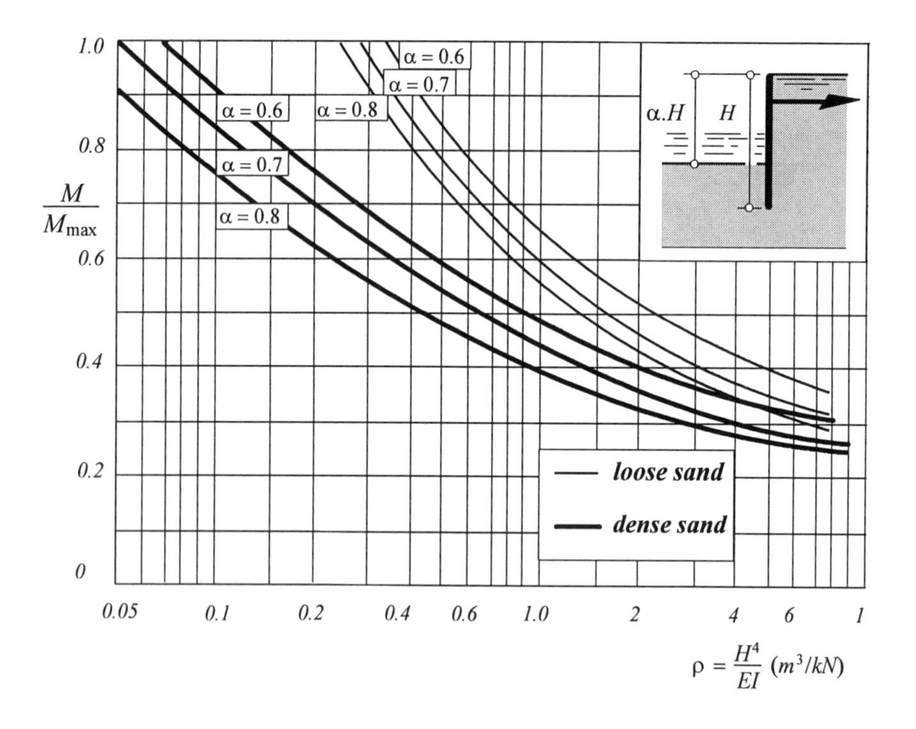

Figure 11.7: Rowe's moment reduction graphs.
(Reproduced by permission of the Institution of Civil Engineers, London.)

Example 11.1

Use the *free earth support method* to check the adequacy of the depth of embedment of the anchored sheet pile cut-off wall driven in the soil conditions depicted in figure 11.8, then determine the appropriate steel section of the pile. The isotropic and homogeneous dense sand through which the pile is driven has a bulk unit weight $\gamma = 19\, kN/m^3$, a saturated unit weight $\gamma_{sat} = 21\, kN/m^3$, and an angle of shearing resistance $\phi' = 38^\circ$; the angle of wall friction on either side of the pile being $\delta = 2\phi'/3$. The pile is assumed to have yielded sufficiently for the active stresses to be fully mobilised behind the wall.

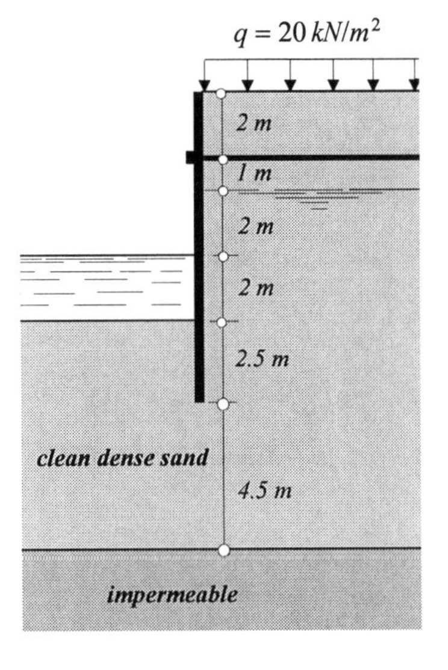

Figure 11.8: Pile dimensions and soil conditions.

First, the *net pressure diagram* needs to be established. The steady state seepage conditions around the pile will cause the effective stresses to increase on the active side and to decrease on the passive side. Hence, it is imperative to calculate the precise nature of water pressure distribution as well as the variation of the hydraulic gradient on each side of the pile. These calculations can easily be made using the Mandel method presented in detail in section 3.5.

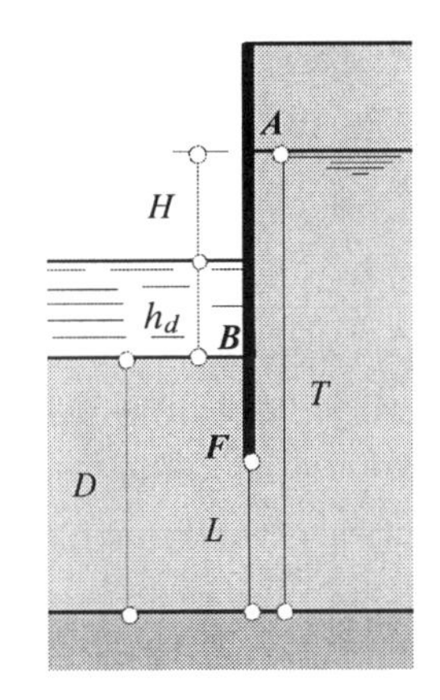

Figure 11.9: Key dimensions.

Referring to figure 11.9, it can be seen that:

$$h_u = 0, \quad h_d = 2\, m, \quad H = 2\, m, \quad L = 4.5\, m, \quad T = 11\, m, \quad D = 7\, m$$

Whence the quantity ξ (equation 3.92, section 3.5.3):

$$\xi = \frac{H}{\ln\left(\frac{T}{L} + \sqrt{\frac{T^2}{L^2} - 1}\right) + \ln\left(\frac{D}{L} + \sqrt{\frac{D^2}{L^2} - 1}\right)}$$

$$= \frac{2}{\ln\left[\frac{11}{4.5} + \sqrt{\left(\frac{11}{4.5}\right)^2 - 1}\right] + \ln\left[\frac{7}{4.5} + \sqrt{\left(\frac{7}{4.5}\right)^2 - 1}\right]} = 0.783\, m$$

and the respective velocity potentials at points *A, F* and *B* (equations 3.91):

$$\Phi_A = \xi \ln\left(\frac{T}{L} + \sqrt{\left(\frac{T}{L}\right)^2 - 1}\right)$$

$$= 0.783 \ln\left[\frac{11}{4.5} + \sqrt{\left(\frac{11}{4.5}\right)^2 - 1}\right] = 1.207\, m$$

$$\Phi_F = 0$$

$$\Phi_B = -\xi \ln\left[\frac{D}{L} + \sqrt{\left(\frac{D}{L}\right)^2 - 1}\right]$$

$$= -0.783 \ln\left[\frac{7}{4.5} + \sqrt{\left(\frac{7}{4.5}\right)^2 - 1}\right] = -0.791\, m$$

Once these three key velocity potentials are known, the variation along the pile of both the hydraulic gradient and porewater pressure can easily be calculated in the following way (refer to section 3.5.3).

- On the *active side* (behind the wall), the velocity potential at any elevation y measured from the top of the impermeable layer, and the corresponding total head are respectively:

$$\Phi_y = \xi \ln\left[\frac{y}{L} + \sqrt{\left(\frac{y}{L}\right)^2 - 1}\right] \quad \text{and} \quad h = \Phi_y - \Phi_A + h_u + T$$

The hydraulic gradient, the porewater pressure, as well as the effective unit weight of sand at any elevation y are thence calculated as follows:

$$i_y = \frac{\Phi_A - \Phi_y}{T - y}, \qquad u = \gamma_w(h - y) \qquad \text{and} \qquad \gamma_y' = \gamma_{sat} - \gamma_w(1 - i_y)$$

The ensuing results on the active side (*i.e.* from point A for which $y = 11\,m$ to point F where $y = 4.5\,m$) are presented in the following table.

$y\,(m)$	$u\,(kN/m^2)$	i_y	$\gamma'\,(kN/m^3)$
11	0	0	11.0
9	18.2	0.09	11.9
7	35.8	0.10	12.0
6	44.1	0.12	12.2
5	51.6	0.14	12.4
4.5	52.9	0.19	12.9

- On the *passive side* (*i.e.* in front of the wall), the velocity potential and the total head are determined using the following (see section 3.5.3):

$$\Phi_y = -\xi \ln\left[\frac{y}{L} + \sqrt{\left(\frac{y}{L}\right)^2 - 1}\right] \qquad \text{and} \qquad h_y = \Phi_y - \Phi_B + h_d + D$$

and the corresponding hydraulic gradient, porewater pressure and effective unit weight of sand along the pile are then calculated as follows:

$$i_y = \frac{\Phi_y - \Phi_B}{D - y}, \qquad u = \gamma_w(h - y), \qquad \gamma_y' = \gamma_{sat} - \gamma_w(1 + i_y)$$

Once applied, these relationships yield the results in the table below on the passive side (*i.e.* between points B where $y = 7\,m$ and F corresponding to $4.5\,m$).

$y\,(m)$	$u\,(kN/m^2)$	i_y	$\gamma'\,(kN/m^3)$
7	20	0	11.0
6	31.7	0.17	9.3
5	44.3	0.21	8.9
4.5	52.9	0.32	7.8

Now that the variations of the porewater pressure as well as the effective unit weight of sand are known on both sides of the pile, the distribution of both *effective active* and *passive pressures* can be calculated using the appropriate *coefficients of earth pressure*. In this respect, a Boussinesq type analysis will be applied through Kerisel and Absi's tables. Accordingly, the coefficients are determined from table 10.1, so that for $\beta = 0,\ \phi' = 38^\circ$, a linear interpolation between the values corresponding to $\phi' = 35^\circ$ and those of $\phi' = 40^\circ$ yields:

$$K_a = 0.247 - \frac{3}{5}(0.247 - 0.202) = 0.22$$

$$K_p = 8 + \frac{3}{5} \times (12 - 8) = 10.4$$

Also, the coefficient of active pressure, applicable to the uniform load q, is found from Kerisel and Absi's tables corresponding to a weightless soil. Thus, it can be seen on p.130 of Kerisel and Absi's book that:

$$\phi' = 35^\circ,\ \Omega = 90^\circ,\ \alpha = 0,\ \delta = 2\phi'/3 \approx 25^\circ \Rightarrow K'_{a1} = 0.250$$

similarly on p.173 (Kerisel and Absi, 1990):

$$\phi' = 40^\circ,\ \Omega = 90^\circ,\ \alpha = 0,\ \delta = 25^\circ \Rightarrow K'_{a2} = 0.203$$

Whence the value corresponding to $\phi' = 38^\circ$:

$$K'_a = 0.25 - \frac{3}{5} \times (0.25 - 0.203) = 0.222$$

Obviously, in this case there is little difference between the values of K_a and K'_a and, accordingly, it is reasonable to assume that, on the active side, the same coefficient $K_a = 0.22$ applies to both the self weight of soil and the uniform load q.

Both procedures (1) and (2) will be used in the following.

(1) *A factor of safety $F = 1$ is applied to K_p*

The stress values are listed in the following tables in which z refers to the depth, on each side of the pile, from the ground surface as depicted in figure 11.10.

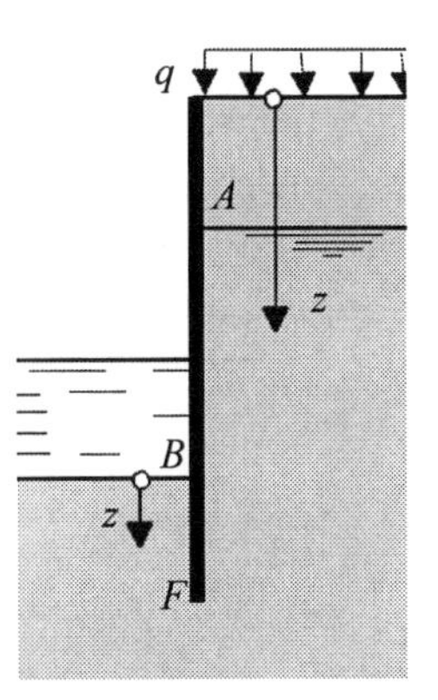

Figure 11.10: Reference system.

The active and passive pressures are calculated in the following way:

- *active effective normal pressure*: $\sigma'_a = K_a\,(\sigma'_v + q)\cos\delta$

$z\,(m)$	$\gamma'\,(kN/m^3)$	$u\,(kN/m^2)$	$\sigma'_v\,(kN/m^2)$	$\sigma'_a\,(kN/m^3)$
0	19	0	0	4
3	19	0	57	15.3
5	11.9	18.2	80.8	20
7	12	35.8	104.9	24.8
8	12.2	44.1	117.1	27.2
9	12	51.6	129.4	29.7
9.5	12.9	52.9	142.3	32.3

- *passive effective normal pressure*: $\sigma'_p = K_p\,\sigma'_v\,\cos\delta$

$z\,(m)$	$\gamma'\,(kN/m^3)$	$u\,(kN/m^2)$	$\sigma'_v\,(kN/m^2)$	$\sigma'_p\,(kN/m^3)$
0	11	20	0	0
1	9.32	31.7	9.3	87.6
2	8.88	44.3	18.2	171.1
2.5	7.84	52.9	22.1	207.9

The *total pressure* as well as the *net total pressure diagrams* can now be drawn as shown in figures 11.11(*a*) and (*b*).

The factor of safety on the passive side, corresponding to the ratio of available passive resistance of the sand to the mobilised one, can easily be calculated by writing that the moments about the anchorage point (point O in figure 11.11*b*) must balance:

$$\Sigma M_{(O)} = 0$$

The detailed calculations are presented in the following table, and reference should be made to the *net pressure diagram* of figure 11.11(*b*).

area (*kN/m*)		*lever arm* (*m*)	*moment* (*kN m/m*)
4×2	$= 8$	1	−8
$0.5 \times (11.5 - 4) \times 2$	$= 7.5$	2/3	−5
11.5×1	$= 11.5$	0.5	5.8
$(15.3 - 11.5) \times 1 \times 0.5$	$= 1.9$	2/3	1.3
15.3×2	$= 30.6$	2	61.2
$(38.2 - 15.3) \times 2 \times 0.5$	$= 22.9$	$1 + (4/3)$	53.4
38.2×2	$= 76.4$	4	305.6
$(40.6 - 38.2) \times 2 \times 0.5$	$= 2.4$	$3 + (4/3)$	10.4
$40.6 \times 0.458 \times 0.5$	$= 9.3$	$5 + (0.458/3)$	47.9
$(48 \times 0.542 \times 0.5)/F$	$= 13/F$	$5.458 + (2 \times 0.542/3)$	$-75.7/F$
$(48 \times 1.5)/F$	$= 72/F$	6.75	$-486/F$
$[0.5 \times (134.1 - 48) \times 1]/F$	$= 43/F$	$6 + (1 \times 2/3)$	$-287/F$
$[(134.1 - 48) \times 0.5]/F$	$= 43/F$	$7 + 0.25$	$-312.1/F$
$[0.5 \times (175.6 - 134.1) \times 0.5]/F$	$= 10.4/F$	$7 + (0.5 \times 2/3)$	$-76.1/F$

Whence

$$\Sigma M_{(O)} = 0 \quad \Rightarrow \quad 472.6 - \frac{1236.9}{F} = 0$$

This yields a factor of safety $F = 2.62$.

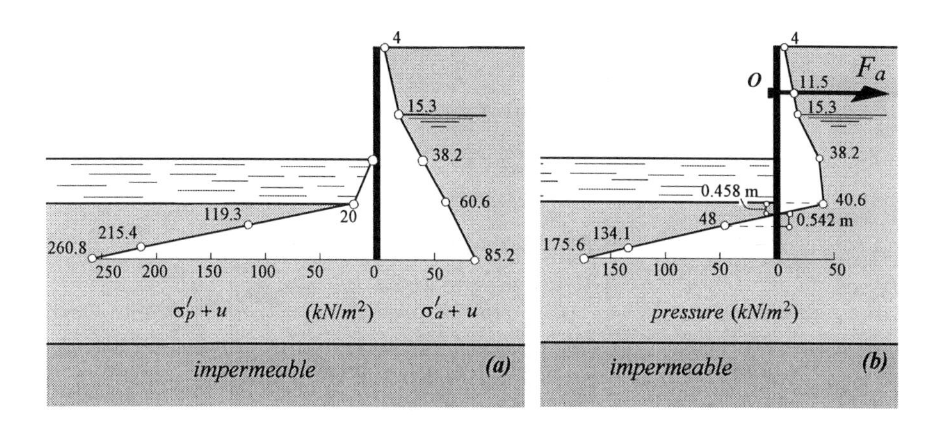

Figure 11.11: (a) Total and (b) net pressure diagrams along both sides of the sheet-pile.

The *anchorage force* can now be found from the equilibrium equation of horizontal forces. Referring to figure 11.11(*b*), it is seen that the sum of the horizontal forces corresponds to the *area* of the net pressure diagram, *augmented* by the anchorage force F_a. Whence the following equilibrium equation:

$$-(4+11.5)\times\tfrac{2}{2}+F_a-(11.5+15.3)\times\tfrac{1}{2}$$

$$-(15.3+38.2)\times\tfrac{2}{2}-(38.2+40.6)\times\tfrac{2}{2}-40.6\times\tfrac{0.458}{2}$$

$$+\tfrac{1}{2.62}\left[48\times 0.542\times\tfrac{1}{2}+(48+134.1)\times\tfrac{1}{2}+(134.1+175.6)\times\tfrac{0.5}{2}\right]=0$$

leading to an anchorage force $F_a = 101.2\,kN/m$.

The precise distribution of shear forces along the pile length is thereafter determined in a straightforward way. The detailed calculations are presented in the following table (in which the depth z is with respect to the ground surface on the active side), and the corresponding shear force diagram is depicted in figure 11.12.

$z\,(m)$	$shear\,force\,(kN/m)$	
0	0	
2^-	$-(4+11.5)\times 2/2$	$=-15.5$
2^+	$-15.5+101.2$	$=85.7$
3	$85.7-(11.5+15.3)\times 0.5$	$=72.3$
5	$72.3-(15.3+32.8)\times 2/2$	$=18.8$
7	$18.8-(38.2+40.6)\times 2/2$	$=-60$
7.458	$-60-(40.6\times 0.458)/2$	$=-69.3$
8	$-69.3+(48\times 0.542)/2$	$=-56.3$
9	$-56.3+(48+134.1)\times 0.5$	$=34.8$
9.5	$34.8+(134.1+175.6)\times 0.5/2$	$=112.2$

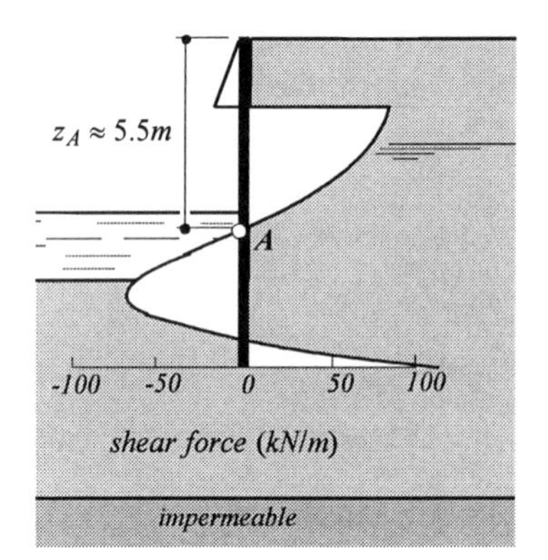

Figure 11.12: Shear force diagram.

The maximum bending moment M_m corresponds to zero shear force, the location of which can easily be obtained from the diagram of figure 11.12 (point A). With reference to figure 11.11(b), it is seen that the *net lateral pressure* at a depth $z_A = 5.5\,m$ is determined using a straightforward linear interpolation and corresponds to a value of $38.5\,kN/m^2$. Evaluating M_m^A at point A, it follows that:

$$M_m^A = -4 \times 2 \times (5.5-1) - (11.5-4) \times \tfrac{2}{2} \times \left(3.5+\tfrac{2}{3}\right)$$

$$+101.2 \times 3.5 - 11.5 \times 1 \times 3 - (15.3-11.5) \times \tfrac{1}{2} \times \left(2.5+\tfrac{1}{3}\right)$$

$$-15.3 \times 2 \times 1.5 - (38.2-15.3) \times \tfrac{2}{2} \times \left(0.5+\tfrac{2}{3}\right)$$

$$-38.2 \times 0.5 \times 0.25 - (38.5-38.2) \times \tfrac{0.5}{2} \times \tfrac{0.5}{3}$$

$$= 169.7\, kN\, m/m$$

The selection of the appropriate steel section can thence be made according to the following: $M_m^A = \sigma_a I/y$. Hence, for $\sigma_a = 180\, N/mm^2$ (mild steel), the section modulus required is:

$$\frac{I}{y} = \frac{M_m^A}{\sigma_a} = \frac{169.7}{180} \times 10^3 = 943\, cm^3/m$$

The nearest pile section satisfying this requirement is a *Larssen LX12* with a section modulus $I/y = 1208\, cm^3/m$. Notice that, in practice, the anchorage force $F_a = 101.2\, kN/m$ calculated previously is increased by about 15% to allow for the possibility of any *horizontal arching effect* or any *stress redistribution* behind the wall. Consequently, the anchor section must be designed to support a horizontal force $F_a^* \approx 116\, kN/m$.

(2) *A factor of safety* $F = 2$ *is used in conjunction with* K_p

In this case, the *active pressures* calculated previously using a factor of safety $F = 1$ on K_p still apply. However, the *passive pressures* are recalculated using the new reduced value of the coefficient of passive pressure $K_p^* = 10.4/2 = 5.2$. The *effective normal passive stresses* are then calculated using the same effective vertical stresses determined previously on the passive side:

$$\sigma_p' = \sigma_v' K_p^* \cos\delta$$

Whence the values listed in the following table:

$z\,(m)$	$\gamma'\,(kN/m^3)$	$u\,(kN/m^2)$	$\sigma'_v\,(kN/m^2)$	$\sigma'_p\,(kN/m^3)$
0	11	20	0	0
1	9.32	31.7	9.3	43.8
2	8.88	44.3	18.2	85.6
2.5	7.84	52.9	22.1	103.9

The corresponding net total pressure diagram is shown in figure 11.13, and the moment equilibrium equation with respect to the anchorage point can now be established. For the sake of clarity, the detailed calculations are once more tabulated as follows.

area (kN/m)		*lever arm (m)*	*moment (kN m/m)*
4×2	$=8$	1	−8
$(11.5-4)\times 0.5\times 2$	$=7.5$	2/3	−5
11.5×1	$=11.5$	0.5	5.8
$(15.3-11.5)\times 1\times 0.5$	$=1.9$	2/3	1.3
15.3×2	$=30.6$	2	61.2
$(38.2-15.3)\times 0.5\times 2$	$=22.9$	$1+4/3$	53.4
38.2×2	$=76.4$	4	305.6
$(40.6-38.2)\times 2\times 0.5$	$=2.4$	$3+4/3$	10.4
$40.6\times 0.91\times 0.5$	$=18.5$	$5+0.91/3$	98
$(4.2\times 0.09\times 0.5)\times 2/F$	$=0.38/F$	$5.91+0.09\times 2/3$	$-2.27/F$
$(4.2\times 1.5)\times 2/F$	$=12.6/F$	$6+1.5/2$	$-85/F$
$(48.6-4.2)\times 0.5\times 2/F$	$=44.4/F$	$6+2/3$	$-296/F$
$(48.6-4.2)\times 0.5\times 2/F$	$=44.4/F$	$7+0.5/2$	$-321.9/F$
$(71.6-48.6)\times 0.25\times 2/F$	$=11.5/F$	$7+0.5\times 2/3$	$-84.3/F$

Accordingly:

$$\Sigma M_{(O)}=0 \quad \Rightarrow \quad 522.7-\frac{789.5}{F}=0$$

This gives $F=1.51$.

This factor of safety represents the ratio of the available passive resistance of the soil *(calculated using a factor of safety of 2 on K_p)* to the mobilised one.

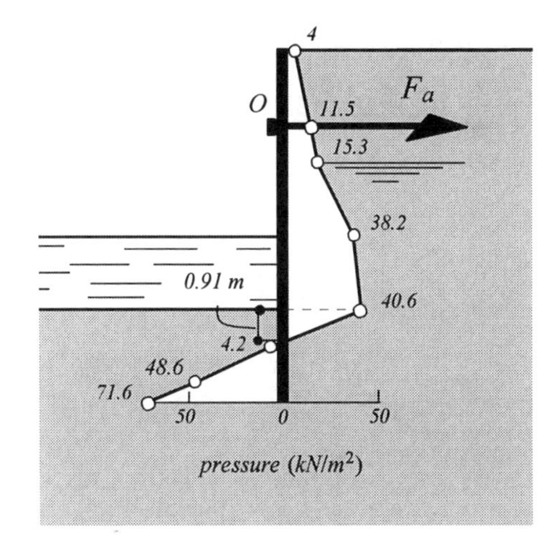

Figure 11.13: Net pressure diagram.

The *anchorage force* is calculated from the equilibrium equation of horizontal forces, established with reference to figure 11.13:

$$F_a = (4+11.5)\times\frac{2}{2} + (11.5+15.3)\times\frac{1}{2}$$

$$+(15.3+38.2)\times\frac{2}{2} + (38.2+40.6)\times\frac{2}{2} + 40.6\times\frac{0.91}{2}$$

$$-\frac{2}{1.5}\times\left[4.2\times\frac{0.09}{2} + (4.2+48.6)\times\frac{1}{2} + (48.6+71.6)\times\frac{0.5}{2}\right]$$

leading to a force $F_a = 104.2\, kN/m$.

The shear force diagram, depicted in figure 11.14, is obtained in precisely the same way, used in conjunction with figure 11.12. Of particular interest, the maximum bending moment M_m^B (corresponding to zero shear force) occurs at a depth $z_B \approx 5.6\, m$. Accordingly:

$$M_m^B = -4\times 2\times(1+3.6)-(11.5-4)\times\left(\frac{2}{3}+3.6\right)+104.2\times 3.6$$
$$-11.5\times(0.5+3.1)-(15.3-11.5)\times\frac{1}{2}\times\left(\frac{1}{3}+2.6\right)$$
$$-15.3\times 2\times 1.6-(38.2-15.3)\times\left(\frac{2}{3}+0.6\right)$$
$$-38.2\times\frac{0.6^2}{2}-(38.9-38.2)\times\frac{0.6^2}{6}=180.2\,kN\,m/m$$

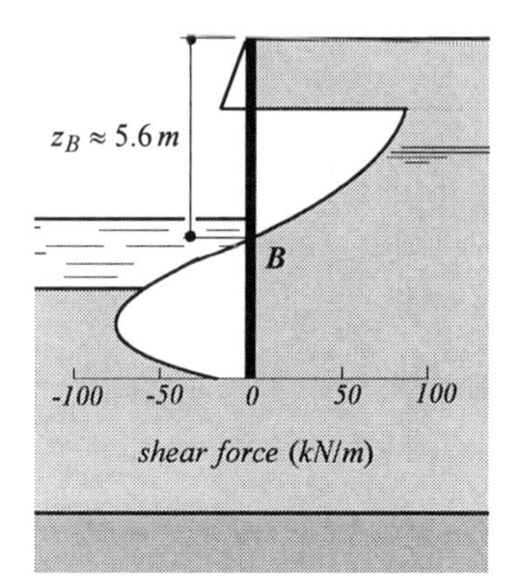

Figure 11.14: Shear force diagram.

Since the pile is embedded in a clean dense sand, the maximum bending moment must be somewhat *reduced* to allow for a degree of flexibility in accordance with equation 11.1:

$$M_c = M_m - \frac{1}{2}(M_m - M_r)$$

where M_r represents the reduced moment obtained from Rowe's method in the following manner.

- An appropriate curve in figure 11.7 corresponding to a ratio $\alpha = 7/9.5 = 0.73$ is selected. Obviously an intermediate curve needs to be pencilled, from which the points $(\rho, M/M_{\max})$ tabulated below are read.

The corresponding values of M^* are calculated by multiplying the different ratios $M/M_{\max}$ by the maximum bending moment $M_m = 180.2\,kN\,m/m$.

$\rho\,(m^3/kN)$	M/M_{max}	M^* (kNm/m)
0.05	0.97	174.8
0.07	0.89	160.4
0.10	0.82	147.8
0.20	0.68	122.5
0.40	0.56	100.9
1.00	0.43	77.5

- The points (ρ, M^*) are thereafter plotted as depicted in figure 11.15.

- The next step consists of using a sheet piling manual to select different pile sections. In this respect, the following characteristics correspond to *Larssen* sections of mild steel with a modulus of elasticity $E = 2.1 \times 10^5\, N/mm^2$ and a working stress $\sigma_a = 180\, N/mm^2$.

Notice that in the following table, I corresponds to the combined moment of inertia of the pile, and I/y represents its section modulus (refer to the piling handbook by British Steel (1997)). Also, $M_t = \sigma_a I/y$ and $\rho_t = H^4/EI$.

pile section	$I\,(cm^4/m)$	$I/y\,(cm^3/m)$	$M_t\,(kNm/m)$	$\rho_t\,(m^3/kN)$
6W	6459	610	109.8	0.6
LX8	12861	830	149.4	0.3
GSP2	8740	874	157.3	0.44
LX12	18723	1208	217.4	0.21

- The points (ρ_t, M_t) are then plotted in figure 11.15, and the *appropriate* pile section corresponds to the point nearest to the curve in the figure, which is in this instance a *Larssen 6W*. Thus, Rowe's method leads to a reduction in the value of the maximum bending moment in the pile from $M_m = 180.2\, kNm/m$ to the value $M_r = 109.8\, kNm/m$ (refer to the table above).

However, a more realistic reduction is calculated from equation 11.1, so yielding the maximum bending moment that must be used to select a pile section:

$$M_c = 180.2 - \frac{1}{2}(180.2 - 109.8) = 145\, kN\, m/m$$

Therefore, a *Larssen LX8* is, according to the table above, a more appropriate section.

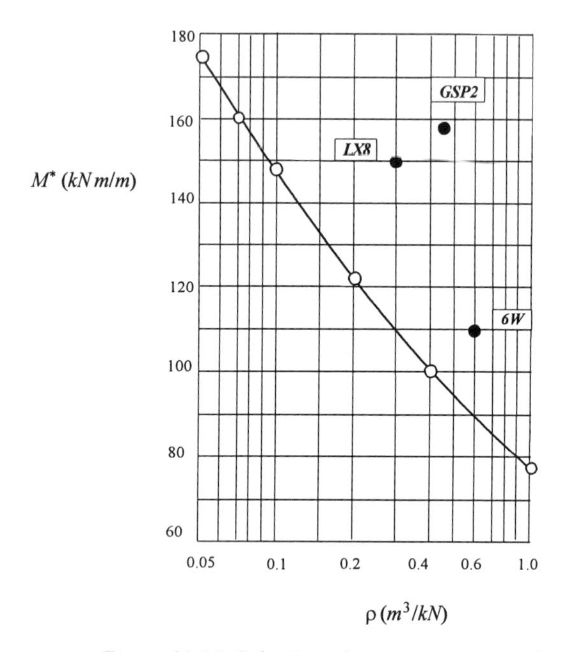

Figure 11.15: Selection of an appropriate steel section.

A quick comparison of the outcome of both procedures (1) and (2) points to the fact that, while the depth of embedment is adequate and the calculated anchorage forces are similar in both cases, the maximum bending moment (which governs the selection of the sheet pile section) obtained from Rowe's method is 17% smaller than that corresponding to the limiting conditions.

Example 11.2

A steel sheet pile is driven into a thick layer of a stiff overconsolidated clay with the following characteristics: $\gamma = 18.5\, kN/m^3$, $\gamma_{sat} = 19.5\, kN/m^3$, $\phi' = 25^\circ$ and $c' = 5\, kN/m^2$.

The (static) water is situated at the same level on each side of the pile which is supported by two lines of anchors as depicted in figure 11.16. Allowing for an angle of wall friction $\delta = 16^\circ$ and a uniform pressure $q = 40\,kN/m^2$ applied on the ground surface on the active side, calculate the *depth of embedment*, the *anchorage forces* and the *steel section* of the pile. The active stress conditions are assumed to have been fully mobilised behind the wall.

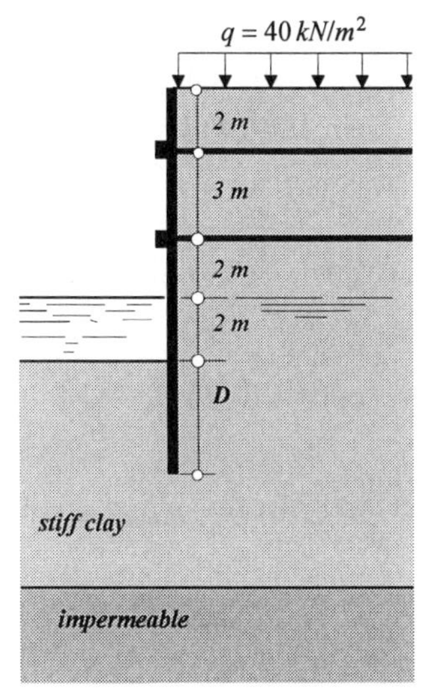

Figure 11.16: Pile dimensions and soil characteristics.

With the use of two lines of anchors, the structure in figure 11.16 becomes statically indeterminate. Such a problem is usually solved in practice using numerical methods such as finite elements. However, an approximate but acceptable solution can be obtained by assuming that the two lines of anchors are reduced to a single fictive line, placed midway between the two actual ones. This assumption allows for the anchorage force F corresponding to the fictive line of anchors to be calculated from the equilibrium equation of horizontal forces. The force F is thereafter spread between the two actual lines of anchors, and the ensuing calculations related to shear forces and bending moments are readily undertaken in the usual manner.

Because the water level is static, only the effective active and passive stress distributions on both sides of the pile need be calculated. Once again, Boussinesq analysis is used to evaluate the coefficients of active and passive earth pressure through the use of Kerisel and Absi's tables. Thus, for $\phi = 25°$, $\delta = 2\phi'/3$, and $\beta = 0$, table 10.1 yields: $K_a = 0.364$, $K_p = 3.7$.

Moreover, the coefficients applicable to the cohesion of the clay and to the uniform load q are found from table 10.2 for $\phi' = 25°, \delta = 2\phi'/3$, $\alpha = 0$ and $\Omega = 90°$: $K'_a = 0.369$, $K'_p = 3.554$.

Using the limiting conditions on the passive side (*i.e.* a factor of safety $F = 1$ in conjunction with K_p), both active and passive normal effective stresses can now be calculated in the following way (refer to section 10.5.4):

$$\sigma'_a = \cos\delta\left(K_a\,\sigma'_v + K'_a\,q\right) - c'\cot\phi\left(1 - K'_a\cos\delta\right)$$

$$\sigma'_p = \sigma'_v\,K_p\cos\delta + c'\cot\phi\left(K'_p\cos\delta - 1\right)$$

σ'_v being the effective vertical stress. Whence the values in the tables below:

Active side

z (m)	σ'_v (kN/m²)	σ'_a (kN/m²)
0	0	174.8
2.0	$18.5 \times 2 = 37$	20.3
3.5	$18.5 \times 3.5 = 64.8$	30
5.0	$18.5 \times 5 = 92.5$	39.7
7.0	$18.5 \times 7 = 129.5$	52.6
9.0	$129.5 + 9.5 \times 2 = 148.5$	59.3
9+D	$148.5 + 9.5D$	$59.3 + 3.325D$

Passive side

z (m)	σ'_v (kN/m²)	σ'_p (kN/m²)
0	0	25.9
D	$9.5D$	$33.782D + 25.9$

Prior to establishing the moment equilibrium equation, the two levels of anchors are replaced by a *fictive* single anchor, whose line of action is situated mid-distance between the two *actual* anchors as shown in figure 11.17. Thus, with reference to the *effective pressure diagram* in the figure, the equilibrium of moments with respect to the fictive anchorage point O is as follows: $\Sigma M_O = 0 \quad \Rightarrow$

$$
\begin{aligned}
&-7.3\times 3.5\times 1.75-(30-7.3)\times \frac{3.5}{2}\times \frac{3.5}{3}+30\times 3.5\times 1.75\\
&+(52.6-30)\times \frac{3.5}{2}\times \frac{2}{3}\times 3.5+52.6\times (2+D)\times \left(3.5+\frac{2+D}{2}\right)\\
&+(59.3-52.6+3.32D)\times \left(\frac{2+D}{2}\right)\times \left[3.5+\frac{2}{3}(2+D)\right]\\
&-25.9\times D\times \left(5.5+\frac{D}{2}\right)-33.78D\times \frac{D}{2}\times \left(5.5+\frac{2D}{3}\right)=0
\end{aligned}
$$

or

$$690.75+183.56D-67.08D^2-10.15D^3=0$$

Whence a depth of embedment $D \approx 3.61\, m$.

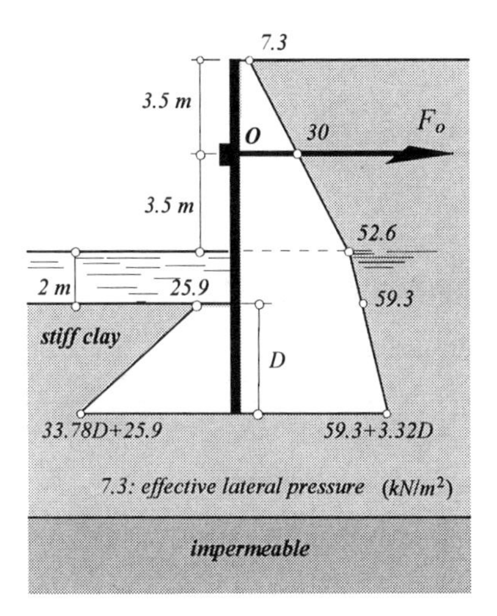

Figure 11.17: Effective lateral pressure diagram.

Next, the net pressure diagram can readily be established as depicted in figure 11.18(a), and the anchorage force F_o can be calculated from the equilibrium equation of horizontal forces. Accordingly:

$$\Sigma F_h = 0 \quad \Rightarrow$$

$$F_o = (7.3 + 52.6) \times \frac{7}{2} + (52.6 + 59.3) \times \frac{2}{2} + 33.4 \times \frac{1.10}{2} - 76.6 \times \frac{2.51}{2}$$
$$= 243.8\ kN/m$$

The anchorage force F_o is then spread equally between the two *actual* anchors as shown in figure 11.18.(*b*) (*i.e.* $F_{a1} = F_{a2} = 121.9\ kN/m$).

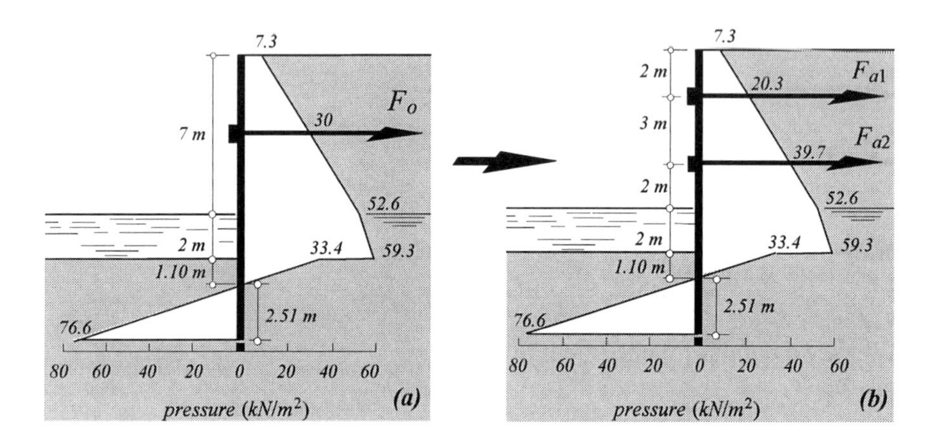

Figure 11.18: Net pressure diagram (a) fictive anchor and (b) actual anchors.

The distribution of shear forces along the pile is calculated in precisely the same way as used in the previous example:

z (*m*)	*shear force* (*kN*/*m*)	
0		0
2^-	$-(7.3 + 20.3) \times 2/2$	$= -27.6$
2^+	$-27.6 + 121.9$	$= 94.3$
5^-	$94.3 - (20.3 + 39.7) \times 3/2$	$= 4.3$
5^+	$4.3 + 121.9$	$= 126.2$
7	$126.2 - (39.7 + 52.6) \times 2/2$	$= 33.9$
9	$33.9 - (52.6 + 59.3) \times 2/2$	$= -78$
10.35	$-78 - 33.4 \times 1.1/2$	$= -96.4$
12.61	$-96.4 + 76.6 \times 2.51/2$	≈ 0

The corresponding shear force and bending moment diagrams are plotted in figure 11.19. Of particular interest are the effects that two lines of anchors have on the shape of the bending moment diagram, as well as the location of the maximum bending moment ($M_{max} = 312\, kN\, m/m$).

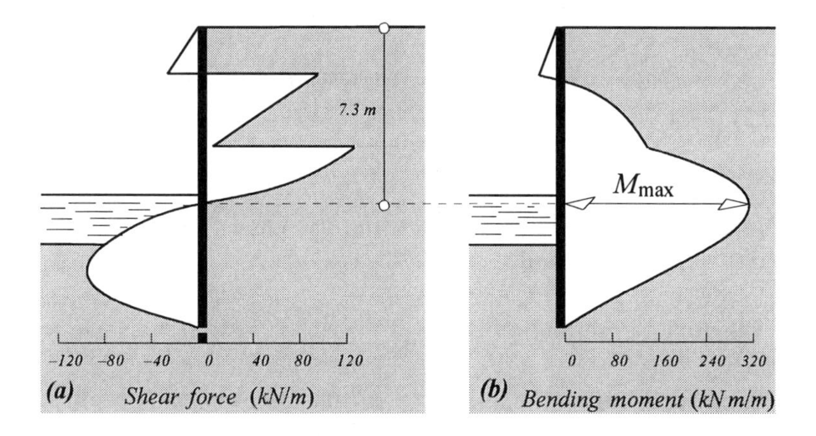

Figure 11.19: (a) Shear force and (b) bending moment diagrams.

Now that all the above quantities have been calculated using the limiting conditions (*i.e.* a factor of safety $F = 1$ on the passive side of the pile), the *actual* characteristics of the pile are such that:

- depth of embedment $D = \sqrt{2} \times 3.61 = 5.10\, m$; hence a total pile length:

$$L = 9 + 5.1 = 14.1\, m$$

- pile steel section: $M_{max} = \sigma_a \dfrac{I}{y} = 312\, kNm/m \quad \Rightarrow$

$$\frac{I}{y} = \frac{M_{max}}{\sigma_a} = \frac{312}{180} \times 10^3 = 1733\, cm^3/m$$

The appropriate steel section corresponds to a *Larssen LX20* with a section modulus $I/y = 2022\, cm^3/m$.

Notice that the calculated anchorage forces must be increased by 25% in the upper line of anchors, and by 15% in the lower one to allow for any stress redistribution behind the wall. The 25% increase in the upper level is justified by the fact that the line of anchors at that level will be subjected, if only temporarily, to a larger horizontal force prior to the installation of the second line. Accordingly, the upper line of anchors must be designed to support a horizontal force $F_{a1} = 121.9 \times 1.25 = 152.4\,kN/m$, *and the lower one must withstand a horizontal force* $F_{a2} = 121.9 \times 1.15 = 140\,kN/m$.

11.4 Design of anchored sheet-piles using the fixed earth support method

11.4.1 Elastic line method

When the depth of embedment D of the pile becomes substantive compared with the height h above dredge level, the *passive pressure* in front of the wall becomes such that it *can no longer be fully mobilised.* Under such circumstances, the deflection of the pile generates an amount of *active pressure* in front of the wall *and passive pressure* behind the wall, due to the anticlockwise rotation of the toe of the pile around the point of fixity C as illustrated in figure 11.20.

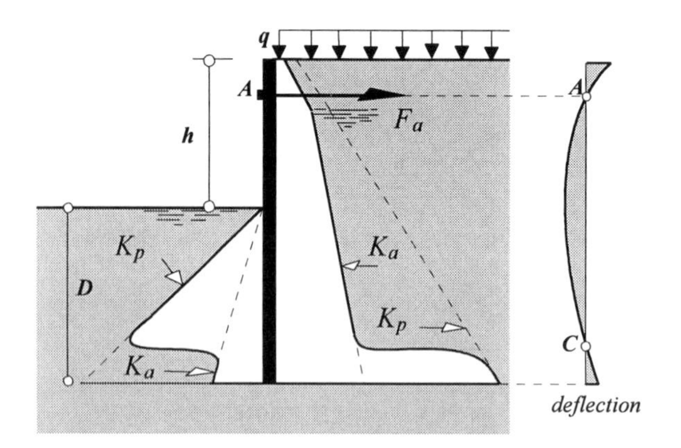

Figure 11.20: Actual lateral pressure distribution for deeply embedded sheet-piles.

The actual pressure distribution can be simplified by *assuming* a depth of embedment D' (smaller than the actual depth D as depicted in figure 11.21) throughout which the passive resistance of the soil is fully mobilised in front of the wall. Behind the wall, the pile is subject to the active thrust *and* to a concentrated load F_c resulting from both active and passive thrusts generated by the anticlockwise rotation of the toe of the pile.

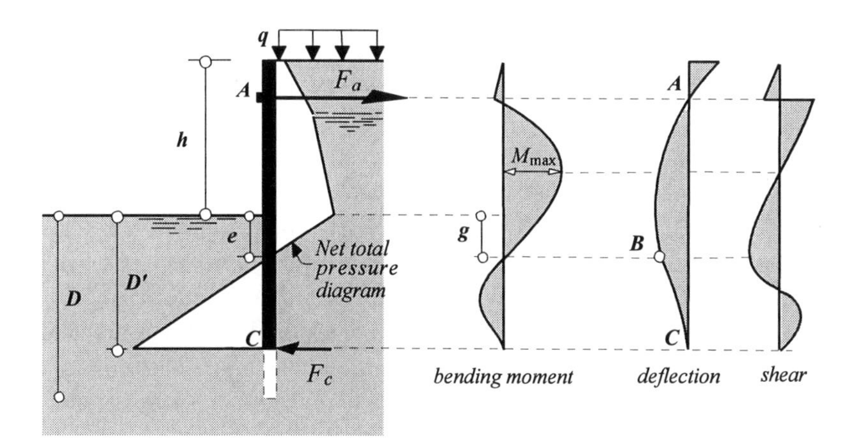

Figure 11.21: Lateral pressure diagram related to the assumed depth of embedment D'.

The *net lateral pressure diagram* can thus be calculated then integrated:

- once to establish the shear forces distribution;
- twice to calculate the bending moments diagram;
- three times to compute the slope;
- four times to determine the deflection.

The associated boundary conditions at point C are as follows:

- the shear force is such that $S_c + F_c = 0$
- the bending moment $M_c = 0$
- the slope is *assumed* to be vertical $y'_c = 0$

If the calculations yield a non-zero deflection at C, then another value of the depth of embedment is adopted, and the entire calculations are repeated following the same procedure. At convergence (*i.e.* when $y_c = 0$), the corresponding depth D' is then increased by 20% to ensure a factor of safety on embedment. This iterative procedure, though not suited for hand calculations, is very easy to program on a computer.

11.4.2 Blum equivalent beam method

This procedure, which is a *variation of the elastic line method,* was developed by Hermann Blum (see for example Blum (1931), (1951)) who undertook an extensive analysis of the behaviour of sheet piles that culminated in the formulation of the *equivalent beam method* which will be detailed shortly. However, the suitability of this method depends on how realistic its *assumptions* are in conjunction with soil conditions in which the pile is driven. In this respect, Blum's method assumes that *the point of zero net pressure and that of zero bending moment occur at the same depth* (that is $e = g$ in figure 11.21); this assumption becomes arguable for non-homogeneous soil conditions (*i.e.* multilayered soils).

Delattre *et al.* (1996) have shown that for *homogeneous* soils, the assumption can be justified provided that the ratio e/h is between 0.1 and 0.2 as depicted in figure 11.22, where e represents the depth below dredge level of the point of zero net pressure, g corresponds to the depth below dredge level of the point of zero bending moment, and h is the height of pile above dredge level (refer to figure 11.21). The error generated by assuming that $e = g$ for ratios $e/h < 0.1$ or > 0.2 (*i.e.* outside the shaded area in figure 11.22) is on the *unsafe side* since the maximum bending moment will be, in this case, underestimated.

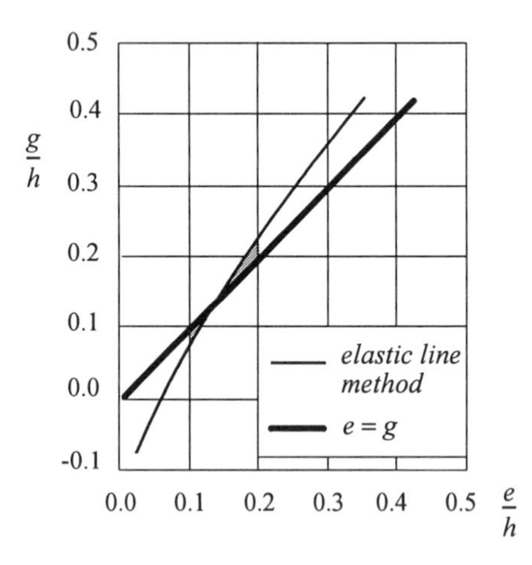

Figure 11.22: Relationship between the depth of zero net pressure and that of zero bending moment. (Reproduced by permission of the Laboratoire Central des Ponts et Chaussées.)

It is therefore recommended to restrict the use of Blum's method to the design of sheet piles, driven into *homogeneous* soils, and for which $0.1 < e/h < 0.2$.

When applicable, the *modus operandi* of the equivalent beam method is as follows.

- Find the *net total pressure diagram* using the limiting conditions (*i.e.* a factor of safety $F = 1$ in conjunction with the coefficient of passive pressure K_p).

- Establish the *point of zero net pressure* (point B, figure 11.23(a)) and its depth e with respect to dredge level.

- Check that the ratio e/h is between 0.1 and 0.2.

- Consider the *two halves* separately (figure 11.23(b).

- Establish the equilibrium equation of horizontal forces in the *upper half.*

- Establish the equilibrium of moments with respect to the anchorage point A in the *upper half.*

- Solve these two equations and calculate the horizontal forces F_a and F_b.

- Establish the equilibrium of moments with respect to point C in the *lower half.*

- Determine the depth of embedment D'.

- Establish the *shear force diagram* and locate the depth z_m corresponding to zero shear force.

- Calculate the *maximum bending moment* $M_{\max}$ occurring at the depth z_m.

- Calculate the total (safe) *depth of embedment* $D = e + 1.2(D' - e)$.

- Increase the anchorage force F_a by 15% to take into account any stress redistribution behind the wall.

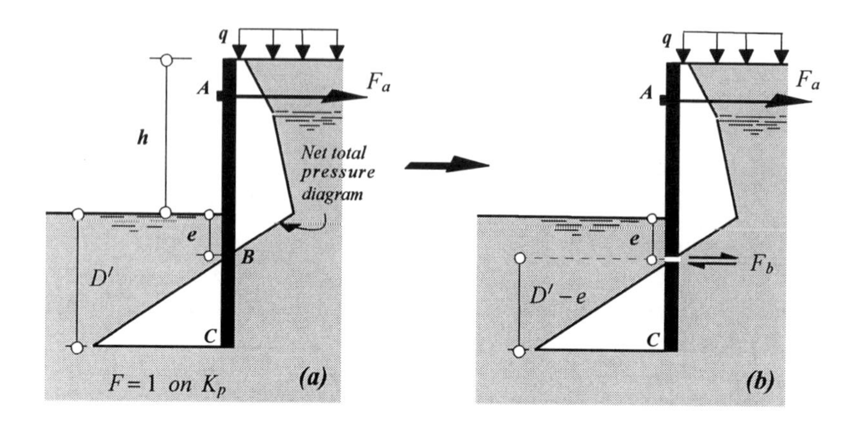

Figure 11.23: (a) Determination of the depth of zero net pressure and (b) separation of the pressure diagram.

Example 11.3

To illustrate the potential pitfalls related to the use of the method, let us apply it to the problem of example 11.1. The *net total pressure diagram* as well as the position of the *zero net pressure* have already been established and are depicted in figure 11.11(*b*). The corresponding ratio *e/h* is in this case: $e/h = 0.458/7 \approx 0.07$.

Now consider the two halves of the pressure diagram as shown in figure 11.24.

Starting with the *upper half*, the equilibrium of horizontal forces is such that:

$$F_a + F_b = (4 + 11.5) \times \frac{2}{2} + (11.5 + 15.3) \times \frac{1}{2}$$

$$+ (15.3 + 38.2) \times \frac{2}{2} + (38.2 + 40.6) \times \frac{2}{2}$$

$$+ 40.6 \times \frac{0.458}{2} = 170.5\, kN/m$$

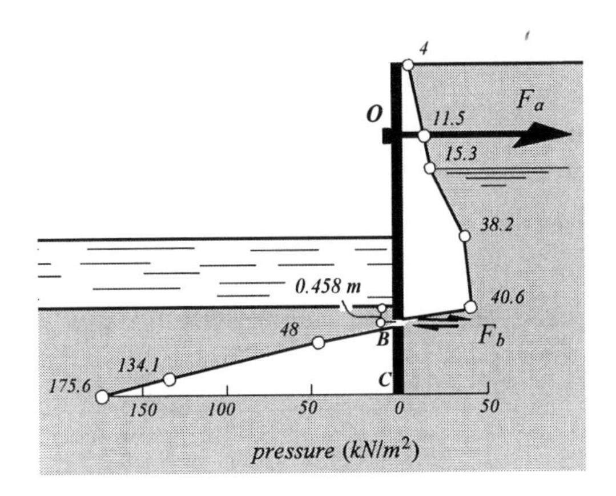

Figure 11.24: Application of the equivalent beam method to the net pressure diagram of example 11.1.

The equilibrium of moments with respect to point O in the upper half implies that:

$\Sigma M_o = 0 \quad \Rightarrow$

$$-2\times 4\times 1-\frac{2}{3}\times(11.5-4)\times\frac{2}{2}+\frac{1}{2}\times 11.5\times 1$$

$$+\frac{2}{3}\times(15.3-11.5)\times\frac{1}{2}+2\times 15.3\times 2$$

$$+\left(1+\frac{2}{3}\times 2\right)(38.2-15.3)\times\frac{2}{2}+4\times 38.2\times 2$$

$$+\left(3+\frac{2}{3}\times 2\right)(40.6-38.2)\times\frac{2}{2}$$

$$+\left(5+\frac{0.458}{3}\right)\times 40.6\times\frac{0.458}{2}-5.458\,F_b=0$$

yielding thus:

$$F_b = 77.2\,kN/m$$

Hence:

$$F_a = 93.3\,kN/m$$

In the *lower half*, the equilibrium of moments with respect to point C yields the factor of safety F on embedment. Whence:

$\Sigma M_c = 0 \quad \Rightarrow$

$$-77.2 \times 2.042 + \frac{1}{F}\Big[48 \times \frac{0.542}{2} \times \left(\frac{0.542}{3} + 1.5\right)$$

$$+48 \times 1.5 \times \frac{15}{2} + (134.1 - 48) \times \frac{1}{2} \times \left(\frac{1}{3} + \frac{1}{2}\right)$$

$$+(134.1 - 48) \times \frac{1}{2} \times \frac{1}{4} + (175.6 - 134.1) \times \frac{1}{2} \times \frac{1}{2} \times \frac{1}{6}\Big] = 0$$

or

$$\frac{124.2}{F} - 157.6 = 0 \qquad \Rightarrow \qquad F = 0.79$$

The *shear force diagram* is determined in precisely the same way used in conjunction with example 11.1, and is shown in figure 11.25.

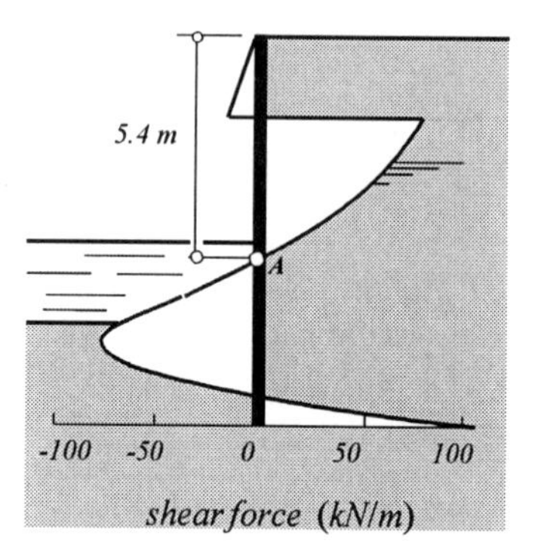

Figure 11.25: Shear force diagram.

The point of zero shear force occurs at a depth $z \approx 5.4\ m$; whence the maximum bending moment M_m, calculated with respect to point A in figure 11.25 is:

$$M_m^A = -4\times 2\times(1+3.4)-(11.5-4)\times\frac{2}{2}\times\left(\frac{2}{3}+3.4\right)$$

$$+93.3\times 3.4-11.5\times 1\times\left(\frac{1}{2}+2.4\right)$$

$$-(15.3-11.5)\times\frac{1}{2}\times\left(\frac{1}{3}+2.4\right)-15.3\times 2.4\times\frac{2.4}{2}$$

$$-(38.2-15.3)\times\frac{2}{2}\times\left(\frac{2}{3}+0.4\right)-(38.2-15.3)\times 0.4\times\frac{0.4}{2}$$

$$= 142.7\,kN\,m/m$$

Comparing these results with the ones calculated with the *free earth support* (with $F = 1$ on K_p) in example 11.1, it emerges clearly that the equivalent beam method:

- produces a smaller anchorage force: $F_a = 93.3\,kN/m$ as opposed to $F_a = 103.8\,kN/m$ for the free earth support method;

- gives a smaller maximum bending moment: $M_{max} = 142.7\,kN\,m/m$ compared with $M_{max} = 178.5\,kN\,m/m$ from the free earth support;

- yields a factor of safety on embedment $F = 0.79$, thus necessitating a depth of embedment much larger than that calculated from the free earth support.

Although these findings do not diverge from the general trend, the difference in results will have been affected to a degree by the fact that, *in this instance,* the ratio e/h was outside the boundaries $0.1 < e/h < 0.2$. Nonetheless, it is seen that the fixed earth support method results in a longer pile with a smaller maximum bending moment, hence a lighter cross-section, which may be harder to drive to the required depth. Under such circumstances, driving can become *the* deciding factor as far as this design method is concerned; whence the need for a cautious approach when using the fixed earth support method to design such structures.

Example 11.4

An anchored sheet pile is driven in a silty sand characterised by a bulk unit weight $\gamma = 19\,kN/m^3$, a saturated unit weight $\gamma_{sat} = 20\,kN/m^3$ and an angle of shearing resistance $\phi' = 30^\circ$. The static water level is situated at a height of 3 m above dredge level as depicted in figure 11.26, and the angles of wall friction are $\delta_a = +2\phi'/3$ on the active side and $\delta_p = -2\phi'/3$ on the passive side.

Use the equivalent beam method, and determine the anchorage force, the depth of embedment of the pile as well as the maximum bending moment.

In order to calculate the lateral pressure distribution on each side of the pile, the coefficients of active and passive pressure need to be known. Using Kerisel and Absi's tables (Kerisel and Absi, 1990), it can be seen that:

- on the active side:

$$\beta/\phi' = 0,\ \delta/\phi' = 0.66,\ \lambda = 0,\ \phi' = 30^\circ \quad \Rightarrow \quad K_a = 0.3 \qquad \text{(p.78)}$$

$$\phi' = 30^\circ,\ \Omega = 90^\circ,\ \alpha = 0,\ \delta_a = 20^\circ \quad \Rightarrow \quad K_a' = 0.304 \qquad \text{(p.132)}$$

- on the passive side:

$$\beta/\phi' = 0,\ \delta/\phi' = -0.66,\ \lambda = 0,\ \phi' = 30^\circ \quad \Rightarrow \quad K_p = 5.3 \qquad \text{(p.14)}$$

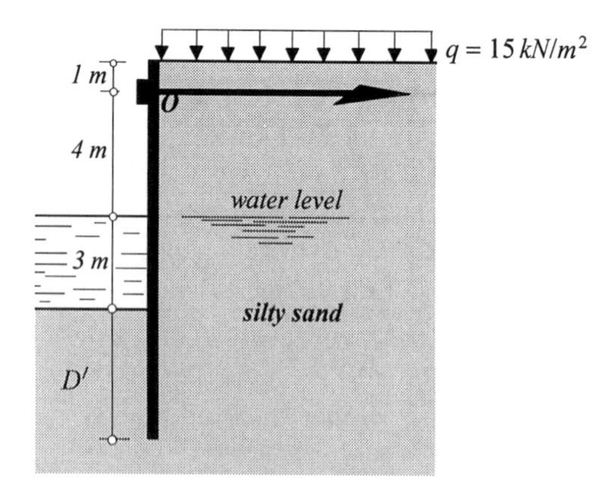

Figure 11.2: Pile dimensions and soil conditions.

Because there is no water seepage, only the *effective* lateral stresses need be calculated in the following way.

- *Active normal effective stress*:

$$\sigma'_{an} = \cos\delta_a\,(K_a\,\sigma'_v + K'_a\,q)$$

- *Passive normal effective stress*:

$$\sigma'_p = K_p\,\sigma'_v\,\cos\delta_p$$

The corresponding *net effective pressure diagram* is shown in figure 11.27, from which the depth of zero net pressure, as well as the ratio *e/h* can easily be calculated:

$$e = \frac{39.6}{46.98} = 0.843\,m \quad \Rightarrow \quad \frac{e}{h} = \frac{0.843}{8} = 0.105$$

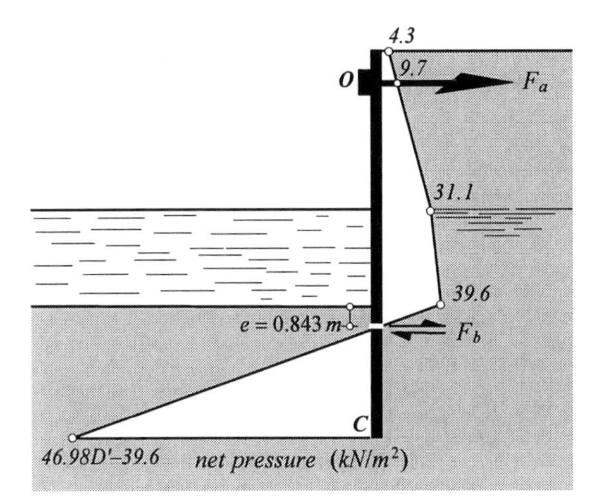

Figure 11.27: Net pressure diagram.

Now consider the equilibrium of moments with respect to the anchorage point in the *upper half* of the pressure diagram:

$\Sigma M_{(o)} = 0 \quad \Rightarrow$

$$-4.3 \times 1 \times 0.5 - (9.7 - 4.3) \times \frac{1}{6} + 9.7 \times 4 \times 2 + (31.1 - 9.7) \times \frac{4}{2} \times \frac{2}{3} \times 4$$

$$+31.1 \times 3 \times \left(\frac{3}{2} + 4\right) + (39.6 - 31.1) \times \frac{3}{2} \times \left(4 + \frac{2}{3} \times 3\right)$$

$$+39.6 \times \frac{0.843}{2} \times \left(7 + \frac{0.843}{3}\right) = 7.843 F_b$$

Hence a shear force: $F_b = 114.7\, kN/m$

Next, the equilibrium of horizontal forces in the same *upper half* is such that:

$$F_a + F_b = (4.3 + 9.7) \times \frac{1}{2} + (9.7 + 31.1) \times \frac{4}{2}$$

$$+(31.1 + 39.6) \times \frac{3}{2} + 39.6 \times \frac{0.843}{2}$$

yielding an anchorage force:

$$F_a = 96.6\, kN/m$$

The depth of embedment is calculated from the equilibrium of moments with respect to point C in the *lower half*. Whence:

$\Sigma M_C = 0 \quad \Rightarrow$

$$(D' - e) F_b = (46.98 D' - 39.6) \frac{(D' - e)^2}{6}$$

or

$$D'^3 - 2.53 D'^2 - 12.51 D' + 11.75 = 0 \quad \Rightarrow \quad D' \approx 4.67\, m$$

Hence, the *actual* depth of embedment of the pile:

$$D = e + 1.2(D' - e) = 0.843 + 1.2 \times (4.67 - 0.843) = 5.43\, m$$

The depth at which the maximum bending moment occurs is determined from the shear force diagram depicted in figure 11.28 and corresponds to the depth of zero shear force (point A). Accordingly, the bending moment at A can be calculated in a straightforward way:

$$M_{\max}^{A} = -4.3 \times 5 \times \left(0.5 + \frac{5}{2}\right) - (31.1 - 4.3) \times \frac{5}{2} \times \left(0.5 + \frac{5}{3}\right)$$

$$-31.1 \times \frac{0.5}{2} \times 0.5 - (32.5 - 31.1) \times \frac{0.5}{2} \times \frac{0.5}{3} + 96.6 \times 4.5$$

$$= 221\ kN\,m/m$$

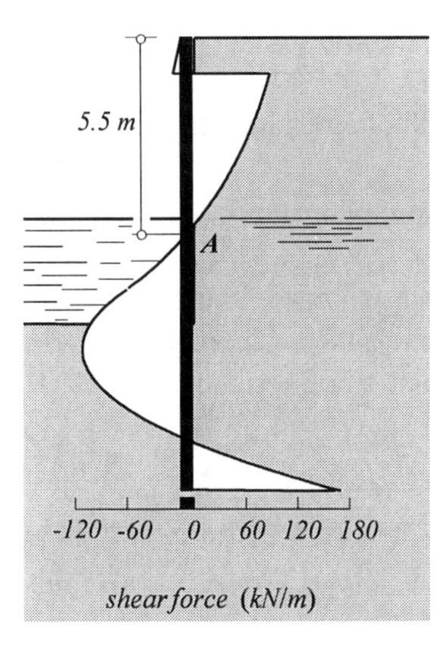

Figure 11.28: Shear force diagram.

If the same structure were designed using the *free earth support method* with a factor of safety $F = 1$ applied to the coefficient of passive pressure, then the net pressure diagram would be identical to that in figure 11.27 without separation (*i.e.* without the shear forces F_b). The critical depth of embedment would then be calculated from the moment equilibrium equation, established with respect to the anchorage point O in the following manner:

$\Sigma M_{(o)} = 0 \quad \Rightarrow$

$$-4.3 \times 1 \times 0.5 - (9.7 - 4.3) \times \frac{1}{6} + 9.7 \times 4 \times 2 + (31.1 - 9.7) \times \frac{4}{2} \times \frac{2}{3} \times 4$$

$$+31.1 \times 3 \times \left(\frac{3}{2} + 4\right) + [39.6 - 31.1] \times \frac{3}{2} \times \left(\frac{2}{3} \times 3 + 4\right)$$

$$+39.6 \times \frac{0.843}{2} \times \left(\frac{0.843}{3} + 7\right)$$

$$-(46.98D' - 39.6)\left(\frac{D' - 0.843}{2}\right)\left[\frac{2}{3}(D' - 0.843) + 7.843\right] = 0$$

or $D'^3 + 9.24D'^2 - 17.7D' - 49.7 = 0 \quad \Rightarrow \quad D' \approx 2.9\,m$

and the actual (safe) depth of embedment:

$$D = D'\sqrt{2} = 4.10\,m$$

Now that the critical depth is known, the precise net pressure diagram can then be drawn, from which the anchorage force is thereafter calculated using the equilibrium equation of horizontal forces. Thus, with reference to figure 11.29:

$$F_a = (4.3 + 9.7) \times \frac{1}{2} + (9.7 + 31.1) \times \frac{4}{2} + (39.6 + 31.1) \times \frac{3}{2}$$

$$+39.6 \times \frac{0.843}{2} - 96.6 \times \frac{2.057}{2} = 112\,kN/m$$

The corresponding shear force diagram is depicted in figure 11.30 which shows that the maximum bending moment occurs at a depth $z = 5.75\,m$ (point B in the figure). Calculating the moment at B, it follows that:

$$M^B_{\max} = -4.3 \times 5 \times \left(0.75 + \frac{5}{2}\right) - (31.1 - 4.3) \times \frac{5}{2} \times \left(\frac{5}{3} + 0.75\right)$$

$$-31.1 \times 0.75 \times \frac{0.75}{2} - (33.2 - 31.1) \times \frac{0.75}{2} \times \frac{0.75}{3}$$

$$+112 \times 4.75 = 291.3\,kN\,m/m$$

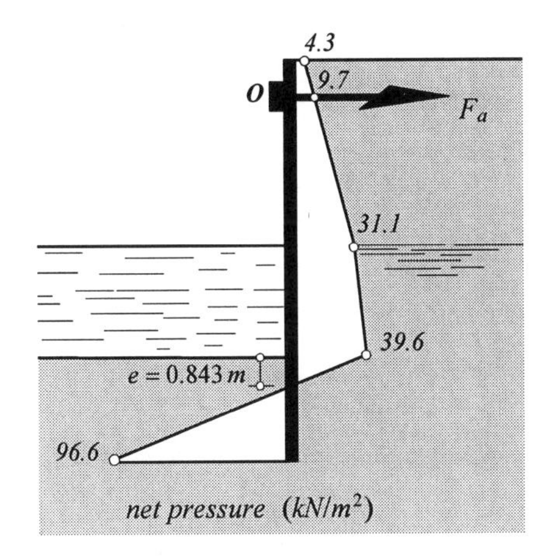

Figure 11.29: Net pressure diagram.

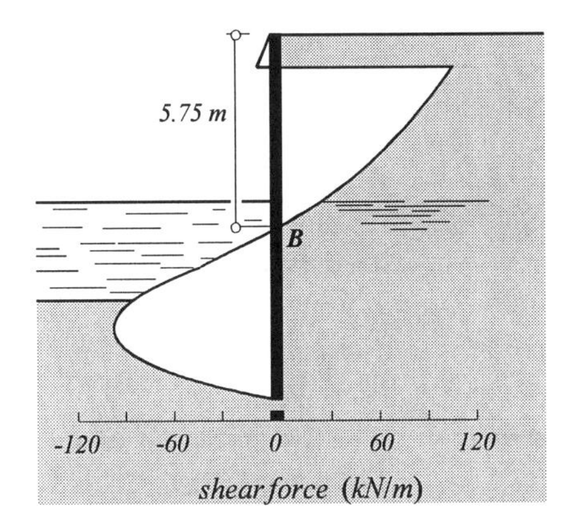

Figure 11.30: Shear force diagram.

These results confirm the findings of the previous example, the *trend* of which is represented diagrammatically in figure 11.31 (Delattre *et al.*, 1996). The figure represents the effect of the angle of shearing resistance of the (cohesionless) soil in which the pile is embedded on the ratios F_a/F_a^B, M_{max}/M_{max}^B and D/D^B where F_a, M_{max} and D are the anchorage force, bending moment and depth of embedment respectively calculated using the *free earth support method*, and F_a^B, M_{max}^B and D^B represent the same respective quantities resulting from the *elastic line method*.

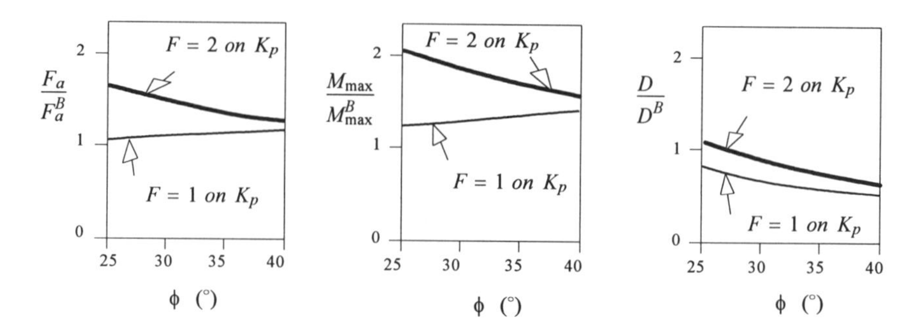

Figure 11.31: Comparison of results obtained using both free earth support and elastic line methods (Delattre et al., 1996). (Reproduced by permission.)

The figure depicts a trend and *should in no way be used for calculation purposes*. Yet, it is clear that the *elastic line method* (*i.e.* the equivalent beam method) underestimates both the anchorage force and the maximum bending moment while, in the meantime, it overestimates the depth of embedment of the pile.

Moreover, the *free earth support method* that consists of using a reduced value of the coefficient of passive pressure K_p always yields an anchorage force and a maximum bending moment in excess of those calculated using the full K_p value.

It is therefore essential that, prior to the selection of a design method, the designer should have a clear idea *vis à vis* the effects of parameters such as the angle of shearing resistance or the ratio e/h presented earlier on the outcome of calculations. Although experience plays an important part in the decision as to which method of design is most suitable, the usefulness of the graphs in figures 11.22 and 11.31 is indubitable.

11.5 Design of cantilever sheet-pile walls

Cantilever sheet pile walls are usually used when the height of retained soil does not exceed 5 *m*. From a behavioural view point, the pile is assumed to rotate about a point of fixity (point C in figure 11.32) so that the resistance to the *active load* is provided by a *passive thrust* developed in front of the wall as well as at the back due to the rotation.

As in the case of anchored sheet piles, the pressure profile in figure 11.32 can be simplified without any loss of accuracy by assuming a depth of embedment D' through which the full passive resistance is mobilised, and by substituting the thrust F_c for the net total pressure below point C as depicted in figure 11.33.

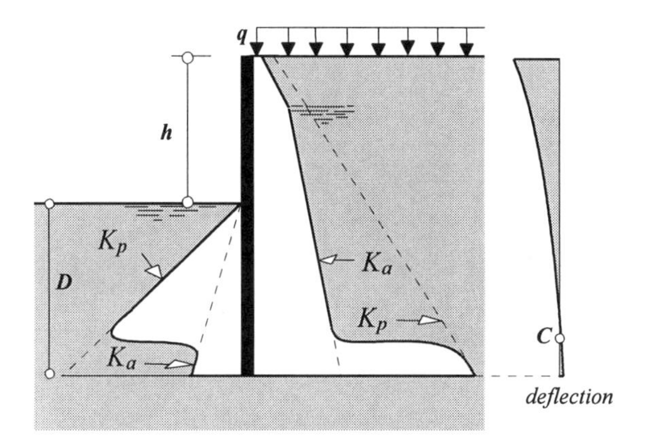

Figure 11.32: Lateral pressure profile around a cantilever sheet-pile wall.

Based on these assumptions, the *elastic line method* can then be applied in precisely the same way used in conjunction with anchored sheet piles (refer to figure 11.21). Alternatively, the *equivalent beam method* can be applied in the knowledge that only the depth of embedment D' and the maximum bending moment need be calculated.

Thus, with reference to figure 11.33, the design of a cantilever sheet pile wall can be undertaken as follows.

- Determine the *net pressure diagram,* taking into account any water seepage when applicable.

- Solve the moment equilibrium equation with respect to point C and determine the critical depth of embedment D'.

- Calculate the actual (safe) *depth of embedment* $D = D'\sqrt{2}$.

- Establish the *shear force diagram.*

- Calculate the *maximum bending moment* in the pile which occurs at the depth corresponding to zero shear force.

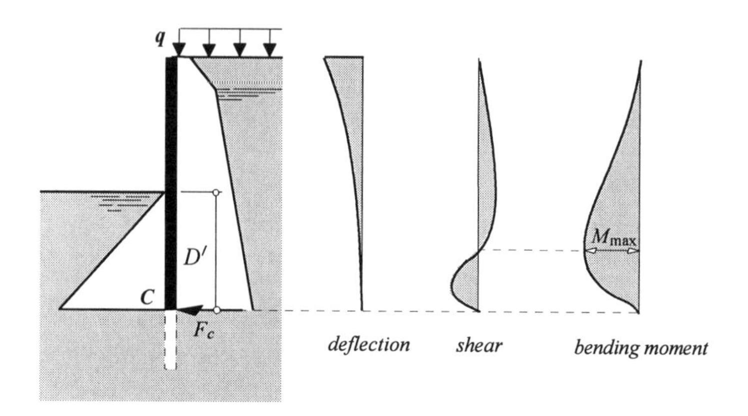

Figure 11.33: Deflection, shear and bending moment diagrams in the case of a cantilever wall.

Notice that for cantilever walls in general, the resistance to any active thrust is entirely due to the passive resistance provided by the soil. Because passive thrusts depend on the angle of shearing resistance, piles driven in clays will require a *deeper* embedment. This can potentially lead to uneconomical design, in which case the use of anchors must be considered.

Example 11.5

A steel sheet pile is driven into a thick layer of a stiff overconsolidated clay overlain by a dense silty sand. The following characteristics apply:

- *sand*: $\gamma = 19\,kN/m^3$, $\phi' = 35°$,
- *clay*: $\gamma_{sat} = 19.5\,kN/m^3$, $\phi' = 25°$, $c' = 5\,kN/m^2$

The (static) water is situated at the same level on each side of the pile as depicted in figure 11.34.

Allowing for an angle of wall friction $\delta_a = +2\phi'/3$ on the active side, $\delta_p = -\phi'/2$ on the passive side, and for a uniform pressure $q = 40\, kN/m^2$ to be applied on the ground surface on the active side. Calculate the depth of embedment and the maximum bending moment of the pile.

The active stress conditions are supposed to have been fully mobilised behind the wall. Also, Boussinesq theory is assumed to apply on both active and passive sides.

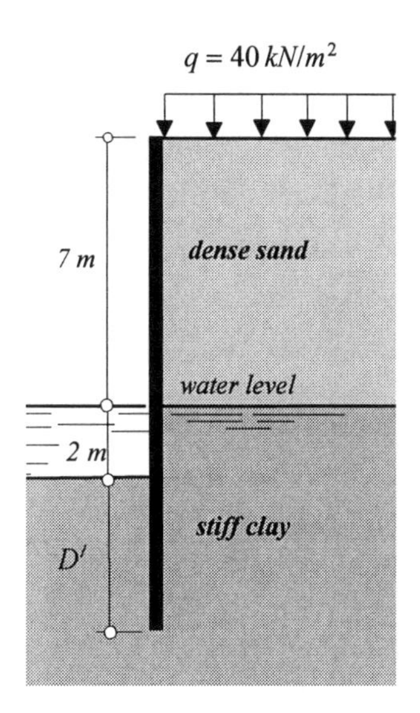

Figure 11.34: Wall dimensions and soil conditions.

Because the water level is static, only the effective active and passive stress distributions on both sides of the pile need be calculated. First, the relevant coefficients of active and passive earth pressure are evaluated using Kerisel and Absi's (1990) tables. Thus:

- *for sand*:

$\phi' = 35°,\ \delta = +2\phi'/3,\ \beta = 0,\ \lambda = 0 \quad \Rightarrow \quad K_{as} = 0.247$ (p.78) (Kerisel & Absi, 1990)

$\phi' = 35°,\ \Omega = 90°,\ \delta = +2\phi'/3,\ \alpha = 0 \quad \Rightarrow \quad K'_{as} = 0.250$ (p.150)

- *for clay on the active side*:

$\phi' = 25°,\ \delta = +2\phi'/3,\ \beta = 0,\ \lambda = 0 \quad \Rightarrow \quad K_{ac} = 0.364$ (p.78)

$\phi' = 25°,\ \Omega = 90°,\ \delta = +2\phi'/3,\ \alpha = 0 \quad \Rightarrow \quad K'_{ac} = 0.369$ (p.120)

- *for clay on the passive side*:

$\phi' = 25°,\ \delta = -\phi'/2,\ \beta = 0,\ \lambda = 0 \quad \Rightarrow \quad K_{pc} = 3.4$ (p. 21)

$\phi' = 25°,\ \Omega = 90°,\ \alpha = +\phi'/2,\ \delta = 0 \quad \Rightarrow \quad K'_{pc} = \frac{1}{0.304} = 3.289$ (p.120)

Note that the coefficients K'_a and K'_p are applicable to the uniform load q and to the cohesion of the clay. Accordingly, if the limiting conditions were used on the passive side (*i.e.* a factor of safety $F = 1$ in conjunction with K_p), both active and passive *normal effective stresses* can be calculated in the following way.

Throughout the sand layer:

$$\sigma'_a = \cos\delta_s\,(K_{as}\sigma'_v + K'_{as}q)$$

Throughout the clay layer:

$$\sigma'_a = \cos\delta_{ac}(K_{ac}\sigma'_v + K'_{ac}q) - c'\cot\phi'\,(1 - K'_{ac}\cos\delta_{ac})$$

$$\sigma'_p = \sigma'_v K_{pc}\cos\delta_{pc} + c'\cot\phi'\,(K'_{pc}\cos\delta_{pc} - 1)$$

σ'_v being the effective vertical stress due to the self weight of soil. The angles of wall friction are such that:

- *for sand*: $\delta_s = \frac{2}{3}\times 35 = 23.3°$
- *for clay on the active side*: $\delta_{ac} = \frac{2}{3}\times 25 = 16.7°$
- *for clay on the passive side*: $\delta_{pc} = -\frac{1}{2}\times 25 = -12.5°$

Whence we obtain the following values:

Active side

z (m)	σ'_v (kN/m^2)		σ'_a (kN/m^2)
0		0	9.2
2	19×2	$= 38$	17.8
7^-	19×7	$= 133$	39.3
7^+	133		53.6
9	$133 + 9.5\times 2$	$= 152$	60.2
$9+D$	$152 + 9.5D$		$60.2+3.312D$

Passive side

z (m)	σ'_v (kN/m^2)	σ'_p (kN/m^2)
0	0	23.7
D	$9.5D$	$23.7 + 31.534D$

The corresponding *effective pressure diagram* is depicted in figure 11.35.

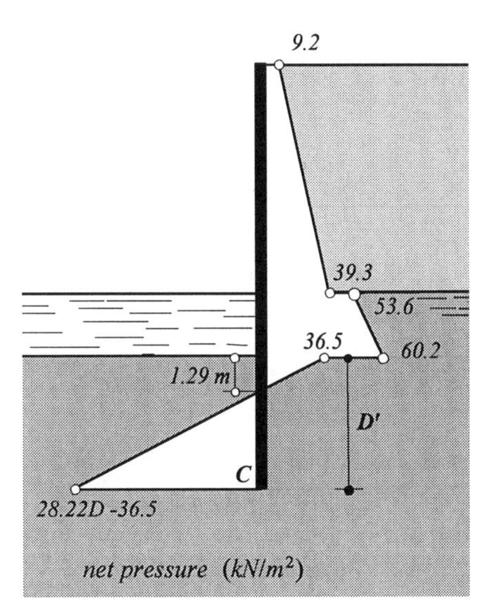

Figure 11.35: Net pressure diagram.

The equilibrium of moments with respect to point C can now be established:

$$\Sigma M_{(c)} = 0 \quad \Rightarrow$$

$$-9.2 \times 7 \times \left(\frac{7}{2} + 2 + D'\right) - (39.3 - 9.2) \times \frac{7}{2} \times \left(\frac{7}{3} + 2 + D'\right)$$

$$-53.6 \times 2 \times (1 + D') - (60.2 - 53.6) \times \frac{2}{2} \times \left(\frac{2}{3} + D'\right)$$

$$-36.5 \times \frac{1.29}{2} \times \left(D' - \frac{1.29}{3}\right) + (28.22D' - 36.5)\frac{(D' - 1.29)^2}{6} = 0$$

or

$$D'^3 - 3.87D'^2 - 60.3D' - 196.2 = 0 \quad \Rightarrow \quad D' \approx 11\,m$$

Hence a safe depth of embedment: $D = D'\sqrt{2} = 15.55\,m$.

The ensuing shear force diagram is depicted in figure 11.36 where the shear force becomes zero at a depth $z = 15\,m$. The corresponding maximum bending moment, calculated at point A is hence:

$$M^A_{\max} = -9.2 \times 7 \times \left(\frac{7}{2} + 8\right) - (39.3 - 9.2) \times \frac{7}{2} \times \left(\frac{7}{3} + 8\right)$$

$$-53.6 \times 2 \times (1 + 6) - (60.2 - 53.6) \times \frac{2}{2} \times \left(\frac{2}{3} + 6\right)$$

$$-36.5 \times \frac{1.29}{2} \times \left(6 - \frac{1.29}{3}\right) + 132.9 \times \left(\frac{6 - 1.29}{2}\right)\left(\frac{6 - 1.29}{3}\right)$$

$$= -2263.4\,kN\,m/m$$

This example gives a clear indication with regard to the cost effectiveness of design methods in conjunction with soil conditions and depth of retained soil, in that the excessive depth of embedment as well as the large maximum bending moment makes the use of a cantilever wall in this instance uneconomical. It also justifies why, in practice, cantilever walls are only used when the height of the retained soil is less than about 5 m.

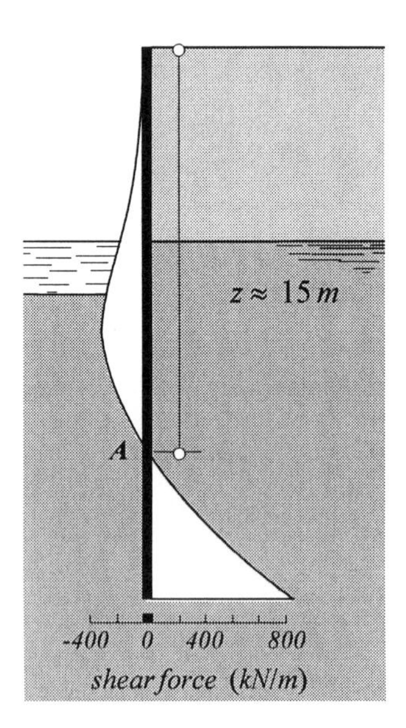

Figure 11.36: Shear force diagram.

Problems

11.1 Consider the anchored sheet pile with the dimensions indicated in figure $p11.1$. Knowing that the soil and loading conditions are the same as those used in example 11.5, and taking advantage of the then established net pressure diagram calculated using a factor of safety $F = 1$ on K_p, determine:

- the safe depth of embedment *D*,
- the anchorage force *F*,
- the maximum bending moment M_{max}.

Ans: $D = 5.97\,m$, $F \approx 187\,kN/m$, $M_{max} \approx 465\ \ kN\,m/m$

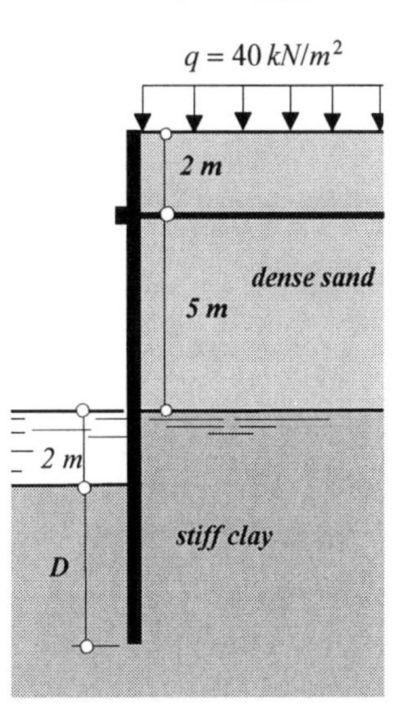

Figure p11.1

11.2 Rework problem $p11.1$, this time using the equivalent beam method.

Ans: $F_a \approx 182\,kN/m$, $D = 7.97\,m$, $M_{max} \approx 320\,kN\,m/m$

11.3 An anchored sheet pile cut-off wall is embedded in the soil strata depicted in figure $p11.3$. The loose sand has a bulk unit weight $\gamma = 19 kN/m^3$ and an effective angle of friction $\phi' = 30°$, and the dense sand is characterised by a saturated unit weight

$\gamma_{sat} = 20 kN/m^3$ and an angle $\phi' = 40°$.

(*a*) Use the free earth support method, in conjunction with the limiting conditions on the passive side (*i.e.* $F = 1$ on K_p), and check the adequacy of the depth of embedment.

(*b*) Calculate the anchorage force F_a and the maximum bending moment M_{max}.

Ans: (*a*) *Factor of safety on passive resistance*: $F = 4.73$
(*b*) $F_a \approx 193\ kN/m$, $M_{max} \approx 234\ kN\,m/m$

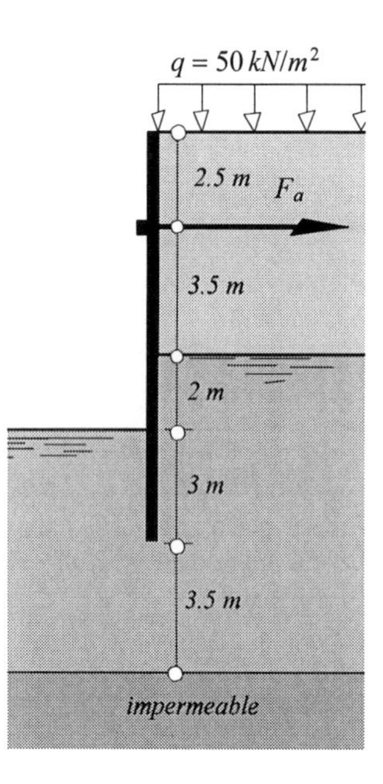

Figure p11.3

11.4 Use the free earth support method with a factor of safety $F = 2$ on K_p in conjunction with the sheet pile cut-off wall in problem *p*11.3 above, then apply Rowe's moment reduction method and estimate the reduced maximum bending moment M_c of the sheet pile.

Ans: $F_a \approx 188\ kN/m$, $M_c = 161.5\ kN\,m/m$

11.5 A diaphragm wall is embedded in a thick layer of a stiff clay overlain by a layer of clean sand (figure *p*11.5). The soil characteristics are as follows:

- sand: $\phi' = 35°$, $\gamma = 18.5\ kN/m^3$
- clay: $\phi' = 23°$, $\gamma = 19\ kN/m^3$, $\gamma_{sat} = 20\ kN/m^3$, $c' = 8\ kN/m^2$

The angle of wall friction can be taken as $\delta = 2\phi'/3$ on both sides of the wall. Using Boussinesq theory to determine the relevant coefficients of active and passive pressure, estimate the critical depth of embedment D' of the wall, and the maximum bending moment $M_{\max}$ as well as the depth at which it occurs.

Ans: $D' = 6.45\ m$, $M_{\max} \approx -633\ kN\,m/m$ *occurring at a depth* $z = 7.8\ m$

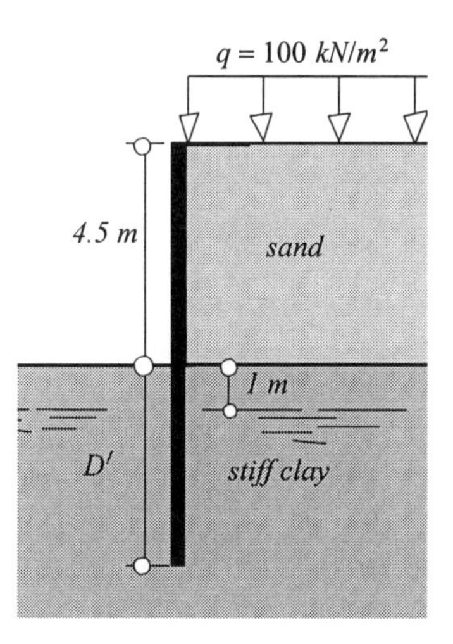

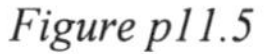
Figure p11.5

References

Blum, H. (1931) *Einspannungsverhältnisse bei Bohlwerken.* Wilhelm Ernest und Sohn Berlin, 27 pp.

Blum, H. (1951) *Beitrag zur Berechnung von Bohlwerken unter Berücksichtigung der Wandverformung, insbesondere bei mit der Tiefe linear zunehmender Widerstandsziffer.* Wilhelm Ernest und Sohn, Berlin, 32 pp.

British Steel (1997) *Piling Handbook,* 7th edn. British Steel Plc.

Delattre, L., Josseaume, H., Mespoulhe, L. and Delmer, T. (1996) *Comparaison des méthodes classiques de dimensionnement des écrans de soutènement ancrés.* Bulletin des Laboratoires des Ponts et Chaussées, 205, pp. 77–90.

Fleming, W. G. K., Weltman, A. J., Randolf, M. F. and Elson, W. K. (1992) *Piling Engineering,* 2nd edn. Wiley, New York.

Kerisel, J. and Absi, E. (1990) *Active and Passive Earth Pressure Tables,* 3rd edn. A.A. Balkema, Rotterdam.

Rowe, P. W. (1952) *Anchored sheet-pile walls.* Proceedings of the Institution of Civil Engineers, London, 1 (1), pp. 27–70.

Rowe, P. W. (1955).*A theoretical and experimental analysis of sheet pile walls.* Proceedings of the Institution of Civil Engineers, January, pp. 32–86.

Rowe, P. W. (1957) *Sheet pile walls in clay.* Proceedings of the Institution of Civil Engineers, London, 7, pp. 629–654.

Rowe, P. W. and Barden, L. (1966) *A new consolidation cell.* Géotechnique, 16 (2), pp. 162–170.

Skempton, A. W. (1951) *The bearing capacity of clays.* Proceedings of the Building Research Congress, London.

Terzaghi, K. (1954) *Anchored bulkheads.* Trans. ASCE, 119.

Tschebotarioff, G. P. (1973) *Foundations, Retaining and Earth Structures,* 2nd edn. McGraw-Hill, New York.

CHAPTER 12

Analysis of the expansion of cylindrical cavities in an infinite soil mass

12.1 The stability of cylindrical cavities: equilibrium equations

Consider the cross-section of a *circular tunnel* subjected to a uniform pressure σ_o at the top as depicted in figure 12.1. Within the circular area surrounding the tunnel opening, both *radial* and *circumferential* (or *hoop*) stresses are assumed to be principal stresses in the absence of any shear stress.

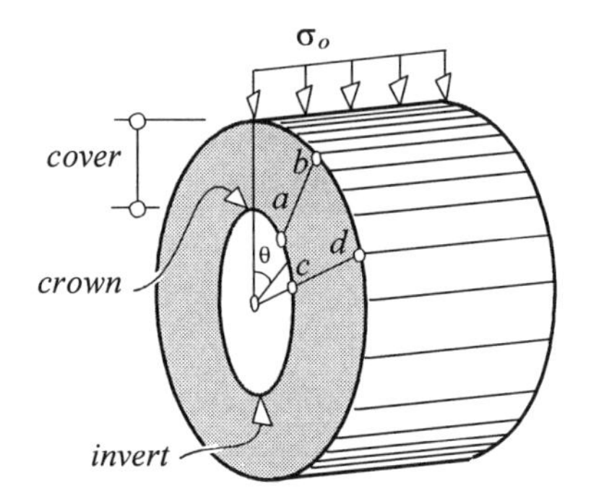

Figure 12.1: Cross-section of a circular tunnel.

Let us now analyse the equilibrium requirements of the curvilinear element *abdc*, whose axis is inclined at an angle θ with respect to the vertical. Using the different dimensions of the element shown in figure 12.2, and neglecting any shear stresses along the sides *ab* and *cd*, then the equilibrium equations can be established in a straightforward way.

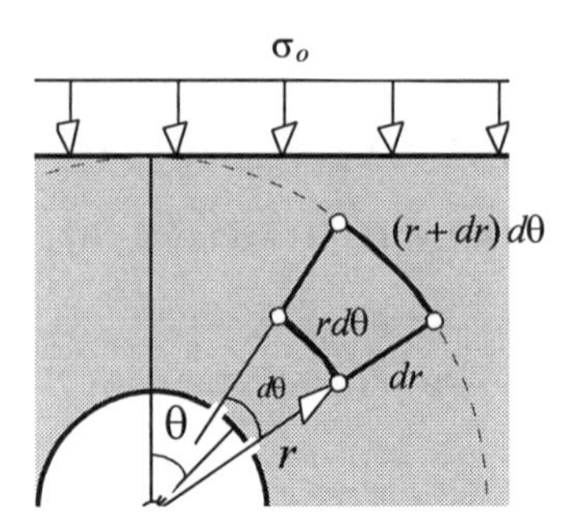

Figure 12.2: Dimensions relating to an elemental cross-section.

Thus, with reference to figure 12.3, the weight w of the element can be evaluated as follows:

$$w \approx \gamma\, r\, d\theta\, dr \tag{12.1}$$

γ being the (relevant) unit weight of soil. Resolving the forces radially in the knowledge that $\sin\frac{d\theta}{2} \approx \frac{d\theta}{2}$, it follows that:

$$(\sigma_r + d\sigma_r)(r + dr)\, d\theta - \sigma_r\, dr\, d\theta - \sigma_\theta\, dr \frac{d\theta}{2}$$
$$-(\sigma_\theta + d\sigma_\theta)\, dr \frac{d\theta}{2} + \gamma\, r\, dr\, d\theta \cos\theta = 0$$

whence:

$$\frac{d\sigma_r}{dr} + \frac{(\sigma_r - \sigma_\theta)}{r} + \gamma \cos\theta = 0 \tag{12.2}$$

Now, resolving perpendicular to the radius:

$$\sigma_\theta\, dr + \gamma\, r\, dr\, d\theta \sin\theta - (\sigma_\theta + d\sigma_\theta)\, dr = 0$$

so that:

$$\frac{d\sigma_\theta}{d\theta} - \gamma\, r \sin\theta = 0 \tag{12.3}$$

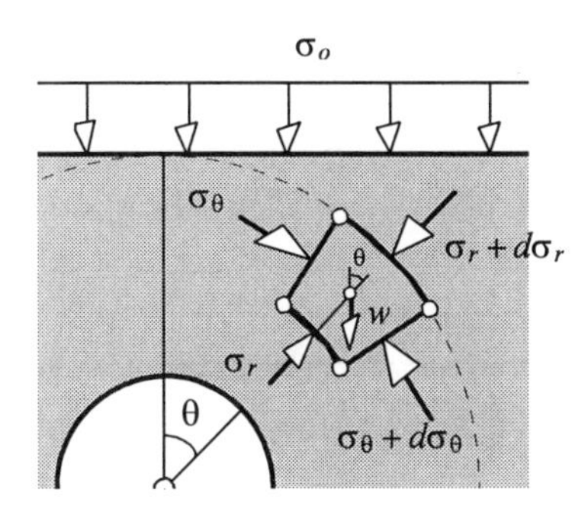

Figure 12.3: Stresses related to the elemental cross-section.

Both equations 12.2 and 12.3 represent the conditions of equilibrium of an element in the vicinity of the tunnel and, as can be seen from figure 12.3, the most unfavourable equilibrium conditions are likely to occur at the *crown* of the tunnel when $\theta = 0$. Under these circumstances, the equilibrium conditions are represented by equation 12.2 which, on substitution for $\theta = 0$, reduces to the following:

$$\frac{d\sigma_r}{dr} = \frac{\sigma_\theta - \sigma_r}{r} - \gamma \tag{12.4}$$

12.2 Stress analysis related to circular tunnels

12.2.1 Circular tunnels in cohesive soils

In the short term, clays derive their strength from their *undrained cohesion* c_u. Hence, in accordance with equation 5.53 in section 5.5.3, the failure criterion in the *passive mode* (*i.e.* $\sigma_\theta > \sigma_r$) under undrained conditions is such that:

$$(\sigma_\theta - \sigma_r) = 2c_u \tag{12.5}$$

Substituting for this quantity in equation 12.4, then integrating between the limits shown in figure 12.4, that is $\sigma_r = \sigma_a$ at $r = a$, and $\sigma_r = \sigma_o$ at $r = R$, it follows that:

$$\int_{\sigma_a}^{\sigma_o} d\sigma_r = \int_a^R \left(\frac{2c_u}{r} - \gamma\right) dr$$

or

$$\sigma_a = \sigma_o + \gamma(R-a) - 2c_u \ln\frac{R}{a} \tag{12.6}$$

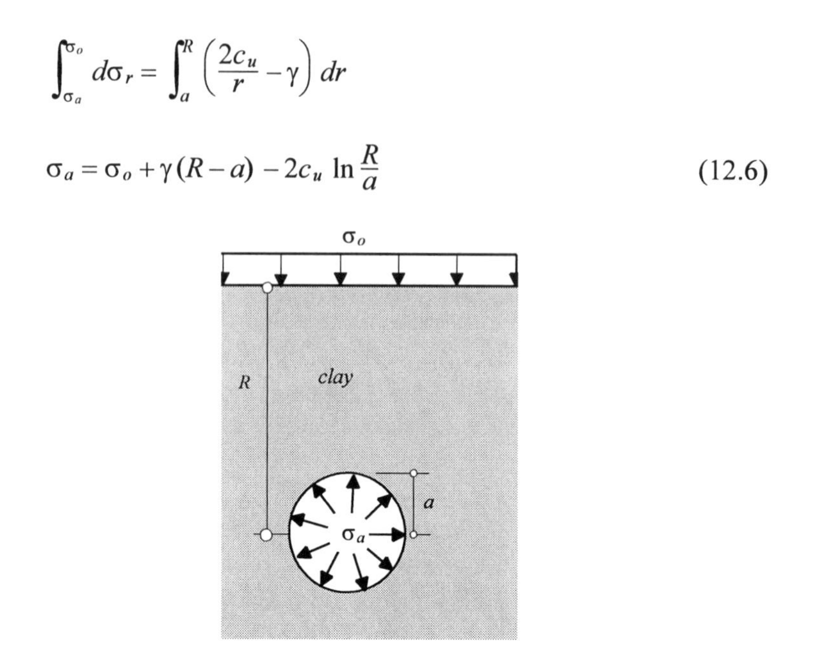

Figure 12.4: Circular tunnels in clays; boundary conditions.

Equation 12.6 yields the pressure σ_a needed in the *short term* to support an unlined tunnel in a clay with an undrained shear strength c_u and a unit weight γ; so that, for example, in the case of a tunnel with a diameter $2a = 4\,m$ and a cover $R - a = 12\,m$, executed in a firm clay characterised by a unit weight $\gamma = 20\,kN/m^3$ and an undrained shear strength $c_u = 60\,kN/m^2$, the support pressure would be, in the absence of any surcharge ($\sigma_o = 0$):

$$\sigma_a = 20 \times 12 - 2 \times 60 \times \ln\frac{14}{2} = 6.5\,kN/m^2$$

This value of (short term) support pressure is in this case surprisingly low given the depth of the tunnel, and reflects a fact known as *stress arching*.

Example 12.1

A circular tunnel is to be excavated unlined in a firm clay having an undrained shear strength $c_u = 50\,kN/m^2$ and a bulk unit weight $\gamma = 20\,kN/m^3$. It is proposed to find the maximum tunnel opening $a_{\max}$ that does not require any support pressure (*i.e.* $\sigma_a = 0$) during excavation, if the ground pressure were $\sigma_o = 60\, kN/m^2$.

Introducing the condition of zero support pressure ($\sigma_a = 0$) into equation 12.6, then rearranging, it is easy to show that:

$$a = \frac{2c_u \ln\left(\frac{R}{a}\right) - \sigma_o}{\gamma\left(\frac{R}{a} - 1\right)}$$

Accordingly, a graphical solution can be achieved by plotting the tunnel radius a *versus* the ratio R/a. The corresponding graph, depicted in figure 12.5 yields a maximum radius $a_{\max} \approx 1.31\, m$, corresponding to a ratio $R/a_{\max} = 4$; whence a distance $R \approx 5.24\, m$ (refer to figure 12.4).

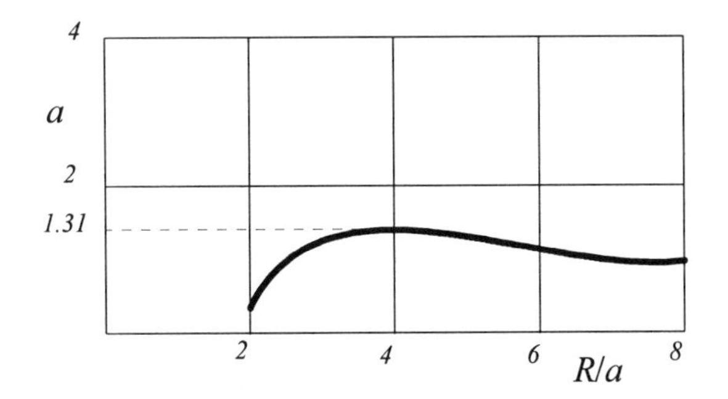

Figure 12.5: Graphical solution to the tunnel opening.

12.2.2 Circular tunnels in cohesionless soils

For soils without cohesion, an effective stress analysis is needed, and for a passive mode of failure, both radial and hoop effective stresses are related at failure through the coefficient of passive earth pressure (equation 10.14, section 10.3):

$$\sigma'_\theta = K_p \sigma'_r \tag{12.7}$$

$$K_p = \frac{1 + \sin\phi'}{1 - \sin\phi'}$$

In the absence of porewater pressure (*i.e.* dry soil), then $\sigma'_\theta \equiv \sigma_\theta$ and $\sigma'_r \equiv \sigma_r$. Thus, substituting for the hoop stress σ_θ from equation 12.7 in the equilibrium equation 12.4, and rearranging:

$$\frac{d\sigma_r}{dr} = \frac{(K_p - 1)\,\sigma_r}{r} - \gamma \tag{12.8}$$

Using the following change of variables $r\xi = \sigma_r$, so that $d\sigma_r = r\,d\xi + \xi\,dr$, then substituting for the quantities σ_r and $d\sigma_r$ in equation 12.8, it is seen that:

$$\frac{dr}{r} = \frac{d\xi}{(K_p - 2)\,\xi\ - \gamma} \tag{12.9}$$

The limits of integration for the above equation are such that (refer to figure 12.4):

$$r = a \quad \Rightarrow \quad \xi = \frac{\sigma_a}{a}, \ \text{and} \ \ r = R \quad \Rightarrow \quad \xi = \frac{\sigma_o}{R}$$

Thus:

$$\int_{\sigma_a/a}^{\sigma_o/R} \frac{d\xi}{(K_p - 2)\,\xi\ - \gamma} = \int_a^R \frac{dr}{r}$$

On integration, it is easy to establish that:

$$\sigma_a = \frac{a}{(K_p - 2)}\left\{\left(\frac{a}{R}\right)^{K_p - 2}\left[(K_p - 2)\left(\frac{\sigma_o}{R} - \frac{\gamma}{(K_p - 2)}\right)\right] + \gamma\right\} \tag{12.10}$$

Example 12.2

A laboratory experiment is undertaken in a small transparent tank to simulate the collapse of a cylindrical cavity. The experiment consists of burying a thin plastic tube in a uniform sand having a unit weight $\gamma = 20\,kN/m^3$ and an angle $\phi' = 37^\circ$. Considering the dimensions indicated in figure 12.6, and assuming the tube radial stiffness is equivalent to $\sigma_a = 1\,kN/m^2$, estimate the maximum static stress σ_o needed to cause collapse of the tube.

The maximum pressure σ_o capable of causing the opening to collapse is calculated from equation 12.10, which in the circumstances is rearranged as follows:

$$\sigma_o = \left\{\left[(K_p - 2)\frac{\sigma_a}{a} - \gamma\right]\left(\frac{R}{a}\right)^{K_p-2} + \gamma\right\}\frac{R}{(K_p-2)}$$

with the coefficient of passive earth pressure: $K_p = \frac{1+\sin 37}{1-\sin 37} = 4.02$.

Substituting for the numerical values into the above equation yields :

$$\sigma_o = \left[\left(\frac{2.02}{0.03} - 20\right) \times \left(\frac{0.5}{0.03}\right)^{2.02} + 20\right] \times \frac{0.5}{2.02} \approx 3448\, kN/m^2$$

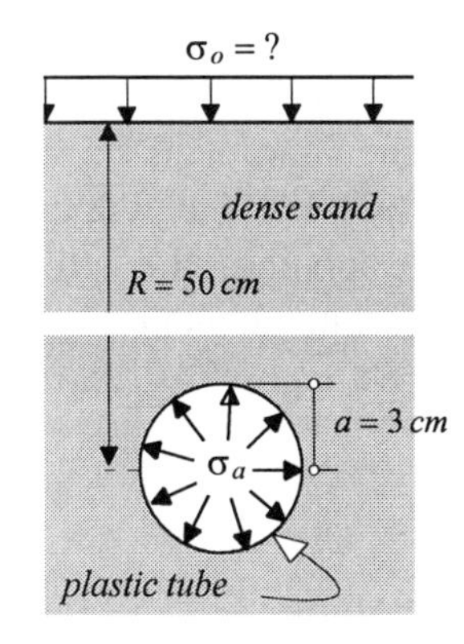

Figure 12.6: Dimensions of opening.

This value indicates plainly that the collapse of very small openings in sands necessitates large static stresses. This is why vibration is used in conjunction with sand compaction, to generate a state nearing *quick conditions* (refer to section 3.2) whereby the shear strength of sand is minimised, thus allowing the solid particles to be optimally positioned with respect to one another, and reducing in the process the void ratio and hence increasing the density of sand.

It is essential to bear in mind that, in practice, the support pressure for cohesionless soils must be provided by structural elements such as concrete or steel linings, since the application of an *air pressure* in this case would increase the total and porewater pressures equally, thus leaving the effective stresses unchanged. More importantly perhaps, *the use of large air pressures for tunnelling in clays at depth can potentially be a health hazard.* Under any circumstances, the air pressure should not exceed 350 kN/m^2 (*i.e.* 3.5 *bar*), under which the maximum duration of work is limited to a mere one hour per day considering that, in this case, the time of decompression (about 50 *min*) is longer than the maximum time per shift (around 30 *min*) then, most importantly, a 6 *hour* rest interval is needed to recover physically (Tschebotarioff, 1973). Remember that an air pressure of 350 kN/m^2 is equivalent to a 35 *m* head of water.

Also, the previous closed-form solutions relates to circular openings executed in simple soil conditions (essentially homogeneous clays and sands), and cannot therefore be extended to more complex soil strata which are likely to be encountered *in situ,* nor can it apply to rocks which are characterised by different constitutive stress–strain relationships and different boundary conditions (see Brady and Brown (1993), Timoshenko and Goodier (1970), Jaeger (1979)). For such complex soil conditions and for different shapes of tunnel opening, the stress analysis is usually undertaken numerically through finite element modelling (see for instance Mestat (1997), Smith and Griffiths (1998), Assadi and Sloan (1991), Zienkiewicz and Taylor (1991)).

12.3 The pressuremeter test analogy

12.3.1 The pressuremeter equipment

The *pressuremeter* was first invented as an *in situ* testing tool in the mid-1950s by Louis Ménard, who then went on to develop a semi-empirical design method for foundations based on pressuremeter test results (Ménard, 1962, 63). The testing equipment and the related design techniques have been continuously refined since and have now become, in France at least, a standard design method for shallow and deep foundations (MELT, 1993).

The equipment itself has undergone considerable technical changes since Ménard's time and different types of pressuremeters are now available, two of which are widely used: the *prebored pressuremeter* (*PBP*) and the

While a PBP such as the one depicted in figure 12.7 is equipped with highly sensitive sensors to detect a borehole wall displacement of 10^{-6} *mm* (or 1 μm), corresponding to a hoop strain $\varepsilon_\theta \approx 0.003\%$, it is obvious that the quality of measurement depends on the degree of soil disturbance which is bound to occur during the preboring of the hole. The disturbance is largely due to the scouring of the cavity wall by the high pressure drilling fluid. In order to *minimise* these effects, the self boring pressuremeter (*SBP*) was developed independently in France (Baguelin *et al.*, 1972, 74) and in Britain (Hughes, 1973). A thorough discussion of the technical details of different types of pressuremeter equipment, as well as the practical aspects related to their use can be found in Clarke (1995).

The pressuremeter test, coupled with the elegant theory of the expansion of cylindrical cavities, can provide the designer with reliable values of soil variables such as the variation with depth of horizontal stresses and shear modulii, as well as the undrained shear strength and porewater pressures (for clays) and the angle of dilation (for sands).

12.3.2 The pressuremeter test in clays

The soil behaviour described in the following analysis is assumed to be elastic–perfectly plastic. Moreover, the ensuing mathematical formulation s based on the analysis by Gibson and Anderson (1961) (also, refer to Baguelin *et al.* (1978)).

Because a pressuremeter test in a clay can be performed very quickly (a est can be practically finished in a matter of minutes), the clay behaviour therefore undrained for the test duration since the excess water pressure ue to the expansion of the borehole remains virtually unchanged. onsequently, the analysis that follows, applicable to clays, is based on tal stresses.

eferring to figure 12.8, the equilibrium of horizontal forces, applied to an ement of soil at a distance *r* from the centre of cavity, having a radial ickness *dr*, requires that:

$$\sigma_r r\, d\theta - (\sigma_r + d\sigma_r)(r + dr)\, d\theta + 2dr\, \sigma_\theta \frac{d\theta}{2} = 0 \tag{12.11}$$

self-boring pressuremeter (*SBP*). Figure 12.7 represents a detailed s
of the *Cambridge Insitu* high pressure dilatometer, a PBP that needs
inserted into a prebored 2.0 *m* long NX-size hole (*i.e.* ≈ 76 *mm* in dia
as shown in the figure. The test consists of applying a radial pressure
soil by inflating a *rubber membrane* and measuring the ensuing
deformation under conditions of axial symmetry and plane strain.

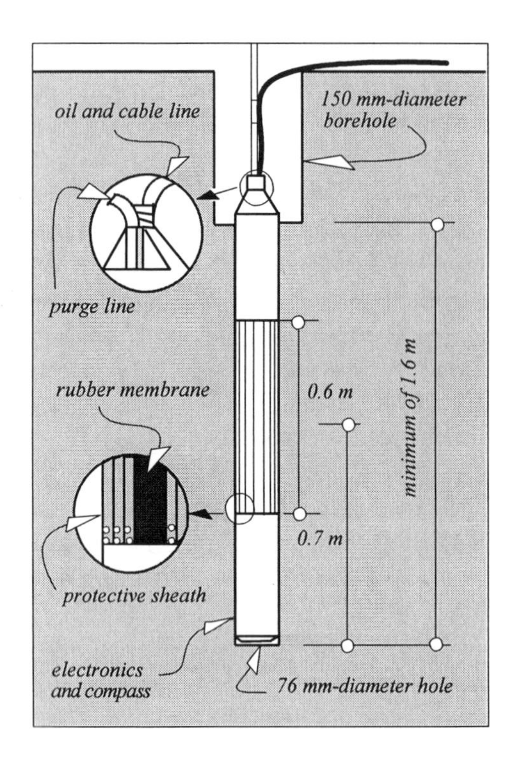

Figure 12.7: Cambridge Insitu high pressure dilatometer.

The rubber membrane is inflated by means of hydraulic o
and the *radial displacement* is averaged from the reading
gauges placed in the same plane about halfway along
height. The *Casing* of the borehole depends on the type of
the depth at which the test is undertaken. A bentonite mud
to prevent the collapse of the borehole walls.

or

$$\frac{(\sigma_r - \sigma_\theta)}{r} + \frac{d\sigma_r}{dr} = 0 \qquad (12.12)$$

Notice that the equilibrium equation 12.12 is similar to equation 12.4 established earlier, except that the weight of the element becomes irrelevant to its equilibrium in the case of a radial stress field such as the one generated by a pressuremeter test.

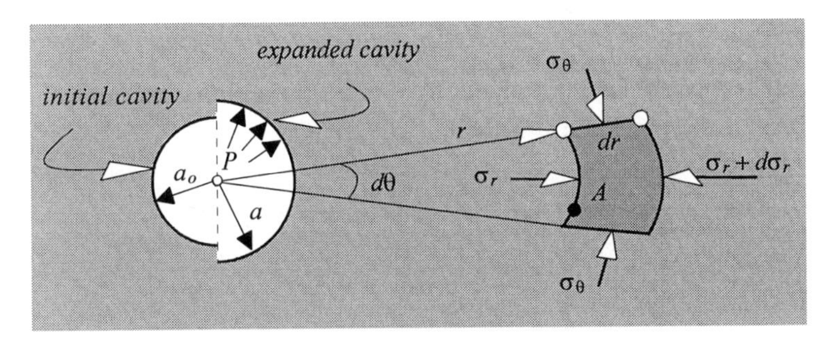

Figure 12.8: Plan view of the stress field applied to an element in the vicinity of an expanded cavity.

A solution to equation 12.12 can be established in conjunction with the elasto–plastic behaviour of the clay; starting first, with the assumed initial linear elastic behaviour.

Prior to testing, the *in situ* stresses are such that: $\sigma_r = \sigma_\theta = \sigma_{ho} = K_o \sigma_v$, where:

σ_{ho} : the total horizontal *in situ* stress
K_o : the coefficient of earth pressure at rest
σ_v : the total vertical *in situ* stress.

This state of stress indicates that, for any deformation to occur, the pressure P applied through the membrane of the pressuremeter (refer to figure 12.7) must exceed σ_{ho} . Accordingly, the radial and circumferential *total* stress increments, resulting from a pressure $P = \sigma_{ho} + \Delta P$ would be such that:

$$\Delta\sigma_r = \sigma_r - \sigma_{ho} \qquad (12.13)$$

$$\Delta\sigma_\theta = \sigma_\theta - \sigma_{ho} \qquad (12.14)$$

Let the displacement undergone by point A in figure 12.8 be w. The principal strains at A can then be expressed in terms of w, so that when *plane strain conditions* are taken into account (*i.e.* $\varepsilon_z = 0$), the following relationships apply:

- *radial strain*: $$\varepsilon_r = \frac{dr - (dr + dw)}{dr} = -\frac{dw}{dr} \tag{12.15a}$$

- *hoop strain*: $$\varepsilon_\theta = \frac{2\pi r - 2\pi (r + w)}{2\pi r} = -\frac{w}{r} \tag{12.15b}$$

Applying Hooke's law of elasticity, and taking into account the *axial symmetry* of the problem, it follows that:

$$\begin{bmatrix} \varepsilon_r \\ \varepsilon_\theta \\ 0 \end{bmatrix} = \begin{bmatrix} -dw/dr \\ -w/r \\ 0 \end{bmatrix} = \frac{1}{E} \begin{bmatrix} 1 & -\nu & -\nu \\ -\nu & 1 & -\nu \\ -\nu & -\nu & 1 \end{bmatrix} \begin{bmatrix} \Delta\sigma_r \\ \Delta\sigma_\theta \\ \Delta\sigma_z \end{bmatrix} \tag{12.16}$$

Moreover, the shear modulus G and the elasticity modulus E of the clay are linked through Poisson's ratio ν as follows:

$$G = \frac{E}{2(1+\nu)} \tag{12.17}$$

Thus, substituting for the different strains from equation 12.15 into the matrix system 12.16, the increments of radial and circumferential stresses can now be determined:

$$\Delta\sigma_r = -\frac{2G}{1-2\nu}\left[(1-\nu)\frac{dw}{dr} + \nu\frac{w}{r}\right] \tag{12.18}$$

$$\Delta\sigma_\theta = -\frac{2G}{1-2\nu}\left[\nu\frac{dw}{dr} + (1-\nu)\frac{w}{r}\right] \tag{12.19}$$

In addition, the quantity $d(\Delta\sigma_r)/dr$ can easily be derived from equation 12.18:

$$\frac{d(\Delta\sigma_r)}{dr} = -\frac{2G}{1-2\nu}\left[(1-\nu)\frac{d^2w}{dr^2} + \frac{\nu}{r}\left(\frac{dw}{dr} - \frac{w}{r}\right)\right] \tag{12.20}$$

Substituting for $\Delta\sigma_r$, $\Delta\sigma_\theta$ and $d(\Delta\sigma_r)/dr$ from equations 12.18, 12.19 and 12.20 respectively into the incremental form of the equilibrium equation 12.12 yields the following second order differential equation:

$$r\frac{d^2w}{dr^2} + \frac{dw}{dr} - \frac{w}{r} = 0 \tag{12.21}$$

for which the boundary conditions are, with reference to figure 12.8:

$$r \to \infty, \qquad w \to 0$$
$$r = a, \qquad \Delta\sigma_r = \Delta P$$

Hence the solution to equation 12.21:

$$w = \frac{a^2}{r}\frac{\Delta P}{2G} \tag{12.22}$$

Now, substituting for w from the latter equation into equations 12.18 and 12.19, then rearranging:

$$\Delta\sigma_r = \frac{a^2}{r^2}\Delta P \tag{12.23}$$

$$\Delta\sigma_\theta = -\frac{a^2}{r^2}\Delta P = -\Delta\sigma_r \tag{12.24}$$

Knowing that $P = \sigma_{ho} + \Delta P$, and inserting $\Delta\sigma_r$ and $\Delta\sigma_\theta$ from equations 12.13 and 12.14 into the above equations, it is straightforward to show that the elastic distribution of both radial and circumferential stresses are as follows:

$$\sigma_r = \frac{a^2}{r^2}P + \sigma_{ho}\left(1 - \frac{a^2}{r^2}\right) \tag{12.25}$$

$$\sigma_\theta = -\frac{a^2}{r^2}P + \sigma_{ho}\left(1 + \frac{a^2}{r^2}\right) \tag{12.26}$$

Moreover, according to equation 12.22, the strain at the borehole wall where the radial displacement is measured is:

$$\varepsilon = \frac{w_{(r=a)}}{a} = \frac{\Delta P}{2G} \tag{12.27}$$

so that at the borehole wall, the *elastic* strain is, in theory, related linearly to the applied pressure P as follows:

$$P = \sigma_{ho} + 2G\,\varepsilon_{(r=a)} \tag{12.28}$$

where G represents the shear modulus of the clay. In practice, however, the clay behaviour is more identifiable with figure 12.9(*b*), which indicates that, *a priori*, the shear modulus cannot be determined from the early elastic response of soil to the applied pressure. This difficulty is overcome by undertaking an unload–reload cycle and measuring G from the corresponding loop as indicated in the figure. *During the unloading phase, care must be taken so as to avoid the collapse of the borehole walls. This may occur when the radial stress becomes a minor stress, in other words, when the amplitude of unloading exceeds* $2c_u$.

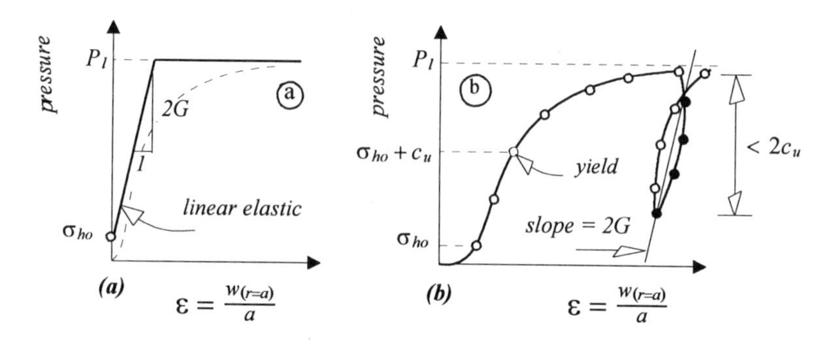

Figure 12.9 : (a) Assumed and (b) experimental aspects of a pressuremeter test in clay.

If the clay were further pressurised at the end of the elastic behaviour, then a plastic annulus with a radius R will start to form around the borehole as soon as the failure criterion is reached, *i.e.*, when:

$$(\sigma_r - \sigma_\theta) = (\Delta\sigma_r - \Delta\sigma_\theta) = 2c_u \tag{12.29}$$

Thus, substituting for σ_r and σ_θ from equations 12.25 and 12.26 into equation 12.29, it is easy to show that, *on yielding*, which first occurs at the borehole wall (*i.e.* $r = a$), the undrained shear strength c_u of the clay is related to the applied pressure P as follows:

$$P = \sigma_{ho} + c_u \tag{12.30}$$

Moreover, at the onset of plasticity, the incremental form of the equilibrium equation still holds:

$$\frac{\Delta\sigma_r - \Delta\sigma_\theta}{r} + \frac{d\Delta\sigma_r}{dr} = 0 \tag{12.31}$$

Whence, substituting for $(\Delta\sigma_r - \Delta\sigma_\theta)$ from equation 12.29, it follows that:

$$\frac{2c_u}{r} + \frac{d\Delta\sigma_r}{dr} = 0 \tag{12.32}$$

On integration:

$$\Delta\sigma_r = -2c_u \ln r + C \tag{12.33}$$

The boundary conditions shown in figure 12.10 are such that at $r = R$ (R being the radius of the plastic annulus):

$$\Delta\sigma_r = \Delta\sigma_\theta + 2c_u \quad \text{(plastic boundary)}$$
$$\Delta\sigma_r = -\Delta\sigma_\theta \quad \text{(elastic boundary)}$$

Hence, combining these two conditions:

$$r = R \rightarrow \Delta\sigma_r = c_u$$

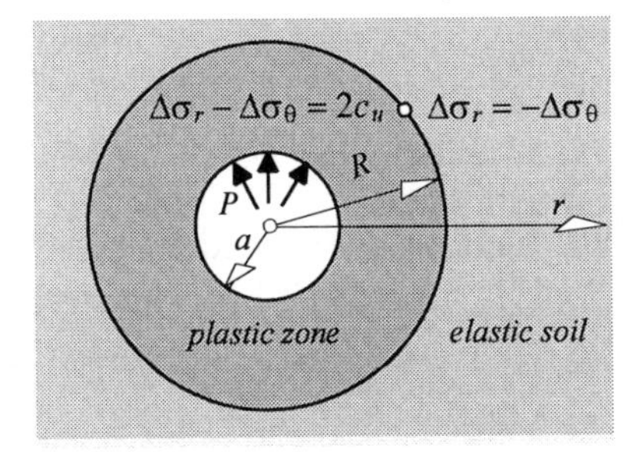

Figure 12.10: Boundary conditions related to the plastic behaviour of clays.

whence the value of the constant C in equation 12.33:

$$C = c_u + 2c_u \ln R \tag{12.34}$$

The radial and circumferential total stress increments in the plastic range are therefore:

$$\Delta\sigma_r = c_u + 2c_u \ln \frac{R}{r} \tag{12.35}$$

$$\Delta\sigma_\theta = -c_u + 2c_u \ln \frac{R}{r} \tag{12.36}$$

The radius R of the plastic annulus can be determined using the boundary condition at $r = a$ (a being the current radius of the borehole):

$$r = a \rightarrow \quad \Delta\sigma_r = \Delta P = P - \sigma_{ho}$$

where P is the applied pressure.

Inserting this condition into equation 12.35, then rearranging:

$$R = a \exp\left(\frac{P - \sigma_{ho} - c_u}{2c_u}\right) \tag{12.37}$$

Finally, substituting for R from the above equation into equations 12.35 and 12.36, it is seen that the variation of both radial and circumferential stresses in the *plastic range* is:

$$\sigma_r = P + 2c_u \ln \frac{a}{r} \tag{12.38}$$

$$\sigma_\theta = P - 2c_u(1 - \ln \frac{a}{r}) \tag{12.39}$$

Notice that the stresses σ_r and σ_θ are related to the stress increments $\Delta\sigma_r$ and $\Delta\sigma_\theta$ through equations 12.13 and 12.14. The stress variations both in the elastic and the plastic range are depicted in figure 12.11.

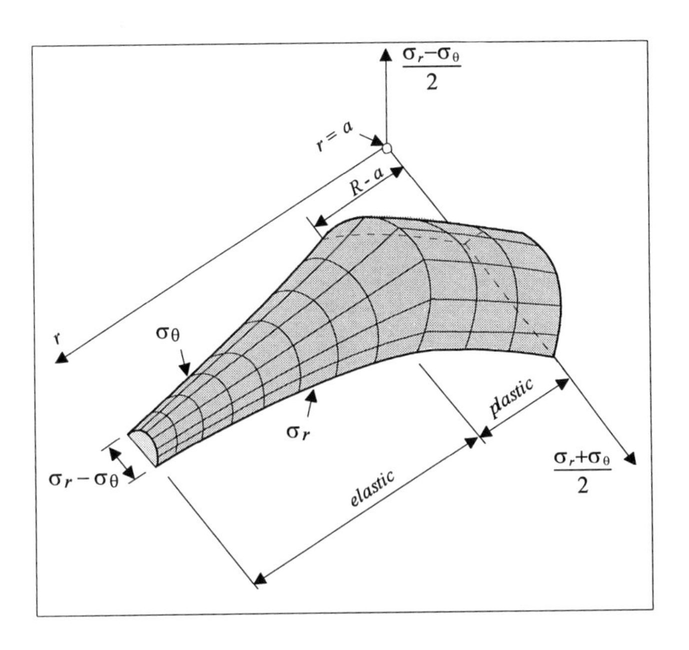

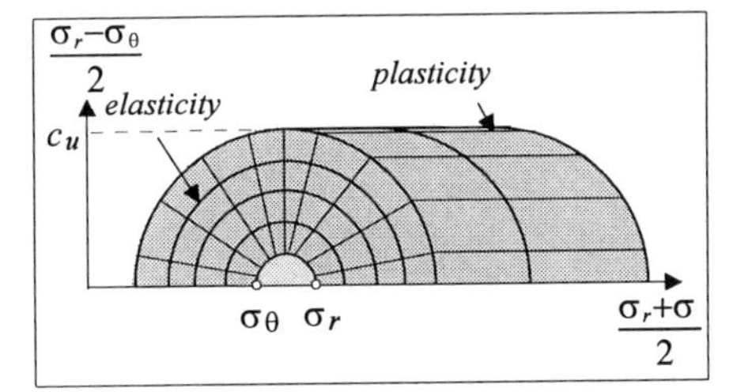

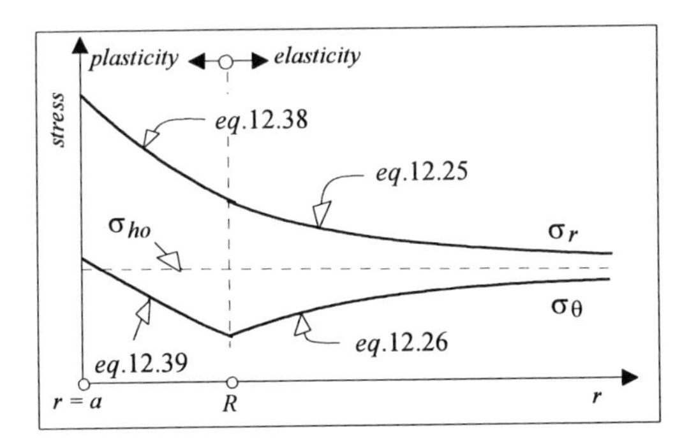

Figure 12.11: Variation of radial and circumferential stresses during a pressuremeter test in a clay.

Because the pressuremeter test is undrained, the radial displacement w related to the formation and expansion of the plastic annulus is such that the volume of the plastic zone remains constant as shown in figure 12.12. Thus, equating the volumes at any two different stages of pressure amounts to:

$$z\pi\left[(R-w)^2 - a_o^2\right] = z\pi\left(R^2 - a^2\right) \tag{12.40}$$

Neglecting the quantity w^2 in the above equation then rearranging:

$$a^2 - a_o^2 = 2Rw \tag{12.41}$$

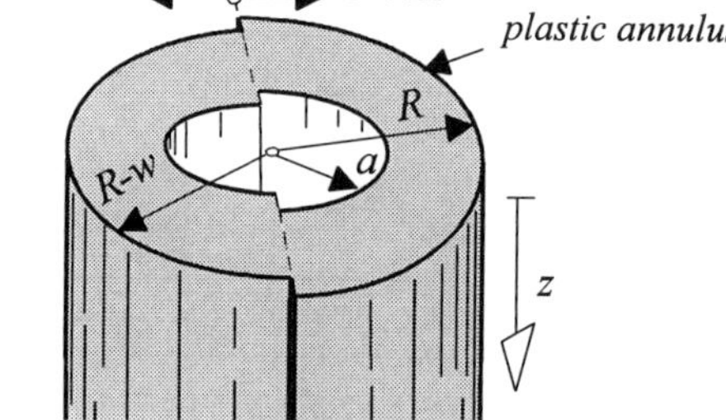

Figure 12.12: Constant volume test.

With reference to figure 12.12, it is clear that both elastic and plastic conditions are simultaneously verified at any point located on the outer boundary of the plastic annulus $r = R$. The elastic condition is obtained from Hooke's law (matrix system 12.16) which yields:

$$w = -\frac{R}{2G}[(1-\nu)\Delta\sigma_\theta - \nu\Delta\sigma_r] \tag{12.42}$$

Since the test is undrained, Poisson's ratio $\nu = 0.5$, and equation 12.42 then reduces to:

$$w = \frac{R}{4G}(\Delta\sigma_r - \Delta\sigma_\theta) \tag{12.43}$$

Now, inserting the plastic condition from equation 12.29 into the above, it follows that:

$$w = \frac{Rc_u}{2G} \tag{12.44}$$

Replacing R by its value from equation 12.37, then substituting for w from equation 12.44 into equation 12.41 yields:

$$\frac{a^2 - a_o^2}{a^2} = \frac{c_u}{G} \exp\left(\frac{\Delta P - c_u}{c_u}\right) \tag{12.45}$$

The left-hand side of equation 12.45 is in fact the *shear strain* expressing the changes in volume of the expanding membrane:

$$\varepsilon_s = \frac{\Delta V}{V} = \frac{a^2 - a_o^2}{a^2} = 1 - \frac{a_o^2}{a^2} \tag{12.46}$$

so that, once rearranged, equation 12.45 becomes:

$$\Delta P = c_u\left(1 + \ln \frac{G}{c_u}\right) + c_u \ln \varepsilon_s \tag{12.47}$$

Note that some authors refer (unjustifiably?) to the strain in equation 12.46 as volumetric strain. This cannot be the case since the actual volume of soil is constant due to the undrained nature of the test.

It can readily be seen from the latter equation that the yielding of clay is characterised by a linear relationship between the applied pressure and the natural logarithm of the shear strain. The corresponding graph is depicted in figure 12.13, where the linear portion representing equation 12.47 and corresponding to the plastic range is characterised by a slope c_u.

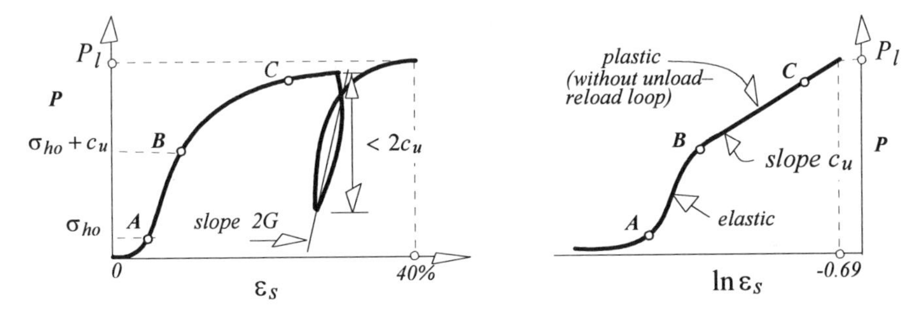

Figure 12.13: Relationship between applied pressure and shear strain during a pressuremeter test in clays.

In practice, the variation in time of the radial displacement at the borehole wall is usually recorded during a pressuremeter test rather than the volume change of the borehole. However, the relationship between the hoop and shear strains can easily be established. Considering the sign convention which stipulates that compressive strains are positive, the hoop strain can be expressed as:

$$\varepsilon_\theta = -\frac{a_o - a}{a_o} = \frac{a}{a_o} - 1 \tag{12.48}$$

thence:

$$\frac{a_o}{a} = \frac{1}{1+\varepsilon_\theta} \tag{12.49}$$

Inserting the above quantity into equation 12.46 yields the relationship between shear and hoop strains:

$$\varepsilon_s = 1 - \frac{1}{(1+\varepsilon_\theta)^2} \tag{12.50}$$

Also, equation 12.47 can be rewritten as follows:

$$\Delta P = P_l + c_u \ln(\varepsilon_s) \tag{12.51}$$

where the *limit pressure* P_l is defined as follows:

$$P_l = c_u\left(1 + \ln\frac{G}{c_u}\right) \tag{12.52}$$

According to equation 12.51, P_l is the *theoretical* pressure needed to induce a shear strain $\varepsilon_s = \Delta V/V = 1$ (or 100%). However, it is an accepted practice to assume that P_l is the pressure under which a hoop strain $\varepsilon_\theta \approx 40\%$ is measured, in which case it is easy to establish from equations 12.50 and 12.51 that P_l corresponds to a shear strain $\varepsilon_s \approx 50\%$, hence the value *–0.69* on the logarithmic scale of figure 12.13.

12.3.3 The Pressuremeter test in sands

A pressuremeter test undertaken in a sandy soil does not generate excess porewater pressures because of the high permeability of such materials.

Consequently, any increase in the applied pressure is transferred to the soil skeleton, and the analysis in this case is made in terms of effective stresses. The behaviour of sand is assumed to be elastic–perfectly plastic.

As far as the elastic behaviour is concerned, the analysis undertaken previously relating to the elastic behaviour of clays is valid for sands, provided that effective stresses are used instead of total stresses. Therefore, only the plastic behaviour will be analysed in what follows.

Once the elastic limit is exceeded, the sand fails and its volume is no longer constant. When expressed in terms of effective stresses, the equilibrium equation 12.12 established earlier becomes:

$$\frac{\sigma'_r-\sigma'_\theta}{r}+\frac{d\sigma'_r}{dr}=0 \qquad (12.53)$$

Moreover, the effective stress ratio at failure (in the active mode) is given by:

$$\frac{\sigma'_\theta}{\sigma'_r}=\frac{1-\sin\phi'}{1+\sin\phi'}=\tan^2\left(\frac{\pi}{4}-\frac{\phi'}{2}\right)=K_a \qquad (12.54)$$

On substitution for σ'_θ from equation 12.54 into equation 12.53, it follows that:

$$-\frac{d\sigma'_r}{\sigma'_r}=(1-K_a)\,\frac{dr}{r} \qquad (12.55)$$

The above equation is then integrated using both boundary conditions at the limit of the failed sand zone ($r=R$) depicted in figure 12.14. Since the effective stress increments are related to the effective horizontal stress *in situ* as follows:

$$\Delta\sigma'_r=\sigma'_r-\sigma'_{ho} \qquad (12.56)$$

$$\Delta\sigma'_\theta=\sigma'_\theta-\sigma'_{ho} \qquad (12.57)$$

the distribution of the radial and circumferential effective stresses with r can be established effortlessly:

$$\sigma'_r = \frac{2\sigma'_{ho}}{1+K_a}\left(\frac{R}{r}\right)^{1-K_a} \tag{12.58}$$

$$\sigma'_\theta = K_a \sigma'_r \tag{12.59}$$

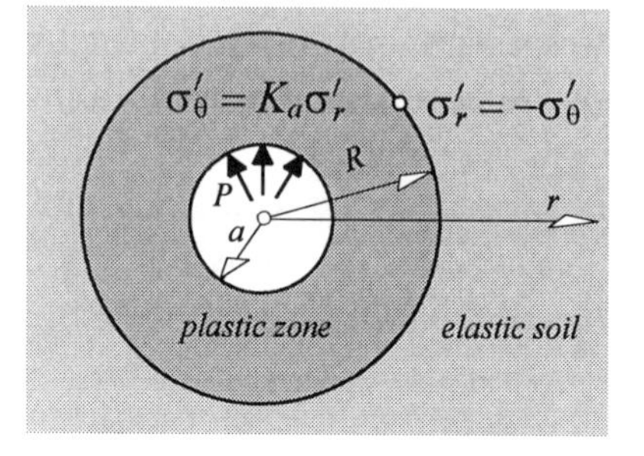

Figure 12.14: Limiting conditions.

where the radius of the plastic annulus R can be determined by using the following boundary condition in conjunction with equation 12.58: $r \to a$ (a being the current radius of the borehole), $\sigma_r = P$ (P is the applied pressure). Thence equation 12.58 yields:

$$R = a\left[\frac{P(1+K_a)}{2\sigma'_{ho}}\right]^{1/(1-K_a)} \tag{12.60}$$

Finally, a substitution for R in equations 12.58 and 12.59 yields the expressions of both effective radial and hoop stresses in the plastic range where $a \le r \le R$:

$$\sigma'_r = P\left(\frac{a}{r}\right)^{1-K_a} \tag{12.61}$$

$$\sigma'_\theta = K_a P\left(\frac{a}{r}\right)^{1-K_a} \tag{12.62}$$

The corresponding patterns of stress changes are shown in figure 12.15.

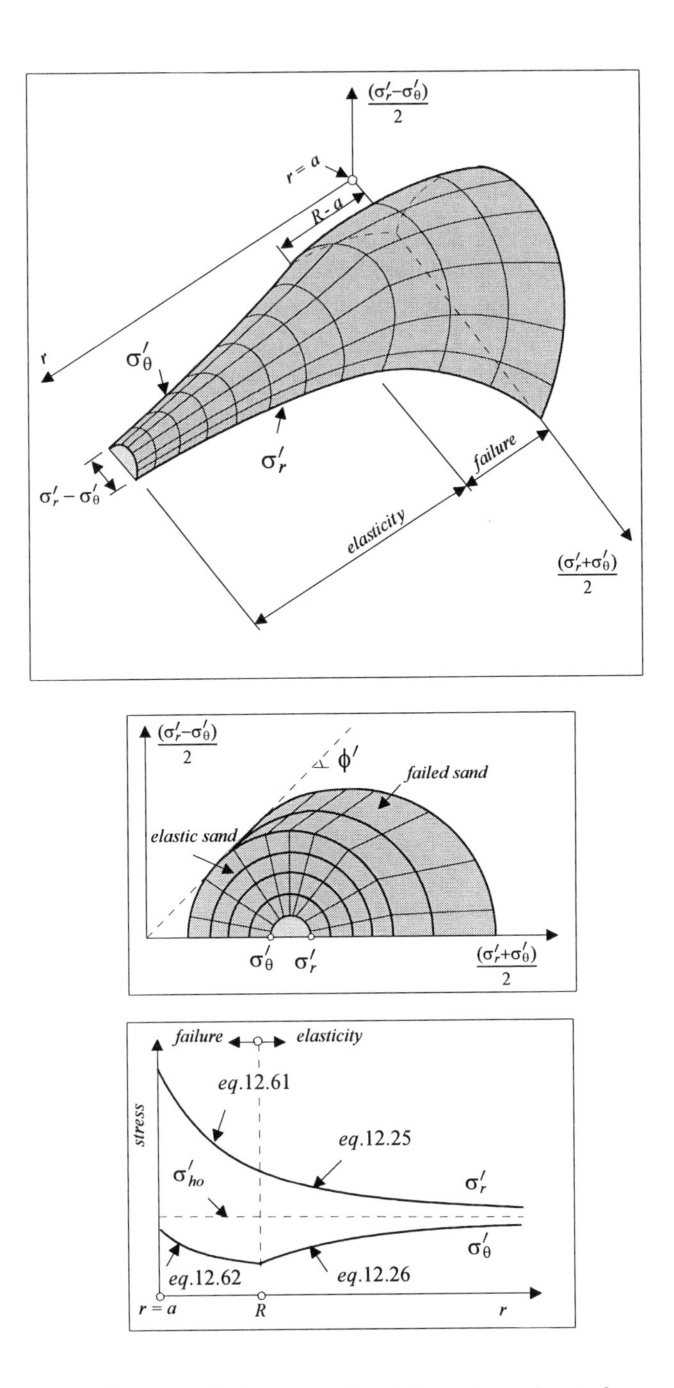

Figure 12.15: Variation of radial and circumferential stresses during a pressuremeter test in a sand.

Because at failure the volume is not constant, *Hughes et al.* (1977) assumed the sand to be failing at a constant dilation rate:

$$\sin \upsilon = -\frac{(\varepsilon_r + \varepsilon_\theta)}{(\varepsilon_r - \varepsilon_\theta)} \tag{12.63}$$

where υ is the *angle of dilation* of sand. They then argued that both dilatancy and internal angle of friction are constant over the strain range of a pressuremeter test once failure has occurred. Under such circumstances, the following relationship based on the Rowe (1962) stress dilatancy theory applies:

$$\frac{1 + \sin \phi'}{1 - \sin \phi'} = \left(\frac{1 + \sin \phi'_c}{1 - \sin \phi'_c}\right)\left(\frac{1 + \sin \upsilon}{1 - \sin \upsilon}\right) = N\left(\frac{1 + \sin \upsilon}{1 - \sin \upsilon}\right) \tag{12.64}$$

ϕ'_c being the critical state angle of shearing resistance.

Next, *Hughes et al.* (1977) integrated the equilibrium equation 12.55 using the boundary condition $\sigma'_r = \sigma'_R$ at the outer boundary $r = R$ in figure 12.14, where σ'_R represents the effective radial stress associated with the onset of failure. On integration, it is easy to show that within the failed area in the figure:

$$\ln\left(\frac{\sigma'_r}{\sigma'_R}\right) = (1 - K_a)\ln(R/r) \tag{12.65}$$

In general, the linear graph corresponding to equation 12.63 has an intercept, and accordingly, the equation can be expressed as follows:

$$-(\varepsilon_r + \varepsilon_\theta) = (\varepsilon_r - \varepsilon_\theta)\sin \upsilon - c \tag{12.66}$$

where the negative intercept c corresponds to a compressive volumetric strain in the absence of any dilatancy. The latter equation can easily be rearranged as follows:

$$-\varepsilon_r = n\varepsilon_\theta - c\frac{(n+1)}{2} \tag{12.67}$$

with:

$$n = \frac{1 - \sin \upsilon}{1 + \sin \upsilon} \qquad (12.68)$$

Substituting for both radial and hoop strains from equations 12.15 into equation 12.67, it is seen that:

$$\frac{dw}{dr} = -n\frac{w}{r} - c\frac{(n+1)}{2} \qquad (12.69)$$

This equation can be solved by making a change of variable similar to that used in conjunction with equation 12.8. Let $\xi = wr^n$ so that:

$$\frac{d\xi}{dr} = \frac{dw}{dr}r^n + n\,w\,r^{n-1}$$

A straightforward integration then yields:

$$\xi = -\frac{c}{2}r^{n+1} + b$$

so that when the change of variable is taken into account, it is seen that:

$$wr^n = -\frac{c}{2}r^{n+1} + b \qquad (12.70)$$

The constant b in equation 12.70 can be found from the outer boundary: $r = R$, $\quad w/R = \varepsilon_R = constant$. Thus:

$$b = R^{(n+1)}\left(\varepsilon_R + \frac{c}{2}\right) \qquad (12.71)$$

The complete solution to equation 12.70 can then be established:

$$\frac{w}{r} = \left(\frac{R}{r}\right)^{n+1}\left(\varepsilon_R + \frac{c}{2}\right) - \frac{c}{2} \qquad (12.72)$$

Substituting for the ratio R/r from equation 12.65 into equation 12.72, then writing at $r = a$, $\quad w/a = \varepsilon_{r=a}$ and $\sigma_r^{\prime} = P$, where P is the applied pressure and $\varepsilon_{r=a}$ is the measured hoop strain:

$$\left(\varepsilon_{r=a}+\frac{c}{2}\right)=\left(P/\sigma'_R\right)^{(n+1)/(1-K_a)}\left(\varepsilon_R+\frac{c}{2}\right)$$

so that:

$$\log\left(\varepsilon_{r=a}+\frac{c}{2}\right)=\frac{n+1}{1-K_a}\log(P)+constant \tag{12.73}$$

The constant c in equation 12.73 is according to *Hughes et al.* (1977) very small, and therefore can safely be discarded. Accordingly, when plotted as the variation of $\log P$ *versus* $\log\varepsilon_\theta$, equation 12.73 yields a linear graph (refer to figure 12.16) with a slope:

$$s=\frac{1-K_a}{n+1}=\frac{(1+\sin\upsilon)\sin\phi'}{1+\sin\phi'} \tag{12.74}$$

On substitution for the quantity N from equation 12.64, the slope s can be related both to the critical state angle ϕ'_c and the angle of dilation υ of the sand in the following way :

$$\sin\phi'=\frac{(N+1)\,s}{(N-1)\,s+2} \tag{12.75}$$

$$\sin\upsilon=\frac{2Ns-N+1}{N+1} \tag{12.76}$$

$$N=\tan^2\left(\frac{\pi}{4}+\frac{\phi'_c}{2}\right) \tag{12.77}$$

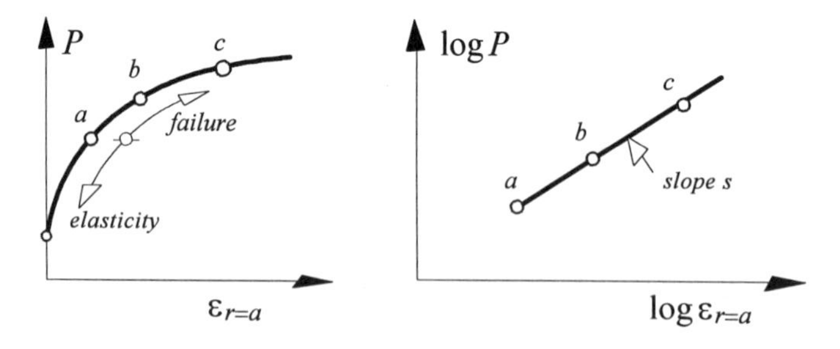

Figure 12.16: Stress–strain relationship at failure during a pressuremeter test in a sand.

12.4 Practical aspects related to the use of the pressuremeter

A successful application of the theoretical basis for the pressuremeter test developed earlier depends, by and large, on how representative are the measured stiffness parameters of the soil. It must be emphasised that, however careful an operator can be, some soil disturbance is bound to occur during testing. Moreover, the results profile depends on the type of equipment used for the test. In this respect, figure 12.17 illustrates typical stress–strain graphs corresponding to the same soil tested using a *prebored PBP* and a *sef-boring SBP* pressuremeters. It has been already mentioned that, in practice, the expansion of cavity during a *PBP* test is limited to a shear strain $\varepsilon_s \approx 50\%$. In contrast, the expansion curve of an *SBP* test tends to have a more realistic form of the initial part of the graph, reflecting the lower levels of disturbance associated with the insertion process. In both tests, the limit pressure for clays which, in theory, corresponds to a ratio $\Delta V/V = 1$, where V represents the expanded volume of the membrane, reflects the assumption that the clay was loaded to a limiting stress condition.

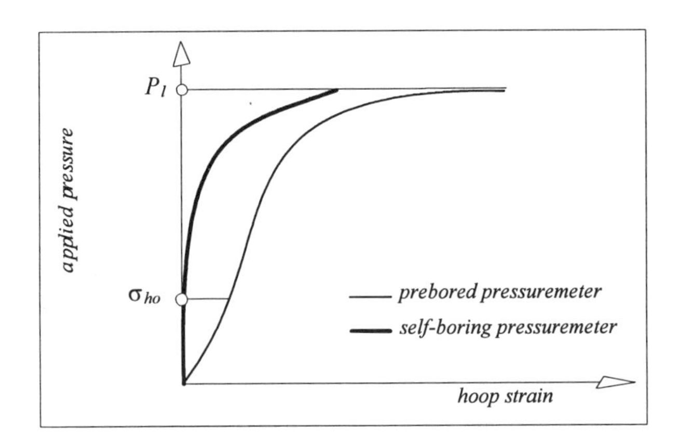

Figure 12.17: Typical PBP and SBP expansion graphs.

Leaving aside the empirical correlation between these graphs, the question that arises at this stage is how representative are these stiffness parameters (that is σ_{ho}, P_l, c_u, ϕ') that have been measured from a pressuremeter test. It is useful to remember that the pressuremeter theory detailed earlier is based on the *assumption* that the soil behaviour is best described as *elastic–perfectly plastic*, and that the expansion of the cavity occurs under conditions of *axial symmetry* and *plane strain* (refer to section 12.3.1). This implies that the pressuremeter membrane (see figure 12.7) is presumed to be *infinitely long*. In practice however, most (self-boring) pressuremeters used commercially are characterised by a *typical ratio* $L/D \approx 6$ (L and D are the membrane length and diameter respectively), and therefore any measured parameter can potentially be affected by the 'end effects' due to the limited length of membrane (*i.e.* the ratio L/D). These effects were examined by Houlsby and Carter (1993), who undertook a finite element simulation of (self-boring) pressuremeter tests in clay. The results suggest that, while the clay shear modulus G is unaffected by the ratio L/D, the undrained shear strength c_u, derived from central strain measurements using a self-boring pressuremeter with a typical ratio $L/D = 6$, may be slightly overestimated. However, such a conclusion should be considered carefully since the finite element analysis in question was undertaken on a linear elastic medium, in other words on an ideal soil.

Furthermore, the assumption related to linear elasticity becomes arguable in the case of tests on clays performed using an *SBP*, for which the initial elastic behaviour is markedly non-linear as illustrated in figure 12.17. Under such circumstances, the clay behaviour is best modelled as *non-linear elastic–perfectly plastic*. Accordingly, applying the previous linear elastic analysis in conjunction with results measured from an SBP test can result in noticeably erroneous stiffness parameters. As an alternative, some authors suggest the use of a power law to describe the variation during the (non-linear) elastic phase of the stiffness with deformation (Bolton and Whittle, 1999):

$$\tau = a\gamma^b \tag{12.78}$$

where the shear stress corresponds to $\tau = (\sigma_r - \sigma_\theta)$, and the shear strain is defined as $\gamma = (\varepsilon_r - \varepsilon_\theta)$, both the parameter a and the power exponent b being soil constants.

It can readily be shown that, were such a power law to be adopted, then the increase in *radial pressure* (*i.e.* $\Delta P = P - \sigma_{ho}$) in the elastic range will be related to the *shear strain* expressed in terms of the *current volume* V (as per equation 12.46) as follows:

$$\Delta P = \frac{a}{b}(\varepsilon_s)^b \qquad (12.79)$$

Thus a plot of the pressure applied at the borehole walls versus the shear strain calculated from equation 12.50 in the space ($\ln \Delta P$, $\ln \varepsilon_s$) yields a line with an intercept *a/b* and a slope *b*.

Figure 12.18 corresponds to results measured from an *SBP* on London clay by Bolton and Whittle (1999). These results were averaged from three unload–reload cycles, where the reversal point in each cycle was taken as an origin. In all three cycles, the clay behaviour was very accurately fitted with a power law type equation, yielding an exponent $b \approx 0.57$ (the corresponding coefficient of correlation in all case was $R^2 = 0.999$).

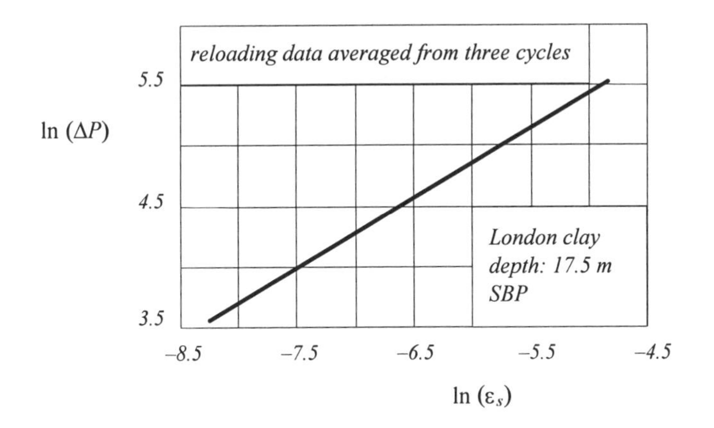

Figure 12.18: SBP data corresponding to three reload cycles, London clay (from Bolton and Whittle (1999), by permission).

In the *plastic range*, the applied pressure becomes related to the undrained shear strength of the clay and to the current volume change as follows:

$$\Delta P = P_l + c_u \ln \varepsilon_s \qquad (12.80)$$

the *limit pressure* being this time defined as:

$$P_l = c_u\left(\frac{1}{b} + \ln \frac{G}{c_u}\right) \qquad (12.81)$$

where the G represents the *secant shear modulus* corresponding to the *yield shear strain* (*i.e.* the shear strain corresponding to the onset of plasticity). Notice that for a *linear* elastic behaviour, the exponent b in equation 12.78 is 1, and equation 12.81 becomes identical to equation 12.52 derived from the Gibson and Anderson theory presented earlier. Similarly, the undrained shear strength c_u of clay is *not* affected by the assumption made in relation to the initial elastic behaviour, since c_u can be obtained using both equations 12.80 and 12.51, and corresponds to the slope of the linear portion of the graph plotted in a semi-logarithmic space as illustrated in figure 12.19 in the case of London clay.

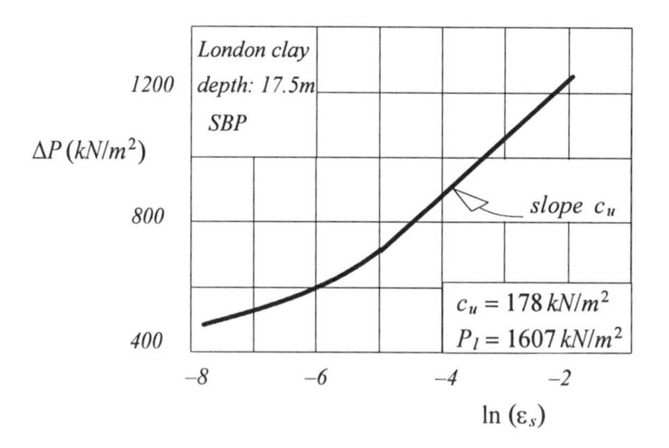

Figure 12.19: Determination of the undrained shear strength of London clay from an SBP test (from Bolton and Whittle (1999), by permission).

These experimental results highlight the need for a cautious approach when interpreting measured data corresponding to a markedly non-linear elastic behaviour. One has to bear in mind that the quality of the outcome of any analysis is a reflection of the assumptions used: unrealistic assumptions yield erroneous results.

Based on these conclusions, and judging from the amount of data amassed through the use of the pressuremeter in different ground conditions and on different soil types thus far, the following facts emerge.

(*a*) although data related to the undrained shear strength c_u of clays measured from pressuremeter tests seem to indicate a considerable scatter, they are heavily dependent on the boundary conditions.

(*b*) An *average* effective angle of friction ϕ'_{av} *can* be measured from *SBP* tests on sands.

(*c*) The total *in situ* horizontal stress σ_{ho} is sensitive to insertion disturbance. However a good estimate of σ_{ho} can be obtained through the use of an appropriate analysis.

(*d*) Provided that the non-linearity during the elastic phase of a pressuremeter test is taken into account, the soil shear modulus G compares favourably with those measured using other *in situ* testing techniques.

(*e*) Although the limit pressure P_l is not constant for a given soil, and is rather intricately related to the stiffness parameters σ_{ho}, G, c_u or ϕ', its value can be reliably determined.

Several semi-empirical methods of design (which generally tend to be grossly conservative), especially concerning shallow and deep foundations, have been developed and are being regularly updated. The details of such methods are out of the scope of this textbook, but can be found in Clarke (1995) for instance. These developments, however, are a clear indication of an increasing reliability of the pressuremeter method and of the willingness of practising engineers to use them with ever growing confidence.

References

Assadi, A. and Sloan, S. W. (1991) *Undrained stability of shallow square tunnels.* ASCE Geotechnical Engineering Division, 117, pp. 1152–1173.

Baguelin, F., Jézéquel, J. F. and Le Méhauté, A. (1972) *Expansion of cylindrical probes in cohesive soils.* Journal of Soil Mechanics, Foundations Division, ASCE, 98 (SM11), pp. 1129–1142.

Baguelin, F., Jézéquel, J. F. and Le Méhauté, A. (1974) *Self-boring placement method of soil characteristics measurement.* Proceedings of Speciality Conference on Subsurface Exploration of Underground Excavation and Heavy Construction, Henniker, pp. 312–332.

Baguelin, F., Jézéquel, J. F. and Shields, D. H. (1978) *The Pressuremeter and*

Foundation Engineering. Trans Tech. Publication, Germany.

Bolton, D. M. and Whittle, R. W. (1999) *A non-linear elastic-perfectly plastic analysis for plane strain undrained expansion tests.* Géotechnique, 49 (1), pp. 133–141.

Brady, B. H. G. and Brown, E. T. (1993) *Rock Mechanics for Underground Mining,* 2nd edn. Chapman & Hall, London.

Clarke, B. G. (1995) *Pressuremeters in Geotechnical Design.* Blackie Academic and Professional, London.

Gibson, R. E. and Anderson, W. F. (1961) *In situ measurement of soil properties with the pressuremeter.* Civil Engineering and Public Works Review, 56, pp. 615–618.

Houlsby, G. T and Carter, J. P. (1993) *The effects of pressuremeter geometry on the results of tests in clay.* Géotechnique, 43 (4), pp. 567–576.

Hughes, J. M. O. (1973) *An instrument for in situ measurement in soft clays.* PhD Thesis, University of Cambridge.

Hughes, J. M. O, Wroth, C. P. and Windle, D. (1977) *Pressuremeter test in sand.* Géotechnique, 27 (4), pp. 455–477.

Jaeger, C. (1979) *Rock Mechanics and Engineering,* 2nd edn. Cambridge University Press.

MELT (1993) *Ministère de l'Equipement, du Logement et des Transports: Règles Techniques de Conception et de Calcul des Fondations des Ouvrages de Génie Civil.* Projet de Fascicule 62, Titre 5, CCTG.

Ménard, L. (1962). *Comportement d'une fondation profonde soumise à des efforts de renversement.* Sol Soils, 3, pp. 9–27.

Ménard, L. (1963) *Calcul de la force portante des fondations sur la base des résultats des essais pressiométriques.* Sol Soils, 5, pp. 9–32.

Mestat, P. (1997) *Maillage d'éléments finis pour les ouvrages de géotechnique: conseils et recommondations.* Bulletin des Laboratoires des Ponts et Chaussées, 212, pp. 39–64.

Rowe, P. W. (1962) *The stress-dilatancy relation for static equilibrium of an assembly of particles in contact.* Proceedings of the Royal Society, A.269, pp. 500–527.

Smith, I. M. and Griffiths, D. V. (1998) *Programming the Finite Element Method.* 3rd edn. John Wiley, New York.

Timoshenko, S. P. and Goodier, J. N. (1970) *Theory of Elasticity,* 3rd edn. McGraw-Hill, International Edition, Singapore.

Tchebotarioff, G. P. (1973) *Foundations, Retaining and Earth Structures,* 2nd edn. McGraw-Hill, New York.

Zienkiewicz, O. C. and Taylor, R. (1991) *The Finite Element Method* (2 volumes), 4th edn. McGraw-Hill, London.

CHAPTER 13

Centrifuge modelling of soil behaviour

13.1 Introduction

Physical modelling of soil behaviour has always played a pivotal role in helping the designer acquire a better understanding of the *actual* behaviour under *similar* stress conditions in the field. In this respect, laboratory tests such as triaxial, consolidation and shear box tests are still used extensively and *do* provide reliable data provided that adequate testing procedures are adhered to. Centrifuge testing of soils constitutes another (recent) development in the field of physical modelling, which has taken off quite rapidly. More and more research laboratories are equipped with centrifuge centres, and the research outcome in this field is becoming widely available so that comparative studies can be undertaken, leading to the development of ever more sophisticated instrumentation equipment.

As in the case of triaxial testing, for instance, a centrifuge test is undertaken on a small size sample of soil, referred to as the *model,* in a way that stress conditions corresponding to a particular event (such as subjecting a pile to a lateral load for example), or to a particular process (as in the case of the execution of an excavation, or the construction of an embankment) are recreated *via* an *inertial acceleration field.* Under such loading conditions, the behaviour of the *model* should be, in theory, a replica of that of the *actual* soil (often referred to as the *prototype*), when subjected to a *similar state of stresses.*

Notwithstanding the practical difficulties related to model preparation and instrumentation, centrifuge modelling is regarded as a valuable means of testing that enhances markedly the understanding of the physical behaviour of soils under complex static or dynamic stress fields (see for instance the paper by Schofield (1980)). Examples vary widely and include (classical) problems such as the stability of slopes, retaining structures, embankments, foundations and tunnels, as well as heat transfer (that is conduction and convection), diffusion (*i.e.* consolidation), seepage, earthquakes, wave loading, contaminant transport, freeze/thaw and the effects of deep mining.

13.2 Modelling principles, stress similarities

The basic principle of centrifuge testing consists of creating stress conditions which are similar to those applied through gravity in the field to a prototype, *in* or *on* a model, with dimensions which are much smaller than those of the prototype. This can be achieved by placing the model in a basket at the end of a centrifuge boom, then subjecting it to an inertial acceleration field. The main features of a centrifuge equipment are depicted in figure 13.1, corresponding to an *Acutronic 680* centrifuge, capable of developing a maximum acceleration of $200g$ (g being the acceleration due to gravity), and whose technical details can be found for instance in the book *Centrifuge 88* edited by Corté (1988).

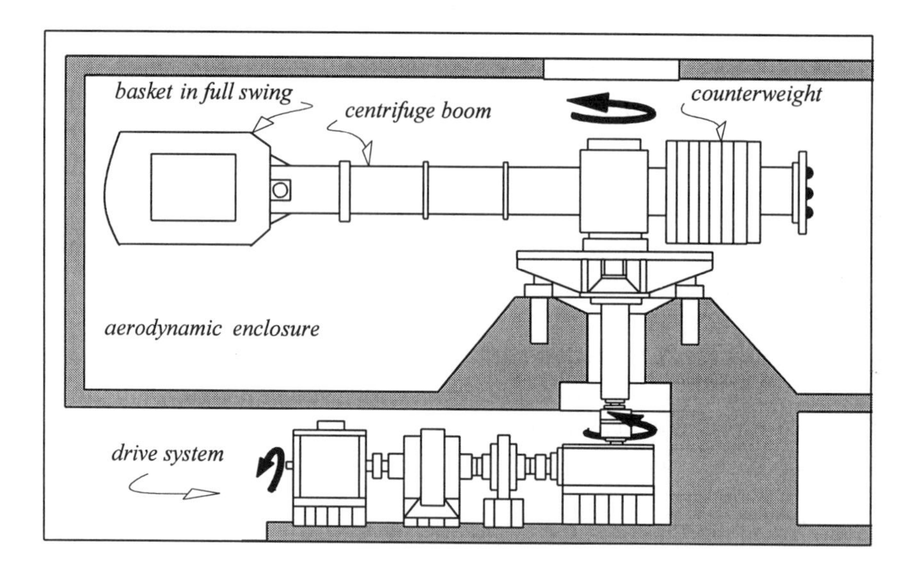

Figure 13.1: Acutronic 680 centrifuge.

In the following analysis, the subscript m refers to the model and the subscript p is used in conjunction with the prototype.

Thus, assuming that all soil properties, including those of the porewater, are identical for the prototype and for the model, and referring to figure 13.2, it is seen that, for a model with a dimension R/n (n being an integer), placed at a distance R (which can be the effective radius of a centrifuge, for instance) from the centre of rotation, and subjected to an acceleration Ng (*i.e.* N times the acceleration due to gravity g), the *centrifugal stress* σ_{vm} is, in theory, *similar* to the vertical stress σ_{vp} due to the self weight of soil in a prototype having a dimension NR/n.

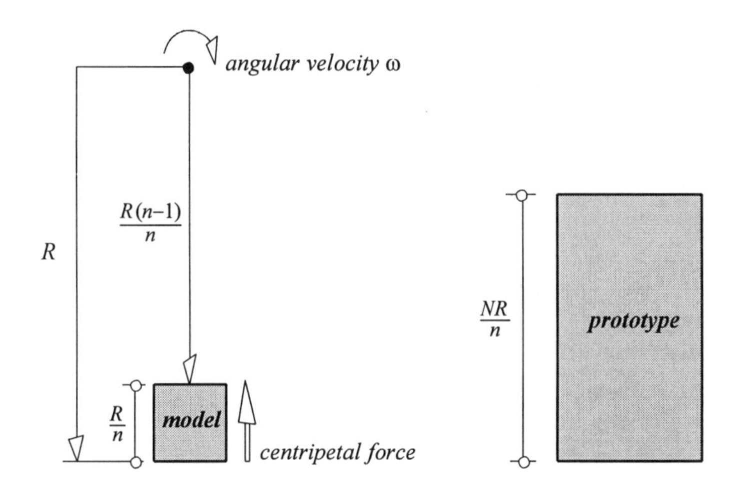

Figure 13.2: Effect of inertial forces on stress similarities between model and prototype.

Accordingly, if the respective dimensions of the model and prototype were:

$$h_m = \frac{R}{n} \quad \text{and} \quad h_p = N\frac{R}{n}$$

then the vertical stress σ_{vm} at a depth h_m in the model, induced by an acceleration Ng will be:

$$\sigma_{vm} = N\rho g h_m \tag{13.1}$$

Similarly, the vertical stress σ_{vp} at depth h_p in the prototype is:

$$\sigma_{vp} = \rho g h_p \tag{13.2}$$

ρ, in both equations, being the soil density.

If the stresses are similar, then clearly the dimensions of the model and prototype are such that:

$$h_m = \frac{h_p}{N} \tag{13.3}$$

In practice however, the similarity between σ_{vm} and σ_{vp} is affected to a certain degree by the negative *centripetal* (or inertial) *forces* applied to the model due to the variation of the angular velocity ω with the radius r.

Referring once more to figure 13.2, it can be seen that the stress σ_{vm} at the base of the model can be evaluated as follows:

$$\sigma_{vm} = \int_{R(n-1)/n}^{R} \rho\omega^2 r\,dr = \rho\omega^2 \frac{R^2}{2}\left(\frac{2n-1}{n^2}\right) \tag{13.4}$$

Moreover, the vertical stress due to gravity at the base of the prototype is:

$$\sigma_{vp} = \rho g N \frac{R}{n} \tag{13.5}$$

Consequently, for these two stresses to be *similar*, equations 13.4 and 13.5 must be equal. Whence:

$$\sigma_{vm} = \sigma_{vp} \quad \Rightarrow \quad \frac{Ng}{\omega^2 R} = 1 - \frac{1}{2n} \tag{13.6}$$

It can be shown that if the stresses σ_{vm} at a depth R/n and σ_{vp} at a depth NR/n are similar, then the maximum stress difference (*i.e.* error) defined as:

$$\Delta S = \frac{\sigma_{vp} - \sigma_{vm}}{\sigma_{vm}} \tag{13.7}$$

occurs at a depth $R/2n$ in the model and $NR/2n$ in the prototype (see for instance Taylor (1995)). Thus, evaluating the stresses at these respective depths, it follows that:

$$\sigma_{vm} = \int_{R(1-1/n)}^{R(1-1/2n)} \rho\omega^2 r\,dr = \rho\omega^2 \frac{R^2}{8n^2}(4n-3) \tag{13.8}$$

and

$$\sigma_{vp} = \rho g N \frac{R}{2n} \tag{13.9}$$

Consequently, it is easy to show that the error corresponding to equation 13.7 is such that:

$$\Delta S = \frac{4nNg}{\omega^2 R(4n-3)} - 1 \tag{13.10}$$

Hence, substituting for the quantity $Ng/\omega^2 R$ from equation 13.6 in equation 13.10 and rearranging, it follows that:

$$\Delta S = \frac{1}{4n-3} \tag{13.11}$$

Moreover, assuming there is similarity of stresses at two-thirds model depth as depicted in figure 13.3, then the graph corresponding to the maximum error ΔS calculated from equation 13.11, and shown in figure 13.4, indicates that for an n value larger than 10, the ensuing error (*i.e.* the shaded area in figure 13.4) is smaller than 3% and therefore negligible.

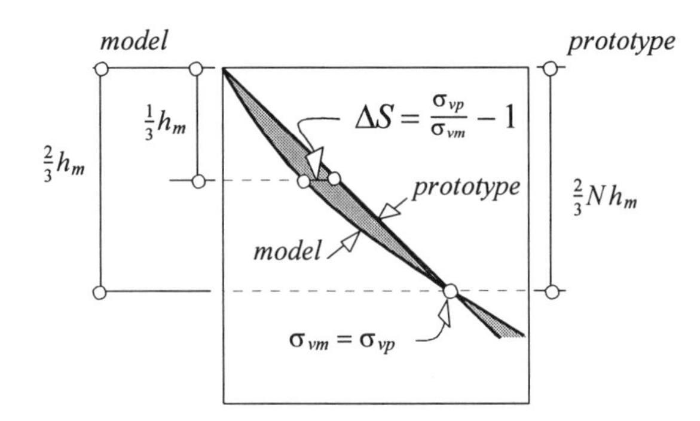

Figure 13.3: Scaling errors due to inertial forces in the model.

In practice, a value $n = 10$ is typical in geotechnical centrifuge testing, so that the model height corresponds to one-tenth of the effective centrifuge radius (measured from the centre of rotation to one-third the model depth).

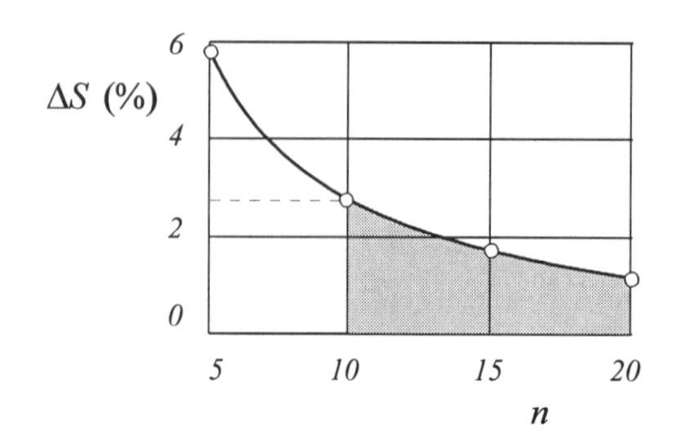

Figure 13.4: Error magnitude related to model dimensions.

13.3 Scale effects

The previous analysis of stress similarities between model and prototype serves as a reminder of the fact that physical (especially centrifuge) modelling of soils is affected by the reduction in dimensions of the tested volume of soil. The *scale effects* are not, however, limited to the model size, since two more phenomena related to centrifuge testing can, in principle, cause the behaviour of the model to be appreciably different from that of the prototype of an identical soil. These two phenomena are associated with (*a*) the particle size and (*b*) the rotational acceleration field.

It has been established previously (refer to equation 13.3) that the stress similarity necessitates the use of a model with dimensions N times smaller than the prototype, with Ng being the acceleration to which the model is subjected in the centrifuge. Moreover, an equivalent dimensional analysis indicates that, for the same soil, the ratio d/L (d being the *average* grain size of the soil, and L a typical boundary dimension) must, in theory, be identical for both the prototype and the model, so that:

$$\frac{d_m}{L_m} = \frac{d_p}{L_p}$$

Substituting for the length $L_m \equiv h_m$ from equation 13.3, and rearranging, it is seen that:

$$d_m = \frac{d_p}{N} \tag{13.12}$$

Equation 13.12 therefore indicates the need to use a model comprising a soil with an average grain size *N*-times smaller than that of the same soil used for the prototype. Manifestly, this conclusion has, in practice, the potential of generating problems since, for instance, a model tested at an acceleration of $100g$, constituted of fine sand that has in the field an average grain size of $d_p = 0.1\,mm$, must have an average grain size of $d_p/100 = 0.001\,mm$ which, according to table 1.1 of section 1.2 corresponds to a clay, with all the implications related to the void ratio, permeability and stress–strain behaviour.

Although no simple answer can be provided as to how the grain size might affect in any appreciable way the behaviour observed during a centrifuge test (see for example Ovesen (1979), Bolton and Lau (1988), Tatsuoka *et al.* (1991)), it seems logical, though, that the grain size effect decays with decreasing grain dimensions, so that while it appears necessary to reduce the grain size in proportion to the centrifuge acceleration when testing a gravel, the adverse effects of not doing so when testing a fine sand, a silt or a clay appear to be minimal. Evidently, these effects need be (or are already in the process of being) investigated thoroughly, and thus an engineering judgement has to be made in conjunction with the soil used for a model and the type of problem to be physically simulated. Therefore some caution must be exercised, so that the above analysis and guidelines are not used in a mechanical way.

Let us now establish the effects of the rotational acceleration field, better known as the *Coriolis effects.* In this respect, it is easier to establish all acceleration components related to a soil element, rotated at a steady velocity V, much in the way that a model contained in a centrifuge basket is subjected to an acceleration Ng. Thus, with reference to figure 13.5, it is seen that the radial co-ordinates of point A, a distance r far from the centre of rotation, are as follows:

$$X = r\cos\theta$$
$$Y = r\sin\theta \qquad (13.13)$$

The components of velocity are therefore:

$$V_x = \frac{dX}{dt} = \cos\theta \frac{dr}{dt} - r\sin\theta \frac{d\theta}{dt}$$
$$V_y = \frac{dY}{dt} = \sin\theta \frac{dr}{dt} + r\cos\theta \frac{d\theta}{dt} \tag{13.14}$$

whence the following accelerations:

$$\frac{d^2X}{dt^2} = \cos\theta \frac{d^2r}{dt^2} - 2\sin\theta \frac{dr}{dt}\frac{d\theta}{dt} - r\sin\theta \frac{d^2\theta}{dt^2} - r\cos\theta \left(\frac{d\theta}{dt}\right)^2$$
$$\frac{d^2Y}{dt^2} = \sin\theta \frac{d^2r}{dt^2} + 2\cos\theta \frac{dr}{dt}\frac{d\theta}{dt} + r\cos\theta \frac{d^2\theta}{dt^2} - r\sin\theta \left(\frac{d\theta}{dt}\right)^2 \tag{13.15}$$

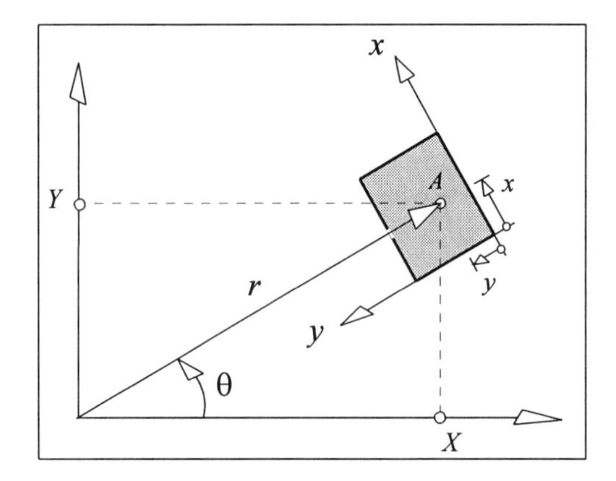

Figure 13.5: Radial co-ordinate system.

Moreover, when the *local co-ordinates system (x, y)* is used, then with reference to figure 13.6, the position of *A* is such that:

$$x = C_1$$
$$y = C_2 - r \tag{13.16}$$

where C_1 and C_2 are constants.

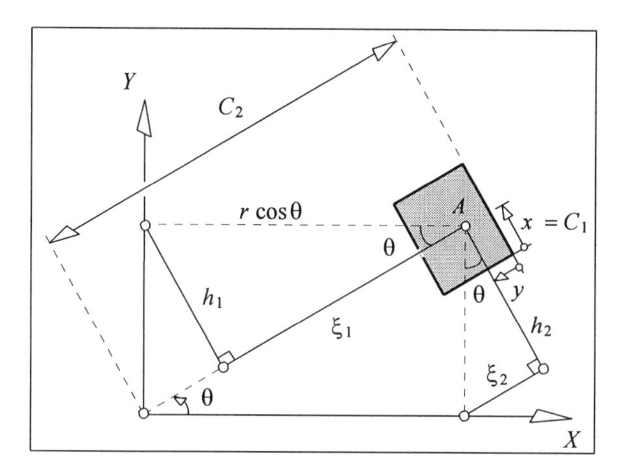

Figure 13.6: Local co-ordinate system.

Referring to figure 13.6, it is seen that:

$$h_1 = r\cos\theta\sin\theta = X\sin\theta$$

$$h_2 = h_1 = Y\cos\theta$$

$$\xi_1 = X\cos\theta$$

$$\xi_2 = Y\sin\theta$$

and

$$r = \xi_1 + \xi_2$$

Accordingly, equations 13.16 can now be expressed as follows:

$$x = C_1 - h_1 + h_2$$

$$y = C_2 - \xi_1 - \xi_2$$

and so, inserting for the quantities h_1, h_2, ξ_1 and ξ_2, it follows that:

$$x = C_1 - X\sin\theta + Y\cos\theta$$

$$y = C_2 - X\cos\theta - Y\sin\theta \tag{13.17}$$

The accelerations, expressed in the local co-ordinates system can then be derived:

$$\frac{d^2x}{dt^2} = -\left(\frac{d^2X}{dt^2}\right)\sin\theta + \left(\frac{d^2Y}{dt^2}\right)\cos\theta$$

$$\frac{d^2y}{dt^2} = -\left(\frac{d^2X}{dt^2}\right)\cos\theta - \left(\frac{d^2Y}{dt^2}\right)\sin\theta$$

Thus, substituting for the quantities derived in equation 13.15 and rearranging:

$$\frac{d^2x}{dt^2} = 2\frac{dr}{dt}\frac{d\theta}{dt} + r\frac{d^2\theta}{dt^2}$$
$$\frac{d^2y}{dt^2} = -\frac{d^2r}{dt^2} + r\left(\frac{d\theta}{dt}\right)^2 \qquad (13.18)$$

Equations 13.18 yield the four components of acceleration:

- *the horizontal shaking*: $\Gamma_h = r\dfrac{d^2\theta}{dt^2}$

- *the vertical shaking*: $\Gamma_v = \dfrac{d^2r}{dt^2}$

- *the inertial acceleration*: $\Gamma_i = r\left(\dfrac{d\theta}{dt}\right)^2 = \omega^2 r = \dfrac{V^2}{r}$

where ω is the angular velocity

- *the Coriolis acceleration*: $\Gamma_c = 2\dfrac{dr}{dt}\dfrac{d\theta}{dt}$

It is noticeable that the first two components Γ_h and Γ_v only apply to *dynamic models* (to simulate quakes for instance). Also, the Coriolis acceleration Γ_c which translates the velocity $v = dr/dt$ of a particle within the model, relative to the velocity of the model centrifuge *in flight* $V = rd\theta/dt$, can be rewritten as follows:

$$\Gamma_c = 2v\frac{d\theta}{dt} = 2v\frac{V}{r} \qquad (13.19)$$

In practice, it is accepted that, for slow moving particles, in other words for a *steady model flight* where no shaking takes place, the error due to Coriolis acceleration is negligible for ratios $\Gamma_c/\Gamma_i < 0.1$, meaning:

$$v < 0.05V \tag{13.20}$$

However, for faster particles, the path of a moving particle within a model becomes curved, with a radius of curvature r_c defined as follows:

$$r_c = \frac{v^2}{\Gamma_c} \tag{13.21}$$

Substituting for Γ_c from equation 13.19, it is seen that:

$$\frac{r_c}{r} = \frac{v}{2V} \tag{13.22}$$

Since r represents the *effective radius of centrifuge* (refer to figure 13.5), equation 13.22 implies that the curvature effect becomes less significant when $r_c > r$. Accordingly, it is suggested that for fast particles (such as in a blast simulation for instance), the errors related to Coriolis acceleration are no longer appreciable as long as the velocities in equation 13.22 are such that:

$$v > 2V \tag{13.23}$$

Thus, both equations 13.20 and 13.23 yield the range *within* which Coriolis effects *must* be taken into account:

$$0.05V < v < 2V \tag{13.24}$$

13.4 Scaling laws

13.4.1 Static models

Length

In accordance with equation 13.3, the stress similarities between a model subjected to an acceleration of Ng and a prototype subjected to the acceleration due to gravity g imply that the model dimensions are N *times smaller* than those of the prototype. Thus, using the (same) subscript m for model and p for prototype, it follows that:

$$L_m = L_p N^{-1} \tag{13.25}$$

Unit weight

The densities of the model and the prototype being identical, the corresponding unit weights are as follows:

$$\gamma_p = \rho g$$

and

$$\gamma_m = \rho N g = N \gamma_p \tag{13.26}$$

Therefore, the unit weight of the model is N *times larger* than that of the prototype.

Hydraulic gradient

With reference to figure 3.4 in section 3.2, it is seen that the hydraulic gradient can be thought of as a pressure gradient, because it is defined as the ratio of the total head loss Δh to the length of the flow path along which Δh is lost through friction, so that at the limit:

$$i = \frac{dh}{dl} \tag{13.27}$$

But the porewater pressure is identical in both the model and the prototype, and therefore:

$$i_p = \frac{dh}{(dl)_p} \qquad \text{and} \qquad i_m = \frac{dh}{(dl)_m}$$

Thus, making use of equation 13.25, it follows that:

$$i_m = \frac{dh}{(dl)_p(1/N)} = N i_p \tag{13.28}$$

Hence, the hydraulic gradient in the centrifuge model is N *times larger* than the one in the prototype.

Seepage velocity

Darcy's law, expressed in terms of seepage velocity in the centrifuge model is as follows:

$$v_m = Ki_m \tag{13.29}$$

Knowing that the model and prototype have identical permeabilities, and substituting for the hydraulic gradient from equation 13.28 into equation 13.29, it follows that:

$$v_m = KNi_p = Nv_p \tag{13.30}$$

which shows that, as in the case of the hydraulic gradient, the seepage velocity is *N times larger* in the centrifuge model than in the prototype.

Seepage time

The seepage time can be expressed as the ratio of the length of the flow path to the seepage velocity. So, for the model:

$$t_m = \frac{L_m}{v_m} \tag{13.31}$$

A straight substitution for L_m and v_m from equations 13.25 and 13.30 respectively into the above equation yields:

$$t_m = \frac{L_p}{N} \times \frac{1}{Nv_p} = \frac{1}{N^2} t_p \tag{13.32}$$

Accordingly, water seeps N^2 *times faster* in the centrifuge model than that in the prototype.

Seepage flow

Using Darcy's law, the flow quantity *per unit length* in the centrifuge model can be expressed as follows:

$$q_m = L_m v_m t_m \tag{13.33}$$

where L_m represents the length across the flow path.

Now, making good use of equations 3.25, 3.30 and 3.32, it is easy to establish that:

$$q_m = \frac{1}{N^2} q_p \tag{13.34}$$

The above equation shows that the flow quantity *per unit length* in the centrifuge model is N^2 *times smaller* than the one in the prototype.

Note that, if the scaling law between *total flow* quantities in a centrifuge model and in a prototype were sought, then one must remember that:

$$Q_m = A_m v_m t_m \tag{13.35}$$

with A_m being the *area* of flow. An approach similar to that used previously leads to the following:

$$Q_m = \frac{1}{N^3} Q_p \tag{13.36}$$

indicating that the *total flow quantity* (in L^3T^{-1}) in a model is N^3 *times smaller* than that occurring in the prototype.

Diffusion problems

The governing parabolic equation of a consolidation process (*i.e.* porewater pressure dissipation with time) can be written as:

$$\frac{\partial^2 u}{\partial z^2} = \frac{1}{c_v} \frac{\partial u}{\partial t} \tag{13.37}$$

The solution to the above equation contains the following time factor written in the case of the centrifuge model as follows:

$$T_v = c_v \frac{t_{cm}}{L_m^2} \tag{13.38}$$

where L represents the drainage path and t is the time, the coefficient of consolidation c_v being identical for both prototype and model. Substituting for t_{cm} and L_m from equations 13.32 and 13.25 respectively, then cancelling appropriate terms:

$$t_{cm} = \frac{1}{N^2} t_p \tag{13.39}$$

This equation, being identical to equation 13.32 in the case of seepage, indicates that the same *stage* of consolidation occurs N^2 *times faster* in a centrifuge model than in a prototype.

Reynolds number

The Reynolds number is of particular interest in centrifuge modelling, because its value is linked to the flow regime. It is well established that for soils, the regime of flow is assumed *laminar* as long as the Reynolds number is kept smaller than 10 (refer to section 3.1). Let us examine the effect that this condition has on the centrifuge model. The dimensionless Reynolds number corresponding to the centrifuge model is defined as:

$$Re_m = v_m \frac{d}{\mu} \tag{13.40}$$

where v_m represents the model seepage velocity, d is the average diameter of soil particles and μ is the kinematic viscosity of water. Since d and μ are identical for both prototype and centrifuge model, equation 13.40 can then be rearranged after introducing the scaling law for seepage velocities from equation 13.30:

$$Re_m = N v_p \frac{d}{\mu} = N \, Re_p \tag{13.41}$$

which implies that the Reynolds number is *N times higher* in a centrifuge model than in a prototype. Consequently, in order to maintain a *laminar regime of flow* in the model, provisions must be made to keep the Reynolds number smaller than 10.

It is of interest to note that, according to equation 13.41, an *identical* Reynolds number for both model and prototype can be achieved by using for the centrifuge model: (*a*) a soil characterised by an average grain diameter that is *N times smaller* than the average diameter in the prototype or (*b*) a fluid *N times more viscous* than water. Clearly the first suggestion is unworkable, however, a variety of oils with differing viscosities can be used as a pore fluid in the model instead of water. One has to bear in mind, though, that the use of a pore fluid other than water may cause the surface properties of the solid particles to change, which may, in turn, affect the behaviour of the model.

13.4.2 Dynamic models

Dynamic events such as quakes can be simulated in a centrifuge, and the scaling laws related to the corresponding models are derived in a manner similar to that used in conjunction with static models. Figure 13.7 depicts a typical horizontal shear wave generated by an earthquake, which is represented by the following differential equation:

$$x = a \sin \omega t \tag{13.42}$$

where x represents the *cyclic motion*, a is the *amplitude of the motion*, and ω corresponds to the *angular velocity*. Obviously, both velocity and acceleration can be derived from equation 13.42:

$$\frac{dx}{dt} = a\omega \cos \omega t \tag{13.43}$$

$$\frac{d^2x}{dt^2} = -a\omega^2 \sin \omega t \tag{13.44}$$

As the quantity $a\omega^2$ in equation 13.44 defines the *magnitude of acceleration*, and because the centrifuge model is subjected to an acceleration N-times larger than that of the prototype, it is clear that:

$$a_m \omega_m^2 = N a_p \omega_p^2 \tag{13.45}$$

also, according to equation 13.25:

$$a_m = \frac{a_p}{N} \tag{13.46}$$

Thus inserting a_m into equation 13.45, and rearranging, it is easy to establish that:

$$\omega_m = N\omega_p \tag{13.47}$$

and because the *angular velocity* is related to the *motion frequency*, it follows that:

$$\omega = 2\pi f \quad \Rightarrow \quad f_m = N f_p \tag{13.48}$$

More importantly perhaps, the quantity $a\omega$ in equation 13.43, which corresponds to the *magnitude of velocity*, is identical for both model and prototype. In fact, substituting for a_m and ω_m from equations 13.46 and 13.47 respectively into equation 13.43, it is seen that:

$$a_m\omega_m \cos(\omega_m t) = a_p\,\omega_p \cos(N\omega_p t) \tag{13.49}$$

Because the velocity magnitude corresponds to the ratio of a length to a time, and knowing that in a centrifuge model the length is reduced by a factor N (see equation 13.25), equation 13.49 therefore implies that the time for dynamic models is reduced by a factor N as opposed to a factor N^2 for seepage or diffusion problems (refer to equations 13.31 and 13.39).

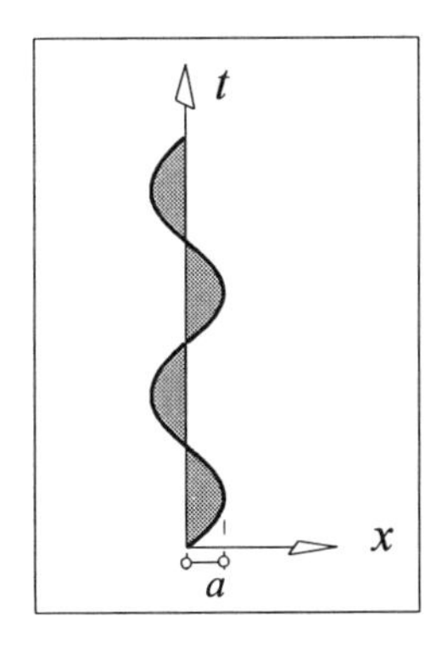

Figure 13.7: Vertically propagating shear wave.

Example 13.1

A model with a volume $V_m = 3 \times 10^{-2}\, m^3$ is subjected to 10 cycles at a frequency $f_m = 100\, Hz$ and with an amliptude $a_m = 1.5\, mm$, while being accelerated in a centrifuge at 100g. Obviously, the duration of shaking is calculated in a straightforward way: $t_m = 10/100 = 0.1s$.

The magnitude of acceleration due to the shaking can be estimated as follows:

$$a_m\omega_m^2 = a_m(2\pi f_m)^2 = 1.5 \times 10^{-3} \times 4\pi^2 \times 10^4$$

$$= 592\, m/s^2 \approx 60.3g$$

This test simulates an earthquake in a prototype, having a volume:

$$V_p = N^3 V_m = 30,000\, m^3$$

subjected to 10 cycles at a frequency: $f_p = \frac{1}{N} f_m = 1\, Hz$, and with an amplitude: $a_p = Na_m = 0.15\, m$.

The duration of the earthquake is such that: $t_p = N t_m = 10\, s$, and the magnitude of acceleration in the prototype is, with reference to figure 13.8:

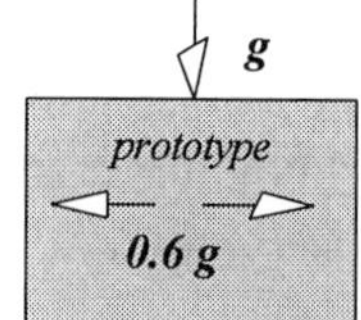

Figure 13.8: Earthquake accelerations.

$$a_p(2\pi f_p)^2 = 0.15 \times 4\pi^2 \times 1 = 5.92\, m/s^2 \approx 0.6\, g$$

Assuming that, during the shaking, the steady centrifuge velocity is $V = 30\, m/s$, then the error due to Coriolis acceleration at maximum velocity of shaking, corresponding to the ratio of the inertial acceleration $\Gamma_i = V^2/r$ to the Coriolis acceleration $\Gamma_c = 2vV/r$ is estimated as follows:

$$v = a_m 2\pi f_m = 1.5 \times 2\pi \times 10^{-1} = 0.94\, m/s$$

whence:

$$\frac{\Gamma_c}{\Gamma_i} = \frac{2v}{V} = \frac{2 \times 0.94}{30} = 0.062$$

Obviously, in this case, the error due to Coriolis acceleration is only 6.2%, and is therefore negligible.

The scaling laws for different physical quantities are summarised in the following table in the case of a model subject to an acceleration $N.g$.

quantity	*scaling law*	*scaling factor*
acceleration	$g_m = Ng_p$	N^{-1}
mass, density	$\rho_m = \rho_p$	1
stress	$\sigma_m = \sigma_p$	1
strain	$\varepsilon_m = \varepsilon_p$	1
velocity	$V_m = V_p$	1
temperature	$\theta_m = \theta_p$	1
length	$L_m = L_p/N$	N
time (static event)	$t_m = t_p/N^2$	N^2
time (dynamic event)	$t_m = t_p/N$	N
displacement, amplitude	$d_m = d_p/N$	N
unit weight	$\gamma_m = N\gamma_p$	N^{-1}
frequency	$f_m = Nf_p$	N^{-1}
hydraulic gradient	$i_m = N i_p$	N^{-1}
seepage velocity	$v_m = Nv_p$	N^{-1}
Reynolds number	$Re_m = NRe_p$	N^{-1}
heat flux	$h_{xm} = Nh_{xp}$	N^{-1}
seepage flow per unit length	$q_m = q_p/N^2$	N^2
total seepage flow	$Q_m = Q_p/N^3$	N^3
diffusion (consolidation)	$t_{cm} = t_{cp}/N^2$	N^2
heat transfer(conduction, convection)	$(\partial\theta/\partial t)_m = (\partial\theta/\partial t)_p/N^2$	N^2

13.5 Practical aspects of centrifuge modelling

Every engineer or researcher would recognise that the outcome of a centrifuge test in geotechnics depends on the meticulous model preparation prior to testing, which is, alas, time consuming. In this respect, the paper by Phillips (1995) constitutes excellent reading as it describes in some detail the potential pitfalls related to model preparation, testing and monitoring.

The one aspect that must always be considered first in relation to centrifuge testing is safety. Centrifuges such as the one depicted in figure 13.1 are powerful machines that must be handled by adequately trained staff. Furthermore, centrifuge testing is unquestionably a multi-disciplinary activity, since the modeller has to be at least conversant in (if not knowledgeable about!) mechanical engineering, electronics equipment, control systems and data acquisition, before even contemplating interpreting the test results from a geotechnical viewpoint. No wonder that such tests are generally intensively resourced, therefore expensive and time consuming.

Manifestly, undertaking a centrifuge test implies spending the majority of time and effort on the careful preparation of the model. This includes the following.

- The selection of the appropriate container depending on the type of soil tested and the type of behaviour to be simulated (static or dynamic) with specific boundary conditions requirements that need to be fulfilled. For instance, modelling a static event such as the consolidation of a clay layer requires the walls of the container to be *ideally* frictionless, whereas the simulation of the creep behaviour of a frozen soil necessitates a strict thermal control within the container.

- The selection of the appropriate soil conditions for the model, in particular, the restrictions related to the flow of pore fluid and the size of the model must always be taken into account.

It was established earlier that the Reynolds number in the model must be kept below 10 for laminar flow conditions to prevail.

This requirement is most likely to be fulfilled through the use of a pore fluid that is *N* times more viscous than water in conjunction with the model (*Ng* being the acceleration to which the model is subjected). Also, in order to minimise the scaling errors due to inertial forces (refer to figures 13.3 and 13.4), the model height must be kept to within one-tenth of the effective centrifuge radius (measured from the centre of rotation to one-third the model depth). On the other hand, the model may have to be protected against any temperature changes or any air movements that can potentially be generated within the aerodynamic enclosure (see figure 13.1) during testing.

- The reconstitution of the model under laboratory conditions must allow for the effective stress profile to be recreated in the case of cohesive soils, or for the change in soil behaviour to be taken into account when interpreting test results. In particular, remoulded samples of clay and silt can be created from a slurry by tamping. Alternatively, the slurry can be consolidated in the centrifuge, in which case, care must be taken not to induce a *differential consolidation* of the slurry mass through the generation of preferential drainage paths due to the high pore pressure within the soil mass (refer to section 5.6). For granular soils, sophisticated techniques such as tamping and pluviation can be used to create a density controlled model.

- Instrumentation and actuation of the model can potentially create problems since neither the instruments, nor the actuator are scaled. Pore pressure, total stress and displacement transducers, as well as thermocouples, that are buried within the model should be placed in a way that minimises any reinforcement effect of the soil. Also, there is a need to insulate strain gauges and lead wires embedded in the model. The restrictive effects on the model behaviour of the actuator (*i.e.* the system that sets off the model to simulate the behaviour of the prototype) must be reduced to a minimum.

- A data acquisition system that can store a large amount of data in a short period of time is needed to conduct a centrifuge test successfully. In this respect, sophisticated systems combining the latest electronics and digital technologies are commercially available.

These points, important though they may be, are only an *aperçu* of the acumen, hard work, and vision required by a modeller to conduct a sophisticated and technically challenging test such as a geotechnical centrifuge test. A thorough analysis which tackles more detailed practical aspects of centrifuge modelling can be found in Phillips (1995).

References

Bolton, M. D. and Lau, C. K. (1988) *Scale Effects Arising from Particle Size. Centrifuge 88* (ed. *J. F. Corté*). A.A. Balkema, Rotterdam, pp. 127–134.

Corté, J. F. (Ed.) (1988) *Centrifuge 88.* A.A. Balkema, Rotterdam.

Ovesen, N. K. (1979) *The scaling law relationship.* Panel discussion. Proceeding of the 7th European conference on Soil Mechanics and Foundation Engineering, Brighton, 4, pp. 319–323.

Phillips, R. (1995) *Centrifuge modelling: practical considerations.* Geotechnical Centrifuge Technology, Blackie, London, pp. 34–60.

Schofield, A. N. (1980) *Cambridge geotechnical centrifuge operations.* Géotechnique, 20, pp. 227–268.

Tatsuoka, F., Okahara, M., Tanaka, T., Tani, K., Morimoto, T. and Siddiquee, M. S. A. (1991) *Progressive failure and particle size effect in bearing capacity of a footing in sand.* ASCE Geotechnical Engineering Congress, Vol. 2 (Geotechnical special publication 27), pp. 788–802.

Taylor, R. N. (1995) *Centrifuges in modelling: principles and scale effects.* Geotechnical Centrifuge Technology, Blackie, London, pp. 19–33.

CHAPTER 14

Finite element modelling in geotechnics

14.1 Finite element modelling

This section aims at presenting the numerical modelling used in conjunction with geotechnical problems from a *practical perspective*. As such, it is not intended to develop in detail the mathematical formalisms of the finite element method, although its working will be explained succinctly. A thorough presentation of such a method of analysis is widely available, and reference should be made to Zienkiewicz and Taylor (1991), Bathe (1982, 1996), Hughes (1987), Stasa (1985), Griffiths and Smith (1991), Smith and Griffiths (1998), Reddy (1993) and Owen and Hinton (1980).

The *finite element method* is one of the most powerful approximate solution method that can be applied to solve a wide range of problems represented by ordinary or partial differential equations. The power of such a method derives from the fact that it can easily accommodate changes in the material stiffness which is evaluated at element level as will be explained shortly. Also, it allows for different boundary conditions to be applied in such a way that an acceptable global approximate solution to the physical problem can be achieved. Considering that closed form solutions cannot be elaborated for a large number of complex physical problems, due to the impossibility of satisfying the boundary conditions related to the corresponding equilibrium equations, the finite element method therefore provides an ideal alternative (approximate) solution method.

In its simplest form, the finite element method consists of:

- dividing a given structure or domain into a number of *elements* (hence the name *finite elements*); this process is known as *discretisation*. Elements are connected by *nodes* at the corners and sometimes at the sides as well;

- modelling the behaviour of the unknown variables at different nodes through the use of appropriate interpolation polynomials, better known as *shape functions*.

The *shape, size, number* and *type* of elements depend on the type of structure or domain, and also on the precision required in the solution, these points being further elaborated below.

In all cases, finite element modelling invariably leads to a matrix formulation that depends on the nature of the physical problem to be solved. Hence, steady-state problems (*i.e.* problems for which the unknown variables are independent of time) always consist of solving (numerically) the following matrix relationship:

$$[K]\{U\} = \{F\} \tag{14.1}$$

where $[K]$ represents the global *stiffness matrix* (that includes the material properties), $\{U\}$ is the vector of *nodal unknown variables* (for instance displacement or water pressure), and $\{F\}$ corresponds to the *vector of applied nodal loads*.

Based on the two basic principles stated above, a finite element procedure consists of the following steps:

(*a*) discretisation and selection of elements;
(*b*) selection of the stress-strain relationships;
(*c*) evaluation of element matrices;
(*d*) assembly of elements matrices and introduction of boundary conditions;
(*e*) solution to nodal unknowns;
(*f*) computation of derived quantities, and *analysis of results.*

(*a*) **Discretisation and elements selection**

The elements used in conjunction with any discretisation process are selected with the view to obtaining sufficiently precise values of the nodal unknowns. Their *shape* and *type* depend therefore on the complexity of the boundaries of the discretised domain, and on the complexity of the physical behaviour to be modelled. In this respect, figure 14.1 illustrates the *deformed shape* of various one-, two-, and three-dimensional elements, each characterised by the *number of nodes* it contains. It is useful to mention at this stage that the number of variables at each node defines the number of *degrees of freedom per node* which are not necessarily identical for all nodes of every element in the mesh. Also, elements for which the *same* shape functions are used to define the unknown variable *and* the geometry of the element (*i.e.* its edges) are known as *isoparametric elements*.

The (deformed) shapes are related to the type of the *shape functions*. Thus, *linear elements* are such that the unknowns vary linearly between any two connected nodes. The edges of a *quadratic element*, on the other hand, are curved since the corresponding shape functions consist of second-order polynomials. *Cubic elements* are generated using third-order polynomials.

While the list of elements depicted in the figure is by no means exhaustive, it contains, however, some of the most widely used elements in finite element modelling in geotechnics. Thus, the three-noded triangle (element 4 in the figure) is the simplest two-dimensional element. Because of its associated linear shape functions, the strains (and consequently the stresses) are constant within the element. For that reason, the element is usually referred to as the *constant strain triangle* (*CST*). Although the *CST* is very easy to program, its use is limited to problems which do not involve the derivatives of the nodal variables such as computational fluid dynamics and seepage applications. Similarly, the same limitations apply to the four-noded quadrangle (element 7 in figure 14.1).

For these reasons, the six-noded triangle and the eight-noded quadrangle (quadratic elements 5 and 8 respectively) are the two-dimensional isoparametric elements *par excellence*, although their performance depends on the integration rule used (see section 14.3) . Also, the 15-noded quadratic (cubic strain) triangle (element 6 with an extra node on each side and three extra interior nodes, figure 14.1) is another quite widely used element which is now incorporated in many finite element packages.

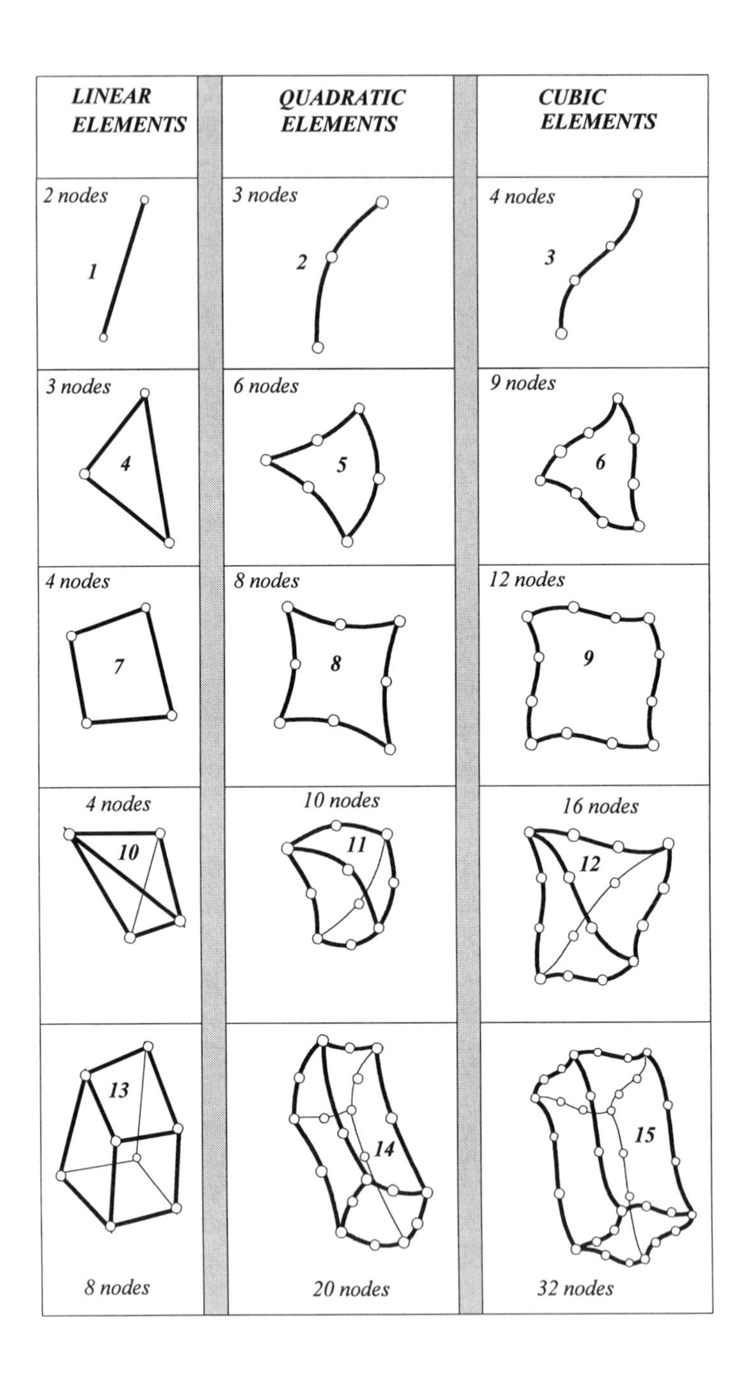

Figure 14.1: Selective types of elements.

These elements are particularly suitable for modelling plane strain and axisymmetric problems. Moreover, the quadratic nature of the associated shape functions makes it possible for domains with curved edges to be discretised in a precise way without the use of an excessive number of elements. It is perhaps worth mentioning that the shape functions associated with the *eight-noded quadrangle* are based on *incomplete* second-order polynomials (the straightforward mathematical details can be found in *any* book on finite elements) and, accordingly, the element is often referred to (in the jargon of finite elements) as a *serendipity quadrangle*.

When it comes to three-dimensional modelling in geotechnics, the two most widely used elements are the ten-noded tetrahedron and the twenty-noded serendipity hexahedron (elements 11 and 14 in figure 14.1). These quadratic 3-D elements can be used successfully to estimate the displacement and stress fields. It must be borne in mind that the coding of mesh generation and output facilities for 3-D finite element analysis can be very complex. Accordingly, 3-D finite element modelling must only be used when neither plane strain nor axisymmetric conditions are suitable, or when the type of project justifies the cost of such a modelling (the case of the simulation of stresses and settlements beneath the foundation of a nuclear power plant for instance).

It should be remembered that a finite element mesh consists generally of a combination of different types of compatible elements (*i.e.* nodes *common* to two elements must have the *same* degrees of freedom, and their shape functions must be characterised by polynomials of the *same* order). This might involve the use of transitional elements characterised by different shape functions on one side (to ensure a correct transition between elements) as illustrated in figure 14.2. More importantly, finite element modelling of soil structure interaction problems (examples include retaining, diaphragm or sheet-pile walls, laterally loaded piles, tunnels) may necessitate the use of interface elements to simulate the friction at the soil–structure interface (refer to figure 14.2).

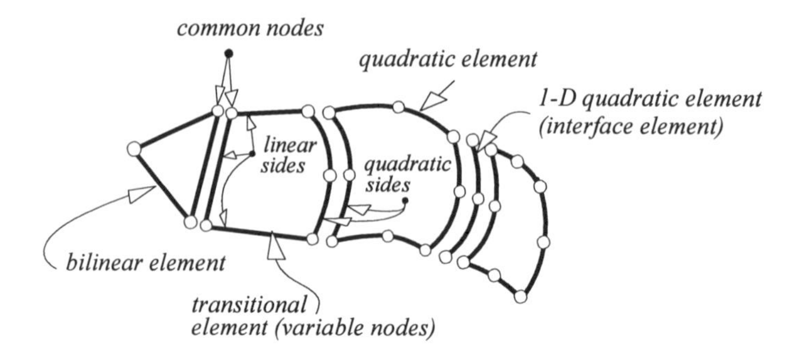

Figure 14.2: Use of transitional and interface elements.

(*b*) Selection of constitutive laws

The predictions based on the outcome of a finite element modelling depend to a large extent on how realistic is the stress–strain relationship used in the calculations. It is a fact that most engineers prefer to use simple linear relationships, and in some cases related to homogeneous isotropic soils, such assumptions can be justified. However, for more complex soil conditions corresponding to layered soils with different stiffness characteristics, the assumption of a unique stress–strain relationship, let alone a linear one, can be *markedly* erroneous. It is essential to realise that finite element modelling, with a however refined mesh and sophisticated elements, will not *per se* offset any shortcomings related to the use of inappropriate stiffness parameters (that is stress–strain relationships), and as such, it cannot be emphasised enough that the numerical analyst must have a good feel, if not a good grasp, of soil behaviour prior to embarking on expensive numerical modelling. Non-linear stress–strain relationships usually lead to iterative procedures in computation, increasing noticeably the cost of calculations. Such cost must therefore be justified at least in terms of obtaining reliable predictions of stresses and deformations.

(*c*) Evaluation of element matrices

In finite element modelling, element matrices are first formulated in terms of local co-ordinates, so that for each element in the mesh, the element stiffness matrix and vector of nodal forces are established using an integral method such as the *Galerkin weighted-residual method,* for instance, thus yielding:

$$[K]^e = \int_{v_e} [B]^T [D][B]\, dv$$

$$\{F\}^e = \int_{v_e} [N]^T \{F_v\}\, dv \qquad (14.2)$$

$[B]^T$ being the *transpose* of the matrix $[B]$ which is derived from the shape functions $[N]$, and $[D]$ corresponds to the matrix of soil properties relating stresses and strains (refer to Hughes (1987), for example).

The numerical integration of the element stiffness matrix is usually undertaken using the *Gauss quadrature method,* whose power is such that the integral in equation 14.2 needs only be evaluated at a few specific points known as *Gauss integration points*. It can be shown that using n points of integration in each direction of the space, this method can integrate exactly up to a $(2n-1)$-order polynomial. For instance, it can readily be shown that in the case of the eight-noded serendipity quadrangle (element 8 in figure 14.2), the integrand $[B]^T[D][B]$ in equation 14.2 is of the *fourth order*. Were the stiffness matrix for this element to be evaluated using nine integration points (that is 3×3 so that $n = 3$), an exact solution *will* be achieved since the use of $n^2 = 9$ integration points can integrate *exactly* up to a *fifth order polynomial* $(2n - 1 = 5)$.

(*d*) Assembly of elements matrices and introduction of boundary conditions

Once all elements' stiffness matrices are evaluated in local co-ordinates, a global stiffness matrix is then determined by first *expanding* each element matrix so that it is expressed in terms of *all* nodal variables of the entire mesh, then *summing* all matrices in a straightforward manner, yielding in the process the now familiar equation 14.1:

$$[K]\{U\} = \{F\}$$

in which the vector $\{U\}$ contains all nodal variables (or degrees of freedom), and $\{F\}$ represents the (expanded) vector containing all nodal forces.

The boundary conditions in terms of nodal forces are inserted into the global vector $\{F\}$ in a straightforward way. However, a prescribed nodal displacement can be introduced in many ways. For example, were the displacement at node i to be prescribed at $u_i = \delta$, then such a boundary condition can be satisfied if the *diagonal coefficient* in the stiffness matrix K_{ii} corresponding to node i is replaced by a large number ($M = 10^6$ for example) *and* the quantity $M\delta$ is substituted for the nodal force F_i.

Notice that the global stiffness matrix is (usually) symmetric and has a dimension of $(m \times m)$ where m represents the total number of degrees of freedom contained in the *entire* mesh. For example a somewhat complex 2-D mesh for a tunnel having, say, 5000 nodes (which is not unusual) with only 2 degrees of freedom per node, yields a global stiffness matrix with a dimension of $(10,000 \times 10,000) = 10^8$. Storing such a colossal matrix in its entirety would undoubtedly cause few problems; fortunately, only a few of the 100 million coefficients in this case are *non-zero*. What is more, the non-zero coefficients are situated in and around the diagonal of the matrix, thus forming a *band*. It is only the width of this band that is of interest, because its storage is done according to a dynamic allocation of space that allows the corresponding coefficients (which may include zeros) to be stored in lines so that they are easily retrieved when required. This method is known as the *skyline storage method*, the details of which can be found in Zienkiewicz and Taylor (1991), for instance.

(*e*) **Solution to nodal unknowns**

This step consists of calculating the vector of nodal unknowns $\{U\}$ from the global equation:

$$[K]\{U\} = \{F\}$$

The obvious solution is to invert the stiffness matrix $[K]$, so that: $\{U\} = [K]^{-1}[F]$. However, given the large size of $[K]$, its inversion is *not practical*. A more effective solution consists of applying the *triangular decomposition method* in conjunction with the stiffness matrix, then using the *forward elimination and back-substitution method* to calculate the unknown nodal variables. Based on the Gaussian elimination method, it can be shown that, provided $[K]$ is not singular (*i.e.* its *determinant* in not zero), it can always be written as the product of two matrices:

$$[K]=[L][S] \tag{14.3}$$

where $[L]$ is a lower triangular matrix in which the diagonal coefficients are unity, and $[S]$ is an upper triangular matrix, so that for example:

$$\begin{bmatrix} 2 & 4 & 8 \\ 4 & 11 & 25 \\ 6 & 18 & 46 \end{bmatrix} = \begin{bmatrix} 1 & 0 & 0 \\ 2 & 1 & 0 \\ 3 & 2 & 1 \end{bmatrix} \begin{bmatrix} 2 & 4 & 8 \\ 0 & 3 & 9 \\ 0 & 0 & 4 \end{bmatrix}$$

$$[K] \quad = \quad [L] \quad\quad [S]$$

Thus, substituting for $[K]$ into the global relationship yields:

$$[L][S]\{U\}=\{F\} \tag{14.4}$$

which can then be solved in a very easy way using the forward elimination and back-substitution method (see Jennings and McKeown (1992), Hoffman (1992)) which consists of the following stages:

- *stage 1*: factorize $[K]$:

$$[K]=[L][S] \tag{14.5a}$$

- *stage 2*: use a forward substitution to solve for $\{Z\}$:

$$[L]\{Z\}=\{F\} \tag{14.5b}$$

- *stage 3*: use a back-substitution to solve for $\{U\}$:

$$[S]\{U\}=\{Z\} \tag{14.5c}$$

Notice that $\{Z\}$ in equations 14.5 represents a vector of intermediate variables.

Such a method of solution, however, may quickly become unsuitable for 3-D finite element analyses, both on grounds of storage capacity and the number of calculations. Under such circumstances, iterative solvers such as the one based on the *pre-conditioned conjugate gradient method* (see Jennings and McKeown (1992)) can be very useful in terms of efficiency and precision.

(*f*) Calculation of derived quantities and analysis of results

Once the primary unknowns in equation 14.4 are determined, the *gradients* or derived quantities can then be established. For instance, the strains are derived from displacements, and the stresses are calculated from the stress–strain constitutive relationships.

Perhaps the most important step in a finite element modelling consists of analysing the outcome of computation. Once more, it is worth reiterating that the finite element method is *only* a method of solution and, as such, it *cannot* make for any deficiency related to soil stiffness parameters used in the calculations. It is naïve to consider a result acceptable simply by virtue of the fact that it was generated through a finite element modelling. The engineer must be in a position to question the very *raison d'être* of such a modelling in the first place. Is it needed ? If yes, then how can one get a *reliable set of soil parameters* (preferably measured *in situ* so as to minimise the effects of disturbance). Furthermore, it is essential that the *soil–structure interaction* mechanisms are well understood before embarking on time consuming expensive numerical modelling. It is also helpful if the working of the finite element method is understood, so that when it comes to the *interpretation of results*, any aberration can be attributed to its appropriate cause(s). In this respect, the following section contains some practical analyses related to the potential pitfalls of finite element modelling.

14.2 Effective stress analysis

Thus far, the term *stress* has been used indiscriminately. However, there is often a need during a finite element analysis of geotechnical problems, to dissociate (numerically) *effective stresses* from *porewater pressures* according to the effective stress principle (an important aspect in geotechnics). Although the specific details of such an analysis is beyond the scope of the present text (reference can be made to Naylor (1974) for example), it is worth mentioning that an effective stress formulation can easily be incorporated in a finite element program. The procedure consists of using an expanded matrix of soil properties relating stresses and strains (matrix $[D\ ]$ in equation 14.2) in the following manner (see Naylor *et al.* (1981)):

$$[D] = [D'\ ] + [m][m]^T [K_f] \tag{14.6a}$$

$$\{\sigma\} = \{\sigma'\} + [m]\{u\} \tag{14.6b}$$

where $[m]^T$ represents the transpose of the column matrix related to the effective stress principle, $[D']$ is the soil skeleton modulus matrix, and $[K_f]$ corresponds to the bulk modulus of the pore fluid element. Note that all coefficients corresponding to shear stress components in $[m]$ are zero, the remaining coefficients being equal to 1.

All calculations involving element matrices, assembly and solution to the global system of equations thus formed are then undertaken as described previously. Once the strains are calculated, the ensuing stress increments are then evaluated using the following relationships:

$$\{\Delta\sigma'\} = [D']\{\Delta\varepsilon\} \tag{14.7a}$$

$$\{\Delta u\} = [K_f]\{\Delta\varepsilon_v\} \tag{14.7b}$$

$\{\Delta u\}$ being the *excess* (*i.e.* load induced) porewater pressure, and $\{\Delta\varepsilon_v\}$ represents the volumetric strain change.

Notice that a *drained analysis* amounts to setting $[K_f]$ to zero. More importantly, the effective stress analysis presented above does not apply to transient flow problems (*i.e.* consolidation problems).

Also, it should be mentioned that the *critical state model* presented in chapter 6 can be (and is often) successfully coupled to finite element programs. In this respect, reference should be made to Britto and Gunn (1986) and Naylor *et al.* (1981).

14.3 Finite element modelling of seepage and consolidation problems

Although seepage problems are *steady state* problems, their finite element solution is nonetheless elaborated in a slightly different way from the one developed earlier, since pressure heads are the corresponding primary unknowns.

Seepage problems are represented by an *elliptic equation,* the two-dimensional form of which has been established earlier (see equation 3.47, section 3.4):

$$k_x \frac{\partial^2 h}{\partial x^2} + k_y \frac{\partial^2 h}{\partial y^2} = 0 \tag{14.8}$$

where k_x and k_y are the coefficients of soil permeability and h corresponds to the total head. Notice that for $k_x \neq k_y$, equation 14.8 is *not* a Laplace equation.

Whilst developing a closed form solution to such an equation can be fraught with difficulties, mainly because in many cases of flow, the corresponding boundary conditions are very difficult to satisfy, the finite element method can be applied in a straightforward manner.

It can be shown that, irrespective of the nature of flow (*i.e.* confined or unconfined), the discretisation of equation 14.8 invariably yields a global relationship:

$$[K]\{H\} = \{Q\} \tag{14.9}$$

which is a familiar matrix relationship, since it is similar to equation 14.1. Here $\{H\}$ is the vector of nodal variables, which in this case correspond to the total heads, and $\{Q\}$ is the vector of nodal flow. It is interesting to mention that solving for total heads implies that each node has only *one degree of freedom*, in other words there is only one unknown quantity at each node.

The global stiffness matrix $[K]$ in equation 14.9 is obtained from the assembly of element matrices, each of which is calculated as follows:

$$[K]^e = \int_{v_e} [B]^T [P][B]\,dv \qquad (14.10)$$

where $[B]^T$ corresponds to the transpose of matrix $[B]$ which is derived from the shape functions, and $[P]$ is the element *permeability matrix* (Naylor *et al.*, 1981):

$$[P] = \begin{bmatrix} k_x & 0 \\ 0 & k_y \end{bmatrix} \qquad (14.11)$$

Notice the diagonal nature of $[P]$.

The numerical solution to equation 14.9 depends on the boundary conditions and therefore on the nature of seepage. Thus, in the case of confined flow as illustrated in figure 14.3 (refer also to section 3.4 for more details related to this type of flow), the corresponding boundary conditions are:

- *along AB*: $H = h_1 + d_1$
- *along CD*: $H = h_2 + d_2$
- *along EF*: $\dfrac{\partial H}{\partial y} = 0$ (impermeable side)

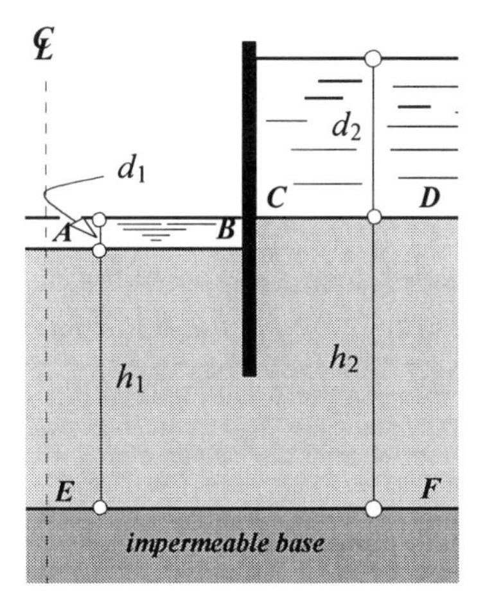

Figure 14.3: Boundary conditions related to confined flow.

For unconfined flow, on the other hand (see figure 14.4), the boundary conditions are:

- *along AB*: $H = h_1$
- *along BC*: $H = y$ *and* $u = 0$ (u being the porewater pressure, refer to section 3.4)
- *along AC*: $\dfrac{\partial H}{\partial y} = 0$ (impermeable side)

in both cases, H is the *total head* expressed with respect to the datum represented by the impermeable base.

The *free surface* in the case of unconfined flow presents an additional problem as its location is not known *a priori*. The precise location can be found using iterative techniques such as the one suggested by Smith and Griffiths (1998), that consists of deforming the mesh until, eventually, its upper surface coincides with the free surface characterised by the boundary condition of a total head identical to the elevation head at any given point.

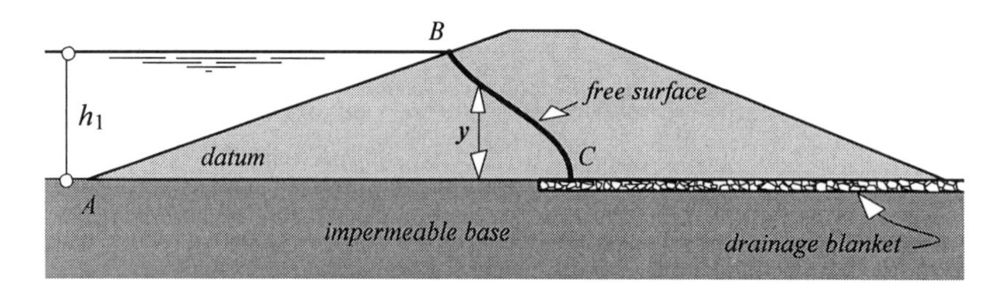

Figure 14.4: Boundary conditions related to unconfined flow.

Time-dependent problems such as consolidation are represented by a *parabolic equation* (refer to the consolidation equation 4.14, section 4.3):

$$c_v \frac{\partial^2 u}{\partial z^2} = \frac{\partial u}{\partial t} \tag{14.12}$$

for which it can readily be shown that finite element modelling yields the following relationship:

$$[C]\left\{\frac{\partial U}{\partial t}\right\} + [K]\{U\} = \{F\} \tag{14.13}$$

where $[C]$ represents the matrix of soil consolidation characteristics. The nodal variables $\{U\}$ this time correspond to the porewater pressure; the remaining symbols having the same meaning as in equation 14.1.

A solution to equation 14.13 can be obtained through a combination of finite difference formulation (to discretise time) and finite elements. Consequently, using a forward difference operator:

$$\left\{\frac{\partial U}{\partial t}\right\} = \frac{1}{\Delta t}\left[\{U\}_{t+\Delta t} - \{U\}_t\right] \tag{14.14}$$

(Δt being the time increment), then inserting the latter quantity into equation 14.13 and rearranging, the following algorithm can easily be established:

$$\{U\}_{t+\Delta t} = \Delta t\,[C]^{-1}\left[\{F\}_t + \left(\frac{1}{\Delta t}[C] - [K]\right)\{U\}_t\right] \tag{14.15}$$

which can then be solved on a step-by-step basis.

14.4 Practical aspects of finite element modelling in geotechnics

Finite element modelling of soil–structure interaction problems needs to be planned and undertaken carefully, so that any anomalies in the results can be spotted and remedied.

The first step consists of discretising the domain, in other words generating a *mesh*. Although some finite element design programs contain mesh-generation pre-processors, it must be remembered that a mesh should always constitute a *compromise* between the *computer capacity to store data and execute calculations* (this capacity can, in some cases, be exceeded even by the standard of very powerful machines), and *reasonable predictions* in terms of outcome such as strains, stresses, porewater pressure etc.

Thus, a reasonable mesh should be refined (*i.e.* formed of smaller elements) near any applied load where the stress and displacement gradients are expected to be large, and around any geometric or material discontinuities (or singularities to use a finite element jargon) such as tunnel openings or excavations. Also, the mesh needs to be refined around the areas where changes in material properties occur, as in the case of different soil layers, or around a soil–structure interface (pile–soil, retaining structure–soil, tunnel–soil etc.). In such cases, interface elements may be needed to simulate friction at the soil–structure interface, particularly when the structure surface is relatively smooth.

Generally, the size of elements used in a mesh depends on the loading conditions and geometric discontinuities stated above. However, numerical evidence seems to indicate that the *aspect ratio* of an element (*i.e.* the ratio of the *largest* to the *smallest* dimensions of an element) must be kept within reasonable limits. In this respect, an aspect ratio smaller than 3 ensures satisfactory results in terms of stresses, unless the soil behaviour is markedly non-linear, in which case an even smaller ratio is required. As far as the *shape* of elements is concerned, and whenever practicable, triangles, quadrangles and hexahedra used to generate a mesh should be as near as possible to equilateral triangles, squares and cubes respectively, so as to avoid any excessive distortion of these elements in the advent of large displacements. Notice that some sophisticated finite element programs offer the possibility of automatically regenerating the mesh while taking into account the level of deformation of different elements.

The size of elements should be *increased progressively* around the refined portions of the mesh in a way that ensures a smooth transition from small to larger size elements. This can be achieved if the ratio of the areas (or volumes in 3-D) of two adjacent elements does not exceed 2.

In any case, it is strongly advised (perhaps one should say it is logical) to avoid using distorted elements such as the ones illustrated in figure 14.5 when generating a mesh. The reason being that, once loaded, each of these elements can potentially deform to the point where the stiffness matrix becomes singular (*i.e.* with a zero *determinant*, so that it cannot be inverted) thus causing numerical instability. As a *guideline*, any angle α within an element must be such that $15^\circ \leq \alpha \leq 165^\circ$. Also, the middle node in a quadratic element should be situated within the middle third of the side [see figure 14.5] (Zienkiewicz and Taylor, 1991).

Transitional elements (refer to figure 14.2 for the principle involved) are mainly used in conjunction with 3-D modelling.

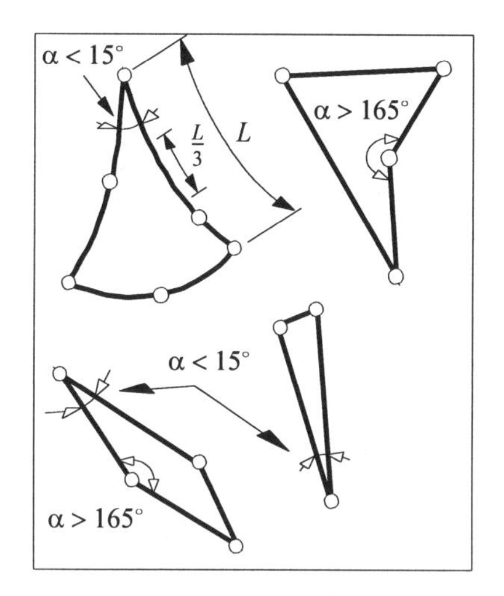

Figure 14.5: Unacceptable elements.

The type of elements used in a finite element modelling depends on the type of problem to be analysed. Nevertheless, one has to bear in mind that the strains and stresses are derived from displacements and, as such, using linear elements (*i.e.* elements characterised by linear shape functions) automatically yields constant strains and stresses within each element, which can be markedly erroneous, particularly in relation to stress distribution. Accordingly, the use of linear elements (refer to figure 14.1) is not advised. Quadratic elements on the other hand, especially the six-noded triangle and the eight-noded quadrangle, are widely used to model different type of problems in geomechanics.

The calculation of strains and stresses is undertaken within each element at specific *integration points*. Often, the integration points used to calculate the strains (and hence the stresses) are fewer than those used to calculate the element stiffness matrix. For example, whilst the stiffness matrix of the *eight-noded serendipity quadrangle* is calculated using 3×3 integration points, numerical evidence shows that, in most cases, satisfactory results in terms of displacements are obtained from a *reduced integration* using 2×2 integration points. In fact, in this case, an exact integration using 3×3 points yields an over-stiff response, so underestimating displacements. Situations might arise whereby a mechanism can occur as a result of a reduced integration using 2×2 points; however, these cases are extreme. As regards the *six-noded triangle*, a happy balance can be achieved using three (non-Gauss) integration points. On the other hand the use of four integration points in conjunction with the *ten-noded tetrahedron* (one of the most widely used 3-D elements) provides satisfactory results in terms of displacements. Interested readers can refer to the paper by Naylor (1994) on integration rules.

Once calculated at the integration points, the stresses are then extrapolated to the element nodes and, because each element has its own stiffness matrix, the magnitude of stresses resulting from each element at a common node such as the one depicted in figure 14.6 is generally different. This stress discontinuity can be smoothed by taking the *average* value at the node, so that the corresponding nodal stress results from the contribution of all adjacent elements.

However, nodal averaging *should not* be used for stress and strain calculations at nodes connecting elements with *different stiffnesses* (for example elements corresponding to different soil layers). Such boundaries are characterised by a *stress discontinuity*, except for the stress component *normal* to the boundary. Numerical evidence shows that the stress field is not affected by the stress discontinuity provided that a refined mesh with smaller size elements around theses boundaries is used.

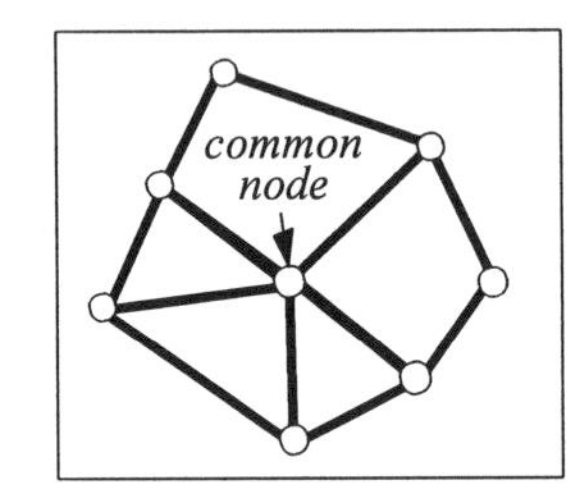

Figure 14.6: Stress calculation at a common node.

When using finite element modelling, special attention must be paid to the degree and type of anisotropy of the soil–structure material. For example the distribution with depth of the vertical stress generated by a uniform surface load within a thick homogeneous isotropic soil layer is *markedly different* in shape and in magnitude from that generated by the same load within a layered anisotropic soil.

As mentioned earlier, geometric or material discontinuities constitute singularities which are unavoidable for a large number of soil–structure interaction problems. This can generate inaccurate stress fields near the singularities (for example, an infinite stress beneath a point load is computed as a finite quantity). However, these effects can be minimised by using a refined mesh around the singularity, so that a right-angle boundary, for instance, can be transformed into a smooth curved boundary.

14.5 Practical aspects of a finite element mesh related to foundations

A finite element mesh related to a given problem in geomechanics must always take into account different aspects evoked previously and linked to elements type and size, and particularly to the nature of the problem (axisymmetry, anisotropy, drainage conditions, soil–structure interaction, geometric singularities, and nature of loading). Consequently, the following *guidelines* should be viewed with a sufficient degree of flexibility, so that local conditions are appropriately considered. As such, they should *in no way* be regarded as rules.

Based on the previous analysis, a finite element (axisymmetric) mesh in the case of a shallow foundation on an isotropic homogeneous soil, with a width B usually includes an area extending to about $5B$ laterally and $8B$ vertically as illustrated in figure 14.7, an area within which most of the stress variations are expected to occur (refer to figures 2.37–2.40, section 2.5).

The conditions imposed at the mesh edges in figure 14.7 allow for a vertical movement (*i.e.* $u = 0$) along the vertical boundaries, while restricting any movement to $u = v = 0$ at the bottom horizontal boundary, where the stresses (and therefore the displacements) are expected to decay. Also, notice how the mesh is refined beneath and around the foundation, with increasing elements size as one moves away from the foundation in each direction.

Were the soil conditions to be different as in the case, for example, of ansiotropic multi-layered and partially submerged soils, then the overall size of mesh as well as the size and type of elements would need to be altered drastically so as to reflect the markedly different characteristics of the problem. Hence the informative nature of the dimensions indicated in figure 14.7.

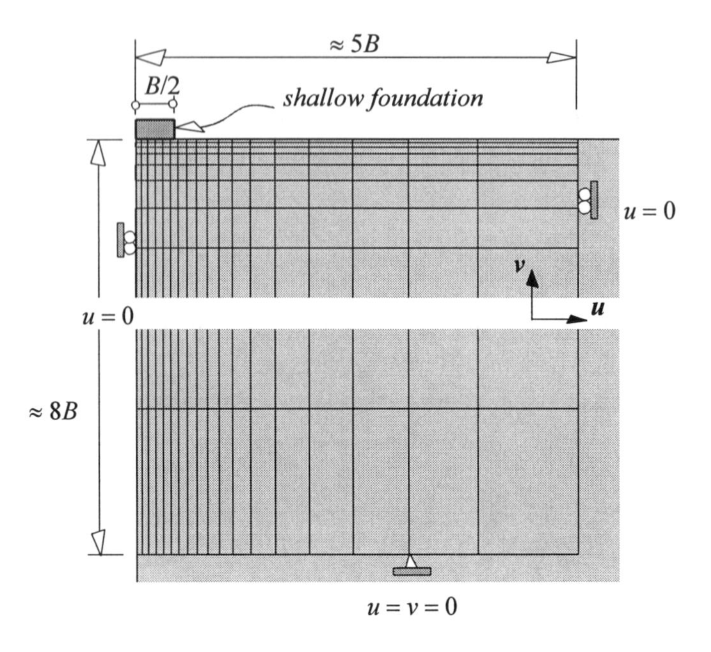

Figure 14.7: Typical (axisymmetric) mesh dimensions for an isolated shallow foundation on isotropic homogeneous soil.

For a deep pile foundation, the mesh depends on the type of loading and the type of pile. Thus, for a *single axially loaded pile* with a length L embedded in an isotropic homogeneous soil, the mesh should ideally cover an area $3L$-deep and about 30-pile diameters wide as depicted in figure 14.8. The typical (axisymmetric) mesh in the figure consists mainly of eight-noded quadrangles and a few six-noded triangles. The size of the elements reflects the stress distribution generated around the pile shaft and tip depicted in figure 9.31 (refer to section 9.5). The boundary conditions are of a similar type to the ones described previously in the case of shallow foundations. However, in this case, there is a need to use interface elements in order to simulate the friction developed along the pile shaft.

It is interesting to notice that the mesh in figure 14.8 is used to simulate the behaviour of an *already* embedded pile. The numerical simulation of pile driving is very complex in nature, not least because soil failure around the pile shaft and tip has to occur every time the pile is struck with the driving tool, so that it can be driven to the required depth, or sometimes until the occurrence of a refusal.

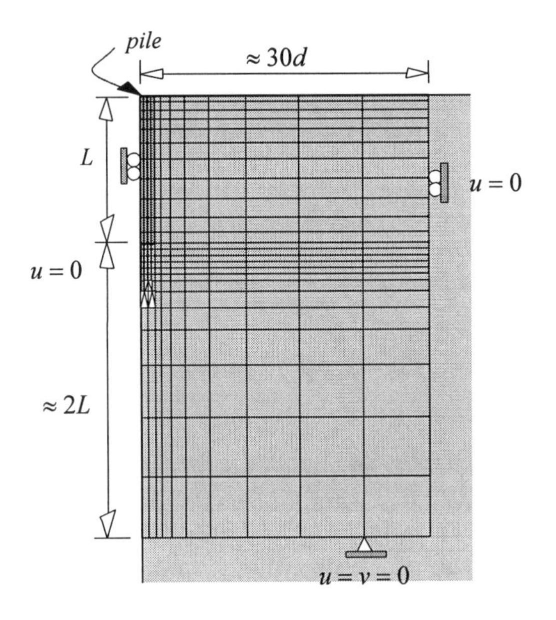

Figure 14.8: Typical (axisymmetric) mesh dimensions for a single axially loaded pile (isotropic homogeneous soil conditions).

A *laterally loaded single pile,* on the other hand, is characterised by a markedly different behaviour. This type of loading occurs especially in conjunction with bridges, flyovers and retaining structures founded on piles, and can be applied either actively or passively as depicted in figure 14.9. In both cases, the solution in terms of displacement, bending moment, shear stress and lateral pressure distribution with depth depends on the soil–pile interaction, in other words on both soil and pile stiffness characteristics.

The soil–pile interaction is depicted in figure 14.10 whereby the resistance to the active load (a combination of a horizontal load and a bending moment applied at the pile head) is provided by the pile stiffness *and* the soil reaction per unit length (that is, the *force per unit length* induced within the soil mass by the active load).

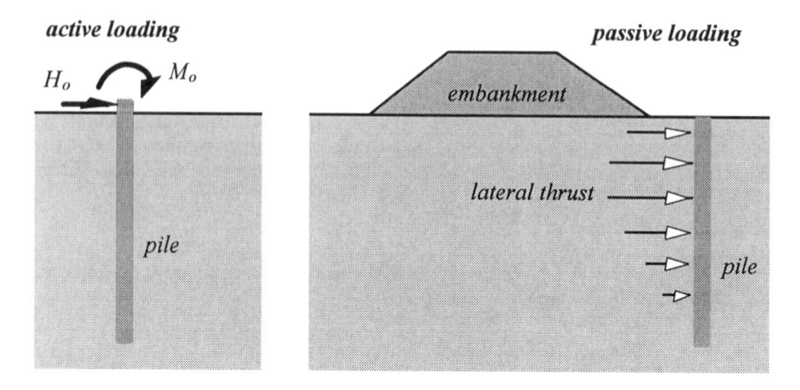

Figure 14.9: Active and passive lateral loading of piles.

While the pile stiffness can be assumed to be constant along the pile shaft, the soil stiffness, on the other hand, depends generally on the depth of embedment. Figure 14.10 shows that at a given depth z, the relationship between the soil reaction P and the lateral displacement y (*i.e.* the pile deflection) is non-linear. Furthermore, the nature of the relationship varies with depth.

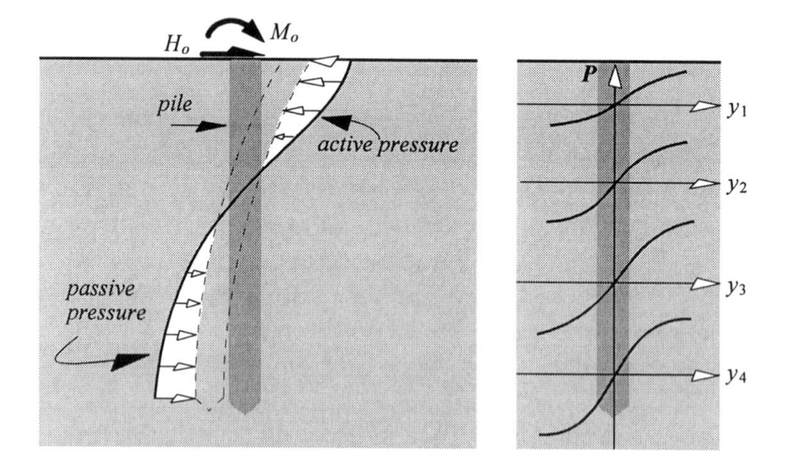

Figure 14.10: Soil–structure interaction along the shaft of an actively loaded single pile.

The profiles of pile deflection, bending moment and soil reaction are also affected by the pile dimensions and the boundary conditions: a long slender pile behaves in a different way from a short rigid pile as illustrated in figures 14.11 and 14.12 in the case of actively loaded piles embedded in an isotropic homogeneous sand. The figures also depict the difference in behaviour between a free head and a fixed head pile.

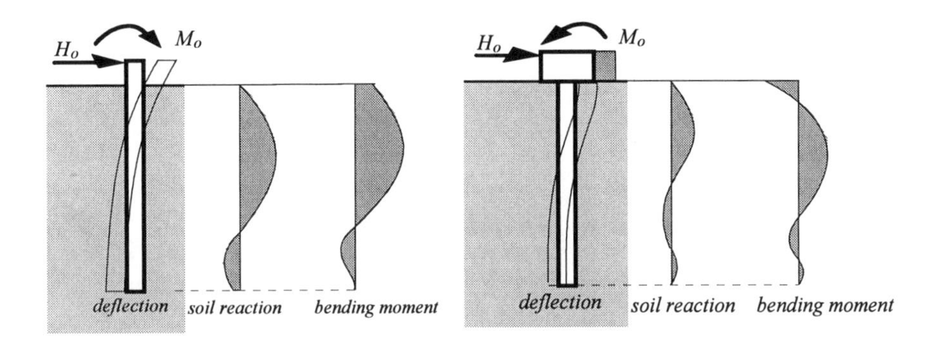

Figure 14.11: Profiles of deflection, soil reaction and bending moment for a long pile in a cohesionless soil.

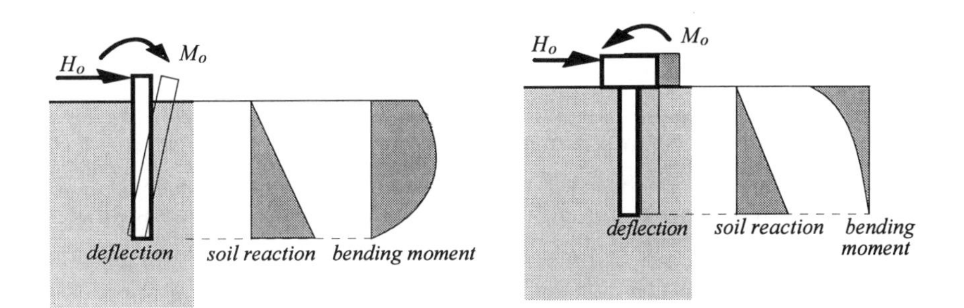

Figure 14.12: Profiles of deflection, soil reaction and bending moment for a short pile in a cohesionless soil.

Accordingly, a finite element mesh in the case of a laterally loaded pile should reflect the soil, pile and loading conditions. Because of the nature of the problem, no axisymmetry can be applied, and the mesh dimensions indicated in figure 14.13 should be applied sensibly.

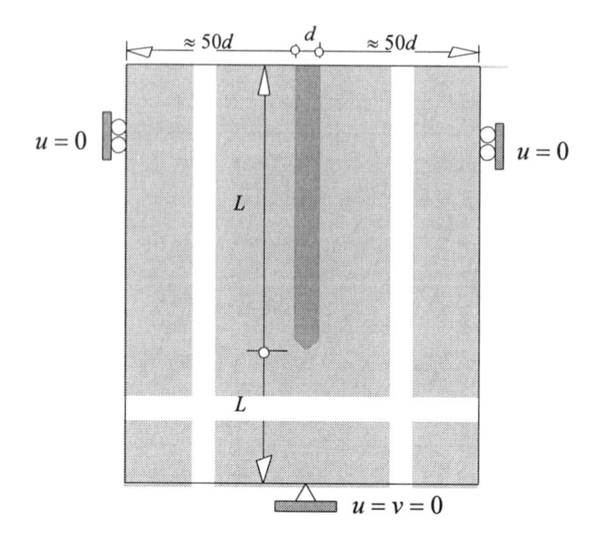

Figure 14.13: Typical mesh dimensions for a single laterally loaded pile (isotropic homogeneous soil conditions).

A pilegroup is much more difficult to model, especially in the absence of axisymmetry (which is generally the case). The difficulties are further illustrated by the fact that several modes of failure can potentially develop (a single pile failure within the group, a row of piles failing at the same time or a block failure en masse, refer to chapter 9). Under such circumstances, a three-dimensional mesh might become a necessity, with all the consequences elicited earlier concerning the potentially very large size of global matrices. However, notwithstanding the somewhat high cost that might be incurred in some cases, 3-D modelling of pilegroups is now becoming an almost routine operation.

Whenever the 3-D problem can be reduced to an axisymmetric problem, then a mesh such as the typical one depicted in figure 14.14 corresponding to a pilegroup embedded in an isotropic soil can be used. The dimensions of the area covered by the mesh extend to about $3L$ in depth, and at least a distance L outside the pile situated at the group edge. Such a mesh should reflect the stress distribution generated within the soil mass by a pilegroup.

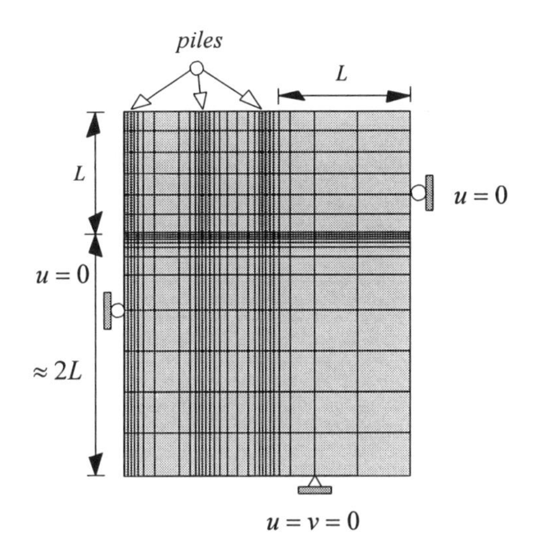

Figure 14.14: Typical (axisymmetric) mesh dimensions for an axially loaded pilegroup in an isotropic soil.

14.6 Finite element mesh related to embankments, retaining structures and tunnels

The finite element modelling of soil loading through an embankment with a height H must be such that the mesh area covers most of the stress increase generated within the soil mass. Also, the area must take into account the possibility of a long term or a short term slope stability failure. Accordingly, for an isotropic homogeneous soil layer, the 2-D mesh should be characterised by a lateral dimension of at least four times the embankment length, and a minimum depth of $5H$ or the depth of stiff substratum, whichever is smaller (see figure 14.15). The mesh should be refined in regions where the maximum stress generated by the loading is expected, and the size of elements should be increased gradually as one moves away from these regions. Note that were the depth of the stiff substratum to apply, then the boundary condition $u = v = 0$ indicated in the figure would correspond to a relatively small change in stiffness between the two layers. If the bottom layer were a rock, for example, overlain by a soft clay, then the boundary condition should be changed to allow for a lateral displacement, in which case $v = 0$.

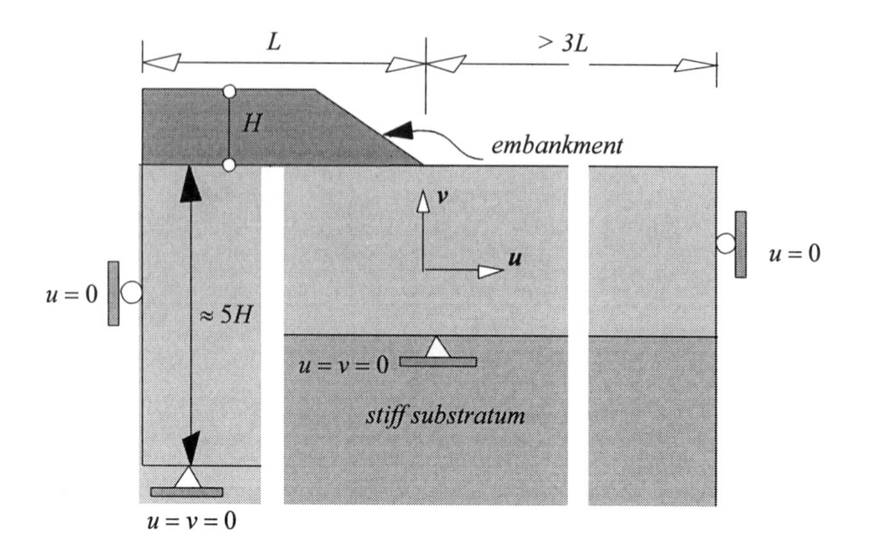

Figure 14.15: Typical mesh dimensions for an isotropic homogeneous soil loaded through an embankment.

As regards a retaining wall with height H and base width L (see figure 14.16), the mesh should be wide enough to include not only the stress changes in the soil mass beneath the wall, but also the potential development of long term active and passive stress failures, as well as the possibility of a deep circular failure. Consequently, for a wall retaining an isotropic homogeneous soil, it is advised to use a mesh with a minimum lateral dimension of $2H$ in front of the wall and $3H$ behind the wall, depending on the type of soil. The mesh should extend a minimum depth of $6L$ or the depth of the stiff substratum, whichever is smaller as illustrated in figure 14.16. Also, interface elements may have to be used if the wall surface is relatively smooth.

The boundary condition related to the stiff substratum indicated in the figure reflects the substantial change in stiffness between the two layers of soil.

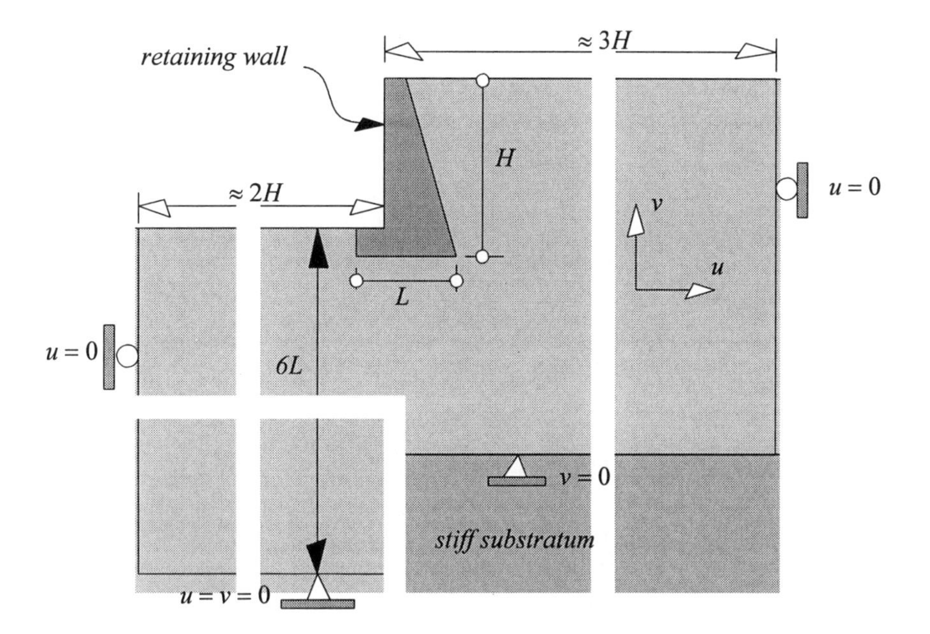

Figure 14.15: Typical mesh dimensions for a wall retaining an isotropic homogeneous soil.

Similar guidelines apply to diaphragm and sheet-pile walls (refer to figure 14.17).

Finite element modelling of an underground cavity or a tunnel requires an engineering judgement as to how such a task can best be undertaken cost-effectively. Thus 3-D modelling is appropriate when there is a variation of soil stratification along the tunnel axis, or when stress conditions near the tunnel face need be studied. On the other hand, an axisymmetrical analysis is suitable in the case of a vertical shaft when the soil stratification varies in the vertical direction only. There are instances in which plane deformations can be assumed to prevail, and therefore only a cross-section of the tunnel needs to be discretised. Under such circumstances, the area covered by the mesh must reflect the expected stress distribution and deformations around the opening.

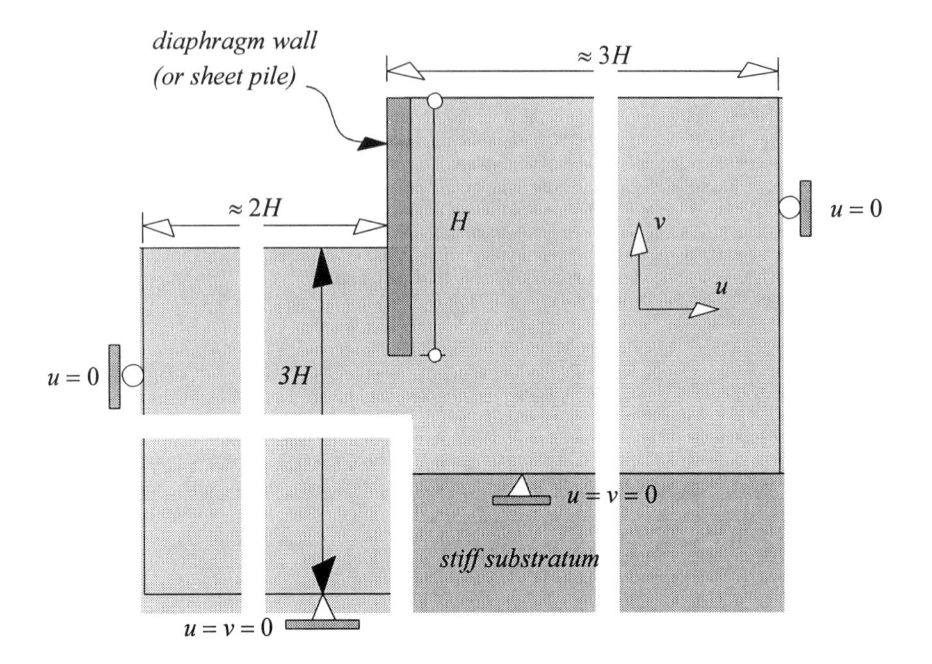

Figure 14.17: Typical mesh dimensions for a diaphragm wall retaining an isotropic homogeneous soil.

As explained earlier in section 12.2, the *stress arching* that occurs around the opening is related to its depth and, accordingly, the dimensions of the mesh area should typically extend to a depth of around $5d$ beneath the *tunnel invert*, and should include the entire height of soil above the *crown*, unless the cover exceeds $10d$, in which case the soil thickness above the crown can be limited to $5d$ (Mestat, 1997). Laterally, the dimension of the mesh area should be extended to about $6d$ from the tunnel axis as illustrated in figures 14.18(*a*) and (*b*). Obviously, these dimensions only represent a guideline, and the engineer must always seek a compromise between reliable predictions in terms of stress–strain distribution, and cost. In so doing, the dimensions advocated above might need to be adjusted to suite the site conditions.

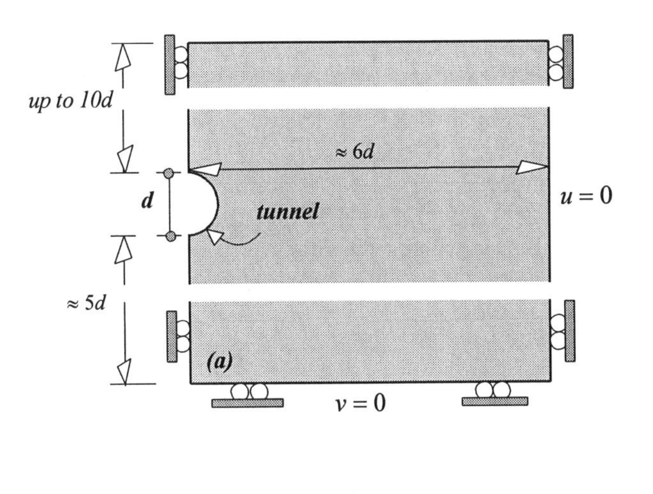

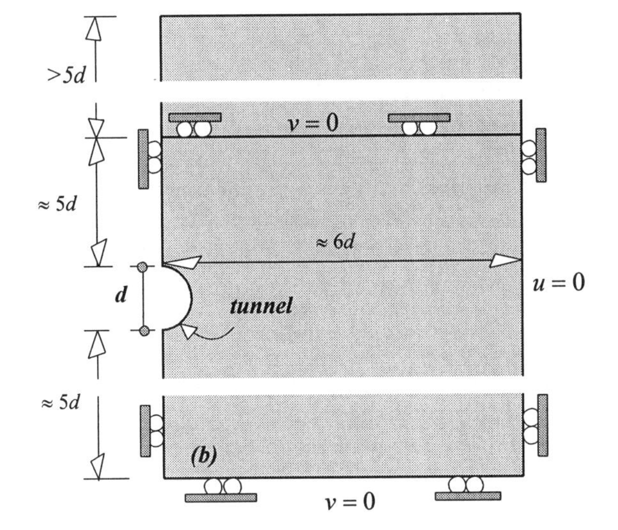

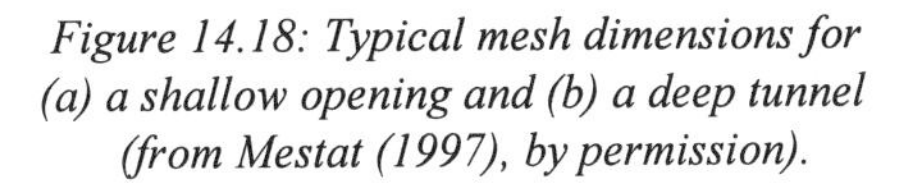

Figure 14.18: Typical mesh dimensions for (a) a shallow opening and (b) a deep tunnel (from Mestat (1997), by permission).

References

Bathe, K. J. (1982) *Finite Element Procedures in Engineering Analysis.* Prentice-Hall, Englewood Cliffs, New Jersey.

Bathe, K. J. (1996) *Numerical Methods in Finite Element Analysis,* 3rd edn. Prentice-Hall, Englewood Cliffs, New Jersey.

Britto, A. and Gunn, M. J. (1986) *Critical State Soil Mechanics via Finite Elements.* Ellis Horwood, Chichester.

Griffiths, D. V. and Smith, I. M. (1991) *Numerical Methods for Engineers.* Blackwell, Oxford.

Hoffman, J. D. (1992) *Numerical Methods for Engineers and Scientists.* McGraw-Hill, New York.

Hughes, T. J. R. (1987) *The Finite Element Method: Linear, Static and Dynamic Finite Element Analysis.* Prentice-Hall, Englewood Cliffs, New Jersey.

Jennings, A. and McKeown, J. J. (1992) *Matrix computation,* 2nd edn. John Wiley & Sons, Chichester.

Mestat, P. (1997) *Maillage d'élements finis pour les ouvrages de géotechnique: conseils et recommandations.* Bulletin de Laboratoires des Ponts et Chaussées, 212, pp. 39–64.

Naylor, D. J. (1974) *Stresses in nearly incompressible materials for finite elements with application to the calculation of excess pore pressures.* International Journal of Numerical Methods in Engineering, 8, pp. 443–460.

Naylor, D. J. (1994) *On integration rules for triangles.* Proceedings of the 3rd European Conference on Numerical Methods in Geotechnical Engineering, Manchester (ed. I. M. Smith), Balkema, Rotterdam.

Naylor, D. J., Pande, G. N., Simpson, B. and Tabb, R. (1981) *Finite Elements in Geotechnical Engineering.* Pineridge Press, Swansea, UK.

Owen, D. R. J. and Hinton, E. (1980) *Finite Elements in Plasticity: Theory and Practice.* Pineridge Press, Swansea.

Reddy, J. N. (1993) *An Introduction to the Finite Element Method.* McGraw-Hill, New York.

Smith, I. M. and Griffiths, D. V. (1998) *Programming the Finite Element Method,* 3rd edn. John Wiley & Sons, New York.

Stasa, F. L. (1985) *Applied Finite Element Analysis for Engineers.* CBS International Edition, New York.

Zienkiewicz, O. C. and Taylor, R. (1991) *The Finite Element Method,* (2 volumes) 4th edn. McGraw-Hill, London.

Subject index

D

M

N

O

P

V

W

Y

Author index